T0327429

Sorghum:

A State of the Art and Future Perspetives

Ignacio A. Ciampitti and P.V. Vara Prasad, editors

Agronomy Monograph 58

American Society of Agronomy
Crop Science Society of America
Soil Science Society of America
5585 Guilford Rd., Madison, WI 53711-5801 USA

agronomy.org • crops.org • soils.org
dl.sciencesocieties.org
SocietyStore.org

Agronomy Monograph Series ISSN: 2156-3276 (online)
ISSN: 0065-4663 (print)

ISBN: 978-0-89118-628-1 (electronic)
ISBN: 978-0-89118-627-4 (print)

doi: 10.2134/agronmonogr58

Printed in the United States of America.

Contents

Foreword ix

Preface xi

1

Genetic Changes in Sorghum 1

Ramasamy Perumal
Passoupathy Rajendrakumar
Frank Maulana
Tesfaye Tesso
Christopher R. Little

2

Sorghum Improvement for Yield 31

Leo Hoffmann, Jr
William L. Rooney

3

Sorghum Genetic Resources 47

Hari D. Upadhyaya
Sangam L. Dwivedi
Yi-Hong Wang
M. Vetriventhan

4

Pedigreed Mutant Library—A Unique Resource for Sorghum Improvement and Genomics 73

Zhanguo Xin
Yinping Jiao
Ratan Chopra
Nicholas Gladman
Gloria Burow
Chad Hayes
Junping Chen,
Yves Emendack
Doreen Ware
John Burke

5

Sorghum Hybrids Development for Important Traits: Progress and Way Forward 97

Are Ashok Kumar

6

Sorghum Breeding from a Private Research Perspective 119

Hugo Zorilla

7

Practical Morphology of Grain Sorghum and Implications for Crop Management 133

Calvin Trostle
Gary Peterson

8

Sorghum Growth and Development 155

Kraig L. Roozeboom
P.V. Vara Prasad

9

Structure and Composition of the Sorghum Grain 173

S.R. Bean
J.D. Wilson
R.A. Moreau
A. Galant
J.M. Awika
R.C. Kaufman
S.L. Adrianos
B.P. Ioerger

10

Sorghum Crop Modeling and Its Utility in Agronomy and Breeding 215

Graeme Hammer
Greg McLean
Al Doherty
Erik van Oosterom
Scott Chapman

11

Drought and High Temperature Stress and Traits Associated with Tolerance 241

P.V.V. Prasad
M. Djanaguiraman
S.V.K. Jagadish
I.A. Ciampitti

12

Water-Use Efficiency 267

Vincent Vadez

13

Genotype × Environment × Management Interactions: US Sorghum Cropping Systems 277

Ignacio A. Ciampitti
P.V.V. Prasad
Alan J. Schlegel
Lucas Haag
Ronnie W. Schnell
Brian Arnall
Josh Lofton

14

The Biology and Control of Sorghum Diseases 297

Christopher R. Little
Ramasamy Perumal

15

Weed Competition and Management in Sorghum 347

Curtis R. Thompson
J. Anita Dille
Dallas E. Peterson

16

Irrigation of Grain Sorghum 361

Danny H. Rogers
Alan J. Schlegel
Johnathon D. Holman
Jonathan P. Aguilar
Isaya Kisekka

17

Future Prospects for Sorghum as a Water-Saving Crop 375

David Brauer
R. Louis Baumhard

18

Sorghum: A Multipurpose Bioenergy Crop 399

P. Srinivas Rao
K.S. Vinutha
G.S. Anil Kumar
T. Chiranjeevi
A. Uma
Pankaj Lal
R.S. Prakasham
H.P. Singh
R. Sreenivasa Rao
Surinder Chopra
Shibu Jose

19

Use of Grain Sorghum in Extruded Products Developed for Gluten-free and Food Aid Applications 425

Sajid Alavi
Sue Ruan
Siva Shankar Adapa
Michael Joseph
Brian Lindshield
Satyanarayana Chilukuri

20

Forage and Renewable Sorghum End Uses 441

Scott Staggenborg

21

Overview of Sorghum Industrial Utilization 463

Guangyan Qi
Ningbo Li
Xiuzhi Susan Sun
Donghai Wang

22

Domestic and International Sorghum Marketing 477

Daniel O'Brien*

23
The Sorghum Industry and Its Market Perspective 503
John Duff
Doug Bice
Ian Hoeffner
Justin Weinheimer

Foreword

Sorghum is a versatile crop said to be the fifth most important cereal grown across the world. Domestication is thought to have occurred in the Sahel region in Africa over 5000 years ago. Today sorghum grain represents an important staple food in some low-income countries and can be used directly as animal feed or as a feedstock for biofuel production. The vegetative components are valuable forage for livestock and a source of sugar and cellulose for biofuel production. Sorghum has tremendous genetic diversity with widely varying phenotypes and photo-period sensitivity. This genetic diversity is still being explored and offers opportunities to further enhance the value of this important crop species. One of the primary characteristics of sorghum of critical importance in the face of climate change is the ability to survive and produce grain under hot and dry conditions that would cause failure for other crop species such as maize. Thus, the importance of sorghum is likely to increase and there is a strong need to maintain and enhance research in support of this crop.

The volume presents a comprehensive coverage of sorghum ranging across genetics, breeding, diseases, production, irrigation, modeling, and utilization. The book will be a valuable resource for researchers, practitioners, and industry representatives alike seeking both technical and practical background information on sorghum. The editors and authors are among the world's thought leaders for the subject matter and have assembled a truly outstanding compendium.

The efforts of the editors, authors, reviewers, members of the book committee, and support staff in the development of this book are greatly appreciated.

Gary Pierzynski
President, American Society of Agronomy
February 2019

Preface / Dedication

Worldwide, sorghum is among the top-five cereal grains in both production and acreage, and a key crop for food security. Sorghum cultivars/hybrids are grown for a range of uses: grain sorghum is a staple food, forage sorghum can be grazed or used for silage, and some are used to produce syrup and for bio-energy. Sorghum is gluten free, and some special grains with high antioxidants and nutritive value are used as a health food. Sorghum is a very versatile crop at scales from subsistence farming to commercial agriculture.

Food and nutritional security goals vary among countries based on their economic development and available resources. In developing countries, the challenge is not only to improve productivity but also to ensure adequate distribution, availability and accessibility nutritious food. Sorghum is a key crop under high-stress environments, ensuring productivity and resilience and access to food when other crops fail.

A critical update on the most recent information on sorghum related to advances in the use of technology, crop modeling, discovery of new traits was our main driver behind leading this project. The last publication that presented a close-to-complete summary for sorghum was published in 2000 (Sorghum: Origin, History, Technology, and Production). Therefore, the concepts and topics presented in this book will provide a needed update and also include new relevant topics for all the sorghum community.

This book is dedicated to all the relevant players working for moving sorghum forward from breeders, biologists, agronomists, farmers all the way to the industry, stakeholders, and consumers.

Lastly, a special dedication to Dr. Richard L. Vanderlip, Emeritus Professor in Crop Production, Department of Agronomy at Kansas State University for his tireless efforts and contributions to sorghum community through his scholarly activities on understanding sorghum growth and development, modeling, agronomy, crop responses to multiple environment and training students and scholars around the world.

Dr. Ignacio A. Ciampitti
Associate Professor, Crop Production and Integrated Farming Systems
Department of Agronomy at Kansas State University.

Dr. P.V. Vara Prasad
Director, Sustainable Intensification Innovation Lab; and
University Distinguished Professor, Crop Ecophysiology and Systems Research
Department of Agronomy at Kansas State University.

Genetic Changes in Sorghum

Ramasamy Perumal,* Passoupathy Rajendrakumar, Frank Maulana, Tesfaye Tesso, and Christopher R. Little

Abstract

Evolution through genetic changes in all crop species is an inevitable process. The available naturally diverse resources in sorghum provide ample material for further genetic improvement using classical and modern breeding tools. The current research priority in sorghum is gene discovery for complex traits related to biotic (pest and diseases) and abiotic (cold, drought, and heat) stresses by integrating high-throughput genotyping (genotyping by sequencing, GBS) and phenotyping facilities with classical breeding methods and advanced computational and statistical tools and marker-assisted breeding. Exploitation of genetic resources through landraces, interspecific and intergeneric crosses, sorghum conversion, and heterosis prediction is discussed in detail in this chapter. The discussion also includes the current status of molecular approaches, including quantitative trait loci (QTL) analysis, mapping studies, marker assisted-selection (MAS), mutagenesis, and epigenetics—for desirable genetic changes and overall sorghum improvement.

Sorghum (*Sorghum bicolor* (L.) Moench) belongs to a member of the tribe Andropogoneae, which is within the grass family, Poaceae, with the other crop species maize (*Zea mays* L.) and sugarcane (*Saccharum* spp.; Clayton and Renvoize, 1986). Naturally established sorghum has a broad genetic base, with a total of 25 species, including five subgenera or sections (*Eu-Sorghum, Chaetosorghum, Heterosorghum, Para-Sorghum,* and *Stiposorghum*). All cultivated (*Sorghum bicolor* subsp. *bicolor*) and wild species (*S. halapense* [Johnsongrass], *S. propinquam, S. xalmum, S. drummondii,* and *S. arundinaceum,* the progenitor of *S. bicolor*) are classified under *Eu-Sorghum* (Harlan and de Wet, 1972; Doggett, 1988). *S. bicolor* ($2n = 2x = 20$), with a genome size of 760 MB, is an annual crop that grows up to 5 m tall and has

---

Abbreviations: CMS, cytoplasmic-nuclear male sterility; EMS, ethyl methanesulfonate; FISH, Fluorescent in situ hybridization; GA, gibberellin; GBS, genotyping by sequencing; GCA, general combining ability; GEBV, genomic estimated breeding value; GWAS, genome-wide association mapping; MAS, marker-assisted selection; NAM, nested association mapping; QTL, quantitative trait loci; rrBLUP, ridge regression best linear unbiased prediction; SCA, specific combining ability; SNP, single-nucleotide polymorphism; SSR, simple sequence repeat; TDMR, tissue-specific differentially methylated region; TILLING, targeting induced local lesions in genomes.

R. Perumal, Agricultural Research Center, Kansas State Univ., Hays, KS 67601; P. Rajendrakumar, ICAR-Indian Institute of Millets Research (formerly the Directorate of Sorghum Research), Rajendranagar, Hyderabad 500030, Telangana, India (rajendra@millets.res.in); F. Maulana, Noble Foundation, 1614N Cedar Loop, Ardmore, OK 73401 (fmaulana@noble.org); T. Tesso, Dep. of Agronomy, Kansas State Univ., Manhattan, KS 66506 (ttesso@ksu.edu); C.R. Little, Dep. of Plant Pathology, Kansas State Univ., Manhattan, KS 66506 (crlittle@ksu.edu). *Corresponding author (perumal@ksu.edu).

doi:10.2134/agronmonogr58.2014.0053

 Sorghum: State of the Art and Future Perspectives, Ignacio Ciampitti and Vara Prasad, editors. Agronomy Monograph 58.

many branching tillers. Information on the genetic diversity and characteristics of cultivars included in world collections is very important for the characterization, management, use, and future collection of exotic germplasm (Perumal et al., 2007). With the naturally available genetic (grain, fodder, sweet, and biofuel types) and geographical diversity from 116 countries, which is accessible as a total of 44,773 germplasms in the US Germplasm Resource Information Network (GRIN), sorghum is considered a unique crop with such diverse uses as food, feed, fodder, fuel, and fiber, is grown globally, and is amenable to genetic change via classical and molecular breeding approaches.

Other well-established sorghum genetic resources that maintain both the accessions and documented information include the online European Search Catalog for Plant Genetic Resources (EURISCO); the plant genetic resource gateway, Genesys; the International Crop Information System (ICIS); the Chinese Crop Germplasm Information System (CGRIS); and the International Crops Research Institute for the Semi-arid Tropics (ICRISAT). These resources provide ample scope for sorghum researchers and farmers interested in overall improvement of the crop with desirable genetic changes.

Genetic Resources and Diversity

Originating as a short-day and photosensitive species in Sudan and Ethiopia, the photoperiod-insensitive sorghum of today evolved by the manipulation of multiple maturity genes through concerted breeding efforts. The diversity of cultivated sorghum in the form of new sorghum types, races, and varieties was created through evolutionary forces such as disruptive selection, geographical isolation, and genetic recombination in the various environments of Northeast Africa and its widespread diversity and is the result of people migrating across the continent (Doggett, 1970, xviii).

Annual types of *S. bicolor* subsp. *bicolor* have thick culms and are classified on the basis of spikelet and panicle morphology alone (Dahlberg, 2000) into five cultivated races—*bicolor, guinea, caudatum, kafir,* and *durra*—and 10 intermediate races (Harlan and de Wet, 1972. The most primitive and heterogeneous race, *bicolor,* has the characteristic open panicles and is distributed globally wherever sorghum is grown. The race *guinea* evolved from *bicolor,* has the same open-panicle type, and originated in West Africa, but it also occurs in East Africa and India (de Wet, 1978; Smith, 1995). The high-yielding semicompact *kafir* race came from southern Africa as a result of population migration. A compact-panicle type, *durra,* a specialized race with a medium-to-large grain (de Wet and Huckabay, 1967), was originally from Ethiopia and Sudan but later spread to maximum acreage in India. *Caudatum* is the most recent and important race in modern agriculture owing to its high-yielding, medium-to-large panicle. The limited distribution of this race is due to the tannin content of the seeds, which have a bitter-flour taste (Stemler et al., 1975). The abovementioned gene pools, along with closely related genera such as sugarcane and maize, are the vital source of genes for economically important traits such as grain and biomass yield and other component traits related to biotic and abiotic stress tolerance, adaptability, and quality. These sources help in the genetic changes and improvement of sorghum through well-strategized breeding and selection programs. The contribution of landraces, cultivated and wild races, sorghum conversion, and interspecific and intergeneric hybridization to the genetic enhancement of sorghum is discussed below.

Landraces

Along with the efforts to genetically improve sorghum via improved breeding lines to enhance sorghum productivity, there is also an immense need to breed for wider adaptation, along with better grain and fodder quality. In this endeavor, sorghum landraces and locally adapted farmers' cultivars, which are reservoirs of valuable genes associated with adaptation and quality, offer an excellent opportunity for the development of superior cultivars having wider adaptability and acceptable quality. The landraces were developed as local farmers exercised selection from the plant population across generations, thereby adapting the genotype to the specific environment in which it grows. Selections were made, especially for yield in a particular environment, tolerance to pests and diseases, superior grain quality, and specific end uses.

Due to its origin in Northeast Africa, many landraces are photoperiod sensitive, having a critical photoperiod of 12 h, which makes growing these landraces difficult in the temperate regions of America and Australia as a summer crop, due to longer day lengths (>13 h) resulting in poor expression of growth-related characters (Reddy et al., 2006). To overcome this hindrance, landraces that normally flower even in day lengths up to 17 h were identified in India and were widely used for the genetic improvement of sorghum worldwide (Rai et al., 1999; Reddy et al., 2006). Landraces are not suitable for cultivation under modern farming methods since they easily lodge due to their tall stature. High-yielding superior cultivars with a shorter plant heights were developed by combining carefully selected high-yielding, tall sorghum landraces and short, photoperiod-insensitive sorghum lines in the crossing programs (Miller, 1980; Rosenow and Dahlberg, 2000).

Before 1949, Chinese farmers were cultivating only local varieties (landraces). However, in 1951, an intensive selection program launched by local farmers resulted in the development and release of some cultivars adapted to local conditions, leading to an increased grain yield of about 10%. With the introduction of 'Combine Kafir 60A' from the USA in 1956, the development of sorghum hybrids became an important priority in China and has led to quantum jumps in grain yield. The vital step in the development of hybrids is the selection of restorer/pollinator lines (R-lines) as male parents, and Chinese sorghum landraces have been used specifically for this purpose. Due to the inherent problems of Chinese sorghum in fertility restoration, the R-lines were developed by hybridization and selection of crosses of non-Chinese sorghum with Chinese sorghum, as well as the composite crosses. As a result, most of the hybrids developed earlier used Chinese sorghum as the R-lines, whereas the hybrids developed later used R-lines that only partially contained the genetic base of Chinese sorghum. Thus, Chinese sorghum forms the major genetic basis of newly developed R-lines (Li and Li, 1998).

Post-rainy-season sorghum, a primary source of food and fodder in India, is grown under residual soil moisture conditions, which is unique to India. The sorghum is photoperiod sensitive, tolerant to terminal moisture stress, and resistant to charcoal rot; usually produces high biomass; and is characterized by lustrous grain having a semicorneous endosperm. All these characters are exemplified best in M 35-1, a selection from a local landrace developed in 1937 at Mohol, Maharashtra, India, that is still grown predominantly during the post-rainy season due to the abovementioned traits as well as its high stable grain and stover yields. Due to a narrow genetic base, exploitation of heterosis in post-rainy-season sorghum was very limited and not successful. However, the superiority of improved-line × landrace crosses over improved-line × improved-line crosses for heterosis due to parental-line diversity was reported by Shinde et al. (1986). Similarly, the hybrids between shoot fly (*Atherigona saccata* Rondani)–resistant A lines (seed parent) and landrace pollinators possess traits such as resistance to shoot fly and terminal drought and pearly white, lustrous, bold grains suited to post-rainy-season adaptation (Reddy et al., 2006), thus highlighting the possible role of landrace-based hybrids in the genetic improvement of post-rainy-season sorghum.

The moisture status, along with the types of soil in which the sorghum is cultivated, influences the productivity of post-rainy-season sorghum in India. Because breeding for wider adaptation was not considered feasible due to variable environmental conditions, development of varieties adapted to specific soil types was suggested (Jirali et al., 2007), and germplasm lines suitable for such conditions were identified (Pawar et al., 2005). Evaluation of 307 landraces collected from the Satpura ranges and the Tapi, Purna, and Godavari river basins, along with 1175 local genotypes, revealed wide variability among the germplasm accessions collected from a district, and they resembled each other in few traits (Patil et al., 2014). Developed by pure line or pedigree breeding in the crosses involving the RSLG collections, the variety Phule Yashoda was released as CSV216R at the national level. The varieties Phule Maulee (shallow to medium soils), Phule Chitra and Phule Suchitra (medium soils), Phule Vasudha (deep soils), Phule Anuradha (shallow soils), and Phule Revati (medium to deep soils) were released between 1999 and 2012 and have been performing well in specific soil situations (Prabhakar et al., 2014).

An attempt was made by Umakanth et al. (2006) to identify the high-yielding cross combinations involving landraces as restorers. Evaluation of 20 hybrids derived from two lines and 10 testers, along with the parents and the checks (CSH 19R and M 35-1), of post-rainy-season sorghum resulted in the identification of 117A × PU 28 as the highest yielder with significant heterosis over the checks as well as the better parent for grain yield and test weight. The hybrid 104A × PU 11 exhibited the highest test weight compared with the checks. The study concluded that the landraces PU 28 and PU 11 are the best restorers for the exploitation of heterosis in future breeding programs.

Sorghum Conversion

Of the available world sorghum germplasm collections, 36,000 are photoperiod sensitive, owing to sorghum's tropical origin. These short-day (a day length of <12 h 20 min is needed to induce flowering) sorghum accessions are late flowering and grow too tall in temperate zones. A sorghum conversion program was hence initiated at the Texas Agricultural Experiment Station (TAES)-USDA in 1963 to introgress day-neutral flowering alleles and dwarf-height genes into exotic photosensitive germplasm backgrounds following classical backcross breeding (Stephens et al., 1967). This scheme was designed to move dwarfing and maturity genes from BTx406, which was used as a seed parent, and it developed a total of 840 sorghum converted lines from diversified tropical accessions to temperately adapted lines (Rosenow et al., 1997). These lines were extensively used by several public and private breeding programs, which developed many outstanding elite parental lines and hybrids around the world and significantly impacted overall sorghum improvement. However, this time-consuming, labor-intensive breeding process with limited resources limits further conversion to introgress numerous key genes tolerant/resistant to biotic and abiotic stresses residing in exotic sources into adapted lines.

In 2009, a renewed public effort was made by the USDA-ARS and NuSeed/MMR Genetics with funding from the National Sorghum Foundation and the United Sorghum Check-Off Program. Accessions from the Sudanese and Ethiopian collections were targeted for conversion. A total of 44 sorghum working groups were represented and included zerazera, caudatum, conspicuum, durra, nandyal, nigricans, guineense, caudatum-kafir, caudatum-guineense, caudatum-nigricans, caudatum-durra, durra-kafir, durra-bicolor, caffrorumbicolor, and durra-dochna, which were used in the partial conversion program. Diversity arrays technology (DarT) markers were integrated into selected $BC_1 F_2$ backcross-derived lines and 220 early-maturing, dwarf-height $BC_1 F_3$ families were further released in 2014 (Klein et al., 2016). These newly developed, partially converted lines helped to minimize several backcross breeding cycles and accelerated the breeding programs. These new germplasms, which were obtained from the USDA-ARS sorghum collection, are currently being used as new genetic resources in temperate zones worldwide. Many new photosensitive, exotic germplasms with different tolerance and resistance backgrounds were recently included in the partial conversion program: IS29091, IS28141, IS29100, and IS15170 for cold tolerance (Kapanigowda et al., 2013a); IS28451 and IS24348 for stalk-rot tolerance (Kapanigowda et al., 2013b); IS7305, IS9745, IS13549, IS16528, IS20632, IS29239, IS30572, and IS31714 for multiple disease resistance to anthracnose, head smut, and downy mildew (Perumal, unpublished data, 2013); and IS7679 and IS20740 for *Potyvirus* spp. (Seifers et al., 2012).

Interspecific Crosses

Most of the wild species of sorghum belong to the tertiary gene pool, and the transfer of genes to the domesticated species becomes difficult due to the presence of sterility barriers (Harlan and de Wet, 1971), predominantly pollen-pistil incompatibilities and the failure to form hybrid embryos (Hodnett et al., 2005). However, in a rare occurrence, a single embryo was formed by the pollination of *S. bicolor* with the pollen of *S. macrospermum*. The embryo was rescued and developed through tissue culture, and the hybridity was confirmed on the basis of cytological and morphological characteristics (Price et al., 2005). When pollinated by alien species, cessation in the growth of prezygotic pollen tubes in the pistils of sorghum is considered to be the primary impediment to interspecific and intergeneric hybridization in sorghum (Hodnett et al., 2005).

As a result of the concerted efforts of the researchers to increase the frequency of hybridization involving wild species for the genetic improvement of cultivated sorghum, an accession was identified that belonged to *S. bicolor* and possessed a recessive gene, *iap* (inhibition of alien pollen), that allowed maize pollen tubes to grow normally through the pistils of *S. bicolor* (Laurie and Bennett, 1989). Using this accession, hybrids between *S. bicolor* ($2n$ = 20; AAB1B1) and *S. macrospermum* ($2n$ = 40; WWXXYYZZ) were obtained from seeds, while those of *S. bicolor*

× *S. angustum* and *S. bicolor* × *S. nitidum* were recovered through embryo rescue and culture (Price et al., 2006). Fluorescent in situ hybridization (FISH) was used to track the introgression of the *S. macrospermum* genome into the cultivated *S. bicolor* (Kuhlman et al., 2006), and because FISH could discriminate between the chromosomes of the parental species, it confirmed the occurrence of recombination through the bivalent formation and allosyndetic pairing. The introgression lines exhibited altered fertility, again a confirmation of the presence of the *S. macrospermum* genome in the backcross progenies. Genomic relationships were adequate to assign AAB_1B_1YYZZ as the genomic constitution for *S. macrospermum,* Y and Z being unknown genomes. The occurrence of allosyndetic recombination indicates that introgression should be possible due to the recovery of viable backcrosses (Kuhlman et al., 2008). The discovery and utilization of the *iap* allele is the first step in the use of unexploited genes in the tertiary gene pool to bring about genetic improvement in yield as well as tolerance to biotic and abiotic stresses resulting from the successful introgression of genes from wild species into cultivated sorghum.

Striga, the most noxious parasitic weed, and its important species, *S. hermonthica* and *S. asiatica,* have a significant negative impact on sorghum production in Africa. Since complete resistance to *Striga* infection does not exist in cultivated sorghum, wild relatives may be a possible reservoir of genes associated with the resistance or tolerance to *Striga* infection. *S. arundinaceum,* a wild relative of sorghum, exhibited tolerance to *S. asiatica* infection, which had a negative effect on host growth and grain production in cultivated sorghum. However, *S. hermonthica* infection had a positive impact on host performance of wild, as well as cultivated, sorghum. Such differences in host-parasite interactions may be due to the timing of parasite attachment to the host and host specificity for these *Striga* species (Gurney et al., 2002).

A strong selection bias toward sorghum midge resistance, stay-green, and other agronomic traits in the sorghum breeding populations in Australia resulted in a decline in genetic diversity, which is the result of selection, genetic drift, and linkage drag (Jordan et al. 1998, 2003). Because of their richness in allelic diversity, the wild relatives of sorghum hold promise for enhancing yield and reducing genetic vulnerability. Since diversification through conventional pedigree breeding may not be successful due to the disruption of favorable linkage blocks, a limited backcrossing approach is a better option to produce useful populations for further selection. Following this method, Jordan et al. (2004) developed backcross (BC_1F_4) progeny comprising 255 lines derived from the cross *S. arundinaceum* (wild species) × 31945-2-2 (elite line), which were further topcrossed to an elite seed parent. Evaluation of the resulting F_1 hybrids in field trials for grain yield led to the identification of a significant number of progeny with grain yield on a par with or greater than that of the recurrent parent. Analysis of this population using molecular markers resulted in the identification of seven genomic regions that are associated with grain yield; two of these regions were derived the favorable alleles from *S. arundinaceum.*

S. propinquum, a wild species of sorghum, is characterized by very large, loose, open panicles and is found especially in southern China: through Thailand, Cambodia, Malaya, and Burma: to the Philippines (Burkill et al., 1966). It differs from *S. bicolor* in about 1.2% of the nucleotides in the coding regions, suggesting a divergence of 1–2 million years between them (Feltus et al., 2004). In a study involving an F_5 recombinant inbred population of *S. bicolor* × *S. propinquum*, about 49 QTL were found to be associated with the enzymatic conversion efficiency of lignocellulosic biomass, of which, 20 QTL were related to the leaf, and 29 to the stem. In addition to these QTL, two QTL associated with biomass crystallinity index (a trait inversely correlated with conversion efficiency) were also identified. The identification of these QTL provides an important lead for identifying specific genes involved in increasing the conversion efficiency of bioenergy feedstocks. Marker-assisted selection using the DNA markers associated with these QTL will be useful for developing high-biomass genotypes and simultaneously achieving increased bioenergy yields (Vandenbrink et al., 2013).

Genetic mapping and QTL analyses were conducted with 248 $F_{2:3}$ plants of the cross 314A (sorghum) × 2002GZ-1 (Sudan grass). The parents exhibited polymorphism for most of the agronomic traits studied. A total of 178 markers (170 AFLP and 8 RAPD) and a mapping population of 248 $F_{2:3}$ plants was used to construct a linkage map, and a total of 98 QTL were identified in two test locations; of these, 26 were detected in both locations. The uneven distribution of these QTL across linkage groups and marker densities disproportional to QTL frequencies indicated the correlation of these QTL with the agronomic traits and the usefulness of the genetic map for the improvement of relevant characters in *S. bicolor* × *S. sudanense* hybrids (Lu et al., 2011).

Intergeneric Crosses

Across the years, the focus of genetic improvement in many crops was based on the population derived from intraspecific hybridization (Kuhlman et al., 2008; Hodnett et al., 2010). Unavailability of donors, especially for tolerance to biotic and abiotic stresses in the primary gene pool, has forced crop breeders to seek such valuable genes in closely related species or genera through interspecific and even intergeneric hybridization, but they have had limited success. Among the members of Poaceae family, sugarcane and maize are considered to be more closely related to sorghum than are the other major cereals. The divergence of sorghum and maize from a common ancestor dates back to 12 million years ago (Gaut et al., 1997; Swigonova et al., 2004), and sorghum and sugarcane may have diverged around 5 million years ago (Sobral et al., 1994). Moreover, sorghum and sugarcane have a similar gene order (Ming et al., 1998) and even produce viable progeny in some cases when crossed (de Wet, 1978). Intergeneric hybridization of sorghum with sugarcane and maize may provide opportunities for trait introgression between genera, leading to genetic improvement in biomass and quality as well as in biotic and abiotic stress tolerance. Even though hybridization between sorghum and sugarcane was reported earlier, the hybrid production frequency was minimal (Gupta et al., 1978), and the hybrids were of little breeding value due to the lack of vegetative vigor and susceptibility to natural incidences of mites (Nair, 1999).

The presence of the *iap* allele in sorghum enhances the success of interspecific and intergeneric hybrids and their progenies. When the sorghum genotype was used with the *iap* allele (Tx3361) as the seed parent, about 14,141 seeds were produced from 252 sorghum × *Saccharum* spp. crosses, which exhibited variation in seed-set frequency, indicating the scope for further success by breeding and selection of sugarcane pollen parents. About 1371 *Sorghum–Saccharum* intergeneric hybrids were successfully developed, which were then advanced for selection of their progenies to develop introgression lines with the potential to be used for sugar or as a biomass feedstock (Hodnett et al., 2010). Variation was observed in the expressivity of the *iap* phenotype when maize pollen was dusted on the stigmas of Tx3361 (*iap iap*) and ATx623 (*Iap Iap*). Maximum adhesion of maize pollen and its germination on the stigmas of Tx3361 was noticed at lower humidity (45%), indicating the role of humidity in pollen adhesion on the stigma and its control of pollen-pistil incompatibility in intergeneric crosses in sorghum (Gill et al., 2014).

Analysis through semiquantitative reverse transcription polymerase chain reaction of the expression of sucrose-related genes, specifically those for sucrose phosphate synthase (*SPS*), sucrose synthase (*SuSy*), and soluble acid invertase (*SAI*), in a *Sorghum* × *Saccharum* hybrid, a *Saccharum* × *Sorghum* hybrid, high- and low-sucrose sugarcane varieties, and sweet and grain sorghum lines revealed higher expression of the *SPS* and *SuSy* genes in high-sucrose varieties, *Saccharum* × *Sorghum* hybrids, and sweet sorghum and lower expression in low-sucrose varieties, *Sorghum* × *Saccharum* hybrids, and grain sorghum. In contrast, the expression of *SAI* was lower in high-sucrose varieties, *Saccharum* × *Sorghum* hybrids, and sweet sorghum, and higher in low-sucrose varieties, *Sorghum* × *Saccharum* hybrids, and grain sorghum. These differences open the possibility for the production of novel hybrids with improved sucrose content and with early maturity (Ramalashmi et al., 2014).

Application and Use of Diverse Sorghum Germplasm

Prediction of Heterosis by Classical Breeding

Hybrid technology is the most significant intervention in agricultural and horticultural crops that has brought about substantial improvements in yield, including for sorghum. Before the 1960s, considerable importance was given to the development of varieties, although the phenomenon of heterosis was demonstrated in sorghum as early as 1927. However, with the discovery of stable cytoplasmic-nuclear male sterility (CMS), hybrids were developed and released for commercial cultivation. A high level of heterosis for grain yield was reported in sorghum and resulted in nearly double the yield for hybrids than for varieties under diverse growing and management conditions (Quinby et al., 1958; Doggett, 1988). Although breeders continue to exploit heterosis to develop superior, high-yielding hybrids in many crop species, there are many outstanding questions about the genetic and molecular bases of heterosis, the foremost being which genes are involved.

One of the prerequisites for producing high-yielding heterotic hybrids is the availability of genetically diverse parental lines. Breeders have to make hundreds of testcrosses and evaluate the F_1 hybrids in different seasons and environments to identify a superior heterotic hybrid. This process is time consuming and resource intensive. Moreover, the results are subjective. The efficiency of hybrid breeding can be increased by developing a simple, rapid, reliable methodology that will help in the identification of superior parental combinations from a large pool of parental lines without field evaluation. Across the years, phenotype-based methods such as per se performance of parents, combining ability, and genetic diversity were used for the prediction of heterosis in many crops. However, with the availability of a wide range of molecular markers, marker-assisted prediction of heterosis based on the molecular diversity among the parental lines has become popular during the last two decades.

Phenotype-Based Prediction

Before the availability of molecular markers, plant breeders mostly relied on the per se performance of parental lines, combining ability, and genetic diversity estimates based on the phenotypic data collected through field evaluation. Per se performance of the parents along with their combining ability can be considered vital criteria in the selection of parents for hybridization in sorghum (Harer and Bapat, 1982). An earlier study revealed a close association of per se performance of hybrids with the heterosis for days to 50% flowering, plant height, leaf area index, brix, panicle length, number of grains per panicle, 100-grain weight, and grain yield per plant, suggesting that per se performance can be considered an important criterion in the selection of the crosses (Premalatha et al., 2006).

Selection of heterotic parents that are good general and specific combiners is necessary for substantial yield improvement in any crop, including sorghum. Usually, plant breeders take only general combining ability (GCA), which accounts for additive genetic variance, into consideration that and ignore the specific combining ability (SCA), or the nonadditive genetic variance. This preference will affect the correlation between combining ability and heterosis, so both GCA and SCA should be considered in the selection of the parental lines. In sweet sorghum, Sandeep et al. (2010) revealed that crosses with higher SCA and heterotic potential can be produced from parents with contrasting GCA effects for bioenergy traits, thus highlighting the predictive power of combining ability.

Plant breeders usually predict heterosis using the extent of genetic diversity between the two parental lines (Zhang et al., 1994; Falconer and Mackay, 1996), and conventionally, the genetic distance between the parental lines has been estimated through multivariate analysis of field evaluation data. But a strong correlation between heterosis and genetic distance between parents has been lacking (Melchinger, 1999; Dixit and Swain, 2000; Singh and Singh, 2004). The absence of perfect correspondence between the level of heterosis and genetic diversity between parents was reported in sorghum by Rani and Rao (2009), who indicated the inability of phenotype-based genetic diversity for the prediction of heterosis. This conclusion has also been supported in other crops by Dave and Joshi, 1995; Singh and Singh, 2004; Shukla and Singh, 2006).

Due to the inconsistency of phenotype-based methods to predict heterosis, there is an immense need to explore the possibility of prediction based on molecular diversity, which is environmentally neutral. Moreover, such an approach will exclude field evaluation of a large number of testcrosses and test only those crosses that are predicted to be heterotic, thereby accelerating the development of superior hybrids.

Marker-Assisted Prediction

Molecular diversity estimates based on DNA marker analysis of the parental gene pool have proved to be useful in the identification of heterotic groups and in assigning inbreds of unknown origin to the established heterotic groups due to the availability of inbreds in large numbers, phenotypic neutrality, and the absence of environmental influences (Hongtrakul et al., 1997; Pejic et al., 1998; Casa et al., 2002). Even though heterotic groups are not as clearly defined in sorghum as in maize, it was shown that marker-assisted prediction of heterosis is improved by using particular linkage groups in models while associating genetic distance with hybrid performance (Jordan et al., 2003). Molecular marker-based prediction of heterosis gained prominence in rice during the 1990s but resulted in inconsistent conclusions (Lee et al., 1989; Godshalk et al., 1990; Smith et al., 1990; Dudley et al., 1991; Peng et al., 1991; Xiao et al., 1996; Liu and Wu, 1998). The main reason for such inconsistency is that heterosis was predicted on the basis of molecular diversity estimated by the use of anonymous DNA markers that may not have been attributed to any gene function. This observation helped shift the focus from the anonymous markers to function-related markers, such as expressed sequence tag–simple sequence repeat (EST-SSR) markers and QTL-linked markers.

In addition to the type of markers, the nature and number of parental lines and the number and distribution of DNA markers may also influence the type of the correlation between molecular diversity among parental lines and heterosis. Even though a weak correlation between DNA marker-based genetic diversity and heterosis was reported in sorghum by means of restriction fragment length polymorphism markers (Jordan et al., 2003), significant correlations have been reported in other crops, including rapeseed (Diers et al., 1996), maize (Smith et al., 1990), sunflower (Cheres et al., 2000), wheat (Smith and Corbellini et al., 2002), and rice (Liu and Wu, 1998; Zhang et al., 2010). Those cases show that the selection of appropriate and effective DNA markers could improve the correlation in sorghum. This possibility is strengthened by the recent identification in sorghum by Ben-Israel et al. (2012) of three heterotic trait loci that exhibit synergistic intralocus effects on grain yield heterosis that is over-dominant in nature. More recently, Rajendrakumar et al. (2013) identified 30 SSR markers that might predict grain yield heterosis in sorghum and that on validation in a new set of experimental hybrids (210) and their parental lines revealed a significant positive correlation of marker polymorphism among parental lines and mid-parent heterosis ($r = 0.48$) as well as better parent heterosis ($r = 0.65$) for grain yield. Phenotypic values of eight traits were used for the selection of yield-related QTL by Lu et al. (2014), and these loci were used for analyzing the correlation between the DNA marker value and grain-yield heterosis. The prediction models for the traits of the hybrids were constructed by stepwise regression, which indicated an average correlation value ($r = 0.65$) between marker value and heterosis while considering dominance and additive effects separately. Such information on the effective loci associated with grain-yield heterosis in sorghum will help molecular breeders exploit them for reliable prediction of heterosis and thereby make them useful in MAS of parental combinations for the development of hybrids. This method could replace the selection of parents through conventional testcrossing.

In recent years, various research groups, using the enormous amount of data generated from the transcriptomic and metabolomic studies in various crops, have attempted to explore novel approaches involving mathematical modeling to improve the prediction of heterosis. Reasonable success was demonstrated in *Arabidopsis*, resulting in the identification of genes related to heterosis, and the allelic combinations of parental inbreds can be used for the prediction of heterosis (Stokes et al., 2006). Significant improvement in the prediction of biomass heterosis was achieved with a combination of genetic markers and the metabolic markers identified via feature selection (Gärtner et al., 2009). The approach also involved metabolite profiles, single-nucleotide polymorphism (SNP) markers, and feature selection (Steinfath et al., 2010). With advances in transcriptomics, all of the transcriptome data can be generated from the parental lines of popular sorghum hybrids to develop effective molecular markers useful for the prediction of heterosis. Such an approach was successfully demonstrated in mice, leading to the prediction of heterosis, even with a limited number of hybrids, as well as performance data (Stokes et al., 2010). Recently, a collection of genes and transcripts related to heterosis and identified from different transcriptomic studies in rice, wheat, and maize were organized in the form of the Heterosis Related Genes Database (HRGD) by Song et al. (2009). The genes and transcripts that show more than 90% homology with sorghum can be targeted for the development of polymerase chain reaction–based markers, which can be employed in the prediction of heterosis.

Quantitative trait loci markers associated with complex traits having significant and large effects are used to build a prediction model. In contrast, in genomic prediction all markers are considered when building a prediction model without a significance test. From the model, the genomic estimated breeding values (GEBVs) are calculated based on the sum of all marker effects (Meuwissen et al., 2001). First, a training population consisting of genotypes that have been both genotyped and phenotyped is used to develop a prediction model that takes genotypic data from a candidate population of untested genotypes to produce GEBVs. To estimate the accuracy of the prediction model, the true breeding value is correlated with the GEBVs using training and validation sets. Stories about the successful use of genomic prediction models in animal breeding (Goddard et al., 2010; Habier, 2010) stimulated the plant breeding community to begin using genomic prediction models to estimate hybrid performance (Lorenzana and Bernardo, 2009; Piepho, 2009). Also, recent advances in next-generation technologies, such as GBS, which can generate a huge amount of genotypic data covering the whole genome at a lower price has further prompted the scientific community to undertake genomic prediction studies to improve selection efficiency. Several prediction models are used in these studies, including the ridge regression best linear unbiased prediction (*rr*BLUP) model (Endelman, 2011).The *rr*BLUP predicts the genotypic value of untested hybrids using effects estimated for each marker (Whittaker et al., 2000; Piepho, 2009).

In maize, simulation studies have shown that grain or biomass yield can be predicted with high accuracy (Albrecht et al., 2011; Zhao et al., 2012). Consequently, such studies can increase the rate of genetic gain since the prediction accuracy of GEBVs is linearly related to the response to selection. Genomic prediction accuracies can be affected by the population size of the training set, trait heritability, pedigree information of the genotypes and linkage disequilibrium between markers with QTL associated with a trait, relatedness of genotypes forming the training and test populations, and the prediction model used. Genotypes from different populations tend to have lower prediction accuracies than those from the same population. For example, a reduction in prediction accuracies as high as 93% was reported for hybrids in training and validation populations that had no common parent shared among them (Gowda et al., 2013). Albrecht et al. (2011) also reported a decrease in prediction accuracies for genotypes in the training population that were not related to those in the test population. Similarly, when trait heritability is low, prediction accuracies tend to be low and vice-versa.

Moreover, different prediction models show variable prediction accuracies for complex traits. For example, *rr*BLUP has been shown to provide more stable and higher prediction accuracies for complex traits (Iwata and Jannink, 2011; Heslot et al., 2012) than other models. This model is very robust and has a low computational load, hence it is well suited for genomic prediction of hybrid performance for complex traits in breeding programs (Piepho, 2009). If successful, these prediction methods will have a significant impact on hybrid breeding by identifying the heterotic groups among the parental gene pool and by reducing the number of testcrosses to be evaluated, apart from saving land, water, input, and labor resources.

Molecular-Marker Breeding

Marker-Assisted QTL Introgression

Selection of effective QTL that contribute at least 20–25% of phenotypic variance for traits of interest, along with the DNA markers associated with them, will rapidly and precisely help improve elite varieties, cultivars, and parental lines of hybrids through marker-assisted QTL introgression. This approach is very significant because quantitative traits, which are not generally targeted for improvement through recurrent backcrossing, can be improved by identifying and introgressing major-effect QTL, or a few QTL with small but additive effects, with a marker-assisted recurrent backcrossing approach, which ensures a speedy recovery of a recurrent genome and also eliminates linkage drag.

Quantitative trait loci associated with drought tolerance (stay-green) and *Striga* resistance are predominantly used for introgression into different farmer-preferred varieties of sorghum in sub-Saharan Africa. This approach has resulted in the successful release of four *Striga*-resistant lines. One or several QTL associated with *Striga* resistance were introgressed from the resistant sorghum line, N13. A *Striga* resistance QTL was introgressed to a farmer-preferred variety in Kenya, Ochuti, in which the backcross derivatives showed lower *Striga* scores than the susceptible variety (Osama, 2013). Similarly, introgression of five QTL associated with *Striga* resistance into the elite farmer-preferred cultivar Tabat was attempted by Gamar and Mohamed (2013), which resulted in the identification of the backcross-derived progenies BC_1S_1-67-H, BC_1S_1-110-H, and BC_1S_1-16-H, all of which possess two QTL from group 5 and exhibit high *Striga* resistance levels, along with normal flowering and superior agronomic performance.

Five QTL associated with *Striga* resistance were introgressed into farmer-preferred sorghum cultivars from Eritrea, Kenya, Mali, and Sudan to validate the marker-assisted backcrossing approach. Field testing selected BC_2 lines possessing zero to five QTL and their parental lines in multiple locations in Mali and Sudan showed that the resistance levels of the best-performing introgression lines were on a par with those of the donor parent, leading to the identification of lines with both high resistance and high grain yield. Evaluation of these introgression lines under the farmer-participatory approach revealed a strong preference for *Striga*-resistant breeding lines that are high yielding, thus indicating the usefulness of marker-assisted introgression of *Striga* resistance into adapted and farmer-preferred sorghum varieties (Muth et al., 2012).

The first products of MAS in sorghum were released for cultivation in sub-Saharan Africa through the development of backcross populations (BC_3S_4) derived by crossing N13 (*Striga* resistant) with three farmer-preferred sorghum cultivars: Tabat, Wad Ahmed, and AG-8 (*Striga* susceptible) by Mohamed et al. (2014). They transferred five QTL identified by Haussmann et al. (2004) using marker-assisted QTL introgression. Twenty resistant lines with two or more major QTL were identified after the marker-assisted foreground and background selection of 31 BC_3S_4 lines with confirmed field resistance to *Striga*. Ten lines were selected and advanced for multilocation testing after the regional evaluation of 20 resistant lines, resulting in the identification of four *Striga*-resistant lines ($T1BC_3S_4$, $AG6BC_3S_4$, $AG2BC_3S_4$, and $W2BC_3S_4$), which were agronomically superior and higher yielding (180–298%) than their recurrent parents.

Stay-green, a well characterized trait associated with post-flowering drought tolerance in sorghum, is characterized by the plant's ability to maintain functional photosynthetic leaf area during the grain-filling stage even under severe post-flowering drought stress. Initial evaluation of partial introgression (BC_3F_3/BC_1F_4 generations) of four stable stay-green QTL (*StgB, Stg1, Stg3,* and *Stg4*) from the donor parent (B 35) into a senescent variety (R 16) revealed higher leaf chlorophyll levels at flowering and a greater leaf area percentage during the later part of grain filling in the majority of introgression lines than in R 16 even though none of the QTL introgression lines exhibited the same level of stay-green as that of B 35. Maintenance of higher relative green leaf area was related to a greater relative grain yield in two of the three post-flowering moisture-deficit environments (Kassahun et al., 2010). Three stay-green QTL (in SBI-01, SBI-07, and SBI-10) from the donor line E36-1 were transferred to a Kenyan-farmer-preferred variety, Ochuti through marker-assisted QTL introgression. Only two QTL (*SBI-07* and *SBI-10*) were transferred into three genotypes, and about 25% of the BC_1F_1 progenies genotyped possessed at least one QTL, because the success rate in QTL introgressions from donor to the recurrent parent depends on the number of plants screened (Ngugi et al., 2013).

Across the years, sorghum research groups around the world have identified QTL for various economically important traits, along with the associated markers, and also mapped them in the sorghum genome. The availability of complete genome of sorghum has enhanced the rate at which these QTL have been mapped, as well as the identification of markers tightly linked to them. Even though many QTL have been identified for traits of interest, it is important to identify large-effect QTL for each trait that contributes at least 20% to the phenotypic variance of a particular trait or to identify a combination of two or more QTL that have a positive effect on the trait. Since the deployment of QTL into elite cultivars is difficult because of the influence of the genetic background on the trait of interest, the successful introgression and effective expression of the trait can be achieved by the deployment of large-effect major QTL.

Nested Association Mapping

QTL mapping approaches based on phenotyping and genotyping of natural populations has also been developed in recent years to complement some of the shortcomings of the classical QTL analysis based on biparental populations (Nordborg and Tavare, 2002). Genome-wide association mapping (GWAS) harnesses the genetic diversity of natural populations to resolve complex trait variation and can be applied to sorghum germplasm that possesses abundant diversity. The association mapping approach offers three advantages over traditional linkage analysis: (i) increased mapping resolution, (ii) increased efficiency (reduced resources), and (iii) the ability to identify large numbers of alleles (Yu and Buckler, 2006). Genome-wide association studies have proven to be a reliable tool to highlight marker-trait associations in a number of crop species, including bioenergy crops (Morris et al., 2013a, 2013b; Evans et al., 2014; Slavov et al., 2014). However, QTL analysis using biparental mapping populations and association mapping panels has limited power to detect the underlying genes of agroclimatic traits due to confounding population structure, low frequency, and allelic heterogeneity. Nested association mapping (NAM) combines both approaches with high statistical power and high mapping resolution and brings together both historic and recent recombination events with high allele richness (Yu et al., 2008)

In the United States, a total of 20 NAM populations (10 from Texas A&M University with a BTx623 seed-parent background and 10 from Kansas State University with an RTx430 seed-parent background) were developed and are currently used for many advanced studies that integrate genomics data with many panicle architecture traits and traits related to cold, drought, and heat tolerance. A high-throughput phenotyping tool with unmanned-hexacopter digital imagery is being used to characterize plant phenotypic traits related to drought tolerance at pre- and post-flowering stages. In Australia, Jordan et al. (2012) developed a NAM population and used DArT markers for QTL analysis. Phenotypic data that were characterized in different environments were subjected to simulation modeling and resulted in gene discovery of quantitative traits and progress in the breeding programs.

Novel and Induced Variation in Sorghum

Mutagenesis has been used as an effective tool by plant breeders for the creation of genetic variability through the use of radiation or chemical mutagens to elucidate morphogenesis, metabolism, and signal transduction pathways in most organisms, including plants (Bentley et al., 2000; Henikoff et al., 2004; Amsterdam and Hopkins, 2006). It has been highly successful for simply inherited traits in which a point mutation will bring about the desired changes in the trait of interest. In sorghum, novel phenotypes have been isolated through mutation (Quinby and Karper, 1942; Gaul, 1964; Sree Ramulu, 1970a, 1970b)—including dwarfism, earliness, high protein digestibility, high lysine, brown midrib, etc.— and have been employed in sorghum improvement (Singh and Axtell, 1973; Quinby, 1975; Ejeta and Axtell, 1985; Oria et al., 2000). An annotated individually pedigreed mutated sorghum library consisting of 6144 pedigreed M_4 seed pools developed from BTx623 by single-seed descent from individual mutagenized seeds (M_1) to the M_3 generation was established to support functional genomic studies (Xin et al., 2008, 2009). It is a reservoir of many useful mutants, such as brown midrib (*bmr*) mutants for biomass with better digestibility and ethanol yield and erect leaf (*erl*) mutants for improved capture of canopy radiation and hence biomass yield (Xin et al., 2009; Saballos et al., 2012; Sattler et al., 2014).

Grain yield is an important trait in any crop that needs genetic improvement, and the seed number per panicle is the trait's key determinant in sorghum and other cereal crops (Richards, 2000; Ashikari et al., 2005; Reynolds et al., 2009; Sreenivasulu and Schnurbusch, 2012) due to its direct relationship with the development of inflorescence architecture and panicle size (Bommert et al., 2005; Sreenivasulu and Schnurbusch, 2012). A stable multiseeded mutant (*msd1*) of sorghum in the BTx623 background was isolated and characterized. Both its sessile and pedicellate spikelets are fertile. In addition to normal seed formation by the sessile spikelets, 75–95% of the pedicellate spikelets in this mutant line can also form viable seeds. This mutant is also characterized by increased length of the panicle branches and a large panicle size. Even though this mutant's individual seeds are smaller than those of the wild type, this reduction was compensated for by increased seed number, leading to a 30–40% increase in the panicle seed weight (Burow et al., 2014). Isolation and characterization of an early-maturing mutant (KFJT-1) obtained by irradiation of sweet sorghum KFJT-CK with carbon ions (80 Gy dose) resulted in improvements in phenology such as greater stalk diameter, higher total biomass yield (78 t ha^{-1}), and a higher sugar content than the wild type, despite the mutant's short stature. This increase in the total biomass yield of the mutant line is due to the greater stalk diameter and internode weight (Cun and Jian, 2014).

Low lignin concentrations in the crop biomass can increase the digestibility of forage for ruminant livestock and improve saccharification yields of biomass for bioenergy. Brown midrib (*bmr*) mutants having brown vascular tissue with altered lignin content were first developed at Purdue University via chemical mutagenesis (Porter et al., 1978), followed by the identification of additional spontaneous brown midrib mutants (Vogler et al., 1994). Both of these groups of *bmr* mutants (*bmr1–bmr28*) show altered cell wall composition, particularly in relation to lignin subunit composition, and some have superior forage quality. Different *bmr* mutants have been characterized that can be introgressed into high-biomass, stay-green sorghum lines (Vermerris et al., 2007). Several of these mutants have resulted in increased yields of fermentable sugars following enzymatic saccharification, which was apparent even when combined with thermochemical pretreatment (Saballos et al., 2008). Among the *bmr* mutants identified, three mutant genes—*bmr6*, *bmr12*, and *bmr18*—have been found to be the most agronomically acceptable (Fritz et al., 1990). Exploitation of the unique interactions of *bmr* alleles, especially in combination, with different genetic backgrounds was suggested by Palmer et al. (2008), and the interactions could be used for the development of high-biomass sorghum. Putative *bmr* mutants isolated from an ethyl methanesulfonate (EMS)–mutagenized population were characterized (Sattler et al., 2014) and resulted in the identification of additional *bmr2*, *bmr6*, and *bmr12* alleles. Six more *bmr* mutants that were not allelic with *bmr2*, *bmr6*, and *bmr12* were identified, and they represented four novel *bmr* loci. A 2-yr field study revealed that most of the *bmr* lines exhibited lower concentrations of lignin in their biomass than the wild type. These new mutants may also affect monolignol biosynthesis like *bmr2*, *bmr6*, and *bmr12* do and so may bring about the desired improvement in bioenergy capacity as well as forage if stacked together.

Cyanogenic glycosides are plant allelochemicals that negatively impact animal and human nutrition since cyanogenic plant species release HCN upon tissue disruption (Conn, 1981; Seigler and Brinker, 1993), making the plants toxic (Oluwole et al., 2000). In sorghum, the cyanogenic glucoside dhurrin is present in higher concentrations in seedlings and young leaves than in mature tissues (Akazawa et al., 1960; Halkier and Møller, 1989; Wheeler et al., 1990; Busk and Moller, 2002). Hence, to protect animals from cyanogenic toxicity, there is a pressing need for the development of sorghum cultivars that produce little or no dhurrin. The cultivar BTx623 was used to create the first targeting induced local lesions in genomes (TILLING) resource in sorghum through EMS mutagenesis (Xin et al., 2008), and the utility of this approach was demonstrated by screening the mutant population for alterations in the genes not associated with cyanogenesis. An acyanogenic forage line (P414L) with a point mutation in the *CYP79A1* gene for cyanogenesis was developed recently by Blomstedt et al. (2012) using a combination of biochemical screening and the TILLING approach. The line exhibited a normal phenotype but had slightly slower growth during the early seedling stages. An EMS-induced mutant that is defective in the release of HCN was isolated by Krothapalli et al. (2013); it accumulated dhurrin but failed to release HCN efficiently on tissue disruption. The causal polymorphism relative to the BTx623 reference genome was identified by integrating SNP data, information on candidate genes associated with cyanogenesis, mutation spectra, and polymorphisms likely to affect phenotypic changes. The acyanogenic phenotype was manifested by the presence of a point mutation in the coding sequence of *dhurrinase2,* which encodes a protein in the dhurrin catabolic pathway and results in a premature stop codon. Recent studies also showed that leaf dhurrin is associated with the stay-green trait in sorghum.

Sorghum contains grain protein in the range of 9–10%, and the quality of protein is poor due to the lower amounts of amino acids, especially lysine. Screening of thousands of Ethiopian world sorghum accessions resulted in the identification of two spontaneous mutants, 1S 11167 and 1S 11758 (Singh and Axtell, 1973). Mohan and Axtell (1975) reported an induced mutant (P721) that had a soft phenotype, a 60% increase in lysine over the wild type, and reduced yield. Conversion lines were developed that behaved differently due to the diverse genetic backgrounds; only a few lines were satisfactory, but they encountered poor acceptance due to soft kernels. However, with the advancements in TILLING approaches, there is ample scope for the development of high-lysine sorghum that has good consumer acceptance but does not compromise grain yield.

A highly digestible sorghum line derived from the mutant P721Q (Mohan, 1975; Weaver et al., 1998) exhibited approximately 10–15% higher protein digestibility when uncooked and about 25% higher digestibility when cooked due to the rearrangement of kafirins, particularly γ-kafirins, which are located on the exterior of the protein bodies. Exposure of the interior α-kafirins and the greater total surface area makes it amenable to hydrolysis by proteolytic enzymes. Soeranto et al. (2001) subjected seeds of the sorghum variety Keris to γ irradiation and caused physiological effects such as abnormality, sterility, and even death of plants in the M_1 generation at high doses. The highest variance in plant height and harvest index was observed at 0.4 kGy, which significantly increased the genetic variability in the M_2 generation. To identify promising lines with desirable agronomic traits, selections were made in the M_2, M_3, and M_4 generations, and one line, ET 20-B, was shown to have higher protein and fat than the control, highlighting the role of mutation breeding in improving the nutritive value of sorghum grain.

A mutant with a phenotype of highly pigmented leaves and low lignin in the stem—*REDforGREEN* (*RG*)—was identified in the M_1 generation of the sweet sorghum cultivar 'Della'. It had been generated through EMS mutagenesis and exhibited dominant inheritance in the M_2 and subsequent generations. *RG* exhibited increased accumulation of lignin in the leaves and lignin depletion in stems, resulting in reduced saccharification efficiency in the leaves and increased saccharification efficiency in the stems compared with the wild type. Low lignin was associated with improved saccharification, and the similarities with the wild type in cellulose content and structure determined by X-ray diffraction analysis, instead of compositional or structural changes in the cellulose, support this correlation. Hence, the red leaf color may be considered a potential marker for improved saccharification efficiency in sorghum (Petti et al., 2013). In addition to tannins, chlorogenic acid, and total phenols, the *RG* mutant also overaccumulates 3-deoxyanthocyanidins in leaf tissue. The pigments luteolinidin and apigeninidin were identified as the main 3-deoxyanthocyanidins in the *RG* mutant through high-performance liquid chromatography, at concentrations of 1768 μg g^{-1} and 421 μg g^{-1}, respectively, compared with the negligible amount in the wild type, indicating a 1000-fold increase in the mutant leaves. These pigments could be developed as functionalized food colors because the 3-deoxyanthocyanidins appear to be acceptable as natural food colorings (Petti et al., 2014).

Mutagenesis, coupled with high-throughput DNA technologies for mutation screening, such as TILLING, high-resolution melt analysis, and EcoTILLING permit breeders to select traits that a few decades ago were difficult to breed. Recent advances in mutation screening techniques allow identification of novel alleles from germplasm collections and mutagenized populations as well as the identification of candidate genes for functional genomics studies and crop improvement.

New Frontiers in Sorghum Improvement

Specialty Traits

In light of increased global food demand and the environmental impact of agricultural expansion caused by unpredictable climatic change and global warming, sorghum, a drought resilient crop, can play a significant role as one of the best food alternatives. Compositionally, sorghum is very similar to the other cereal grains and can be used almost interchangeably with maize. Protein and starch digestibility are lower in sorghum than in maize. Although sorghum is used as a food grain in Asia and Africa, in the Western Hemisphere it is used as a feed or industrial crop (Dykes et al., 2005). Recently, the value of sorghum has been recognized because of its unique grain traits—inherent bioactive compounds and gluten-free quality—and the crop is becoming more importance in the dietary and specialty food markets. Researchers have reported that sorghum, more specifically, sorghum bran, contains phytonutrients in the form of phenolic compounds that offer bioactive properties important to human health (Hagerman et al., 1998; Dykes et al., 2005; Gu et al., 2007). In vitro testing of sorghum has confirmed its high antioxidant capacity, which is associated with mitigating cardiovascular (Klopfenstein et al., 1981), inflammatory (Burdette et al., 2010), and cancer-related diseases (Yang et al., 2009; Hargrove et al., 2011). However, the research on sorghum's phytonutritive properties has been restricted to a few commercial hybrids and advanced lines, and the majority of sorghum accessions have not been evaluated because sorghum has not received the research emphasis given to other crops such as rice, wheat, and maize. Technologies that are routinely used to enhance management and productivity in other crops are still far from deployment in sorghum. The tannin composition of polyphenolic compounds was much higher (23–62 mg/g) in sumac and black-seeded sorghum varieties than in antioxidant-rich blueberries (5 mg/g) and pomegranate (2.0–3.5 mg/g) juice. Sorghum has thus proved to be the richest and cheapest source of antioxidants and is gaining more importance in the food and beverage industries. Seed from the black-seeded pollinator RTx3362 (Rooney et al., 2013a) and parents ATx3363 seed BTx3363 (Rooney et al., 2013b) was released from Texas A&M University and recommended for the production of hybrids rich in 3-deoxyanthocyanins and for the development of new parents with this trait.

To improve the nutritional, bioethanol, and bioindustrial value in sorghum, Taylor et al. (2006) highlighted the need for research in the following areas: (i) protein and starch digestibility, (ii) sorghum endosperm matrix protein and cell wall components; (iii) starch gelatinization temperature and β-amylase activity, and (iv) preprocessing to recover valuable by-products such as the kafirin prolamin proteins and the pericarp wax.

Epigenetics

Throughout the years, plant breeders have altered crops genetically to improve their performance; however, they have also speculated that nongenetic, or epigenetic, variations could play vital roles in the regulation of gene expression, operating through mechanisms such as DNA methylation, histone modifications, and RNA interference. Such epigenetic variations in crop plants have been implicated in heterosis (Groszmann et al., 2011; Shivaprasad et al., 2012), time of flowering and maturity (Schmitz and Amasino, 2007; Heo and Sung, 2011), inbreeding depression (Cheptou and Donohue, 2013), and genotype × environment interactions (Dooner and Weil, 2007; Smith et al., 2012). During the last decade, a few studies in sorghum have demonstrated the epigenetic control of tissue-specific gene expression, culm bending, fertility restoration, and hybrid performance. Still, there is no clear evidence for the use of epigenetic variations for crop improvement.

Although the role of cytosine methylation in mammalian development was known, its role in plant development was not clear until recently, when it became evident from studies in *Arabidopsis* and barley that established the essentiality of cytosine methylation during specific developmental stages (Xiao et al., 2006; Berdasco et al., 2008; FitzGerald et al., 2008; von Wettstein, 2009). Identification and characterization of tissue-specific, differentially methylated regions (TDMRs) are an important step in determining the extent of cytosine methylation and understanding its role in plant development. Enriched TDMRs in sorghum are specific to the endosperm, whereas the embryo, leaf, root, young inflorescence, anther, and ovary tissues show distinct patterns, though similar levels, of cytosine methylation. The correlation between the expression pattern of some TDMRs and their tissue-specific methylation state indicated the role of DNA methylation in the regulation of tissue-specific or preferential gene expression (Zhang et al., 2011). Gibberellin (GA) deficiency in sorghum leads to a loss-of-function mutation in four genes (*SbCPS1, SbKS1, SbKO1, SbKAO1*), resulting in severe dwarfism and abnormal culm bending. These effects demonstrated in sorghum by using mutant lines in which application of GA alleviated bending of the culm and dwarfism, whereas application of uniconazole ((β*E*)-β-[(4-chlorophenyl)methylene]-α-(1,1-dimethylethyl)-1*H*-1,2,4-triazole-1-ethanol; a GA biosynthesis inhibitor) induced dwarfism and abnormal culm bending in the wild type depending on the concentration (Ordonio et al., 2014).

Plants respond to changing environmental conditions through physiological or developmental changes, and there is growing evidence that epigenetic mechanisms, especially DNA methylation, are responsible for the responses of plants and animals to various environmental signals (Feil and Fraga, 2012). The role of epigenetic mechanisms in maintaining the stability of fertility restoration in new types of male-sterile cytoplasms (A_4, 9E, M35) was demonstrated in sorghum by Elkonin et al. (2006). They concluded that fertility restoration was governed by a dominant gene whose expression was dependent on water availability during the critical stage of anther and pollen formation. The fertility restorer gene exhibited stable expression in self-pollinated progenies of restored hybrids but such expression failed to transmit to testcross hybrids cultivated under arid climatic conditions. In a by Elkonin and Tsvetova (2012), higher water availability during panicle and pollen formation significantly enhanced male fertility of the 9E CMS–based F_1 hybrid, whereas less-pronounced effects were noticed in F_2 populations. However, fertility was reverted when male-sterile F_1 plants were transferred from the dry plot to the greenhouse. Moreover, stable inheritance of male fertility was observed for three cycles of self-pollination in the progenies of these revertants. These studies provide evidence for the epigenetic regulation of fertility regulation in sorghum. In sorghum, the analysis of the level and pattern of DNA methylation (Wang et al., 2010) by methylation-sensitive amplified polymorphism in a pair of reciprocal F_1 hybrids and their parental inbreds treated with low-dose laser irradiation revealed that irradiation induced significant alterations in DNA methylation level and pattern. The alteration frequency in F_1 hybrids was higher than in the parental inbreds, thus demonstrating the role of epigenetics in the generation of heritable variations.

Methylation-sensitive amplified polymorphism analysis in inbred sorghum lines revealed persistent and moderate changes in cytosine methylation levels at the early seedling stage under 150 M Al^{3+} toxicity and low pH (4.0), with more changes in the cytosine-N-guanine methylation levels than in the cytosine-guanine methylation levels. Analysis of polymorphisms through the basic local alignment search tool (BLAST) identified possible polymorphisms that could have a major role in the regulation of gene expression associated with stress resistance since they were present in the regulatory regions of Al-tolerance genes (Kimatu et al., 2011). In another study, Agboola et al. (2012) assessed changes in the methylation levels and patterns induced by a combined effect of B (boron) and salt toxicity. They concluded that at the constant concentration of high NaCl (100 mM), hypermethylation was increased at a B concentration of 300 mM, while hypomethylation was induced at 400 mM, implicating the role of epigenetic alterations in the expression and repression of stress-responsive genes.

The epigenetic control of gene expression was recently demonstrated by de la Rosa Santamaria et al. (2014) using the gene *MSH1* (*MutS Homolog1*), which is present in every plant species and encodes a mitochondrial and plastid-localized protein (Abdelnoor et al., 2003; Xu et al., 2011). The gene's level of expression appears to be influenced by environmental stress (Shedge et al., 2010; Xu et al., 2011).

Developmental reprogramming in *Arabidopsis msh1* mutants (MSH1-dr) is characterized by dwarfing, variegation, delay in maturity transition and flowering, altered branching, and woody growth with aerial rosettes at short day length, which is associated with drastic changes in the expression of genes, particularly those involved in organelle and stress-response functions (Xu et al., 2012). Similarly, a substantial phenotypic variation (primarily epigenetic) induced by crossing a MSH1-dr line with the wild type was observed in sorghum by de la

Rosa Santamaria et al. (2014). The MSH1-dr lines were maintained as transgene nulls on elite inbred backgrounds, which would help to achieve the MSH1-enhanced growth phenomenon in hybrid breeding. The crossing of the MSH1-dr transgene null with its wild type generates maximum variation in the F_2 population and, followed by selection, results in a genetically uniform population with enhanced vigor and productivity by the F_4. This result is a clear indication that untapped epigenetic mechanisms have immense potential for the genetic improvement of crops.

The majority of the agronomically important traits in sorghum, which are partitioned as genetic (G), environmental (E) and gene × environment interactions (G×E), are quantitatively inherited. With the recent studies in many other crops besides sorghum, the focus has shifted in analyzing the epigenetic mechanisms as the nongenetic component in the expression of different traits. According to Springer (2013), the relative contribution of genetic and epigenetic variations toward the manifestation of different quantitative traits may be ascertained by dissecting their relative contributions to gene expression levels. The potential for this variation to be captured through breeding programs may be determined after understanding the stability and heritability of the epigenetic variation. Detailed studies on the role of epigenetic mechanisms in sorghum will help unravel the complex mechanisms involved in the manifestation of heterosis, tolerance to biotic and abiotic stresses, and plant development.

Conclusions

Tremendous genetic changes in sorghum have resulted through a variety of methods: advances in quantitative genetics, exploration of novel approaches involving mathematical heterosis prediction modeling, mutagenesis, DNA marker technology, transcriptomic and metabolomics studies, genomics integrated with the development of accelerated partial sorghum converted lines, double haploid development, GWAS in NAM populations via integration of high-throughput phenotyping facilities and GBS-based SNP data, refinement of phenotyping and computational/statistical analysis, availability of cost-effective genotyping platforms, and sequencing of the complete genome of more than 50 sorghum genotypes. These tools, in conjunction with the large number of unexploited naturally occurring, diverse genetic resources in sorghum, will help identify many novel traits and genes for marker-assisted breeding, acceleration of classical breeding programs, and overall sorghum improvement in the years to come.

Acknowledgments

This publication is contribution no. 16-363-B from the Kansas Agricultural Experiment Station, Manhattan, KS.

References

Abdelnoor, R.V., R. Yule, A. Elo, A.C. Christensen, G. Meyer-Gauen, and S.A. Mackenzie. 2003. Substoichiometric shifting in the plant mitochondrial genome is influenced by a gene homologous to MutS. Proc. Natl. Acad. Sci. USA 100:5968–5973. doi:10.1073/pnas.1037651100

Agboola, R.S., J.N. Kimatu, Y.C. Liao, and B. Liu. 2012. Morphological and cytosine DNA methylation changes induced by a combined effect of boron (B) and salt toxicity in *Sorghum bicolor* inbred line. Afr. J. Biotechnol. 11:10874–10881.

Akazawa, T., P. Miljanich, and E.E. Conn. 1960. Studies on cyanogenic glycoside of *Sorghum vulgare*. Plant Physiol. 35:535–538. doi:10.1104/pp.35.4.535

Albrecht, T., V. Wimmer, H.J. Auinger, M. Erbe, C. Knaak, M. Ouzunova, H. Simianer, and C.C. Schon. 2011. Genome-based prediction of testcross values in maize. Theor. Appl. Genet. 123:339–350. doi:10.1007/s00122-011-1587-7

Amsterdam, A., and N. Hopkins. 2006. Mutagenesis strategies in zebrafish for identifying genes involved in development and disease. Trends Genet. 22:473–478. doi:10.1016/j.tig.2006.06.011

Ashikari, M., H. Sakakibara, S. Lin, T. Yamamoto, T. Takashi, A. Nishimura, E.R. Angeles, Q. Qian, H. Kitano, and M. Matsuoka. 2005. Cytokinin oxidase regulates rice grain production. Science 309:741–745. doi:10.1126/science.1113373

Ben-Israel, I., B. Kilian, H. Nida, and E. Fridman. 2012. Heterotic trait locus (HTL) mapping identifies intra-locus interactions that underlie reproductive hybrid vigor in *Sorghum bicolor*. PLoS One 7:e38993. doi:10.1371/journal.pone.0038993

Bentley, A., B. MacLennan, J. Calvo, and C.R. Dearolf. 2000. Targeted recovery of mutations in *Drosophila*. Genetics 156:1169–1173.

Berdasco, M., R. Alcazar, M.V. Garcıa-Ortiz, E. Ballestar, A.F. Fernandez, T. Roldan-Arjona, A.F. Tiburcio, T. Altabella, N. Buisine, H. Quesneville, A. Baudry, L. Lepiniec, M. Alaminos, and M.F. Fraga. 2008. Promoter DNA hypermethylation and gene repression in undifferentiated *Arabidopsis* cells. PLoS One 3:e3306. doi:10.1371/journal.pone.0003306

Blomstedt, C.K., R.M. Gleadow, N. O'Donnell, P. Naur, K. Jensen, T. Laursen, C.E. Olsen, P. Stuart, J.D. Hamill, B.L. Møller, and A.D. Neale. 2012. A combined biochemical screen and TILLING approach identifies mutations in *Sorghum bicolor* L. Moench resulting in acyanogenic forage production. Plant Biotechnol. J. 10:54–66. doi:10.1111/j.1467-7652.2011.00646.x

Bommert, P., N. Satoh-Nagasawa, D. Jackson, and H. Hirano. 2005. Genetics and evolution of inflorescence and flower development in grasses. Plant Cell Physiol. 46:69–78. doi:10.1093/pcp/pci504

Burdette, A., P.L. Garner, E.P. Mayer, J.L. Hargrove, D.K. Hartle, and P. Greenspan. 2010. Anti-inflammatory activity of select sorghum (*Sorghum bicolor*) brans. J. Med. Food 13:879–887. doi:10.1089/jmf.2009.0147

Burkill, I.H., W. Birtwistle, F.W. Foxworthy, J.B. Scrivenor, and J.G. Watson. 1966. A dictionary of the economic products of the Malay Peninsula. Vol. 1, 2nd ed. A.H. Government of Malaysia and Singapore, Kuala Lumpur, Malaysia.

Burow, G., Z. Xin, C. Hayes, and J. Burke. 2014. Characterization of a multi-seeded (*msd1*) mutant of sorghum for increasing grain yield. Crop Sci. 54:2030–2037. doi:10.2135/cropsci2013.08.0566

Busk, P.K., and B.L. Moller. 2002. Dhurrin synthesis in sorghum is regulated at the transcriptional level and induced by nitrogen fertilization in older plants. Plant Physiol. 129:1222–1231. doi:10.1104/pp.000687

Casa, A.M., S.E. Mitchell, O.S. Smith., J.C. Register III, S.R. Wessler, and S. Kresovich. 2002. Evaluation of *Hbr* (MITE) markers for assessment of genetic relationships among maize (*Zea mays* L.) inbred lines. Theor. Appl. Genet. 104:104–110. doi:10.1007/s001220200012

Cheptou, P.O., and K. Donohue. 2013. Epigenetics as a new avenue for the role of inbreeding depression in evolutionary ecology. Heredity 110:205–206. doi:10.1038/hdy.2012.66

Cheres, M.T., J.F. Miller, J.M. Crane, and S.J. Knapp. 2000. Genetic distance as a predictor of heterosis and hybrid performance within and between heterotic groups in sunflower. Theor. Appl. Genet. 100:889–894. doi:10.1007/s001220051366

Clayton, W.D., and S.A. Renvoize. 1986. Genera Graminum: Grasses of the world. Kew Bull. Additional Ser. 13. Royal Botanic Gardens, Kew, London, p. 338–345.

Conn, E.E. 1981. Cyanogenic glycosides. In: P.K. Stumpf and E.E. Conn, editors, The biochemistry of plants: A comprehensive treatise. Vol. 7 of Secondary plant products. Academic Press, New York, NY. p. 479–499. doi:10.1016/B978-0-12-675407-0.50022-1

Corbellini, M., M. Perenzin, M. Accerbi, P. Vaccino, and B. Borghi. 2002. Genetic diversity in bread wheat, as revealed by coefficient of parentage and molecular markers, and its relationship to hybrid performance. Euphytica 123:273–285. doi:10.1023/A:1014946018765

Cun, D.X., and L.W. Jian. 2014. Evaluation of KTJT-1, an early-maturity mutant of sweet sorghum acquired by carbon ions irradiation. Nuc. Sci. Tech. 25:020305.

Dahlberg, J.A. 2000. Classification and characterization of sorghum. In: C.W. Smith and R.A. Frederiksen, editors, Sorghum: Origin, history, technology and production. John Wiley & Sons, Hoboken, NJ. p. 99–130.

Dave, R.V., and P. Joshi. 1995. Divergence and heterosis for fodder attributes in pearl millet. Indian J. Genet. Plant Breed. 55:392–397.

De la Rosa Santamaria, R., M.R. Shao, G. Wang, D.O.N. Liu, H. Kundariya, Y. Wamboldt, I. Dweikat, and S.A. Mackenzie. 2014. MSH1-induced non-genetic variation provides a source of phenotypic diversity in *Sorghum bicolor*. PLoS One 9:e108407. doi:10.1371/journal.pone.0108407

de Wet, J.M.J. 1978. Systematics and evolution of sorghum sect. Sorghum (Gramineae). Am. J. Bot. 65:477–484. doi:10.2307/2442706

de Wet, J.M.J., and J.P. Huckabay. 1967. The origin of *Sorghum bicolor*. II. Distribution and domestication. Evolution 21:787–802. doi:10.2307/2406774

Diers, B.W., P.B.E. McKetty, and T.C. Osborn. 1996. Relationship between heterosis and genetic distance based on restriction fragment length polymorphism markers in oilseed rape (*Brassica napus* L). Crop Sci. 36:79–83. doi:10.2135/cropsci1996.0011183X003600010014x

Dixit, U.N., and D. Swain. 2000. Genetic divergence and heterosis in sesame. Indian J. Genet. Plant Breed. 60:213–219.

Doggett, H. 1970. Sorghum. Tropical Agriculture Series. Longmans, London.

Doggett, H. 1988. Sorghum. Longman Scientific and Technical. London.

Dooner, H.K., and D.F. Weil. 2007. Give-and-take: Interactions between DNA transposons and their host plant genomes. Curr. Opin. Genet. Dev. 17:486–492. doi:10.1016/j.gde.2007.08.010

Dudley, J.W., M.A. Saghai Maroof, and G.K. Rufener. 1991. Molecular markers and grouping of parents in maize breeding programs. Crop Sci. 31:718–723. doi:10.2135/cropsci1991.0011183X003100030036x

Dykes, L., L.W. Rooney, R.D. Waniska, and W.L. Rooney. 2005. Phenolic compounds and antioxidant activity of sorghum grains of varying genotypes. J. Agric. Food Chem. 53:6813–6818. doi:10.1021/jf050419e

Ejeta, G., and J. Axtell. 1985. Mutant gene in sorghum causing leaf "reddening" and increased protein concentration in the grain. J. Hered. 76:301–302.

Elkonin, L.A., V.V. Kozhemyakin, and O.P. Kibalnik. 2006. Genetic and epigenetic regulation of male fertility restoration in the 9E, A_4 and M35 CMS-inducing cytoplasms of sorghum. Acta Agron. Hung. 54:281–289. doi:10.1556/AAgr.54.2006.3.2

Elkonin, L.A., and M.I. Tsvetova. 2012. Heritable effect of plant water availability conditions on restoration of male fertility in the "9E" CMS-inducing cytoplasm of sorghum. Front. Plant Sci. 3:91. doi:10.3389/fpls.2012.00091

Endelman, J.B. 2011. Ridge regression and other kernels for genomic selection with R package rrBLUP. Plant Genome 4:250–255. doi:10.3835/plantgenome2011.08.0024

Evans, L.M., G.T. Slavov, E. Rodgers-Melnick, J. Martin, P. Ranjan, W. Muchero, A.M. Brunner, W. Schackwitz, L. Gunter, J.-G. Chen, G.A. Tuskan, and S.P. DiFazio. 2014. Population genomics of *Populus trichocarpa* identifies signatures of selection and adaptive trait associations. Nat. Genet. 46:1089–1096. doi:10.1038/ng.3075

Falconer, D.S., and T.F.C. Mackay. 1996. Introduction to quantitative genetics. 4th ed. Longman, London.

Feil, R., and M.F. Fraga. 2012. Epigenetics and the environment: Emerging patterns and implications. Nat. Rev. Genet. 13:97–109.

Feltus, F.A., J. Wan, S.R. Schulze, J.C. Estill, N. Jiang, and A.H. Paterson. 2004. A SNP resource for rice genetics and breeding based on subspecies *indica* and *japonica* genome alignments. Genome Res. 14:1812–1819. doi:10.1101/gr.2479404

FitzGerald, J., M. Luo, A. Chaudhury, and F. Berger. 2008. DNA methylation causes predominant maternal controls of plant embryo growth. PLoS One 3:e2298. doi:10.1371/journal.pone.0002298

Fritz, J.O., K.J. Moore, and E.H. Jaster. 1990. Digestion kinetics and cell wall composition of brown midrib sorghum × sudangrass morphological components. Crop Sci. 30:213–219. doi:10.2135/cropsci1990.0011183X003000010046x

Gamar, Y.A., and A.H. Mohamed. 2013. Introgression of *Striga* resistance genes into a *Sudanese* sorghum cultivar, Tabat, using marker assisted selection (MAS). Greener J. Agric. Sci. 3:550–556.

Gärtner, T., M. Steinfath, S. Andorf, J. Lisec, R.C. Meyer, T. Altmann, L. Willmitzer, and J. Selbig. 2009. Improved heterosis prediction by combining information on DNA and metabolic markers. PLoS One 4:e5220. doi:10.1371/journal.pone.0005220

Gaul, H. 1964. Mutations in plant breeding. Radiat. Bot. 4:155–232. doi:10.1016/S0033-7560(64)80069-7

Gaut, B.S., L.G. Clark, J.F. Wendel, and S.V. Muse. 1997. Comparisons of the molecular evolutionary process at *rbcL* and *ndhF* in the grass family (Poaceae). Mol. Biol. Evol. 14:769–777. doi:10.1093/oxfordjournals.molbev.a025817

Gill, J.R., W.L. Rooney, and P.E. Klein. 2014. Effect of humidity on intergeneric pollinations of *iap* (*inhibition of alien pollen*) sorghum [*Sorghum bicolor* (L.) Moench]. Euphytica 198:381–387. doi:10.1007/s10681-014-1113-5

Goddard, M.E., B.J. Hayes, and T.H. Meuwissen. 2010. Genomic selection in livestock populations. Genet. Res. 92:413–421. doi:10.1017/S0016672310000613

Godshalk, E.B., M. Lee, and K.R. Lamkey. 1990. Relationship of RFLP to single cross hybrid performance of maize. Theor. Appl. Genet. 80:273–280. doi:10.1007/BF00224398

Gowda, M., Y. Zhao, H.P. Maurer, E.A. Weissmann, T. Wurschum, and J.C. Reif. 2013. Best linear unbiased prediction of triticale hybrid performance. Euphytica 191:223–230. doi:10.1007/s10681-012-0784-z

Groszmann, M., I.K. Greaves, N. Albert, R. Fujimoto, C.A. Helliwell, E.S. Dennis, and W.J. Peacock. 2011. Epigenetics in plants-vernalisation and hybrid vigour. Biochim. Biophys. Acta 1809:427–437. doi:10.1016/j.bbagrm.2011.03.006

Gu, L., S.E. House, L. Rooney, and R.L. Prior. 2007. Sorghum bran in the diet dose dependently increased the excretion of catechins and microbial-derived phenolic acids in female rats. J. Agric. Food Chem. 55:5326–5334. doi:10.1021/jf070100p

Gupta, S.C., J.M.J. de Wet, and J.R. Harlan. 1978. Morphology of *Saccharum-Sorghum* hybrid derivatives. Am. J. Bot. 65:936–942. doi:10.2307/2442680

Gurney, A.L., M.C. Press, and J.D. Scholes. 2002. Can wild relatives of sorghum provides new sources of resistance or tolerance against *Striga* species? Weed Res. 42:317–324. doi:10.1046/j.1365-3180.2002.00291.x

Habier, D. 2010. More than a third of the WCGALP presentations on genomic selection. J. Anim. Breed. Genet. 127:336–337. doi:10.1111/j.1439-0388.2010.00897.x

Hagerman, E., K.M. Ried, G.A. Jones, K.N. Sovik, N.T. Ritchard, P.W. Hartzfeld, and T.L. Riechel. 1998. High molecular weight plant polyphenolics (tannins) as biological antioxidants. J. Agric. Food Chem. 46:1887–1892. doi:10.1021/jf970975b

Halkier, B.A., and B.L. Møller. 1989. Biosynthesis of the cyanogenic glucoside dhurrin in seedlings of *Sorghum bicolor* (L.) Moench and partial purification of the enzyme system involved. Plant Physiol. 90:1552–1559. doi:10.1104/pp.90.4.1552

Harer, P.N., and D.R. Bapat. 1982. Line × tester analysis of combining ability in grain sorghum. J. Maharashtra Agric. Univ. 7:230–232.

Hargrove, J.L., P. Greenspan, D.K. Hartle, and C. Dowd. 2011. Inhibition of aromatase and *a*-amylase by flavonoids and proanthocyanidins from *Sorghum bicolor* bran extracts. J. Med. Food 14:799–807. doi:10.1089/jmf.2010.0143

Harlan, J.R., and J.M.J. de Wet. 1971. Toward a rational classification of cultivated plants. Taxon 20:509–517.

Harlan, J.R., and J.M.J. de Wet. 1972. A simplified classification of cultivated sorghum. Crop Sci. 12:172–176. doi:10.2135/cropsci1972.0011183X001200020005x

Haussmann, B.I.G., D.E. Hess, G.O. Omanya, R.T. Folkertsma, B.V.S. Reddy, M. Kayentao, H.G. Welz, and H.H. Geiger. 2004. Genomic regions influencing resistance to the parasitic weed *Striga hermonthica* in two recombinant inbred populations of sorghum. Theor. Appl. Genet. 109:1005–1016. doi:10.1007/s00122-004-1706-9

Henikoff, S., B.J. Till, and L. Comai. 2004. TILLING. Traditional mutagenesis meets functional genomics. Plant Physiol. 135:630–636. doi:10.1104/pp.104.041061

Heo, J.B., and S. Sung. 2011. Encoding memory of winter by noncoding RNAs. Epigenetics 6:544–547. doi:10.4161/epi.6.5.15235

Heslot, N., H.P. Yang, M.E. Sorrells, and J.L. Jannink. 2012. Genomic selection in plant breeding: A comparison of models. Crop Sci. 52:146–160. doi:10.2135/cropsci2011.06.0297

Hodnett, G.L., B.L. Burson, W.L. Rooney, S.L. Dillon, and H.J. Price. 2005. Pollen-pistil interactions result in reproductive isolation between *Sorghum bicolor* and divergent *Sorghum* species. Crop Sci. 45:1403–1409. doi:10.2135/cropsci2004.0429

Hodnett, G.L., A.L. Hale, D.J. Packer, D.M. Stelly, J. da Silva, and W.L. Rooney. 2010. Elimination of a reproductive barrier facilitates inter-generic hybridization of *Sorghum bicolor* and *Saccharum*. Crop Sci. 50:1188–1195. doi:10.2135/cropsci2009.09.0486

Hongtrakul, V., G.M. Huestis, and S.J. Knapp. 1997. Amplified fragment length polymorphisms as a tool for DNA fingerprinting sunflower germplasm: Genetic diversity among oilseed inbred lines. Theor. Appl. Genet. 95:400–407. doi:10.1007/s001220050576

Iwata, H., and J.L. Jannink. 2011. Accuracy of genomic selection prediction in barley breeding programs: A simulation study based on the real single nucleotide polymorphism data of barley breeding lines. Crop Sci. 51:1915–1927. doi:10.2135/cropsci2010.12.0732

Jirali, D.I., B.D. Biradar, and S.S. Rao. 2007. Performance of rabi sorghum genotypes under receding moisture conditions in different soil types. Karnataka J. Agric. Sci. 20:603–604.

Jordan, D.R., D. Butler, B. Henzell, J. Drenth, and C.L. McIntyre. 2004. Diversification of Australian sorghum using wild relatives. In: R.A. Fischer, editor, New directions for a diverse planet. Proceedings of the 4th International Crop Science Congress, Brisbane, Australia. 26 September–1 October. CSSA. Madison, WI.

Jordan, D.R., E. Mace, A. Cruickshank, and R.G. Henzell. 2012. Development and use of a sorghum nested association mapping population. Abstracts, International Plant and Animal Genome Conference, San Diego, CA. 14–18 January.

Jordan, D.R., Y.Z. Tao, I.D. Godwin, R.G. Henzell, M. Cooper, and C.L. McIntyre. 1998. Loss of genetic diversity associated with selection for resistance to sorghum midge in Australian Sorghum. Euphytica 102:1–7. doi:10.1023/A:1018311908636

Jordan, D.R., Y. Tao, I.D. Godwin, R.G. Henzell, M. Cooper, and C.L. McIntyre. 2003. Prediction of hybrid performance in grain sorghum using RFLP markers. Theor. Appl. Genet. 106:559–567.

Kapanigowda, M.H., R. Perumal, R.M. Aiken, T.J. Herald, S.R. Bean, and C.R. Little. 2013a. Analysis of sorghum [*Sorghum bicolor* (L.) Moench] lines and hybrids in response to early season planting and cool conditions. Can. J. Plant Sci. 93:773–784. doi:10.4141/cjps2012-311

Kapanigowda, M., R. Perumal, D. Maduraimuthu, M.R. Aiken, P.V.V. Prasad, T. Tesso, and C.R. Little. 2013b. Genotypic variation in sorghum [*Sorghum bicolor* (L.) Moench] exotic germplasm collection for drought and disease tolerance. Springerplus 2:650. doi:10.1186/2193-1801-2-650

Kassahun, B., F.R. Bidinger, C.T. Hash, and M.S. Kuruvinashetti. 2010. Stay-green expression in early generation sorghum [*Sorghum bicolor* (L.) Moench] QTL introgression lines. Euphytica 172:351–362. doi:10.1007/s10681-009-0108-0

Kimatu, J.N., M. Diarso, C. Song, R.S. Agboola, J. Pang, X. Qi, and B. Liu. 2011. DNA cytosine methylation alterations associated with aluminium toxicity and low pH in *Sorghum bicolor.* Afr. J. Agric. Res. 6:4579–4593.

Klein, R.R., F.R. Miller, S. Bean, and P.E. Klein. 2016. Registration of 40 converted germplasm sources from the reinstated Sorghum Conversion Program. J. Plant Reg. 10:57–61. doi:10.3198/jpr2015.05.0034crg

Klopfenstein, C.F., E. Varriano-Marston, and R.C. Hoseney. 1981. Cholesterol-lowering effect of sorghum diet in guinea pigs. Nutr. Rep. Int. 24:621–627.

Krothapalli, K., E.M. Buescher, X. Li, E. Brown, C. Chapple, B.P. Dilkes, and M.R. Tuinstra. 2013. Forward genetics by genome sequencing reveals that rapid cyanide release deters insect herbivory of *Sorghum bicolor*. Genetics 195:309–318. doi:10.1534/genetics.113.149567

Kuhlman, L.C., B.L. Burson, P.E. Klein, R.R. Klein, D.M. Stelly, H.J. Price, and W.L. Rooney. 2008. Genetic recombination in *Sorghum bicolor* × *S. macrospermum* interspecific hybrids. Genome 51:749–756. doi:10.1139/G08-061

Kuhlman, L.C., B.L. Burson, P.E. Klein, D.M. Stelly, and W.L. Rooney. 2006. Interspecific sorghum breeding using *S. macrospermum*. Paper presented at: Proceedings of the ASA-CSSA-SSA 2006 International Meetings, Indianapolis, IN. 12–16 November.

Laurie, D., and M.D. Bennett. 1989. Genetic variation in *Sorghum* for the inhibition of maize pollen tube growth. Ann. Bot. 64:675–681.

Lee, M., E.B. Godshalk, K.R. Lamkey, and W.L. Woodman. 1989. Association of restriction length polymorphism among maize inbreds with agronomic performance of their crosses. Crop Sci. 29:1067–1071. doi:10.2135/cropsci1989.0011183X002900040050x

Li, Y., and C. Li. 1998. Genetic contribution of Chinese landraces to the development of sorghum hybrids. Euphytica 102:47–57. doi:10.1023/A:1018374203792

Liu, X.C., and J.L. Wu. 1998. SSR heterogenic patterns of parents for marking and predicting heterosis in rice breeding. Mol. Breed. 4:263–268. doi:10.1023/A:1009645908957

Lorenzana, R.E., and R. Bernardo. 2009. Accuracy of genotypic value predictions for marker-based selection in bi-parental plant populations. Theor. Appl. Genet. 120:151–161. doi:10.1007/s00122-009-1166-3

Lu, X.P., D.D. Liu, S.Y. Wang, F.G. Mi, P.A. Han, and E.S. Lu. 2014. Genetic effects and heterosis prediction model of *Sorghum bicolor* × *S. sudanense* grass. Acta Agron. Sin. 40()3:466–475. doi:10.3724/SP.J.1006.2014.00466

Lu, X.P., J.F. Yun, C.P. Gao, and A. Surya. 2011. Quantitative trait loci analysis of economically important traits in *Sorghum bicolor* × *S. sudanense* hybrid. Can. J. Plant Sci. 91:81–90. doi:10.4141/cjps09112

Melchinger, A.E. 1999. Genetic diversity and heterosis. In: J.G. Coors and S. Pandey, editors, The genetics and exploitation of heterosis in crops. CSSA, Madison, WI. p. 99–118.

Meuwissen, T.H.E., B.J. Hayes, and M.E. Goddard. 2001. Prediction of total genetic value using genome-wide dense marker maps. Genetics 157:1819–1829.

Miller, F.R. 1980. The breeding of sorghum. Texas Agric. Exp. Stn. Bull. 1451:128–136.

Ming, R., S.C. Liu, Y.R. Lin, J. da Silva, W. Wilson, D. Braga, A. van Deynze, T.F. Wenslaff, K.K. Wu, P.H. Moore, W. Burnquist, M.E. Sorrells, J.E. Irvine, and A.H. Paterson. 1998. Detailed alignment of *Saccharum* and *Sorghum* chromosomes: Comparative organization of closely related diploid and polyploid genomes. Genetics 150:1663–1682.

Mohamed, A., R. Ali, O. Elhassan, E. Suliman, C. Mugoya, C.W. Masiga, A. Elhusien, and C.T. Hash. 2014. First products of DNA marker-assisted selection in sorghum released for cultivation by farmers in sub-Saharan Africa. J. Plant Sci. Mol. Breed. 3, art. 3. doi:10.7243/2050-2389-3-3

Mohan, D.D. 1975. Chemically induced high lysine mutants in *Sorghum bicolor* (L.) Moench. Ph.D. diss., Purdue Univ., West Lafayette, IN.

Mohan, D.D., and J.D. Axtell. 1975. Diethyl sulphate-induced high-lysine mutants in sorghum. In: Proceedings of the 9th Biennial Grain Sorghum Research and Utilization Conference, Lubbock, TX.

Morris, G.P., P. Ramu, S.P. Deshpande, C.T. Hash, T. Shah, H.D. Upadhyaya, O. Riera-Lizarazu, P.J. Brown, C.B. Acharya, S.E. Mitchell, J. Harriman, J.C. Glaubitz, E.S. Buckler, and S. Kresovich. 2013a. Population genomic and genome-wide association studies of agro-climatic traits in sorghum. Proc. Natl. Acad. Sci. USA 110:453–458. doi:10.1073/pnas.1215985110

Morris, G.P., D.H. Rhodes, Z. Brenton, P. Ramu, V.M. Thayil, S. Deshpande, C.T. Hash, C. Acharya, S.E. Mitchell, E.S. Buckler, J. Yu, and S. Kresovich. 2013b. Dissecting genome-wide association signals for loss-of-function phenotypes in sorghum flavonoid pigmentation traits. G3: Genes, Genomes, Genet. 3:2085–2094. doi:10.1534/g3.113.008417

Muth, P., O. Elhassan, A. Mohammed, H.F.W. Rattunde, A. Toure, and B.I.G. Haussmann. 2012. Effect of marker-assisted backcrossing to introgress resistance to *Striga hermonthica* into African sorghum varieties. In: Eric Tielkes, editor, Resilience of agricultural systems against crises. Tropentag 2012: International Research on Food Security, Natural Resource Management and Rural Development, Georg-August-Universität, Göttingen. 19–21 September. Cuvilier Verlag, Gottingen, Germany.

Nair, N.V. 1999. Production and cyto-morphological analysis of intergeneric hybrids of *Sorghum* × *Saccharum*. Euphytica 108:187–191. doi:10.1023/A:1003633015836

Ngugi, K., W. Kimani, D. Kiambi, and E.W. Mutitu. 2013. Improving drought tolerance in *Sorghum bicolor* L. Moench: Marker-assisted transfer of the stay-green quantitative trait loci (QTL) from a characterized donor source into a local farmer variety. Int. J. Scientific Res. Knowledge 1:154–162. doi:10.12983/ijsrk-2013-p154-162

Nordborg, M., and S. Tavare. 2002. Linkage disequilibrium: What history has to tell us. Trends Genet. 18:83–90. doi:10.1016/S0168-9525(02)02557-X

Oluwole, O.S.A., A.O. Onabolu, H. Link, and H. Rosling. 2000. Persistence of tropical ataxic neuropathy in a Nigerian community. J. Neurol. Neurosurg. Psychiatry 69:96–101. doi:10.1136/jnnp.69.1.96

Ordonio, R.L., Y. Ito, A. Hatakeyama, K.O. Shinohara, S. Kasuga, T. Tokunaga, H. Mizuno, H. Kitano, M. Matsuoka, and T. Sazuka. 2014. Gibberellin deficiency pleiotropically induces culm bending in sorghum: An insight into sorghum semi-dwarf breeding. Sci. Rep. 4. doi:10.1038/srep05287

Oria, M.P., B.R. Hamaker, J.D. Axtell, and C.P. Huang. 2000. A highly digestible sorghum mutant cultivar exhibits a unique folded structure of endosperm protein bodies. Proc. Natl. Acad. Sci. USA 97:5065–5070. doi:10.1073/pnas.080076297

Osama, K.S. 2013. Marker assisted introgression of *Striga* resistance into farmer preferred sorghum variety (Ochuti). M.S. thesis. College of Agriculture and Veterinary Sciences, University of Nairobi, Kenya.

Palmer, N.A., S.E. Sattler, A.J. Saathoff, D. Funnell, J.F. Pedersen, and G. Sarath. 2008. Genetic background impacts soluble and cell wall-bound aromatics in *brown midrib* mutants of sorghum. Planta. 229:115. doi:10.1007/s00425-008-0814-1

Patil, J.V., S.P. Reddy, K.B. Prabhakar, A.V. Umakanth, S. Gomashe, and K.N. Ganapathy. 2014. History of post-rainy season sorghum research in India and strategies for breaking the yield plateau. Indian J. Genet. Plant Breed. 74(3):271–285. doi:10.5958/0975-6906.2014.00845.1

Pawar, K.N., B.D. Biradar, J. Shamarao, and M.R. Ravikumar. 2005. Identification of germplasm sources for adaptation under receding soil moisture situations in post-rainy sorghum. Agric. Sci. Dig. 25:56–58.

Pejic, I.P., M. Ajamone-Marsan, M. Morgante, V. Kozumplick, P. Castiglioni, G. Taramino, and M. Motto. 1998. Comparative analysis of genetic similarity among maize inbred lines detected by RFLPs, RAPDs, SSRs and AFLPs. Theor. Appl. Genet. 97:1248–1255. doi:10.1007/s001220051017

Peng, J.Y., S.S. Virmani, and A.W. Julfiquar. 1991. Relationship between heterosis and genetic distance in rice. Oryza 28:129–133.

Perumal, R., K. Renganayaki, M.M. Menz, S. Katile, J. Dahlberg, C.W. Magill, and W.L. Rooney. 2007. Genetic diversity among sorghum races and working groups based on AFLP and SSRs. Crop Sci. 47:1375–1383. doi:10.2135/cropsci2006.08.0532

Petti, C., A.E. Harman-Ware, M. Tateno, R. Kushwaha, A. Shearer, A.B. Downie, M. Crocker, and S. DeBolt. 2013. Sorghum mutant RG displays antithetic leaf shoot lignin accumulation resulting in improved stem saccharification properties. Biotechnol. Biofuels 6:146. doi:10.1186/1754-6834-6-146

Petti, C., R. Kushwaha, M. Tateno, A.E. Harman-Ware, M. Crocker, J. Awika, and S. DeBolt. 2014. Mutagenesis breeding for increased 3-deoxyanthocyanidin accumulation in leaves of *Sorghum bicolor* (L.) Moench.: A source of natural food pigment. J. Agric. Food Chem. 62(6):1227–1232. doi:10.1021/jf405324j

Piepho, H.P. 2009. Ridge regression and extensions for genome-wide selection in maize. Crop Sci. 49:1165–1176. doi:10.2135/cropsci2008.10.0595

Porter, K.S., J.D. Axtell, V.L. Lechtenberg, and V.F. Colenbrander. 1978. Phenotype, fiber composition, and *in vitro* dry matter disappearance of chemically induced brown midrib (*bmr*) mutants of sorghum. Crop Sci. 18:205–208. doi:10.2135/cropsci1978.0011183X001800020002x

Prabhakar, K.B., J.V. Patil, and S.P. Reddy. 2014. *Rabi* sorghum improvement: Past, present and future. Karnataka J. Agric. Sci. 27:433–444.

Premalatha, N., N. Kumaravadivel, and P. Veerabadhiran. 2006. Heterosis and combining ability for grain yield and its components in sorghum [*Sorghum bicolor* (L.) Moench]. Indian J. Genet. Plant Breed. 66:123–126.

Price, H.J., G.L. Hodnett, B.L. Burson, S.L. Dillon, and W.L. Rooney. 2005. A *Sorghum bicolor* × *S. macrospermum* hybrid recovered by embryo rescue and culture. Aust. J. Bot. 53:579–582. doi:10.1071/BT04213

Price, H.J., G.L. Hodnett, B.L. Burson, S.L. Dillon, D.M. Stelly, and W.L. Rooney. 2006. Genome dependent interspecific hybridisation of *Sorghum bicolor* (Poaceae). Crop Sci. 46:2617–2622. doi:10.2135/cropsci2005.09.0295

Quinby, J.R. 1975. The genetics of sorghum improvement. J. Hered. 66:56–62.

Quinby, J.R., and R.E. Karper. 1942. Inheritance of mature plant characters in sorghum: Induced by radiation. J. Hered. 33:323–327.

Quinby, J.R., N.W. Kramer, J.C. Stephens, K.A. Lahir, and R.E. Karper. 1958. Grain sorghum production in Texas. Bull. 912. Texas Agric. Exp. Stn., College Station.

Rai, K.N., D.S. Murty, D.J. Andrews, and P.J. Bramel-Cox. 1999. Genetic enhancement of pearl millet and sorghum for the semi-arid tropics of Asia and Africa. Genome 42:617–628. doi:10.1139/g99-040

Rajendrakumar, P., K. Hariprasanna, I. Jaikishan, R. Madhusudhana, and J.V. Patil. 2013. Potential of microsatellite marker polymorphism in the prediction of grain yield heterosis in sorghum [*Sorghum bicolor* (L.) Moench]. In: S. Rakshit et al., editors, Compendium of papers and abstracts: Global consultation on millets promotion for health and nutritional security. Society of Millets Research, Directorate of Sorghum Research, Hyderabad, India. p. 329.

Ramalashmi, K., P.T. Prathima, K. Mohanraj, and N.V. Nair. 2014. Expression profiling of sucrose metabolizing genes in *Saccharum, Sorghum* and their hybrids. Appl. Biochem. Biotechnol. 174:1510–1519. doi:10.1007/s12010-014-1048-2

Rani, K.J., and S.S. Rao. 2009. Relationship between heterosis and genetic divergence in rabi sorghum [*Sorghum bicolor* (L.) Moench]. Res. Crops 10:319–322.

Reddy, B.V.S., S. Ramesh, and P.S. Reddy. 2006. Sorghum genetic resources, cytogenetics and improvement. In: R.J. Singh and P.P. Jauhar, editors, Cereals. Vol. 2 of Genetic resources, chromosome engineering, and crop improvement. CRC Taylor and Francis, Boca Raton, FL. p. 309–363.

Reynolds, M., M.J. Foulkes, G.A. Slafer, P. Berry, M.A.J. Parry, J.W. Snape, and W.J. Angus. 2009. Raising yield potential in wheat. J. Exp. Bot. 60:1899–1918. doi:10.1093/jxb/erp016

Richards, R.A. 2000. Selectable traits to increase crop photosynthesis and yield of grain crops. J. Exp. Bot. 51:447–458. doi:10.1093/jexbot/51.suppl_1.447

Rooney, W.L., J. Awaika, and L. Dykes. 2013a. Registration of Tx3362 sorghum germplasm J. Plant Reg. 7:104–107. doi:10.3198/jpr2012.04.0262crg

Rooney, W.L., O. Portillo, and C.M. Hayes. 2013b. Registration of ATx3363 and BTx3363 black sorghum germplasms. J. Plant Reg. 7:342. doi:10.3198/jpr2013.01.0006crg

Rosenow, D.T., and J.A. Dahlberg. 2000. Collection, conversion and utilisation of sorghum. In: C.W. Smith and R.A. Frederiksen, editors, Sorghum: Origin, history, technology and production. John Wiley, Sons, New York, NY. p. 309–328.

Rosenow, D.T., J.A. Dahlberg, J.C. Stephens, F.R. Miller, D.K. Barnes, G.C. Peterson, J.W. Johnson, and K.F. Schertz. 1997. Registration of 63 converted sorghum germplasm lines from the sorghum conversion program. Crop Sci. 37:1399–1400. doi:10.2135/cropsci1997.0011183X003700040090x

Saballos, A., S.E. Sattler, E. Sanchez, T.P. Foster, Z. Xin, C. Kang, J.F. Pedersen, and W. Vermerris. 2012. Brown midrib2 (*bmr2*) encodes the major 4-coumarate:coenzyme A ligase involved in lignin biosynthesis in sorghum [*Sorghum bicolor* (L.) Moench]. Plant J. 70:818–830. doi:10.1111/j.1365-313X.2012.04933.x

Saballos, A., W. Vermerris, L. Rivera, and G. Ejeta. 2008. Allelic association, chemical characterization and saccharification properties of brown midrib mutants of sorghum [*Sorghum bicolor* (L.) Moench]. BioEnergy Res. 1:193–204. doi:10.1007/s12155-008-9025-7

Sandeep, R.G., M.R. Gururaja Rao, S. Ramesh, Chikkalingaiah, and H. Shivanna. 2010. Parental combining ability as a good predictor of productive crosses in sweet sorghum [*Sorghum bicolor* (L.) Moench]. J. Appl. Nat. Sci. 2:245–250.

Sattler, S.E., A. Saballos, Z. Xin, D.L. Funnell-Harris, W. Vermerris, and J.F. Pedersen. 2014. Characterization of novel sorghum brown midrib mutants from an EMS-mutagenized population. G3: Genes, Genomes, Genet. 4:2115–2124. doi:10.1534/g3.114.014001

Schmitz, R.J., and R.M. Amasino. 2007. Vernalization: A model for investigating epigenetics and eukaryotic gene regulation in plants. Biochim. Biophys. Acta 1769:269–275. doi:10.1016/j.bbaexp.2007.02.003

Seifers, D.L., R. Perumal, and C.R. Little. 2012. New sources of resistance in sorghum [*Sorghum bicolor* (L.) Moench] germplasm are effective against a diverse array of *Potyvirus* spp. Plant Dis. 96:1775–1779. doi:10.1094/PDIS-03-12-0224-RE

Seigler, D.S., and A.M. Brinker. 1993. Characterisation of cyanogenic glycosides, cyanolipids, nitroglycosides, organic nitro compounds and nitrile glucosides from plants. In: P.G. Waterman, P.M. Dey, and J.B. Harborne, editors, Alkaloids and sulphur compounds. Vol. 8 of Methods in plant biochemistry. Academic Press, London. p. 51–131.

Shedge, V., J. Davila, M.P. Arrieta-Montiel, S. Mohammed, and S.A. Mackenzie. 2010. Extensive rearrangement of the *Arabidopsis* mitochondrial genome elicits cellular conditions for thermotolerance. Plant Physiol. 152:1960–1970. doi:10.1104/pp.109.152827

Shinde, V.K., K.G. Nandanwankar, and S.S. Ambekar. 1986. Heterosis and combining ability for grain yield in rabi sorghum. Sorghum Newsl. 26:19.

Shivaprasad, P.V., R.M. Dunn, B.A. Santos, A. Bassett, and D.C. Baulcombe. 2012. Extraordinary transgressive phenotypes of hybrid tomato are influenced by epigenetics and small silencing RNAs. EMBO J. 31:257–266. doi:10.1038/emboj.2011.458

Shukla, S., and S.P. Singh. 2006. Genetic divergence in relation to heterosis in opium poppy (*P. somniferum* L.). J. Med. Arom. Plant Sci. 28:4–8.

Singh, R., and J.D. Axtell. 1973. High lysine mutant gene that improves protein quality and biological value of grain sorghum. Crop Sci. 13:535–539. doi:10.2135/cropsci1973.0011183X001300050012x

Singh, S.P., and M. Singh. 2004. Multivariate analysis in relation to genetic improvement in *Cuphea procumbens*. J. Genet. Breed. 58:105–112.

Slavov, G.T., R. Nipper, P. Robson, K. Farrar, G.G. Allison, M. Bosch, J.C. Clifton-Brown, I.S. Donnison, and E. Jensen. 2014. Genome-wide association studies and prediction of 17 traits related to phenology, biomass and cell wall composition in the energy grass *Miscanthus sinensis*. New Phytol. 201:1227–1239. doi:10.1111/nph.12621

Smith, A.M., C.N. Hansey, and S.M. Kaeppler. 2012. TCUP: A novel *h*AT transposon active in maize tissue culture. Front. Plant Sci. 3:6. doi:10.3389/fpls.2012.00006

Smith, B.D. 1995. The emergence of agriculture. Scientific American Library, New York.

Smith, O.S., J.S.C. Smith, S.L. Bowen, R.A. Tenborg, and S.J. Wall. 1990. Similarities among a group of elite maize inbreds as measured by pedigree, F_1 grain yield, grain yield heterosis and RFLPs. Theor. Appl. Genet. 80:833–840. doi:10.1007/BF00224201

Sobral, B.W.S., D.P.V. Braga, E.S. LaHood, and P. Keim. 1994. Phylogenetic analysis of chloroplast restriction enzyme site mutations in the Saccharinae Griseb. subtribe of the Andropogoneae Dumort. tribe. Theor. Appl. Genet. 87:843–853.

Soeranto, H., T.M. Nakanishi, and M.T. Razzak. 2001. Mutation breeding in sorghum in Indonesia. Radioisotopes 50:169–175. doi:10.3769/radioisotopes.50.169

Song, S., Y. Huang, X. Wang, G. Wei, H. Qu, W. Wang, X. Ge, S. Hu, G. Liu, Y. Liang, and J. Yu. 2009. HRGD: A database for mining potential heterosis-related genes in plants. Plant Mol. Biol. 69:255–260. doi:10.1007/s11103-008-9421-6

Springer, N.M. 2013. Epigenetics and crop improvement. Trends Genet. 29:241–247. doi:10.1016/j.tig.2012.10.009

Sree Ramulu, K. 1970a. Induced systematic mutations in sorghum. Mutat. Res., Fundam. Mol. Mech. Mutagen. 10:77–80. doi:10.1016/0027-5107(70)90149-1

Sree Ramulu, K. 1970b. Sensitivity and induction of mutations in sorghum. Mutat. Res., Fundam. Mol. Mech. Mutagen. 10:197–206. doi:10.1016/0027-5107(70)90116-8

Sreenivasulu, N., and T. Schnurbusch. 2012. A genetic playground for enhancing grain number in cereals. Trends Plant Sci. 17:91–101. doi:10.1016/j.tplants.2011.11.003

Steinfath, M., T. Gartner, J. Lisec, R.C. Meyer, T. Altmann, L. Willmitzer, and J. Selbig. 2010. Prediction of hybrid biomass in *Arabidopsis thaliana* by selected parental SNP and metabolite markers. Theor. Appl. Genet. 120:239–247. doi:10.1007/s00122-009-1191-2

Stemler, A., J.R. Harlan, and J.M.J. De Wet. 1975. *Caudatum sorghums* and speakers of Chari-Nile languages in Africa. J. Afr. Hist. 16:161–183. doi:10.1017/S0021853700001109

Stephens, J.C., Miller, F.R. and Rosenow, D.T. 1967. Conversion of alien sorghum to early combine types. Crop Sci. 7:396. doi: 10.2135/cropsci1967.0011183X000700040036x

Stokes, D., F. Fraser, C. Morgan, M. Carmel, C.M. O'Neill, R. Dreos, A. Magusin, S. Szalma, and I. Bancroft. 2010. An association transcriptomics approach to the prediction of hybrid performance. Mol. Breed. 26:91–106. doi:10.1007/s11032-009-9379-3

Stokes, D., C. Morgan, C.M. O'Neill, F. Fraser, and I. Bancroft. 2006. Transcriptome-based predictive modeling of heterosis. Abstracts, Heterosis in Plants, Max Planck Institute, Potsdam-Golm, Germany. 18–20 May.

Swigonova, Z., J. Lai, J. Ma, W. Ramakrishna, V. Llaca, J.L. Bennetzen, and J. Messing. 2004. Close split of sorghum and maize genome progenitors. Genome Res. 14:1916–1923. doi:10.1101/gr.2332504

Taylor, J.R.N., T.J. Schober, and S.R. Bean. 2006. Novel food and non-food uses for sorghum and millets. J. Cereal Sci. 44:252–271. doi:10.1016/j.jcs.2006.06.009

Umakanth, A.V., S.S. Rao, and S.V. Kuriakose. 2006. Heterosis in landrace hybrids of post-rainy sorghum [*Sorghum bicolor* (L.) Moench]. Indian J. Agric. Res. 40:147–150.

Vandenbrink, J.P., V. Goff, H. Jin, W. Kong, A.H. Paterson, and F.A. Feltus. 2013. Identification of bioconversion quantitative trait loci in the interspecific cross *Sorghum bicolor* × *Sorghum propinquum*. Theor. Appl. Genet. 126:2367–2380. doi:10.1007/s00122-013-2141-6

Vermerris, W., A. Saballos, G. Ejeta, N.S. Mosier, M.R. Ladisch, and N.C. Carpita. 2007. Molecular breeding to enhance ethanol production from corn and sorghum stover. Crop Sci. 47(Supplement 3):S142–S153. doi:10.2135/cropsci2007.04.0013IPBS

Vogler, R., G. Ejeta, K. Johnson, and J. Axtell. 1994. Characterization of a new brown midrib sorghum line. In: 2009 Agronomy abstracts. ASA, Madison, WI. p. 124.

von Wettstein, D. 2009. Mutants pave the way to wheat and barley for celiac patients and dietary health. In: Q.Y. Shu, editor, Induced plant mutations in the genomics era. Food and Agriculture Organization of the United Nations, Rome. p. 187–190.

Wang, H., Q. Feng, M. Zhang, W. Sha, and B. Liu. 2010. Alteration of DNA methylation level and pattern in sorghum (*Sorghum bicolor* L.) pure-lines and inter-line F_1 hybrids following low-dose laser irradiation. J. Phytochem. Photobiol. B 99:150–153. doi:10.1016/j.jphotobiol.2010.03.011

Weaver, C.A., B.R. Hamaker, and J.D. Axtell. 1998. Discovery of grain sorghum germplasm with high uncooked and cooked in vitro protein digestibilities. Cereal Chem. 75:665–670. doi:10.1094/CCHEM.1998.75.5.665

Wheeler, J.L., A.C. Mulcahy, J.J. Walcott, and G.G. Rapp. 1990. Factors affecting the hydrogen cyanide potential of forage sorghum. Aust. J. Agric. Res. 41:1093–1100. doi:10.1071/AR9901093

Whittaker, J.C., R. Thompson, and M.C. Denham. 2000. Marker-assisted selection using ridge regression. Genet. Res. 75:249–252. doi:10.1017/S0016672399004462

Xiao, W., K.D. Custard, R.C. Brown, B.E. Lemmon, J.J. Harada, R.B. Goldberg, and R.L. Fischer. 2006. DNA methylation is critical for *Arabidopsis* embryogenesis and seed viability. Plant Cell 18:805–814. doi:10.1105/tpc.105.038836

Xiao, J., J. Li, L. Yuan, S.R. McCouch, and S.D. Tanksley. 1996. Genetic diversity and its relationships to hybrid performance and heterosis in rice as revealed by PCR-based markers. Theor. Appl. Genet. 92:637–643. doi:10.1007/BF00226083

Xin, Z., M.L. Wang, N.A. Barkley, G. Burow, C. Franks, G. Pederson, and J. Burke. 2008. Applying genotyping (TILLING) and phenotyping analyses to elucidate gene function in a chemically induced sorghum mutant population. BMC Plant Biol. 8:103. doi:10.1186/1471-2229-8-103

Xin, Z., M.L. Wang, G. Burow, and J. Burke. 2009. An induced sorghum mutant population suitable for bioenergy research. BioEnergy Res. 2:10–16. doi:10.1007/s12155-008-9029-3

Xu, Y.Z., de la Rosa Santamaria R., K.S. Virdi, M.P. Arrieta-Montiel, F. Razvi, S. Li, G. Ren, B. Yu, D. Alexander, L. Guo, X. Feng, I.M. Dweikat, T.E. Clemente, and S.A. MacKenzie. 2012. The chloroplast triggers developmental reprogramming when MUTS HOMOLOG1 is suppressed in plants. Plant Physiol. 159:710–720. doi:10.1104/pp.112.196055

Xu, Y.Z., M.P. Arrieta-Montiel, K.S. Virdi, W.B. de Paula, J.R. Widhalm, G.J. Basset, J.I. Davila, T.E. Elthon, C.G. Elowsky, S.J. Sato, T.E. Clemente, and S.A. Mackenzie. 2011. MutS HOMOLOG1 is a nucleoid protein that alters mitochondrial and plastid properties and plant response to high light. Plant Cell 23(9):3428–3441. doi:10.1105/tpc.111.089136

Yang, L., J.D. Browning, and J.M. Awika. 2009. Sorghum 3-deoxyanthocyanins possess strong phase II enzyme inducer activity and cancer cell growth inhibition properties. J. Agric. Food Chem. 57:1797–1804. doi:10.1021/jf8035066

Yu, J., and E.S. Buckler. 2006. Genetic association mapping and genome organization of maize. Curr. Opin. Biotechnol. 17:155–160. doi:10.1016/j.copbio.2006.02.003

Yu, J., J.B. Holland, M.D. McMullen, and E.S. Buckler. 2008. Genetic design and statistical power of nested association mapping in maize. Genetics 178:539–551. doi:10.1534/genetics.107.074245

Zhang, Q.F., Y.J. Gao, S.H. Yang, R.A. Ragab, M.A. Saghai Maroof, and Z.B. Li. 1994. A half-diallel analysis of heterosis in elite hybrid rice based on RFLP and microsatellites. Theor. Appl. Genet. 89:185–192.

Zhang, M., C. Xu, D. von Wettstein, and B. Liu. 2011. Tissue-specific differences in cytosine methylation and their association with differential gene expression in sorghum. Plant Physiol. 156:1955–1966. doi:10.1104/pp.111.176842

Zhang, T., X.L. Ni, K.F. Jiang, H.F. Deng, Q. He, Q.H. Yang, L. Yang, X.Q. Wan, Y.J. Cao, and J.K. Zheng. 2010. Relationship between heterosis and parental genetic distance based on molecular markers for functional genes related to yield traits in rice. Rice Sci. 17:288–295. doi:10.1016/S1672-6308(09)60029-9

Zhao, Y., M. Gowda, W. Liu, T. Wurschum, H.P. Maurer, F.H. Longin, N. Ranc, and J.C. Reif. 2012. Accuracy of genomic selection in European maize elite breeding populations. Theor. Appl. Genet. 124:769–776. doi:10.1007/s00122-011-1745-y

Sorghum Improvement for Yield

Leo Hoffmann, Jr and William L. Rooney*

A C_4 grass with center of origin in Africa, sorghum (*Sorghum bicolor* L. Moench) is adapted to a diverse set of environments ranging from arid and semiarid to tropical and temperate regions throughout the world. As a species, sorghum is classified into five distinct races; bicolor, caudatum, durra, guinea and kafir, and an array of intermediate classes (Harlan and de Wet, 1972). The genetic diversity and wide adaptation provide current sorghum improvement programs with a wide array of unique phenotypes and genotypes to use in future breeding efforts.

Sorghum ranks fifth among cereal grain crops in area planted and total production (FAOSTAT, 2014). This ranking is as a grain crop; however, it is equally important, if not more so, as a forage and fuel crop. Depending on the end use, yields of value can be either grain, or biomass, or a combination of the two.

Yields in sorghum are highly variable and influenced by environment, genotype and their interaction. According to FAOSTAT (2014), the worldwide sorghum grain yield average was 1457 kg ha^{-1} in 2013. However, there was significant variation based on country of production. For example, the top producing countries were in excess of 3000 kg ha^{-1}. When compared to the documented yield potential of grain sorghum which is as high as 15,000 kg ha^{-1} (Rooney, 2004), these averages indicate that actual yields are only 20% of the yield potential of the crop. Consequently, there are not only opportunities to improve the genetic yield potential of the crop, but to identify the environmental constraints that produce such a low actual to potential yield.

The purpose of this chapter is to discuss the challenges of breeding sorghum targeting yield improvement through both offensive and defensive breeding strategies. Within this context, offensive breeding refers to the genetic improvement of yield potential, while defensive breeding refers to the incorporation of traits designed to mitigate losses due to both abiotic and biotic stresses.

Historical Perspective: From Domestication to Hybrids

The center of origin for sorghum is Eastern Africa, and much of the diversity within the species still exists within this region (Dahlberg, 2000). By all accounts, sorghum was domesticated in the same area when humans identified and selected plants of value. As a self-pollinated species, these new variants likely arose from

---

Abbreviations: QTL, quantitative trait loci.

Department of Soil and Crop Sciences, Texas A&M University, College Station, TX 77843-2474. *Corresponding author (wlr@tamu.edu)

doi:10.2134/agronmonogr58.2014.0055

 Sorghum: State of the Art and Future Perspectives, Ignacio Ciampitti and Vara Prasad, editors. Agronomy Monograph 58.

outcrossing and/or mutations that were subsequently made genetically homozygous as the plants continued to self-pollinate over generations. If the trait was of value (either naturally or to those producing it), it was maintained. This form of "improvement" was utilized and moved with the crop as it expanded around the world. Only in the early 20th century were the principles of basic genetics applied to sorghum improvement.

The movement and development of sorghum in the United States demonstrates a similar scenario. While it is not clearly documented, the first sorghums introduced to the United States were likely associated with the West African slave trade. These sorghums were tall, late to anthesis, and had the open-panicle characteristic of Guinea-race sorghums typical in Western Africa. However, these types were of little value for grain production in temperate environments; they functionally disappeared without contributing to further sorghum development in the United States.

Grain sorghum grown in the United States today traces its lineage back to the late 19th century when Durra and Kafir race sorghums were brought to North America (Smith and Frederiksen, 2000; Vinall et al., 1936). Like previous introductions, early sorghums were typically tall, and late (usually due to photoperiod sensitivity) and they easily lodged. All of these factors resulted in marginal grain yield potential. Farmers made the initial selections for adaptation (earlier maturity) height (tall to short) and panicle shape, consequently early sorghum crop scientists introduced selected cultivars with improved grain harvesting characteristics, notably shorter and earlier, with slightly improved and stable yields. (Smith and Frederiksen, 2000).

The application of genetic principles to sorghum in the early 20th century allowed sorghum agronomists and breeders to directly improve the crop. The first sorghum cultivars produced from intentional crosses were 'Chiltex' and 'Premo', released in 1923 by Vinall and Cron (Quinby and Martin, 1954). Since that time, sorghum improvement programs have systematically modified plant height, maturity, biotic and abiotic stress tolerance as well as yield by crossing and selecting pure-line cultivars from the segregating progeny. These pure-line cultivars of sorghum were used until an economically feasible method of producing hybrid seed was developed.

Heterosis in sorghum was documented long before hybrids were commercially available (Quinby and Karper, 1946), but the absence of any obvious means to produce a hybrid minimized work in the area. However, interest in hybrid sorghum was further increased by the development of hybrid corn and its acceptance by corn producers; it was not until a cytoplasmic male sterility system was identified in sorghum that the prospect of hybrid sorghums was economically feasible (Stephens and Holland, 1954). The system was integrated and adopted quickly using standard cultivars as parental lines with first hybrids sold in 1956 and by 1960, hybrids were planted on over 90% of grain sorghum production area.

The development of hybrid sorghum transitioned the sorghum industry from a few publicly-supported breeding programs to numerous private industry breeding programs. In all situations, sorghum breeding programs increased in both size and scope of research. Offensive traits such as grain yield under high inputs (water, fertilizer) were prioritized, as was the protection of yield from either biotic or abiotic stresses. These factors formed the basis of the modern sorghum breeding programs that are now working to improve the crop. These efforts led to substantial improvements in yield potential, grain and forage quality and the protection of this potential through abiotic and biotic stress tolerances.

Rates of Yield Gain

Yield gains over the period of systematic improvement document the preceding history and can be roughly defined into pre- and post-hybrid phases. From 1908 to 1955, national average yields increased 300% from 500 to 1500 kg ha^{-1} primarily on adaptation and selection (Monk et al., 2014). The second phase of genetic improvement effectively captured the heterotic potential as well as the increase in breeding efforts and from 1955 to 1980, improved genetics increased yield approximately 2% annually (Miller and Kebede, 1984). Since that time, sorghum improvement rates have slowed due several different factors, including changes in agronomic management (less irrigation) and shifts to less productive ground, and reduced research spending due to industry consolidation (Monk et al., 2014).

Current Breeding Efforts

Whether the cultivar grown is a true breeding pure line (inbred) or hybrid, most sorghum breeding programs rely on the production of segregating populations from which new inbred lines are derived. Where pure-line cultivars are grown, the potential of these new lines is evaluated; if hybrids are the ultimate goal, inbred line development is combined with a testcrossing program to identify those lines with superior general and specific combining ability.

For either an inbred or hybrid cultivar, inbred lines must be produced. While specific approaches vary and may include breeding methods such as pedigree, bulk, and single-seed descent (Acquaah, 2012), but they all allow for self-pollination. In general, specific breeding crosses are made between lines for the specific improvement of certain traits (Rooney, 2004). F_1 progeny are self-pollinated to produce an F_2 population. From the F_2 generation, until a generation in which sufficient uniformity exists (defined by the breeder), breeders use various methods of selection for agronomic, disease-tolerance, and stress-tolerance traits. The appropriate generation for the selection of specific traits is dependent on the heritability of the trait and the production environments. Traits with higher heritability (maturity, height, grain color, etc.) are generally selected in the early generations. To speed the process, there is interest in developing doubled haploid technology, which would result in the production of inbred lines more quickly than self-pollination through multiple generations. Haploids have long been documented in sorghum (Brown, 1943), but specific and effective systems that are applicable in breeding programs have yet to be identified and developed.

Most hybrid sorghum breeding programs divide their inbred development into seed parents (*A*/*B* lines) and pollinator (*R* lines) parents. Separating the two groups maintains heterosis that exists either naturally or has been developed over sixty years of systematic breeding for hybrid performance. Testcrossing of new inbred lines is typically initiated between the F_3 and F_5 generations, with F_4 being the most common. Pollinator parents are immediately available for testcrossing, while the new seed parents (*B*-lines) must be male-sterilized, which typically involves backcrossing the new line to an *A*-line (Rooney, 2004). The *B*-line is the recurrent parent, and with enough backcrosses an *A*-line is produced that is isocytoplasmic to the *B*-line. This process adds approximately two to four years to the development of new *A*-lines.

In testcrossing, new lines are hybridized to standard tester lines from the opposite heterotic group to identify the lines with the best general combining ability. Those lines are then advanced for additional testing to identify unique hybrid combinations with superior specific combining ability. Elite hybrids have high yield, good stability and agronomic characteristics (height, maturity, etc.), and abiotic and biotic stress tolerances sufficient for the area of adaptation and are candidates for commercialization.

Essential to any sorghum breeding program is access to genetic diversity. Sorghum is an incredibly diverse species with variation for most traits of importance to sorghum production. That diversity is critical to sorghum improvement. Duncan et al. (1991) estimated that almost every sorghum hybrid currently grown in the United States had germplasm derived from exotic sorghum accessions in their pedigree wherein an exotic parent is a genotype that is not adapted to the production region. Currently, large collections of sorghum germplasm are curated at the International Center for Research in the Semi-Arid Tropics (ICRISAT) and the USDA Germplasm Resources Information Network (GRIN), which hold approximately 30,000 and 40,000 sorghum accessions, respectively. One limitation to the utilization of most of the accessions is the lack of adaptation to the target environment. The most common form of poor adaptation is photoperiod sensitivity, wherein a photoperiod sensitive genotype will not flower under the long days in temperate environments. To mitigate this issue, sorghum geneticists have used several approaches. First, the sorghum conversion program converted exotic sorghums from photoperiod sensitive to photoperiod insensitive (Stephens et al., 1967). This program was highly successful, producing over 700 converted lines, many of which have been used in applied breeding programs (Dahlberg et al., 1998; Rosenow et al., 1997). Jordan et al. (2011) utilized unadapted sorghum germplasm through a modified backcross program combined with agronomic desirability selection to produce breeding lines with agronomic potential and additional genetic diversity and variation. This system has been effective in tapping into exotic sorghum germplasm for a variety of different traits that enhance or protect productivity and quality.

Trait-Based Breeding Efforts

Because sorghum is grown across a wide range of environments, the optimum ideotype varies among these production regions. Regardless, there are common traits of importance to all producers and in all environments. First and foremost, successful hybrids are high yielding and stable in their production across environments. Because water is commonly limited in sorghum production environments, drought tolerance is a required trait as well. These two traits demonstrate the challenge of integrating both offensive (inherent yield potential) and defensive (protection of yield potential from abiotic or biotic stresses) traits. In sorghum, history demonstrates that overemphasis on either the offensive trait or the defensive traits can be problematic. For example, pauses in the productivity of sorghum hybrids has been influenced by the initial absence of resistance to a disease or insect followed by an overemphasis on breeding for resistance which slowed efforts in breeding for yield (Monk et al., 2014). This has also been demonstrated in the initial transgenic soybeans from the late 1990s (Elmore et al., 2001).

Consequently, a balanced approach to improvement of both defensive and offensive traits is necessary.

Defensive Breeding

Defensive traits are defined as traits that protect the inherent yield potential or quality attributes of a crop. For the most part, these genes and/or traits do not increase the yield potential of the crop, but they reduce or minimize losses when the stress occurs. In some situations, the presence of a defensive trait has had a negative effect on performance in the absence of the stress to which it provides protection; or in other situations, breeding for stress tolerance has reduced breeding for increased yield (Monk et al., 2014). Therefore, improvement programs must identify and prioritize which stresses are common in their target area and breed tolerance to those stresses.

Sorghum production encounters both abiotic and biotic forms of stress. The most prevalent form of abiotic stress is drought. Worldwide, water stress, primarily in the form of drought, is the single largest factor in reducing sorghum productivity. Other forms of abiotic stresses include temperature (heat and/or cold), and soil fertility and/or composition (specifically soil pH, micronutrients, or fertility). Among the biotic stresses, both insects and plant pathogens can reduce yield and quality. However, their presences are target-area specific and often related to endemic weather conditions in the region of production. A brief summary of breeding effort for the major stress tolerances is provided below.

Drought

While sorghum is well-known for its drought tolerance, there is significant variation among sorghum in the expression of this trait. Drought stress causes reductions in plant height, plant biomass, delays in panicle development, kernel size, kernel number and increase in lodging, stalk rot (*Macrophomina phaseolina*), and ultimately grain yield loss (Rosenow et al., 1983). The two types of drought stress response are generally recognized in sorghum as pre- and post-flowering drought stress (Rosenow and Clark, 1981). As described, pre-flowering drought stress resistance is important prior to flowering, and post-flowering drought stress is important after flowering. Sources of resistance have been identified for both and they are independent traits. In most cases, germplasm with tolerance of one type; germplasm with both pre- and post-flowering drought tolerance are rare (Rosenow et al., 1983).

The most important drought tolerance trait in sorghum is stay-green. This post-flowering drought tolerance delays senescence and essentially keeps the plant alive (Rosenow and Clark, 1981). Stay-green, or versions thereof, have been incorporated into sorghum hybrids grown around the world. Borrell et al. (2000) reported that there is no detriment to sorghum yield by having stay-green when fields were irrigated, but under dry land (drought emulation) hybrids with confirmed stay-green had significantly higher yields than hybrids without.

The stay-green trait is well characterized with several different approaches to indirectly phenotype for its presence. These include visual screening under drought stress, leaf chlorophyll (Xu et al., 2000) and dhurrin concentrations (Hayes et al., 2016), and use of quantitative trait loci (QTL) identified on chromosomes 2 (*Stg3*), 3 (*Stg1* and *Stg2*), and 5 (*Stg4*) of sorghum (Sanchez et al., 2002).

Another approach to improve drought tolerance includes evaluation of root system size and branching (Mace et al., 2012; Singh et al., 2010). Masi and

Maranville (1998) proposed the utilization of fractal geometry to predict root system structure for different genotypes of sorghum. Once fractal models were built, they could be applied to improve the efficiency of root phenotyping. More recently, the use of ground penetrating radar (GPR) techniques to estimate root mass and structure (Zhu et al., 2011). The goal is to use these methods to develop molecular markers and QTLs associated with root traits that can be deployed in MAS or genomic selection systems.

Improvement for drought tolerance can also be achieved by capitalizing on epicuticular wax on leaf and stalk. Literature showed that increase levels of wax can reduce in evapotraspiration in sorghum fields (Ebercon et al., 1977; Jordan et al., 1984). Awika (2012) found inverted correlation between wax load in and canopy temperatures especially under drought emulated environments.

Soil Fertility and pH

Sorghum is typically produced in marginal production environments where some form of nutrient stress is encountered. Depending on the location and soil type, these nutrient deficiencies include aluminum toxicity in acid soils, salinity toxicity, and iron chlorosis on alkaline soils. As such research to characterize the basic response of the sorghum plant and to identify sources of tolerance has been completed.

Estimates indicate that approximately half of cropland soils in the World are acidic, and are a major impediment for agriculture in countries or crops with low inputs and where lime to adjust soil pH is not available. When sorghum is grown on soils with a pH < 5, aluminum (Al^{3+}) toxicity and low phosphorus (P) become significant problems and reduce sorghum yield by inhibiting root growth (Caniato et al., 2011; Leiser et al. 2012, 2014; Samac and Tesfaye, 2003). Efforts to breed sorghum to tolerate Al^{+3} have been successful, transferring tolerance from several sources (SC283, CMS225 and SC566) in elite breeding lines (Magalhaes et al., 2004). The genes associated with this tolerance have been identified as Alt_{sb} and *sbMATE* (Kochian et al., 2009; Leiser et al., 2014; Magalhaes et al. 2004, 2007).

As a species, sorghum genotypes exhibit moderate tolerance to saline soils and sorghum is more tolerant to saline soils at germination and seedling growth than any other stage (Francois et al., 1984). The effect of saline soils was more pronounced on grain yield than on vegetative biomass yield. Yang et al. (1990) reported that *Sorghum halepense* (L.) Pers. was more tolerant than *Sorghum bicolor* (L.) Moench and that *S. halepense* may be useful for increasing the salinity tolerance of grain sorghum.

Sorghum is highly susceptible to iron chlorosis which occurs in high-pH, calcareous soils where the availability of soluble iron in the soil is reduced. Because of soil variability, plants in susceptible areas of the field are chlorotic and slower growing, which results in uneven crop growth and maturity, reduces yield and introduces management challenges. Genetic variation for iron chlorosis was reported in multiple studies in both inbred lines and hybrids (Esty et al., 1980; McKenzie et al., 1984). Most of these studies are completed in laboratory assays and while differences in genotypes have been described under field conditions, breeding for field level tolerance has not solved this problem. Consequently, foliar applications of iron remain the most effective method to alleviate the deficiency (Withee and Carlson, 1959).

Insects

While the number of pathogens and insects that cause economic damage to sorghum is modest, those that do can have devastating effects on crop productivity and/or quality. Among the insects, headworm [*Helicoverpa zea* (Boddie)], midge [*Contarinla sorghicola* (Coquillett)], greenbug [*Schizaphis graminum* (Rondani)] and sugarcane aphid (*Melanaphis sacchari*) are insect pests in that category. Breeding for tolerance of these pests has had varying success.

Losses due to greenbug vary from 0 to 80% of potential yield, depending of the interaction biotype, environment, host (Radchenko and Zubov, 2007). Several biotypes (B, C, E, F, I, H, and K) of greenbug have been identified, forcing breeders to identify new sources of resistance. From that screening, numerous sources of resistance were made available and successfully used in breeding programs (Dixon et al., 1991; Harvey et al. 1991, 1997; Nagaraj et al., 2005; Peterson et al., 1984; Radchenko and Zubov, 2007). Between 1970 and 1975, sorghum improvement programs devoted a large proportion of their total breeding program to greenbug resistance. While this was very successful, this overemphasis on a single trait is often identified as the reason for a reduced rate of improvement in yield during the same time (Monk et al., 2014).

Sorghum midge causes damage when midge larvae consume the developing sorghum seed resulting in loss of grain development. Genetic tolerance to midge has been identified in sorghum and the mechanisms of resistance are often different and range from antibiosis to structural inhibition (Johnson et al., 1973; Page, 1979). Using these sources, breeding for midge resistance has been possible but successful deployment has varied across production regions. For example, breeding for midge resistance has been very successful in Australia, but cultural and chemical control methods are more effective in the Southern United States. The differences are primarily due to the insect life cycle and environment in which the pressure occurs. Quantitative train loci for midge resistance have been identified with two QTL associated to antixenosis (non-preference) and another QTL for antibiosis (Tao et al., 2003). For midge management, the use of cultural, chemical and genetic resistance has all been important.

In the case of headworm, genetic resistance has not been identified, so avoidance and chemical control have been the only mechanisms of control. Transgenic approaches might be effective, but economic and gene flow issues have stopped the deployment of this approach.

More recently, the sugarcane aphid has become a significant pest of sorghum. Sources of resistance have been identified and some, but not all are also resistant to the greenbug (Mbulwe et al., 2016; Armstrong et al., 2015). To date, control has been primarily chemical, but as tolerance and/or resistance are bred into elite germplasm, genetic tolerance should eventually become the primary mechanism of control.

Diseases

Common diseases that result in economic losses in sorghum yield include anthracnose (*Colletotricum sublineolum*), downy mildew (*Peronosclerospora sorghi*), smuts (*Sporisorium reilianum* and *S.cruenta*) and grain molds (caused by *Fusarium* sp., *Alternaria* sp., and *Curvularia* sp.). In addition, plant lodging caused by Charcol rot (*Macrophomina phaseoli*), Fusarium Stalk Rot (*Fusarium ssp.)* and anthracnose

(*Colletotrichum graminicola*) are likely the most important diseases in grain sorghum given the severity and potential of yield losses they can inflict. Breeding for tolerance of these pests have had varying success.

Most phenotypic screening for disease resistance is completed at the field in environments conducive to the disease. In some cases, fields may be inoculated to ensure the presence of the disease. Inoculation provides a fair distribution of the infectious agent through the field and increases efficiency of selection. In other situations, greenhouse screenings supplement or replace the field screenings, especially at the initial phases of selection.

Control of anthracnose is almost exclusively via genetic resistance. Numerous sources of resistance have been identified, but resistance is often environment specific due to race differences in the pathogens at these different locations (Mehta et al., 2005). In most cases the genetics of resistance is controlled by different, but highly heritable genes (Mehta et al., 2005). In some cases, QTL have been identified for the resistance and marker-assisted breeding has been used in breeding (Burrell et al., 2015, Mehta et al., 2005; Patil et al., 2017).

Downy mildew is an example of a plant disease mitigated using genetic and/or chemical control. Soon after its emergence as economically damaging disease, its impact was mitigated by effective genetic resistance. However, the pathogen has often shifted pathotypes resulting in new outbreaks of the disease (Isakeit and Jaster, 2005). Resistance to this pathotype was more difficult to breed; concurrently, chemical seed treatments provided suitable protection and soon reduced the emphasis on genetic resistance. Eventually, pathotypes with resistance to chemical control emerged and at that point genetic resistance became important as well. Currently, both methods are used to control the disease.

One challenge in screening for downy mildew is the inconsistency of expression; inoculum pressure and distribution through the field varies from year to year. To mitigate this problem, infection rates are monitored by the continual inclusion of susceptible checks to monitor disease pressure and maintain inoculum. Downy mildew resistance is considered is often multigenetic, and has high level interaction to environment and host (Rooney et al., 2002). Utilization of markers and breeding for target environment are the most efficient to approach this disease (Perumal et al., 2008).

Soon after hybrids were introduced and widely grown, sorghum head smut expanded as a significant problem because several early hybrids were quite susceptible to the disease. Control of this trait is primarily via genetic resistance and significant variation for resistance exists in sorghum germplasm (Oh and Magill, 1994). Like downy mildew, the inconsistency in expression of head smut requires consistent soil inoculum and continual inclusion of susceptible checks to monitor disease pressure and maintain inoculum.

The grain mold and weathering complex describes the diseased appearance of sorghum grain resulting from infection by one or more parasitic fungi (Williams and Rao, 1981). The trait is of significant economic impact when grain maturity occurs during rainy, wet or humid weather. Like many other diseases, the primary form of control is genetic resistance. Over the years, an array of screening and selection methodologies have been developed for grain mold, and significant information has been published on the identification of resistance, the mechanism of resistance, the inheritance of resistance, and the potential for future improvements (Bandyopadhyay et al., 1988, Castor and Frederiksen, 1980). From this work, numerous traits (kernel hardness, tannins, grain color) have been associated with increased grain mold resistance, but no specific trait confers complete resistance to the disease (Esele et al., 1993). The germplasm sources of grain mold resistance in sorghum have been identified which contain those various kernel traits that influence the trait. Each of these mechanisms and the sources of resistance have been very important in the development of new germplasm with even better grain mold resistance.

Many of the traits associated with grain mold and weathering tolerance are qualitatively inherited (Rooney, 2004) but their contributions to total grain mold resistance usually accounts for only a portion of the total variation for the trait. Thus, grain mold resistance commonly is considered a quantitative trait, and general and specific combining ability components of variation for grain mold resistance are highly significant (Dabholkar and Baghel, 1980). Several QTL for grain mold resistance have been identified (Klein et al., 2001; Rami et al., 1998) Thus, breeding for stable resistance to grain mold is difficult because grain mold resistance mechanisms differ by environment, and resistance mechanisms are controlled by different genetic loci.

Lodging and Stalk Rot Resistance

Stalk rots and stalk lodging are probably the most important stress in any type of sorghum production because of their immediate impact on yield; lodged plants that cannot be harvested directly reduce yield and quality is reduced even if the grain, forage, and/or biomass is harvestable. Several different types of stalk rots and lodging occur in sorghum and are associated with different production environments. Of all the stalk rots, the most common is charcoal rot [caused by *Macrophomena phaeseolina* (Tassi) Goid.] and it is typically associated with post-flowering drought conditions. The most effective means of control of charcoal rot is either avoidance of drought conditions or introgression of stay-green to maintain the stalk health of the crop as the crop is matures. *Fusarium* stalk rot (caused by *Fusarium moniloforme sensu lato*) and anthracnose stalk rot [caused by *Colletotrichum graminicola* (Ces.) Wils.] are not as common as charcoal rot but they are important in specific regions. For Fusarium and Anthracnose stalk rot, screening methodologies involve pathogen inoculation and visual evaluation of infection (Rosenow, 1984, Tesso et al., 2005). For both Fusarium and anthracnose sources of resistance have been identified and utilized in breeding programs (Bandara et al., 2015; Bramel-Cox et al., 1988).

Offensive Breeding

Sorghum yield potential depends on type of cultivar (inbred line or hybrid), the specific genotype and its adaptation of the environment, and the environment. Sorghum grain yield consists of several different interrelated yield components. Those yield components include but are not limited to seed number per panicle, panicle number per area, seed size, grain fill duration, plant leaf architecture, seed nutrient composition, flowering time, plant height and yield potential and correlations among them are well documented (Burow et al., 2014; Miller and Kebede, 1984; Morgan et al., 2002; Murray et al., 2008; Maman et al., 2004; Yang et al., 2010).

Yield Components

Seed size varies significantly in sorghum ranging from 1.41 to 5.5g 100 seed^{-1} (Han et al., 2015, Tuinstra et al., 2001) and the trait is moderately to highly heritable. Seed size is important from both a processing and utilization perspective, but in terms of yield potential there is little correlation between seed size and yield. Multi-seeded mutants have been proposed as a mechanism to increase grain yield by increasing the number of seed in a panicle. Twin seeded sorghum was reported and hybrids developed in the mid-1970s (Miller, 1976), but comparable yields and processing issues ended interest in this concept. More recently, a multi-seeded (tri-seed) mutant was described in which seed size was reduced but seed number tripled resulting in up to 37% higher yield than the wild-type (Burow et al., 2014). However, recent testing of these lines combinations indicated that while seed number increased, the total grain yield was reduced (Tolk and Schwartz, 2017).

Another method to increase yield involves panicle size and panicle number per unit area. Often these two traits are inversely related, but that interaction is further complicated by tillering ability of sorghum hybrids. In corn, increasing plant density has resulted in more ears and/or acre, and is the primary reason for yield increases (Duvick, 1984). A similar approach might be possible, but tiller in sorghum often compensates for planting density. However, sorghum plants with upright leaves and shade tolerance would likely increase yield potential (Morgan et al., 2002, Xin et al., 2009).

Breeding for root architecture would be a possible means to improve sorghum to sustain yield per plant while increase plant populations. In studies at the University of Queensland, genetic variability for root shape and root angle were detected among sorghum genotypes (Singh et al., 2010) and some structures were correlated to water use on deeper layers of the soil profile as adaptation to water limited environments (Mace et al., 2012).

Plant Height

The general tendency in all crops is to select genotypes that are short as possible to facilitate mechanized harvesting, reduce lodging, and to improve water use efficiency. In sorghum, there is a positive correlation between height and grain yield (Rooney, 2004). Consequently, sorghum breeders must choose between a taller hybrid with outstanding yield performance, or an average height sorghum with an average yield. This issue exemplifies how yield potential is limited by pragmatism and market demand.

New Breeding Approaches for Yield Improvement

As exemplified at the beginning of the chapter, world demand for cereals in general, food or feed is increasing making plant breeding an essential tool to prevent food security crises in the future. In the past, hybrid vigor, maturity and dwarfing genes were the primary approaches to increase adaptation and yield. Today with the integration of informatics and genome sequencing analysis (bioinformatics) the amount of data generated that can be used in breeding is enormous. It is safe to say that contemporary sorghum breeding programs must have broader disciplinary interaction. To match the speed and efficiency of genotyping, phenotyping must be done faster and more comprehensively, with multiple traits observed simultaneously to allow for their integration into bioinformatics and plant breeding decisions.

High-throughput phenotyping provides the potential to evaluate genotypes faster and for traits heretofore impossible to measure using traditional approaches. Initial studies in this area are promising; height can be measured effectively (Shi et al., 2016) and traits previously mapped using traditional approaches have been mapped to the same genomic location using data collected from remote sensing (Tanger et al., 2017). However, this area is new and must be developed both in the form of the appropriate screening methodology and the hardware and software available to complete the phenotyping.

In corn breeding programs, the adoption and deployment of doubled haploid technology has revolutionized the inbred line development process. This process is successful because inducer lines were selected that increased the frequency of haploid progeny high enough to justify the effort to produce doubled haploid (Röber and Geiger, 2005). There is interest in deployment of the same technology, but to date, haploid induction frequencies have been too low to justify use in sorghum. Even if the frequencies increase, temporal sterility systems must be developed so that the same system will work in sorghum.

Finally, genomic selection and gene editing has emerged as means to increase the efficiency of breeding through improved selection and trait modification. Significant resources have been devoted to the development of genomic resources in sorghum (U.S. Department of Energy ARPA-E TERRA projects) and many groups have used those resources to characterize both germplasm and traits of economic importance. Many of these resources have been used to develop marker assisted selection programs for traits of specific importance, but to date, genomic selection programs have not been specifically described in sorghum (Bernardo and Yu, 2007; Nakaya and Isobe, 2012).

In summary, utilization of all assets available is a must for a breeding program determined to have a holistic and comprehensive approach on yield improvement. Decline in cost of molecular analysis are to its lowest, allowing outsource of services of this matter to be contract, making it accessible even to the smallest breeding programs. Therefore, phenotyping faster and in larger numbers became the new challenge for breeding programs. The use of bioinformatics is what is going to revolutionize the dynamics of how plant breeding is done.

References

Acquaah, G. 2012. Principles of plant genetics and breeding, 2nd ed. Wiley-Blackwell, New York. doi:10.1002/9781118313718

Armstrong, J.S., W.L. Rooney, G.C. Peterson, R.T. Villenueva, M.J. Brewer, and D. Sekula-Ortiz. 2015. Sugarcane aphid (Hemiptera: Aphididae): Host range and sorghum resistance including cross-resistance from greenbug sources. J. Econ. Entomol. 108:576–582. doi:10.1093/jee/tou065

Awika, H.O. 2012. Determining genetic overlap between staygreen, leaf wax and canopy temperature depression in sorghum RILs. M.S. thesis. Texas A&M University. College Station, TX.

Bandara, Y.M.A.Y., R. Perumal, and C.R. Little. 2015. Integrating resistance and tolerance for improved evaluation of sorghum lines against Fusarium stalk rot and charcoal rot. Phytoparasitica 43:485–499. doi:10.1007/s12600-014-0451-0

Bandyopadhyay, R., L. Mughogho, and K.P. Rao. 1988. Sources of resistance to sorghum grain molds. Plant Dis. 72:504–508. doi:10.1094/PD-72-0504

Bernardo, R., and J. Yu. 2007. Prospects for genomewide selection for quantitative traits in maize crop science. Crop Sci. 47:1082–1090. doi:10.2135/cropsci2006.11.0690

Borrell, A., G. Hammer, and R.G. Henzell. 2000. Does maintaining green leaf area in sorghum improve yield under drought? II. Dry matter production and yield. Crop Sci. 40:1037–1048. doi:10.2135/cropsci2000.4041037x

Bramel-Cox, P.J., I.S. Stein, D.M. Rodgers, and L.E. Claflin. 1988. Inheritance of resistance to Macrophomina phaseolina (Tassi) Goid. and Fusarium moniliforme Sheldom in sorghum. Crop Sci. 28:37–40. doi:10.2135/cropsci1988.0011183X002800010009x

Brown, M.S. 1943. Haploid plants in sorghum. J. Hered. 34:163–166. doi:10.1093/oxfordjournals.jhered.a105274

Burow, G., Z. Xin, C. Hayes, and J. Burke. 2014. Characterization of a Multiseeded (msd1) Mutant of Sorghum for Increasing Grain Yield. Crop Sci. 54:2030–2037. doi:10.2135/cropsci2013.08.0566

Burrell, A.M., A. Sharma, N.Y. Patil, S.D. Collins, W.F. Anderson, and W.L. Rooney. 2015. Sequencing of an anthracnose-resistant sorghum genotype and mapping of a major QTL reveal strong candidate genes for anthracnose resistance. Crop Sci. 55:790–799. doi:10.2135/cropsci2014.06.0430

Caniato, F.F., C.T. Guimarães, M. Hamblin, C. Billot, J.-F. Rami, B. Hufnagel, L.V. Kochian, J. Liu, A.A.F. Garcia, C.T. Hash, P. Ramu, S. Mitchell, S. Kresovich, A.C. Oliveira, G. de Avellar, A. Borem, J.-C. Glaszmann, R.E. Schaffert, J.V. Magalhaes. 2011. The Relationship between population structure and aluminum tolerance in cultivated sorghum. PLoS One 6:e20830. doi:10.1371/journal.pone.0020830

Castor, L., and R. Frederiksen. 1980. Fusarium head blight occurrence and effects on sorghum yield and grain characteristics in Texas. Plant Dis. 64:1017–1019. doi:10.1094/PD-64-1017

Dabholkar, A., and S. Baghel. 1980. Inheritance of resistance to grain mould of sorghum. Indian J. Genet. Plant Breed. 40:472–475.

Dahlberg, J. 2000. Classification and characterization of sorghum. Sorghum: Origin, history, technology, and production: 99-130.

Dahlberg, J., D. Rosenow, G. Peterson, L. Clark, F. Miller, and A. Sotomayor-Rís. 1998. Registration of 40 converted sorghum germplasms. Crop Sci. 38:564–565. doi:10.2135/crop sci1998.0011183X003800020090x

Dixon, A.G.O., P.J. Bramel-Cox and T.L. Harvey. 1991. [*Sorghum bicolor* (L.) Moench] Complementarity of genes for resistance to greenbug [*Schizaphis graminum* (Rondani)], biotype E, in sorghum. Theor. Appl. Genet. 81:105–110. doi:10.1007/BF00226119

Duncan, R., P. Bramel-Cox, and F. Miller. 1991. Contributions of introduced sorghum germplasm to hybrid development in the USA. Use of plant introductions in cultivar development part 1: 69-102. Crop Science Society of America and American Society of Agronomy, Madison, WI.

Duvick, D.N. 1984. Genetic contributions to yield gains of U.S. hybrid maize, 1930 to 1980. In: W.R. Fehr, editor, Genetic contributions to yield gains of five major crop plants. Crop Science Society of America, American Society of Agronomy, Madison, WI. p. 15-47.

Ebercon, A., A. Blum, and W.R. Jordan. 1977. A rapid colorimetric method for epicuticular wax context of sorghum leaves. Crop Sci. 17:179–180. doi:10.2135/cropsci1977.0011183X001700010047x

Elmore, R.W., F.W. Roeth, L.A. Nelson, C.A. Shapiro, R.N. Klein, S.Z. Knezevic, and A. Martin. 2001. Glyphosate-resistant soybean cultivar yields compared with sister lines. Agron. J. 93:408–412. doi:10.2134/agronj2001.932408x

Esele, J.P., R.A. Frederiksen, and F.R. Miller. 1993. The association of genes controlling caryopsis traits with grain mould resistance in sorghum. Phytopathology 83:490–495. doi:10.1094/Phyto-83-490

Esty, J.C., A.B. Onken, L.R. Hossner, and R. Matheson. 1980. Iron use efficiency in grain sorghum hybrids and parental lines. Agron. J. 72:589–592. doi:10.2134/agronj1980.000 21962007200040004x

FAOSTAT. 2014. World agriculture: Towards 2015/2030 summary report. Food and Agriculture Organization of the United States. Rome, Italy. http://faostat.fao.org/site/567/ DesktopDefault.aspx?PageID = 567#ancor (22 Dec. 2017).

Francois, L.E., T. Donovan, and E.V. Maas. 1984. Salinity effects on seed yield, growth, and germination of grain sorghum. Crop Sci. 76:741–744.

Han, L., J. Chen, E.S. Mace, Y. Liu, M. Zhu, N. Yuyama, D.R. Jordan, and H. Cai. 2015. Fine mapping of qGW1, a major QTL for grain weight in sorghum. Theor. Appl. Genet. 128:1813–1825. doi:10.1007/s00122-015-2549-2

Harlan, J.R., and J.M.J. de Wet. 1972. A simplified classification of cultivated sorghum1. Crop Sci. 12:172–176. doi:10.2135/cropsci1972.0011183X001200020005x

Harvey, T.L., K.D. Kofoid, T.J. Martin, and P.E. Sloderbeck. 1991. A new greenbug virulent to e-biotype resistant sorghum. Crop Sci. 31:1689–1691. doi:10.2135/cropsci1991.0011183X003100060062x

Harvey, T.L., G.E. Wilde, and K.D. Kofoid. 1997. Designation of a new greenbug, Biotype K, injurious to resistant sorghum. Crop Sci. 37:989–991. doi:10.2135/cropsci1997.0011183X003700030047x

Hayes, C.M., B.D. Weers, M. Thakran, G. Burow, Z. Xin, Y. Emendack, W.R. Rooney, and J.E. Mullet. 2016. Discovery of a dhurrin QTL in sorghum: Co-localization of dhurrin biosynthesis and a novel stay-green QTL. Crop Sci. 56:104–112. doi:10.2135/cropsci2015.06.0379

Isakeit, T., and J. Jaster. 2005. Texas has a new pathotype of Peronosclerospora sorghi, the cause of sorghum downy mildew. Plant Dis. 89:529. doi:10.1094/PD-89-0529A

Johnson, J.W., D.T. Rosenow, and G.L. Teetes. 1973. Resistance to the sorghum midge in converted exotic sorghum cultivars. Crop Sci. 13:754–755. doi:10.2135/cropsci1973.0011183X001300060051x

Jordan, D.R., E.S. Mace, A.W. Cruickshank, C.H. Hunt, and R.G. Henzell. 2011. Exploring and exploiting genetic variation from unadapted sorghum germplasm in a breeding program. Crop Sci. 51:1444. doi:10.2135/cropsci2010.06.0326

Jordan, W.R., P.J. Shouse, A. Blum, F.R. Miller, and R.L. Monk. 1984. Environmental physiology of sorghum. II. Epicuticular wax load and cuticular transpiration1. Crop Sci. 24:1168–1173. doi:10.2135/cropsci1984.0011183X002400060038x

Klein, R.R., R. Rodriguez-Herrera, J.A. Schlueter, P.E. Klein, Z.H. Yu, and W.L. Rooney. 2001. Identification of genomic regions that affect grain-mould incidence and other traits of agronomic importance in sorghum. Theor. Appl. Genet. 102:307–319. doi:10.1007/s001220051647

Kochian, L., J. Liu, J.V. de Magalhaes, C.T. Guimaraes, R.E. Schaffert, and V.M.C. Alves. 2009. Sorghum aluminum tolerance gene, SbMATE. U.S. Patent Full-Text and Image Database. U.S. Patent and Trademark Office, Washington, D.C.

Leiser, W., H.F. Rattunde, H.P. Piepho, E. Weltzien, and A. Diallo. 2012. Selection strategy for sorghum targeting phosphorus-limited environments in West Africa: Analysis of multi-environment experiments. Crop Sci. 52:2517. doi:10.2135/cropsci2012.02.0139

Leiser, W.L., H.F. Rattunde, E. Weltzien, and N. Cisse. 2014. AM, Diallo, AO Tourè, JV Magalhaes, and BI Haussmann 2014. Two in one sweep: Aluminum tolerance and grain yield in P-limited soils are associated to the same genomic region in West African sorghum. BMC Plant Biol. doi:10.1186/s12870-014-0206-6

Mace, E.S., V. Singh, E.J. Van Oosterom, G.L. Hammer, C.H. Hunt, and D.R. Jordan. 2012. QTL for nodal root angle in sorghum (Sorghum bicolor L. Moench) co-locate with QTL for traits associated with drought adaptation. Theor. Appl. Genet. 124:97–109. doi:10.1007/s00122-011-1690-9

Magalhaes, J., D. Garvin, Y. Wang, M. Sorrells, P. Klein, and R. Schaffert. 2004. Comparative mapping of a major aluminum tolerance gene in sorghum and other species in the poaceae. Genetics 167:1905–1914. doi:10.1534/genetics.103.023580

Magalhaes, J.V., J. Liu, C.T. Guimaraes, U.G.P. Lana, V.M.C. Alves, and Y.-H. Wang. 2007. A gene in the multidrug and toxic compound extrusion (MATE) family confers aluminum tolerance in sorghum. Nature Genetics 39: 1156-1161. http://www.nature.com/ng/journal/v39/n9/suppinfo/ng2074_S1.html (verifed 27 Dec. 2017).

Maman, N., S.C. Mason, D.J. Lyon, and P. Dhungana. 2004. Yield components of pearl millet and grain sorghum across environments in the Central Great Plains. Crop Sci. 44:2138. doi:10.2135/cropsci2004.2138

Masi, C.E.A., and J.W. Maranville. 1998. Evaluation of sorghum root branching using fractals. J. Agric. Sci. 131:259–265. doi:10.1017/S0021859698005826

Mbulwe, L., G.C. Peterson, J.S. Armstrong, and W.L. Rooney. 2016. [Melanaphis sacchari (Zehntner)] Registration of sorghum germplasm Tx3408 and Tx3409 with tolerance to sugarcane aphid. J. Plant Reg. 10:51–56. doi:10.3198/jpr2015.04.0025crg

McKenzie, D.B., L.R. Hossner, and R.J. Newton. 1984. Sorghum cultivar evaluation for iron chlorosis resistance by visual scores. J. Plant Nutr. 7:677–685. doi:10.1080/01904168409363232

Mehta, P.J., C.C. Wiltse, W.L. Rooney, S.D. Collins, R.A. Frederiksen, and D.E. Hess. 2005. Classification and inheritance of genetic resistance to anthracnose in sorghum. Field Crops Res. 93:1–9. doi:10.1016/j.fcr.2004.09.001

Miller, F.R. 1976. Twin-seeded sorghum hybrids: Fact or fancy? Proceedings of the Annual Corn and Sorghum Research Conference. American Seed Trade Association, Washington, D.C.

Miller, F.R., and Y. Kebede. 1984. Genetic contributions to yield gains in sorghum, 1950 to 1980. In: W.R. Fehr, editor, Genetic contributions to yield gains of five major crop plants. Crop Science Society of America and American Society of Agronomy, Madison, WI. p. 1–14.

Monk, R., C. Franks, and J. Dahlberg. 2014. Sorghum. In: S. Smith, B. Diers, J. Specht, and B. Carver, editors, Yield gains in major U.S. field crops. American Society of Agronomy, Inc., Crop Science Society of America, Inc., and Soil Science Society of America, Inc., Madison, WI. p. 293-310.

Morgan, P.W., S.A. Finlayson, K.L. Childs, J.E. Mullet, and W.L. Rooney. 2002. Opportunities to improve adaptability and yield in grasses. Crop Sci. 42:1791–1799. doi:10.2135/cropsci2002.1791

Murray, S.C., A. Sharma, W.L. Rooney, P.E. Klein, J.E. Mullet, and S.E. Mitchell. 2008. Genetic improvement of sorghum as a biofuel feedstock: I. QTL for stem sugar and grain nonstructural carbohydrates. Crop Sci. 48:2165–2179. doi:10.2135/cropsci2008.01.0016

Nagaraj, N., J.C. Reese, M.R. Tuinstra, C.M. Smith, P. St. Amand, and M.B. Kirkham. 2005. Molecular mapping of sorghum genes expressing tolerance to damage by greenbug (Homoptera: Aphididae). J. Econ. Entomol. 98:595–602. doi:10.1093/jee/98.2.595

Nakaya, A., and S.N. Isobe. 2012. Will genomic selection be a practical method for plant breeding? Ann. Bot. (Lond.). doi:10.1093/aob/mcs109

Oh, B.R.A.F., and C.W. Magill. 1994. Identification of molecular markers linked to head smut resistance gene (Shs) in sorghum by RFLP and AFLP analyses. Phytopathology 84:830–833. doi:10.1094/Phyto-84-830

Page, F.D. 1979. Resistance to sorghum midge (Contarinia sorghicola Coquillet) in grain sorghum. Aust. J. Exp. Agric. Anim. Husb. 19:97–101. doi:10.1071/EA9790097

Patil, N.Y., R.R. Klein, C. Williams, S.D. Collins, J.E. Knoll, A.M. Burrell, W.F. Anderson, W.L. Rooney, and P.E. Klein. 2017. Identification of quantitative trait loci associated with anthracnose resistance in Sorghum. Crop Sci. 57:877–890. doi:10.2135/cropsci2016.09.0793

Perumal, R., P. Nimmakayala, S.R. Erattaimuthu, E.-G. No, U.K. Reddy, and L.K. Prom. 2008. Simple sequence repeat markers useful for sorghum downy mildew (Peronosclerospora sorghi) and related species. BMC Genet. 9:77. doi:10.1186/1471-2156-9-77

Peterson, G.C., J.W. Johnson, G.L. Teetes, and D.T. Rosenow. 1984. Registration of Tx2783 greenbug resistant sorghum germplasm line. Crop Sci. 24:390. doi:10.2135/cropsci1984.0011183X002400020062x

Quinby, J.R., and R.E. Karper. 1946. Heterosis in sorghum resulting from the heterozygous condition of a single gene that affects duration of growth. Am. J. Bot. 33:716–721. doi:10.2307/2437497

Quinby, J.R., and J.H. Martin. 1954. Sorghum improvement. Adv. Agron. 6:305–359. doi:10.1016/S0065-2113(08)60388-0

Radchenko, E.E., and A.A. Zubov. 2007. Genetic diversity of sorghum in greenbug resistance. Russ. Agric. Sci. 33:223–225. doi:10.3103/S1068367407040039

Rami, J.-F., P. Dufour, G. Trouche, G. Fliedel, C. Mestres, and F. Davrieux. 1998. Quantitative trait loci for grain quality, productivity, morphological and agronomical traits in sorghum (Sorghum bicolor L. Moench). Theor. Appl. Genet. 97:605–616. doi:10.1007/s001220050936

Röber, F.K., and H.H. Geiger. 2005. In: vivo haploid induction in maize-performance of new inducers and significance of doubled haploid lines in hybrid breeding. Maydica 50:275–283.

Rooney, W.L. 2004. Sorghum improvement—Integrating traditional and new technology to produce improved genotypes. Advances in Agronomy. Academic Press, Waltham, MA. p. 37-109.

Rooney, W.L., S.D. Collins, R.R. Klein, P.J. Mehta, R.A. Frederiksen, and R. Rodriguez-Herrera. 2002. Breeding sorghum for resistance to anthracnose, grain mold, downy mildew, and head smuts. In: J.F. Leslie, editor, Sorghum and millets diseases. Wiley-Blackwell, Hoboken, NJ. p. 273–279.

Rosenow, D. 1984. Breeding for resistance to root and stalk rots in Texas. In: Mughogho, L.K., and G. Rosenberg, editors, Sorghum root and stalk rots: A critical review. ICRISAT, Patancheru, India. p. 209–217.

Rosenow, D., J. Dahlberg, J. Stephens, F. Miller, D. Barnes, and G. Peterson. 1997. Registration of 63 converted sorghum germplasm lines from the sorghum conversion program. Crop Sci. 37:1399–1400. doi:10.2135/cropsci1997.0011183X003700040090x

Rosenow, D.T., and L.E. Clark. 1981. Drought tolerance in sorghum. Proceedings of Annual Corn and Sorghum Conference 36: 18-30.

Rosenow, D.T., J.E. Quisenberry, C.W. Wendt, and L.E. Clark. 1983. Drought tolerant sorghum and cotton germplasm. Agric. Water Manage. 7:207–222. doi:10.1016/0378-3774(83)90084-7

Samac, D., and M. Tesfaye. 2003. Plant improvement for tolerance to aluminum in acid soils– a review. Plant Cell Tissue Organ Cult. 75:189–207. doi:10.1023/A:1025843829545

Sanchez, A.C., P.K. Subudhi, D.T. Rosenow, and H.T. Nguyen. 2002. Mapping QTLs associated with drought resistance in sorghum (Sorghum bicolor L. Moench). Plant Mol. Biol. 48:713–726. doi:10.1023/A:1014894130270

Shi, Y., J.A. Thomasson, S.C. Murray, N.A. Pugh, W.L. Rooney, S. Shafian, N. Rajan, A. Ibrahim, G. Rouze, C.L.S. Morgan, H.L. Neely, A. Rana, M.V. Bagavathiannan, J. Henrickson, E. Bowden, J. Valasek, J. Olsenholler, M.P. Bishop, R. Sheridan, E.B. Putman, S. Popescu, T. Burks, D. Cope, B.F. Mccutchen, D.D. Baltensperger, R.V. Avant, M. Vidrine, and C. Yang. 2016. Unmanned aerial vehicles for high-throughput phenotyping and agronomic research. PLoS One 11(7):E0159781. 10.1371/journal.pone.0159781

Singh, V., E. van Oosterom, D. Jordan, C. Messina, M. Cooper, and G. Hammer. 2010. Morphological and architectural development of root systems in sorghum and maize. Plant Soil 333:287–299. doi:10.1007/s11104-010-0343-0

Smith, C.W., and R.A. Frederiksen. 2000. Sorghum: Origin, history, technology, and production. John Wiley & Sons. New York.

Stephens, J.C., and R.F. Holland. 1954. Cytoplasmic male sterility for hybrid sorghum seed production. Agron. J. 46:20–23. doi:10.2134/agronj1954.00021962004600010006x

Stephens, J.C., F.R. Miller, and D.T. Rosenow. 1967. Conversion of alien sorghums to early combine genotypes. Crop Sci. 7:396. doi:10.2135/cropsci1967.0011183X000700040036x

Tanger, P., S. Klassen, J.P. Mojica, J.T. Lovell, B.T. Moyers, M. Baraoidan, M.E.B. Naredo, K.L. McNally, J. Poland, D.R. Bush, H. Leung, J.E. Leach, and J.K. McKay. 2017. Field-based high throughput phenotyping rapidly identifies genomic regions controlling yield components in rice. Sci. Rep. 7:42839. doi:10.1038/srep42839

Tao, Y.Z., A. Hardy, J. Drenth, R.G. Henzell, B.A. Franzmann, and D.R. Jordan. 2003. Identifications of two different mechanisms for sorghum midge resistance through QTL mapping. Theor. Appl. Genet. 107:116–122. doi:10.1007/s00122-003-1217-0

Tesso, T.T., L.E. Claflin, and M.R. Tuinstra. 2005. Analysis of stalk rot resistance and genetic diversity among drought tolerant sorghum genotypes contribution No. 04-051-J from the Kansas Agric. Exp. Stn. Crop Science 45:645–652. doi:10.2135/cropsci2005.0645

Tolk, J.A., and R.C. Schwartz. 2017. Do more seeds per panicle improve grain sorghum yield? Crop Sci. 57:490–496. doi:10.2135/cropsci2016.04.0245

Tuinstra, M.R., G.L. Liang, C. Hicks, K.D. Kofoid, and R.L. Vanderlip. 2001. Registration of KS 115 Sorghum Registration by CSSA. Crop Sci. 41:932–933. doi:10.2135/cropsci2001.413932x

Vinall, H.N., J.H. Martin, and J.C. Stephens. 1936. Identification, history, and distribution of common sorghum varieties. US Dep. of Agriculture, Washington, D.C.

Williams, R.J., and K.N. Rao. 1981. A review of sorghum grain moulds. Trop. Pest Manage. 27:200–211. doi:10.1080/09670878109413652

Withee, L.V., and C.W. Carlson. 1959. Foliar and soil applications of iron compounds to control iron chlorosis of grain sorghum. Agron. J. 51:474–476. doi:10.2134/agronj1959.00021962005100080010x

Xin, Z., M. Wang, G. Burow, and J. Burke. 2009. An induced sorghum mutant population suitable for bioenergy research. BioEnergy Res. 2:10–16. doi:10.1007/s12155-008-9029-3

Xu, W., D.T. Rosenow, and H.T. Nguyen. 2000. Stay green trait in grain sorghum: Relationship between visual rating and leaf chlorophyll concentration. Plant Breed. 119:365–367. doi:10.1046/j.1439-0523.2000.00506.x

Yang, Y.W., R.J. Newton, and F.R. Miller. 1990. Salinity tolerance in sorghum. I. Whole plant response to sodium chloride in S. bicolor and S. halepense. Crop Sci. 30:775–781. doi:10.2135/cropsci1990.0011183X003000040003x

Yang, Z., E. Yang, D. van Oosterom, A. Jordan, G. Doherty and G.L. Hammer. 2010. Genetic variation in potential kernel size affects kernel growth and yield of sorghum. Crop Sci. 50:685. doi:10.2135/cropsci2009.06.0294

Zhu, J., P.A. Ingram, P.N. Benfey, and T. Elich. 2011. From lab to field, new approaches to phenotyping root system architecture. Curr. Opin. Plant Biol. 14:310–317. doi:10.1016/j.pbi.2011.03.020

Sorghum Genetic Resources

Hari D. Upadhyaya,* Sangam L. Dwivedi, Yi-Hong Wang, and M. Vetriventhan

Abstract

Sorghum is a smart food, feed, and bioenergy crop, adapted worldwide to temperate and tropical climates. Sustained gains in plant breeding rely on variation in crop gene pool. Over 231,000 sorghum accessions are conserved globally, with the International Crops Research Institute for the Semi-Arid Tropics (ICRISAT) genebank holding 39,234 accessions from 93 countries. Representative subsets in the form of core and mini core collections and reference sets have been developed and used to identify new sources of variations for stress resistance, phenology, seed yield and quality, and for bioenergy traits. Ethnolinguistic diversity, human migration, and social boundaries have shaped the abundant diversity among landraces, which significantly impacted conservation and distribution of on-farm diversity. Sorghum is a genomic resources rich crop. Genome-wide association mapping unraveled many significant marker–trait associations: flowering, plant height, tillering, culm length, inflorescence architecture, number of panicles, panicle length, seed weight, and stalk sugar; resistance to anthracnose, grain mold, and rust; increased seed yield under P-deficient or Al-toxic soils; or seed quality, with many of these markers comapped on the same linkage groups previously reported as harboring quantitative trait loci (QTL) or candidate genes associated with some of these traits. The sequencing of diverse sorghum germplasm and its comparison with a reference genome has revealed substantial untapped diversity, which offers a great opportunity for the genetic improvement of sorghum. Targeting Induced Local Lesions IN Genome (TILLING) populations unraveled new variations not previously known in sorghum. Genomic tools and information are being applied to harness new sources of variations by using *S. halepense* and *S. propinquum* to adapt sorghum to extreme climatic conditions, both in tropical and temperate regions.

Sustained gains in plant breeding rely on variation in crop gene pool. Natural variations which evolved over millions of years are the key to the success in crop breeding. Globally, over 7 million germplasm accessions are stored across

Abbreviations: BIN, biological nitrification inhibition; EST, expressed sequence tags; GP, gene pool; HI, harvest index; ICRISAT, International Crops Research Institute for the Semi-Arid Tropics; INDELS, insertions and deletions; SNP, single nucleotide polymorphism; SPAD, soil plant analysis development; SSR, simple sequence repeat; TE, transpiration efficiency; TILLING, Targeting Induced Local Lesions IN Genome; QTL, quantitative trait loci; WE, water extraction.

H.D. Upadhyaya, S.L. Dwivedi (s.dwivedi@cgiar.org), and M. Vetriventhan (m.vetriventan@cgiar.org), ICRISAT, Patancheru Post Office, Pin Code 502324, Telangana, India; H.D. Upadhyaya, also at Dep. of Agronomy, Kansas State Univ., Manhattan, KS 66506, and UWA Inst. of Agriculture, Univ. of Western Australia, Crawley WA 6009, Australia; and Y.-H. Wang, Dep. of Biology, Univ. of Louisiana at Lafayette, 300 E. St. Mary Blvd., 108 Billeaud Hall, Lafayette, LA 70504 (yxw9887@louisiana.edu).
*Corresponding author (h.upadhyaya@cgiar.org).

doi:10.2134/agronmonogr58.2014.0056.5

1750 genebanks (Food and Agricultural Organization, 2010). Genebanks play an important role in conserving and safeguarding this priceless genetic diversity for present and future generation's uses. Sorghum is the fifth most important cereal grain, grown in 73 countries (production > 10,000 t); Africa, the Americas, and Asia together during the 2009 to 2013 period on average contributed 95% of the world sorghum production (58.75 m t) (Table 1). The ICRISAT has one of the largest collections of sorghum germplasm, both cultivated and wild and weedy relative types; however, to date the use of this germplasm resource in sorghum improvement has been very limited. The low use of germplasm in most crops including sorghum is due to nonavailability of reliable information on traits of economic importance that show high genotype × environment interaction, due to lack of accurate and precise large-scale multilocation evaluation of germplasm collections; nonavailability of information needed by the breeder for genetically diverse, trait-specific, and agronomically desirable germplasm in genebank databases; linkage load of many undesirable genes and assumed risks; restricted access to the germplasm collections due to limited seed quantities, particularly of wild relatives and unadapted landraces and regulations governing international exchange; the enhanced role of nonadditive genetic variation when diverse exotic germplasm is used by the breeders; lack of robust and cost-effective tools to facilitate efficient utilization of exotic germplasm in plant breeding programs; and limited exposure to available germplasm and recirculation of the same germplasm in breeding programs (Dwivedi et al., 2009).

Sorghum is an important food, feed, and bioenergy crop. It is also a good source of fodder. Sorghum grains are gluten free and therefore an alternative food for those who suffer from Celiac disease (Dahlberg et al., 2011). Worldwide, sorghum production is constrained by many abiotic and biotic stresses. Drought, heat, salinity, and aluminum toxicity are the major abiotic stresses. The biotic stresses include anthracnose [*Colletotrichum graminicola* (Ces.) GW Wilson], charcoal rot and/or stalk rot [*Macrophomina phaseolina* (Tassi) Goid], downy mildew (*Peronosclerospora sorghi* [Wetson and Uppal (Shaw)]), grain mold (caused by a complex of several fungal diseases), leaf blight [*Exserohilum turcicum* (Pass.) KJ Leonard and EG Suggs], rust (*Puccinea purpurea* Cooke), and several viral diseases, while head bug (*Calocoris angustatus* Leth), midge (*Contarinia sorghicola* Coq.), stem borer [*Chilo partellus* (Swinhoe) and *Sesamia inferens* (Walker)], and shoot fly [*Atherigona soccata* (Rondani)] are the major pests of sorghum. These stresses often occur in combinations that cause substantial yield losses to sorghum production (House, 1985; Sharma, 1993; Kumar et al., 2011). Further, the evidence to date suggests an overall negative effect of global warming on agricultural biodiversity, food production, and food quality (Dwivedi et al., 2013). Discovering climate-resilient traits in germplasm collections and their use in crop improvement should be taken up urgently to tailor crop cultivars adapted to climate change and variability.

Genomic and bioinformatic tools are now available in many crops to enhance the quality, efficiency, and cost-effectiveness of genebank operations; for example, large-scale genotyping and targeted resequencing has the potential to significantly advance the rational conservation, characterization, and utilization of crop genetic resources, providing new information on germplasm collections. This will enable the genebank curators to more effectively target future collection missions, manage seed multiplication, develop in-house quality control procedures, and disseminate

Table 1. Global sorghum production (5-yr average, 2009–2013) by region and countries.†

Region and country	Production	Percent of the global production
	million t	%
	Sorghum production by region	
Africa	24.29	41.34
Americas	21.63	36.82
Asia	9.84	16.75
Europe	0.85	1.45
Oceania	2.14	3.64
Total	58.75	100
	Sorghum production by country	
United States	8.02	13.70
Nigeria	6.58	11.20
Mexico	6.55	11.20
India	6.44	11.00
Ethiopia	3.76	6.40
Sudan	3.57	6.10
Argentina	3.49	5.90
Australia	2.14	3.60
China	2.04	3.50
Brazil	1.88	3.20
Burkina Faso	1.78	3.00
Total	46.25	78.80

† Source: http://faostat.fao.org.

information about the value of germplasm from end-user's perspective (McCouch et al., 2012). Genomics science in sorghum has advanced considerably in the last decade, both in terms of assays and marker technologies, genetic maps, mapping and cloning quantitative trait loci (QTL) associated with agronomically beneficial traits, and genome sequencing. At the same time, there has been substantial investment in developing high-throughput phenotyping platforms for rapid and accurate assessment of phenotypic traits, including resistance to abiotic stress, with such facilities also created at ICRISAT, Patancheru, India. More importantly, resequencing of diverse germplasm and its sequence comparisons with a reference genome provides researchers opportunities to discover markers and whether such differences in phenotypes relate to the differences in sequence variations between accessions. This chapter provides an overview of the current status of the sorghum germplasm collection maintained at the ICRISAT genebank: the formation of representative subsets—core/mini core collections or genotype-based reference sets—capturing appropriate levels of diversity (~80%) present in the entire collection; assessing the impact of mating systems on gene flow and diversity; ethnolinguistic diversity impacting conservation and distribution of on-farm diversity; discovering climate-resilient traits by using representative germplasm collections; and on population structure and diversity, marker–trait associations and identifying multitrait genetically diverse and agronomically beneficial germplasm to support breeding and genomics in sorghum.

Origin, Taxonomy, Domestication, and Gene Pools

Sorghum belongs to the family Poaceae, tribe Andropogoneae, subtribe Sorghinae, and genus *Sorghum* Moench (Clayton and Renvoize, 1986). Dillon et al. (2007)

detailed the origin, taxonomy, domestication, and gene pools of sorghum. Sub-Saharan Africa and more particularly Ethiopia, Sudan, and Chad are considered the primary center of origin and domestication of sorghum species, while India and China are secondary centers of diversity (de Wet, 1978; Doggett, 1988). The domestication of sorghum commenced around 4000–3000 BC. The improved sorghum types disseminated through movement of people and trade routes into other regions of Africa, India, the Middle East, and into the Far East. By the time sorghum moved into the United States during the late 1800s to early 1900s, it had numerous improved forms that were created through the movement of people, disruptive selection, geographic isolation, and genetic recombination. Early domestication of sorghum was associated with changing the small-seeded, shattering open panicles to larger, nonshattering seeds and more compact panicles. These changes brought a greater number of branches within the inflorescence, decreased the internode length of the rachis, and increased the seed size.

The genus *Sorghum* has 25 species grouped into 5 taxonomic subgenera: *Eusorghum, Chaetosorghum, Heterosorghum, Parasorghum,* and *Stiposorghum. Eusorghum* contains all the domesticated and cultivated sorghum races and varieties of *S. bicolor* subsp. *bicolor,* as well as the wild and weed species *S. halepense* (L) Pers. (Johnson grass), *S. propinquum* (Kunth) Hitchc, *S.* × *almum* Parodi, *S.* × *drummondi* (Steud) Millsp. and Chase, and *S. arundinaceum* (Desv.) Stapf., the known progenitor of *S. bicolor.* All the *S. bicolor* subsp. *bicolor* are annuals, $2n = 2x = 20$ chromosomes, and classified into 5 basic (*bicolor, caudatum, durra, guinea, kafir*) and 10 intermediate (*guinea-bicolor, caudatum-bicolor, kafir-bicolor, durra-bicolor, guinea-caudatum, guinea-kafir, guinea-durra, kafir-caudatum, durra-caudatum, kafir-durra*) races, all recognized by spikelet–panicle morphology, and can be linked to their specific environments and the nomadic peoples that first cultivated them (Harlan and de Wet, 1972).

Sorghum has three gene pools (GPs): primary (GP 1), secondary (GP 2), and tertiary (GP 3). Gene pool 1 consists of *S. bicolor* complex, including a wild diploid *S. propinquum* (Kunth) Hitchc. The species in this gene pool easily intercross and produce fertile hybrids. Gene pool 2 includes *S. halepense* (L) Pers., a tetraploid perennial with well-developed creeping rhizomes. Some of the species in this gene pool can be crossed with primary gene pool species to produce fertile hybrids, indicating that gene transfer between two gene pools (GP 1 and GP 2) is possible; however, it is usually difficult to achieve. The species from subgenera *Chaetosorghum, Heterosorghum, Parasorghum,* and *Stiposorghum* constitute GP 3 species, and these do not cross readily with GP 1 species. Hybrids produced, if any, are invariably sterile; special techniques are needed to effect gene transfer to the GP 1 species (reviewed in Upadhyaya et al., 2014a). Using low-copy nuclear and plastid genes and representative samples from genera *Sorghum, Cleistachne sorghoides,* and those from outgroups, Liu et al. (2014) estimated the crown age of *Sorghum* (including *Cleistachne sorghoides*) and subg. *Sorghum* as 12.7 and 8.6 million years ago, respectively. Further, they proposed a new subgeneric classification of *Sorghum* Moench into three distinct subgenera: subg. *Chaetosorghum* with two sections (*Chaetosorghum* and *Heterosorghum*), each with a single species; subg. *Parasorghum* with 17 species, and subg. *Sorghum* with nine species, and named *Cleistachne sorghoides* as *S. sorghoides.*

Global Sorghum Genetic Resources

Germplasm Stored in Genebank

Worldwide, 231,725 sorghum accessions, both cultivated (98.3%) and wild and weedy relatives (1.7%), are preserved in genebanks. The bulk of the cultivated germplasm come from South Asia, North America, sub-Saharan Africa, South America, and East Asia, while southeastern Asia and Oceania constitute the major wild and weedy relatives of sorghum. Sub-Saharan Africa and North America each constitute about 5% of wild species collections (Table 2). Representation by major sorghum producing countries revealed that India, the United States, China, Brazil, Ethiopia, and Mexico have the largest collection of cultivated germplasm, while India, Australia and the United States dominate the wild and weedy relatives of sorghum. Some genebanks hold large proportions of these germplasm; for example, the ICRISAT genebank at Patancheru, India, and the USDA-ARS at Griffin, GA, each share ~16% of the cultivated germplasm collections, whereas the National Bureau of Plant Genetic Resources (NBPGR) genebank at Delhi, India, has the largest collection (66.6%) of sorghum wild relatives (Fig. 1). Ethiopia is the primary center of diversity for sorghum. The Institute of Biodiversity Conservation (IBC) genebank in Ethiopia contains 4.2% of the global collection of cultivated germplasm, while the International Livestock Research Institute (ILRI) genebank in Ethiopia manages 1.3% of sorghum wild and weedy relatives. The global collection of cultivated sorghum is represented by 5 races and 10 intermediate races, which are fairly represented in some genebanks; for example, in the ICRISAT collection, *caudatum, durra, guinea, durra-caudatum, guinea-caudatum, durra-bicolor,* and *caudatum-bicolor* are more fairly represented than that of other intermediate races (Fig. 2). Some genebanks have unique collection of wild relatives not commonly found in other genebanks, notably *S. halepense* (Sudangrass) accessions at ATTC genebank in Albania, USDA-ARS in Griffin, GA, NPGBI-SPII in Iran, and ICRISAT in India (Fig. 3). *S. halepense* is a noxious and invasive weed species; however, it is a source of perenniality,

Table 2. Status of sorghum genetic resources, both cultivated and wild, conserved across regions.†

Region	Number of accessions	Percent of germplasm conserved across 168 genebanks
		%
Cultivated		
Southern Asia	58,566	25.3
North America	43,560	18.8
Sub-Saharan Africa	38,332	16.5
South America	28,035	12.1
Eastern Asia	24,144	10.4
Europe	16,011	6.9
Central America	12,729	5.5
Oceania	4141	1.8
Wild		
Southeastern Asia	3194	79.6
Oceania	346	8.6
North America	199	5.0
Sub-Saharan Africa	189	4.7
Europe	73	1.8

† Source: http://apps3.fao.org/wiews.

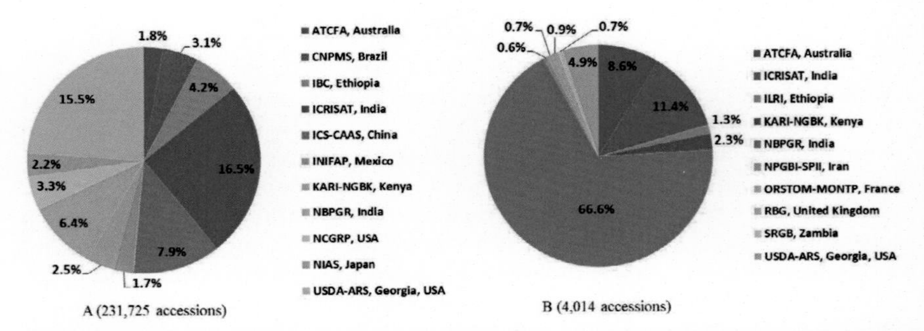

Fig. 1. Status of (A) cultivated and (B) wild sorghum germplasm accessions (%) conserved by major genebanks (http://apps3.fao.org/wiews).

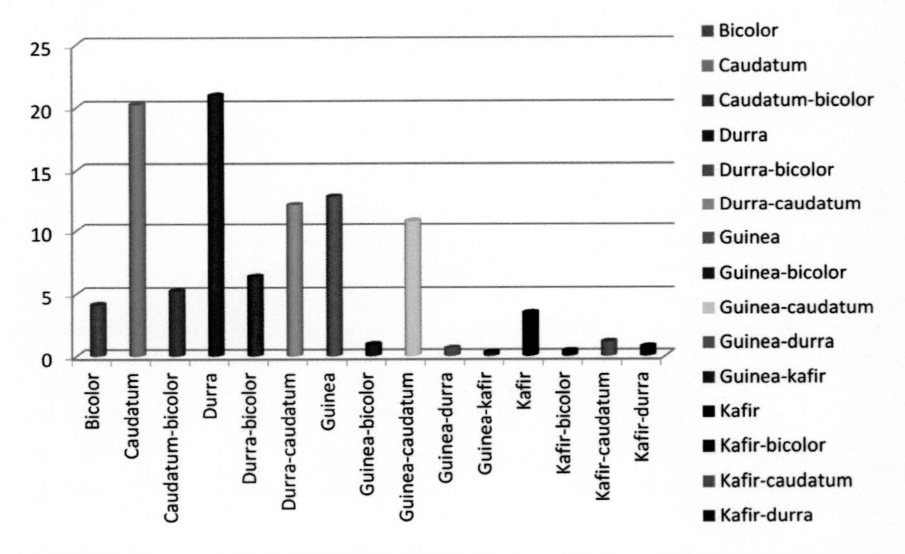

Fig. 2. Status of cultivated sorghum germplasm accessions (%) conserved at the International Crops Research Institute for the Semi-Arid Tropics (ICRISAT) by 5 races and 10 intermediate races (www.icrisat.org).

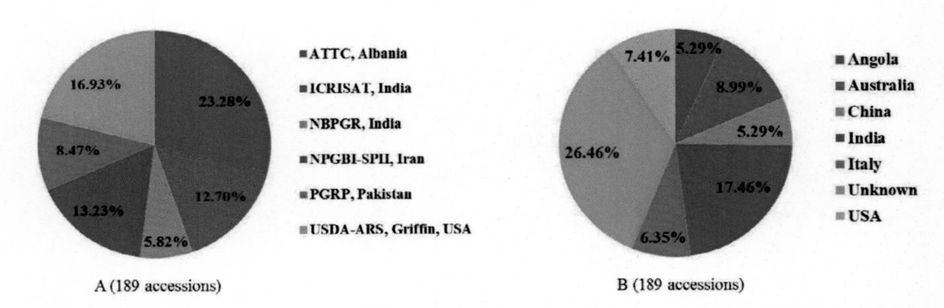

Fig. 3. ***Sorghum halepense*** **accessions (%) by (A) institute and (B) country conserved in major genebanks globally (http://apps3.fao.org/wiews).**

high-biomass production, and extreme resistance to temperature stress (heat and cold tolerance). Another species with similar characteristics is *S. propinquum*, but only six accessions are available in databases across genebanks. The native Australian sorghum species have great diversity in seed size and shape, the nature of endosperm, the distribution of protein bodies throughout the endosperm, and shape and size of the starch granules (Shapter et al., 2008, 2009a, 2009b); all these have potential to impact the nutritional quality of the grains.

Traits and Factors Shaping Landraces Diversity

Phenotypic Diversity

Landraces form the substantial number of sorghum germplasm preserved in national and international genebanks; for example, 86% of the sorghum accessions (33,307) maintained at ICRISAT are landraces from 87 countries (http://www.icrisat.org/what-we-do/crops/sorghum/Project1/pfirst.asp). Previous reports have shown a high level of diversity for morpho-agronomic traits among sorghum landraces, highlighting their central role in subsistence agriculture (Zongo et al., 1993; Appa Rao et al., 1996; Ayana and Bekele, 1998, 1999, 2000; Grenier et al., 2004; Geleta and Labuschagne, 2005; Manzelli et al., 2005). Abundant diversity among sorghum landraces is shaped by ethnic, linguistic, customary, end-use, and sociological factors, in addition to edaphic, temporal, and spatial factors. Ethiopia and Sudan are the primary center of diversity in sorghum. In Ethiopia, farmers from Amhara National Regional State refer to discrete sorghum types by different names; for example, the name Wotet-begunche refers to the sorghum landrace having mature seeds with a milky taste, Chibite to compact head, Cherekit to white seed color, Minchiro to loose drooping panicles, Marchuke and Maar-Beshenbeko to honey, Shilme to a head with different colored seeds, Gubete to soft roasted seeds, Ahyo and Wofaybelash to bird resistance or Mera, Mognayfere, Minchiro, and Ckerekit as *Striga* tolerant landraces (Benor and Sisay, 2003). Those from North Shewa and South Welo in Ethiopia named the landraces based on midrib color, seed color, seed size, glume color, glume hairiness, and seed shape (Teshome et al., 1997). Dissan farmers from southern Mali named the varieties based on maturity, yield, and taste. For example, Boboka is known for its good taste and adaptation to rainfed conditions, Segatono for its resistance to *Striga*, and Kalo Saba for early maturity (Lacy et al., 2006). On-farm sorghum diversity in Ethiopia is also reported to be shaped by field altitude, field size, soil pH, P, available N, organic matter, exchangeable K, and clay content (Mekbib et al., 2009). Likewise, a higher level of diversity was reported in the intermediate altitudes than in the lowlands and highlands of Ethiopia. Varietal mixture is one of the strategies adopted by the farmers to maintain on-farm genetic diversity (Teshome et al., 1999; Mekbib et al., 2009). Eastern province is the major sorghum growing zone in Kenya, and landraces in this region are continuously maintained based on cultural preferences and traditional practices adopted by the farmers. For example, landraces from Mbeere region had diverse seed color compared with those from Kitui, Mutomo, and Makueni regions, with many of these landraces unique in their adaptation, food quality, grain yield, quality of harvested products, biotic stress resistance, and postharvest processing (Muui et al., 2013).

Sudan is another important country with abundant racial diversity in sorghum, often differentiated by region-specific phenological traits. For example, landraces from Gezira-Gedarif region are short in stature, early maturing, less sensitive to changes in photoperiod, and adapted to low rainfall and mechanized cultivation. They possess long, narrow, and compact panicles. In contrast, those sorghum from the Equatorial region are taller and late maturing, with small and light kernel types. Kassala region landraces are often difficult to thresh, while those from the Blue Nile region are agronomically superior, many with white kernels, poorly covered, and easy to thresh. Sorghum landraces from the Upper Nile region have loose panicles with poorly covered kernels and adapted to high rainfall in Southern Sudan. More importantly, Sudanese landraces, although differentiated by regions, possess a high level of within-region diversity (Grenier et al., 2004).

Tribal communities in Adilabad district of Andhra Pradesh, India, have cultivated sorghum landraces since ancient times. Adilabad is the fifth largest district in Andhra Pradesh, inhibited predominantly by tribal populations. The tribal population is dominated among others by Gond, Lambada, and Kolam. Sorghum is one of the main crops grown by these communities, and the landraces are named by their phenotypic diversity. As examples, Vubiri patti jonna refers to midrib color; Erra jonna and Pachcha jonna to grain color; Chinna jonna, Chinna boda jonna, and Pedda jonna to grain size; Tekedari jonna to glume color; Leha jonna to glume hairiness; Pandemutte jonna and Pelala jonna to earhead shape; and Sevata jonna to grain shape. Likewise, they named a local landrace as Sai jona, a trust reposed on the holy saint Sai Baba, which never fails to give some assured yield (Pandravada et al., 2013). Some of the other nomenclature adopted by the Indian farmers cultivating sorghum include that of Moli jowar from Madhya Pradesh that fetches higher prices because of attractive grains; the irungu cholam and mathappu cholam from Tamil Nadu, with the former having the best quality porridge and the latter for preparing a jelly like food called khali; the beed maldandi and bidri from Maharashtra produce the best quality jowar roti or uncleavened bread; Kodumurugu jola, allin jola, and allur jola from Karnataka for making laddus, papads, and pops, respectively. Likewise, the Pachcha jonna from Andhra Pradesh, the Barmuda local, Deshi Chari and Gudli local from Rajasthan, and Bendri dagdi and Khondya from Maharashtra are excellent fodder types. Several landraces such as Valsangh maldandi local, Vadgaon dagdi maldani, Tongralingaon maldandi, Tongralingaon dagdi, Sultanpur local dagdi, Sultanpur maldandi, Harni jogdi, Chungi maldandi, Musti local, Chungi kuch-kachi, Baddi jowar and Chakur maldandi from Maharashtra, or Sai jonna from Andhra Pradesh are considered by the farmers as drought tolerant landraces (Elangovan et al., 2009). Clearly, several factors contributed to landrace diversity, necessitating the need to further document this diversity based on farmers perceptions, agroecological factors, and end-use patterns from the regions not explored.

Genotypic Diversity

Molecular profiling of landraces showed large allelic diversity. For example, substantial allelic diversity exists among landraces (30 accessions and 10 simple sequence repeats [SSRs]) from Botswana; however, these differences could not be related to diversity in agroecological regions, ethnicities, and races (Motlhaodi et al., 2014). The sub-Sahelian landraces (124 accessions and 29 SSRs) from Burkina Faso showed the highest gene diversity, and the seed color was the principal factor delineating this diversity (Barro-Kondombo et al., 2010). The Ethiopian landraces (137 accessions and 20 SSRs) preserved in the USDA-ARS genebank showed absence of population structure (Fst = 0.10), though they clustered into three groups. This reveals the significance of this germplasm as an important resource for association mapping (Cuevas and Prom, 2013). The Sudanese landraces (95 accessions and 39 SSRs) showed high within population variability and clustered into three groups independent of their geographic origins (Gamar et al., 2013). The Somalian landraces (9 accessions and 5 SSRs) showed high gene diversity, and within accessions high diversity reveals no selection pressure and/or a continuous gene flow between populations (Manzelli et al., 2007). The Ugandan landraces (241 accessions and 21 SSRs) showed higher variation within races and agroecologies than among races and agroecological zones, which indicated significant gene flow (Mbeyagala et al., 2012). The Zambian landraces (27 accessions and 10 SSRs) showed substantial diversity, clustered into four groups based on ethnic differences (Ng'uni et al., 2011). The landraces adapted to clay soils in dry season (after cessation of rains) in northern Cameroon (Lake Chad Basin) showed significant levels of differentiation (Fst, 0.52), named as Baburi (early sown) and Muskuwaari (late sown) types, and clearly grouped into two clusters (Soler et al., 2013). Burow et al. (2012) grouped the landraces (159 accessions and 41 SSRs) from northeast China into eight clusters and reported a number of genetically distinct accessions with agronomically beneficial traits for use in sorghum breeding. Clearly, limited studies on assessment of sorghum landraces from the center of diversity, preserved in genebanks, delineated the accessions into distinct groups, and in many cases this differentiation could be related to ethnic, regional, and racial diversity.

Forming and Assessing Diversity in Representative Subsets

Forming Representative Subsets

Forming a representative germplasm set that captures the entire diversity of a given species conserved in a genebank is a way to economize genebank operations and promote the use of plant genetic resources in crop improvement. Core (Frankel, 1984) and mini core (Upadhyaya and Ortiz, 2001) collections sample representative variability from the entire collection. Conventional core (based on morpho-agronomic traits data) collections in sorghum consists of 3475 accessions (Prasad Rao and Ramnatha Rao, 1995), 2247 landraces (Grenier et al., 2001), or 3011 accessions (Dahlberg et al., 2004), while mini core collections consist of 242 accessions (Upadhyaya et al., 2009). More recently, Billot et al. (2013) reported a genotype-based reduced subset, named as reference set, which consists of 384 accessions, representing the basic and intermediate races, and a few wild and weedy relatives of sorghum. These reduced subsets are sufficiently diverse to serve as a useful panel for discovering new sources of variations and for mining allelic variations associated with agronomically beneficial traits for use in the breeding and genomics of sorghum.

Phenotypic Diversity in Representative Subsets

An extensive evaluation of a sorghum reference set (384 accessions) for five season–locations revealed large variability (as measured by Shannon–Weaver diversity index) for plant height, 100-seed weight, grain yield, panicle weight, and for chlorophyll content as measured by soil plant analysis development (SPAD) chlorophyll meter reading (Billot et al., 2013). Further, the qualitative traits such as mid rib color (white), plant pigmentation, glume color (black and purple), and seed color (white, purple, and brown), which scored in greater proportion in the reference set, could be used as markers to select accessions of desired end-use types as these were associated with fodder quality, resistance to grain mold, or for seed preference to food and industrial (beverage) uses (Seetharam, 2011). Drought tolerant accessions identified were predominantly from the race *caudatum* as these were early flowering types, had shortest plant height, and high panicle weight and grain yield (Seetharam, 2011).

Upadhyaya et al. (2014b) evaluated a sorghum mini core collection for Brix (a measure of sugar concentration) content for two seasons under irrigated and terminal drought-stressed conditions. Accessions IS 13294, 13549, 23216, 23684, 24139, 24939, and 24953 recorded 14–15% greater Brix concentration than the best control, IS 33844 (Brix, 12.4%); however, these were lower yielding. IS 1004, 4698, 23891, and 28141 produced 12–23% greater seed yield over IS 33844 (35–42 g/plant) and had almost the same Brix concentration (~13%). The current research at ICRISAT also led to the identification of seed mineral dense (Fe and Zn) accessions from the sorghum mini core collection, some with significantly greater seed yield and specific adaptation to irrigated and/or drought-stressed environments (Upadhyaya et al., 2016a).

Genotypic Diversity in Representative Subsets

Billot et al. (2013) undertook the largest ever study to dissect the population structure and diversity in a composite collection (3367 accessions, representing five races and 10 intermediate races, and wild relatives) using 41 genomic SSRs. This study revealed that accessions from central and eastern Africa had the largest number of alleles, and cultivated races were structured around geographic regions and race within regions, detecting 13 groups of variable size with peripheral groups in western Africa, southern Africa and eastern Asia. Further, except for the race *kafir*, there was little correspondence between races and marker-based groups. *Bicolor*, *caudatum*, *durra*, and *guinea* types were each dispersed in three groups or more. Wild and weedy relatives, in contrast, were very diverse and scattered among cultivated types, indicating large gene flow between the two types. Ramu et al. (2013) further analyzed this reference set by using expressed sequence tags (EST)–SSRs to validate the genetic structure unraveled by Billot et al. (2013). They also found that genotypes were grouped primarily based on race within geographic origins. Accessions derived from African continents contributed 88.6% allelic diversity, confirming the African origin of sorghum. *Guinea margaritiferum* subrace from West Africa formed a separate cluster in close proximity to wild accessions, suggesting that this group represents an independent domestication event. Likewise, *guineas* from India and West Africa formed two distinct clusters. *Kafir* race accessions formed the most homogeneous group. Furthermore, high subpopulation genetic variance (72%) suggests that germplasm lines included in the set are highly diverse.

Using 265,000 single nucleotide polymorphism (SNP) data on 971 sorghum accessions (which also included a sorghum reference set), Morris et al. (2013) showed structured populations along both morphological types and geographical origins, which provided insights into historical processes of crop diffusion within and across the agroclimatic zones of Africa and Asia. For example, *caudatum* types showed the least population structure; *durra* types diffused widely across Africa and Asia but were restricted to regions with semiarid and desert climates; *guinea* types diffused over long distances from western Africa to southeastern Africa and eastern India but restricted to tropical savanna climates; *kafir* types predominated in southern Africa showed the strongest pattern of population subdivision, phenotypic divergence, and genetic bottlenecking resulting from a shift to a contrasting agroclimatic zones; *bicolor* types were not notably clustered (Morris et al., 2013).

Assessing In Situ Genetic Diversity

An understanding of geographical, environmental, and social patterns of genetic diversity on spatial and temporal scales is the key to the sustainable in situ management of genetic resources. Moreover, understanding patterns of in situ landrace diversity aids in deciphering evolutionary forces shaping its diversity, which has implications on the conservation of genetic resources and their use in crop improvement programs. This aspect of assessing in situ diversity has been lacking in the past in most crops, including sorghum. Barnaud et al. (2007) assessed the spatial patterns of planting and farmers' perceptions of 21 landraces in northern Cameroon by using SSRs, which delineated the accessions into four clusters which correspond to functionally and ecologically distinct groups of landraces. This differentiation among landraces was substantial and significant (Fst 0.36), which could relate to historical factors, variation in breeding systems, and farmers practices, affecting patterns of genetic variation among these landraces. Deu et al. (2008) conducted a genetic analysis of 484 sorghum varieties collected from 79 villages evenly distributed across Niger by using 28 SSRs. They detected a high level of SSR diversity that varied between eastern and western Niger, with a lower allelic richness in the eastern part of the country. Such differences could be related to botanical races, geographical distribution, and ethnic groups. More recently, Adugna (2014) assessed in situ diversity and genetic structure of eight cultivated sorghum landrace populations (160 individual plants) representing three diverse geographic regions from Ethiopia by using 7 phenotypic traits and 12 highly polymorphic SSRs. They reported a very high percentage of rare alleles (78 alleles, 63%), averaging 10.25 alleles per polymorphic locus. Neighbor-joining and STRUCTURE analysis grouped the 160 samples from 8 populations differently with high divergence (Fst, 0.40), which indicated low levels of gene flow. Further, they could relate 54.44% of the variation to be within populations, 32.76% among populations within a region, and 12.8% among the regions of origin. Populations from Wello displayed a close relationship with the remote Gibe and Metekel populations, indicating that this variation followed a human migration pattern. In an earlier study, Shewayrga et al. (2008) reported loss of diversity in sorghum landraces from Wello region, Ethiopia, which was largely due to a switch in cultivation from sorghum to tef [*Eragrostis tef* (Zucc.) Trotter].

The dynamics of crop diversity is also influenced by farming systems adapted by sorghum farmers. Rabbi et al. (2010) used 16 polymorphic SSR markers and studied the genetic architecture of landraces and modern sorghum cultivars from two contrasting agroecosystems in eastern Sudan and western Kenya. The structure analysis detected two contrasting patterns. Landraces from western Kenya formed widely overlapping clusters, indicating weak genetic differentiation, whereas those from eastern Sudan formed clearly distinguishable groups. Likewise, modern varieties from Sudan displayed high homogeneity, whereas those from western Kenya were very heterogeneous. In western Kenya, the farmland is highly fragmented and farmers plant different sorghum varieties in the same fields, which increases the likelihood of intervariety gene flow. This could explain the low genetic differentiation between the landraces and heterogeneity of the modern varieties from western Kenya.

These observations support the role of farmers and farming systems in shaping on-farm genetic variation in sorghum.

Allelic Richness in Wild–Weedy Relatives vis-à-vis Crop Cultigens

Wild relatives of sorghum harbor genes for adaptation and resistance to pests and diseases (Venkateswaran et al., 2014). The information on allelic richness and diversity, unlike cultivated germplasm, between wild and cultivated species are limited in sorghum. Using 104 accessions including cultivated (98 landraces) and wild (31 accessions) relatives and 98 SSRs, Casa et al. (2005) found that landraces retained 86% of the allelic diversity of the wild sorghums, showed weak evidence of population structure (Fst, 0.13) between cultivated and wild as well as among cultivated sorghum accessions, and mapped seven SSR loci in or near-genomic regions associated with domestication related QTLs in sorghum. Wild sorghum is widely distributed in Kenya and displays extensive phenotypic and genetic diversity. Unlocking diversity through use of DNA markers provide genetic proximity between wild and cultivated germplasm. Mutegi et al. (2012) detected a relatively greater number of alleles (257) in cultivated than wild (238 alleles) gene pools by using 24 SSRs; however, wild sorghum gene pools harbored significantly more genetic diversity than its domesticated counterpart, revealing that domestication was accompanied by a genetic bottleneck in sorghum. Further, they claimed that the observed genetic proximity may have arisen primarily because of historical and/or contemporary gene flow between the two congeners, with farmers' practices explaining differences in interregional gene flow. In another study using 18 SSRs on 62 wild sorghum populations collected from 4 main sorghum growing regions in Kenya, Muraya et al. (2011b) showed that wild sorghum is highly variable, with the coastal region displaying the highest diversity; gene flow is not restricted to populations within the same geographic region, and a weak regional differentiation among populations reflect human intervention in shaping wild sorghum genetic structure through seed-mediated gene flow. Thus, the sympatric occurrence of wild and cultivated sorghums coupled with extensive seed-mediated gene flow and high outcrossing rates of wild sorghum suggest a potential crop-to-wild gene flow and vice versa across regions.

Mating Systems Impacting Wild-Crop Gene Flow and Diversity

A mating system is a major evolutionary force that largely shapes the genetic structure and variability within and among populations. Knowledge of mating systems and their consequences for gene flow aid judicial conservation and use of genetic resources in crop improvement. Sorghum is predominantly a self-pollinated crop (Doggett, 1988), although reported to be up to 40% outcrossing (Barnaud et al., 2008). In sorghum, the relative roles of natural and human selection in facilitating or constraining gene flow are not fully understood. In a study involving Duppa farmers (who predominantly cultivate sorghum for sustenance) and sorghum weed–crop complex in northern Cameroon, Barnaud et al. (2009) claimed that farmers exert conscious and unconscious selection to manage landrace diversity in relation to the weedy relatives of sorghum. Farmers are well versed at practicing selection against weedy types for cultivation that favor gene flow to domesticated sorghum types. Other studies on crop–wild gene flow also report occurrence of putative crop–wild hybrids in fields where both cultivated and wild sorghum occurred intermixed with and adjacent to cultivated sorghum, largely because the flowering period of wild and weedy complex overlap with those of cultivated sorghum (Tesso et al., 2008), with most of the gene flow occurring from cultivated to wild sorghum, leading to introgression of crop genes into wild sorghum (Muraya et al., 2011a). This may have risk, in the long run, on the very existence of sorghum wild relatives in nature.

Little is known about the extent and patterns of gene flow and genetic diversity between cultivated sorghum and its wild related taxa under natural conditions in Africa. Mutegi et al. (2012) examined the magnitude and dynamics of crop–wild gene flow and genetic variability in a crop–wild–weedy complex by using 10 polymorphic SSRs on 110 cultivated sorghum and 373 wild sorghum individuals under traditional farming system in Kenya. They detected asymmetric gene flow with higher rates from crop-to-wild forms than vice versa, and the two congeners that they found retained substantial phenotypic distinctness but were genetically not different; however, both the congeners had many rare alleles, which may be linked to important adaptive traits. Further, isolation by distance in cultivated or wild sorghum had no impact on gene dispersal in the two conspecifics. This knowledge on the amount and organization of genetic diversity within crop–wild–weedy complex can assist conservationists and/or crop genetic enhancer's as they focus on conserving and exploiting truly distinct groups in sorghum improvement. The discovery of rare alleles should be investigated if these could be related to adaptive traits to exploit in breeding programs for enhancing sorghum adaptation to extreme environmental variations. Further, when studying the adaptive value of hybrid generations (F_1, F_2, and F_3) from wild × cultivated cross, Muraya et al. (2012) report that progenies had no serious fitness penalties, with some hybrids as fit as their respective wild parents across the three generations evaluated, thus wild × cultivated hybrids may act as an avenue for introgression.

Ethnolinguistic Diversity Impacting Conservation and Distribution of On-Farm Diversity

Farmers are the custodians and end-users of plant genetic resources. Differences in their ethnic, linguistic, cultural, socioeconomic, family structure, environmental stress leading to animal and human migration, economic factors, and, above all, informal seed exchange represent a key mechanism in the dynamics of crop genetic diversity (Delêtre et al., 2011; Pautasso et al., 2013). The indigenous knowledge–based sorghum classification and naming has been a long tradition in Ethiopia. Mekbib (2007) employed a range of methodologies to gather information on folk taxonomy of naming sorghum varieties in Ethiopia and concluded that farmers use a range of morphological, response to abiotic and biotic stresses, and end-use related traits in folk taxonomy of sorghum to classify their gene pool into distinct groups. The most important morphological traits used by farmers in describing the taxonomy of sorghum are panicle-related traits (as also reflected in classification of cultivated plants developed by Harlan and de Wet [1971]). Thus, the integrated folk–formal taxonomy is imperative for management and utilization of on-farm genetic resources (Mekbib, 2007). The role of social boundaries in sorghum evolution and diversification that use 18 SSR makers and 15 morphological descriptors on 297 individual plants derived from seeds collected under 16 named landraces was assessed between three ethnolinguistic groups (Chuka, Mbeere, and Tharaka) in eastern Kenya. Labeyrie et al. (2014) found that the spatial distribution of landrace names and the overall genetic spatial patterns were significantly correlated with ethnolinguistic partition. However, the genetic structure inferred from molecular markers did not discriminate the short-cycle landraces despite their morphological distinctness, which highlight the effects of the social organization of communities on the diffusion of seed, practices, and variety of nomenclature (Pautasso et al., 2013; Labeyrie et al., 2014). Crop diversity patterns result not only from an interaction between genetic and environmental factors, G × E, but from a three-way interaction, G × E × S (social boundaries) (Leclerc and d'Eeckenbrugge, 2012).

Caudatum-based sorghum cultivation in some parts of Africa (north of lake Kyoga in Uganda; south of 12° north latitude in Sudan; between 9.5° and 12° north latitude in Chad; as far south as 9.5° north latitude in Cameroon; northeastern Nigeria) has played a key role in the life of inhabitants speaking Chari–Nile languages. *Caudatum* sorghums are adapted under drought-stressed conditions. The correspondence of the distributions of *caudatum* sorghums and Chari-Nile speakers over such a vast area suggest that *caudatum* sorghums are very much part of the agriculture carried to every place the Chari–Nile-speaking inhabitants settled (Stemler et al., 1975). More recently, Westengen et al. (2014) examined the genetic structure of sorghum and detected three major populations associated with the distribution of ethnolinguistic groups in Africa. The codistribution of the central sorghum population and the Nilo-Saharan language family demonstrate a close and causal relationship between the distribution of sorghum and languages in the region between the Chari and Nile rivers. The southern sorghum population is associated with the Bantu languages of the Niger-Congo language family, while the northern sorghum population is distributed across early Niger-Congo and Afro-Asiatic language family areas with dry agroclimatic conditions. Further, the genetic structure within the central sorghum population is associated with language group expansions within the Nilo-Saharan language family. The western Nilotic ethnolinguistic group (Pari people) provides a window into the social and cultural factors involved in generating and maintaining continent-wide diversity patterns. Westengen et al. (2014) further claimed that the age–grade system, a cultural institution important for the expansive success of this ethnolinguistic group in the past, plays a central role in the management of sorghum landraces and continues to underpin the resilience of their traditional seed system. Thus, these observations support the "farming-language co-dispersal hypothesis," which proposes that farming and language families have moved together through population growth and migration (Diamond and Bellwood, 2003; Jobling et al., 2013).

Enhancing Sorghum Adaptation to Climate Change and Variability

Climate-Resilient Germplasm

Climate models predict increased intensity of drought and heat stress as well as a threat to altered pests and diseases due to global warming, which will have adverse impact on food production and quality (Dwivedi et al., 2013). Plant genetic resources are the most sought after biological resource to adapt to climate change and variability. Developing crop cultivars that resist stress, and at the same time use resources efficiently, are needed to minimize production losses associated with climate change. Discovering and utilizing climate-resilient germplasm (traits) in crop improvement is the way forward for developing climate-ready crops. A number of germplasm resistance to abiotic and biotic stresses are reported (reviewed in Upadhyaya et al. [2014a]). Seed yield under drought-stressed conditions in sorghum is a function of harvest index (HI), transpiration efficiency (TE), and water extraction (WE). Vadez et al. (2011) reported a 10-fold difference in seed yield and HI, a twofold difference in TE, and 1.25-fold difference in WE under drought-stressed conditions among 149 reference set accessions. Accessions from *durra* landrace had the highest WE capacity, whereas those from *caudatum-bicolor* and *caudatum-durra* races had poor WE capacity. Likewise, *durra, caudatum,* and *caudatum-guinea* races had the highest TE, whereas the *guinea* race had the lowest TE. Kapanigowda et al. (2013) reported large variation for drought tolerance–related traits among exotic germplasm and advanced breeding lines evaluated for postflowering drought tolerance. Chlorophyll content (SPAD chlorophyll meter reading), seed yield, and HI under dryland conditions decreased by 9, 44, and 16%, respectively, compared with irrigated conditions. Genotype RT × 7000 and PI 475432 had higher leaf temperature and grain yield, while PI 570895 had lower leaf temperature and higher grain yield under dryland conditions. Genotypes with increased grain yield and optimum leaf temperature were IS 1212, PI 510898, and PI 533946, while those with decreased grain yield but with optimum leaf temperature under dryland conditions were IS 14290, 12945, and 1219. Likewise, genotypes IS 14290, 12945, and 1219 showed greater susceptibility to drought with lower grain yield and HI along with late maturity. Taken together, these accessions are ideal genetic resources to be used in breeding and genomics for drought tolerance in sorghum. Kholova et al. (2014) note that limited transpiration rates under high vapor deficit if combined with enhanced WE capacity at the root level had shown a positive effect on production in sorghum. Further, variability in leaf development (smaller canopy size, later plant vigor, or increased leaf appearance rate) also increases grain yield under severe drought, although it causes a stover yield trade-off under milder stress. They thus conclude that grain yield in sorghum under water-limited conditions is largely determined by the amount of water availability after anthesis, and more so with stress severity.

Djanaguiraman et al. (2014) reported heat stress tolerant (DK-54-00, Pioneer 84G62, and SC 1019) and sensitive (DK 28-E, SC 15, and TX 7078) sorghum germplasm lines. They further claimed that heat stress tolerant genotypes had relatively less oxidative damage in leaves and pollen grains than sensitive genotypes.

Phosphorous (P) deficiency and aluminum (Al) toxicity are serious constraints to sorghum production in West Africa. Leiser et al. (2014) identified a genomic region on chromosome 3 in sorghum associated with a major Al tolerance gene, *SbMATE,* colocated in this region. The *SbMATE*-specific SNPs showed very high associations to grain yield in P-deficient soils. Furthermore, Caniato et al. (2014) reported SNPs with the strongest association signal among highly Al tolerant accessions, which may provide researchers greater opportunity to identify accessions adapted to P-deficient and Al-toxic soils in diverse germplasm by using allele mining strategies.

Nitrification refers to the oxidation of ammonia to nitrate (NO_3^-), which is lost through leaching which can contaminate ground water, leading to adverse impacts on the environment and human health (Galloway et al., 2008; Schlesinger, 2009). Certain plant species release nitrification inhibitors, such as sorgoleone, termed as biological nitrification inhibition (BNI), from their roots (Subbarao et al., 2006, 2007). Sorgoleone production, which inhibit nitrification, and BNI activity released by roots are closely associated. The evidence to date suggest that sorghum genotypes release varying quantity of sorgoleone which suppresses soil nitrification. Genetic differences for sorgoleone release and its functional link with BNI in sorghum germplasm have been found. A sorghum genotype GDLP 34-5-5-3 released higher quantities of sorgoleone into the rhizosphere, which had higher BNI activity, and greatly suppressed soil nitrification compared with IS 41245 (Tesfamariam et al., 2014). This approach has potential for genetic improvement of sorghum BNI capacity and the deployment of such germplasm-cultivars in low-nitrogen sorghum production systems.

Wild and Weedy Relatives as Resources to Identifying New Variations

Wild and weedy relatives of crop species including sorghum harbor genes for resistance to stresses, conferring greater adaptive advantage under adverse environments (Dwivedi et al., 2008). A global initiative is under way to collect, conserve, evaluate, and make available this diversity in a form that plant breeders can readily use (a prebreeding product) to produce varieties adapted to the new climatic conditions that farmers, particularly in the developing world, are already encountering (Dempewolf et al., 2014). Recently, Venkateswaran et al. (2014) consolidated a list of wild sorghum species with reported resistance to pests and diseases, which include resistance to downy mildew, shoot fly, spotted stem borer, and midge. Some wild sorghum races such as vigratum, arundinaceum, and verticilliflorum showed ecologically wide adaptation. They can be found in African rainforests where cultivated sorghums are poorly adapted, or thrive in deserts because of their greater drought and heat tolerance (reviewed in Venkateswaran et al. [2014]). Wild and weedy sorghum relatives offer immense opportunities to mine and transfer agronomically beneficial alleles to cultivated sorghum.

Johnsongrass [*S. halepense* (L.) Pers], a tetraploid ($2n = 2\times = 40$) and highly invasive weed species, originated from hybridization between *S. propinquum* (Kunth.) Hitchc. and *S. bicolor* ssp. *arundinaceum* followed by chromosome doubling (de Wet, 1978). *S. halepense* is native to western Asia and adapted to very cold climates (Warwick et al., 1986), whereas *S. propinquum* is adapted to the moist habitats in southeast Asia (Paterson et al., 1995), and both are perennials and rhizomatous (Chittenden et al., 1994). Perennial sorghum cropping systems offer agroecological benefits such as increased soil organic carbon, reduced soil erosion, low input use, and higher net energy return (Costanza et al., 1997; Lewandowski et al., 2003). Overwintering ability is a critical component of perennial sorghum systems in temperate regions and relates directly to the production of rhizomes (Paterson et al., 1995; Washburn et al., 2013a). The genus *Sorghum* is a model plant for dissecting the molecular basis of rhizomatousness (Paterson et al., 1995). *S. propinquum* is cross-compatible with *S. bicolor*, whereas *S. halepense* and *S. bicolor* are very difficult to intermate (de Wet, 1978). Rhizome formation is a quantitatively inherited trait (Paterson et al., 1995), and two QTLs on linkage group SBI-01 for overwintering have been found. One QTL (*Overwintering 2011B*) located within ~14.5 Mb region overlaps directly to a genomic region associated with rhizome growth, while the other QTL for overwintering differ from *Overwintering 2011B*, indicating that two unique genes on SBI-01 are involved in overwintering (Washburn et al., 2013b). More importantly, an earlier study reported no trade-off between allocation of resources to rhizomes vs. seeds supporting the possibility of combining high seed yield and production of rhizomes in the development of perennial grain sorghum adapted to temperate conditions (Piper and Kulakow, 1994). Knowledge gained from this study should aid in the development of cultivated perennial sorghums adapted to extreme cold conditions in temperate regions (Cox et al., 2002). The overwintering sorghum types could then be used for improvement in grain, forage, and biofuel sorghum production.

S. propinquum, like *S. halepense*, also has some interesting traits (other than rhizomes) of applied agronomic value. For example, a *S. bicolor* × *S. propinquum* were cross segregated for traits related to plant architecture, growth and development, reproduction, and life history (Kong et al., 2013). The evaluation of this population in environments (hot tropical) to which *S. bicolor* adapted is expected to map new QTLs for flowering that previously eluded detection. Evaluations in temperate (cold) environments may unravel QTLs for traits that were previously below the significance threshold, and may also provide more variation relevant to temperate latitudes adaptation. Terminal drought and heat stress remains the most important abiotic constraints to sorghum production in Africa and Asia, whereas adaptation to extreme cold is needed to expand sorghum adaptation in the Americas. Currently, a USAID-supported project is investigating the potential of *S. halepense* and *S. propinquum* to enhance cultivated sorghum adaptation to severe drought and heat stress conditions in the tropics or extend adaptation to temperate regions with extreme cold conditions, both for feed, grain, and bioenergy crops.

Applying TILLING to Create New Variations

Targeting Induced Local Lesions IN Genome (TILLING) is a powerful reverse genetic technique that employs a mismatch-specific endonuclease to detect single base pair allelic variation in a target gene by using high-throughput assay. Its advantages over other reverse techniques include its adaptability to virtually any organism, its facility for high throughput, and its independence of genome size, reproductive system, or generation time (Gilchrist and Haughn, 2005). TILLING has been successfully used for the detection of both induced and natural variation in pathogenic bacteria and plant and animal species (reviewed in Dwivedi et al. [2007]). In sorghum, ethyl methane sulfonate (EMS)-mutanized population consisting of 1600 lines was generated from the inbred line BT × 623, and when phenotyped, numerous phenotypes with altered morphological and agronomic traits were discovered (Xin et al., 2008). Further, when a subset of 768 mutant lines were analyzed by TILLING by using four target genes, they identified five mutations, one per 526 kb genomic region. Two mutations were detected in the gene encoding caffeic acid O-methyltransferase (*COMT*), while the other two mutant lines segregated for brown midrib (*bmr*) phenotype, a trait associated with altered lignin content and increased digestibility. Sorghum contains the cyanogenic glucoside dhurrin (HCNp), consumption of which can cause a significant problem for animal and human because of its ability to release toxic hydrogen cyanide, sometimes causing the death of grazing animals. By using a TILLING approach, mutant lines exhibiting lower or negligible HCNp compared with nonmutated parent lines were reported in sorghum (Blomstedt et al., 2012). These examples clearly demonstrate the effectiveness of a TILLING approach to create new variations, not previously known in crop plants.

Genomic Resources Promoting Conservation and Utilization of Genetic Resources

Variation in genome sequences among the germplasm lines, both wild and cultivated types, can be used to guide crop improvement. Today, the sorghum research community has access to numerous genomic resources, including DNA markers (SSR, diversity arrays technology [DArT], and SNP) (Mace et al., 2008, 2013; Li et al., 2009; Singhal et al., 2011; Billot et al., 2012; Evans et al., 2013; Morris et al., 2013), high-density genetic maps (Bowers et al., 2003; Mace et al., 2009; Kong et al., 2013), and the sequenced genomes (Paterson et al., 2009; Mace et al., 2013). Evans et al. (2013) reported ~2.8 million SNPs and ~300,000 insertions and deletions (INDELS) that distinguish sorghum genotype Tx7000 and/or BTx642 from the reference genotype, BTx623. Further, comparison of genome sequences of these genotypes revealed greater than 100-fold variation in the density of DNA polymorphisms. BTx642 (also known as B35) is stay-green and resistant to terminal drought (Tuinstra et al., 1997; Crasta et al., 1999), whereas Tx7000 is non-stay-green germplasm and susceptible to terminal drought. BTx642 has been used for the development of commercial hybrids (Henzell et al., 2001).

An SSR kit (http://sat.cirad.fr/sat/sorghum_ssr_kit) is available for diversity analysis of sorghum germplasm. It contains information on 48 technically robust sorghum SSRs and 10 DNA controls. It can be used to calibrate sorghum SSR genotyping data acquired with different technologies to genetic diversity references (Billot et al., 2012). More recently, a robust SNP array platform has been reported which enables screening for genome-wide and trait-linked polymorphisms in genetically diverse *S. bicolor* populations (Bekele et al., 2013). This array of the whole-genome sequences with 6× to 12× coverage, from five genetically diverse *S. bicolor* genotypes (three sweet sorghums and two grain sorghums) aligned to the sorghum reference genome. From this resource, they selected 2124 informative SNPs aligned with 50% allele frequencies and evenly spaced throughout the six *S. bicolor* genomes. Further, they selected 876 SNPs associated with early-stage chilling tolerance. When further testing this array on additional 564 sorghum genotypes, they validated 2620 robust and polymorphic sorghum SNPs, underlining the efficiency of this array technology for whole-genome SNP selection and screening, with diverse applications, genetic mapping, genome-wide association, and genomic selection (Bekele et al., 2013).

The analysis of the genomes of populations growing in contrasting environments will unravel the genes subject to natural selection in adaptation to climate variations. The whole-genome sequencing of such populations should define the numbers and types of genes associated with climate adaptation. This strategy is most easily applied to species for which high-quality reference genome sequences are available and where populations of wild relatives are found growing in diverse environments or across environmental gradients. Such an approach will also guide germplasm curators in preserving populations with novel alleles. Crop genetic enhancers can use such genetic resources to enrich cultigen gene pools (Henry, 2013, 2014). Many of sorghum's wild relatives are found in harsh and diverse agroecologies where some naturally outcross with cultivated types. These can produce populations which are ideal genetic resources to mine alleles associated with adaptation to environmental stresses, and when such alleles are introgressed into cultivated types, populations with enhanced adaptation to adverse climatic conditions originate to effect selection of desired types.

To date a number of significant marker–trait associations, using representative subsets or diversity panels and SSR/SNP markers, have been reported in sorghum: days to flowering, plant height, tillering, culm length, number of panicles, panicle length, seed weight, and maturity, inflorescence architecture, stalk sugar content (Brix), and resistance to anthracnose, grain mold, and rust. Many of these markers are comapped on the same linkage groups previously reported as harboring QTL or a candidate gene associated with tillering, plant height, maturity, and resistance to anthracnose, rust, and grain mold (reviewed in Upadhyaya et al. [2014a]). Sorghum's sensitivity to low soil temperatures adversely impacts seed germination. Three SNP loci associated with seed germination and one for seedling vigor at low temperature stress (12°C) were reported (Upadhyaya et al., 2016b). Further, genome-wide association mapping, involving 300 sorghum accessions, 10 seed quality traits, and 1290 SNPs, resulted in eight significant marker–trait associations, of which one SNP (*starch synthase IIa*) was associated with seed hardness and two SNPs (*starch synthase IIb* and *pSB1120*) with starch content (Sukumaran et al., 2012). Sorghum production in sub-Saharan Africa is seriously constrained by phosphorous (P) deficiency and aluminum (Al) toxicity. Using 187 diverse sorghum accessions and a large panel of SNPs (220,934), Leiser et al. (2014) identified a genomic region on chromosome 3 highly associated to grain yield. They further showed that a major tolerance gene in sorghum, *SbMATE*, was colocated in this region, and *SbMATE*-specific SNPs showed very high associations to grain yield production, especially in P-deficient or Al-toxic soils. These reports clearly indicate that once SNPs associated with useful variations are identified, they can be used as genetic tags for discovering new sources of variation from the germplasm collections.

Resequencing 44 sorghum lines from the primary gene pool (representing taxonomic, geographic, and end-use diversity), the progenitor (*S. bicolor* subsp. *verticilliflorum*) and the allopatric Asian species *S. propinquum* resulted a 16 to 45 × genome coverage, 8 million high-quality SNPs, 1.9 million INDELS, and specific gene loss and gain events in *S. bicolor* (Mace et al., 2013). They further showed that sorghum possesses a diverse primary gene pool but with decreased diversity in both landrace and improved groups, and a great untapped pool of diversity in *S. propinquum*. More importantly, they found that sorghum has strong racial structure, a complex domestication history involving at least two distinct domestication events, and modern cultivated sorghum evolved from a limited sample of racial variation. Thus, discovery of untapped diversity in the other races of *S. bicolor* but also in *S. propinquum* offer a great opportunity for the genetic improvement of sorghum and other grass species.

Outlook

Great diversity, both in cultivated and wild relatives, exists in sorghum germplasm collections preserved worldwide. Abundant genetic and genomic resources are available in sorghum. Researchers are using the representative subsets (core and mini core collections or reference sets) to identify new sources of variations, assess population structure and diversity, and identify marker–trait associations. Many of the markers comapped on the same linkage groups previously reported as harboring QTLs or candidate genes are associated with agronomically beneficial traits. The untapped diversity unraveled through genome sequencing of a diverse set of germplasm offers a great opportunity for the genetic improvement of sorghum. The genomes of *S. halepense* and *S. propinquum* are amenable to introgress agronomically beneficial alleles to enhance sorghum adaptation to extreme climatic conditions.

References

Adugna, A. 2014. Analysis of in situ diversity and population structure in Ethiopian cultivated *Sorghum bicolor* (L.) landraces using phenotypic traits and SSR markers. SpringerPlus 3:212.

Appa Rao, S., K.E. Prasada Rao, M.H. Mengesha, and V.G. Reddy. 1996. Morphological diversity in sorghum germplasm in India. Genet. Resour. Crop Evol. 43:559–567. doi:10.1007/BF00138832

Ayana, A., and E. Bekele. 1998. Geographical pattern of morphological variation in sorghum (*Sorghum bicolor* L. Moench) germplasm from Ethiopia and Eritrea: Qualitative characters. Hereditas 129:195–205. doi:10.1111/j.1601-5223.1998.t01-1-00195.x

Ayana, A., and E. Bekele. 1999. Multivariate analysis of morphological variation in sorghum (*Sorghum bicolor* L. Moench) germplasm from Ethiopia and Eritrea. Genet. Resour. Crop Evol. 46:273–284. doi:10.1023/A:1008657120946

Ayana, A., and E. Bekele. 2000. Geographical pattern of morphological variation in sorghum (*Sorghum bicolor* L. Moench) germplasm from Ethiopia and Eritrea: Quantitative characters. Euphytica 115:91–104. doi:10.1023/A:1003998313302

Barnaud, A., M. Deu, E. Garine, J. Chantereau, J. Bolteu, E.O. Koïda, D. McKey, and H.I. Joly. 2009. A weed-crop complex in sorghum: The dynamics of genetic diversity in a traditional farming system. Am. J. Bot. 96:1869–1879. doi:10.3732/ajb.0800284

Barnaud, A., M. Deu, E. Garine, D. McKey, and H.I. Joly. 2007. Local genetic diversity of sorghum in a village in northern Cameroon: Structure and dynamics of landraces. Theor. Appl. Genet. 114:237–248. doi:10.1007/s00122-006-0426-8

Barnaud, A., G. Triguero, D. Mckey, and H.I. Joly. 2008. High outcrossing rates in fields with mixed sorghum landraces: How are landraces maintained? Heredity 101:445–452. doi:10.1038/hdy.2008.77

Barro-Kondombo, C., F. Sagnard, J. Chantereau, M. Deu, K. vom Brocke, P. Durand, E. Gozé, and J.D. Zongo. 2010. Genetic structure among sorghum landraces as revealed by morphological variation and microsatellite markers in three agroclimatic regions of Burkina Faso. Theor. Appl. Genet. 120:1511–1523. doi:10.1007/s00122-010-1272-2

Bekele, W.A., S. Wieckhorst, W. Freidt, and R.J. Snowdon. 2013. High-throughput genomics in sorghum: From whole-genome resequencing to a SNP screening array. Plant Biotechnol. J. 11:1112–1125. doi:10.1111/pbi.12106

Benor, S., and L. Sisay. 2003. Folk classification of sorghum (*Sorghum bicolor* L. (Moench)) landraces and its ethnobotanical implication: A case study in Northeastern Ethiopia. Etnobiologia 3:29–41.

Billot, C., P. Ramu, S. Bouchet, J. Chantereau, M. Deu, L. Gardes, J.-L. Noyer, J.-F. Rami, R. Rivallan, Y. Li, P. Lu, T. Wang, R.T. Folkertsma, E. Arnaud, H.D. Upadhyaya, J.-C. Glaszmann, and C.T. Hash. 2013. Massive sorghum collection genotyped with SSR markers to enhance use of global genetic resources. PLoS ONE 8:e59714. doi:10.1371/journal.pone.0059714

Billot, C., R. Rivallan, M.N. Sall, D. Fonceka, M. Deu, J.-C. Glaszmann, J.-E. Noyer, J.-F. Rami, A.-M. Risterucci, P. Wincker, P. Ramu, and C.T. Hash. 2012. A reference microsatellite kit to assess for genetic diversity of *Sorghum bicolor* (Poaceae). Am. J. Bot. 99:e245–e250. doi:10.3732/ajb.1100548

Blomstedt, C.K., R.M. Gleadow, N. O'Donnell, P. Naur, K. Jensen, T. Laursen, C.E. Olsen, P. Stuart, J.D. Hamill, B.L. Møller, and A.D. Neale. 2012. A combined biochemical screen and TILLING approach identifies mutations in *Sorghum bicolor* L. Moench resulting in acynogenic forage production. Plant Biotechnol. J. 10:54–66. doi:10.1111/j.1467-7652.2011.00646.x

Bowers, J.E., et al. 2003. A high-density genetic recombination map of sequence-tagged sites for sorghum, as a framework for comparative structural and evolutionary genomics of tropical grains and grasses. Genetics 165:367–386.

Burow, G., C.D. Franks, Z. Xin, and J.J. Burke. 2012. Genetic diversity in a collection of Chinese sorghum landraces accessed by microsatellites. Am. J. Plant Sci. 3:1722–1729. doi:10.4236/ajps.2012.312210

Caniato, F.F., M.T. Hamblin, C.T. Guimaraes, Z. Zhang, R.E. Schaffert, L.V. Kouchian, and J.V. Magalhaes. 2014. Association mapping provides insights into the origin and the fine structure of the sorghum aluminum tolerance locus, Alt_{SB}. PLoS ONE 9:e87438. doi:10.1371/journal.pone.0087438

Casa, A.M., S.E. Mitchell, M.T. Hamblin, H. Sun, J.E. Bowers, A.H. Paterson, C.F. Aquadro, and S. Kresovich. 2005. Diversity and selection in sorghum: Simultaneous analyses using simple sequence repeats. Theor. Appl. Genet. 111:23–30. doi:10.1007/s00122-005-1952-5

Chittenden, L.M., K.F. Schertz, Y.R. Lin, R.A. Wing, and A.H. Paterson. 1994. A detailed RFLP map of *Sorghum bicolor* × *S. propinquum*, suitable of high-density mapping, suggest ancestral duplication of *Sorghum* chromosomes or chromosomal segments. Theor. Appl. Genet. 87:925–933. doi:10.1007/BF00225786

Clayton, W.D., and S.A. Renvoize. 1986. Genera granium grasses of the world. Kew Bulletin Addition Ser. XIII. Royal Botanic Gardens, Kew, London, UK, p. 338–345.

Costanza, R., R. d'Arge, R. de Groot, S. Farber, M. Grasso, B. Hannon, K. Limburg, S. Naeem, R.V. O'Neill, J. Paruelo, R.J. Raskin, P. Sutton, and M. van den Belt. 1997. The value of world's ecosystem services and natural capital. Nature 387:253–260. doi:10.1038/387253a0

Cox, T.S., M. Bender, C. Picone, D.L. van Tassel, J.B. Holland, E.C. Brummer, B.E. Zoeller, A.H. Paterson, and W. Jackson. 2002. Breeding perennial grain crops. Crit. Rev. Plant Sci. 21:59–91. doi:10.1080/0735-260291044188

Crasta, O.R., W.W. Xu, D.T. Rosenow, J. Mullet, and H.T. Nguyen. 1999. Mapping of post-flowering drought resistance traits in grain sorghum: Association between QTLs influencing premature senescence and maturity. Mol. Gen. Genet. 262:579–588. doi:10.1007/s004380051120

Cuevas, H.E., and L.K. Prom. 2013. Assessment of molecular diversity and population structure of the Ethiopian sorghum (*Sorghum bicolor* L. (Moench)) germplasm collection maintained by the USDA-ARS National Plant Germplasm System using SSR markers. Genet. Resour. Crop Evol. 60:1817–1830. doi:10.1007/s10722-013-9956-5

Dahlberg, J.A., J. Berenji, V. Sikora, and D. Latković. 2011. Assessing sorghum (*Sorghum bicolor* (L.) Moench) germplasm for new traits: Food, fuel and unique uses. Maydica 56:85–92.

Dahlberg, J.A., J.J. Burke, and D.T. Rosenow. 2004. Development of a sorghum core collection: Refinement and evaluation of a subset from Sudan. Econ. Bot. 58:556–567. doi:10.1663/0013-0001(2004)058[0556:DOASCC]2.0.CO;2

Delêtre, M., D.B. McKey, and T.R. Hodkinson. 2011. Marriage exchanges, seed exchanges, and the dynamics of manioc diversity. Proc. Natl. Acad. Sci. USA 108:18249–18254. doi:10.1073/pnas.1106259108

Dempewolf, H., R.J. Eastwood, L. Guarino, C.K. Khoury, J.V. Müller, and J. Toll. 2014. Adapting agriculture to climate change: A global initiative to collect, conserve, and use crop wild relatives. Agroecol. Sust. Food Syst. 38:369–377.

Deu, M., F. Sagnard, J. Chantereau, C. Calatayud, D. Hérault, C. Mariac, J.-L. Pham, Y. Vigouroux, I. Kapran, P.S. Troare, A. Mamadou, B. Gerard, J. Ndjeunga, and G. Bezançon. 2008. Niger-wide assessment of in situ sorghum genetic diversity with microsatellite markers. Theor. Appl. Genet. 116:903–916. doi:10.1007/s00122-008-0721-7

de Wet, J.M. 1978. Systematics and evolution of sorghum sect. *Sorghum* (Gramineae). Am. J. Bot. 65:477–484. doi:10.2307/2442706

Diamond, J., and P. Bellwood. 2003. Farmers and their languages: The first expansions. Science 300:597–603. doi:10.1126/science.1078208

Dillon, S.L., F.M. Shapter, R.J. Henry, G. Cordeiro, L. Izquierdo, and L.S. Lee. 2007. Domestication to crop improvement: Genetic resources for *Sorghum* and *Saccharum* (Andropogoneae). Ann. Bot. (Lond.) 100:975–989. doi:10.1093/aob/mcm192

Djanaguiraman, M., P.V. Vara Prasad, M. Murugan, R. Perumal, and U.K. Reddy. 2014. Physiological differences among sorghum (*Sorghum bicolor* L Moench) genotypes under high temperature stress. Environ. Exp. Bot. 100:43–54. doi:10.1016/j.envexpbot.2013.11.013

Doggett, H. 1988. Sorghum. 2nd ed. Longmans, London.

Dwivedi, S.L., J.H. Crouch, D.J. Mackill, Y. Xu, M.W. Blair, M. Ragot, H.D. Upadhyaya, and R. Ortiz. 2007. The molecularization of public sector crop breeding: Progress, problems, and prospects. Adv. Agron. 95:163–318. doi:10.1016/S0065-2113(07)95003-8

Dwivedi, S.L., K. Sahrawat, H. Upadhyaya, and R. Ortiz. 2013. Food, nutrition and agrobiodiversity under global climate change. Adv. Agron. 120:1–128. doi:10.1016/B978-0-12-407686-0.00001-4

Dwivedi, S.L., H.D. Upadhyaya, and C.L.L. Gowda. 2009. Approaches to enhance the value of genetic resources in crop improvement. In: Strengthening information on plant genetic resources in Asia. FAO, Bangkok, Thailand. p. 71–77.

Dwivedi, S.L., H.D. Upadhyaya, H.T. Stalker, M.W. Blair, D.J. Bertioli, S. Nielen, and R. Ortiz. 2008. Enhancing crop gene pools with beneficial traits using wild relatives. Plant Breed. Rev. 30:179–230.

Elangovan, M., Prabhakar, V.A. Tonapi, and D.C.S. Reddy. 2009. Collection and characterization of Indian sorghum landraces. Indian J. Plant Genetic Resour. 22:173–181.

Evans, J., R.F. McCormick, D. Morishige, S.N. Olson, B. Weers, J. Hilley, P. Klein, W. Rooney, and J. Mullet. 2013. Extensive variation in the density and distribution of DNA polymorphism in sorghum genomes. PLoS ONE 8:e79192. doi:10.1371/journal.pone.0079192

Food and Agricultural Organization. 2010. The second report on the state of the world's plant genetic resources for food and agriculture. FAO, Rome.

Frankel, O.H. 1984. Genetic prospective of germplasm conservation. In: W. Arber, K. Limensee, P.J. Peacock, and P. Stralinger, editors, Genetic manipulation: Impact on man and society. Cambridge Univ. Press, Cambridge, UK. p. 161–170.

Galloway, J.N., A.R. Townsend, J.W. Erisman, M. Bekunda, Z. Cai, J.R. Freney, L.A. Martinelli, S.P. Seitzinger, and M.A. Sutton. 2008. Transformation of the nitrogen cycle: Recent trends, questions and potential solutions. Science 320:889–892. doi:10.1126/science.1136674

Gamar, Y.A., D. Kiambi, M. Kairichi, M. Kyallo, and M.H. Elgada. 2013. Assessment of genetic diversity and structure of Sudanese sorghum accessions using simple sequence repeat (SSRs) markers. Greener J. Plant Breeding Crop Sci. 1:16–24.

Geleta, N., and M.T. Labuschagne. 2005. Qualitative trait variation in sorghum (*Sorghum bicolor* L. (Moench) from eastern highlands of Ethiopia. Biodivers. Conserv. 14:3055–3064. doi:10.1007/s10531-004-0315-x

Gilchrist, E.J., and G.W. Haughn. 2005. TILLING without a plough: A new method with applications for reverse genetics. Curr. Opin. Plant Biol. 8:211–215. doi:10.1016/j.pbi.2005.01.004

Grenier, C., P.J. Bramel, J.A. Dahlberg, A. El-Ahmadi, M. Mahmoud, G.C. Peterson, D.T. Rosenow, and G. Ejeta. 2004. Sorghums of the Sudan: Analysis of regional diversity and distribution. Genetic Resour. Crop Evol. 51:489–500. doi:10.1023/B:GRES.0000024155.43149.71

Grenier, C., P. Hamon, and P.J. Bramel-Cox. 2001. Core collection of sorghum: II. Comparison of three random sampling strategies. Crop Sci. 41:241–246. doi:10.2135/cropsci2001.411241x

Harlan, J.R., and J.M.J. de Wet. 1971. Toward a rational classification of cultivated plants. Taxon 20:509–517. doi:10.2307/1218252

Harlan, J.R., and J.M.J. de Wet. 1972. A simplified classification of cultivated sorghum. Crop Sci. 12:172–196. doi:10.2135/cropsci1972.0011183X001200020005x

Henry, R.J. 2013. Sequencing of wild crop relatives to support the conservation and utilization of plant genetic resources. Plant Genetic Resour. 12(S1):S9–S11. doi:10.1017/S1479262113000439

Henry, R.J. 2014. Genomic strategies for germplasm characterization and the development of climate resilient crops. Front. Plant Sci. 5:68. doi:10.3389/fpls.2014.00068

Henzell, R.G., B.R. Hare, D.R. Jordan, D.S. Fletcher, A.N. McCosker, G. Bunker, and D.S. Persley. 2001. Sorghum breeding in Australia: Public and private endeavors. In: A.K. Borrell, R.G. Henzell, and P.N. Vance, editors. Proceedings of the 4th Australian Sorghum Conference, Kooralbyn, Australia. 5–8 Feb. Aust. Inst. of Agric. Sci., Melbourne.

House, L.R. 1985. A guide to sorghum breeding. 2nd ed. Int. Crops Res. Inst. for the Semi-Arid Tropics, Patancheru, India.

Jobling, M.A., E. Hollox, M. Hurles, T. Kivisild, and C. Tyler-Smith. 2013. Human evolutionary genetics. 2nd ed. Taylor and Francis, New York.

Kapanigowda, M.H., R. Perumal, M. Djanaguiraman, R.M. Aiken, T. Tesso, P.V. Vara Prasad, and C.R. Little. 2013. Genotypic variation in sorghum [*Sorghum bicolor* (L.) Moench] exotic germplasm collections for drought and disease resistance. SpringerPlus 2:650. doi:10.1186/2193-1801-2-650

Kholova, J., T. Murugesan, S. Shakti, S. Malayee, R. Baddam, G.L. Hammer, G. McLean, S. Deshpande, C.T. Hash, P.Q. Craufurd, and V. Vadez. 2014. Modelling the effect of plant water traits on yield and stay-green expression in sorghum. Funct. Plant Biol. 41:1019–1034. doi:10.1071/FP13355

Kong, W., H. Jin, C.D. Franks, C. Kim, R. Bandopadhyay, M.K. Rana, S.A. Auckland, V.H. Goff, L.K. Rainville, G.B. Burow, C. Woodfin, J.J. Burke, and A.H. Paterson. 2013. Genetic analysis of recombinant inbred lines for *Sorghum bicolor* × *Sorghum propinquum*. Genes, Genomes. Genetics 3:101–108.

Kumar, A.A., B.V.S. Reddy, H.C. Sharma, C.T. Hash, P.S. Rao, B. Ramaiah, and P.S. Reddy. 2011. Recent advances in sorghum genetic enhancement research at ICRISAT. Am. J. Plant Sci. 2:589–600. doi:10.4236/ajps.2011.24070

Labeyrie, V., M. Deu, A. Barnaud, C. Calatayud, M. Buiron, P. Wambugu, S. Manel, J.-C. Glaszmann, and C. Leclerc. 2014. Influence of ethnolinguistic diversity on the sorghum genetic patterns in subsistence farming systems in eastern Kenya. PLoS ONE 9:e92178. doi:10.1371/journal.pone.0092178

Lacy, S.M., D.A. Cleveland, and D. Soleri. 2006. Farmer choice of sorghum varieties in Southern Mali. Hum. Ecol. 34:331–353. doi:10.1007/s10745-006-9021-5

Leclerc, C., and G.C. d'Eeckenbrugge. 2012. Social organization of crop genetic diversity. The G × E × S interaction model. Diversity 4:1–32. doi:10.3390/d4010001

Leiser, W.L., H.F.W. Rattunde, E. Weltzien, N. Cisse, M. Abdou, A. Diallo, A.O. Tourè, J.V. Magalhaes, and B.I.G. Haussmann. 2014. Two in one sweep: Aluminum tolerance and grain yield in P-limited soils are associated to the same genomic region in West African sorghum. BMC Plant Biol. 14:206. doi:10.1186/s12870-014-0206-6

Lewandowski, I., J.M.O. Scurlock, E. Lindvall, and M. Christou. 2003. The development and current status of perennial rhizomatous grasses as energy crops in the US and Europe. Biomass Bioenergy 25:335–361. doi:10.1016/S0961-9534(03)00030-8

Li, M., N. Yuyama, L. Luo, M. Hirata, and H. Cai. 2009. In silico mapping of 1758 new SSR markers developed from public genomic sequences for sorghum. Mol. Breed. 24:41–47. doi:10.1007/s11032-009-9270-2

Liu, Q., H. Liu, J. Wen, and P.M. Peterson. 2014. Infrageneric phylogeny and temporal divergence of *Sorghum* (Andropogoneae, Poaceae) based on low-copy nuclear and plastid sequences. PLoS ONE 9:e104933. doi:10.1371/journal.pone.0104933

Mace, E.S., J.-F. Rami, S. Bouchet, P.E. Klein, R.R. Klein, A. Kilian, P. Wenzl, L. Xia, K. Halloran, and D.R. Jordan. 2009. A consensus genetic map of sorghum that integrates multiple component maps and high-throughput diversity array technology (DArT) markers. BMC Plant Biol. 9:13. doi:10.1186/1471-2229-9-13

Mace, E.S., et al. 2013. Whole-genome sequencing reveals untapped genetic potential in Africa's indigenous cereal crop sorghum. Nat. Commun. 4:2320.

Mace, E.S., L. Xa, D.R. Jordan, K. Halloran, D.K. Parh, E. Huttner, P. Wenzl, and A. Kilian. 2008. DArT markers: Diversity analysis and mapping in *Sorghum bicolor*. BMC Genomics 9:26. doi:10.1186/1471-2164-9-26

Manzelli, M., S. Benedettelli, and V. Vecchio. 2005. Agricultural biodiversity in Northwest Somalia– an assessment among selected Somali sorghum (*Sorghum bicolor* (L.) Moench) germplasm. Biodivers. Conserv. 14:3381–3392. doi:10.1007/s10531-004-0545-y

Manzelli, M., L. Pileri, N. Lacerenza, S. Benedettelli, and V. Vecchio. 2007. Genetic diversity assessment in Somali sorghum (*Sorghum bicolor* (L.) Moench) accessions using microsatellite markers. Biodivers. Conserv. 16:1715–1730. doi:10.1007/s10531-006-9048-3

Mbeyagala, E.K., D.D. Kiambi, P. Okori, and R. Edema. 2012. Molecular diversity among sorghum (*Sorghum bicolor* (L.) Moench) landraces in Uganda. Int. J. Bot. 8:85–95. doi:10.3923/ijb.2012.85.95

McCouch, S.R., K.R. McNally, W. Wang, and R.S. Hamilton. 2012. Genomics of gene banks: A case study in rice. Am. J. Bot. 99:407–423. doi:10.3732/ajb.1100385

Mekbib, F. 2007. Infra-specific folk taxonomy in sorghum (*Sorghum bicolor* (L.) Moench) in Ethiopia: Folk nomenclature, classification, and criteria. J. Ethnobiol. Ethnomed. 3:38. doi:10.1186/1746-4269-3-38

Mekbib, F., A. Bjørnstad, L. Sperling, and G. Synnevåg. 2009. Factors shaping on-farm genetic resources of sorghum (*Sorghum bicolor* L. (Moench) in the center of diversity, Ethiopia. Int. J. Biodiversity Conserv. 1:45–59.

Motlhaodi, T., M. Geleta, T. Bryngelsson, M. Fatih, S. Chite, and R. Ortiz. 2014. Genetic diversity in ex-situ conserved sorghum accessions of Botswana as estimated by microsatellite markers. Aust. J. Crop Sci. 8:35–43.

Morris, G.P., P. Ramu, S.P. Deshpande, C.T. Hash, T. Shah, H.D. Upadhyaya, O. Riera-Lizarazu, P.J. Brown, C.B. Acharya, S.E. Mitchell, J. Harriman, J.C. Glaubitz, E.S. Buckler, and S. Kresovich. 2013. Population genomic and genome-wide association studies of agroclimatic traits in sorghum. Proc. Natl. Acad. Sci. USA 110:453–458. doi:10.1073/pnas.1215985110

Muraya, M.M., H.H. Geiger, S. de Villiers, F. Sagnard, B.M. Kaanyenji, D. Kiambi, and H.K. Parzies. 2011a. Investigation of pollen competition between wild and cultivated sorghums (*Sorghum bicolor* (L.) Moench) using simple sequence repeats markers. Euphytica 178:393–401. doi:10.1007/s10681-010-0319-4

Muraya, M.M., S. de Villiers, H.K. Parzies, E. Mutegi, F. Sagnard, B.M. Kaanyenji, D. Kiambi, and H.H. Geiger. 2011b. Genetic structure and diversity of wild sorghum populations (*Sorghum* spp.) from different eco-geographical regions of Kenya. Theor. Appl. Genet. 123:571–583. doi:10.1007/s00122-011-1608-6

Muraya, M.M., H.H. Geiger, F. Sagnard, L. Toure, P.C.S. Traore, S. Togola, S. de Villiers, and H.K. Parzies. 2012. Adaptive value of wild × cultivated sorghum (*Sorghum bicolor* (L.) Moench) hybrids in generations F_1, F_2, and F_3. Genet. Resour. Crop Evol. 59:83–93. doi:10.1007/s10722-011-9670-0

Mutegi, E., F. Sagnard, M. Labuschagne, L. Herselman, K. Semagn, M. Deu, S. de Villers, B.M. Kanyenji, C.N. Mwongera, P.C.S. Traore, and D. Kiambi. 2012. Local scale patterns of gene flow and genetic diversity in a crop-wild-weedy complex of sorghum (*Sorghum bicolor* (L.) Moench) under traditional agricultural field conditions in Kenya. Conserv. Genet. 13:1059–1071. doi:10.1007/s10592-012-0353-y

Muui, C.W., R.M. Muasya, and D.T. Kirubi. 2013. Identification and evaluation of sorghum (*Sorghum bicolor* (L.) Moench) germplasm from Eastern Kenya. African J. Agric. Res. 8:4573–4579.

Ng'uni, D., M. Geleta, and T. Bryngelsson. 2011. Genetic diversity in sorghum (*Sorghum bicolor* (L.) Moench) accessions of Zambia as revealed by simple sequence repeats (SSRs). Hereditas 148:52–62. doi:10.1111/j.1601-5223.2011.02208.x

Pandravada, S.R., N. Sivaraj, N. Sunil, R. Jairam, Y. Prasanthi, S.K. Chakrabarty, P. Ramesh, I.S. Bisht, and S.K. Pareek. 2013. Sorghum landraces patronized by tribal communities in Adilabad district, Andhra Pradesh. Indian J. Traditional Knowledge 12:465–471.

Paterson, A.H., et al. 2009. The *Sorghum bicolor* genome and the diversification of grasses. Nature 457:551–556. doi:10.1038/nature07723

Paterson, A.H., K.F. Schertz, Y.-R. Lin, S.-C. Liu, and Y.L. Chang. 1995. The weediness of wild plants: Molecular analysis of genes influencing dispersal and persistence of Johnsongrass, *Sorghum halepense* (L.) Pers.). Proc. Natl. Acad. Sci. USA 92:6127–6131. doi:10.1073/pnas.92.13.6127

Pautasso, M., et al. 2013. Seed exchange networks for agrobiodiversity conservation. A review. Agron. Sustainable Dev. 33:151–175. doi:10.1007/s13593-012-0089-6

Piper, J.K., and P.A. Kulakow. 1994. Seed yield and biomass allocation in *Sorghum bicolor* and F_1 and backcross generations of *S. bicolor* × *S. halepense* hybrids. Can. J. Bot. 72:468–474. doi:10.1139/b94-062

Prasad Rao, K.E., and V. Ramnatha Rao. 1995. Use of characterization data in developing a core collection of sorghum. In: T. Hodgkin et al., editors, Core collection of plant genetic resources. John Wiley, Chichester, UK. p. 109–111.

Rabbi, I.Y., H.H. Geiger, B.I.G. Haussmann, D. Kiambi, R. Folkerstma, and H.K. Parzies. 2010. Impact of farmers' practices and seed systems on the genetic structure of common sorghum varieties in Kenya and Sudan. Plant Genetic Resour. 8:116–126. doi:10.1017/S147926211000002X

Ramu, P., C. Billot, J.-F. Rami, S. Senthilvel, H.D. Upadhyaya, L.A. Reddy, and C.T. Hash. 2013. Assessment of genetic diversity in the sorghum reference set using EST-SSR markers. Theor. Appl. Genet. 126:2051–2064. doi:10.1007/s00122-013-2117-6

Schlesinger, W.H. 2009. On the fate of anthropogenic nitrogen. Proc. Natl. Acad. Sci. USA 106:203–208. doi:10.1073/pnas.0810193105

Seetharam, K. 2011. Phenotypic assessment of sorghum (*Sorghum bicolor* L. Moench) germplasm reference set for yield and related traits under postflowering drought conditions. Ph.D. diss., Tamil Nadu Agricultural Univ., Coimbatore, Tamil Nadu, India.

Sharma, H.C. 1993. Host plant resistance to insects in sorghum and its role in integrated pest management. Crop Prot. 12:11–34. doi:10.1016/0261-2194(93)90015-B

Shapter, F.M., M.P. Dawes, L.S. Lee, and R.J. Henry. 2009a. Aleurone and sub-aleurone morphology in native Australian wild cereal relatives. Aust. J. Bot. 57:688–696. doi:10.1071/BT07086

Shapter, F.M., P. Eggler, L.S. Lee, and R.J. Henry. 2009b. Variation in granule bound starch synthase I (GBSSI) loci amongst Australian wild cereal relatives (Poaceae). J. Cereal Sci. 49:4–11. doi:10.1016/j.jcs.2008.06.013

Shapter, F.M., L.S. Lee, and R.J. Henry. 2008. Endosperm and starch granule morphology in wild cereal relatives. Plant Genetic Resour. 6:85–97. doi:10.1017/S1479262108986512

Shewayrga, H., D.R. Jordan, and I.D. Godwin. 2008. Genetic erosion and changes in distribution of sorghum (*Sorghum bicolor* L. (Moench)) landraces in north-eastern Ethiopia. Plant Genetic Resour. 6:1–10. doi:10.1017/S1479262108923789

Singhal, D., P. Gupta, P. Sharma, N. Kashyap, S. Anand, and H. Sharma. 2011. In silico single nucleotide polymorphisms (SNP) mining of *Sorghum bicolor* genome. Afr. J. Biotechnol. 10:580–583.

Soler, C., A.-A. Saidou, T.V.C. Hamadou, M. Pautasso, J. Wencelius, and H.H.I. Joly. 2013. Correspondence between genetic structure and farmers' taxonomy—A case study from dry-season sorghum landraces in Northern Cameroon. Plant Genetic Resour. 11:36–49. doi:10.1017/S1479262112000342

Stemler, A.B.L., J.R. Harlan, and J.M.J. deWet. 1975. Caudatum sorghums and speakers of Chari-Nile languages in Africa. J. Afr. Hist. 16:161–183. doi:10.1017/S0021853700001109

Subbarao, G.V., T. Ishikawa, O. Ito, K. Nakahara, H.Y. Wang, and W.L. Berry. 2006. A bioluminescence assay to detect nitrification inhibitors released from plant roots: A case study with *Brachiaria humidicola*. Plant Soil 288:101–112. doi:10.1007/s11104-006-9094-3

Subbarao, G.V., M. Rondon, O. Ito, T. Ishikawa, I.M. Rao, K. Nakahara, C. Lascano, and W.L. Berry. 2007. Biological nitrification inhibition (BNI)—Is it a wide spread phenomenon? Plant Soil 294:5–18. doi:10.1007/s11104-006-9159-3

Sukumaran, S., W. Wang, S.R. Bean, J.F. Pedersen, S. Kresovich, M.R. Tuinstra, T.T. Tesso, M.T. Hamblin, and J. Yu. 2012. Association mapping for grain quality in a diverse sorghum collection. Plant Gen. 5:126–135. doi:10.3835/plantgenome2012.07.0016

Tesfamariam, T., H. Yoshinaga, S.P. Deshpande, P. Srinivasa Rao, K.L. Sahrawat, Y. Ando, K. Nakahara, C.T. Hash, and G.V. Subbarao. 2014. Biological nitrification inhibition in sorghum: The role of sorgoleone production. Plant Soil 379:325–335. doi:10.1007/s11104-014-2075-z

Teshome, A., B.R. Baum, L. Fahrig, J.K. Torrance, T.J. Arnason, and J.D. Lambert. 1997. Sorghum (*Sorghum bicolor* L. (Moench)) landrace variation and classification in North Shewa and South Welo, Ethiopia. Euphytica 97:255–263. doi:10.1023/A:1003074008785

Teshome, A., L. Fahrig, J.K. Torrance, J.D. Lambert, T.J. Arnason, and B.R. Baum. 1999. Maintenance of sorghum (*Sorghum bicolor*, Poaceae) landrace diversity by farmers' selection in Ethiopia. Econ. Bot. 53:79–88. doi:10.1007/BF02860796

Tesso, T., I. Kapran, C. Grenier, A. Snow, P. Sweeney, J. Pedersen, D. Marx, G. Bothma, and G. Ejeta. 2008. The potential of crop-to-wild gene flow in sorghum in Ethiopia and Niger: A geographic study. Crop Sci. 48:1425–1431. doi:10.2135/cropsci2007.08.0441

Tuinstra, M.R., E.M. Grote, P.B. Goldsbrough, and G. Ejeta. 1997. Genetic analysis of post-flowering drought tolerance and components of grain development of *Sorghum bicolor* (L.) Moench. Mol. Breeding 3:439–448. doi:10.1023/A:1009673126345

Upadhyaya, H.D., and R. Ortiz. 2001. Mini core subset for capturing diversity and promoting utilization of chickpea genetic resources in crop improvement. Theor. Appl. Genet. 102:1292–1298. doi:10.1007/s00122-001-0556-y

Upadhyaya, H.D., R.P.S. Pundir, S.L. Dwivedi, C.L.L. Gowda, V.G. Reddy, and S. Singh. 2009. Developing a mini core collection of sorghum for diversified utilization of germplasm. Crop Sci. 49:1769–1780. doi:10.2135/cropsci2009.01.0014

Upadhyaya, H.D., S. Sharma, S.L. Dwivedi, and S.K. Singh. 2014a. Sorghum genetic resources: Conservation and diversity assessment for enhanced utilization in sorghum improvement. In: Y.-H. Wang, H.D. Upadhyaya, and C. Kole, editors, Genetics, genomics and breeding of sorghum. CRC Press, Boca Raton, FL. p. 28–55.

Upadhyaya, H.D., S.L. Dwivedi, P. Ramu, S.K. Shailesh, and S. Singh. 2014b. Genetic variability and effect of postflowering drought on stalk sugar content in sorghum mini core collection. Crop Sci. 54:2120–2130. doi:10.2135/cropsci2014.01.0040

Upadhyaya, H.D., S.L. Dwivedi, S. Singh, K.L. Sahrawat, and S.K. Singh. 2016a. Genetic variation and post flowering drought effects on seed iron and zinc in ICRISAT sorghum mini core collection. Crop Sci. 56:374–384.

Upadhyaya, H.D., Y.-H. Wang, D.V.S.S.R. Sastry, S.L. Dwivedi, P.V.V. Prasad, A.M. Burrell, R.R. Klein, G.P. Morris, and P.E. Klein. 2016b. Association mapping of low temperature germinability and seedling vigor in sorghum under controlled low-temperature conditions. Genome 59:137–145.

Vadez, V., L. Krishnamurthy, C.T. Hash, H.D. Upadhyaya, and A.K. Borrell. 2011. Yield, transpiration efficiency, and water-use variations and their interrelationships in the sorghum reference collection. Crop Pasture Sci. 62:645–655. doi:10.1071/CP11007

Venkateswaran, K., M. Muraya, S.L. Dwivedi, and H.D. Upadhyaya. 2014. Wild sorghums—Their potential use in crop improvement. In: Y.-H. Wang, H.D. Upadhyaya, and C. Kole, Genetics, Genomics and breeding of sorghum. CRC Press, Boca Raton, FL. p. 56–88.

Warwick, S.T., D. Phillips, and C. Andrews. 1986. Rhizome depth: The critical factor in winter survival of *Sorghum halepense* (L.) Pers. (Johnsongrass). Weed Res. 26:381–387. doi:10.1111/j.1365-3180.1986.tb00721.x

Washburn, J.D., D.K. Whitmire, S.C. Murray, B.L. Burson, T.A. Wickersham, J.J. Heitholt, and R.W. Jessup. 2013a. Estimation of rhizome composition and overwintering ability in perennial *Sorghum* spp. using near-infrared spectroscopy (NIRS). BioEnergy Res. 6:822–829. doi:10.1007/s12155-013-9305-8

Washburn, J.D., S.C. Murray, B.L. Burson, R.R. Klein, and R.W. Jessup. 2013b. Targeted mapping of quantitative trait locus regions for rhizomatousness in chromosome SBI-01 and analysis of overwintering in a *Sorghum bicolor* × *S. propinquum* population. Mol. Breeding 31:153–162. doi:10.1007/s11032-012-9778-8

Westengen, O.T., M.A. Okongo, L. Onek, T. Berg, H.D. Upadhyaya, S. Birkeland, S.D.K. Khalsa, K.H. Ring, N.C. Stenseth, and A.K. Brysting. 2014. Ethnolinguistic structuring of sorghum genetic diversity in Africa and the role of local seed systems. Proc. Natl. Acad. Sci. USA 111:14,000–14,105.

Xin, Z., M.L. Wang, N.A. Barkley, G. Burow, C. Franks, G. Pederson, and J. Burke. 2008. Applying genotyping (TILLING) and phenotyping analyses to elucidate gene function in a chemically induced sorghum mutant population. BMC Plant Biol. 8:103. doi:10.1186/1471-2229-8-103

Zongo, J.D., P.H. Gouyon, and M. Sandeier. 1993. Genetic variability among sorghum accessions from the Sahelian agroecological region of Burkina Faso. Biodivers. Conserv. 2:627–636. doi:10.1007/BF00051963

Pedigreed Mutant Library—A Unique Resource for Sorghum Improvement and Genomics

Zhanguo Xin,* Yinping Jiao, Ratan Chopra, Nicholas Gladman, Gloria Burow, Chad Hayes, Junping Chen, Yves Emendack, Doreen Ware, and John Burke

Abstract

Sorghum is a versatile crop used for food, feed, fodder, and biofuel. Because of its resilience to environmental stresses and low soil fertility, sorghum is becoming increasingly important in meeting the growing need for food and energy in the face of declining arable land and fresh water resources. Low grain yield and poor nutritional quality are two major limiting factors for sorghum production. New resources are needed to overcome these limits. We have developed a pedigreed mutant library in BTx623, the inbred line for generating the reference genome sequence. The library consists of 6400 M_4 seed pools, each of which was derived from an independent M_1 seed through single-seed descent. The mutant library displays a wide range of visible phenotypes, and many are beneficial traits that have potential to be used as breeding materials to improve grain yield and quality. A selection of 256 lines was sequenced to 16x coverage of the whole genome. More than 1.8 million canonical ethyl methanesulfonate (EMS)–induced guanine/cytosine to adenosine/thymine single nucleotide polymorphisms (SNPs) were discovered, covering 94% of the genes in the sorghum genome. Over 57% of the genes carry large effect mutations that may alter the function of the gene product. In comparison with natural variations, 97.5% of the EMS-induced mutations are novel. Thus, the pedigreed mutant library can serve as a unique resource for selection of novel agronomic traits and identifying their causal mutations for sorghum improvement and for elucidating gene function through isolation of mutant series for the traits of interest.

A well-annotated induced-mutant library from an inbred line with uniform genetic background and a sequenced genome provides important resources

---

Abbreviations: EMS, ethyl methanesulfonate; EW, epicuticular wax; HSP, heat-shock protein; HT, high temperature; NGS, next-generation sequencing; PCR, polymerase chain reaction; QTL, quantitative trait loci; SNP, single nucleotide polymorphism; TILLING, targeting induced local lesions in genomes.

Z. Xin, R. Chopra (Ratan.Chopra@ars.usda.gov), N. Gladman (Nicholas.Gladman@ars.usda.gov), G. Burow (gloria.burow@ars.usda.gov), C. Hayes (chad.hayes@ars.usda.gov), J. Chen (junping.chen@ars.usda.gov), Y. Emendack (yves.Emendack@ars.usda.gov), J. Burke (john.burke@ars.usda.gov), Plant Stress & Germplasm Development Unit, Cropping Systems Research Lab., USDA-ARS, 3810 4th St., Lubbock, TX 79415; Y. Jiao (yjiao@cshl.edu) Plant Stress & Germplasm Development Unit, Cropping Systems Research Lab., USDA-ARS, 3810 4th St., Lubbock, TX 79415, and Cold Spring Harbor Laboratory, 1 Bungtown Rd., Cold Spring Harbor, NY 11724; D. Ware (ware@cshl.edu), Cold Spring Harbor Lab., 1 Bungtown Rd., Cold Spring Harbor, NY 11724, and Plant, Soil and Nutrition Research Unit, USDA-ARS, 538 Tower Rd., Ithaca NY 14853. *Corresponding author (Zhanguo.xin@ars.usda.gov).

doi:10.2134/agronmonogr58.2014.0057

for isolating useful traits for breeding and for identifying the causal mutations with modern genomic tools, such as next-generation sequencing (NGS) techniques. To support functional genomic studies in sorghum [*Sorghum bicolor* (L.) Moench] and accelerate sorghum improvement, we developed a pedigreed sorghum mutant library in the elite inbred line BTx623, which has been used for generating the reference genome sequence (Paterson et al., 2009). Ethyl methanesulfonate was used as a mutagen to generate the mutant population because of the chemical's high rate of success in inducing a broad spectrum of mutations in many plant species (Greene et al., 2003). Individual EMS-treated sorghum seeds (M_1) were advanced to the M_3 generation by single-seed descent. M_4 seeds pooled from 10 randomly selected panicles from each M_3 plot were deposited into the library for public distribution and screening for beneficial traits for sorghum improvement. A significant advantage of the pedigreed mutant library is that most of the induced mutations can be preserved, including recessive lethal mutations and mutations that significantly reduce the ability of the mutated plants to survival.

The mutant library displays a diverse array of phenotypes that are distinct from the unmutated BTx623 plant. Many mutant phenotypes, such as multiseededness (*msd*) and erect leaf (*erl*), have potential as breeding resources to enhance sorghum biomass and grain yield. An effective strategy based on NGS has been developed to identify the causal mutations for these beneficial traits. Based on these causal mutations, perfect molecular markers can be designed to accelerate the introgression of these beneficial traits into elite lines through marker-assisted selection. Recently, we have sequenced 256 lines that were randomly selected from the pedigreed mutant library. Initial analysis of the 3.1 Tb of sequencing data discovered more than 1.8 million canonical EMS-induced mutations (Jiao et al., 2016). Of these 256 core lines, over 97.5% of the mutations are novel and not present in the natural sorghum collection (Mace et al., 2013; Morris et al., 2013). These sequenced mutant lines will be an important resource for identifying gene mutations that underlie important traits and for elucidating gene functions through in silico evaluation of mutants that harbor disruptive mutations in genes of interest.

A Historic Collection of Sorghum Mutants and Breeding Lines

We have recently rescued a collection of sorghum genetic stocks amassed during many years of arduous work by Dr. Keith Schertz, a USDA-ARS sorghum geneticist (Xin et al., 2013). This collection possesses a rich diversity of morphological, growth, and developmental traits (Table 1). This historic collection is significant for cereal crop research because it includes accessions and lines that were reported as early as the 1930s (Karper and Stephens, 1936). Although the collection has many interesting and important phenotypes, only a few lines have been used in current sorghum research, largely due to the lack of proper documentation. The collection includes natural variation and induced mutations. The genetic background of some of these lines is not clear. However, a number of the entries have enough information to decipher the genetic background, such as the irradiated RTx7078 and those with common names or sorghum accession numbers. With NGS mapping-by-sequencing, whether the original genetic background is known or not, it is still possible to identify the causal gene for the observed phenotypes in the collection. For example, the original mutant can be

Table 1. Summary of various sorghum mutants collected by Keith Schertz.

Mutation	Number of lines	Phenotype and range of variation
Seedling	98	Albino, pale green, yellow, variegation
Leaf	90	Spotty, necrotic, golden, yellow, striped
Reproduction	35	Homeotic, female sterility, twin-seed, etc.
Male sterile	20	*ms1*, *ms2*, *ms3*, *ms7*
Maturity	51	Days to flowering: 44–97
Height	31	43 to 188 cm
Grain	21	Endosperm structure, lysine, seed coat, etc.
Bloomless	23	Sparse bloom to bloomless
Stem	12	Zigzag, lazy, tenuous
Tiller	10	Monoculm, multiple tillers
Brown midrib	8	Brown-colored midvein
Glume color	6	Various color
Liguleless	4	No ligule
Translocation	10	Various chromosome translocation line
Linkage	14	With visible markers for linkage analysis
Unclassified	74	No remarkable phenotype
Total	507	

outcrossed to BTx623. Pooled mutants selected from the F_2 population can be subjected to NGS. If the genetic background of the original mutants is sufficiently distant from that of BTx623, the causal gene can be identified through the ShoreMap strategy (Schneeberger et al., 2009). If the genetic background is very close to BTx623, the causal gene can be identified through an isogenic mapping-by-sequencing strategy (Abe et al., 2012). For mutants with genetic backgrounds that are moderately diverse from that of BTx623, a combination of both strategies can be used to identify the causal gene. The key is that the mutant phenotypes need to segregate as Mendelian traits in the F_2 population and can be unambiguously identified. Efforts are underway to cross all historically important mutants or genetic stocks, such as height and maturity classes, to a cytoplasmic male sterile BTx623. This collection will allow the identification of most of the genes important for the adaptation of sorghum from a tropical crop to US environments.

Pedigreed Sorghum Mutant Library

A well-categorized mutant library generated from an inbred line with a uniform genetic background and having detailed annotation of phenotypes provides a powerful resource to isolate independent alleles of mutants with agronomically interesting and relevant traits. Mutagenesis has long been used as a complementary approach to breed sorghum with novel phenotypes (Quinby and Karper, 1942; Gaul, 1964). Many mutants with unique phenotypes have been selected from mutant populations treated with various mutagens, such as X-rays and γ-irradiation, EMS, methyl methanesulfonate, diethyl sulfate, *N*-nitroso methyl urea, *N*-nitroso ethyl urea, or combinations of chemical and irradiation mutagens (Quinby and Karper, 1942; Sree Ramulu, 1970a,b; Sree Ramulu and Sree Rangasamy, 1972). Many beneficial mutations, including dwarfing, early flowering, high protein digestibility, high lysine, and others, have been widely used in sorghum breeding (Singh and Axtell, 1973; Quinby, 1975; Ejeta and Axtell, 1985; Oria et al., 2000). Those early efforts in mutagenesis often targeted specific traits of interest through bulk mutagenesis, but most of the early populations are no longer available. We started a systematic approach to develop a pedigreed mutant library. The library consists of pedigreed M_4 seeds derived from individual M_1 seeds; thus, each line is considered an independent mutation event. Since the M_4 seeds are pooled from 10 individual M_3 plants without selection, most mutations, including recessive lethal mutations, are preserved in the library. The library posses a wide range of phenotypes and will serve as an important resource for rapid discovery of genes through mapping-by-sequencing with NGS technologies. It can also be used as a reverse genetic resource to isolate mutant series of genes to establish their functions by targeting induced local lesions in genomes (TILLING; Henikoff et al., 2004).

The completion of the sorghum genome sequence and annotated transcriptome has made it possible to study gene function on a genome-wide scale and to compare gene function with that in other plants (Paterson, 2008; Paterson et al., 2009). Complimentary to the genome sequence, a systematic mutant library that contains multiple mutations for all genes in the sorghum genome is urgently needed to deduce gene functions. Xin et al. (2008) reported a modest population of 768 pedigreed EMS-mutagenized lines of BTx623, the inbred line used for sorghum genome sequencing. Genomic DNA was prepared using leaf samples collected from the M_2 plants that were used to produce M_3 seeds (Fig. 1). Phenotypes are annotated at the M_3 generation to ensure that any mutant phenotype observed in a family is descended from a single mutagenesis event (a single germ cell), which is represented by a M_2 plant used to prepare genomic DNA. Following phenotype annotation, 10 M_3 panicles are bulked as M_4 seeds, which are deposited in the library and will be distributed to end users on request. Currently the library has been expanded to 6400 M_4 seed pools, each derived from an independent M_1 seed.

The mutant library displays an incredible diversity of phenotypes in plant and leaf types, roots, tillers, panicles, and seeds that are distinct from unmutated BTx623 (Fig. 2). More than 20% of the lines segregated for albino plants, a result that signifies a high mutational load. In addition, many lines segregated for altered leaf size, shape, color, and angle and necrosis (Xin et al., 2008). Although these mutants may not have direct application for the improvement of sorghum, they do furnish materials for elucidating fundamental biological processes that

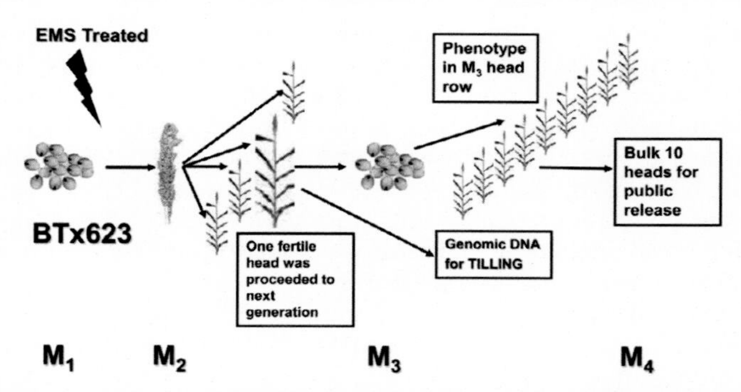

Fig. 1. Procedure to develop the pedigreed mutant library. BTx623 seeds were treated with ethyl methane sulfonate (M_1). The M_1 seeds were propagated to M_3 through single-seed descent. Ten M_3 panicles were pooled as the M_4 seed pools, which were deposited in the pedigreed mutant library. EMS, ethyl methanesulfonate.

provide knowledge required for sorghum improvement. For example, leaves are crucial for photosynthesis, but genes that mediate leaf size, shape, and angle are poorly understood in sorghum and other crop plants. The large collection of mutants with altered leaf color, morphology, and angle provide key materials for identifying the genes that control leaf formation and for designing ideal

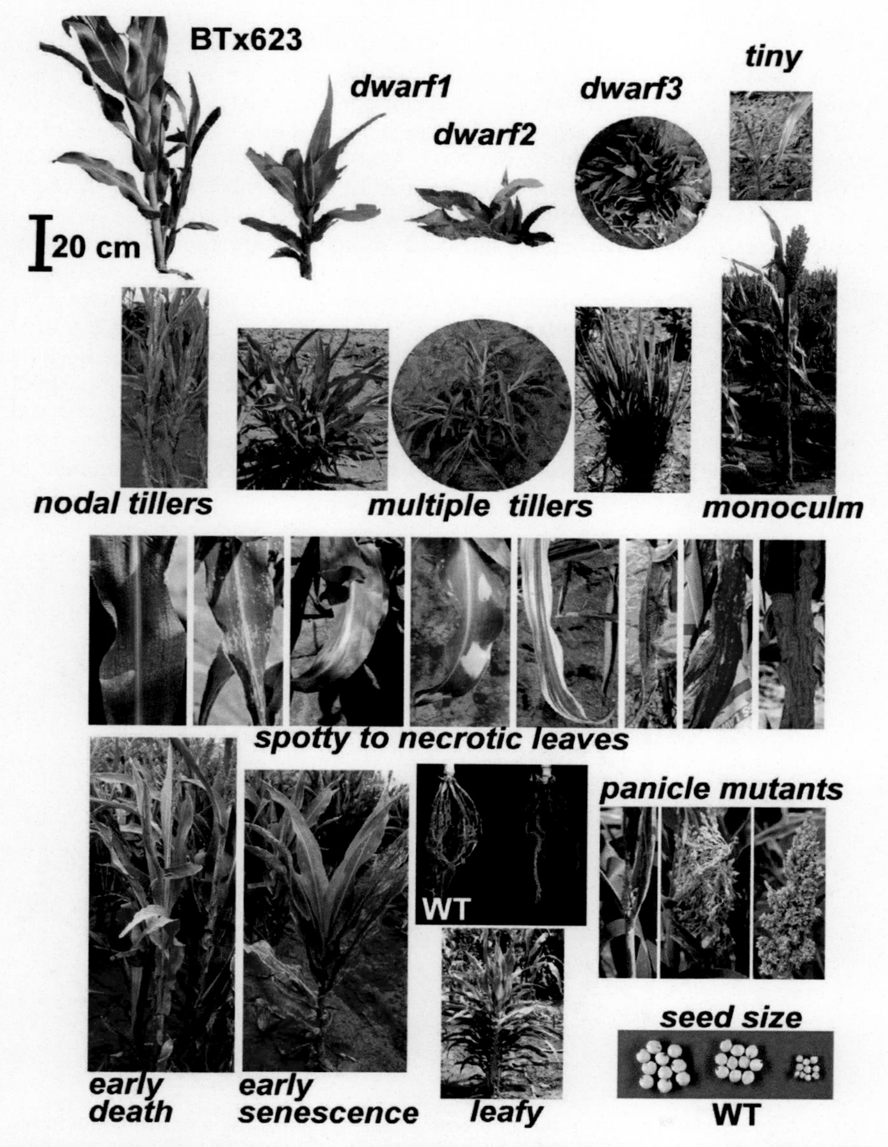

Fig. 2. A gallery of selected mutant phenotypes. Only a fraction of selected mutant phenotypes are presented to illustrate the diversity of phenotypes observed from the sorghum mutant library. WT, wild type.

leaf architecture to maximize light capture and photosynthesis. Through careful inspection of the mutant library under field conditions, many beneficial traits that have potential to improve sorghum biomass, grain yield, and quality were isolated. Several traits from the mutant collection are discussed below.

Selected Potential Traits

Multiseeded Mutants

Grain yield is determined by the number of plants per acre, seed number per plant, and seed weight. Among the yield components, seed number per panicle is a major determinant of grain yield in sorghum and other cereals (Saeed et al., 1986; Duggan et al., 2000; Richards, 2000; Ashikari et al., 2005; Reynolds et al., 2009). Seed number per panicle is determined by the number and length of primary and secondary flower branches and the fertility of spikelets. The sorghum panicle consists of a main rachis on which many primary branches are developed. Secondary branches, and then sometimes tertiary branches, are developed from the primary branches (Brown et al., 2006; Burow et al., 2014). The main inflorescence and primary, secondary, and tertiary branches all end with a terminal triplet spikelet, which consists of one sessile fertile and two sterile pedicellate spikelets (Walters and Keil, 1988). Below the terminal spikelet, one or more spikelet pairs can develop, and these adjacent spikelet pairs consist of one sessile and one pedicellate spikelet. In the wild-type BTx623 and other sorghum accessions, only the sessile spikelets can develop into seeds (Fig. 3A). The development of

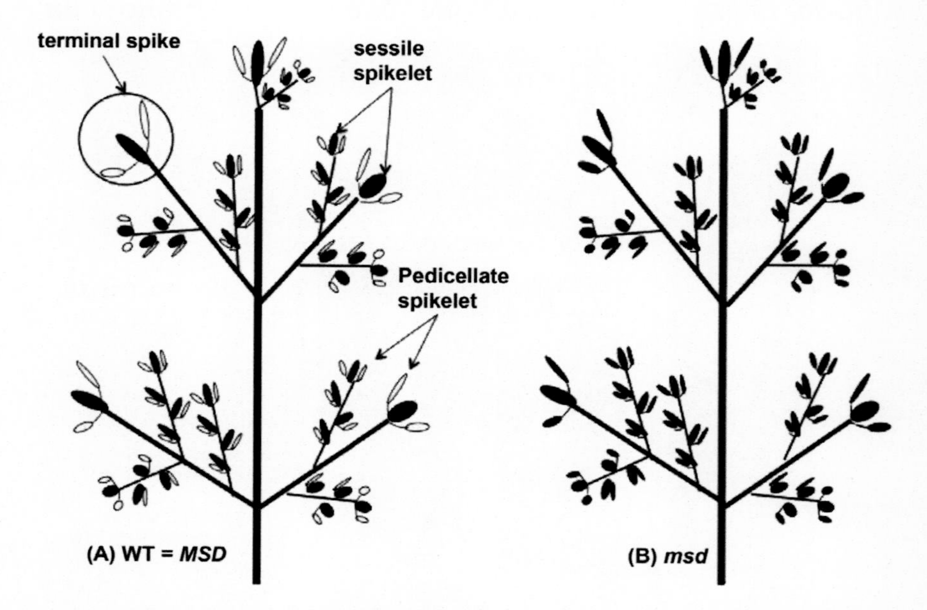

Fig. 3. A comparison between the wild-type and *msd* panicle. The drawing illustrates that in the wild-type BTx623 (left), only the sessile spikelets (directly attached to a flower branch) are fertile (filled ovals). The pedicellate spikelets (attached to a flower branch through a short pedicel) are infertile (open ovals). In the *msd* mutants, both types of spikelets are fertile.

pedicellate spikelets is arrested at various stages in different sorghum lines. In some lines, the pedicellate spikelets can develop anthers and shed viable pollen, but few lines can develop ovaries (Karper and Stephens, 1936). The pedicellate spikelets eventually abort.

Recently we isolated and characterized a novel class of sorghum mutants, referred to as multiseeded (*msd*) mutants, in which the developmental arrest of the pedicellate florets was released (Fig. 3B). In this class of mutants, all spikelets, both sessile and pedicellate, developed into perfect flowers and produced seeds. In addition, the *msd* mutants have more and longer primary inflorescence branches than the wild type. Consequently, the *msd* mutants have the potential to produce three times the seed number and twice the seed weight per panicle compared with the wild-type BTx623 (Burow et al., 2014). Over 40 independent *msd* mutants have been isolated from the mutant library. All *msd* mutants display similar features except for some variations in panicle shape, such as compactness, width, and length. Genetic analyses indicate that all *msd* mutants are recessive and have three distinctive structural changes: an increased number of primary branches, enlarged size of the primary branches, and fertile sessile and pedicellate spikelets (Fig. 4). Genetic complementation is underway to determine the number of loci that controls this trait.

Fig. 4. Morphological features of the *msd* mutant panicle. Three coordinated changes in the panicle morphology are shown: (A) an increased number of the primary inflorescence branches, (B) increased size of the primary and secondary branches, (C) full fertility of both sessile and pedicellate spikelets, and (D) enlarged overall panicle size.

Erect Leaf Mutants

Leaf inclination angle is an important architectural trait of canopy improvement to maximize solar radiation capture and increase total biomass and grain production. Duvick and Cassman (1999) analyzed 10 morphological and agronomical parameters in 36 maize (*Zea mays* L.) hybrids released from 1936 to 1991 for traits contributing to the genetic gains in grain yield. The leaf angle score of new hybrids displayed an improvement of 122% over the old ones, the greatest change among the plant traits examined (Duvick and Cassman, 1999). Modern maize hybrids have a much more acute (erect) leaf angle than older hybrids, which allows the hybrids to be planted at a higher density to capture more solar radiation per unit land area (Duvick and Cassman, 1999). Erect leaf mutants in rice (*Oryza sativa* L.) have been shown to have increased biomass and grain yield (Sakamoto et al., 2006).

Compared with modern maize hybrids, sorghum exhibits an open canopy with such wide leaf angles that the leaves are almost parallel to the ground. A liguleless sorghum mutant with an erect leaf angle has been reported previously (Singh and Drolsom, 1973). Another erect mutant has been identified from historic sorghum genetic stocks obtained from Western Blackhull kafir. It has a short leaf and compact canopy (Burow et al., 2013; Xin et al., 2013). However, these two mutants have not been used to improve sorghum leaf angle.

Leaf inclination angle is a complex trait that may be under the control of many QTLs in maize and sorghum (Tian et al., 2011; Li et al., 2015; Truong et al., 2015). Although a few major leaf angle genes, such as *osdwarf4* and *leaf inclination2*, have been identified in rice (Sakamoto et al., 2006; Zhao et al., 2010), the vast majority of the genes that control leaf angle have not been identified. We have isolated a series of erect leaf mutants (Fig. 5, Table 2). Among the 6400 M_3 field plots in the field, more than 50 segregated for upright leaves (Xin et al., 2009). Eleven of

Fig. 5. Leaf architecture of BTx623 and *erl1-1* and *erl1-2* mutants. At physiological maturity, typical BTx623 and *erl1-1* and *erl1-2* mutants were transplanted from field to a studio and photographed.

Table 2. Erect leaf mutants confirmed at the M_4 generation.

Line	Height	Angle of second leaf†	Width of second leaf	Length of second leaf	Length of head	Seed weight per head
	cm	°		cm		g
BTx623	163	45	7.5	56	32	78
M2P1374	96	82	9	65	34	7.6
M2P0514	101	80	6	56	25	10.2
MUT841	121	70	7.4	74.5	35	32.1
MUT1008	112	70	8.5	43	21	56.6
M2P0630	122	70	8.7	56	28	73.4
MUT1169	131	65	7	48	32	60.9
M2P0819	122	65	7	55	25	44
M2P0684	124	60	6	65	26	28.7
M2P0784	128	60	7	60	29	38.4
25M2–0552	113	60	7	49	23	12.3
20M2–0024	138	60	8.5	54	35	74

† Leaf angle was measured at full bloom on the leaf below the flag leaf. A leaf parallel to the ground has an angle of zero, and a leaf perpendicular to ground has an angle of 90 degrees.

these mutants were confirmed in the M_4 generation. Several mutants have similar or slightly bigger panicles than the wild type. These *erl* mutants may prove to be useful for improving sorghum biomass and grain production based on the yield gains in maize hybrids brought about by the more acute leaf angle (Dhugga, 2007).

Due to the improved leaf inclination angles, the post–Green Revolution maize hybrid yield continues to increase while sorghum yield remains flat (Duvick and Cassman, 1999; Dhugga, 2007; Truong et al., 2015). Since sorghum posses the same efficient C_4 photosynthesis pathway as maize, sorghum should have the potential to produce similar increases in biomass and grain yield as maize (Wang et al., 2009). It would be interesting to test if the erect leaf mutants isolated from the mutant library could help increase sorghum biomass production and narrow the yield gap between these two important C_4 crops (Zhu et al., 2010).

Brown Midrib Mutants

A distinct example of beneficial mutants detected is that of the brown midrib mutants isolated from C_4 cereals such maize, sorghum, and millet through natural or induced mutations (Sattler et al., 2010). This mutant phenotype is typified by a distinctive brownish colored mid veins of leaves, which can be easily identified in the field. A typical *bmr* sorghum mutant is shown in Fig. 6. Some mutants accumulate reddish brown to yellow pigment in the stalk, root, and stem pith. The *bmr* mutation is associated with reduced lignin content, better digestibility for livestock, and a higher conversion efficiency of sorghum stover to ethanol than the wild type (Vermerris et al., 2007).

Sorghum *bmr* mutants were first isolated by Porter et al. (1978) from a diethyl sulfate–mutagenized population. Twenty-eight sorghum *bmr* mutants representing four loci (*bmr2*, *bmr6*, *bmr12*, and *bmr19*) have been isolated from various sources, including natural mutation (Sattler et al., 2010). Three of the loci have been cloned by candidate gene approaches. The *bmr6* harbors a mutation in the cinnamyl alcohol dehydrogenase gene, and *bmr12* harbors a mutation in the caffeic *O*-methyl transferase gene (Bout and Vermerris, 2003; Saballos et al., 2009; Sattler et al., 2009). The *bmr2* harbors a mutation in the 4-coumarate:CoA ligase

Fig. 6. A collection of the brown midrib (*bmr*) mutant. A comparison of the midrib between BTx623 and a *bmr* mutant is shown in the two top left photos. The top right photo shows a *bmr* mutant growing in the field. The bottom photo displays a series of *bmr* mutants from the mutant library observed under field conditions, with the wild type on the far left.

gene (Saballos et al., 2012). These three enzymes are involved in the biosynthesis of monolignols, the precursors for lignin biosynthesis. Among these four loci, *bmr2* and *bmr19* are represented by a single locus, indicating that saturation mutagenesis has not been achieved (Sattler et al., 2010). Moreover, both *bmr6* and *bmr12*, which are the main sources for commercial *bmr* forage sorghum, are complete knockout mutations. To identify additional *bmr* mutants and to isolate non-knockout alleles of *bmr6* and *bmr12*, we initiated a systematic approach to isolate additional *bmr* mutants. Thirty independent *bmr* mutants have been isolated (Xin et al., 2009). A selection of the *bmr* mutants is shown in Fig. 6. Complementation analysis showed that in addition to many alleles of the previously known *bmr* loci, there are six novel mutants that could not complement the previously known loci. These six mutants represent four new loci that had not yet been reported (Sattler et al., 2014). These novel mutants and the new alleles of the previously known loci provide new genetic resources for improving the digestibility of forage sorghum and the conversion efficiency of stover to ethanol while minimizing the effect of *bmr* mutations on biomass production and lodging. Since the new *bmr* mutants are isolated from the sequenced inbred line BTx623, these mutants can be used for discovering additional genes of lignin biosynthesis through mapping-by-sequencing.

Bloomless Mutants

Sorghum produces and deposits copious amount of epicuticular wax (EW) in the form of readily visible large flakes of wax, known as bloom, on the aerial parts of the plants (Fig. 7; Ebercon et al., 1977; Jenks et al., 1992). The EW layer has been proposed as a barrier to transpiration through the cuticular layer, adding to the reputation of sorghum as one of the most water-use-efficient crops with tolerance

Fig. 7. A comparison of a bloomless (*blm*) mutant with unmutated BTx623. The aerial surface of BTx623 (left) is covered with a thick layer of white wax particles. The aerial surface of the *blm* mutants (right) is vibrant green because there is no wax layer.

to drought. This notion is supported by a study showing that cuticular transpiration is negatively correlated with EW load in detached leaves in 38 near-isogenic lines with EW loads ranging from 0.03 to 0.1 g m^{-2} (Jordan et al., 1984). An EW load of more than 0.07 g m^{-2} can provide an effective barrier to water loss through the cuticle. Furthermore, water use efficiency is positively correlated with EW load under both irrigated and unirrigated conditions in the field (Premachandra et al., 1994). In a greenhouse study, cuticular wax has been shown to reduce cuticle transpiration and increase transpiration efficiency (Burow et al., 2008, 2009).

All plant species accumulate some EW layers on their aerial surface, but only sorghum accumulates a copious thick layer of wax on the abaxial and adaxial leaf surfaces, sheath, and peduncle (Peters et al., 2009). The accumulation of EW in sorghum is considered to be under the control of many genes. From about 4000 segregating plants in the M_2 population, Peters et al. (2009) isolated 38 bloomless (*blm*) mutants. Complementation analysis with previously known *blm* mutants discovered 19 loci that mediate EW deposition in sorghum. From 6400 pedigreed mutant lines, we isolated 120 independent *blm* mutants. These newly isolated *blm* mutants, together with the previously isolated ones, provide a rich resource for identification of genes involved in EW biosynthesis, transport, deposition, and regulation in sorghum through forward genetics of mapping-by-sequencing. Elucidation of the pathways and regulation of EW deposition in sorghum may provide genetic arsenals for improving water use efficiency and abiotic stress tolerance in other cereals like maize, rice, and wheat.

A Mutant with Altered Abiotic Stress Tolerance

High-temperature (HT) tolerance in plants is complex, involving multiple pathways and many biological, biochemical, and physiological processes, such as a reduced photosynthesis rate, altered leaf temperature, chlorophyll fluorescence, and changes in enzymatic and nonenzymatic antioxidants (Iba, 2002; Larkindale et al., 2005; Burke and Chen, 2006; Kotak et al., 2007; Djanaguiraman et al., 2014). Several mechanisms of HT tolerance have emerged from studies in model organisms and crop plants. Induction of heat-shock proteins (HSPs) is a well-known mechanism by which plants acquire enhanced tolerance to extreme HTs in response to a brief exposure to sublethal elevated temperatures (Gurley, 2000; Hong and Vierling, 2000; Nieto-Sotelo et al., 2002). Many HSPs function as molecular chaperones and enable organisms to survive brief exposures to extreme HTs by preventing aggregation of denatured polypeptides and repairing misfolded proteins (Vierling, 1991; Wang et al., 2004; Lee et al., 2005). In maize, induction of HSP101, upregulation of a chloroplast elongation factor (EF-Tu), and enhanced thermostability of rubisco activase have all been implicated in HT tolerance (Bhadula et al., 2001; Crafts-Brandner and Salvucci, 2002; Nieto-Sotelo et al., 2002; Vargas-Suárez et al., 2004; Momcilovic and Ristic, 2007). The maintenance of proper membrane stability and fluidity through adjustments to membrane lipid composition and fatty-acid saturation levels is another well-characterized adaptive response to stresses associated with high and low temperatures (Marcum, 1998; Alfonso et al., 2001; Sung et al., 2003; Falcone et al., 2004; Chen et al., 2006a). Other responses, such as the production of antioxidants and modification of protein properties or enzyme activities, also play roles in protecting plants from extreme temperature stresses (Larkindale and Huang, 2004; Burke and Chen, 2006; Chen et al., 2006b; Wang et al., 2006).

Crops in the field are often exposed to a combination of HT and drought stresses. The mechanisms for tolerance to such stresses under field conditions are expected to be more complex than those observed in model organisms under controlled laboratory conditions. Using two well-studied maize inbred lines that have different HT tolerances under field conditions, Chen et al. (2012) observed that all known mechanisms of HT tolerance, such as expression of HSPs, are very similar between the two lines under normal or HT stress. But the two lines differ significantly in phosphatidic acid, a minor phospholipid that acts as a signal molecule in plants, before and after an HT stress. Because of the complexity of HT stresses under field conditions, very little is known about the mechanisms of HT tolerance. The sorghum mutant library may provide a useful resource to fill this critical gap. From 2000 mutant lines, more than 100 segregated for a wide range of HT sensitive phenotypes under field conditions after a heat wave, including leaf firing, panicle blast, etc. (Fig. 8). These phenotypes are similar to the heat-sensitive phenotypes observed in field-grown maize (Chen et al., 2012).

Sorghum is considered as one of the most HT tolerant crops. Because maize and sorghum shared a common ancestor as recently as 11 million years ago (Paterson et al., 2004), the HT-sensitive mutants may serve as tools to dissect the

Fig. 8. Heat-sensitive phenotypes observed in the mutant population under field conditions. Under field conditions in Lubbock, TX, both leaves and panicles of BTx623 are healthy and show no heat injury phenotypes (left). Both leaf firing and panicle blast, typical heat sensitive phenotypes of maize under the same environment, were observed in many lines of the sorghum mutant library (right).

mechanism of HT tolerance of maize under field conditions. For example, if the sorghum heat-sensitive mutation is mapped close to a maize HT tolerance QTL, the sorghum mutant may be used to clone the sorghum gene through mapping-by-sequencing approaches. Then, the maize homology of the sorghum gene can be tested to determine if it contributes to maize HT tolerance. In addition to HT-sensitive mutants, mutants that germinate better under cold temperatures were also identified. Systematic screening of these mutants under various abiotic stresses is necessary to take full advantage of the available mutant library to identify genes of stress tolerance.

Identification of Causal Mutations Underlying Agronomic Traits

Many important signal transduction components, developmental regulators, and biochemical pathways have been discovered through forward genetic analysis of mutants in model plants (http://theArabidopsisbook.org). The process begins with mutagenesis of a pure (inbred) line with either a physical or chemical mutagen followed by isolation of mutants with relevant phenotypes. The mutant of interest is then outcrossed to another accession that has extensive DNA polymorphism to the line used for mutagenesis. Linkage of the mutant phenotype with DNA markers is analyzed using as many markers as possible until the mutation is mapped to a very small region flanked by two DNA markers. The process, called map-based cloning, depends on two critical factors to succeed: a large segregating mapping population and dense DNA markers, to provide the resolution power to delimit the mutation into a small region. Usually, over 1000 individual F_2 individuals with the accurate phenotype need to be analyzed with thousands of DNA markers to narrow the mutation to a region harboring only a few genes (Jander et al., 2002). The entire region is sequenced to identify the gene that carries the expected mutation. To confirm the identity of the gene, the wild-type gene is introduced into the mutant through transformation to determine if the gene can restore the mutant to the wild-type phenotype. This last step can be bypassed if two or more independent mutant alleles are identified that carry unique mutations in the same gene.

Next-generation sequencing (NGS) techniques provide sizable amounts of high-resolution genotypic data quickly (Metzker, 2010), which can replace the expensive and time-consuming linkage analysis of conventional map-based cloning (Schneeberger and Weigel, 2011). Three general approaches have been developed that use NGS to clone genes represented by relevant phenotypes. The first strategy is represented by mapping-by-sequencing (ShoreMap; Schneeberger et al., 2009; Hartwig et al., 2012) or next-generation mapping (Austin et al., 2011). This approach is similar to conventional map-based cloning. It starts with development of a mapping population by outcrossing a mutant to a divergent line. Homozygous mutants or wild type (if the mutation is dominant) are selected from the segregating F_2 population. Genomic DNA from the selected lines is pooled in equal amounts for NGS analysis following the bulk-segregant analysis principle (Michelmore et al., 1991). For the current Illumina Hi-Seq 2000 machine, one sequencing lane can produce more than 30 Gb of high-quality sequencing data, which is sufficient to provide more than 40× coverage of the sorghum genome (~730 Mb). Sufficient resolution occurs at 15× coverage, when most homozygous SNPs and short insertion/deletion markers that exist between the two lines will be captured (Jiao et al., 2016). Thus, one sequencing lane can roughly accommodate three mapping populations. Because only homozygous F_2 mutants are selected for whole-genome sequencing, the SNPs surrounding the causal mutation site are expected to be predominantly of the mutant genotype. The further away a SNP is from the causal mutation, the more genotype from the divergent parent will appear. For loci that are unlinked to the mutation, we would expect the appearance of SNPs from both parents at an approximate ratio of 50:50. Since the estimated frequency of SNPs between any two divergent sorghum lines is more than 2 SNP/kb (Nelson et al., 2011), it should be possible to narrow the causal gene to a single or a few candidate genes by NSG if a sufficient number of homozygous mutants (>100) is used. With the ShoreMap strategy, developing the mapping population is probably the most time-consuming step; once sufficient homozygous mutants are selected from the mapping F_2 population, the sequencing and bioinformatics analysis take merely a few days to complete (Schneeberger and Weigel, 2011).

The second approach is isogenic mapping-by-sequencing, a variation of the ShoreMap technique, designed to map mutant phenotypes for which the degree of manifestation of the phenotype may be different in different genetic backgrounds (Abe et al., 2012; Hartwig et al., 2012; Zhu et al., 2012). For example, the presence of leaf angle QTLs in different lines may affect the expression of leaf angle of the *erl* mutants. In the isogenic mapping-by-sequencing approach, the mutant is crossed to a wild type of the same genotype as the mutant to avoid interference from a divergent genetic background. Figure 9 illustrates the process of identifying a causal gene mutation using *msd1* as an example. The *msd1* mutant is crossed to BTx623. After confirming that the F_2 population segregates at a ratio of *msd1* mutants to wild-type at approximately 1 to 3, leaf samples from 50 homozygous *msd1* mutants are collected. The genomic DNA from homozygous *msd1* mutants is pooled for NGS. Each F_2 individual may harbor hundreds of mutations unrelated to the *msd1* phenotype. Because only homozygous *msd1* mutants are selected, the causal mutation for the phenotype is expected to be 100% composed of mutated SNPs, whereas unlinked background mutations are expected to

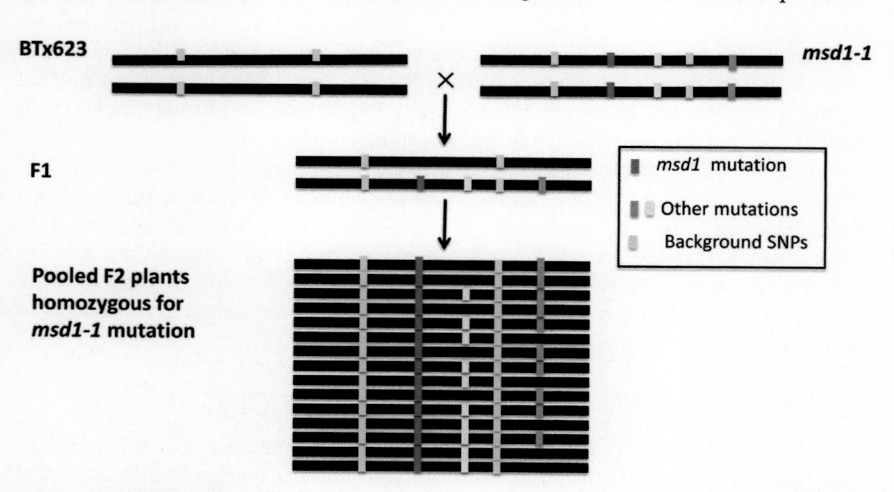

Fig. 9. Gene identification by the isogenic mutant mapping method. A homozygous mutant with the desired phenotype (*msd*) is crossed to the wild type, BTx623. Homozygous mutants from the F_2 segregating population are selected. The genomic DNA from the selected mutants is pooled equally and subjected to next-generation sequencing. The causal single-nucleotide polymorphism (SNP) for the phenotype, as well as background SNPs present in both the mutant and the wild-type BTx623, will exhibit 100% mutated SNPs. Other strategies are needed to distinguish the causal SNP from the background SNPs.

comprise around 50% mutated SNPs and 50% wild-type SNPs. Mutations close to the causal mutation will vary from 50 to 100%, depending on the genetic distance to the causal mutation. Thus, it is possible to determine the causal mutation from the background mutations. This approach is particularly useful in mapping subtle agronomic traits that can be modified by genetic background (Abe et al., 2012; Zhu et al., 2012). One complication of this approach is that the background mutations preexisted in the wild type before mutagenesis. In that case, BTx623 will also display 100% mutated SNPs. Knowledge of the biological processes contributing to the trait of interest may help rule out some mutations. While challenging, it is possible to identify the causal mutation by this approach alone. Definitive evidence for the identification of the causal gene mutation can be obtained if independent alleles of the same locus are found.

The third approach is direct sequencing of the mutants if multiple independent alleles exist. If each allele of the locus carries a unique mutation in a gene, this gene will be the probable candidate for the causal gene. Theoretically, this approach should work, but it has not been reported (Schneeberger and Weigel, 2011). The presence of many mutations in individual mutants makes it possible that two independent mutants may harbor unique mutations in genes that are not related to the phenotype of interest. A combination of the three approaches would also help to identify the causal gene or genes. For example, the *MSD1* gene was originally narrowed to 11 candidate genes by the isogenic mutant-mapping approach. The sequencing of another allele of the *msd1* mutant pointed to a single gene that has unique mutations in two independent *msd1* alleles.

With the continual improvement in NGS techniques and decrease in cost, any of three approaches will become more affordable. It is expected that cloning of the genes represented by well-defined and relevant mutants will become routine in the near future. The key bottleneck will be identification of the mutants of interest and the development of mapping populations. In contrast to the mutant populations developed in model plants, the sorghum mutant library can help in isolating many traits of agronomic and nutritional importance.

Using a combination of isogenic mutant mapping and direct sequencing of independent alleles of a mutant locus, in the last few years we have identified two genes that confer the *msd* phenotype when mutated. With the continual increase in sequencing data output and decrease in sequencing cost, we expect to identify more causal gene mutations that contribute to biomass and grain yield, quality, and abiotic stress tolerance. Molecular markers will be developed accordingly to accelerate sorghum improvement.

Mutation-Indexed Mutant Library

The development of reverse genetic techniques to detect small insertions or deletions (indels) and point mutations has revived interest in chemically inducing mutant populations for many plant and animal species to identify genes involved in signal transduction and growth, development, and adaptive processes (McCallum et al., 2000a; Till et al., 2003; Wienholds et al., 2003; Gilchrist and Haughn, 2005; Winkler et al., 2005; Gilchrist et al., 2006; Xin et al., 2008; Tsai et al., 2011). TILLING is a technique that can be efficiently used for detecting small indels and point mutations (McCallum et al., 2000a,b). TILLING is particularly suitable for genome-wide analysis of a large mutant population induced by mutagenesis and can be used in a high-throughput format to make links between characterized mutant genotypes and the resulting phenotypes. TILLING mainly includes the following steps: development of a mutant population with a reasonable number of individuals by mutagenesis; collection of leaf tissue from M_2 plants and extraction of DNA; pooling of DNA samples from mutants by one or multiple dimensions; design of polymerase chain reaction (PCR) primers covering regions of interest from sequence database; PCR amplification of targeted regions of interest; heteroduplex formation through denaturation and reannealing; cleavage of the heteroduplex by endonucleases that can recognize single base pair mismatch, such as CelI; and separation and detection of cleaved fragments by electrophoresis (Till et al., 2003). The TILLING process is long and requires meticulous optimization for each species and amplicon. A subset of 768 lines from the sorghum mutant library was selected to conduct a pilot TILLING (Xin et al., 2008). Despite extensive optimization of the conditions for cleavage, separation, and detection of the method, only 1/526 kb mutation was found in four amplicons, a mutation rate that is much lower than expected from the frequency of phenotypes observed in the mutant library and the whole-genome sequencing (Xin et al., 2008; Jiao et al., 2016).

During the process of identifying the causal mutations for the *msd* phenotype through whole-genome sequencing, we found that the average mutation rate for 17 of the independent *msd* mutants is about 10/Mb (Y. Jiao and Z. Xin, unpublished data, 2014). Based on this mutation rate, whole-genome sequencing of 250 lines will produce an average of three mutations in the exon of each gene. Thus, a minicore of 256 lines was randomly selected for whole-genome sequencing on the Illumina HiSeq2000 to 15× coverage. Preliminary analysis of the 3.3 Tb of sequencing data uncovered more than 1.8 million canonical EMS-induced mutations (Jiao et al., 2016). The average mutation rate is 11.2/Mb, or about one SNP per 344 bp in the 256 lines. All 10 chromosomes were evenly covered with mutations (Fig. 10). Table 3 summarizes the effect of the SNPs. About 236,000 (6.2%) of the SNPs are located in 30,294 genes, which covers about 92% of sorghum genes. A total of 111,850 nonsynomynous SNPs are located in the exons of 25,605 (77.5%) genes, with an average of four mutations per gene. There were a total of 8043 stop-gained or splice site mutations that potentially produce a knockout mutation of the genes. This large-scale, sequenced mutant library will be a useful forward and reverse genetic resource for in silico identification of genes underlying important traits and for mutant series of interesting genes to elucidate their function in sorghum. Although this sequenced subset of mutants can provide mutations for many genes in the sorghum genome, it is far from sufficient to provide knockout mutations for each gene. For functional genomics studies that require complete knockout alleles, TILLING the rest of mutant lines in addition to the 256 sequenced lines is necessary. Based on the mutation frequency of the 256 sequenced lines, TILLING an additional 5000 lines will produce at least 60 independent mutations per gene and an average of three knockout mutations per gene.

Perspectives

New sequencing technologies that have higher throughput and lower cost are continually being developed. It is now already possible to map and identify causal genes represented by interesting mutants through whole-genome sequencing of

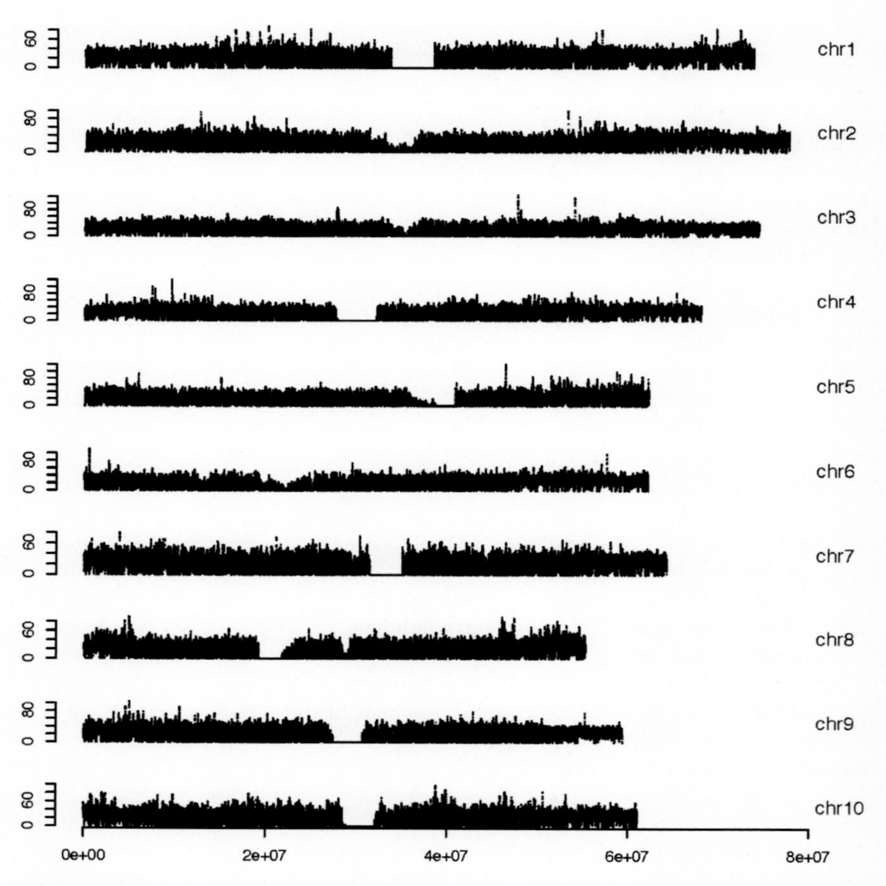

Fig. 10. Distribution of single nucleotide polymorphism (SNPs) among the 10 sorghum chromosomes of sorghum. The number of SNPs within a slide window of 10 kb was plotted against the base position of the chromosomes. The slide window was moved by 1 kb each step.

the bulked F_2 segregants from a backcrossed population. It will soon become routine to rapidly identify causal genes through mapping-by-sequencing. With a genome size of about 730 Mb and a completed genome sequence in the reference inbred line BTx623, sorghum is suitable for mapping-by-sequencing strategies developed in model plants Arabidopsis and rice. The mutant library we have developed, the historic collection of sorghum mutants and genetic stocks, and the new sorghum mutant populations developed at Purdue University and the University of Queensland (Clifford Weil and David Jordan, personal communication, 2012), will serve as critical resources for the rapid discovery of genes critical for improving sorghum as a food, feed, and biofuel crop. It is expected that sequencing technologies will continue to improve and provide larger amounts of higher-quality sequence data at more affordable prices. The key is to develop screen technologies that can identify mutants with the potential to improve sorghum's yield, adaptation to environment, and quality.

Table 3. Functional annotation of the SNPs.

Type	Mutation SNPs		Sorghum genes affected by mutation	
	Number	Percentage total SNPs	Number	Percentage covered
Non_synonymous_coding	111,850	2.94	25,605	77.52
Stop_gained	6015	0.16	4134	12.52
Splice_site_donor†	880	0.02	643	1.95
Splice_site_acceptor‡	1148	0.03	846	2.56
Splice_site_region§	10,008	0.26	5613	16.99
Start_gained	6891	0.18	4634	14.03
Start_lost	152	0.00	139	0.42
Utr_5_prime	38,945	1.02	14,602	44.21
Utr_3_prime	60,909	1.60	19,568	59.24
Sub Total	236,798	6.21	30294	92
Upstream¶	781,853	20.56	31,418	95.11
Downstream¶	769,911	20.24	31,196	94.44
Intron	297,143	7.81	21,726	65.77
Synonymous_coding	70,781	1.86	22,681	68.66
Intergenic	1,646,651	43.30	—	—

† Defined as two bases before exon start, except for the first exon

‡ Defined as two bases after coding exon end, except for the last exon.

§ Is within the region of the splice site, either within one to three bases of the exon or three to eight bases of the intron.

¶ 5,000 bases upstream or downstream of a gene.

Acknowledgments

We thank the United Sorghum Checkoff Program for financial support to develop the sorghum mutant library and Hale Hughes and Lan Liu-Gitz for technical support.

References

Abe, A., S. Kosugi, K. Yoshida, S. Natsume, H. Takagi, H. Kanzaki, et al. 2012. Genome sequencing reveals agronomically important loci in rice using MutMap. Nat. Biotechnol. 30:174–178. doi:10.1038/nbt.2095

Alfonso, M., I. Yruela, S. Almarcegui, E. Torrado, M.A. Perez, and R. Picorel. 2001. Unusual tolerance to high temperatures in a new herbicide-resistant D1 mutant from *Glycine max* (L.) Merr. cell cultures deficient in fatty acid desaturation. Planta 212:573–582. doi:10.1007/s004250000421

Ashikari, M., H. Sakakibara, S. Lin, T. Yamamoto, T. Takashi, A. Nishimura, et al. 2005. Cytokinin oxidase regulates rice grain production. Science 309:741–745. doi:10.1126/science.1113373

Austin, R.S., D. Vidaurre, G. Stamatiou, R. Breit, N.J. Provart, D. Bonetta, et al. 2011. Next-generation mapping of Arabidopsis genes. Plant J. 67:715–725. doi:10.1111/j.1365-313X.2011.04619.x

Bhadula, S.K., T.E. Elthon, J.E. Habben, T.G. Helentjaris, S. Jiao, and Z. Ristic. 2001. Heat-stress induced synthesis of chloroplast protein synthesis elongation factor (EF-Tu) in a heat-tolerant maize line. Planta 212:359–366. doi:10.1007/s004250000416

Bout, S., and W. Vermerris. 2003. A candidate-gene approach to clone the sorghum *Brown midrib* gene encoding caffeic acid *O*-methyltransferase. Mol. Genet. Genomics 269:205–214.

Brown, P.J., P.E. Klein, E. Bortiri, C.B. Acharya, W.L. Rooney and S. Kresovich. 2006. Inheritance of inflorescence architecture in sorghum. Theor. Appl. Genet. 113: 931–942. doi:10.1007/s00122-006-0352-9.

Burke, J.J., and J. Chen. 2006. Changes in cellular and molecular processes in plant adaptation to heat stress. In: B. Huang, editor, Plant-environment interactions. 3rd. ed. CRC Press, Boca Raton, FL. p. 27–46. doi:10.1201/9781420019346.ch2

Burow, G., C. Franks, Z. Xin, and J. Burke. 2013. Developmental and genetic characterization of a short leaf mutant of sorghum [*Sorghum bicolor* (L.) Moench]. Plant Growth Regul. 71:271–280. doi:10.1007/s10725-013-9827-2

Burow, G., Z. Xin, and J. Burke. 2014. Characterization of multi-seeded (*msd1*) mutants of sorghum for increasing grain yield. Crop Sci. doi:10.2135/cropsci2013.08.0566

Burow, G.B., C.D. Franks, V. Acosta-Martinez, and Z. Xin. 2009. Molecular mapping and characterization of *BLMC*, a locus for profuse wax (bloom) and enhanced cuticular features of sorghum (*Sorghum bicolor* L. Moench.). Theor. Appl. Genet. 118:423–431. doi:10.1007/s00122-008-0908-y

Burow, G.B., C.D. Franks, and Z. Xin. 2008. Genetic and physiological analysis of an irradiated bloomless mutant (epicuticular wax mutant) of sorghum. Crop Sci. 48:41–48. doi:10.2135/cropsci2007.02.0119

Chen, J., J.J. Burke, Z. Xin, C. Xu, and J. Velten. 2006a. Characterization of the Arabidopsis thermosensitive mutant atts02 reveals an important role for galactolipids in thermotolerance. Plant Cell Environ. 29:1437–1448. doi:10.1111/j.1365-3040.2006.01527.x

Chen, J., J.J. Burke, J. Velten, and Z. Xin. 2006b. FtsH11 protease plays a critical role in Arabidopsis thermotolerance. Plant J. 48:73–84. doi:10.1111/j.1365-313X.2006.02855.x

Chen, J., W. Xu, J. Velten, Z. Xin, and J. Stout. 2012. Characterization of maize inbred lines for drought and heat tolerance. J. Soil Water Conserv. 67:354–364. doi:10.2489/jswc.67.5.354

Crafts-Brandner, S.J., and M.E. Salvucci. 2002. Sensitivity of photosynthesis in a C4 plant, maize, to heat stress. Plant Physiol. 129:1773–1780. doi:10.1104/pp.002170

Dhugga, K.S. 2007. Maize biomass yield and composition for biofuels. Crop Sci. 47:2211–2227. doi:10.2135/cropsci2007.05.0299

Djanaguiraman, M., P.V. Vara Prasad, M. Murugan, R. Perumal, and U.K. Reddy. 2014. Physiological differences among sorghum (*Sorghum bicolor* L. Moench) genotypes under high temperature stress. Environ. Exp. Bot. 100:43–54. doi:10.1016/j.envexpbot.2013.11.013

Duggan, B.L., D.R. Domitruk, and D.B. Fowler. 2000. Yield component variation in winter wheat grown under drought stress. Can. J. Plant Sci. 80:739–745. doi:10.4141/P00-006

Duvick, D.N., and K.G. Cassman. 1999. Post-Green Revolution trends in yield potential of temperate maize in the North-Central United States. Crop Sci. 39:1622–1630. doi:10.2135/cropsci1999.3961622x

Ebercon, A., A. Blum, and W.R. Jordan. 1977. A rapid colorimetric method for epicuticular wax contest of sorghum leaves. Crop Sci. 17:179–180. doi:10.2135/cropsci1977.0011183X001700010047x

Ejeta, G., and J. Axtell. 1985. Mutant gene in sorghum causing leaf "reddening" and increased protein concentration in the grain. J. Hered. 76:301–302.

Falcone, D.L., J.P. Ogas, and C.R. Somerville. 2004. Regulation of membrane fatty acid composition by temperature in mutants of Arabidopsis with alterations in membrane lipid composition. BMC Plant Biol. 4:17. doi:10.1186/1471-2229-4-17

Gaul, H. 1964. Mutations in plant breeding. Radiat. Bot. 4:155–232. doi:10.1016/S0033-7560(64)80069-7

Gilchrist, E.J., and G.W. Haughn. 2005. TILLING without a plough: A new method with applications for reverse genetics. Curr. Opin. Plant Biol. 8:211–215. doi:10.1016/j.pbi.2005.01.004

Gilchrist, E.J., N.J. O'Neil, A.M. Rose, M.C. Zetka, and G.W. Haughn. 2006. TILLING is an effective reverse genetics technique for *Caenorhabditis elegans*. BMC Genomics 7:262. doi:10.1186/1471-2164-7-262

Greene, E.A., C.A. Codomo, N.E. Taylor, J.G. Henikoff, B.J. Till, S.H. Reynolds, et al. 2003. Spectrum of chemically induced mutations from a large-scale reverse-genetic screen in Arabidopsis. Genetics 164:731–740.

Gurley, W.B. 2000. HSP101: A key component for the acquisition of thermotolerance in plants. Plant Cell 12:457–460. doi:10.1105/tpc.12.4.457

Hartwig, B., G.V. James, K. Konrad, K. Schneeberger, and F. Turck. 2012. Fast isogenic mapping-by-sequencing of ethyl methanesulfonate-induced mutant bulks. Plant Physiol. 160:591–600. doi:10.1104/pp.112.200311

Henikoff, S., B.J. Till, and L. Comai. 2004. TILLING. Traditional mutagenesis meets functional genomics. Plant Physiol. 135:630–636. doi:10.1104/pp.104.041061

Hong, S.W., and E. Vierling. 2000. Mutants of *Arabidopsis thaliana* defective in the acquisition of tolerance to high temperature stress. Proc. Natl. Acad. Sci. USA 97:4392–4397. doi:10.1073/pnas.97.8.4392

Iba, K. 2002. Acclimative response to temperature stress in higher plants: Approaches of gene engineering for temperature tolerance. Annu. Rev. Plant Biol. 53:225–245. doi:10.1146/annurev.arplant.53.100201.160729

Jander, G., S.R. Norris, S.D. Rounsley, D.F. Bush, I.M. Levin, and R.L. Last. 2002. Arabidopsis map-based cloning in the post-genome era. Plant Physiol. 129:440–450. doi:10.1104/pp.003533

Jenks, M.A., P.J. Rich, P.J. Peters, J.D. Axtell, and E.N. Ashworth. 1992. Epicuticular wax morphology of bloomless (*bm*) mutants in *Sorghum bicolor.* Int. J. Plant Sci. 153:311–319. doi:10.1086/297034

Jiao, Y., J.J. Burke, R. Chopra, G. Burow, J. Chen, B. Wang, et al. 2016. A sorghum mutant resource as an efficient platform for gene discovery in grasses. Plant Cell 28:1551–1562. doi:10.1105/tpc.16.00373

Jordan, W.R., P.J. Shouse, A. Blum, F.R. Miller, and R.L. Monk. 1984. Environmental physiology of sorghum. II. Epicuticular wax load and cuticular transpiration. Crop Sci. 24:1168–1173. doi:10.2135/cropsci1984.0011183X002400060038x

Karper, R.E., and J.C. Stephens. 1936. Floral abnormalities in sorghum. J. Hered. 27:183–194.

Kotak, S., J. Larkindale, U. Lee, P. von Koskull-Doring, E. Vierling, and K.D. Scharf. 2007. Complexity of the heat stress response in plants. Curr. Opin. Plant Biol. 10:310–316. doi:10.1016/j.pbi.2007.04.011

Larkindale, J., J.D. Hall, M.R. Knight, and E. Vierling. 2005. Heat stress phenotypes of Arabidopsis mutants implicate multiple signaling pathways in the acquisition of thermotolerance. Plant Physiol. 138:882–897. doi:10.1104/pp.105.062257

Larkindale, J., and B. Huang. 2004. Thermotolerance and antioxidant systems in *Agrostis stolonifera*: Involvement of salicylic acid, abscisic acid, calcium, hydrogen peroxide, and ethylene. J. Plant Physiol. 161:405–413. doi:10.1078/0176-1617-01239

Lee, U., C. Wie, M. Escobar, B. Williams, S.W. Hong, and E. Vierling. 2005. Genetic analysis reveals domain interactions of Arabidopsis Hsp100/ClpB and cooperation with the small heat shock protein chaperone system. Plant Cell 17:559–571. doi:10.1105/tpc.104.027540

Li, C., Y. Li, Y. Shi, Y. Song, D. Zhang, E.S. Buckler, et al. 2015. Genetic control of the leaf angle and leaf orientation value as revealed by ultra-high density maps in three connected maize populations. PLoS One 10:e0121624. doi:10.1371/journal.pone.0121624

Mace, E.S., S. Tai, E.K. Gilding, Y. Li, P.J. Prentis, L. Bian, et al. 2013. Whole-genome sequencing reveals untapped genetic potential in Africa's indigenous cereal crop sorghum. Nat. Commun. 4:2320. doi:10.1038/ncomms3320

Marcum, K.B. 1998. Cell membrane thermostability and whole-plant heat tolerance of Kentucky bluegrass. Crop Sci. 38:1214–1218. doi:10.2135/cropsci1998.0011183X003800050017x

McCallum, C.M., L. Comai, E.A. Greene, and S. Henikoff. 2000a. Targeted screening for induced mutations. Nat. Biotechnol. 18:455–457. doi:10.1038/74542

McCallum, C.M., L. Comai, E.A. Greene, and S. Henikoff. 2000b. Targeting induced local lesions IN genomes (TILLING) for plant functional genomics. Plant Physiol. 123:439–442. doi:10.1104/pp.123.2.439

Metzker, M.L. 2010. Sequencing technologies—The next generation. Nat. Rev. Genet. 11:31–46. doi:10.1038/nrg2626

Michelmore, R.W., I. Paran, and R.V. Kesseli. 1991. Identification of markers linked to disease-resistance genes by bulked segregant analysis: A rapid method to detect markers in specific genomic regions by using segregating populations. Proc. Natl. Acad. Sci. USA 88:9828–9832. doi:10.1073/pnas.88.21.9828

Momcilovic, I., and Z. Ristic. 2007. Expression of chloroplast protein synthesis elongation factor, EF-Tu, in two lines of maize with contrasting tolerance to heat stress during early stages of plant development. J. Plant Physiol. 164:90–99. doi:10.1016/j.jplph.2006.01.010

Morris, G.P., P. Ramu, S.P. Deshpande, C.T. Hash, T. Shah, H.D. Upadhyaya, et al. 2013. Population genomic and genome-wide association studies of agroclimatic traits in sorghum. Proc. Natl. Acad. Sci. USA 110:453–458. doi:10.1073/pnas.1215985110

Nelson, J.C., S. Wang, Y. Wu, X. Li, G. Antony, F.F. White, et al. 2011. Single-nucleotide polymorphism discovery by high-throughput sequencing in sorghum. BMC Genomics 12:352. doi:10.1186/1471-2164-12-352

Nieto-Sotelo, J., L.M. Martinez, G. Ponce, G.I. Cassab, A. Alagon, R.B. Meeley, et al. 2002. Maize HSP101 plays important roles in both induced and basal thermotolerance and primary root growth. Plant Cell 14:1621–1633. doi:10.1105/tpc.010487

Oria, M.P., B.R. Hamaker, J.D. Axtell, and C.P. Huang. 2000. A highly digestible sorghum mutant cultivar exhibits a unique folded structure of endosperm protein bodies. Proc. Natl. Acad. Sci. USA 97:5065–5070. doi:10.1073/pnas.080076297

Paterson, A.H. 2008. Genomics of sorghum. Int. J. Plant Genomics, article ID 362451. doi:10.1155/2008/362451

Paterson, A.H., J.E. Bowers, R. Bruggmann, I. Dubchak, J. Grimwood, H. Gundlach, et al. 2009. The *Sorghum bicolor* genome and the diversification of grasses. Nature 457:551–556. doi:10.1038/nature07723

Paterson, A.H., J.E. Bowers and B.A. Chapman. 2004. Ancient polyploidization predating divergence of the cereals, and its consequences for comparative genomics. Proc. Natl. Acad. Sci. U.S.A. 101: 9903–9908. doi:10.1073/pnas.0307901101.

Peters, P.J., M.A. Jenks, P.J. Rich, J.D. Axtell, and G. Ejeta. 2009. Mutagenesis, selection, and allelic analysis of epicuticular wax mutants in sorghum. Crop Sci. 49:1250–1258. doi:10.2135/cropsci2008.08.0461

Porter, K.S., J.D. Axtell, V.L. Lechtenberg, and V.F. Colenbrander. 1978. Phenotype, fiber composition, and in vitro dry matter disappearance of chemically induced brown midrib (*bmr*) mutants of sorghum. Crop Sci. 18:205–208. doi:10.2135/cropsci1978.0011183X001800020002x

Premachandra, G.S., D.T. Hahn, J.D. Axtell, and R.J. Joly. 1994. Epicuticular wax load and water-use efficiency in bloomless and sparse-bloom mutants of *Sorghum bicolor* L. Environ. Exp. Bot. 34:293–301. doi:10.1016/0098-8472(94)90050-7

Quinby, J.R. 1975. The genetics of sorghum improvement. J. Hered. 66:56–62.

Quinby, J.R., and R.E. Karper. 1942. Inheritance of mature plant characters in sorghum: Induced by radiation. J. Hered. 33:323–327.

Reynolds, M., M.J. Foulkes, G.A. Slafer, P. Berry, M.A.J. Parry, J.W. Snape, et al. 2009. Raising yield potential in wheat. J. Exp. Bot. 60:1899–1918. doi:10.1093/jxb/erp016

Richards, R.A. 2000. Selectable traits to increase crop photosynthesis and yield of grain crops. J. Exp. Bot. 51:447–458. doi:10.1093/jexbot/51.suppl_1.447

Saballos, A., G. Ejeta, E. Sanchez, C. Kang, and W. Vermerris. 2009. A genomewide analysis of the cinnamyl alcohol dehydrogenase family in sorghum [*Sorghum bicolor* (L.) Moench] identifies *SbCAD2* as the *Brown midrib6* gene. Genetics 181:783–795. doi:10.1534/genetics.108.098996

Saballos, A., S.E. Sattler, E. Sanchez, T.P. Foster, Z. Xin, C. Kang, et al. 2012. *Brown midrib2* (*Bmr2*) encodes the major 4-coumarate:coenzyme A ligase involved in lignin biosynthesis in sorghum (*Sorghum bicolor* (L.) Moench). Plant J. 70:818–830. doi:10.1111/j.1365-313X.2012.04933.x

Saeed, M., C.A. Francis, and M.D. Clegg. 1986. Yield component analysis in grain sorghum. Crop Sci. 26:346–351. doi:10.2135/cropsci1986.0011183X002600020028x

Sakamoto, T., Y. Morinaka, T. Ohnishi, H. Sunohara, S. Fujioka, M. Ueguchi-Tanaka, et al. 2006. Erect leaves caused by brassinosteroid deficiency increase biomass production and grain yield in rice. Nat. Biotechnol. 24:105–109. doi:10.1038/nbt1173

Sattler, S.E., D.L. Funnell-Harris, and J.F. Pedersen. 2010. Brown midrib mutations and their importance to the utilization of maize, sorghum, and pearl millet lignocellulosic tissues. Plant Sci. 178:229–238. doi:10.1016/j.plantsci.2010.01.001

Sattler, S.E., A.J. Saathoff, E.J. Haas, N.A. Palmer, D.L. Funnell-Harris, G. Sarath, et al. 2009. A nonsense mutation in a cinnamyl alcohol dehydrogenase gene is responsible for the sorghum *brown midrib6* phenotype. Plant Physiol. 150:584–595. doi:10.1104/pp.109.136408

Sattler, S.E., A. Saballos, Z. Xin, D.L. Funnell-Harris, W. Vermerris, and J.F. Pedersen. 2014. Characterization of novel Sorghum brown midrib mutants from an EMS-mutagenized population. G3: Genes, Genomes, Genet. 4:2115–2124. doi:10.1534/g3.114.014001

Schneeberger, K., S. Ossowski, C. Lanz, T. Juul, A.H. Petersen, K.L. Nielsen, et al. 2009. SHOREmap: Simultaneous mapping and mutation identification by deep sequencing. Nat. Methods 6:550–551. doi:10.1038/nmeth0809-550

Schneeberger, K., and D. Weigel. 2011. Fast-forward genetics enabled by new sequencing technologies. Trends Plant Sci. 16:282–288. doi:10.1016/j.tplants.2011.02.006

Singh, R., and J.D. Axtell. 1973. High lysine mutant gene (*hl*) that improves protein quality and biological value of grain sorghum. Crop Sci. 13:535–539. doi:10.2135/cropsci1973.0011183X001300050012x

Singh, S.P., and P.N. Drolsom. 1973. Induced recessive mutations affecting leaf angle in *Sorghum bicolor*. J. Hered. 64:65–68.

Sree Ramulu, K. 1970a. Induced systematic mutations in sorghum. Mutat. Res. 10:77–80. doi:10.1016/0027-5107(70)90149-1

Sree Ramulu, K. 1970b. Sensitivity and induction of mutations in sorghum. Mutat. Res. 10:197–206. doi:10.1016/0027-5107(70)90116-8

Sree Ramulu, K., and S.R. Sree Rangasamy. 1972. An estimation of the number of initials in grain sorghum using mutagenic treatments. Radiat. Bot. 12:37–43. doi:10.1016/S0033-7560(72)90015-4

Sung, D.-Y., F. Kaplan, K.-J. Lee, and C.L. Guy. 2003. Acquired tolerance to temperature extremes. Trends Plant Sci. 8:179–187. doi:10.1016/S1360-1385(03)00047-5

Tian, F., P.J. Bradbury, P.J. Brown, H. Hung, Q. Sun, S. Flint-Garcia, et al. 2011. Genome-wide association study of leaf architecture in the maize nested association mapping population. Nat. Genet. 43:159–162. doi:10.1038/ng.746

Till, B.J., T. Colbert, R. Tompa, L.C. Enns, C.A. Codomo, J.E. Johnson, et al. 2003. High-throughput TILLING for functional genomics. Methods Mol. Biol. 236:205–220.

Truong, S.K., R.F. McCormick, W.L. Rooney, and J.E. Mullet. 2015. Harnessing genetic variation in leaf angle to increase productivity of *Sorghum bicolor*. Genetics 201:1229–1238. doi:10.1534/genetics.115.178608

Tsai, H., T. Howell, R. Nitcher, V. Missirian, B. Watson, K.J. Ngo, et al. 2011. Discovery of rare mutations in populations: TILLING by sequencing. Plant Physiol. 156:1257–1268. doi:10.1104/pp.110.169748

Vargas-Suárez, M., A. Ayala-Ochoa, J. Lozano-Franco, I. García-Torres, A. Díaz-Quiñonez, V.F. Ortíz-Navarrete, et al. 2004. Rubisco activase chaperone activity is regulated by a post-translational mechanism in maize leaves. J. Exp. Bot. 55:2533–2539. doi:10.1093/jxb/erh268

Vermerris, W., A. Saballos, G. Ejeta, N.S. Mosier, M.R. Ladisch, and N.C. Carpita. 2007. Molecular breeding to enhance ethanol production from corn and sorghum stover. Crop Sci. 47:S-142–S-153. doi:10.2135/cropsci2007.04.0013IPBS

Vierling, E. 1991. The roles of heat shock proteins in plants. Annu. Rev. Plant Physiol. Plant Mol. Biol. 42:579–620. doi:10.1146/annurev.pp.42.060191.003051

Walters, D.R., and D.J. Keil. 1988. Vascular plant taxonomy. 4th ed. Kendall/Hunt Pub. Co., Dubuque, IA.

Wang, P., W. Duan, A. Takabayashi, T. Endo, T. Shikanai, J.Y. Ye, et al. 2006. Chloroplastic NAD(P)H dehydrogenase in tobacco leaves functions in alleviation of oxidative damage caused by temperature stress. Plant Physiol. 141:465–474. doi:10.1104/pp.105.070490

Wang, W., B. Vinocur, O. Shoseyov, and A. Altman. 2004. Role of plant heat-shock proteins and molecular chaperones in the abiotic stress response. Trends Plant Sci. 9:244–252. doi:10.1016/j.tplants.2004.03.006

Wang, X., U. Gowik, H. Tang, J.E. Bowers, P. Westhoff, and A.H. Paterson. 2009. Comparative genomic analysis of C4 photosynthetic pathway evolution in grasses. Genome Biol. 10:R68. doi:10.1186/gb-2009-10-6-r68

Wienholds, E., F. van Eeden, M. Kosters, J. Mudde, R.H. Plasterk, and E. Cuppen. 2003. Efficient target-selected mutagenesis in zebrafish. Genome Res. 13:2700–2707. doi:10.1101/gr.1725103

Winkler, S., A. Schwabedissen, D. Backasch, C. Bokel, C. Seidel, S. Bonisch, et al. 2005. Target-selected mutant screen by TILLING in *Drosophila*. Genome Res. 15:718–723. doi:10.1101/gr.3721805

Xin, Z., G.B. Burow, C. Woodfin, C.D. Franks, R.R. Klein, K.F. Schertz, et al. 2013. Registration of a diverse collection of sorghum genetic stocks. J. Plant Reg. 7:119–124. doi:10.3198/jpr2011.09.0502crgs

Xin, Z., M.L. Wang, N.A. Barkley, G. Burow, C. Franks, G. Pederson, et al. 2008. Applying genotyping (TILLING) and phenotyping analyses to elucidate gene function in a chemically induced sorghum mutant population. BMC Plant Biol. 8:103. doi:10.1186/1471-2229-8-103

Xin, Z., M. Wang, G. Burow, and J. Burke. 2009. An induced sorghum mutant population suitable for bioenergy research. BioEnergy Res. 2:10–16. doi:10.1007/s12155-008-9029-3

Zhao, S.Q., J. Hu, L.B. Guo, Q. Qian, and H.W. Xue. 2010. Rice leaf inclination2, a VIN3-like protein, regulates leaf angle through modulating cell division of the collar. Cell Res. 20:935–947. doi:10.1038/cr.2010.109

Zhu, X.-G., S.P. Long, and D.R. Ort. 2010. Improving photosynthetic efficiency for greater yield. Annu. Rev. Plant Biol. 61:235–261. doi:10.1146/annurev-arplant-042809-112206

Zhu, Y., H.G. Mang, Q. Sun, J. Qian, A. Hipps, and J. Hua. 2012. Gene discovery using mutagen-induced polymorphisms and deep sequencing: Application to plant disease resistance. Genetics 192:139–146. doi:10.1534/genetics.112.141986

Sorghum Hybrids Development for Important Traits: Progress and Way Forward

Are Ashok Kumar*

Abstract

Sorghum is one of the most important cereal crops grown in the semiarid tropics (SAT) of Africa, Asia, Australia and Americas for its food, feed, fodder and fuel value. While impressive progress has been made in increasing productivity over recent years, global sorghum production is still constrained by several biotic and abiotic stresses. Hybrid technology played a major role in improving sorghum productivity in more than half of sorghum growing areas, particularly in grain, dual-purpose and forage sorghum in the Americas, Asia and Australia, but it has yet to make a dent in Sub-Saharan Africa. It is important to take this technology to Sub-Saharan Africa, which has highest sorghum area and largest scope for enhancing productivity. In addition, sorghum has high potential for postrainy adaptation, grain micronutrient density, sweet stalk and high mass and for niche ecologies like rice–fallow areas. The need of the hour is to breed specific hybrids for these traits with adaptation to various geographies. This chapter describes the sorghum floral biology, cultivar options, breeding methods, hybrid development and trait focus for increasing genetic gain through deployment of heterotic hybrids.

Sorghum (Sorghum bicolor (L.) Moench) is a predominantly self-pollinating (although cross-pollination exceeds 5% at times) diploid ($2n = 2x = 20$) belonging to Gramineae family with a genome size of 730 Mb, about 25% the size of maize or sugarcane. It is a C4 plant with higher photosynthetic efficiency and abiotic stress tolerance (Nagy et al., 1995; Reddy et al., 2009b). Its small genome makes it an attractive model for studying functional genomics of C4 grasses taking advantage of synteny. The tolerance to water-deficit stress makes sorghum especially important in dry regions such as northeast Africa (its center of diversity), India, and the southern plains of the United States (Paterson et al., 2009). In addition to its significance as a food and fodder crop, sorghum plays an important role in supplying micronutrients at low cost (Parthasarathy Rao et al., 2006, 2010). In addition, the genetic variation in the partitioning of carbon into sugar stores versus cell wall mass, its perenniality and associated features such as tillering and stalk reserve retention, makes it an attractive system for the study of

---

Abbreviations: bmr, brown-midrib; LGP, length of growing period; SAT, semiarid tropics; SC, Sorghum conversion

Keywords: Sorghum, hybrids, breeding, yield, nutritional quality, high biomass, stress tolerance

International Crops Research Institute for the Semi-Arid Tropics (ICRISAT) Patancheru-502324, Telangana, India. *Corresponding author (a.ashokkumar@cgiar.org)

doi:10.2134/agronmonogr58.2014.0059

perennial cellulosic biomass crops (Paterson et al., 1995). Sorghum is one among the climate resilient crops that can better adapt to climate change conditions (Cooper et al., 2009; Reddy et al., 2011). This chapter deals with the biology of sorghum and cultivar options, need for hybrids development, important new traits like higher Zn and Fe, postrainy season adaptation where in no commercial hybrids available thus far and approaches used in improving parents and hybrids.

Sorghum plays a major role in global food security. It is the fifth most important cereal crop globally after rice, wheat, maize and barley and is the dietary staple of more than 500 million people in over 30 countries, primarily in the developing world. On an average it is grown on 40 million ha area in more than 90 countries in Africa, Asia, Oceania, and the Americas (Hariprasanna and Rakshit, 2016). Among these, the United States, Nigeria, India, Mexico, Sudan, China, and Argentina are the major sorghum producers globally. Sorghum accounts for 6% of the global coarse cereals production in the world and is particularly well suited to hot and dry agroecologies in the world and responds well to high input management. In spite of its higher adaptability, stress-tolerance and wider distribution, global sorghum productivity is low (1.4 t ha^{-1}) with wide variation in different parts of the world (Ashok Kumar et al., 2013c; Sekoli and Morojele 2016).

Sorghum is consumed in different forms for various end uses. Its grain is mostly used directly for food purposes (55%), and is consumed in the form of porridges (thick or thin) and flat breads. It is also an important feed grain (33%), especially in Australia and the Americas. It is among the highly traded coarse cereals globally. The United States and Australia are the largest sorghum exporters while China, Japan and Mexico are the largest importers, mainly for feed use. Its stover (crop residue after grain harvest) is an important feed source in mixed crop–livestock systems prevalent in semiarid tropics. Of late, sweet sorghum with sugar-rich juicy stalks is emerging as an important biofuel crop (Vermerris et al., 2007; Reddy et al., 2008). Sorghum grain is a rich source of micronutrients, particularly Fe and Zn (Ashok Kumar et al. (2011, 2015) and is also a rich and cheap source of starch. Thus, sorghum is a unique crop with multiple uses as food, feed, fodder, fuel and fiber. It is generally grown in rainy season (spring/summer) across many countries but in Maharashtra and Karnataka states in India and Lake Chad areas of Africa it is grown in both rainy and postrainy seasons (Reddy et al., 2009a). It is possible to grow sorghum year-round in some tropical countries like India by using appropriate hybrids.

Floral Biology, Mode of Reproduction and Artificial Hybridization

An understanding of crop floral biology, pollination control mechanisms and seed development is essential for designing effective breeding strategies and suitable breeding methods for systematic genetic improvement. Breeding methods and procedures to be employed for genetic enhancement of any crop species is largely determined by its mode of reproduction (Aruna and Audilakshmi, 2008). Sorghum is a short day plant, and blooming is hastened by short days and long nights. However, varieties differ in their photoperiod sensitivity (Quinby and Karper 1947; Cuevas, 2016). In traditional varieties, reproductive stage is initiated when daylengths return to 12 h. Floral primordial initiation takes place 30 to 40 d after germination. Usually, the initiation is 15 to 30 cm above the ground when the plants are about 50 to 75 cm tall (House, 1980). Floral initiation marks the end of the vegetative phase. The time required for transformation from the vegetative primordial to reproductive primordial is largely influenced by the genotype and the environment. The grand growth period in sorghum follows the formation of a floral bud and consists largely of cell enlargement. Hybrids take less time to reach panicle initiation, more days to expand the panicle, and a longer grain filling period than their corresponding parents (Maiti 1996). This is one of the factors enabling exploitation of heterosis through hybrid development and their commercialization.

The natural cross pollination in sorghum varies from 0.6 to 30% depending on the genotype, panicle type, wind direction and velocity (Maiti 1996). Inflorescence is a raceme, consisting of one to several spikelets. The spikelets usually occur in pairs, one being sessile and the second borne on a short pedicel, except the terminal sessile spikelet, which is accompanied by two pedicelled spikelets. Anthesis starts with the exertion of complete panicle from the boot leaf. In general flowers begin to open 2 d after complete emergence of the panicle. The sorghum head begins to flower from the tip and proceeds successively downward. Anthesis takes place first in the sessile spikelets. It takes about 6 d to complete anthesis with a maximum flowering in first three or four days. Anthesis flowering starts in the morning hours, and frequently occurs just before or just after sunrise, but delayed on cloudy damp mornings. Maximum flowering is observed between 0600 and 0900 h but varies from place to place depending on aerial temperature. Because all heads in a field do not flower at the same time, pollen is usually available for a period of 10 to 15 d. At the time of flowering, the glumes open and all the three anthers fall free, while the two stigmas protrude, each on a stiff style. The anthers dehisce when they are dry and pollen is blown into the air. Pollen in the anthers remains viable several hours after pollen shedding. Flowers remain open for 30 to 90 min. Dehiscence of the anthers for pollen diffusion takes place through the apical pore. The pollen drifts to the stigma, where it germinates; the pollen tube, with two nuclei, grows down the style, to fertilize the egg and form a $2n$ nucleus (Aruna and Audilakshmi, 2008). Stigmas get exposed before the anthers dehisce subjecting to cross pollination. Artificial pollination for crossing purposes should start soon after normal pollen shedding is completed during morning hours.

Sorghum is among breeders' friendly crops as it is amenable for both crossing and selfing quite easily. For selfing, after panicle exertion, bagging should be done by snipping off the flowered florets at the tip. Crossing is done by emasculation of selected panicles and dusting of pollen from identified plants. Hand emasculation is the most commonly practiced in sorghum. Because of this ease in crossing, hybridization is most commonly followed in sorghum for trait improvement.

Cultivar Options in Sorghum

The crop improvement methods applicable in both self and cross pollinated crops can be conveniently employed for developing improved sorghum cultivars. This is the reason why one can find sorghum pure line varieties, hybrids and populations as cultivar options in different parts of the world. However, sorghum hybrids are superior to pure lines and populations for yield and other important agronomic traits. The discovery of cytoplasmic-nuclear male sterility system (CMS) helped to produce hybrids on mass scale using three-line system (A– male sterile, B– male fertile and R– fertility restorer) breeding for commercial hybrids development (Stephens and Holland, 1954) and revolutionized sorghum production across the world (Reddy et al., 2009a).

Sorghum improvement deals with production of new cultivars that are superior to existing ones for the traits of interest. Availability of genetic variability, knowledge about heritability and inheritance, effective phenotyping methodologies are needed for success of any crop improvement program. In fact, the efficiency of phenotyping and its robustness decides the success of the program in terms of producing a tangible product or technology. In sorghum, a large collection of global germplasm is available at ICRISAT (~40,000 accessions) which is also replicated in Svalbard Global Seed Vault with characterization information available for various morphological, agronomic and adaptive traits. Inheritance of major traits is well studied and phenotyping techniques have been developed for efficient selection and/or screening for major traits of interest. There is continuous exchange of material and information across research groups. As a result, a large number of sorghum cultivars were developed and commercialized across the world for traits of interest. For example, during the period 1976 to 2016, a total of 242 sorghum cultivars were developed from the ICRISAT-bred sorghum material and released in 44 countries (Ashok Kumar et al., 2011). Recent records indicate that this number has increased to 270 cultivars. The list is quite exhaustive if we consider cultivars developed by other centers in all sorghum-growing countries. Most sorghum breeding programs focuses on hybrid parents' development so as to develop heterotic hybrids. The R-lines developed in the process form good candidates for open pollinated varieties (OPVs) development. Agronomically promising partial restorers also form OPVs for commercialization. Pure line selection, mass selection, pedigree and backcross are commonly employed breeding methods for OPVs development in sorghum.

Open pollinated varieties are the cultivar options in sorghum in Western and Central Africa and Eastern and Southern Africa and postrainy sorghum areas in India. One of the reasons hybrids are not popular in these regions is that heterosis is limited and seed industry is not well established to take hybrids to the farmers' doorstep. Sometimes stringent quality considerations by farmers and marketing limit the adoption of hybrids.

Hybrid Development

Heterosis in sorghum was demonstrated by Conner and Karper (1927), but its commercial exploitation was possible only after the discovery of a stable and heritable cytoplasmic–nuclear male sterility (CMS) mechanism and its designation as A_1 *(milo)* (Stephens and Holland, 1954). Utilization of this system required development of a CMS line (A-line), its maintainer line (B-line) and fertility restorer (R-line) which is relatively convenient compared to genetic male sterility (GMS) based hybrids development. Since then a large number of hybrids have been developed and released/marketed for commercial cultivation in Asia, the Americas, Australia and Africa. The hybrids have contributed significantly to increased grain and forage yields in several countries. The grain productivity increased by 47% in China and by 50% in India from the 1960s to the 1990s (FAO 1960–1996), which corresponds well with the adoption of hybrids in these countries. Adoption of the first commercial hybrid (CSH 1) in India ranges over much of the rainy season sorghum area, while local varieties are confined to fairly narrow specific environmental niches. This stands testimony to the wide adaptability of hybrids over varieties (House et al., 1997). Currently, over 95% of the sorghum area is adopted to the hybrids in the United States, Australia and China and 85% rainy season sorghum area in India (Reddy et al., 2006).

In the prehybrid era of early 1960s, the average sorghum productivity was low with 0.49 t ha^{-1} in India, 0.66 t ha^{-1} in China, 0.76 t ha^{-1} in sub-Saharan Africa, 1.48 t ha^{-1} in Australia, and 2.8 t ha^{-1} in the United States. In the United States, Northern and Central America, where commercial hybrids were exploited, there was a 40% increase in productivity from early 1960s to early 1990s. A similar trend was noticed globally including China and India. However, the productivity remained static in Africa from 1960s to early 1990s which can be directly attributed to nonexploitation of heterosis through hybrid development (Reddy et al., 2006; Rattunde et al., 2013).

Trait Focus in Hybrids

Rainy Season: High Yielding Grain and Dual-Purpose Types

This is the most important sorghum adaptation globally grown from May and/or June to August and/or September with more than 30 million ha area across various continents falling under this adaptation. A variety of sorghum belonging to different races (basic or intermediate races), and grain pericarp colors (red, brown, white, etc.) are grown for a variety of end uses in more than 90 sorghum growing countries. Nearly 50% of this area is hybrid. For development of new cultivars, crop improvement programs must focus on target environments, consumer preferences and seed supply chain. For example, medium height dual–purpose hybrids with bold white seeds are preferred in India for both food and feed uses whereas grain types with red pericarp are preferred for food and brewing purposes in East Africa while tall, long-duration guinea sorghums are preferred in West Africa for food. Similarly, short to medium height grain hybrids are preferred in the United States, South America and Australia for mechanical harvesting for use as animal feed (Rooney 2014). In grain sorghum, plant height, leaf sheath anthocyanin pigmentation, days to 50% flowering, maturity duration, panicle exertion, panicle size, glume coverage, grain number, grain size and color and grain threshability are major selection criteria in addition to the grain yield. In dual purpose types, apart from grain yield, stover yield and quality are also important while in energy sorghums, the biomass yield, juice yield and ^{0}Bx% are major selection criterion (Rooney 2014). The important biotic constraints in rainy season sorghum include shoot fly, stem borer, midge, aphids, grain mold, anthracnose, charcoal rot, striga and among abiotic constraints, drought and cold predominates (Gressel et al., 2004; Reddy et al., 2009b).

Grain yield is the most important trait in sorghum breeding as in other crops; however stover yield is equally important, considering its feed value. Breeding for grain yield improvement is performed by selecting genotypes directly for grain yield and its component traits like panicle length, number of grains per panicle and grain size. For higher yield, genotypes with a plant height of around 1.5 m are desirable, which are amenable for mechanical harvesting with medium maturity duration (100–120 d). Longer duration types give higher yields but the length of growing period (LGP) in most sorghum growing areas does not allow for breeding long duration types, with the exception of West Africa. If we reduce the crop duration, it is likely that the yield goes down. Therefore, breeder has to first fix the plant height and maturity duration for a given environment. However, in the context of climate change, longer duration types need to be maintained in the breeding program considering the fact that when temperatures increase by 2 °C, the longer-duration types behave as medium duration types and produce higher yields (Cooper et al., 2009; Haussmann et al., 2012). Another important consideration in sorghum improvement is photoperiod sensitivity. It is the ability of a genotype to mature at a given period in the calendar year irrespective of its planting date. It is feasible to identify the photoperiod-sensitive genotypes by planting them in different dates (at 15 or 30 d interval) and recording the days for 50% blooming and days to maturity in the genotypes. Genotypes that take less time for flowering when planted late in the growing season can be considered photoperiod-sensitive. In West Africa and postrainy sorghum in India, photoperiod sensitivity is a key trait for improvement (Rattunde et al., 2013). Among the component traits, long panicles, bold and higher number of grains per panicle and higher test weight contribute for grain yield and most of these traits have high heritability, enabling the plant breeder to improve for these traits through simple selection. The gap between flag leaf sheath and panicle base should be minimal to have good grain filling. Also, less glume coverage on grains results in higher threshability. Grain size can be visually judged and grain color can be selected as per the consumer and/or market preference in the given adaptation (House 1980; Reddy et al., 2009b; Haussmann et al., 2012). Improving cultivars for various processing needs and the stover with high digestibility is critical for the success of hybrids (Hash et al., 2003; Blümmel et al., 2003).

Grain and Stover Yield

In areas where sorghum stover is important as animal feed, breeding dual-purpose types is the best choice. Heterosis for grain and stover yield is high in sorghum and therefore hybrid development should be targeted. A heterosis of 30 to 40% for grain yield is reported compared to the best varieties (Pfeiffer et al., 2010; Hayes and Rooney 2014; Ringo, 2015; Mindaye et al., 2016). Hybrid parents' development is critical for exploiting heterosis and therefore genetic and cytoplasmic diversification of hybrid parents is a major breeding objective. Population improvement is also followed (Rooney 2004; Morris et al., 2013).

Quality of grain and stover is as important as grain yield. This is even more so in the postrainy season sorghum grown predominantly in India on 4.0 million ha area where consumers prefer bold, lustrous white grain types, which are generally available only in landrace varieties (Reddy BVS et al., 2012). The narrow genetic base of these landraces is more challenging to improve for postrainy season adaptation. Similarly heterosis is low when both parents are derived from landraces. A more practical method for developing postrainy season hybrids is by using rainy season adapted lines (mostly *caudatum* types) as females and landrace varieties as pollinators (Reddy et al., 2012). While improving the stover yield, one has to keep in the mind the stover digestibility and protein content (Seetharama et al., 1990; Reddy PS et al., 2012). The stover yields have to be recorded on oven dried samples after harvesting the grains and for stover quality. Indirect selection using NIRS is the most practical method (Montes et al., 2009).

Height and Maturity

Plant height is a major consideration in sorghum improvement. In fact it is one the criterion used for classifying sorghums as grain sorghums, dual-purpose sorghums, sweet sorghums and forage sorghums. In sorghum, four loci are known to be involved in the control of plant height. These genes are assigned *Dw1, Dw2, Dw3,* and *Dw4*. Tallness is partially dominant to dwarfness. The zero dwarf type (dominant [DW-] at all loci) may reach a height of 4 m. The change from four to three dominant genes may result in a height change of 50 cm or more. If genes at one or more of the loci are recessive, the difference in height resulting from the recessive condition at an additional locus may have a lesser effect in reducing plant height. The difference between a 3-dwarf and a 4-dwarf type may be only 10 or 15 cm (House 1980). Breeders have to keep in mind these facts while selecting genotypes with appropriate height.

In sorghum, factors at four loci have been identified to influence maturity, *Ma1, Ma2, Ma3,* and *Ma4*. Generally tropical types are dominant (Ma-) at all four of these loci, and a recessive condition (*mama*) at any one of them will result in more temperate zone adaptation that takes more time for maturity (Quinby, 1967). Most sorghum improvement programs target medium maturity types (crop duration less than 120 d) as they yield high, however the targeted maturity is to be decided based on the length of growing period (LGP) of the target area. In general, tropical sorghums take 35 to 40 d from flowering to maturity. The parents and hybrids should be developed keeping this in to consideration. The grain is to be harvested at the stage of physiological maturity. The hilum turns dark at physiological maturity, and this is an important criterion for harvesting (House, 1980).

Most temperate-adapted sorghum cultivars are photoperiod-insensitive and dwarfed for grain production. Classical segregation studies predict that temperate adaptedness involves four major loci each for maturity and dwarfing. Two major maturity loci, Ma1 (PRR37) and Ma3 (phytochrome B), and a single major dwarfing locus, Dw3 (PGP1/br2), have been cloned. Sorghum conversion (SC) lines essentially are exotic varieties that have been introgressed with early maturity and dwarfing QTL from a common, temperate-adapted donor using a minimum of four backcrosses. In a study, partially-isogenic populations were generated by crossing five diverse SC lines to their corresponding exotic progenitor (EP) lines to assess the phenotypic effects of individual introgressions from the temperate-adapted donor and 192 F3 lines from the five populations were phenotyped for plant height and maturity. Subsets of 109–175 F3 lines were genotyped using Illumina genotyping-by-sequencing (GBS) and used for QTL analysis. QTL models explained 62.31 to 88.16% of the phenotypic variation for height and maturity in these partially isogenic populations. Nearly all variation was accounted for by the linked Ma1/Dw2 loci on chromosome 6 and the Dw3 and Dw1 loci on chromosomes 7 and 9, respectively. The Dw1 locus fractionated into linked QTL for height and maturity, and a novel height QTL on chromosome 3 was discovered (Higgins, 2013). In another study association mapping of the 39 markers with 242 accessions from the sorghum mini core collection identified five markers associated with maturity or height. All were clustered on chromosomes 6, 9, and 10 with previously mapped height and maturity markers or QTLs. One marker associated with both height and maturity was 84 kb from recently cloned Ma1 (Upadhyaya et al., 2012).

Postrainy Season: Dual-Purpose Types with High Yield and Quality

Postrainy season is a unique adaptation to India (approximately 4.0 million ha) where the crop is grown from September and/or October to January and/or February with residual and receding moisture in black soils. Grain is preferred for food use in India owing to its bold globular lustrous nature. However, sensory evaluation in the sorghum-eating population showed no differences between the flat breads made from rainy season (but matured under rain-free condition) and postrainy season sorghums (Ashok Kumar et al., 2013c). The stover from postrainy crop is the most important animal feed, particularly in summer. In addition to grain and stover yield, photoperiod sensitivity, temperature insensitivity and grain color, size and luster, and stover quality are the major selection criterion. Restoration percentage (judged by percent seed set under bagging in hybrids) is another important criterion for developing postrainy sorghum hybrids. The restoration percentage in hybrids should be > 80%. In postrainy sorghum, open-pollinated varieties are the cultivar choice as of now, but there is good scope for hybrid development using the white grained rainy season adapted lines as female parents and land race restorers as pollinators. While terminal drought is the major production constraint, shoot fly, aphids and charcoal rot play havoc with postrainy season production (Ashok Kumar et al., 2011).

Some hybrids have been developed for postrainy adaptation in India. However they are not adopted by farmers as they lack heterosis or traits preferred by farmers. Recently 75 elite hybrids were evaluated along with three checks (8712–hybrid check, Parbhani Moti and M 35–1, OPV checks to compare the hybrids quality) with farmers participation at across four postrainy testing locations in India (ICRISAT–Patancheru, VNMKV–Parbhani, MPKV–Rahuri and CRS–Solapur) during 2013 postrainy season. Twenty elite hybrids with a grain yield of 3.6 to 4.9 t ha-1 were superior to the hybrid check, 8712 (3.5 t ha-1) with comparable grain size and luster. Of the twenty, eight hybrids showed restoration percentage of 80% or more, indicating their superiority. The hybrids matched preferred OPVs, Parbhani Moti and M35–1, for grain quality with significantly higher grain yield over them. The hybrids were at par with checks for animal feed quality.

Shoot Fly Resistance

Shoot fly is a major problem in late-sown crop in in most sorghum growing regions in Africa and Asia. Shoot fly resistance is a complex trait with large genetic variability. The interlard-fishmeal technique has been used for screening to develop shoot fly resistant parents and hybrids (Nwanze 1997). Resistant sources in desirable agronomic backgrounds like ICSV 702, ICSV 705, ICSV 708, PS 21318, PS 30715–1 and PS 35805 as well as germplasm source IS 18551, IS 923, IS 1057, IS 1071, IS 1082, IS 1096, IS 2394, IS 4663, IS 5072, IS 18369, IS 4664, IS 5470 and IS 5636 were used in crossing programs. Following a trait-based pedigree breeding approach, a large number of shoot fly–resistant seed parents for both rainy season (e.g., ICSA//B-409 to ICSA-/B-436 and ICSA-/B- 29001 to 29006) and postrainy season adaptation (e.g., ICSA-/B-437 to ICSA-/B-463) were developed (Reddy et al., 2004; Riyazaddin et al., 2015). High yielding, shoot fly–resistant parents were developed and heterotic hybrids produced using these parents; the need for having shoot fly resistance in both female and male parents for producing shoot fly resistant high yielding hybrids was demonstrated (Ashok Kumar et al., 2008). On comparing the A1 and A2 CMS systems for shoot fly resistance, no significant differences were observed between A1 and A2 cytoplasms, indicating the opportunity for cytoplasmic diversification (Reddy et al., 2009b).

In sorghum, quantitative trait loci (QTL) governing various component traits contributing for shoot fly resistance have been identified and mapped in the parent IS 18551 (Satish et al., 2009; Ashok Kumar et al., 2014; Kiranmayee et al., 2016). The QTL have been transferred to two cultivated backgrounds, namely BT×623 and 296B at ICRISAT, and these lines are used to transfer shoot fly resistance in to elite sorghum varieties and parents. Phenotyping of 20 QTL introgression lines (BC2F3) was developed using QTL introgressed BT×623 and 296B as donors, along with their respective elite recurrent and QTL donor parents with shoot fly resistant and susceptible checks showed significant differences among the introgression lines for shoot fly resistance. The mean of dead hearts scored in introgression lines was significantly fewer compared with that of their recurrent parents. Introgression lines 6018–15, 5135–8 and 6018–5 with QTL for reduced oviposition showed lesser percentage of plants with shoot fly eggs, 10%, 22% and 25% less respectively compared to the recurrent parent (ICSB 29004) carrying 39% shoot fly eggs. Similarly, the line 6026–13 carrying QTL for trait oviposition nonpreference and dead hearts percentage, showed 5% shoot fly eggs and 10% shoot fly dead hearts compared to the recurrent parent ICSB 29004 showing 20% shoot fly dead hearts. All the entries carrying QTL for trait glossiness showed glossy phenotype. These lines are being validated for use in shoot fly resistance improvement programs for commercialization. Two hybrid combinations (ICSA 434 × M 35–1-19 and ICSA 445 × ICSV 702) showed significantly higher yield and lower shoot fly dead hearts percentage compared to best check CSH 16 (Ashok Kumar et al., 2008).

Grain Mold Resistance

Sorghum grain mold is a major production constraint in Asia and parts of Africa and grain weathering in the United States, particularly under inclement weather conditions (Tuinstra, 2008). The white grain medium duration genotypes are more prone to grain mold attack as their grain development coincides with heavy rainfall particularly in Western Africa and in India. A complex of pathogenic and saprophytic fungi causes grain mold, and the major fungi associated with early infection events are Fusarium spp., Curvularia lunata, Alternaria alternata and Phoma sorghina (Thakur et al., 2003, 2006). Grain mold damage resulting from early infection causes reduced kernel development, discoloration of grains, colonization and degradation of endosperm, and decreased grain density, germination and seedling vigor (Thakur et al., 2006). Several species of Fusarium associated with grain mold complex have been shown to produce mycotoxins, such as fumonisins and trichothecenes that are harmful to human and animal health (Thakur et al., 2006; Sharma et al., 2011). Phenotyping for grain mold reaction is done under field conditions during rainy season. No artificial inoculation is required since sufficient natural inoculum of mold fungi are present during the rainy season over sorghum fields in India for natural field epiphytotic conditions (Bandyopadhyay et al., 1988; Thakur et al., 2007). The test lines are sown in the first half of June so that grain maturing stages coincided with periods of frequent rainfall in August and/or September. To enhance mold development, high humidity (> 90% RH) is provided through sprinkler irrigation of test plots twice a day for 30 min each between 10 AM and 12 PM, and between 4 and 6 PM on rain-free days from flowering to physiological maturity. The visual panicle grain mold rating (PGMR) is taken at the prescribed physiological maturity using a progressive 1 to 9 scale, where 1 = no mold infection, 2 = 1–5%, 3 = 6–10%, 4 = 11–20%, 5 = 21–30%, 6 = 31–40%, 7 = 41–50%, 8 = 51–75% and 9 = 76-100% molded grains on a panicle to categorize the test entries into resistant (1.0–3.0 score), moderately resistant (3.1–5.0 score), susceptible (5.1–7.0 score) and highly susceptible (> 7.0 score) reaction types (Thakur et al., 2006). The resistant (IS 14384) and susceptible (Swarna) checks are invariably included for comparison. A greenhouse screening method has been developed at ICRISAT Patancheru that facilitates screening sorghum lines against individual mold pathogen under controlled conditions (Thakur et al., 2007).

Resistance to grain mold is a polygenic trait and both additive and non-additive gene action has been reported. To develop grain mold-resistant hybrids, at least one parent should possess grain mold resistance (Kumar et al., 2011b). Hard grain and colored glumes contribute to grain mold resistance in white grain types and red grain possesses better grain mold resistance than white grain types. In a study at ICRISAT, 156 grain mold tolerant and/or resistant lines were identified by screening 13,000 photoperiod-insensitive sorghum germplasm lines (Bandyopadhyay et al., 1988). Several grain mold resistant accessions (IS 2815, IS 21599, IS 10288, IS 3436, IS 10646, IS 10475 and IS 23585) have been used in breeding to develop restorer lines varieties and hybrid parents. White/chalky white-grained mold resistant accessions such as IS 20956, -21512, -21645 IS 2379 and -17941 have been selected from the sorghum mini-core collection (Sharma et al., 2010). In a trait-specific breeding program, a number of grain mold resistant lines with maintainer reaction have been converted into male-sterile lines. Fifty-eight seed parents with A1 cytoplasm with white, red and brown grain have been developed. Also, the grain mold resistant accession IS 9470 with A1 (milo), A2, A3, and A4 (maldandi), and IS 15119 with A3 and A4 (maldandi) cytoplasms have been developed and charactrerized characterized (Kumar et al., 2011a). It was showed that there are no cytoplasmic differences between A1 and A2 cytoplasms for grain mold resistance and it is feasible to develop white pericarp grain mold resistant high yielding sorghum hybrids with stable performance by using improved grain mold resistant hybrid parents, at least one of the parents being resistant to grain mold (Reddy et al., 2009b; Kumar et al., 2008b). More recently, some test hybrids developed using mold resistant advanced hybrid parents (A- and R-lines) have shown promising results for mold resistance and grain yield at ICRISAT (Thakur et al., 2007; Kumar et al., 2011a). For identifying QTL for grain mold resistance, two mapping populations (RILs) were developed (296 B × PVK 801; PVK 801 × 296 B) and the phenotyping of these populations for grain mold resistance is in progress at ICRISAT.

In spite of good progress, the identification and characterization of resistance genes for sorghum grain mold have been elusive targets. Thus, a strategy to look for the presence of common resistance gene motifs is the correct course of action (Little et al., 2012). As grain mold is caused by multiple fungi, with the current ability to identify and track alleles of genes that confer resistance to a specific pathogen with easily screened PCR-based markers, the task of combining multiple genes into cultivars with broad grain mold resistance will be greatly simplified. However, due to the many host genes and fungal species involved, it seems likely that selection of locally adapted varieties will remain the most effective breeding tool for some time (Little et al., 2012). In a study on hybrids, grain density and germination percentage were found to be useful in measuring grain mold resistance and the resistant hybrids contained higher grain density and germination percentage compared to susceptible hybrids (Rao et al., 2013).

Micronutrient Density

Breeding crops for enhanced nutrition is termed biofortification, and is one of the major breeding objectives in sorghum. Biofortification, wherever feasible, complements the ongoing efforts to address hidden hunger, which is rampant in Sub-Saharan Africa and South Asia (Meenakshi et al., 2010; Bouis et al., 2011). It is one of the cheapest and most sustainable options to combat the malnutrition in predominantly sorghum eating populations (Parthasarathy Rao et al., 2006; Ashok Kumar et al., 2010a). Earlier efforts in sorghum indicated that there exits limited variability for b-carotene content, but large variability for Fe and Zn concentration in sorghum grain (Reddy et al., 2005; Ashok Kumar et al., 2009, 2012). After assessing grain Fe in Zn in popular sorghum cultivars, the baselines for Fe (30 ppm) and Zn (20 ppm) were fixed. Based on sorghum grain consumption levels, nutrient retention in grain storage and processing, and nutrient bioavailability, ICRISAT targeted 60 ppm Fe and 32 ppm Zn concentration in grain for addressing micronutrient malnutrition in populations who depend predominantly on sorghum for their nutrient requirements (Ashok Kumar et al., 2015). A total of 2267 core germplasm accessions were screened and promising donors were identified for use in breeding programs (Ashok Kumar et al., 2009). Significant positive association between grain Fe and Zn concentration was observed. However no significant association between grain Fe and Zn contents and agronomic traits were observed (Reddy PS et al., 2010). In sorghum, the grain Fe and Zn are quantitatively inherited traits and show continuous variation. While grain Zn is predominantly controlled by additive gene action both additive and non-additive gene action play a major role in controlling grain Fe concentration (Ashok Kumar et al., 2013a). This mean one need to improve both the parents for Fe and Zn so as to develop high Fe and Zn hybrids.

A total of 66 commercial sorghum cultivars developed by public- and private-sector partner organizations in India were assessed for grain Fe and Zn contents and promising cultivars identified; some of them are in leading hybrids (Ashok Kumar et al., 2010). Recently, promising parents with high Fe and Zn concentration were identified and new hybrids were developed using them (Ashok Kumar et al., 2015). A biofortified variety ICSR 14001 completed the multilocation testing and currently under on-farm testing for commercialization. Two new hybrids ICSA 29007 × ICSR 89058 and ICSA 101 × ICSR 196 were found promising over two years of testing in three locations.

Forage Sorghum

Sorghum as forage is popular in Americas, Europe, China, Australia and India. Forage sorghums, by definition, include annual grain sorghum (*Sorghum bicolor* L. Moench) and sudangrass (*Sorghum bicolor* × *Sorghum sudanense* (Piper) Stapf.) hybrids grown for green fodder or forage. *Sorghum bicolor* is primarily cultivated for for food and animal feed and the dry fodder and/or stover is used as forage. The other species of sorghum grown for annual forage is *Sorghum sudanense* (Piper) Stapf. Forage sorghum has an advantage over maize as it requires less water and inputs, is drought tolerant, is multi-cut due to high regeneration ability and is well adapted to range of soil and climatic conditions. Under good management, forage sorghum makes excellent hay for supplemental feeding during times of inadequate forage production. One of the greatest advantages of sorghum is that farmers can adopt various management options depending on the growing conditions to maximize its productivity.

Hybrids are popular cultivar option in forage sorghum. Brown midrib types with high forage yield are preferred to increase the digestibility (Oliver et al., 2004). Some of the constituents that affect palatability, acceptability and animal performance include protein and lignin content, lignin type and chemistry, mineral content, plant morphology, antinutritional components such as HCN, anatomical components and forage digestibility (Hanna 1993). Earlier studies indicated that green forage sorghum is more succulent and has higher dry matter content than maize with other chemical constituents being comparable. The crude protein in sorghum stover is more than that of wheat, rice and pearl millet straw (Singh et al., 1974; Ashok Kumar et al., 2011). Lignin concentrations in brown-midrib (bmr) mutants are consistently lower than their normal counterparts in sorghum (Cherney et al., 1988; Astigarraga et al., 2014). In vitro digestibility and the tendency for improved animal performance of bmr genotypes has been reported to be higher than normal genotypes. Most digestible, slowly and partially digestible tissues in sorghum leaves are degraded by fiber-digesting bacteria to a greater extent in bmr mutants than in normal genotypes. Geneticists are now attempting to incorporate the bmr trait into a range of backgrounds in sorghum (Cherney et al., 1991). Recently, effects of the six new bmr mutants on enzymatic saccharification of lignocellulosic materials were determined, but the amount of glucose released from the stover was similar to wild-type in all cases. Like bmr2, bmr6, and bmr12 which are widely used, these mutants may affect monolignol biosynthesis and may be useful for bioenergy and forage improvement when stacked together or in combination with the three previously described bmr alleles (Sattler et al., 2014; Umakanth et al., 2014).

High yielding Intraspecific grain hybrids and varieties provide grain and stover during rainy season in most parts of the world. Multi-cut forage sorghum hybrids are grown during spring/summer. Unlike grain sorghum, sudangrass inter-specific hybrids tiller profusely, produce succulent stems, high leaf to stem ratio, regrow quickly, withstand multi-cuts, are low in hydrocyanic acid (HCN), tannins. Single and three-way inter-specific hybrids have been developed by public and private sectors. However, three-way cross hybrids are predominantly cultivated, as private seed industry largely supplies the hybrid seed. Red-grained three-way hybrids are cultivated, though there is no difference between white- and red-grained forage hybrids. There is a need to dispel the farmers' perceptions with regards to grain color as it limits the exploitation of the rich genetic potential of sorghum. However the companies are now switching to single-cross hybrids with the ease in seed production. The A3 cytoplasm can be used to diversify the cytoplasm base than currently used A1 cytoplasm only for hybrids development (Aruna et al., 2013).

The forage quality improvement revolves around increasing feed intake and digestibility, and reducing anti-quality attributes (Smith et al., 1997). In vitro dry matter digestibility (IVDMD), a measure of digestibility, is under genetic control and is correlated with crude protein (CP), Neutral Detergent Fiber (NDF) and

Acid Detergent Fiber (ADF) and water soluble carbohydrates (WSC) (Snyman and Henda, 1996; Marsalis et al., 2010); Corral-Luna et al., 2013). It is important to incorporate insect resistance and leaf disease resistance in the parents keeping in view the production constraints in crop growing areas. Similarly the seed farmers should aim for higher yields in forage hybrid seed production plots so that large-scale commercialization is feasible. Promising forage sorghum hybrid parents for high green fodder and stover yields and quality traits with high tillering ability and ratoonability were developed at ICRISAT (Ashok Kumar et al., 2011). One of the successful examples in recent times is CSH 24 MF, a popular multicut, high-yielding, insect resistant forage hybrid with high digestibility suitable for forage cultivation in India. The female parent ICSA 467 is stem borer resistant and high yielding line and the male parent Pant Chari 6 is a high tillering, fast regenerating cultivar with sweet and juicy stem. To meet the seed demand by farmers, ICAR-IIMR licensed seed production of this hybrid to 11 seed companies.

Sweet and High Biomass Sorghum

Sweet sorghum is a multipurpose crop that yields food, fodder and fuel (Reddy et al., 2008; Vermerris 2011). It is being used for syrup and ethanol production globally. Biomass sorghums are produced as a source of lignocellulosic biofuels. They can grow very tall due to photoperiod sensitivity under long day conditions; in this case the plants continue to grow vegetatively until harvest (Rooney et al., 2007). Breeding for sweet sorghum is similar to dual-purpose sorghum but with additional traits like green stalk yield, soluble solids concentration (0Bx) and juice volume should be recorded so as to compute the estimated ethanol and/or sugar yield for a given genotype. Open pollinated varieties were developed initially in many countries for the traits improvement. Keller, Recova, sugar grace, SSV 74, SSV 84, M81-E are among popular sweet sorghum cultivars. Hybrids are the cultivar options in sweet sorghum, as hybrids are high-yielding, early flowering and less photoperiod-sensitive compared to the varieties (Reddy et al., 2008). Hybrids are amenable for planting across various planting dates to meet the feedstock requirements of distilleries for ethanol production (Rao et al., 2013). Continuous availability of sweet sorghum, transport and storing much mass and minimizing the postharvest loss of fermentable sugars are fundamental to exploiting sweet sorghum as a bioenergy crop (Anami et al., 2015). QTLs have been identified for architectural traits contributing to bioenergy in sorghum that includes roots, leaves, plant height, flowering time, stem sugar content, and bioconversion efficiency haven identified and have been put to use (Vermerris et al., 2007; Srinivas et al., 2009; Mace et al., 2012; Vandenbrink et al., 2013; Nagaraja Reddy et al., 2013; Li et al., 2014; Anami et al., 2015)

A number of sweet sorghum hybrids have been developed based on improved parental lines were commercialized. CSH 22 SS, one of the popular hybrids for ethanol production, was the first sweet sorghum hybrid released in India using the ICRISAT–bred female parent ICSA 38. Strategic research indicated that ethanol production in India using sweet sorghum is cost-effective and its cultivation gives 23% additional income to farmers compared to the grain sorghum (Reddy et al. 2008). There are minimal food–fuel tradeoffs in sweet sorghum but season-specificity exists (Murray et al., 2008a). Earlier work showed that a combination of centralized and decentralized models for supply chain management is critical for commercial ethanol production (Ashok Kumar et al., 2010b). While sweet sorghum ethanol production is technically feasible, commercial viability depends on policy support it receives from various countries (Ashok Kumar et al., 2013b; Basavaraj et al., 2015). Sweet sorghum, when fed directly as forage, was found to have high daily intake and higher digestibility in large ruminants (cows and buffalo (Blümmel et al., 2009). No significant differences were observed in the intake or body weight of animals when bagasse and stripped leaves feed blocks were used to feed the ruminants, indicating that sweet sorghum bagasse (after extraction of juice) can be used as animal feed.

Energy sorghum is a genetic model for the design of C4 grass bioenergy crops. High biomass yield, efficient nitrogen recycling, and preferential accumulation of stem biomass with low nitrogen content contributes to energy sorghum's elevated nitrogen use efficiency. Sorghum's integrated genetics–genomics–breeding platform, diverse germplasm, and the opportunity for annual testing of new genetic designs in controlled environments and in multiple field locations is aiding fundamental discovery and accelerating the improvement of biomass yield and optimization of composition for biofuels production. Recent advances in wide hybridization of sorghum and other C4 grasses could allow the deployment of improved genetic designs of annual energy sorghum in the form of wide-hybrid perennial crops. The current trajectory of energy Sorghum genetic improvement indicates that it will be possible to sustainably produce biofuels from C4 grass bioenergy crops that are cost competitive with petroleum-based transportation fuels (Mullet et al., 2014).

Rice–Fallow Sorghum

Rice–fallow sorghum as a niche ecology emerged in last five years, but is spreading rapidly in India, particularly in the coastal states like Andhra Pradesh (AP). In the coastal belt, rice is the major crop in rainy (kharif) season. After rice harvest, farmers used to sow pulses in the last week of December under zero-tillage to avail the residual soil moisture. However, delayed rice sowing and increased viral disease incidence on the pulses made the system vulnerable and uneconomical in the recent past. Therefore in the last 15 yr, this system was replaced by rice–maize cropping, and the farmers could harvest nearly 5 tons of maize per hectare. In recent years, as a consequence of climate change, water scarcity, the rice–maize system became uneconomical. Because of this, farmers have been looking for alternative crops to replace maize, and found sorghum to be a highly suitable crop for this system for increasing their productivity with grain yields in farmers' fields reaching up to 5.7 t ha^{-1} (Mishra et al., 2011).

Private sector seed companies are active in supplying the hybrids adapted to this ecology. Major breeding targets for this region include early maturity, medium height with long panicles, juicy stems with high ^{0}Bx, and high digestibility. Optimal utilization of inputs is critical to maximize the profits (Chapke et al. 2011a, 2011b). One of the popular hybrids, CSH-16, developed for rainy season adaptation is performing well in these ecologies and increasing farmers incomes.

The Way Forward

Global sorghum productivity has shown substantial improvement in last 60 yr and a large proportion of this can be directly attributed to adoption of hybrid cultivars. Hybrids not only increase productivity but stabilize production and bring higher returns to growers through new market opportunities. Sorghum hybrids have helped the seed industry to firmly establish themselves and strengthen the seed chain. However OPVs are still reigning in more than 50% sorghum area, particularly in Sub-Saharan Africa. There is an urgent need to develop adapted hybrids and institutions to deliver the improved hybrids to these farmers. Declining area, limited export opportunities, and limited funding to research are some of the major challenges to sorghum researchers. However, its high productivity and resilience makes sorghum one of the best crops for bioenergy. More efforts are needed in this direction. The dairy and meat sectors are growing at high rate in fast developing economies like Ethiopia, China, India, Uzbekistan, Laos and Philippines. There is need to develop highly digestible forage sorghums to capture the feed markets in these countries. Genetic and cytoplasmic diversification of trait based parents development should be given high priority in sorghum improvement research. It is important to complement crop improvement with genomic tools to increase the breeding efficiency for enhanced genetic gains. More investment is needed for sorghum research and in disseminating the health and environmental benefits associated with its utilization.

References

Anami, S.E., L. Zhang, Y. Xia, Y. Zhang, Z. Liu, and H. Jing. 2015. Sweet sorghum ideotypes: Genetic improvement of the biofuel syndrome. Food and Energy Security. 4(3):159–177. doi:10.1002/fes3.63

Aruna, C., P.K. Shrotria, S.K. Pahuja, A.V. Umakanth, B. Venkatesh Bhat, A. Vishala Devender, and J.V. Patil. 2013. Fodder yield and quality in forage sorghum: Scope for improvement through diverse male sterile cytoplasms. Crop Pasture Sci. 63(11-12):1114–1123.

Aruna, C., and S. Audilakshmi. 2008. Breeding methods in sorghum. In: B.V.S. Reddy, S. Ramesh, A. Ashok Kumar, C.L.L Gowda, editors, Sorghum improvement in the new millennium. Patancheru 502 324. International Crops Research Institute for the Semi-Arid Tropics. Andhra Pradesh, India. p. 28-30.

Ashok Kumar, A., K. Anuradha, B. Ramaiah, H. Frederick, W. Rattunde, P. Virk, W.H. Pfeiffer and S. Grando. 2015. Recent advances in sorghum biofortification research. Plant Breed. Rev. 39:89–124.

Ashok Kumar, A., B.V.S. Reddy, M. Blümmel, S. Anandan, Y.R. Reddy, C.R. Reddy, P.S. Rao, P.S. Reddy, and B. Ramaiah. 2010b. On-farm evaluation of elite sweet sorghum genotypes for grain and stover yields and fodder quality. Anim. Nutr. Feed Technol. 10S:69–78.

Ashok Kumar, A., B.V.S. Reddy, B. Ramaiah, K.L. Sahrawat, and W.H. Pfeiffer. 2012. Genetic variability and character association for grain iron and zinc contents in sorghum germplasm accessions and commercial cultivars. Eur. J. Plant Sci. Biotech. 6(1):66–70.

Ashok Kumar, A., V.S. Belum Reddy, B. Ramaiah, K.L. Sahrawat, and W.H. Pfeiffer. 2013a. Gene effects and heterosis for grain iron and zinc concentration in sorghum [Sorghum bicolor (L.) Moench]. Field Crops Res. 146:86–95. doi:10.1016/j.fcr.2013.03.001

Ashok Kumar, A., B.V.S. Reddy, B. Ramaiah, P. Sanjana Reddy, K.L. Sahrawat, and H.D. Upadhyaya. 2009. Genetic variability and plant character association of grain Fe and Zn in selected core collections of sorghum germplasm and breeding lines. J. SAT Agric. Res. 7(1):1-4 http://oar.icrisat.org/1991/

Ashok Kumar, A., B.V.S. Reddy, K.L. Sahrawat, and B. Ramaiah. 2010a. Combating micronutrient malnutrition: Identification of commercial sorghum cultivars with high grain iron and zinc. J. SAT Agric. Res. 8(1):1-5. http://ejournal.icrisat.org/volume8/Sorghum_Millets/Combating_micronutrient.pdf (verified 14 Sept. 2017).

Ashok Kumar, A., B.V.S. Reddy, B. Ramaiah, and R. Sharma. 2011b. Heterosis in white-grained grain mold resistant sorghum hybrids. J. SAT Agric. Res. (9). http://ejournal.icrisat.org/Volume9/Sorghum_Millets/Heterosis.pdf (verified 14 Sept. 2017).

Ashok Kumar, A., B.V.S. Reddy, H.C. Sharma, C.T. Hash, P. Srinivasa Rao, and B. Ramaiah. 2011a. Recent advances in sorghum genetic enhancement research at ICRISAT. Am. J. Plant Sci. 2:589–600. doi:10.4236/ajps.2011.24070

Ashok Kumar, A., B.V.S. Reddy, H.C. Sharma, and B. Ramaiah. 2008a. Shoot fly (Atherigona soccata) resistance in improved grain sorghum hybrids. J. SAT Agric. Res. (6). http://ejournal.icrisat.org/Volume6/Sorgum_Millet/Ashok_Shootfly.pdf (verified 14 Sept. 2017).

Ashok Kumar, A., B.V.S. Reddy, R.P. Thakur, and B. Ramaiah. 2008b. Improved sorghum hybrids with grain mold resistance. J. SAT Agric. Res. 6(1):1–4. http://oar.icrisat.org/2695/1/improvedsorghum6(1).pdf (verified 14 Sept. 2017).

Ashok Kumar, A., S. Gorthy, H.C. Sharma, Y. Huang, R. Sharma, and B.V.S. Reddy. 2014. Understanding genetic control of biotic stress resistance in sorghum for applied breeding. In: Y. Want, H.D. Upadhyaya and C. Kole, editors, Genetics, genomics and breeding of sorghum. CRC Press, Taylor & Francis Group. p. 198-225.

Ashok Kumar, A., C. Ravinder Reddy, J.V. Patil, and V.S. Belum Reddy. 2013b. Sweet sorghum for ethanol: A new beginning. In: B.V.S. Reddy, A. Ashok Kumar, C.R. Reddy, P. Parthasarathy Rao and J.V. Patil, editors, Developing a sweet sorghum ethanol value chain. Patancheru 502 324. International Crops Research Institute for the Semi-Arid Tropics. Andhra Pradesh, India. p. 1-22.

Ashok Kumar, A., H.C. Sharma, R. Sharma, M. Blummel, P.S. Reddy, and B.V.S. Reddy. 2013c. Phenotyping in sorghum [Sorghum bicolor (L.) Moench]. In: S.K. Panguluri and A. Ashok Kumar, editors, Phenotyping for plant breeding. Springer New York. p. 73-109.

Astigarraga, L., A. Bianco, R. Mello, and D. Montedónico. 2014. Comparison of brown midrib sorghum with conventional sorghum forage for grazing dairy cows. Am. J. Plant Sci. 5:955–962. doi:10.4236/ajps.2014.57108

Bandyopadhyay, R., L.K. Mughogho, and K.E.P. Rao. 1988. Sources of resistance to sorghum grain molds. Plant Dis. 72:504–508. doi:10.1094/PD-72-0504

Basavaraj, G., P. Parthasarathy Rao, C. Ravinder Reddy, A. Ashok Kumar, D.S. Mazumdar, Y. Ramana Reedy, P. Srinivasa Rao, S.M. Karuppan Chetty, and B.V.S. Reddy. 2015. Sweet sorghum: A smart crop to meet the demands of food, fodder, fuel and feed. (In: R.D. Christy, C.A. da Silva, M. Mhlanga, E. Mabaya and K. Tihanyi, editors, Innovative institutions, public policies and private strategies for agro-enterprise development. Food and Agriculture Organization and World Scientific Publishing Co. Rome, Italy. p. 169-187.

Blümmel, M., E. Zerbini, B.V.S. Reddy, C.T. Hash, F. Bidinger, and A.A. Khan. 2003. Improving the production and utilization of sorghum and pearl millet as livestock feed: Progress towards dual-purpose genotypes. Field Crops Res. 84:143–158. doi:10.1016/S0378-4290(03)00146-1

Blümmel, M., S.S. Rao, S. Palaniswami, L. Shah, and B.V.S. Reddy. 2009. Evaluation of sweet sorghum (Sorghum bicolor (L) Moench) used for bio-ethanol production in the context of optimizing whole plant utilization. Anim. Nutr. Feed Technol. 9:1–10.

Bouis, H.E., C. Hotz, B. McClafferty, J.V. Meenakshi, and W.H. Pfeiffer. 2011. Biofortification: A new tool to reduce micronutrient malnutrition. Food Nutr. Bull. 32(Supplement 1):S31–S40. doi:10.1177/15648265110321S105

Chapke, R.R., B. Mondal, and J.S. Mishra. 2011a. Resource-use Efficiency of Sorghum (Sorghum bicolor) Production in Rice (Oryza sativa)-fallows in Andhra Pradesh, India. J. Hum. Ecol. 34(2):87–90.

Chapke, R.R., S. Rakshit, J.S. Mishra, and J.V. Patil. 2011b. Factors associated with sorghum cultivation under rice-fallows. Indian Research Journal of Extension Education 11(3):67–71.

Cherney, J.H., D.J.R. Cherney, D.E. Akin, and J.D. Axtell. 1991. Potential of brown–midrib, low lignin mutants for improving forage quality. Adv. Agron. 46:157–198. doi:10.1016/S0065-2113(08)60580-5

Cherney, J.H., J.D. Axtell, M.M. Hassen, and K.S. Anliker. 1988. Forage quality characterization of a chemically induced brown–midrib mutant in pearl millet. Crop Sci. 28:783–787. doi:10.2135/cropsci1988.0011183X002800050012x

Conner, A.B., and R.E. Karper. 1927. Hybrid vigor in sorghum. Texas Experiment Stations Bulletin no. 359. Texas A&M University, College Station, TX.

Cooper, P.J.M., K.P.C. Rao, P. Singh, J. Dimes, P.S. Traore and K. Rao. 2009. Farming with current and future climate risk: Advancing a 'hypothesis of hope' for rain-fed agriculture in the semi-arid tropics. Journal of SAT Agricultural Research, 7. p. 1-19.

Corral-Luna, A., D. Dominguez-Diaz, M.R. Murphy, F.A. Rodriguez-Almeida, G.Villalobos, and J.A. Ortega-Gutierrez. 2013. Relationship between chemical composition, in vitro dry matter, neutral detergent fiber, digestibility and in vitro gas production of Corn and Sorghum silages. J. Anim. Vet. Adv. 12(20):1524–1529.

Cuevas, H.E., C. Zhou, H. Tang, P.P. Khadke, S. Das, Y.-R. Lin, Z. Ge, T. Clemente, H.D. Upadhyaya, and C. Thomas Hash. 2016. The evolution of photoperiod-insensitive flowering in sorghum, a genomic model for panicoid grasses. Mol. Biol. Evol. 33(9):2417–2428. doi:10.1093/molbev/msw120

Gressel, J., A. Hanafi, G. Head, W. Marasas, A.B. Obilana, J. Ochanda, T. Souissi, and G. Tzotzos. 2004. Major heretofore intractable biotic constraints to African food security that may be amenable to novel biotechnological solutions. Crop Prot. 23:661–689. doi:10.1016/j.cropro.2003.11.014

Hanna, W.W. 1993. Improving forage quality by breeding. In: D.A. Buxton, R. Shibles, R.A. Forsberg, et.al, editors, International Crop Science I. CSSA, Madison, WI. p. 671-675.

Upadhyaya, H.D., Y.-H. Wang, S. Sharma, and S. Singh. 2012. Association mapping of height and maturity across five environments using the sorghum mini core collection. Genome 55(6):471–479. doi:10.1139/g2012-034

Hariprasanna, K., and S. Rakshit. 2016. Economic importance of sorghum. In: S. Rakshit and Y.H. Wang, editors. The sorghum genome. Springer International Publishing AG, Cham, Switzerland. p. 1-25.

Hash, C.T., A.G. Bhasker Raj, S. Lindup, A. Sharma, C.R. Beniwal, R.T. Folkertsma, V. Mahalakshmi, E. Zerbini, and M. Blümmel. 2003. Opportunities for marker-assisted selection (MAS) to improve the feed quality of crop residues in pearl millet and sorghum. Field Crops Res. 84:79–88. doi:10.1016/S0378-4290(03)00142-4

Haussmann, B.I.G., H. Fred Rattunde, E. Weltzien-Rattunde, P.S.C. Traoré, K. vom Brocke, and H.K. Parzies. 2012. Breeding strategies for adaptation of pearl millet and sorghum to climate variability and change in West Africa. J. Agron. Crop Sci. 198(5):327–339. doi:10.1111/j.1439-037X.2012.00526.x

Hayes, C.M. and W.L. Rooney. 2014. Agronomic performance and heterosis of specialty grain sorghum hybrids with a black pericarp. Euphytica 196: 459 -466. doi:10.1007/s10681-013-1047-3

Higgins, R. 2013. Genetic dissection of sorghum height and maturity variation using sorghum converted lines and their exotic progenitors. M.S. Thesis. University of Illinois at Urbana-Champaign, Urbana, IL.

House, L.R. 1980. A guide to sorghum breeding. ICRISAT, Patancheru, India.

House, L.R., B.N. Verma, G. Ejeta, B.S. Rana, I. Kapran, A.B. Obilana. 1997. Developing countries breeding and potential of hybrid sorghum. In: Proceedings of the International Conference on Genetic Improvement of Sorghum and Pearl Millet, Lubbock, TX, 22–27 Sept. 1996. Collaborative Research Support Program on Sorghum and Pearl millet 97-5. ICRISAT, Patancheru, India. p. 84-96.

Kiranmayee, K.N.S.U., P.B.K. Kishor, C.T. Hash, S.P. Deshpande. 2016. Evaluation of QTLs for shoot fy (Atherigona soccata) resistance component traits of seedling leaf blade glossiness and trichome density on sorghum (Sorghum bicolor) chromosome SBI-10L. Trop. Plant Biol. 9(1):12–28. doi:10.1007/s12042-015-9157-9

Li, R., Y. Han, P. Lv, R. Du, and G. Liu. 2014. Molecular mapping of the brace root traits in sorghum (Sorghum bicolor L. Moench). Breed. Sci. 64:193–198. doi:10.1270/jsbbs.64.193

Little, C.R., R. Perumal, T. Tesso, L.K. Prom, G.N. Odvody, C.W. Magill. 2012. Sorghum pathology and biotechnology-A fungal disease perspective: Part I. Grain mold, head smut, and ergot. The European Journal of Plant Science and Biotechnology. p. 10-26.

Mace, E., V. Singh, E. Van Oosterom, G. Hammer, C. Hunt, and D. Jordan. 2012. QTL for nodal root angle in sorghum (Sorghum bicolor L. Moench) co-locate with QTL for traits associated with drought adaptation. Theor. Appl. Genet. 124:97–109. doi:10.1007/s00122-011-1690-9

Maiti, R. 1996. Sorghum science. Oxford & IBH Publishing Co. Pvt. Ltd. New Delhi, India. p. 352.

Marsalis, M.A., S.V. Angadi, and F.E. Contreras-Govea. 2010. Dry matter yield and nutritive value of corn, forage sorghum, and BMR forage sorghum at different plant populations and nitrogen rates. Field Crops Res. 116:52–57. doi:10.1016/j.fcr.2009.11.009

Meenakshi, J.V., N.L. Johnson, V.M. Manyong, H. Degroote, J. Javelosa, and D.R. Yanggen. 2010. How cost-effective is biofortification in combating micronutrient malnutrition? An ex- ante assessment. World Development Report 38:64–75. doi:10.1016/j.worlddev.2009.03.014

Mindaye, T.T., E.S. Mace, I.D. Godwin, and D.R. Jordan. 2016. Heterosis in locally adapted sorghum genotypes and potential of hybrids for increased productivity in contrasting environments in Ethiopia. The Crop Journal. 4(6):479–489 doi:10.1016/j.cj.2016.06.020

Mishra, J.S., B. Subbarayudu, R.R. Chapke, and N. Seetharama. 2011. Evaluation of sorghum (Sorghum bicolor) cultivars in rice (Oryza sativa)-fallows under zero-tillage. Indian J. Agric. Sci. 81(3):277–279.

Montes, J.M., V. Mirdita, K.V.S.V. Prasad, M. Blummel, B.S. Dhillon, and A.E. Melchinger. 2009. A new near infrared spectroscopy sample presentation unit for measuring feeding quality of maize stover. J. Near Infrared Spectrosc. 17:195–201. doi:10.1255/jnirs.841

Morris, G.P., P. Ramu, S.P. Deshpande, C.T. Hash, T. Shah, H.D. Upadhyaya, O. Riera-Lizarazu, P.J. Brown, C.B. Acharya, S.E. Mitchell, J. Harriman, J.C. Glaubitz, E.S. Buckler, and S. Kresovich. 2013. Population genomic and genome-wide association studies of agro-climatic traits in sorghum. Proc. Natl. Acad. Sci. USA 110(2):453–458. doi:10.1073/pnas.1215985110

Mullet, J., D. Morishige, R. McCormick, S. Truong, J. Hilley, B. McKinley, R. Anderson, S.N. Olson, and W. Rooney. 2014. Energy sorghum–a genetic model for the design of C4 grass bioenergy crops. J. Exp. Bot. 65(13):3479–3489. doi:10.1093/jxb/eru229

Murray, S.C., A. Sharma, W.L. Roone, P.E. Klein, J.E. Mullet, and S.E. Mitchell. 2008a. Genetic improvement of sorghum as a biofuel feedstock I: Quantitative loci for stem sugar and grain nonstructural carbohydrates. Crop Sci. 48:2165–2179. doi:10.2135/cropsci2008.01.0016

Nagy, Z., Z. Tuba, F. Zsoldus, and L. Erdei. 1995. CO_2 exchange and water retention responses of sorghum and maize during water and salt stress. J. Plant Physiol. 145:539–544. doi:10.1016/S0176-1617(11)81785-2

Nwanze, K.F. 1997. Screening for resistance to sorghum shoot fly. In: H.C. Sharma, F. Singh, and K.F. Nwanze, editors, Plant resistance to insects in sorghum (Patancheru 502 324). International Crops Research Institute for the Semi-Arid Tropics, Andhra Pradesh, India. p. 35–37.

Oliver, A.L., R.J. Grant, J.F. Pendersen, and J. O'Rear. 2004. Comparison of Brown Midrib-6 and -18 forage sorghum with conventional sorghum and corn silage in diets of lactating dairy cows. Faculty Papers and Publications in Animal Science. Paper 738. http://digitalcommons.unl.edu/animalscifacpub/738

Parthasarathy Rao, P., B.S. Birthal, V.S. Reddy Belum, K.N. Rai, and S. Ramesh. 2006. Diagnostics of sorghum and pearl millet grains-based nutrition in India. International Sorghum and Millets Newsletter 47:93–96.

Parthasarathy Rao, P., G. Basavaraj, W. Ahmad, and S. Bhagavatula. 2010. An analysis of availability and utilization of sorghum grain in India. J. SAT Agric. Res. 8:1–8.

Paterson, A.H., K.F. Schertz, Y.R. Lin, S.C. Liu, and Y.L. Chang. 1995. The weediness of wild plants—molecular analysis of genes influencing dispersal and persistence of johnsongrass, Sorghum halepense (l) pers. Proc. Natl. Acad. Sci. USA 92:6127–6131. doi:10.1073/pnas.92.13.6127

Paterson, A.H., J.E. Bowers, R. Bruggmann, I. Dubchak, J. Grimwood, and H. Gundlach. 2009. The Sorghum bicolor genome and the diversification of grasses. Nature 457:551–556. doi:10.1038/nature07723

Pfeiffer T. W, M. J. Bitzer, J. J. Toy and J. F. Pedersen. 2010. Heterosis in sweet sorghum and selection of a new sweet sorghum hybrid for use in syrup production in Appalachia. Crop Science. 50: 1788–1794.

Quinby, J.R. 1967. The maturity genes of sorghum. In: A.G. Norman, editor, Advances in agronomy. Vol. 19. Academic Press, New York. p. 267–305.

Quinby, J.R., and R.E. Karper. 1947. The effect of short photoperiod on sorghum varieties and first generation hybrids. J. Agric. Res. 75:295–300.

Thirumala Rao, V., P. Sanjana Reddy, R.P. Thakur, and B.V.S. Reddy. 2013. Physical kernel properties associated with grain mold resistance in sorghum (Sorghum bicolor (L.) Moench). Int. J. Plant Breed. Genet. 7:176–181. doi:10.3923/ijpbg.2013.176.181

Rao, S.S., J.V. Patil, P.V.V. Prasad, D.C.S. Reddy, J.S. Mishra, A.V. Umakanth, B.V.S. Reddy, and A.A. Kumar. 2013. Sweet sorghum planting effects on stalk yield and sugar quality in semi-arid tropical environment. Agron. J. 105:1458–1465. doi:10.2134/agronj2013.0156

Rattunde, H.F.W., E. Weltzien, B. Diallo, A.G. Diallo, M. Sidibe, A.O. Touré, A. Rathore, R.R. Das, W.L. Leiser and A. Touré. 2013. Yield of photoperiod-sensitive sorghum hybrids based on guinea-race germplasm under farmers' field conditions in Mali. Crop Science 53: 2454-2461. doi:10.2135/cropsci2013.03.0182

Reddy, B.V.S., H.C. Sharma, R.P. Thakur, S. Ramesh, F. Rattunde, and M. Mgonja. 2006. Sorghum hybrid parents research at ICRISAT–Strategies, status, and impacts. SAT E-Journal. 2(1): 1-24.

Reddy, B.V.S., P. Sanjana Reddy, A.R. Sadananda, E. Dinakaran, A. Ashok Kumar, S.P. Deshpande, P. Srinivasa Rao, H.C. Sharma, R. Sharma, L. Krishnamurthy, and J.V. Patil. 2012. Postrainy season sorghum: Constraints and breeding approaches. J. SAT Agric. Res. 10:1–12.

Reddy, P.S., J.V. Patil, S.V. Nirmal and S.R. Gadakh. 2012. Improving post-rainy season sorghum productivity in medium soils: Does ideotype breeding hold a clue? Curr. Sci. 102(6):904–908.

Reddy, P.S., B.V.S. Reddy, A.A. Kumar, S. Ramesh, K.L. Sahrawat, and P.V. Rao. 2010b. Association of grain Fe and Zn contents with agronomic traits in sorghum. Indian J. Plant. Genet. Resour. 23(3):280–284.

Reddy, B.V.S., S. Ramesh, P.S. Reddy, and A.A. Kumar. 2009a. Comparison of A1 and A2 cytoplasmic male-sterility for combining ability in sorghum (Sorghum bicolor (L.) Moench). Indian J. Genet. Plant Breed. 69(3):199–204.

Reddy, B.V.S., A.A. Kumar, and P.S. Reddy. 2010a. Recent advances in sorghum improvement research at ICRISAT. Kasetsart J. (Nat. Sci.). 44: 499-506.

Reddy, B.V.S., A.A. Kumar, S. Ramesh, and P.S. Reddy. 2011. Breeding sorghum for coping with climate change. In: S.S. Yadav, B. Redden, J.L. Hatfield and H. Lotze-Campen, editors, Crop adaptation to climate change. John Wiley & Sons, Inc., Ames, IA. 2011. p. 326-339.

Reddy, B.V.S., P. Rao, and U.K. Deb. 2004. Global sorghum genetic enhancement processes at ICRISAT. In: M.C.S. Bantilan, U.K. Deb, C.L.L. Gowda, B.V.S. Reddy, A.B. Obilana, and R.E. Evenson, editors, Sorghum genetic enhancement: Research process, dissemination and impacts. International Crops Research Institute for the Semi-Arid Tropics, Patancheru 502 324, Andhra Pradesh, India. p. 65–102.

Reddy, B.V.S., S. Ramesh, A.A. Kumar, S.P. Wani, R. Ortiz, H. Ceballos, and T.K. Sreedevi. 2008. Bio-fuel crops research for energy security and rural development in developing countries. BioEnergy Res. 1:248–258. doi:10.1007/s12155-008-9022-x

Reddy, B.V.S., S. Ramesh, and T. Longvah. 2005. Prospects of breeding for micronutrients and carotene-dense sorghums. International Sorghum and Millets Newsletter 46:10–14.

Reddy, B.V.S., S. Ramesh, P.S. Reddy, and A.A. Kumar. 2009b. Genetic enhancement for drought tolerance in sorghum. Plant Breed. Rev. 31:189–222.

Nagaraja Reddy, R., R. Madhusudhana, S.M. Mohan, D.V.N. Chakravarthi, S.P. Mehtre, N. Seetharama, and J.V. Patil. 2013. Mapping QTL for grain yield and other agronomic traits in post-rainy sorghum Sorghum bicolor (L.) Moench. Theor. Appl. Genet. 126:1921–1939. doi:10.1007/s00122-013-2107-8

Ringo, J., A. Onkware, M. Mgonja, S. Deshpande, A. Rathore, E. Mneney, and S. Gudu. 2015. Heterosis for yield and its components in sorghum (Sorghum bicolor L. Moench) hybrids in dry lands and sub-humid environments of East Africa. Aust. J. Crop Sci. 9(1):9–13.

Riyazaddin, M.D., P.B. Kavi Kishor, A. Ashok Kumar, B.V.S. Reddy, R.S. Munghate and H.C. Sharma. 2015. Mechanisms and diversity of resistance to sorghum shoot sly, Atherigona soccata. Plant Breed. 134(4):423–436. doi:10.1111/pbr.12276

Rooney, W.l. 2004. Sorghum Improvement– Integrating traditional and new technology to produce improved genotypes. Adv. Agron. 83:37–109. doi:10.1016/S0065-2113(04)83002-5

Rooney, W.L. 2014. Sorghum. In: D.L. Karlen, editor, Cellulosic energy cropping systems. John Wiley & Sons, Ltd, Chichester, UK. p. 109–129, doi:10.1002/9781118676332.ch7.

Rooney, W.L., J. Blumenthal, B. Bean, and J.E. Mullet. 2007. Designing sorghum as a dedicated bioenergy feedstock. Biofuels Bioprod. Biorefin. 1:147–157. doi:10.1002/bbb.15

Satish, K., G. Srinivas, R. Madhusudhana, P.G. Padmaja, R. Nagaraja Reddy, S. Murali Mohan, and N. Seetharama. 2009. [Sorghum bicolor (L.) Moench] Identification of quantitative trait loci for resistance to shoot fly in sorghum. Theor. Appl. Genet. 119:1425–1439. doi:10.1007/s00122-009-1145-8

Sattler, S.E., A. Saballos, Z. Xin, D.L. Funnell-Harris, W. Vermerris, and J.F. Pedersen. 2014. Characterization of novel sorghum brown midrib mutants from an EMS-mutagenized population. G3 Genes. Genomes. Genetics 4:2115–2124.

Seetharama, N., S. Singh and B.V.S Reddy. 1990. Strategies for improving rabi sorghum productivity. Proceedings of the Indian National Science Academy B56 (5&6): 455-467.

Sekoli, M.M.S., and M.E. Morojele. 2016. Sorghum productivity trends and growth rate for Lesotho. Global Journal of Agricultural Research 4(1):52–57.

Sharma, R., R.P. Thakur, S. Senthilvel, S. Nayak, S.V. Reddy, V.P. Rao, and R.K. Varshney. 2011. Identification and characterization of toxigenic Fusaria associated with sorghum grain mold complex in India. Mycopathologia 171:223–230. doi:10.1007/s11046-010-9354-x

Sharma, R., V.P. Rao, H.D. Upadhyaya, V.G. Reddy, and R.P. Thakur. 2010. Resistance to grain mold and downy mildew in a mini-core collection of sorghum germplasm. Plant Dis. 94:439–444. doi:10.1094/PDIS-94-4-0439

Singh, J., T.R. Raju, L.L. Relwani, A. Kumar, and A.K. Mehta. 1974. Forage yield and chemical composition of different strains of pearl millet. Indian J. Agric. Res. 8:179–184.

Smith, K.F., K.F.M. Reed, and J.Z. Foot. 1997. An assessment of the relative importance of specific traits for the genetic improvement of nutritive value in dairy pasture. Grass Forage Sci. 52:167–175. doi:10.1111/j.1365-2494.1997.tb02347.x

Snyman, L.D., and W.J. Henda. 1996. Effect of maturity stage and method of preservation on the yield and quality of forage sorghum. Anim. Feed Sci. Technol. 57(1–2):63–73. doi:10.1016/0377-8401(95)00846-2

Srinivas, G., K. Satish, R. Madhusudhana, R.N. Reddy, S.M. Mohan, and N. Seetharama. 2009. Identification of quantitative trait loci for agronomically important traits and their association with genic-microsatellite markers in sorghum. Theor. Appl. Genet. 118:1439–1454. doi:10.1007/s00122-009-0993-6

Stephens, J.C., and P.F. Holland. 1954. Cytoplasmic male sterility for hybrid sorghum seed production. Agron. J. 46:20–23. doi:10.2134/agronj1954.00021962004600010006x

Thakur, R.P., B.V.S. Reddy, and K. Mathur. 2007. Screening techniques for sorghum diseases. Information Bulletin No. 76, International Crops Research Institute for the Semi-Arid Tropics, Patancheru 502324, Andhra Pradesh, India.

Thakur, R.P., B.V.S. Reddy, S. Indira, V.P. Rao, S.S. Navi, X.B. Yang, and S. Ramesh. 2006, Sorghum grain mold. Information Bulletin No. 72. International Crops Research Institute for the Semi-Arid Tropics, Patancheru 502324, Andhra Pradesh, India.

Thakur, R.P., V.P. Rao, S.S. Navi, T.B. Garud, G.D. Agarkar, and B. Bhat. 2003. Sorghum grain mold: Variability in fungal complex. International Sorghum and Millet Newsletter 44:104–108.

Tuinstra, M.R. 2008. Food-grade sorghum varieties and production considerations: A review. J. Plant Interact. 3(1):69–72. doi:10.1080/17429140701722770

Umakanth, A.V., B.V. Bhat, M. Blummel, C. Aruna, N. Seetharama, and J.V. Patil. 2014. Yield and stover quality of brown mid-rib mutations in different genetic backgrounds of sorghum. Indian J. Anim. Sci. 84(2):181–185.

Vandenbrink, J.P., R.N. Hilten, K. Das, A.H. Paterson, and F. Alex Feltus. 2013. Quantitative models of hydrolysis conversion efficiency and biomass crystallinity index for plant breeding. Plant Breed. 132:252–258. doi:10.1111/pbr.12066

Vermerris, W. 2011. Survey of genomics approaches to improve bioenergy traits in maize, sorghum and sugarcane. J. Integr. Plant Biol. 53(2):105–119. doi:10.1111/j.1744-7909.2010.01020.x

Vermerris, W., A. Saballos, G. Ejeta, N.S. Mosier, M.R. Ladisch, and N.C. Carpita. 2007. Molecular breeding to enhance ethanol production from corn and sorghum stover. Crop Sci. 47:S-142–S-153. doi:10.2135/cropsci2007.04.0013IPBS

Sorghum Breeding from a Private Research Perspective

Hugo Zorrilla*

Abstract

For centuries sorghum has primarily been grown in the semiarid tropical and subtropical areas under poor management and highly variable environmental conditions. One of the main goals for the private sorghum breeder is to deliver stable, high-yielding hybrids to comply with the sorghum grower's demands. The hybrids have to provide profits to the growers as well as the seed providers. Sorghum breeders continuously work to improve the following traits: yield, maturity, dry down, seed quality and/or treatments, seed size, seed color, root and stalk quality, drought tolerance, heat tolerance, water use efficiency, cold (emergence) tolerance, seedling vigor, postfreeze lodging tolerance, disease tolerance, insect tolerance, and herbicide tolerance. This chapter focuses on grain sorghum breeding from the private research perspective. It will explore agronomic and genetic contributions to sorghum yield gains, the beginning of the sorghum hybrid industry, sorghum breeding sources, and sorghum grower expectations. The activities of a typical sorghum breeding station, breeding steps, and time required to deliver farmers' needs will change as private research switches from traditional sorghum breeding to programs based on modern molecular technologies (quantitative trait loci mapping, marker-assisted selection, and marker-assisted backcrossing). Thanks to the gene sequencing and publication of the sorghum genome (Paterson et al., 2009), new and improved sorghum products can be delivered to farmers faster (5–6 yr) by using the new technologies.

Sorghum [*Sorghum bicolor* (L.) Moench] is among the most drought and heat tolerant of cereal crops. The FAOSTAT's (2013) ranking of food and agricultural commodities production placed sorghum as the fifth major cereal crop in terms of production after maize (*Zea mays* L.), paddy rice (*Oriza sativa* L.), wheat (*Triticum aestivum* L.), and barley (*Hordeum vulgare* L.). For centuries sorghum has primarily been grown in the semiarid tropical and subtropical areas under poorly managed and highly variable environmental conditions. Major sorghum producers are the United States, Mexico, Argentina, and Australia (Monk et al., 2014).

There are several uses of a sorghum plant:

- Grain: Food (dietary and gluten-free) in Africa and India. Animal feed as energy and protein source in the United States, South America, and China. Industrial products (alcohol, beer, adhesives, etc.).

Abbreviations: ALS, Acetolactate synthase; MAS, marker-assisted selection; NSP, National Sorghum Producers; QTL, quantitative trait loci.

Hugo Zorrilla, Retired Sorghum Plant Breeder, 3444 Treesmill Dr., Manhattan, KS 66503. Received 22 July 2015. Accepted 27 July 2016. *Corresponding author(hzorrilla@cox.net).

doi:10.2134/agronmonogr58.2014.0060

 Sorghum: State of the Art and Future Perspectives, Ignacio Ciampitti and Vara Prasad, editors. Agronomy Monograph 58.

- Stalks: Fodder, building materials, and biofuels.
- Whole plant: Silage forage source and/or animal feed as pasture, green-chop or hay.

Sorghum's drought and heat tolerance make it a good choice for humankind to be grown in this era of global warming, as the lack of rain (drought) is causing water control issues by governments due to lower water-reservoirs, declining soil water tables, and the desire to optimize water usage for human population and the agro-industry.

Sorghum Planted and Grain Harvested in the United States from 1981 to 2016

United States grain sorghum production data for the last 36 yr (1981 to 2016) of all sorghum planted vs. sorghum grain harvested is shown in Fig. 1 (USDA Crop Production Report April 2014 and January 2017). During those 36 yr, averages of 4.1 million hectares (10.1 million acres) were planted and 3.6 million hectares (8.8 million acres) of sorghum grain harvested to render an 87% harvest-to-planted ratio. In 1985, the United States passed major farm bill legislation (farm bill payments, insurance policies, and set aside acreages) that negatively impacted the sorghum seed industry since many of the acres used by sorghum were planted with corn, soybeans, and cotton because of commodity prices, etc. After 1985, sorghum was relegated to marginally productive land where drought and heat stress were prevalent (Monk et al., 2014). Prior to 1985, sorghum investments were high with a considerable number of public and private research programs. After 1985, public research support was minimized and private research programs were consolidated because of the changing economic environment (farm bill, new technologies, commodity prices, etc.). Private sorghum research investments are proportional to the sorghum area planted: the higher the area planted (market share) the larger the research investments. Between 1993 and 1995, sorghum acreage again declined as herbicide and insect-tolerant traits were added to corn, soybeans, and cotton through biotechnology. In the mid-1990s, the sorghum industry decided not to pursue the GMO Roundup Ready technology to minimize the potential impact of Johnson grass [*Sorghum halepense* (L.)] acquiring tolerance.

The lowest area planted occurred in 2010, where only 2.2 million hectares (5.4 million acres) were planted and only 2.0 million hectares (4.9 million acres) of sorghum grain harvested. The grain harvest-to-planted ratio for that year was 91%, and the average yields were 4522 kg ha^{-1} (72.0 bu ac^{-1}). Finally, in 2016, 2.7 million hectares (6.7 million acres) were planted and 2.5 million hectares (6.2 million acres) of grain harvested with a 93% harvest-to-planted ratio and 4892 kg ha^{-1} (77.9 bu ac^{-1}) as overall yield (USDA Crop Production Report April 2014 and January 2017).

Sorghum yield averages in the United States from 1981 to 2016 are shown in Fig. 2. The yields fluctuated from 3077 kg ha^{-1} (49 bu ac^{-1}) to 4892 kg ha^{-1} (77.9 bu ac^{-1}) with an overall average of 4000 kg ha^{-1} (64 bu ac^{-1}) covering 36 yr (USDA, 2014, 2017).

High commodity prices were responsible for the above yield fluctuations, as well as new technologies for corn, soybeans, and cotton, which relegated sorghum

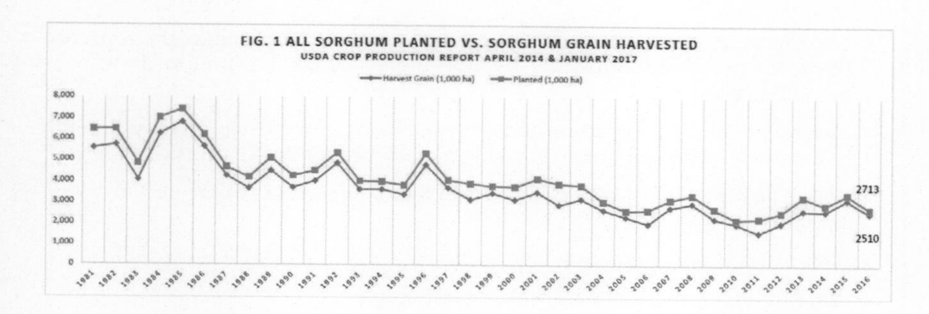

Fig. 1. All sorghum planted vs. sorghum grain harvested (USDA, 2017).

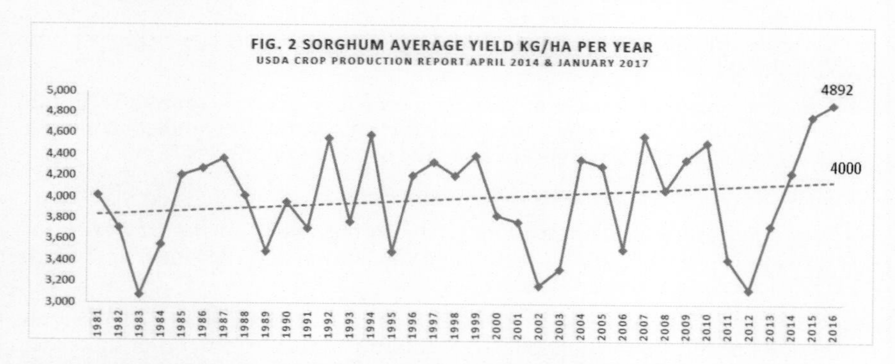

Fig. 2. Sorghum average yield per acre (bu) per year (USDA, 2017).

to poor marginally productive land where drought, heat stress, diseases, insects, and poor management are prevalent.

Sorghum Yield Gains

Sorghum yield gain is a dynamic process due to plant breeding (genetics) and diverse agronomic practices. Their specific contributions are difficult to assess because the majority of the harvested data comes from a mixture of poor environmental growing conditions (poor soils, low soils fertility, lack of rain, and lower management).

Several of the agronomic contributions to improving grain yield are listed below:

- Improved plant establishment (seed quality) due to fungicide, insecticide, and herbicide seed treatments that protect the germinating seed and seedlings.
- Optimum plant population for the right environment.
- The right hybrid placement to match the environment.
- Better weed control (pre-emergent herbicides).

- Minimum tillage that allows sorghum growers to manage the soil profile moisture more efficiently by optimizing plant population as well as fertilization.
- Larger, better, and faster equipment for planting, fertilizing, spraying, and harvesting with their corresponding electronics for data collection.
- Better and faster information management (communication and/or extension) for growers.
- Better water management where available.
- Good safety and environmental stewardship.

Besides, agronomic contributions, some of the genetic contributions to yield include the following:

- Hybrid vigor (Conner and Karper 1927), replacing older open-pollinated varieties and so increasing yield and agronomic performance (Karper and Quinby 1937).
- Use and implementation of different breeding sources (public sorghum breeders) into commercial programs providing variability and contributing to increase heterosis or hybrid vigor (Karper and Quinby 1947).
- Genetic improvements in tolerance to diseases.
- Improved drought tolerance due to the introgression of the stay-green or no senescence trait (pre-flowering and post-flowering) (Rosenow and Clark, 1995; Jordan and Monk, 1980).
- Miller and Kebede (1984) compared hybrids produced in the 1950s with hybrids produced in the 1980s. In their study, the newer hybrids showed a 48% yield increase with increases in seed numbers per panicle, plant weight and plant height, and leaf area at four different growth stages. These new hybrids showed a better stay-green response and retained the green leaf area until full maturity, while the old hybrids began to lose the green leaf area shortly after flowering. Assefa and Staggenborg (2010, 2011) indicated that 60 to 65% of yield gain was due to genetic improvement. Furthermore, they compared commercial hybrids released during a 50-yr span in a greenhouse study under differential water regimes and concluded that the total water usage was similar for all hybrids, but for the newer hybrids the total biomass produced was greater, indicating a net increase in water use efficiency with time.

The yield potential for sorghum is about 2.5 to 3.8 times the national average 4,000 kg ha^{-1} (64 bu ac^{-1}) when grown under good management conditions (such as rain-fed, limited to full irrigation, the right hybrid, the right placement, good quality seed/fungicide, insecticide and herbicide treatments, optimum plant population, and good management practices).

The National Sorghum Producers (NSP) Yield Contest provides sorghum growers the chance to compete every year at the national level for maximum yield under a variety of good management conditions. The National and State yield winners are recognized every year as the best sorghum growers in the United States by the NSP at an award banquet in conjunction with the Commodity Classic. The NSP, The United Sorghum Checkoff Program and The Private Sorghum Seed Companies provide incentives to the growers' who participate in the contest (fees, cash, traveling...). You can follow the NSP yield test results at https://www.sorghumgrowers.com/yield-contest (verified 18 May 2017).

The annual yield and management contests result sponsored by the NSP reported (state first place winners) for 2014, yields from 9,546 kg ha^{-1} (152 bu ac^{-1}) to a record 15,448 kg ha^{-1} (246 bu ac^{-1}). For 2015, yields from 10,802 kg ha^{-1} (172 bu ac^{-1}) to a record 15,072 kg ha^{-1} (238 bu ac^{-1}) and for 2016, yields from 11,053 kg ha^{-1} (176 bu ac^{-1}) to 13,062 kg ha^{-1} (208 bu ac^{-1}) under a variety of good management conditions.

There are tremendous challenges for the public and private sorghum breeders as well as the sorghum industry to close the above yield gaps.

The Beginning of the Sorghum Hybrid Industry

Sorghum, a C4 plant, is a self-pollinated crop (female and male are in the same spikelet) and it is classified as diploid ($2n = 20$) with less than 15% of outcrossing (Doggett, 1988).

Sorghum hybrids and hybrid seed production are possible thanks to the discovery by Stephens and Holland in 1954 of the male sterility resulting from the interaction of cytoplasmic (Milo) and nuclear genes (Kafir) named "cytoplasmic-genetic male sterility" (A_1 cytoplasm). This was a historical event:"The Beginning of the Hybrid Sorghum Industry".

Production of Hybrid Sorghum Seed

I. Parental seed:

- A line = male sterile Female
- B line = male fertile female, maintainer
- A × B = A line/male sterile, increasing the A line
- R line = male fertile, restorer
- A × R = sorghum hybrid, fertile
- The B/female and R/male lines are recognized as the two distinctive breeding patterns for sorghum based on their fertility restoration or lack of.

II. Adequate wind for pollinations to occur.

III. Blooming–nick time for male and female.

IV. Isolations by space or time.

V. Good environmental conditions for harvest yield and seed quality.

History of Sorghum Hybrid Seed Development

Quinby in 1971 presented a paper "Hybrid Sorghum……A Triumph of Research" to the Pioneer Hi-Bred Int., Inc. Plant Breeding Division. He summarized the history of the hybrid sorghum seed development as follows:

- 1878-1880 – Milo and Kafir sorghum varieties were introduced into the United States.
- 1879-1890 – Sorghum growers selected short and temperately adapted varieties.
- 1925 – Heterosis or hybrid vigor in sorghum was identified.
- 1914 and 1932 – Producing sorghum hybrid seed was possible by a. Hand emasculation and pollination. b. Hot water emasculation and pollination.

- The 1930s – Sorghum growers' field selected short varieties from natural mutations which allowed combine harvest.
- 1929, 1935 and 1943 – Male sterility identified to produce sorghum seed. a. Genetic male-sterility. b. Other genetic-male sterility sources.
- 1952 – Cytoplasmic-genetic male sterility was discovered in sorghum.
- 1954 – Technology was recognized to produce hybrid sorghum seed.
- 1955 – Field sorghum hybrid seed production demonstrated by 200 growers.
- 1956 – Ten thousand acres of hybrid seed produced in Texas.
- 1957 – Hybrid seed covers 15% of the sorghum U. S. acreage.
- 1960 – Sorghum growers adopted the production technology for hybrid seed that led to the production of parental seed (A, B and R), and hybrid seed covers 95% of sorghum acreage.

He mentioned that the sorghum growers quickly adopted the hybrid production technology because of the previous and successful experiences from the hybrid corn growers. It took 4 yr to take over the U. S. hybrid sorghum seed acreage compared with 25 yr for hybrid corn seed to take over the U. S. corn acreage.

Quinby (1974) described and recognized 25 yr of sorghum research efforts by the United States Department of Agriculture and the Texas Agricultural Experimental Stations, starting with finding sorghum genetic male sterility and ending with the discovery of cytoplasmic-genetic male sterility by Stephens and Holland (1954). Quinby mentioned that the early discovery of cytoplasmic-genetic male sterility in onions by Jones and Clark in 1943 and their corresponding inheritance studies heavily influenced the search for cytoplasmic-genetic male sterility in sorghum.

Quinby also recognized the participation of the private research activities of the DeKalb Agricultural Association since 1949. Their first hybrid (using genetic male sterility) was sold in 1956. Cytoplasmic-genetic male sterility hybrids were sold by DeKalb in 1957. DeKalb's successes accelerated the release of parental sorghum lines and motivated other seed companies to start sorghum breeding programs.

Sorghum breeding sources

The introduction of sorghum breeding sources is the key factor for creating and maintaining diversity in breeding programs. There are plenty of sources available to breeders depending on their needs and objectives:

1. Sorghum World Collections (Plant Introduction, Indian Sorghum)
    - International Crop Research Institute for the Semi-Arid Tropics (ICRISAT)
    - The National Seed Storage Laboratory (NSSL), Fort Collins, CO
    - USDA–ARS Plant Genetic Resources Conservation Unit (PGRCU), Griffin, GA
2. University Sorghum Releases (A, B and R lines, breeding populations, tropical conversions)
    - Texas A&M University
    - Kansas State University

- University of Nebraska-Lincoln
- Purdue University
- Oklahoma State University

3. National and/or International Institutions with Sorghum Breeding Programs
4. Private Sorghum Breeding Programs
    - Exchange within private breeding programs (proprietary)
    - Seed alliances (use of parental lines and paying royalties)
5. Sorghum Foundation Seed Suppliers
    - Licensing parental lines and hybrids through catalogs

Sorghum Conversion Program

To take advantage of the tremendous genetic diversity (yield potential, maturity, heterotic patterns, cytoplasmic sources, drought and heat tolerance, water use efficiency, good stalks and roots, insect and disease tolerances, nutritional values, etc.) from the sorghum world collections (the majority of accessions are tall, late, and photoperiod sensitive, meaning they will not flower under temperate environments like that of the United States) a Tropical Conversion Program was established in 1963. These collaborative project included the USDA tropical location at Mayaguez, Puerto Rico, and the Texas A&M temperate locations at Chillicothe and Lubbock.

The breeding scheme will take the photoperiod-sensitive tall and late accessions and convert them to short early photoperiod insensitive by using backcrossing (takes 3–5 cycles) where the short early photoinsensitive type is the non-recurrent parent. The crossing and backcrossing are done at Mayaguez, Puerto Rico, and the short early segregates are selected at Lubbock, TX, and send back to Mayaguez for backcrossing and selfing (Stephens et al., 1967). Texas A&M releases like TX623 A&B, TX2752 A&B, and TX 430, to name a few, are examples of a successful conversion program. The breeding scheme was proven to be very useful, and public and private sorghum breeders started their own conversion programs in the 1970s.

Sorghum breeding sources (Introductions) from countries like Australia, Africa, and India cannot enter the United States directly without going through the USDA quarantine program in Saint Croix. These breeding sources are planted in Saint Croix and the USDA will certify and release the sources, making sure that they do not carry new diseases and/or insects to the United States.

Prior to 1995, exchanges of ideas as well as sorghum breeding sources were free (Freedom to Operate [FTO]) and it was required that the originator's idea be recognized as well as the seed provider. After 1995, these exchanges between the public and private sector (a key factor for sorghum improvement) were minimized because of the implementation of the Plant Variety Protection Act and Patents. Today, material exchanges between public and private sectors depend on the Material Transfer Agreement (MTA) contracts, followed by contracts that usually involve paying royalties to the creator of the source being used.

Sorghum Growers' Demands

One of the main goals for the private sorghum breeder is to deliver high-yielding (stable) hybrids to comply with sorghum growers' demands. The hybrids have to provide profits for the growers as well as the seed providers. The higher the sale volumes (market shares) for the hybrids, the higher the profits which positively cause an increase for sorghum research investments.

The agronomy, sales, and marketing groups will pass along the sorghum growers' needs to the sorghum research breeding teams so that they can focus and deliver the right sorghum products to the right customer environment.

To deliver hybrids that satisfy growers' demand, breeders continuously work to improve the following traits:

1. Yield, maturity choices, dry down, seed quality (seed treatments), seed size, seed colors (Red, White & Brown), roots and stalk quality [*Macrophomina phaseolina* (Tassi, Goid) and *Fusarium moliniforme* (Sheldon)], drought tolerance, heat tolerance, water use efficiency, cold (emergence) tolerance, seedling vigor, and post-freeze lodge tolerance.

2. Disease tolerances to downy mildew (*Peronosclerospora sorghi* (Weston and Uppal)], head smut [*Sporisorium reilianum* (Kuhn, Langdon and Fullerton)], gray leaf spot [*Cercospora sorghi* (Ellis and Everhart)], anthracnose [*Colletotrichum graminicola* (Cesati, Wilson)], leaf blights [*Exserohilum turcicum* (Leo and Sug.)], rust [*Puccinia purpurea* (Cooke)], sooty stripe [*Ramulispora sorghi* (Ellis and Everhart, Olive and Lefebvre)], bacterial stripe [*Pseudomonas andropogonis* (E. F Smith, Stapp)], bacterial streak [*Xanthomonas holcicola* (Elliot, Starr and Burkholder)], ergot [*Claviceps africana* (Freder., Mantle and De Milliano)], viruses and/or Maize Dwarf Mosaic, and Sugarcane Mosaic.

3. Insect tolerances to green bugs [*Schizaphis graminum* (Rondaniz)], chinch bugs [*Blissus leucopterus* (Say)], midge [*Contarinea sorghicola* (Coquillet)], and sugarcane aphids [*Melanaphis sacchari* (Zehntner)]. Some insecticide seed treatments will protect the sorghum plants from green bugs and chinch bugs up to 35 days after planting.

4. Herbicide tolerance to Acetolactate Synthase (ALS) inhibiting herbicides started in sorghum in 2007 with the release of 18R lines and 16B lines with genetic tolerance to ALS herbicides (post emergency weed control) by the Kansas State University Research Foundation–Kansas State University (Tuinstra and Al-Khatib, 2006). The ALS tolerance was identified in shattercane (*Sorghum bicolor* spp. *bicolor)* field populations. DuPont Crop Protection licensed the trait from KSURF–KSU to develop and register the herbicide and offered to the sorghum seed companies the chance to license the trait. Government approval, stewardship protocols, and the release of commercial tolerant hybrids are targeted by 2015 or 2016.

A Typical Private Sorghum Breeding Station

A typical sorghum breeding station is strategically located (areas of adaptation) where there is a potential sorghum market. To be able to deliver the customer sorghum demands, a research program needs the following:

1. Dependable annual budgets: breeding programs require a long commitment, which is usually inadequate compared with other major crops like corn,

soybeans, cotton, and wheat. This lack of ample resources forces sorghum breeders to manage their budgets in a more efficient and effective manner.

2. Have a good breeding strategy (pedigree system or other recurrent selection procedures) as well as good breeding materials diverse and variable enough to impact hybrid vigor or heterosis (elite inbreds, release inbreds, tropical conversion sources, world collections, good breeding populations, etc.). There are more than forty thousand individual accessions in the world collections (70–85% are photoperiod sensitive) that need evaluation so they can be classified as B or R sources, yield (heterosis) potential, drought and heat tolerance, water use efficiency, disease and insect tolerances, as well as nutritional values. Breeders need to be aware of the specific traits needed by sorghum growers. Breeding programs should be sorted and/or grouped by maturities (areas of adaptation) with the proper maturity checks.
3. Research personnel: 1 to 2 plant-molecular breeders, 1 to 2 research associates, 6 to 8 part-time employees and adequate administrative support. About 75% of the budget goes to pay employees.
4. Office space for personnel: seed preparation lab, cold room (good and precise inventory), buildings to store field equipment (tractor, planter, combine, trucks and trailers, tillage equipment, sprayer, etc.), drying facilities, and adequate work areas.
5. Data and communication equipment: office computers, electronic data collection devices for planting, harvesting, and field notes. A software package (field plot designs and statistical analysis) that will link all the yield test locations data over the years (3–5) to be able to make good hybrid decisions.
6. A home irrigated field nursery as well as dry land satellite nurseries to be able to screen breeding sources for root and stalk strength, and disease and insects prevalent at each area of adaptation. Make sure the station budgets for home and winter nurseries covers the use of creating quantitative trait loci (QTL) associated with breeding populations, the marker-assisted selection (MAS) needs for field sampling and laboratory DNA analysis, as well as the information management support.
7. Three to five uniform dry land yield test locations at farmers' fields to ensure good data collection, complemented by two to three observation locations to be able to screen hybrids for root and stalk strength, and diseases and insects prevalent at the area of adaptation. Yield test program should be sorted and/or grouped by maturities with the proper maturity and yield checks.
8. One to two winter nursery locations with greenhouse facilities (tropical environment like Puerto Rico, Hawaii, Mexico, or Chile) to advance one to three breeding cycles per year (accelerating genetic gains), as well as increasing hybrid seed for yield testing back home.
9. A research station should also develop proprietary manuals (plant breeding objectives) that include detailed protocols and/or instructions to use for planting time, flowering (pollination) time, harvest time, for home nurseries, winter nurseries, and yield test locations. Breeders have to be familiar with the parental seed and production seed protocols as well as safety and environmental stewardship protocols.

10. Station personnel have to be familiar with the annual phytosanitary field inspections as well as the site- specific sorghum quarantinable diseases (stem and bulb nematode, bacterial leaf, ergot, head smut, periconia root rot, viruses, etc.) to be able to send and/or receive seeds from winter locations as well as other international breeding projects.
11. Ten to fifteen percent of the station effort should be dedicated to trait introgression (improving elite sources), tropical conversion, population improvement, visiting public and private breeders to exchange and/or find new ideas, attending seminars, field days, visiting the sorghum sales–marketing demonstration plots, attending annual meetings, and visiting with agronomy, entomology, pathology, physiology, statisticians, biotechnology, and food scientists to improve and/or update personal knowledge that will end up benefiting the sorghum growers.

Breeding Steps

Public breeders may release inbreds with specific traits (disease, insect, quality, drought tolerance, herbicide tolerance, etc.) and limited combining ability information. Private breeders will move the above traits to proprietary elite sources as well as find their combining ability in hybrid combinations with the corresponding maturity and yield checks.

Between 1957 and 1995, a conventional breeding program required about 10 to 12 years to develop a hybrid and make it a commercial product. After 1995 (Table 1 and Table 2), some programs adopted various "Fast-Track" pedigree systems to reduce the amount of time used to develop parental lines. This system (budget and time driven) basically moved yield and specific traits for sorghum by using two to three selection cycles per year complemented with greenhouse-field screening at the tropical nursery locations, which allowed elite males (*R*) and elite females (A & B) to reach commercial status in 6 to 7 yr, respectively.

Making the initial crosses is the most important breeding step and requires good planning that depends on actual or previously collected data focusing on customer needs. During the selection process, the breeder (to comply with proprietary production standards) also will make sure that the female (A) is a good sterile line that produces enough seeds per acre (profitable) and the male (*R*) is a good restorer as well as a good pollinator.

The top cross stage is the beginning of yield testing at several locations to sort out the general (2–4 testers) combining ability of the breeding sources. There is a high cost associated with the YLD T3 to YLD T5 stages or years 4, 5, and 6 (see Tables 1 and 2), because of the field isolations needed by the parental seed group and the production group to be able to increase parental lines and produce hybrid seed for yield testing by research, sales, and marketing (commercial product).

Nowadays (since 2007) sorghum products can be delivered in 5 to 6 yr by using MAS, marker-assisted backcrossing, embryo rescue, greenhouse screenings and two to three selection cycles per year between home locations and tropical nursery locations. Sorghum breeders are taking advantage of the molecular breeding to create products with specific traits (insect and diseases continue to evolve) and move them faster to the marketplace.

Sorghum growers are slow to change compared with corn and soybean growers. Once they learn how to manage a sorghum product (this takes 3 yr) they

Table 1. Fast-track male breeding steps.†

Years	Dates	Station	Generations	Activities	Research, sales, marketing
1	Oct.	Puerto Rico	F0 making crosses (R × R)	plastic bag and/or emasculation	research
	Feb.	Puerto Rico	F1 ID crosses (trait or morphological markers)	Bulk 2 to 5 heads $cross^{-1}$	research
	June	Home	F2	grow F2s	research
2	Oct.	Puerto Rico	F3	grow F3s	research
	Feb.	Puerto Rico	F4 making TC/T1 seed		research
	June	home	F5 YLD TC/T1	3 loc; 2 to 3 reps loc^{-1}	research
3	June	home	F6 YLD T2 BB increase (10 to 20)	6+ locs; 2 to 3 reps loc^{-1}	research
4	June	home	YLD T3 PSG making hybrid seed PSG increasing inbred back-up BB (20)	6+ locs; 2 to 3 reps loc^{-1}	research PSG isolation PSG isolation
5	June	home	YLD T4 advance trials PG increasing hybrid seed	6+ locs; 2 to 3 reps loc^{-1} 20 to 30 locs	Research sales and/or marketing PG Isolation
6	June	home	YLD T5 commercial advance trials PG increasing hybrid seed	6+ locs; 2 to 3 reps loc^{-1} 20 to 30 locs	research sales and/or marketing PG isolation

† BB, breeder's bulk; PG, production group; PSG, parental seed group; TC, top cross; T1, testing stage; YLD, yield.

Table 2. Fast-track female breeding steps.†

Years	Dates	Station	Generations	Activities	Research, sales, marketing
1	Oct.	Puerto Rico	F0 making crosses (B × B)	plastic bag and/or emasculation	research
	Feb.	Puerto Rico	F1	bulk 2 to 5 heads per cross	research
	June	home	F2	grow F2s	research
2	Oct.	Puerto Rico	F3	grow F3s	research
	Feb.	Puerto Rico	F4 sterilizable (B × B) × A		research
	June	home	(F3) BC0		research
3	Oct.	Puerto Rico	(F4) BC1		research
	Feb.	Puerto Rico	(F5) BC2 make TC seed		research
	June	home	(F6) BC3 YLD TC/T1	3 locs; 2–3 reps loc^{-1}	research
4	June	home	YLD T2 BB increase (10 through 20# A & B)	6+ locs; 2 to 3 reps loc^{-1}	research
5	June	home	YLD T3 PSG making hybrid seed PSG increasing A & B back-up BB (20 A & B)	6+ locs; 2 to 3 reps loc^{-1}	research PSG isolation PSG Isolation
6	June	home	YLD T4 advanced trials PG increasing hybrid seed	6+ locs; 2 to 3 reps loc^{-1} 20 to 30 locs	research sales/marketing PG isolation
7	June	home	YLD T5 commercial advance trials PG increasing hybrid seed	6+ locs; 2 to 3 reps loc^{-1} 20 to 30 $locs^{-1}$	research sales/marketing PG isolation

† BB, breeder's bulk; BC0, back cross; PG, production group; PSG, parental seed group; TC, top cross; T1, testing stage; YLD, yield.

usually stay with it for a long time. The life cycle for a sorghum product is 8 to 10 yr. However, some sorghum products remain in the marketplace for 20 to 25 yr.

As new technologies arise and become available for sorghum, breeder's schemes get updated. The use of molecular tools (marker assistant selection, marker-assisted backcrossing, and gene sequencing), as well as improvement in phenotyping capacity, provide a tremendous advantage in increasing the response to selection, improving genetic gain, and speeding up the delivery of a high-yielding stable product to sorghum growers faster than previously accomplished.

Keys to successful private research

- Adequate financial support over years.
- The research focused on customer needs.
- Breeders with considerable hybrid experience and well trained over years.
- A very diverse germplasm base that includes proprietary and public breeding sources.
- Research centers located in potential market areas at national and international levels.
- Extensive home breeding nurseries and tropical nurseries to speed up product delivery.
- Extensive hybrid testing programs to match the corresponding customers' areas of adaptations.
- A strong relationship (communications) between research, agronomy, sales, and marketing to monitor customer needs as well as market changes (due to biotic and abiotic threats, environmental changes, as well as market share trends).
- Leveraging modern technologies (planting and harvesting equipment, field plots designs, field and greenhouse screenings, embryo culture, statistical analysis, MAS, MABC, DNA and gene sequencing analysis, yield monitoring, and stalk and root lodging) to improve sorghum breeding efficiencies.
- Providing and maintaining a very competitive array of high-yielding and stable products (specific traits) that will benefit (profits) the sorghum growers at their corresponding environmental conditions.

Breeders' challenges

- Continue to close the yield gaps between traditional and more favorable environmental conditions. Private breeders will continue to develop proprietary elite inbreds, test public released inbreds, and evaluate them in hybrid combinations.
- Leveraging new technologies from corn and other crops as the cost of molecular markers is getting cheaper with time. The double haploid (DH) technology for sorghum (finding inducers–mutations) will be a tremendous breeding tool. High-yielding (stable) hybrids with specific traits (adaptation areas) can be delivered to sorghum growers faster (3–5 yr).

- Continue to take advantage of the sorghum world collections to sort out traits (yield, heterotic patterns, new cytoplasm's, biotic and abiotic stress, etc.) to maximize sorghum improvements to benefit sorghum growers.
- Deliver high-yielding and stable products with wider areas of adaptation to reduce sorghum production costs worldwide (improving profitability).
- Improve knowledge to be able to balance traditional breeding (forward breeding) with the molecular breeding (new technologies). Make sure to "field verify" molecular breeding (genotyping) with traditional breeding (phenotyping). Support and promote public and private plant breeding education by providing theoretical and practical training.
- Continue to improve stalk, root lodging, drought and heat tolerance, disease and insect tolerance (sugarcane aphid challenge since 2013), as well as herbicide tolerance by maximizing the use of all the molecular technologies available.
- Continue to improve the sorghum grain nutritional value for the benefit of humankind (dietary source and gluten-free source).
- Promote sorghum as the food and feed for the future because of the global challenges (human population and agro-industry water usage).
- Take advantage of the monetary support available through the USDA, NSP, United Sorghum Checkoff Program, the Sorghum and Millet Innovation Lab, as well as the private seed industry by submitting proposals that will benefit and/or impact the sorghum growers at the national and/or international levels.
- Finally, to be able to deliver the right product to the right environment (sorghum grower needs) requires a tremendous work-group effort that involves public and private sorghum breeding, agronomy, entomology, phytopathology, physiology, biotechnology, statistics, food sciences, as well as sales and marketing.

References

Assefa, Y., and S.A. Staggenborg. 2010. Grain sorghum yield with hybrid advancement and changes in agronomic practices from 1957 through 2008. Agron. J. 102:703–706. doi:10.2134/agronj2009.0314

Assefa, Y., and S.A. Staggenborg. 2011. Phenotypic changes in grain sorghum over the last five decades. J. Agron. Crop Sci. 197:249–257. doi:10.1111/j.1439-037X.2010.00462.x

Conner, A.B., and R.E. Karper. 1927. Hybrid vigor in sorghum. Texas Agric. Exp. Sta. Bul. 359:1–23.

Doggett, H. 1988. Sorghum. 2nd ed. Tropical Agriculture Series. Longman Scientific & Technical, London.

FAOSTAT. 2013. Food and agricultural commodities production: Commodities by regions. http://faostat3.fao.org/browse/rankings/commodities_by_regions/E (accessed 8 June 2015) (verified 18 May 2017).

Jordan, W.R., and R.L. Monk. 1980. Enhancement of drought resistance of sorghum: Progress and limitations. In: Proceedings of the 35th Annual Corn and Sorghum Research Conference. Am. Seed Trade Assoc. Chicago, IL. p. 185–204.

Karper, R.E., and J.R. Quinby. 1937. Hybrid vigor in sorghum. J. Heredity. 28:83–91.

Karper, R.E., and J.R. Quinby. 1947. Sorghum-Its production, utilization and breeding. Econ. Bot. 1:355–371. doi:10.1007/BF02858895

Miller, F.R., and Y. Kebede. 1984. Genetic contributions to yield gains in sorghum, 1950 to 1980. In: W.R. Fehr, editor, Genetic contributions to yield gains in five major crop plants. CSSA Spec. Publ. 7. ASA and CSSA, Madison, WI.

Monk, R., C. Franks, and J. Dahlberg. 2014. Sorghum. In: S. Smith, B. Diers, J. Specht, and B. Carver, editors, Yield gains in major U. S. Field Crops. CSSA Spec. Publ. 33. CSSA, Madison, WI. p. 293–310. doi:10.2135/cssaspecpub33.c11

Paterson, A.H., J.E. Bowers, R. Bruggmann, I. Dubchak, J. Grimwood, H. Gundlach, G. Haberer, U. Hellsten, T. Mitros, A. Poliakov, J. Schmutz, M. Spannagl, H. Tang, X. Wang, T. Wicker, A.K. Bharti, J. Chapman, F.A. Feltus, U. Gowik, I.V. Grigoriev, E. Lyons, C.A. Maher, M. Martis, A. Narechania, R.P. Otillar, B.W. Penning, A.A. Salamov, Y. Wang, L. Zhang, N.C. Carpita, M. Freeling, A.R. Gingle, C.T. Hash, B. Keller, P. Klein, S. Kresovich, M.C. McCann, R. Ming, D.G. Peterson, M. Rahman, D. Ware, P. Westhoff, K.F.X. Mayer, J. Messing, and D. Rokhsaret. 2009. The Sorghum bicolor genome and the diversification of grasses. Nature 457:551–556. doi:10.1038/nature07723

Quinby, J.R. 1971. A triumph of research…Sorghum in Texas. Texas A&M Univ. Press, College Station, TX. p. 1–19.

Quinby, J.R. 1974. Origin of hybrid Sorghum. Sorghum improvement and genetic growth. Texas A&M Univ. Press, College Station, TX. p. 9–17.

Rosenow, D.T., and L.E. Clark. 1995. Drought and lodging resistance for a quality sorghum crop. In: Proceedings of the 50th Annual Corn and Sorghum Research Conference. Am. Seed Trade Assoc., Chicago, IL. p. 82–96.

Stephens, J.C., and R.F. Holland. 1954. Cytoplasmic male sterility for hybrid sorghum seed production. Agron. J. 46:20–23. doi:10.2134/agronj1954.00021962004600010006x

Stephens, J.C., F.R. Miller, and D.T. Rosenow. 1967. Conversion of alien sorghums to early combine genotypes. Crop Sci. 7:396. doi:10.2135/cropsci1967.0011183X000700040036x

Tuinstra, M.R., and K. Al-Khatib. 2006. Acetolactate synthase herbicide resistant sorghum. US Provisional Patent application 60/873,529. Date issued: 7 December.

United States Department of Agriculture (USDA). 2014. Crop Production Historical Track Records, April 2014. National Agricultural Statistics Service (NASS). Washington, D.C. http://usda.mannlib.cornell.edu/usda/nass/htrcp//2010s/2014/htrcp-04-11-2014.pdf (verified 18 May 2017).

United States Department of Agriculture (USDA). 2017. Crop Production Historical Track Records, April 2017. National Agricultural Statistics Service (NASS). Washington, D.C. http://usda.mannlib.cornell.edu/usda/current/htrcp/htrcp-04-13-2017.pdf (verified 18 May 2017).

Practical Morphology of Grain Sorghum and Implications for Crop Management

Calvin Trostle* and Gary Peterson

Morphology of sorghum plants is described in detail in numerous publications and reference works dating back 100 years (Artschwager, 1948; Miller, 1916). This discussion mostly emphasizes "field" morphology—morphological features of interest to a farmer, crop consultant, agronomist, or educator—and which may factor in sorghum crop management. These morphological traits can be observed without the aid of expensive equipment, a microscope, or chemical tests. Simple observations using a shovel, sharp knife or blade, or a small 10X handheld magnifying glass can reveal numerous practical aspects of sorghum morphology.

Some characteristics of sorghum morphology may dictate or respond to specific crop management practices (Cothren et al., 2000). These include tillering, rooting, and degree of grain fill. Other morphological traits are innate to the specific hybrid or germplasm line, including plant and grain color, glume traits, and panicle type. Specific sorghum morphological traits that may influence crop management are highlighted below as *'Morphology driven sorghum management considerations.'*

A crop physiologist, sorghum breeder or taxonomist would note that there are several diverse types of sorghums including durra, feterita, hegari, kafir, kaoliang, milo (not to be confused with the term milo referring to grain sorghum in general that is used in some regions), and shallu (Bennett et al., 1990). Due to the myriad of crosses that have occurred to develop sorghum breeding lines and hybrids, some morphological traits that might have distinguished these sorghum types from each other may not be reliable today.

Root

Basic sorghum root growth and development is reviewed in Assefa et al. (2014) and Singh et al. (2010). The first root of grain sorghum to emerge from the seed, the radicle, develops vertically into the soil and becomes the primary root (Fig. 1). Contrary to what others have written about grain sorghum or what may be commonly believed, there are no other seminal, or lateral, roots that emerge from

---

Abbreviations: GPD, growing point differentiation.

C. Trostle, Texas A&M AgriLife Extension Service; G. Peterson, Texas A&M AgriLife Research, Texas A&M University System, Lubbock, Texas 79403
*Corresponding author (ctrostle@ag.tamu.edu)

doi:10.2134/agronmonogr58.2014.0061

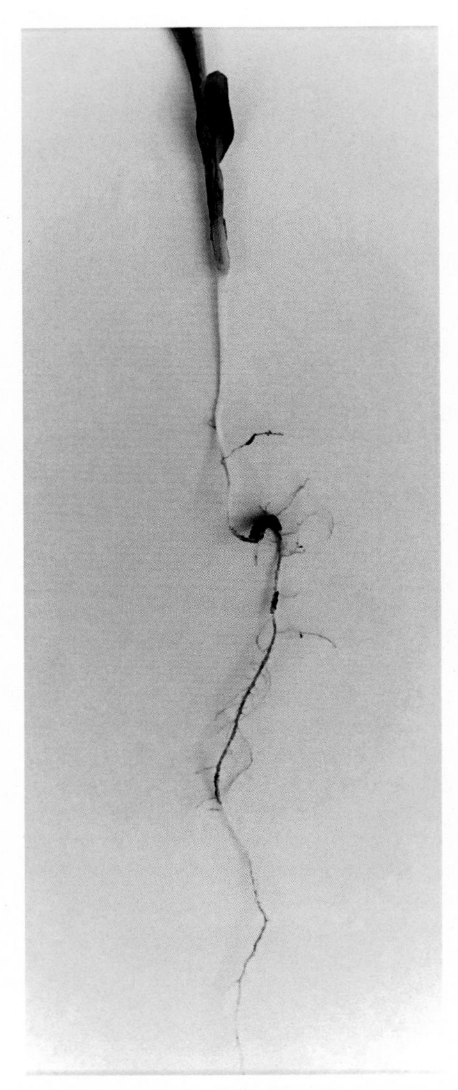

Fig. 1. Grain sorghum seedling at leaf stage two (V2) with single primary root which has developed from the seed radicle.

the sorghum seed (in contrast corn generally has three or more lateral, or seminal, roots from the seed (Assefa et al., 2014). There is some discrepancy among authors in how the term 'seminal root'—either the initial embryonic, or primary, root and any additional roots developing from the seed—is used. Some authors refer only to this first root from the seed in contrast to later roots (crown, nodal) that emerge from the base of the stem. Sorghum has only one seminal root, the primary root (Fig. 2A & 2B) (Blum et al., 1977). Other authors, however, speak of the seminal root *system* (Roozeboom and Prasad, 2018; this publication), which at a minimum implies there is more than one root. Again, sorghum *does not* develop a traditional seminal root from the seed (other than the primary root), and this discrepancy is found in both the literature and more general publications.

During this initial growth germinating and emerging grain sorghum seedlings rely on two structures to assist emergence. The first, the mesocotyl, is a root-like underground stem feature that grows upward from the seed (Fig. 3A&B). Mesocotyl length adjusts to seeding depth to place the base of the future crown below the soil line. The coleoptile is a tube-like plant structure growing up from the mesocotyl (Fig. 3A&B). The coleoptile accomplishes initial emergence at the soil surface where it stops growing once it encounters light. This generates hormonal feedback from the coleoptile to the mesocotyl whose growth then ceases (Vanderhoef and Briggs, 1978). The coleoptile, which is similar to a leaf (but no collar thus not a true leaf), enables the first leaf to emerge above the soil line.

Grain sorghum root morphology is further characterized by crown, or nodal, roots (Fig. 4) and brace roots. Beginning with landmark sorghum root studies by Miller (1916) sorghum is long noted for its fine, fibrous root system (Fig. 5). Crown roots emerge from the base of the developing stem, usually from the first four to six nodes (Fig. 6; Assefa et al., 2014). Crown roots provide the backbone to sorghum growth and development due to their extensive role in water and nutrient uptake.

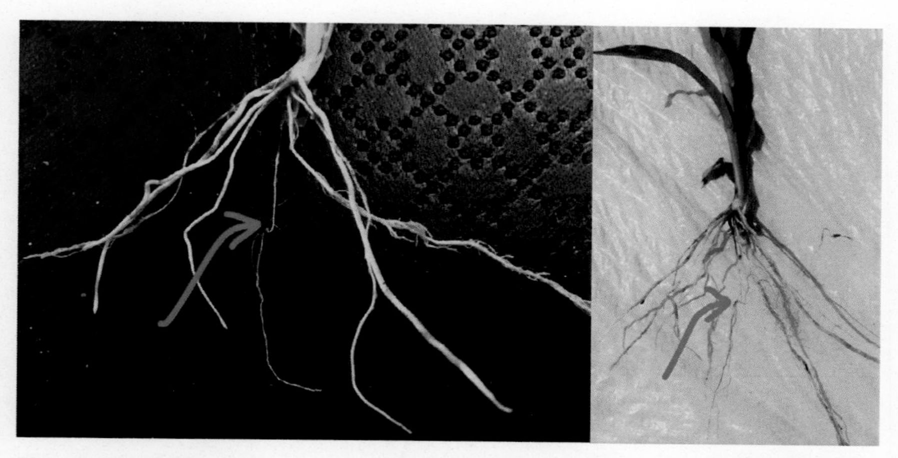

Figs. 2A&B. Grain sorghum roots at leaf stage five (V5). Several crown or nodal roots have developed but the single primary root (pointer) remains the sole root to originate from the seed (no additional seminal roots).

In contrast brace roots emerge from nodes usually above the soil line, commonly nodes seven and eight and possibly node nine (Fig. 7A&B). For some sorghum plants roots emerging from node six and perhaps even node five may appear and function more like brace roots as they have a thick tubular appearance (Fig. 6). Also, since these roots are exposed to light they may develop chlorophyll thus appear green. This might occur when seeds germinate at a shallow depth and the base of the crown is near the soil line or perhaps erosion has removed some soil from around the developing plant, potentially exposing lower nodes. The presence of a well-developed brace root system may have a substantial bearing on the success of a grain sorghum crop due to their essential contribution to standability. Sorghum growers should consider how planting practices and management of soil around the base of the sorghum stalk can impact sorghum production. This leads to our first consideration of sorghum morphology and sorghum management.

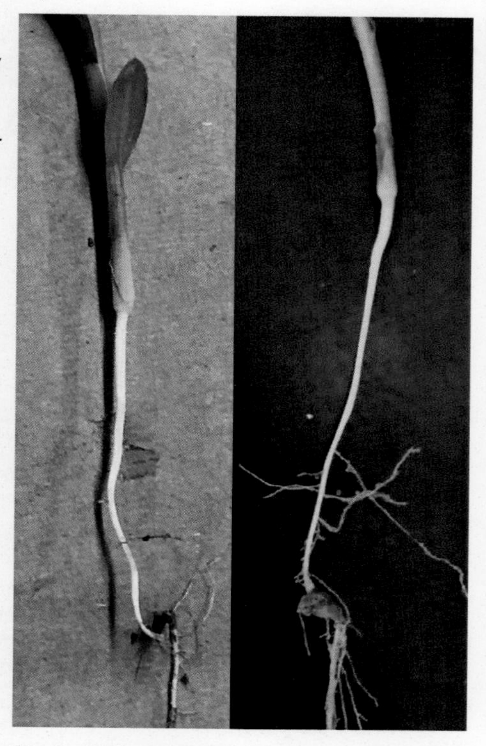

Fig. 3. Grain sorghum seedling mesocotyl and coleoptile which assist emergence of seedling from the soil. Leaf 1 (rounded tip) sheath extends to base of coleoptile.

Fig. 4. Development of nodal, or crown, root system in grain sorghum (nodes 1 to 4). Growth ranges from single primary root with initial crown roots (left, V3) to initiation of brace roots (right, V9).

Fig. 5. (left) Fibrous root system of grain sorghum.

Fig. 6. (right) Crown (nodal) root development from base of sorghum stalk (knife tip likely points to node 5). Due to possible shallow location of the crown, thick tubular roots from node 5 and even node 4 appear similar to brace roots.

Morphology-driven Sorghum Management Consideration #1

Sorghum planting pattern may affect later brace root development and standability pending sorghum planting in a furrow, flat, or on an elevated bed.

Row-crop sorghum is planted in a range of planting patterns from furrow planting to flat to planting on an elevated bed. Due to the increase of reduced tillage and no-till systems grain sorghum is more likely planted on flat ground than in the past.

In drier conditions it is more likely brace roots may have trouble penetrating dry or hard soil at the base of the plant. Some sorghum historically has been planted so that the soil line above the seed is at a depressed position relative to the surrounding soil. In some cases, in semiarid regions sorghum may even be planted 'in the hole,' in a furrow that may be 10–20 cm deep, which enhances potential access to moisture. In this situation (and also for flat-planted sorghum) where mechanical cultivation is still widely used, it is common for cultivation to throw soil around the base of plants about 40 cm or taller. This practice buried small weeds in the row and placed the lower nodes (up to node 9), which produce brace roots, in the soil (Fig. 8A&B). Sorghum managed in this fashion is less likely to root lodge producing better standability.

Fig. 7. Typical sorghum brace root development from slightly elongated nodes just above the soil line.

In contrast, western and southern areas of the U.S. grain sorghum belt where sorghum is grown in semiarid dryland conditions and wind erosion is a major concern, many row crop fields—especially cotton—are planted (in local terms, listed) on elevated 'beds' that leave a furrow of 15–25 cm deep (Fig. 9). At cotton planting the top of the bed is knocked off to plant the seed in moist soil that is warmer than soil temperatures in flat ground. This practice, however, when extended to grain sorghum is often detrimental to root development and sorghum standability (Trostle et al., 2010). When sorghum is planted in this elevated position (Fig. 10A&B) and brace roots emerge, these roots are compromised as the brace roots:

- Encounter soil that is both warmer and drier, and individual brace root penetration is not assured (Fig. 11, Fig. 12A&B). This condition limits brace

Figs. 8A&B. Soil cultivated around the base of sorghum plants buries brace roots (A); example of large brace roots that developed in the soil of an individual plant (B).

Fig. 9. In semi-arid regions of the U.S. sorghum belt sorghum and other crops are often planted and grown in an elevated position, which may curtail normal sorghum brace root anchorage.

Fig. 10. View of bed-planted sorghum preharvest (A) and furrow view postharvest (B) demonstrating the elevated position (top of the bed) that sorghum and other crops like cotton are sometimes planted in to prevent bare soil wind erosion. In this field, surface residues after harvest have been retained (10B, in contrast to Fig. 9) as the field has been relisted for the next crop (Texas).

roots' ability to penetrate the soil—roots will not move into or through dry soil (whereas soil in a furrow is likely cooler and more moist).

- If cultivation is attempted, it is more difficult to cultivate soil to the elevated base of the sorghum plants, because soil must be moved upward and remain there.

Morphology-driven Sorghum Management Consideration #2

Location of Sorghum Roots Near the Soil Surface and N Fertilizer Application

Use a shovel to dig a 20-cm deep cross-section of a plant's root profile crosswise to the direction of the planted row of sorghum. This will indicate where roots are present near the soil surface. This may become important when a farmer applies sidedress nitrogen fertilizer with a knife applicator or coulter rig (Nakayama and van Bavel, 1963). Farmers should ensure that only a minimal amount of the grain sorghum root system is severed. If needed, sidedress N applications must be made further from the planted row. Checking to see the location of the root system is also helpful as the same hybrid planted in the same soil type but in a different year may have a somewhat different rooting pattern, particularly in wet vs. dry years.

Root Maladies in Sorghum Due to Herbicide Damage

Sorghum root injury in the field is most likely due to in-season application of growth regulator herbicides dicamba and 2,4-D, which produce similar symptoms, or from higher concentrations of residual soil applied 'yellow' herbicides, particularly trifluralin (Smith and Scott, 2004; Frederiksen and Odvody, 2000).

For dicamba or 2,4-D herbicide sev

eral symptoms may involve roots, leaves, (curling, twining, or epinasty), and stalks (curvature, or goose necking). A possible root symptom will be a lack of crown root development (2,4-D, Fig. 13A; dicamba, Fig. 13B) or deformed roots (2, 4-D, Fig. 14, which also shows leaning in the stalk). Also, dicamba and 2,4-D may lead to symptoms of fusing several brace roots into one mass (Fig. 15), which then usually do not grow normally and will not penetrate the soil thus standability is compromised (Al-Khatib, 2017).

Fig. 11. Brace root elongation in sorghum on top of a listed ridge where due to drier, warmer soil, some brace roots (left) have extended elongation, not yet able to penetrate the soil.

Seedling growth inhibitors such as trifluralin, pendimethalin, or ethalfluralin may also lead to swollen or clubbed root tips in crown roots and brace roots, and these roots may cease to grow or have trouble penetrating the soil. These conditions most likely occur when previous herbicides have not degraded sufficiently over time to a safe concentration level. These herbicides may also cause these injury symptoms when applied post emergence onto exposed roots. Symptoms might also become pronounced due to dry conditions, overapplication of herbicide, or land preparation that

Fig. 12. Individual sorghum brace roots unable to penetrate soil due to soil compaction, hot and dry conditions drying and warming surface soil, or possible herbicide carryover.

Fig. 13. Root injury from 2,4-D with untreated control (right, A) and dicamba (B) led to little crown root development and thus lodged sorghum plants.

Fig. 14. (left) Sorghum root deformation and leaning due to 2,4-D injury. (Photo courtesy Brent Bean, United Sorghum Checkoff Program.)
Fig. 15. (right) Fusing of brace roots due to too-high 2,4-D exposure and subsequent injury. (Photo courtesy Curtis Thompson, Kansas State Univ.)

concentrates herbicide residues in the seed zone (e.g., a rolling cultivator use to prepared elevated beds for planting).

Most of the literature and pictorial examples of these condition in annual warm-season grains is for corn (*Zea mays* L.), which behaves similarly to sorghum. For resources in learning about sorghum symptomology consult with local and regional weed scientists with state or federal agencies, or with staff from chemical companies.

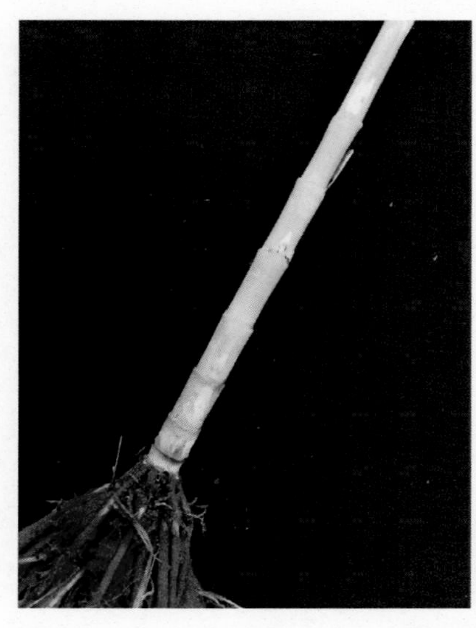

Fig. 16. Grain sorghum stalked stripped of leaves and leaf sheaths during heading to reveal internodes and their length.

Stalk and Initial Reproductive Development of Panicle

The sorghum stalk (culm) develops as early growth adds additional nodes. The first internode is visible between nodes four and five—this is internode four (5 to 6 mm in length) and is located below the soil line. Internodes up the stalk are progressively longer as the stalk elongates (Fig. 16). One leaf is attached to each node, in an alternating pattern on each side of the stalk, via a leaf sheath which enfolds the stalk. Above the topmost internode is the peduncle which supports the panicle. Rapid peduncle growth after boot stage is essential to ensure the peduncle elevates the panicle above the flag leaf sheath and canopy to facilitate harvest (Roozeboom and Prasad, 2018).

A key growth and development stage in grain sorghum is growing point differentiation (GPD; Fig. 7 of Roozeboom and Prasad, 2018), the point at which the growing point switches from producing another leaf to initiating and developing the panicle (Vanderlip, 1993). This development is hidden from view without splitting the stalk to find the growing point (early development visible, Fig. 17). This contrasts with small grains such as wheat, barley, rye, and oats that, like sorghum, are also members of the grass family *Poaceae* (formerly *Gramineae*). In sorghum there is nothing externally visible on the plant to indicate GPD is occurring. But in small grains, GPD can be closely tracked in a field by finding the first node in an individual stem (Miller, 1999), either visually or by feeling stem. (If nodes are found occasionally in a stem, then it is likely the rest of the stems across the field are at or just beginning GPD). In contrast, for grain sorghum we must rely on 1) calendar estimates and hybrid maturity for gauging of leaf number as an estimate of GPD, or 2) slice stems to find the growing point, which is not easy for many producers to identify small growing points with accuracy. Differentiation typically initiates about 30 to 40 d after germination, depending on environmental conditions (especially growth rates as related to the accumulation of growing degree days). It will be complete for an individual growing point in no more than 7 d (Vanderlip, 1993), then the developing panicle can be observed increasing in size as it moves up the stalk (Fig. 18A&B).

Fig. 17. Split sorghum stalk showing slight expansion of upper internodes and the location of the growing point, which has differentiated to early development of the panicle.

Figs. 18A&B. Subsequent progressive development of the sorghum panicle in the lower stalk (8 mm in 18A). Further panicle development (~50 mm, 18B) fosters clear definition of internodes to left indicating continued elongation will occur to exsert the panicle from the eventual boot.

Understanding the implications of this important stage is key for optimizing management of grain sorghum production. First, GPD is sensitive to environmental impacts and sorghum management activities. The determination of spikelet number and seeds per spikelet is a key component of yield potential. Factors such as drought stress, delayed nitrogen fertilizer applications or irrigation, and improper

Fig. 19. Primary tiller development from crown nodes at base of sorghum stalk.

timing of herbicides (especially growth regulator type herbicides like dicamba or 2,4-D), insecticides and even fungicides may damage the developing panicle (head) and thus artificially cap potential grain yield early in the life of the plant. Poor conditions at GPD cannot be fully compensated for in later growth (Gerik et al., 2003).

Morphology-driven Sorghum Management Consideration #3

Keys for optimizing grain sorghum production due to growing point differentiation.

Sorghum growers are often unaware of the connection between three sorghum management keys of production practices and growing point differentiation—optimum timing of management is the purpose of capitalizing on and preserving grain sorghum's yield potential:

- Timing of irrigation, if available. When conditions are dry irrigation applied at or just before GPD will preserve the development of yield potential by setting higher spikelet number and thus potential seeds per panicle. Irrigation applied after GPD is still important for yield potential, but it will not affect potential spikelet number (panicle size). An objection raised by farmers in rainfed conditions is that, without irrigation, there is nothing they can do about moisture supply during GPD—not true: foremost, reduced seeding rates, which should be considered anyway for lower rainfall and potentially droughty conditions afford more moisture per plant, which will enhance potential panicle size; likewise, soil moisture conservation practices (reduced tillage, maintenance of crop residues on the soil surface) will enhance the desired outcome of GPD. In addition, planting date can often be managed to coincide with when rainfall is more likely to occur.
- Timing of in-season nitrogen fertilizer application. Likewise, as nitrogen fertilizer is often applied after planting, most U.S. grain sorghum production guidelines recommend ensuring that the bulk of N is applied in advance or during GPD (Trostle et al., 2010). This is similar to small grains production for the same reason: providing adequate N nutrition during GPD aids in ensuring sorghum panicle development is not limited thus setting higher grain yield potential earlier in the growth of the sorghum plant. Many studies demonstrate that timely irrigation during sorghum's boot state provides the greatest potential yield increase. But producers who have irrigation available should recognize that excessively dry conditions around GPD can lower yield potential whereby no amount of irrigation later in the boot stage can recapture the yield potential that may be lost during GPD.
- Herbicide injury potential. Growth regulator herbicides dicamba and 2,4-D may injure the growing point, which is particularly sensitive during GPD. Many producers do not realize the connection between growth stage and label restrictions for several herbicides. Common labeling instructs growers

to apply over-the-top applications of dicamba or 2,4-D no later than approximately 20 cm tall (this corresponds to about leaf stage 5 or five fully formed leaves). This stage of growth is approximately 4 wk after emergence. At this point further applications of dicamba or 2,4-D in sorghum up to 38 cm tall—approximately leaf stage 7 to 8—should use a drop nozzle or hooded sprayer to minimize contact of the herbicide to leaves and especially deposition in the whorl. This practice reduces the exposure of sensitive GPD tissues to herbicide and the potential subsequent injury in the developing panicle, which is above the soil line, but still well down in the stalk. Significant dicamba or 2,4-D injury may lead to blanking or sterility in the panicles with little grain development.

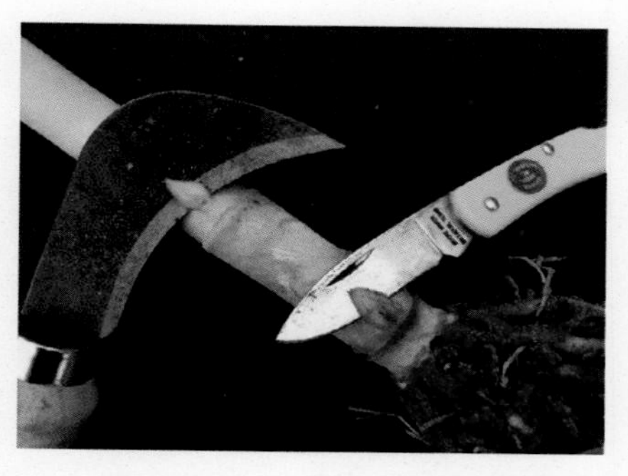

Fig. 20. Axillary buds near base of sorghum stalk which potentially initiate new tiller development after cutting for forage (most likely in sorghum/sudan) or may initiate late-season growth near or after maturity (grain sorghum).

Fig. 21. Tiller initiation from both basal and above ground nodes with regrowth after forage harvest in sorghum/sudan.

The summation of management of grain sorghum around GPD is all about developing and retaining higher yield potential. The maximum potential number of spikelets and florets per spikelet—a major component of maximum yield potential—represent up to 70% of sorghum's final grain yield, and is a result based on the maximum number of possible seeds determined during GPD over a period of about 7 to 10 d (which includes tillers that trail main panicle development). Subsequent environmental conditions such as drought and heat during flowering will influence how many florets per spikelet flower and fertilize then develop seed.

Tillers

Most productive tillers in sorghum, whether for grain or forage production, originate from the same four to six nodes (Fig. 19) noted above where crown roots develop (Assefa et al., 2014). Each node has one axillary bud that may develop a productive basal tiller. These buds on the lowest nodes are difficult to see even with magnification. Tiller initiation is generally a function of the interaction between environment and genetics. Cooler conditions are likely to generate more tillers than the same hybrid planted in warmer conditions one to 2 mo later. In addition, plant density (derived from seeding rate) may also likely influence the ability of a sorghum plant to tiller. Increasing plant population within a planted row will decrease the amount of tillering that will occur. In production environments where drought potential is

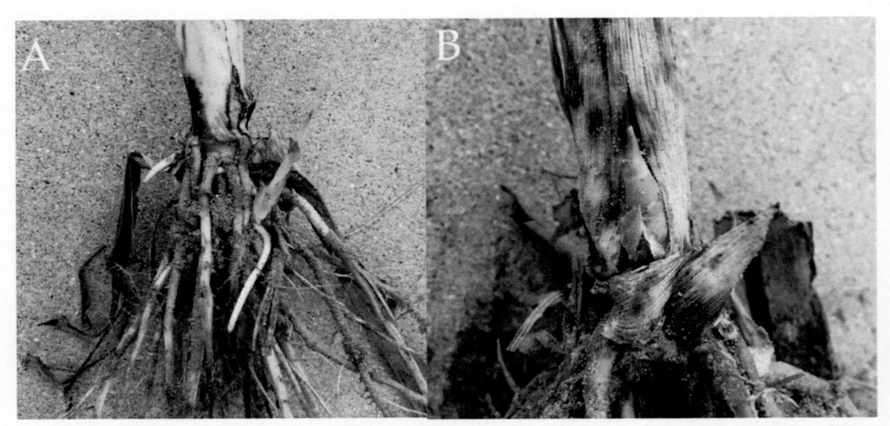

Fig. 22. Late-season tiller development from base nodes in crown (A) and just above soil line (B) in sorghum.

Fig. 23. Late-season basal tiller development in sorghum after maturity.

Fig. 24. Upper node tiller development in sorghum producing a 'sucker' head.

significant, lower tillering hybrids may be favorable in that substantial tillering early in the season may result in excessive vegetative and reproductive growth that the plant cannot sustain during moisture stress and may lower yield potential.

All nodes in a sorghum plant have one axillary bud that potentially can develop a tiller. Each axillary bud may or may not develop a tiller. Unknown to many sorghum workers and farmers is the presence of these potential buds (Fig. 20). When sorghum is cut as forage, including forage sorghums and especially sorghum/sudans, buds potentially produce regrowth tillers (Fig. 21). This valuable trait enables grazing and multiple hay cuts for sorghum family forages, especially those with a sudan [*Sorghum sudanense* (Piper) Stapf] background.

Depending on environmental conditions, some tillers in sorghum—both near the base of the stalk and from nodes above ground—may initiate unproductive tillers later in the season. If a node in the crown (first ~6 nodes) has already produced a tiller it will not produce another tiller, but axillary buds can still generate tillers in the crown and in nodes just above the soil line (Fig. 22A&B). Late in the season, often near or even after harvest, tiller growth from the base of the stalk may occur (Fig. 23) but is of no consequence to the crop.

A drawback to tillering in grain sorghums is late tiller development from nodes above the soil line (Fig. 24). In contrast to early basal tillers, which may contribute significantly to grain yield, these late tillers are often initiated as the panicles from basal tillers are approaching grain maturity. The effect has a genetic component and may be influenced by cooler late-season weather. These 'sucker tillers' may interfere with timely grain harvest due to either significant new foliage or if time is sufficient immature grain in late tillers (Fig. 25). Harvest aids such as sodium chlorate or glyphosate may be used to either desiccate or kill the plant if the main crop panicles are mature (Gigax and Burnside, 1976). This harvest aid drying effect stops growth of sucker tillers and reduces interference with harvest. Otherwise, some sucker tiller development may delay harvest until weather terminates the crop. This delay often places the sorghum crop at greater risk for lodging with no significant additional grain yield benefit. Texas A&M AgriLife observations note that sucker tiller development tends to be more common on true early-season hybrids, especially when planted late in the season.

Fig. 25. Substantial sucker tiller and head development in sorghum poses possible harvest issues requiring a use of harvest aid or delay in harvesting due to immature panicles when primary panicles are harvestable.

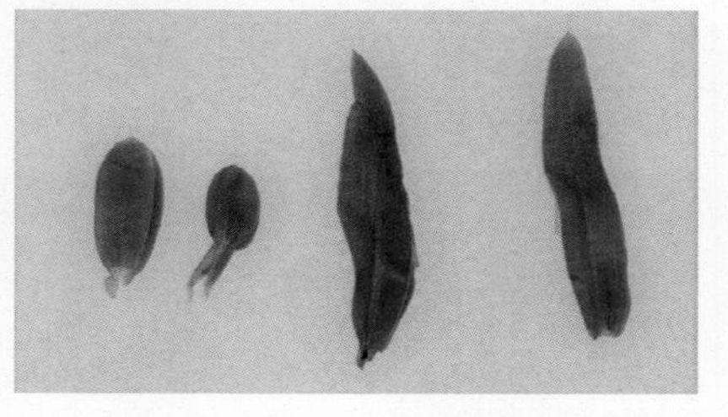

Fig. 26. Rounded leaf tip of leaf one in grain sorghum (left) compared to pointed tips of leaf two.

Leaves

The first grain sorghum leaf emerging through the coleoptile is distinct from all other sorghum leaves due to a round leaf tip (Fig. 26). This is a convenient reference to establish leaf number. Subsequent leaves have a pointed tip. For each newly expanding sorghum leaf, once the collar forms the leaf is at maximum size (Gerik et al., 2003). Sorghum leaf staging is numbered based on the most recent fully matured leaf even though two or three younger leaves are emerging from or visible down in the whorl.

Examination of the base of the sorghum plant up to six weeks after emergence may still identify the brown shriveled remains of the first leaf. This is helpful in determining accurate leaf number. Researchers needing to track leaf number later in the life of a sorghum plant should mark lower leaves in some fashion for reference. Alternatively, node five may be located by splitting the base of the plant and finding the node above the first visible internode which is about 5 to 6 mm long. The fifth leaf originates just above this internode at node five, so sometimes this is helpful in determining plant leaf number if you can properly identify node five.

Sorghum leaves alternate from one side of the plant to the other. This also makes counting leaf number easy once your starting point is determined. Each alternating leaf is attached to a single node progressively up the stalk of the plant.

Grain sorghum hybrids that retain a higher degree of green foliage (and stalks) are often desirable for two reasons. First, this staygreen trait can prolong sorghum's ability to fill grain when water is deficient thus countering potentially

Fig. 27. Sorghum plant color of purple/red (left two stalks) vs. tan plants as expressed in mature sorghum stalks and leaf sheaths (A). Red/purple plant color and red grain (left) and tan plant color with a crème/lemon-yellow grain (B).

detrimental drought conditions (Borrell et al., 2014). Second, the staygreen trait maintains stalk integrity longer vs. hybrids that senesce sooner. This reduces potential lodging after physiological maturity and before harvest.

Plant Color

Pigmentation in sorghum stalks and leaves determines one of three plant colors—purple, red, or tan (Fig. 27A&B). For purple and red pigmentation, the color is seldom distinct and frequently looks like a mixture of both colors. It is also possible to have a dark-purple (some sorghum breeders call it black-purple). The tan color is non-pigmented. For practical purposes in hybrid classification all grades of purple/red are classified as purple. The hybrids that are traditionally grown are all purple pigmentation, but tan plant color, which is often important in the food industry where tan glumes are desired, is commonly coupled with white or lemon-yellow grain (see below). Plant pigmentation will also be apparent on the leaves due to foliar disease or insecticide phytoxicity ('burn').

Plant pigmentation is generally only of interest to breeders who may use the trait as indication of genetic composition or to produce "food grade" type sorghum hybrids which must avoid the dark speckling in food products that are sometimes viewed by grain brokers or consumers as an impurity or contaminant. The presence of pronounced pigmentation, particularly for red and purple plant color, may confound the identification of plant diseases of the leaves and stalk.

Panicle (Head)

Sorghum panicles exhibit a range of appearance depending on the hybrid and growing conditions. Panicles can range from compact to semi-open (Fig. 28A) to open (loose) (Fig. 28B) depending on the length of the primary branches of each spikelet.

Compact panicles have short primary branches and loose panicles have long primary branches. Panicle length and seed number per panicle are dependent on genetics and environment during the floral initiation period. Adverse conditions prior to, during, and just after GPD can lead to pollen sterility, poor grain set, and aborted kernels. Sorghum management includes timing of planting to minimize potential heat and drought stress from about two weeks prior to two weeks after flowering.

Morphology-driven Sorghum Management Consideration #4

Environment and Insect Feeding Potential May Influence Hybrid Choice of Panicle Type

Loose panicle hybrids that are fast growing and dry down rapidly are generally more preferred in areas with higher humidity and greater panicle feeding insect pressure. Tight panicles may enable insects and worms to better hide from feeding birds. Likewise, tight panicles also reduce insecticide penetration for feeding worms and other insects that may move into the panicle (e.g., stinkbugs, sugarcane aphids). Furthermore, tight panicles are more likely to develop grain molds in humid environments due to poor drying within the panicle. Most producers probably don't consider panicle type, but accept whatever hybrid is best adapted or recommended for their area regardless of panicle type (or plant or grain color).

◊

Flowering of grain sorghum panicles begins from the tip and moves downward, completing flowering in four to seven days for typical size sorghum panicles. The larger the panicle, the more likely producers should check grain maturity as defined by black layer at the bottom of the panicle (as well as on tillers, which may be later to bloom by several days vs. the main panicle). Seed companies, breeders, and agronomists commonly refer to half bloom in sorghum as a relative measure of crop growth and development or maturity. When applied to an individual panicle, half bloom is when flowering has occurred half way down from the top of the panicle. In a population setting (plot, field) half bloom is the point at which half of the panicles in the field are at any stage of bloom.

Fig. 28. Sorghum panicle types range from compact to semi-open (A) to open or loose (B).

Fig. 29. Basic grain color types in grain sorghum ranging from white (left two panicles), lemon-yellow (center-left) and red (four panicles of varying grain color expression, right).

Grain

Grain traits include endosperm type, seed color, and the presence of glumes, which enfold the seed.

Grain color may be a preference among some producers based on their historical production or experience. Some farmers want red grain, some may want cremes. Different grain types may weather better in the field under wetter conditions and delayed harvest.

Grain Endosperm

Seed endosperm characteristics reflect two types of the starchy material in sorghum. Normal endosperm is colorless (white). Yellow endosperm sorghum has a high level of carotenoid pigments. Yellow endosperm and lemon-yellow pericarp are unrelated characteristics.

Grain Color

Many colors have been associated with sorghum grain including red, white, brown, bronze, yellow, orange, gray, pink, and salmon pink. These colors are the visual expression or phenotype resulting from the interaction of genes for pericarp color, thickness, and presence or absence of the testa, and spreader genes. Technically, the genetics of sorghum grain allow for only three colors: red, lemon yellow, and white (colorless) (Fig. 29). If present, beneath the pericarp is a layer that can be highly pigmented, called the testa. The testa layer is usually brown. Tannins are present with the pigmented testa and produce a bitter taste, but almost all commercial grain sorghum hybrids now lack a testa.

Grain color is not a significant factor nutritionally. The specific color derives from a combination of three variables: i) pericarp, or seed coat, color, and thickness, ii) whether a testa, or sub-layer, is present under the pericarp, and iii) endosperm color and texture (Bean, 2017).

In the distant past the presence of a testa was associated with reduced bird feeding, but due to high tannins and their undesirability in the animal feed and food industries, tannin grain sorghum hybrids are largely absent from U.S. production. Only a few grain sorghum hybrids are sold in the United States with a testa. An intentional testa exception is the commercialized Onyx sorghum (Hayes and Rooney, 2014) which develops a black color (though not related to tannin but due to exposure to sunlight) and is used in specialty food products. Some hybrid trials with Texas A&M AgriLife since the late 1990s note that bird feeding is sometimes less on specific crème or white grains, but it is not expected that the differential is attributed to tannins due to their general absence from commercial grain sorghum hybrids.

The pericarp color, thickness, presence or absence of the testa, and endosperm all interact for the visual expression of grain color. The term 'bronze' sorghum refers to grain with a thin red pericarp, no testa, and yellow endosperm. If the pericarp was thick the yellow endosperm color would not be apparent. A true yellow endosperm sorghum has thin pericarp with no testa and appears yellow because of the endosperm color.

Fig. 30. Grain color is an unreliable indicator of grain development and maturity. The same panicles are presented with reddish color (A) facing mid-day and afternoon sun (in this case south and west, northern hemisphere) vs. the opposite side of the panicle, which has less color development and appears less mature.

Morphology-driven Sorghum Management Consideration #5

Mistake: Associating developing grain color with maturity.

Too often farmers may draw incorrect conclusions about grain development and maturity based on observations from the edge if the field without checking for physiological grain maturity. Producers recognize that among different grain sorghum hybrids there is variation in mature grain color, and not all grain will achieve a pronounced red or orange color (let alone a crème or white grain) that might be assumed to represent maturity.

Our observation is that when the seed in the panicle begins tinting toward its final color, the seed is most likely still at milk stage. As a sorghum field further develops color, readily observed from the edge of the field or while driving by, the sorghum grain tends to be intermediate between milk stage and soft dough

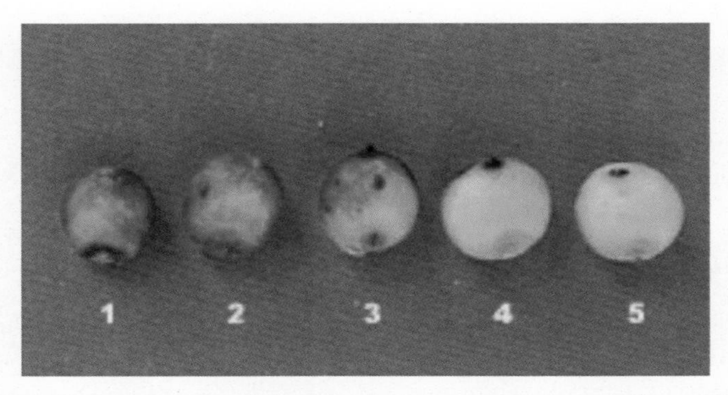

Fig 31. Increasing development of physiological maturity (right to left) as defined by black layer at the base of the sorghum kernel (bottom in the picture) which is attached to the plant. The black layer is initially visible in stage 3 and becomes more distinguishable as the seed loses moisture. The fading black dot at the top of the seed is a remnant of floral structure attachment during fertilization and should not be confused with black layer. (Photo used by permission of Texas A&M AgriLife Extension Service.)

(we call this 'mealy' or 'gel' stage; pressing the seed with finger and thumbnail no longer produces any liquid, but seed contents are squeezed out). Once the sorghum panicle is fully colored from top to bottom, the grain is typically in the late soft to hard dough stage.

Also, the direction which you observe a maturing sorghum field's grain color may lead one to a different conclusion about the development of a field. Typically, sorghum panicles color more quickly on the sides facing the sun (Fig. 30A&B).

Some producers erroneously use field grain color alone as a gauge for one last possible irrigation if available, but grain color is not a reliable means for deciding to irrigate again. Producers must check grain stage of development and probe for remaining soil moisture. Although final irrigation is more appropriately gauged in response to physiological growth and development and not morphology, the latter can potentially mislead producers in interpreting the stage of growth. Irrigation termination may be evaluated once sorghum grain reaches the hard dough stage or when the grain color changes throughout most of the sorghum panicle (keep in mind, though, that sorghum tillers may lag the primary panicles by up to a week, so assessment should reflect total grain development and maturation). At this time, the decision to terminate irrigation should be made based on stored soil moisture and anticipated rainfall. High yielding grain sorghum can still potentially use five or more inches of water from late soft dough to harvest. Some plant use of available soil moisture is not attributed directly to final grain yield but may be important in maintaining stalk strength in high-yielding environments. While the plant remains green, some water is required to maintain stalk integrity thus reducing lodging potential.

Mature grain is never defined by grain color, but by physiological black layer development, which can vary by up to seven days in a larger panicle, and up to seven days between a primary panicle vs. a tiller panicle. Growers can readily determine when sorghum grain is physiologically mature, when black layer appears at the bottom of the sorghum seed (enfolded by a pair of glumes) as it is attached in the spikelet. It is a distinct dark black spot. For an individual seed, the black color transitions from no color (Fig. 31, #5) to faint black, to eventual fully black (Fig. 31, #1) (Stichler and Livingston, 2003). A sorghum crop will transition toward full physiological maturity as an increasing proportion of kernels develop black layer. Black layer should not be confused with the opposite end of the kernel where floral structures were attached the seed ovule at fertilization. This often also leaves a black dot on the seed as well, but it is visible well before maturity. Thus, when picking seeds from a panicle for black layer determination be sure to evaluate the base of the kernel attached to the plant. As Stichler and Livingston (2003) note, with practice producers can disregard grain color to determine relative maturity of not only seeds but grain sorghum crops as a whole.

◊

Glume Traits

Glume color will also influence the appearance of grain color. Pigmented (purple or red) plants have purple or red glumes. Most tan plants will have tan or straw-colored glumes. The effect of glume color on the grain is through the glumes 'staining' of the grain at maturity. This is a consideration only if a food type (tan plant, white grain) sorghum is grown. This sorghum should have tan or straw color glumes to reduce the staining of the grain at maturity. Flour processed from white food type grain will be a consistent white color.

How to Distinguish Young Sorghum Plants from Corn

Before grain sorghum and corn (*Zea mays* L.) seedlings and plants develop readily identifiable external physiological characteristics that distinguish one from the other, a few morphological characteristics in the first 30 d or so are helpful in identifying these two plants. For newly emerged seedlings, as noted above, the first leaf has a characteristic rounded tip and may look similar in sorghum and corn, but this first leaf in corn tends to be more ovoid and much larger, sometimes triple or more in size (Fig. 32A&B). It is easy, however, to dig not only just seedlings, but likely larger plants up to ~10-leaf stage (40–50 cm tall) and follow the primary root (the first root from the seed, which grows vertically into the soil) to the seed remnant. Usually on careful observation, even if not initially obvious, the remaining portion of the seed can be identified unless it has simply rotted away or fungus has consumed it. If the seed remnants are gone, corn tends to have several seminal roots growing

Figs. 32A&B. Small first leaf on grain sorghum (A) at 40 days (V7). First leaf on upper plant (A) is deteriorated. Corn first leaf (B) planted nearby in the same field on the same day expressing larger ovoid first leaves.

laterally from where the seed was in addition to the single primary root. Also, sorghum leaf margins in general are always serrated (sawtooth) whereas corn leaf edges are smoother (Hannaway and Meyers, 2004). If this is not readily identified visually then a low magnification lens will reveal the difference.

Summary

This discussion focuses on practical grain sorghum morphology that is of interest to growers, consultants, and agronomists. Small-scale morphological characteristics that are too small to see without magnification aids are discussed in other publications. The objective is to describe some key morphological and physiological traits of sorghum that may influence how sorghum is managed and how sorghum management may be improved. These traits include sorghum rooting response to planting pattern and how roots are best able to penetrate soil leading to increased standability. Sorghum seeded in an elevated position may lead to poor root brace, root penetration of surface soil, and compromised standability. The location of roots in response to field planting pattern and the individual year's growth and development is a factor in minimizing root damage from side-dress fertilizer applications in the soil. The development of the sorghum growing point and when differentiation to panicle development occurs is a management consideration for mid-season nitrogen application, irrigation if available, and exclusion of certain herbicides and other pesticides which may damage the growing point and thus diminish yield potential. Growing point differentiation has no external sign to indicate its development status thus examination of split sorghum stalks is emphasized to better manage these crop inputs. Sorghum panicle type may be a management consideration where effective insect management is needed, or grain is at a disadvantage due to poor drying conditions which can foster disease development. Finally, producers that conduct late season "drive by" management of the crop for final irrigation (if available) or maturity may improperly gauge grain development if decisions are based solely on grain color. An improved understanding of these morphological traits and potential management—which do not require additional inputs—will help sorghum retain its value and lead to improved production. This is especially important where sorghum competes with other crops (especially corn) which often have higher crop value and more seed defense traits for pests or herbicide tolerance.

References

Al-Khatib, K. 2017. Herbicide damage. Univ. of California, Davis, CA. http://herbicidesymptoms.ipm.ucanr.edu/HerbicideDamage/ (retrieved 16 June 2018).

Artschwager, E. 1948. Anatomy and morphology of the vegetative organs of Sorghum vulgare. USDA Technical Bulletin No. 957, Washington, D.C.

Assefa, Y., K.L. Roozeboom, C.R. Thompson, A.J. Schlegel, L. Stone, and J.E. Lingenfelser. 2014. Corn and grain sorghum morphology, physiology and phenology, p. 3-14. Academic Press, Waltham, MA.

Bean, B. 2017. Sorghum grain color. United Sorghum Checkoff Program, Lubbock, TX http://www.sorghumcheckoff.com/news-and-media/newsroom/2017/12/29/sorghum-grain-color/ (Retrieved 16 June 2018).

Bennett, W.F., B.B. Tucker, and A.B. Maunder. 1990. Modern grain sorghum production. Iowa State Univ. Press, Ames, IA.

Blum, A., G.F. Arkin, and W.R. Jordan. 1977. Sorghum root morphogenesis and growth I. Effect of maturity genes. Crop Sci. 17:149–153. doi:10.2135/cropsci1977.0011183X001700010039x

Borrell, A.K., J.E. Mullet, B. George-Jaeggli, E.J. van Oosterom, G.L. Hammer, P.E. Klein, and D.R. Jordan. 2014. Drought adaptation of stay-green sorghum is associated with canopy development, leaf anatomy, root growth, and water uptake. J. Exp. Bot. 65:6251–6263. doi:10.1093/jxb/eru232

Cothren, J.T., J.E. Matocha, and L.E. Clark. 2000. Integrated crop management for sorghum. In: C.W. Smith and R.A. Frederiksen, editors, Sorghum: Origin, history, technology, and production. Wiley & Sons, New York.

Frederiksen, R.A., and G.N. Odvody. 2000. Compendium of sorghum diseases, 2nd edn. Amer. Phytopathological Soc., St. Paul, MN.

Gerik, T., B. Bean, and R. Vanderlip. 2003. Sorghum growth and development, B-6137. Texas Coop. Ext., College Station, TX.

Gigax, D.R., and O.C. Burnside. 1976. Chemical desiccation of grain sorghum. Agron. J. 68:645–649. doi:10.2134/agronj1976.00021962006800040028x

Hannaway, D.B., and D. Myers. 2004. Forage fact sheet: Sorghum. http://forages.oregonstate.edu/php/fact_sheet_print_grass.php?SpecID524&use5Forage

Hayes, C., and W. Rooney. 2014. Agronomic performance and heterosis of specialty grain sorghum hybrids with a black pericarp. Euphytica 196:459–466. doi:10.1007/s10681-013-1047-3

Miller, E.C. 1916. Comparative study of the root systems and leaf areas of corn and sorghum. J. Agric. Res. 6:311–332.

Miller, T. 1999. Growth stages of wheat: Identification and understanding improve crop management. SCS-1999-10. Texas A&M AgriLife Extension Service, College Station, TX.

Nakayama, F.S., and C.H.M. van Bavel. 1963. Root activity distribution patterns of sorghum and soil moisture conditions. Agron. J. 55:271–274. doi:10.2134/agronj1963.00021962005500030020x

Roozeboom, K., and P.V.V. Prasad. 2018. Sorghum growth and development. (this publication). In: I. Ciampitti and P.V.V. Prasad, editors, Sorghum: State of the art and future perspectives. Amer. Soc. Agronomy, Madison, WI.

Singh, V., E.J. Van Oosterom, J. Jordan, C.D. Messina, M. Cooper, and L. Graeme. 2010. Morphological and architectural development of root systems in sorghum and maize. Plant Soil 333:287–299. doi:10.1007/s11104-010-0343-0

Smith, K., and B. Scott. 2004. Weed control in grain sorghum. In: L. Espinoza and J. Kelley, editors, Arkansas grain sorghum production handbook, MP297. Cooperative Ext. Serv., Univ. of Arkansas, Fayetteville, AR.

Stichler, C., and S. Livingston. 2003. Harvest aids in sorghum. L-5435. Texas A&M AgriLife Ext. Serv., College Station, TX.

Trostle, C., B. Bean, N. Kenny, T. Isakeit, P. Porter, R. Parker, D. Drake, and T. Baughman. 2010. West Texas production guide. United Sorghum Checkoff Program, Lubbock, TX.

Vanderhoef, L.N., and W.R. Briggs. 1978. Red light-inhibited mesocotyl elongation in maize seedlings. I. The auxin hypothesis. Plant Physiol. 61:534–537. doi:10.1104/pp.61.4.534

Vanderlip, R.L. 1993. How a sorghum plant develops. S-3. Agricultural Experiment Station and Cooperative Extension Service, Kansas State University, Manhattan, KS.

Sorghum Growth and Development

Kraig L. Roozeboom* and P.V. Vara Prasad

Abstract

Understanding sorghum growth and development is essential for comprehending sorghum plant response to environmental stresses and making sound production management decisions. Sorghum development has been separated into three major divisions: vegetative (GS-1), reproductive (GS-2), and grain fill (GS-3), with about a third of the life cycle spent in each. Vegetative growth can be subdivided into emergence (S0), third leaf collar (S1), and fifth leaf collar (S2). Reproductive growth begins when the terminal meristem forms panicle structures rather than additional leaves at growing point differentiation (S3). Reproductive growth can be subdivided into flag leaf appearance (S4) and boot (S5) stages. Reproductive growth transitions to grain filling at half bloom (S6) when plant height and leaf area are at their maximum, and about half the total aboveground dry matter has been accumulated. Grain fill is subdivided into soft dough (S7) and hard dough (S8) and ends at physiological maturity (S9) when dry matter accumulation has ceased. A visual indicator is the appearance of the abscission layer (black layer) opposite the embryo. Although genetics and environmental conditions can modify the timing dramatically, it takes roughly 10 d for a medium maturity, commercial hybrid growing in the central Great Plains to progress from one stage to another. The sorghum plant possesses a vast capacity for adapting to its environment via tiller production, adjustments in the number of potential kernels per head, seed set, and changes in kernel size. As a result, sorghum yields tend to be relatively stable across years, environments, and management systems.

Sorghum growth is defined as an increase in plant dry matter over time. The goal of grain sorghum production is to increase accumulation of dry matter in the form of grain. Development is the progression of the sorghum plant through its life cycle, from seed to grain. Knowledge of how sorghum proceeds from one stage to another, how it accumulates dry matter along the way, and how it partitions that dry matter to tissue and grain is foundational for understanding how a sorghum plant is likely to respond to environmental factors and management practices. Applying this understanding to specific hybrids planted in a specific environment across a specific landscape forms the essence of managing a sorghum crop for efficient production.

Unless otherwise indicated, information in this chapter is derived from the 2016 revision of Kansas State University Research and Extension publication S-3, *Sorghum Growth and Development* (Vanderlip, 1993). Plant materials used to

---

Kraig L. Roozeboom, Dep. of Agronomy, Kansas State Univ., 2004 Throckmorton Plant Sciences Center, 1712 Claflin Rd., Manhattan, KS 66506. P.V. Vara Prasad, Feed the Future Sustainable Intensification Innovation Lab., 108 Waters Hall, 1603 Old Claflin Place, Kansas State Univ., Manhattan, Kansas 66506 (vara@ksu.edu). *Corresponding author (kraig@ksu.edu).

doi:10.2134/agronmonogr58.2014.0062

generate the images illustrating the various growth stages were field grown at Manhattan, KS, in 2009 and 2010 using best management practices for weed control, fertility, seeding rate, etc., in a no-till cropping system. As a result, leaves often exhibit wind, insect, or disease damage as would be evident in a typical production field (although severely damaged plants were avoided). Plants were removed from the field and taken indoors to facilitate uniform, consistent light conditions for photographs. Two commercial hybrids (DEKALB DKS54–00 representing the medium-late [ML] maturity class, and Pioneer P86G32 representing the medium-early [ME] maturity class) were seeded in multiple blocks to generate plant material. Ten plants of each hybrid were harvested every 7 d, separated into leaves, stems, panicle, and grain and dried and weighed to quantify dry matter accumulation (Fig. 1). Leaf area and plant height were quantified as well (Fig. 2). Values from both hybrids were standardized as a percent of total accumulation for each hybrid maturity (ME and ML) or in some cases were combined to represent predicted accumulation for a medium maturity hybrid (M). Samples were analyzed for nutrient concentration by the Kansas State University Soil Testing Laboratory to estimate nutrient accumulation.

Plant Structures and Terminology

Although sorghum plants are complex organisms, the main plant organs (seed, leaf, stem, and panicle) can be used to track development (Fig. 3 and 4). The first structure to appear outside the seed at germination is the radicle, which develops into the seminal root system. Eventually the nodal roots grow from the base of the stem, above the coleoptile node. The mesocotyl is the root-like stem structure between the seed and the base of the coleoptile, the first plant structure to emerge at the soil surface. The true leaves appear soon after emergence, but the remnant of the coleoptile often can be seen attached to the base of young sorghum seedlings. True leaves can be identified by the presence of a collar, the tissue at the transition between the leaf blade and leaf sheath. The peduncle is the stem structure that supports the panicle (Fig. 4). Elongation of this structure pushes the panicle from the sheath of the final leaf (flag leaf) just before flowering.

Development and Growth Stages

This chapter uses the staging scheme presented by Vanderlip and Reeves (1972) that divides development of the sorghum plant into 10 distinct stages with clearly identifiable characteristics (Table 1). These stages align with the three growth stages defined by Eastin (1972) as indicated in Table 1. Each of these stages marks an important developmental milestone indicating continued progress of the sorghum plant through its life cycle. In most cases this development is associated with concurrent growth (Fig. 1). Although the time needed to reach each stage depends on both genetic and environmental factors (Hammer et al., 1989), the time to each stage presented in Table 1 represents the approximate number of days required to reach each stage for a medium maturity hybrid grown at Manhattan, KS.

Tropical sorghum varieties and some forage sorghum hybrids are short-day plants that require a long night to initiate floral development (Prasad and Staggenborg, 2009). Most commercial grain sorghum hybrids have been converted to be

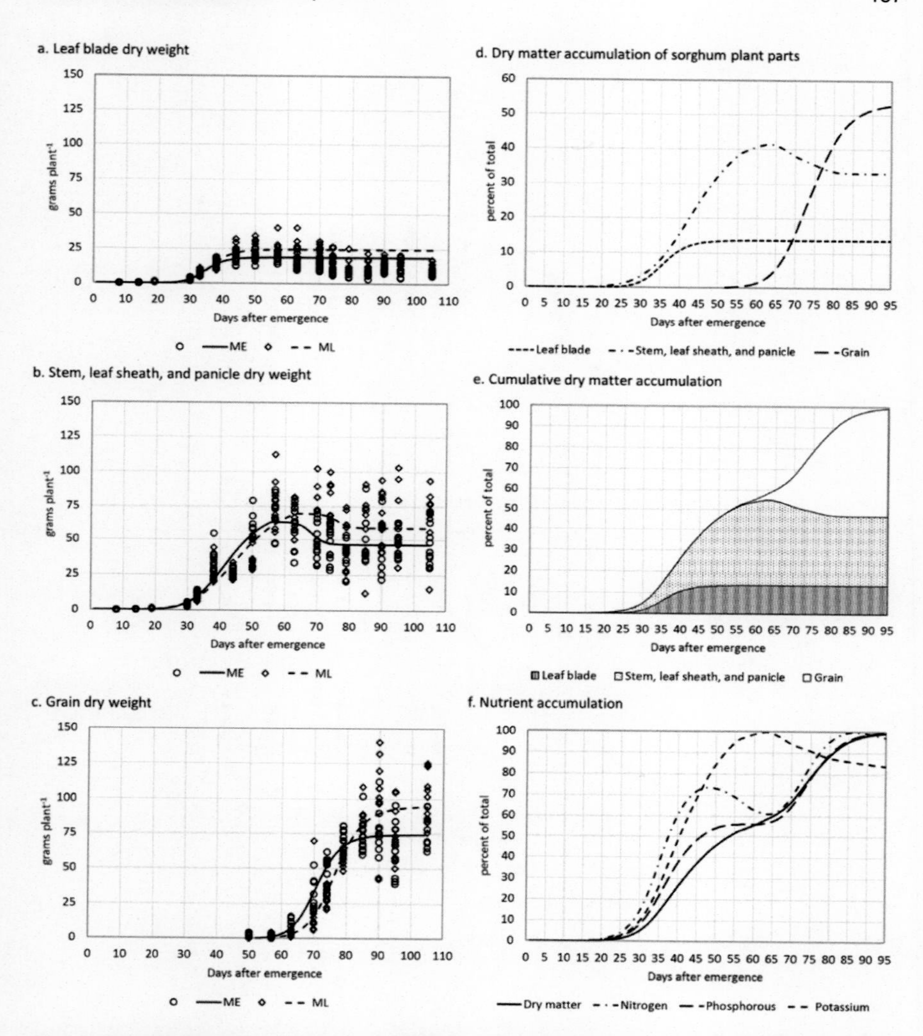

Fig. 1. Dry matter accumulation of medium-early (ME) and medium-late (ML) hybrids (a, b, c) and estimated dry matter and nutrient accumulation for a medium maturity hybrid (d, e, f).

insensitive to photoperiod, and development is driven primarily by temperature in the absence of water stress. The SORKAM growth model uses thermal time to calculate duration of growth stages, but cardinal temperatures change depending on growth stage (Rosenthal et al., 1989). Gerik et al. (2003) and others (Kansas Mesonet, 2016) use a 10°C base temperature and 38°C upper limit to calculate sorghum growing degree units (GDU). Because of its inherent ability to tolerate temperature and moisture extremes, sorghum often is grown in stress-prone environments that can alter the phenology of the sorghum plant, often with different effects depending on hybrid (Hammer et al., 1989; Peacock and Heinrich, 1984).

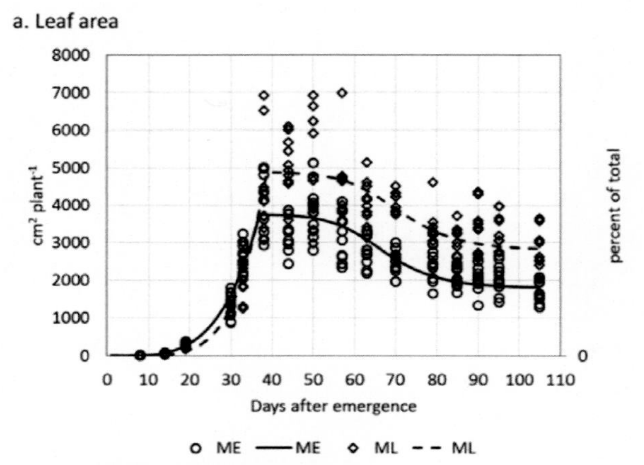

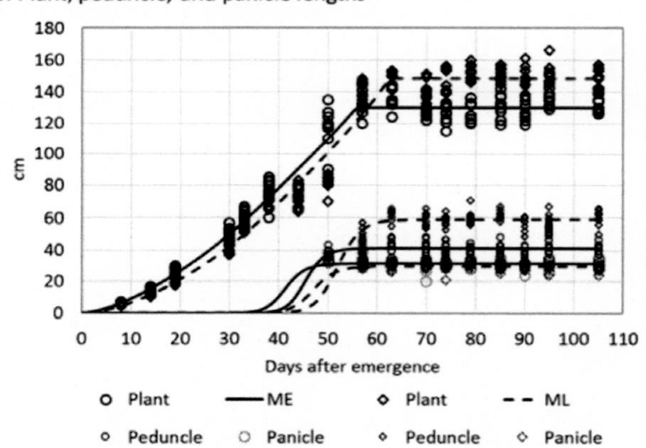

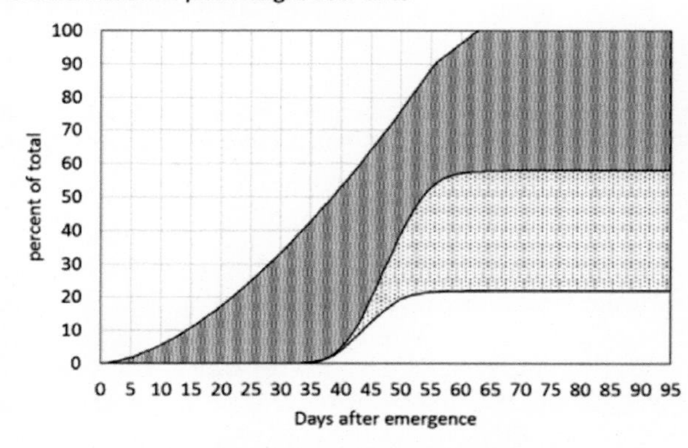

Fig. 2. Leaf area and plant height development of medium-early (ME) and medium-late (ML) hybrids (a, b) and estimated distribution of plant height for a medium maturity hybrid (c).

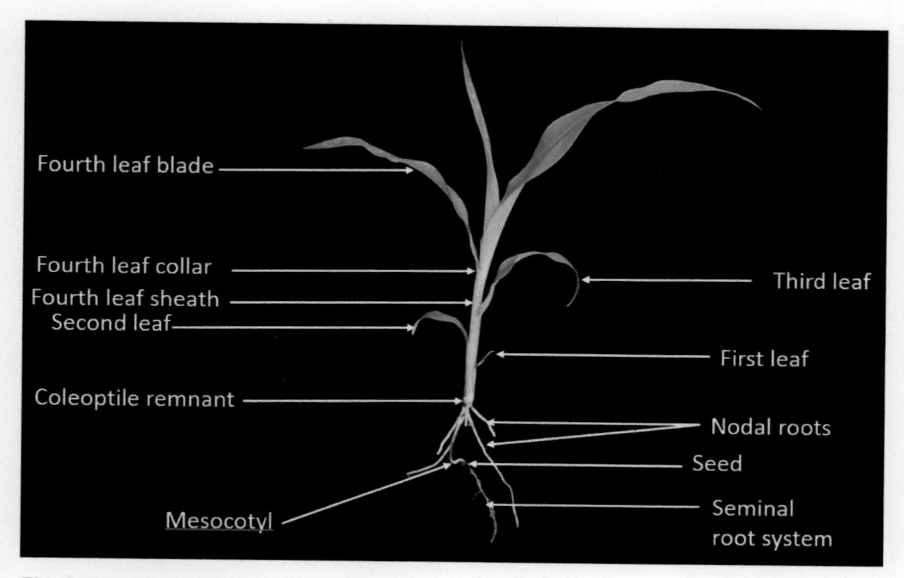

Fig. 3. Seedling sorghum plant parts.

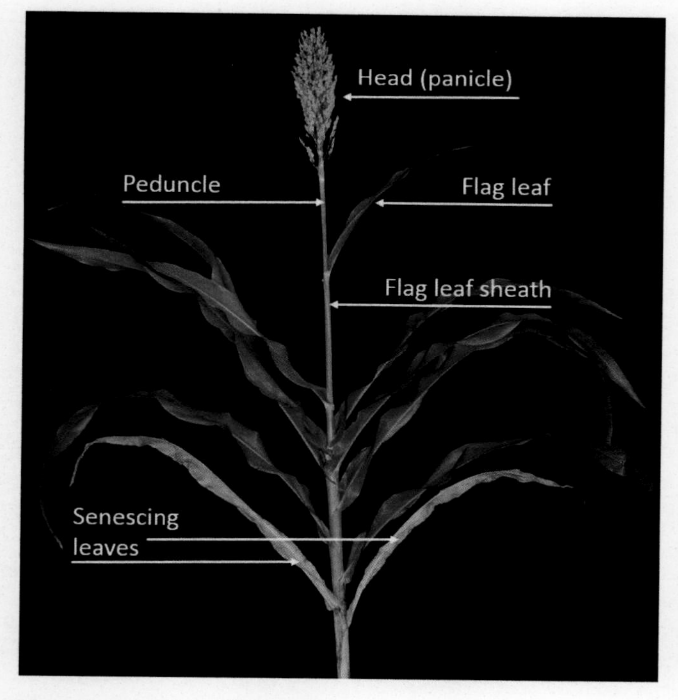

Fig. 4. Postflowering sorghum plant parts.

Table 1. Grain sorghum developmental stages and indicators.

Growth stage†	Developmental stage‡	Approximate time to each stage		Stage name	Indicator(s)
		Days§	GDU¶		
GS-1 Vegetative (germination to panicle initiation)	0	0	55–110	emergence	coleoptile or first leaf visible at soil surface
	1	10	140–200	third leaf	collar of third leaf visible
	2	20	220–330	fifth leaf	collar of fifth leaf visible
	3	30	390–640	growing point differentiation	7 to 10 leaf collars (Panicle initiation)
GS-2 Reproductive (panicle development)	4	40	530–780	flag leaf visible	final leaf beginning to emerge from whorl
	5	50	670–915	boot	panicle is located in flag leaf sheath
	6	60	800–1060	half bloom	half of plants blooming (anthesis)
GS-3 Grain fill (bloom to maximum dry matter accumulation)	7	70	970–1220	soft dough	seed crushes easily but no liquid
	8	85	1220–1470	hard dough	seed resists crushing
	9	95	1420–1670	physiological maturity	black layer formation

† Eastin (1972).

‡ Vanderlip and Reeves (1972).

§ Approximate days from emergence required to reach each stage for a medium maturity hybrid grown at Manhattan, KS, with normal weather conditions.

¶ Range of GDU required from planting to reach each stage for the typical range of hybrid maturities planted at Manhattan, KS. Daily GDU = (max. temp. + min temp.)/2–10 where max. temp. >38 = 38 and min. temp. < 10 = 10 (Gerik et al., 2003).

Freezing temperatures or extreme drought can terminate dry matter accumulation before plants are physiological mature (Staggenborg et al., 1999). As a result, sorghum development often can be difficult to predict based on GDU alone. Even so, GDU provides a useful basis for comparison that accounts for variations in daily temperatures. For the purposes of this discussion, timing of development is related to an approximate number of days and range of GDU (daily GDU = [(max. temp. + min. temp.)/2] − base temp., where min. temp. < 10°C = 10, max. temp. > 38°C = 38, and base temp. = 10) that captures the requirements for the typical range of hybrid maturities planted in the central Great Plains.

Seed

Like most commercially important grain crops, sorghum relies on seeds to establish each new generation. The seed should represent both the best available genetic potential for a given growing environment and management system as well as the best seed quality as measured by germination and purity. Yields of similar-maturity hybrids can vary by more than 20% in trials where all hybrids receive the same management (Lingenfelser et al., 2014), and the difference can be greater across maturity classes. Sorghum seed is relatively small compared with most other grain crops with seed weight ranging from less than 20 g to more than 40 g 1000 seeds^{-1} in unconditioned seed lots. Under field conditions, seed weight is determined by environmental conditions during grain filling (Prasad et al., 2008) and genetics (Tuinstra et al., 2001; Yang et al., 2010). Although seed conditioned for commercial sale meets germination and purity standards, seed size can differ from one seed lot to the next. It is essential to plant a target number of seeds per hectare rather than a target weight of seed per hectare . For example, if two seed lots have seed weights of 20 and 30 g 1000 seeds^{-1}, the resulting number of seeds per hectare for each is 250,000 and 151,515, respectively, if both seed lots were planted at a rate of 5 kg ha^{-1}. Given the need to match plant density with environmental and management resources, a change in the number of seeds per hectare of this magnitude is unacceptable. To achieve the desired plant density, germination rate (based on laboratory analysis of the planted seed lot) and expected field emergence rate (based on producer experience, soil and residue characteristics, and environmental conditions) should be considered when selecting the seeding rate.

Germination

Germination of the sorghum seed is a complex process that begins with the seed imbibing water held in the surrounding soil matrix and culminates when the radicle breaks through the seed coat. A number of factors affect the rate at which germination proceeds, but two of the most important are temperature and water availability (Anda and Pinter, 1994). The literature varies regarding the base temperature required for sorghum growth. Hammer et al. (1989) reported base temperatures for different hybrid groups adapted to temperate, subtropical, or tropical environments ranging from 12.1 to 14°C from emergence to growing point differentiation and 10 to 11°C from growing point differentiation to half bloom. Lafarge et al. (2002) used 11°C as a base temperature to quantify thermal time for sorghum development in Australia. The SORKAM grain sorghum growth model uses an even lower base temperature when calculating thermal time during early sorghum development (Rosenthal et al., 1989). Although these reports support a base temperature for sorghum of 10°C, Brar and Stewart (1994) reported no germination at 10°C across several hybrids. Prompt germination requires substantially higher temperatures. Brar and Stewart (1994) reported 7 d to germination at 15.5°C, and 1 d to germination at 37.5°C averaged across several hybrids in laboratory conditions. Although hybrids differ in their germination response to temperature, the optimal temperature range reported in the literature is 21 to 35°C (Brar and Stewart, 1994; Mortlock and Vanderlip, 1989; Peacock and Heinrich, 1984). A soil temperature of 18 to 21°C often is recommended to promote rapid germination and emergence in temperate environments (Roozeboom et al., 2010).

Even if two seed lots have similar germination, it is common for field establishment to differ greatly from laboratory germination (Vanderlip et al., 1973). A poor relationship between seed size and seedling emergence under field conditions often is observed across sorghum hybrids. If seed sizes within a seed lot are considered, both extremely small and extremely large seed have lower than average establishment capability (Abdullahi and Vanderlip, 1972; Maranville and Clegg, 1977).

Stage 0: Emergence

Emergence occurs when the plant, usually the coleoptile, is visible at the soil surface. Time required from planting to emergence can vary from 3 to 10 d or more (55 to 110 GDU), depending on soil temperature, moisture, and planting depth. Vigor of the emerging seedling also can affect emergence rate and is a function

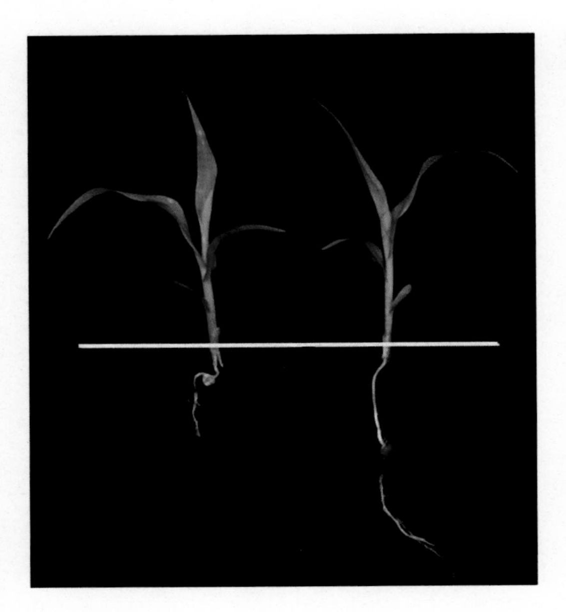

Fig. 5. Sorghum seed depth and mesocotyl length.

of both seed characteristics (e.g., size and composition) and genetic factors. Until green leaf tissue appears above the soil surface and becomes active photosynthetically, the emerging seedling depends on the seed reserves for energy and most nutrients. The length of the mesocotyl adjusts depending on seed depth to place the base of the coleoptile and eventually the nodal roots, at a relatively consistent depth (Fig. 5). This is accomplished via a hormonal signal that causes the mesocotyl to stop elongating when light is detected by the coleoptile (Vanderhoef and Briggs, 1978).

True leaves break through the coleoptile soon after emergence, often the same day. True leaves can be distinguished from the coleoptile by the presence of the distinct structures of blade, collar, and sheath. The first true leaf can be identified by its short blade with a slightly rounded tip compared with the relatively longer leaf blade and more pointed tip of subsequent leaves (Fig. 5). Assuming optimal temperatures, the collar of the second leaf is visible 3 or 4 d after appearance of the first leaf collar, and the third leaf collar appears 3 to 4 d after that.

Stage 1: Three-Leaf Stage

Depending on temperature, the third leaf collar should be visible about 8 to 10 d (140 to 200 GDU) after emergence. Although the plant is about 5% of its final height (Fig. 2c), less than 1% of the total aboveground dry matter (Fig. 1e) and total leaf area (Fig. 2a) has been accumulated. The plant is composed mostly of leaf blades and sheaths at this stage. No elongation of the stem has occurred so the growing point is still below the soil surface. Nodal roots are just beginning to appear. Eventually these roots will develop into the main support structure and means of nutrient and water acquisition for the growing plant. However, until the nodal roots become established, the sorghum seedling relies on the branching radicle and seminal roots. With optimal temperatures, the fourth leaf collar is visible 4 or 5 d after the plant reaches the three-leaf stage. The first four nodes remain compressed so the growing point is still below the soil surface, protecting it from temperature extremes and physical damage.

Stage 2: Five-Leaf Stage

Approximately 2 to 3 wk after emergence (220 to 330 GDU), the sorghum plant has five fully expanded leaves, although the next two leaves have emerged from the whorl and are expanding rapidly (Fig. 6). At this point the nodal root system is beginning to take over as the primary source of support, nutrients, and water. The stem is beginning to elongate so the growing point is starting to separate from the base of the stem, although it is usually still below the soil surface. Although only about 1% of the total dry matter has been accumulated (Fig. 1e), the plant is poised to grow rapidly with more noticeable stem elongation and leaf area development. Dry matter accumulation will soon accumulate at a nearly constant rate until maturity. Close to 20% of the plant height has been achieved (Fig. 2b and 2c). Sorghum plants at this stage are becoming more robust and will begin to out compete emerging weeds or late-emerging sorghum plants. New leaf collars will be visible every 3 to 4 d with favorable temperatures and adequate moisture.

Up to this point, loss of the leaf area does not necessarily translate into large yield losses. With favorable growing conditions, the sorghum plant can recover rapidly and may have relatively minor grain yield loss. Heading likely will be delayed, but at the five-leaf stage, less than 10% of the leaf area has developed (Fig. 2a), leaving most of the leaf area available to support continued growth and development (Vanderlip et al., 1977). This assumes that remaining leaves can physically push through the damaged leaves and leaf sheaths.

Fig. 6. Dissected Stage 2 sorghum plant.

All nodes, except at the base of the peduncle, have axillary buds. Yet, tillers typically develop only from the first five or six nodes. Hybrids differ in their propensity to generate tillers. Tiller emergence from the lower nodes is determined by photosynthate availability in the main stem and by changes in light quality resulting from competition from neighboring plants for upper nodes (Lafarge et al., 2002). Continued development, seed set, and grain production of emerged tillers also depends on photosynthate availability at tiller emergence. Although it may differ with hybrid, Lafarge et al. (2002) showed that tiller emergence was most consistent at the third node across varying plant densities, followed by the fourth, second, and fifth nodes in that order. Tillers also emerged at a relatively low frequency of about 25% from the first node regardless of plant density, but emerged from the sixth node only when plant density was very low. Grain production from tillers depended on plant density, but, after the main stem, the third-node tiller typically produced the most grain with rapidly decreasing contributions from tillers above and below, following the pattern of tiller emergence. This appears to be a stable pattern in that it was observed across plant densities that resulted in tiller numbers between 0.2 and 4.9 per plant and contributions to grain yield from tillers of 5 to 78% for tiller emergence frequency, tiller fertility frequency, and grain number per tiller. Few tillers survived and produced grain as plant density increased. One driver of tiller development and survival is temperature, probably via its effect on photosynthate availability at tiller emergence (Lafarge et al., 2002). Later planted sorghum exposed to higher temperatures typically produces fewer fertile tillers than earlier planted sorghum in the temperate United States (Maiga, 2011; Roozeboom et al., 2013).

Stage 3: Growing Point Differentiation

About 1 mo after emergence (390–640 GDU), depending on growing conditions and hybrid maturity, the growing point of the sorghum plant changes from producing leaves to producing the panicle (Fig. 7). This marks a fundamental developmental change that is reflected in movement of the plant from vegetative (Growth Stage 1) to reproductive (Growth Stage 2) growth. This change is not outwardly visible without dissecting the plant and using magnification, but usually occurs during the period when the seventh to tenth leaf collars appear. Because the growing point is now producing panicle structures, no new leaves will be initiated on the main stem. Although the total number of leaves has been determined, only about one third of the total leaf area has fully expanded (Fig. 2a). Hybrid maturity and prevailing temperatures determine the timing of this shift from vegetative to reproductive growth, with earlier hybrids and warmer temperatures hastening the transition (Rosenthal et al., 1989; Hammer et al., 1989).

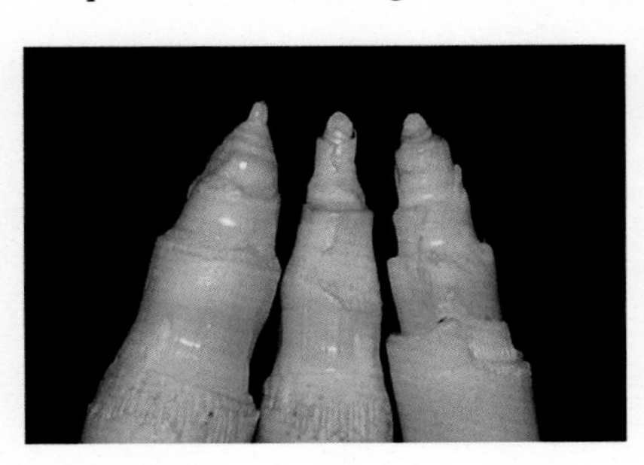

Fig. 7. Dissected sorghum stems showing panicle forming soon after reaching Stage 3.

At growing point differentiation, the sorghum plant is about one third of the way to physiological maturity in terms of time, but has accumulated only about 5 to 10% of its total dry matter (Fig. 1a, 1b, 1d, and 1e). Nutrients are accumulating more rapidly in plant tissues than dry matter at this stage (Fig. 1f). Elongation of the stem has pushed the growing point, now the developing panicle, up through the enclosing leaf sheaths so that it is several centimeters above the soil surface. As a result, the sorghum plant is likely to suffer greater yield reductions from physical damage or environmental stresses because a greater portion of the potential leaf area has been developed, and the panicle is susceptible to either direct injury or reductions in potential grain number due to limited photosynthate availability. Plant height is now about a third of the eventual maximum (Fig. 2 b and 2c), and will increase rapidly with favorable temperatures and moisture availability.

Stage 4: Flag Leaf Visible

Rapid stem and leaf growth after growing point differentiation results in the appearance of the flag (final) leaf in the whorl about 35 to 45 d after emergence (530 to 780 GDU). Dissection of the plant may be required to determine if the emerging leaf is in fact the flag leaf. Although only about 25% of total dry matter has been accumulated at this stage (Fig. 1), all but the last three to four leaves are fully expanded. With leaf area approaching maximum (Fig. 2a), light interception also is approaching maximum, fueling rapid growth and nutrient uptake. Nutrient uptake continues to outpace dry matter accumulation with close to 60% of total N, 50% of total K, and 40% of total P already in the plant (Fig. 1f).

With expansion of the stem diameter, the first few leaves likely have been lost by this stage. Carefully splitting the stalk and finding the fifth node (the first to separate from the first four compressed nodes) and its associated leaf can help in determining leaf number. Counting down from the flag leaf is another way to quantify leaf number if the lower leaves are lost, but those leaf numbers are not associated with the growth stages described above. Counting leaves in this manner can be useful for quantifying canopy development as the last several leaves are expanding.

Stage 5: Boot

Once the collar of the flag leaf is visible, the sorghum plant is entering the boot stage (Fig. 8), at 45 to 55 d after emergence (670 to 915 GDU). At this stage, all leaves have fully expanded, resulting in maximum leaf area (Fig. 2a) and light interception. The panicle is approaching full size (Fig. 2b and 2c) but is still completely enclosed within the flag leaf sheath. The potential number of grains in the panicle has been determined so that, in some ways, yield potential can only decrease after this point, depending on success of grain set and grain filling. Tillers that will not produce grain begin to gradually die off starting at this point and continue to do so until plant maturity (Lafarge et al., 2002).

Peduncle elongation is beginning (Fig. 2b and 2c) and will result in exsertion of the panicle from the flag leaf sheath. Extent of exsertion is regulated by the interaction of genetic and environmental factors. With adequate moisture and optimal temperatures, the panicle should fully emerge from the boot in commercial hybrids. However, with drought stress and nonoptimal temperatures, panicle exsertion can be inhibited, often depending on a hybrid's ability to maintain peduncle elongation in the face of these stresses. Incomplete or slow

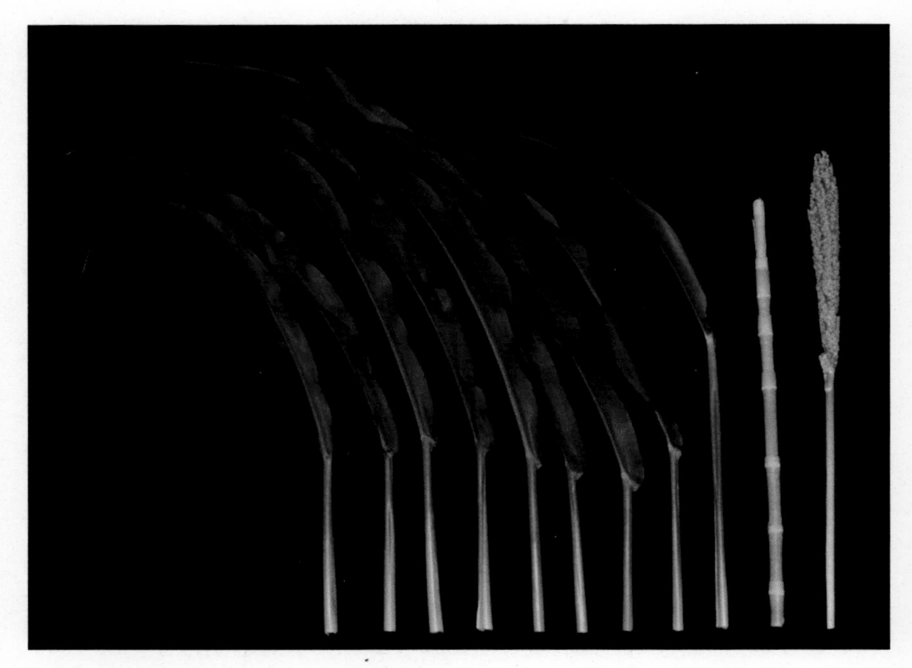

Fig. 8. Dissected Stage 5 sorghum plant.

exsertion can negatively affect pollination, kernel set, and grain size and slows harvest because more leaves must pass through the combine.

Stage 6: Half Bloom

At approximately 2 mo after emergence (55 to 65 d or 800 to 1060 GDU), the sorghum plant will be at the half bloom stage. On an individual plant basis, half bloom is defined as when flowering (defined here as appearance of the anthers) has progressed halfway down the panicle. When applied to a field or section of a field, half bloom usually is defined as when one half of the plants in the area under consideration are in some stage of bloom. Flowering of an individual head in commercial hybrids occurs over a period of 3 to 5 d or longer depending on temperature. Anthers first appear at the top of the panicle with a new flush of anthers moving down the panicle in a lower band each day. All later stages of grain development (soft dough, hard dough, and physiological maturity) follow this same pattern, with grains at the base of the head being the last to reach each stage. Anthesis occurs roughly from midnight to 10 AM each day, peaking between 6 and 8 AM (Prasad and Staggenborg, 2009).

Rapid elongation of the peduncle after the boot stage usually pushes the panicle through the flag leaf sheath and exposes the entire panicle before flowering begins. However, environmental factors, such as temperature and moisture status, and genetic factors, such as strength of exsertion and drought tolerance, can result in flowering commencing before the entire panicle has emerged. Although internode length has been reduced in modern sorghum hybrids compared with wild relatives, length of the peduncle is similar. With normal elongation of the peduncle, the panicle is placed well above the leaf canopy, facilitating combine harvest. The plant has reached maximum height at the end of flowering, (Fig. 2b and 2c).

As with most crops, the number of grains per unit area is a major determinant of final grain yield for sorghum (Heinrich et al., 1983), making successful pollination and grain set essential for maximizing yield. Prasad et al. (2008) demonstrated that the period surrounding half bloom is a time of particular vulnerability to temperature stress. High-temperature episodes consisting of 10 d of 40–30°C day–night temperatures before or after flowering reduced grain set by 22 to 54%, with the greater reductions associated with high temperatures at and immediately after flowering. Singh et al. (2015) reported similar results and demonstrated that genetic variability exists in susceptibility to yield reductions caused by high temperatures before and after anthesis. Elevated night temperatures have been linked to reductions in floret initiation during panicle development (Peacock and Heinrich, 1984). Elevated night temperatures after panicle emergence caused a number of negative physiological changes in leaves and pollen grains (Prasad and Djanaguiraman, 2011). Low minimum night temperatures before and during anthesis also can result in reduced pollen viability or even male sterility (Peacock and Heinrich, 1984; Prasad and Staggenborg, 2009). Reductions in seed set caused by temperature extremes, especially nighttime temperatures, during this critical period will have a substantial influence on grain yield, regardless of potential head size and environmental conditions during the remainder of grain development.

Approximately half of the total dry matter has been accumulated by the half bloom stage (Fig. 1). With all leaves fully expanded and the panicle completely formed (Fig. 9), this results in a grain harvest index (portion of total aboveground dry matter allocated to grain) of close to 50% if grain set and grain fill are not hindered. Harvest index values in typically stressful production environments

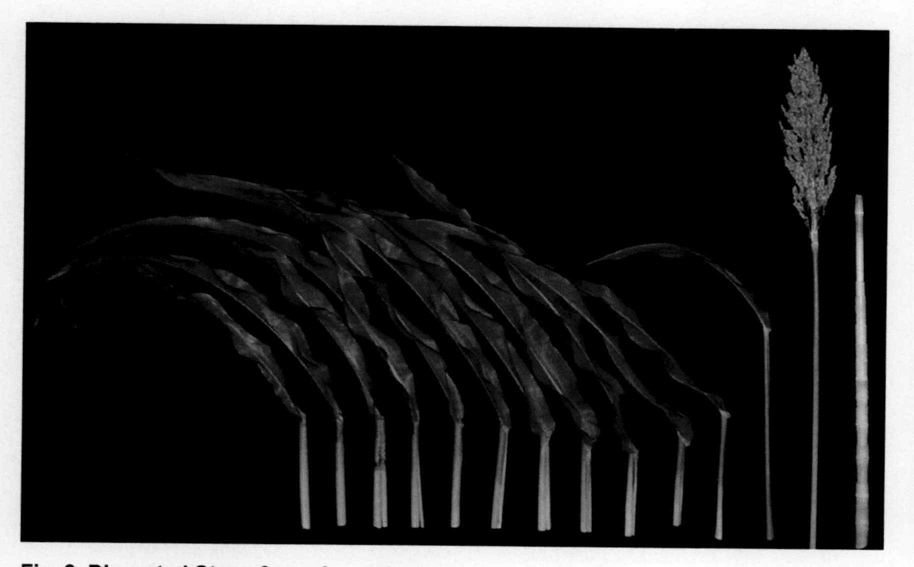

Fig. 9. Dissected Stage 6 sorghum plant.

are usually less than 50% but can be greater in optimal conditions (van Duivenbooden et al., 1995). Loss of leaf area has the greatest negative effect on yield at this time compared with leaf loss before or after, as no new leaves will be formed and essentially no grain fill has occurred. Nutrient accumulation still exceeds dry matter formation with about 95% of the total K, 65% of the total N, and 55% of the total P taken up at half bloom.

Stage 7: Soft Dough

Completion of flowering marks the beginning of grain fill (Growth Stage 3; Eastin, 1972), with substantial increases in water content and grain volume during the first several days after anthesis. Approximately 10 d after half bloom (65 to 75 d or 970 to 1220 GDU after emergence) the grain has reached the soft dough stage. The volume of individual kernels has been determined, but dry matter will accumulate in the grain for many days.

Yang et al. (2010) identified three distinct phases of grain fill based on rates of change in kernel water content, volume, and dry weight. All three parameters increase rapidly during the first phase, and rates of increase in individual grain volume and water content are maximized. During the second phase, dry matter accumulates at its maximum rate, volume increases at a slightly reduced rate compared with phase one, and water content is fairly constant. Dry matter accumulation continues, but at a slower rate, while volume stabilizes, and water content decreases during the third phase. The end of the third phase is defined as the point when dry matter accumulation ceases and corresponds with physiological maturity. Water content and volume both decline after maturity.

Genetic control of grain size appears to be mediated by extending the time in phase one so that individual grains reach a larger volume, providing a larger receptacle for assimilation on a per grain basis. This translates into a longer total duration of grain fill and heavier grains. Sorghum breeding lines have been developed to maximize grain fill period, which results in greater grain weight and yield (Tuinstra et al., 2001). Although individual grains can be two- or threefold heavier in these lines compared with normal sorghum lines, yields generally do not increase by the same magnitude because grain number per plant is reduced (Yang et al., 2010). The relatively soft consistency of the grain during the early stages of filling makes it particularly susceptible to insect pests during this stage. This type of insect feeding is particularly damaging because they are directly reducing economic product.

Nutrient uptake continues, but almost all K is already in the plant (Fig. 1f). Approximately 30% of N and 35% of P must still be taken up by the plant, illustrating the importance of adequate season-long supplies of these nutrients. Although stem weight continues to increase for a short time after bloom, it soon begins to decrease as grain filling takes priority for current and stored carbohydrates (Fig. 1). The loss in stem weight can account for up to 10% or moreof grain dry matter. Lower leaves continue to senesce during this period, as illustrated by the loss of leaf area in Fig. 2a. The extent of lower leaf loss depends on environmental, management, and genetic factors. Both drought stress and N deficiency increase rate of lower leaf loss. Hybrids with the ability to maintain green leaf area during grain filling have canopy and root development characteristics that reduce canopy size at flowering. This reduces preflowering drought stress and allows them to better maintain grain fill rate and duration, even with some degree of drought stress (Borrell et al., 2000, 2014).

Stage 8: Hard Dough

From soft to hard dough, grain filling continues at a rapid pace for 10 to 15 d so that close to 90% of grain dry matter has been accumulated (Fig. 1c) at this stage (80 to 90 d or 1220 to 1470 GDU after emergence). Weight loss from the stem has leveled off, but leaf loss can continue. Susceptibility to insect feeding on the grain decreases as the kernel hardens. Nutrient uptake is approaching completion, but substantial yield loss can occur if severe drought stress or frost terminate grain filling before physiologic maturity (Staggenborg and Vanderlip, 1996). Given the relatively consistent duration of grain fill across maturities for most commercial hybrids, managing sorghum planting and hybrid maturity so heading occurs early enough to avoid late season drought or early frost is critical for allowing the sorghum plant to fully fill grain (Shroyer et al., 1987).

Stage 9: Physiological Maturity

This stage is marked by maximum accumulation of dry matter in the grain (Fig. 1c–1e) and is reached 90 to 100 d after emergence (1420 to 1670 GDU). An important visual indicator is formation of the black layer, a dark spot at the base of the grain opposite of the embryo. Formation of this abscission layer indicates that the vascular connection between the panicle and the kernel has been severed, preventing additional dry matter accumulation. Grain moisture content at physiological maturity can range from 25 to 35% depending on hybrid and environmental conditions (Gigax and Burnside, 1976; Vanderlip and Reeves, 1972; Yang et al., 2010). About one third of the leaf area has been lost (Fig. 2a) along with more than 15% of K and a small portion of N (Fig. 1f), either through leaf loss or leaching from senesced leaves.

Grain fill duration, the time from flowering until physiological maturity, can be modified by several factors, but generally represents about one third of the time from planting to physiological maturity. With similar rates of dry matter accumulation for most commercial hybrids, duration of grain filling becomes an important determinant of grain yield. If soil moisture and temperature are not limiting, longer grain fill duration, whether due to genetic or environmental factors, usually results in greater grain yield.

At this stage the sorghum plant is still active unless it is killed by severe drought or freezing temperatures. With favorable temperature and moisture, tillers may appear at upper nodes even after harvest.

Conclusions

Sorghum is a remarkably resilient plant, with the ability to compensate for changing resource availability (Heinrich et al., 1983), to tolerate high temperatures, and to produce grain in water-limited environments (Stone et al., 2006). With adequate water, temperatures, and nutrients, within a month of emergence the sorghum plant has built the framework of its photosynthetic infrastructure and transitions to reproductive development, although vegetative growth continues as leaves expand and the stem elongates. After another month, the plant has essentially reached full stature and is dedicating all resources to grain production, to the point of remobilizing resources from vegetative plant parts. Assuming growth and development has gone well and depending on environmental conditions during grain fill, physiological maturity will be reached in about another month. This is followed by a period of grain drying to ultimately reach harvest maturity, allowing for combine harvest and safe storage of grain. Properly identifying the growth stages of the sorghum plant and understanding how dry matter and nutrients are accumulating at those stages facilitates management of a sorghum crop to maximize production and will provide the basis for understanding deviations from normal development that may result from environmental stresses, management decisions, diseases, or insects.

Acknowledgments

Richard Vanderlip authored the first version of S-3, "How a sorghum plant develops," and deserves credit for the framework used here to describe sorghum development. Bob Holcombe, Kansas State University, Department of Communications and Agricultural Education, took the digital photographs used to illustrate various growth stages. The United Sorghum Checkoff Program and K-State Center for Sorghum Improvement provided funding for generating dry matter and nutrient accumulation data and digital images for the most recent revision of S-3.

References

Abdullahi, A., and R.L. Vanderlip. 1972. Relationships of vigor tests and seed source and size to sorghum seedling establishment. Agron. J. 64:143–144. doi:10.2134/agronj1972.00021962006400020004x

Anda, A., and L. Pinter. 1994. Sorghum germination and development as influenced by soil temperature and water content. Agron. J. 86:621–662. doi:10.2134/agronj1994.00021962008600040008x

Borrell, A.K., G.L. Hammer, and R.G. Henzell. 2000. Does maintaining green leaf area in sorghum improve yield under drought? II. Dry matter production and yield. Crop Sci. 40:1037–1048. doi:10.2135/cropsci2000.4041037x

Borrell, A.K., J.E. Mullet, B. George-Jaeggli, E.J. van Oosterom, G.L. Hammer, P.E. Klein, and D.R. Jordan. 2014. Drought adaptation of stay-green sorghum is associated with canopy development, leaf anatomy, root growth, and water uptake. J. Exp. Bot. 65:6251–6263. doi:10.1093/jxb/eru232

Brar, G.S., and B.A. Stewart. 1994. Germination under controlled temperature and field emergence of 13 sorghum cultivars. Crop Sci. 34:1336–1340. doi:10.2135/cropsci1994.0011183X003400050036x

Eastin, J.D. 1972. Photosynthesis and translocation in relation to plant development. In: N.G. Prasada Rao and L.R. House, editors, Sorghum in seventies. Oxford and IBH, New Delhi, India.

Gerik, T., B. Bean, and R. Vanderlip. 2003. Sorghum growth and development. Texas Agricultural Experiment Station and Cooperative Extension. Texas A&M Univ. Syst., College Station, TX.

Gigax, D.R., and O.C. Burnside. 1976. Chemical desiccation of grain sorghum. Agron. J. 68:645–649. doi:10.2134/agronj1976.00021962006800040028x

Hammer, G.L., R.L. Vanderlip, G. Gibson, L.J. Wade, R.G. Henzell, D.R. Younger, J. Warren, and A.B. Dale. 1989. Genotype-by-environment interaction in grain sorghum. II. Effects of temperature and photoperiod on ontogeny. Crop Sci. 29:376–384. doi:10.2135/cropsci1989.0011183X002900020029x

Heinrich, G.M., C.A. Francis, and J.D. Eastin. 1983. Stability of grain sorghum yield components across diverse environments. Crop Sci. 23:209–212. doi:10.2135/cropsci1983.0011183X002300020004x

Kansas Mesonet. 2016. Kansas Mesonet agriculture growing degree days. Dep. of Agron., Kansas State Univ.. http://mesonet.k-state.edu/agriculture/growingdegrees/ (accessed 2 July 2016).

Lafarge, T.A., I.J. Broad, and G.L. Hammer. 2002. Tillering in grain sorghum over a wide range of population densities: Identification of a common hierarchy for tiller emergence, leaf area development and fertility. Ann. Bot. (Lond.) 90:87–98. doi:10.1093/aob/mcf152

Lingenfelser, J., D. Jardine, J. Whitworth, and M. Knapp. 2014. 2014 Kansas performance tests with grain sorghum hybrids. SRP1113. Kansas State Univ. Agric. Exp. Stn. and Coop. Ext. Serv., Manhattan, KS.

Maiga, A. 2011. Effect of planting practices and nitrogen management on grain sorghum production. Ph.D. thesis, Kansas State Univ., Manhattan, KS.

Maranville, J.W., and M.D. Clegg. 1977. Influence of seed size and density on germination, seedling emergence, and yield of grain sorghum. Agron. J. 69:329–330. doi:10.2134/agronj1977.00021962006900020032x

Mortlock, M.Y., and R.L. Vanderlip. 1989. Germination and establishment of pearl millet and sorghum of different seed qualities under controlled high-temperature environments. Field Crops Res. 22:195–209. doi:10.1016/0378-4290(89)90092-0

Peacock, J.M., and G.M. Heinrich. 1984. Light and temperature responses in sorghum. In: S.M. Virmani and M.V.K. Sivakumar, editors, Agrometeorology of sorghum and millet in the semi-arid tropics: Proceedings of the International Symposium, Patancheru, India. 15–20 Nov. 1982. Int. Crops Res. Inst. for the Semi-Arid Tropics, Patancheru, India. p. 143–158.

Prasad, P.V.V., and M. Djanaguiraman. 2011. High night temperature decreases leaf photosynthesis and pollen function in grain sorghum. Funct. Plant Biol. 38(12):993–1003. doi:10.1071/FP11035

Prasad, P.V.V., S.R. Pisipati, R.N. Mutava, and M.R. Tuinstra. 2008. Sensitivity of grain sorghum to high temperature stress during reproductive development. Crop Sci. 48:1911–1917. doi:10.2135/cropsci2008.01.0036

Prasad, P.V.V., and S.A. Staggenborg. 2009. Growth and production of sorghum and millets, soils, plant growth and crop production. In: H. Verheye, editor, Encyclopedia of life support systems (EOLSS) developed under the auspices of the UNESCO. Eolss Publishers, Oxford, UK.

Roozeboom, K., D. Ruiz Diaz, D. Jardine, C. Thompson, R.J. Whitworth, and D.H. Rogers. 2013. Kansas Sorghum Management 2013. MF3046. Agric. Exp. Stn. and Coop. Ext. Serv., Kansas State Univ., Manhattan, KS.

Roozeboom, K., S. Staggenborg, D. Shoup, K. Martin, S. Duncan, and B. Olson. 2010. Planting. In: J. Dahlberg, E. Roemer, J. Casten, G. Kilgore, and J. Vorderstrasse, editors, Central and eastern plains production handbook. United Sorghum Checkoff Program, Lubbock, TX. p. 22–30.

Rosenthal, W.D., R.L. Vanderlip, B.S. Jackson, and G.F. Arkin. 1989. SORKAM: A grain sorghum crop growth model. Research Center Program and Model Documentation. MP-1669. Texas Agric. Exp. Stn., College Station, TX.

Shroyer, J.P., R.L. Vanderlip, D.L. Bark, J.A. Schaffer, and T.L. Walter. 1987. Probability of sorghum maturing before freeze. AF-162. Agric. Exp. Stn. and Coop. Ext. Serv., Kansas State Univ., Manhattan, KS.

Singh, V., C.T. Nguyen, E. van Oosterom, D. Jordan, S. Chapman, G. McLean, B. Zheng, and G. Hammer. 2015. Heat stress effects on grain sorghum productivity-biology and modelling, in "Building Productive, Diverse and Sustainable Landscapes," Proceedings of the 17th ASA Conference, Hobart, Australia. 20–24 Sept. 2015.

Staggenborg, S.A., R.L. Vanderlip, G.J. Roggenkamp, and K.D. Kofoid. 1999. Methods of simulating freezing damage during sorghum grain fill. Agron. J. 91:46–53. doi:10.2134/agronj1999.00021962009100010008x

Staggenborg, S.A., and R.L. Vanderlip. 1996. Sorghum grain yield reductions caused by duration and timing of freezing temperatures. Agron. J. 88:473–477. doi:10.2134/agronj1996.00021962008800030019x

Stone, L.R., A.J. Schlegel, A.H. Khan, N.L. Klocke, and R.M. Aiken. 2006. Water supply: Yield relationships developed for study of water management. J. Nat. Resour. Life Sci. Educ. 35:161–173.

Tuinstra, M.R., G.L. Liang, C. Hicks, K.D. Kofoid, and R.L. Vanderlip. 2001. Registration of KS 115 sorghum. Crop Sci. 41:932–933. doi:10.2135/cropsci2001.413932x

Vanderhoef, L.N., and W.R. Briggs. 1978. Red light-inhibited mesocotyl elongation in maize seedlings. I. The auxin hypothesis. Plant Physiol. 61:534–537. doi:10.1104/pp.61.4.534

Vanderlip, R.L. 1993. How a sorghum plant develops. S-3. Agric. Exp. Stn. and Coop. Ext. Serv., Kansas State Univ., Manhattan, KS.

Vanderlip, R.L., J.D. Ball, P.J. Banks, F.N. Reece, and S.J. Clark. 1977. Flaming grain sorghum to delay flowering. Crop Sci. 17:902–905. doi:10.2135/cropsci1977.0011183X001700060021x

Vanderlip, R.L., F.E. Mockel, and H. Jan. 1973. Evaluation of vigor tests for sorghum seed. Agron. J. 65:486–488. doi:10.2134/agronj1973.00021962006500030039x

Vanderlip, R.L., and H.E. Reeves. 1972. Growth stages of sorghum [Sorghum bicolor, (L.) Moench.]. Agron. J. 64:13–16. doi:10.2134/agronj1972.00021962006400010005x

van Duivenbooden, N., C.T. de Wit, and H. van Keulen. 1995. Nitrogen, phosphorus and potassium relations in five major cereals reviewed in respect to fertilizer recommendations using simulation modelling. Fert. Res. 44:37–49. doi:10.1007/BF00750691

Yang, Z., E.J. van Oosterom, D.R. Jordan, A. Doherty, and G.L. Hammer. 2010. Genetic variation in potential kernel size affects kernel growth and yield of sorghum. Crop Sci. 50:685–695. doi:10.2135/cropsci2009.06.0294

Structure and Composition of the Sorghum Grain

S.R. Bean,* J.D. Wilson, R.A. Moreau, A. Galant, J.M. Awika, R.C. Kaufman, S.L. Adrianos, and B.P. Ioerger

Abstract

Sorghum has a high degree of genetic diversity, and, as such, sorghum grain composition and structure can vary widely. Such variability can be of great benefit in supplying a diversity of uses but can also be a negative when uniformity is desired. Despite sharing similarities to other cereals such as maize and the millets, sorghum has several unique attributes related to the chemistry and composition of the grain. Since grain composition is linked to utilization, this chapter focuses on the basic composition and structure of major grain components and also how such attributes relate to new, upcoming, or potential uses of sorghum.

Sorghum ranks fifth in the world with respect to overall grain production (with the first four grains being maize, rice, wheat and barley) and accounted for 2.2% of total worldwide grain production in 2013 (Food and Agriculture Organization, 2014). Regional production may vary; for example, sorghum ranks third in grain production (behind wheat and maize) in the United States (Food and Agriculture Organization, 2014) and in parts of the semiarid tropics. Sorghum is a primary crop and basic food staple for large numbers of people (Belton and Taylor, 2004).

Use of sorghum grain can vary widely by location. As mentioned above, sorghum grain is a food staple for millions of people in many parts of the world and is consumed in a wide range of traditional food and beverage products (Murty and Kumar, 1995; Rooney and Waniska, 2000). In other parts of the world, sorghum is primarily an animal feed, while in the United States sorghum grain is increasingly used for the production of ethanol for the biofuels market (Wang et al., 2008a; Jessen, 2010).

Abbreviations: DDGS, distillers dried grains with solubles; HPLC, high-performance liquid chromatography; SDS-PAGE, sodiumdodecylsulfate polyacrylamide electrophoresis.

S.R. Bean, J.D. Wilson (jeff.d.wilson@ars.usda.gov), A. Galant (a.l.galant@gmail.com), R.C. Kaufman (rhett.kaufman@bayer.com), S.L. Adrianos (sherry.adrianos@ars.usda.gov), and B.P. Ioerger (brian.ioerger@ars.usda.gov), Center for Grain and Animal Health Research, USDA-ARS, 1515 College Ave., Manhattan, KS 66502; A. Galant, currently at Busch Agricultural Resources, 2101 26th St. S, Moorhead, MN 56560; R.C. Kaufman, currently at Bayer CropScience LP, 925 County Rd. 378, Beaver Crossing, NE 68313 (rhett.kaufman@bayer.com); R.A. Moreau (robert.moreau@ars.usda.gov), Eastern Regional Research Center, USDA-ARS, 600 E. Mermaid Ln., Wyndmoor, PA 19038; and J.M. Awika (jawika@ag.tamu.edu), Soil and Crop Science Dep. and Nutrition and Food Science Dep., Texas A&M Univ., 429 Heep Center, College Station, TX 77843. *Corresponding author (srbean@ksu.edu, scott.bean@ars.usda.gov).

doi:10.2134/agronmonogr58.2014.0081

 Sorghum: State of the Art and Future Perspectives, Ignacio Ciampitti and Vara Prasad, editors. Agronomy Monograph 58.

Grain chemistry and structure is an important factor in governing the end-use quality of any cereal grain and is important for all types of grain utilization, including food, feed, and fuel. Sorghum is similar in some aspects to other grains, especially maize. However, sorghum also varies considerably in some aspects from other cereal grains. For example, despite similarities between maize and sorghum proteins, sorghum has much lower protein digestibility than maize (Duodu et al., 2003). Some sorghum germplasm is also very unique in terms of phenolic content and composition (Awika and Rooney, 2004).

Sorghum is also increasingly being incorporated into gluten-free foods in Western countries (Taylor et al., 2006). "Specialty" sorghum germplasm with unique phenolic compounds with potential health-promoting properties are being developed and studied (Awika and Rooney, 2004). Biomaterial uses of sorghum proteins such as bioplastic films (Taylor et al., 2005), micro-encapsulation agents (Elkhalifa et al., 2009; Taylor et al., 2009), and adhesives have also been reported (Li et al., 2011). The role of sorghum proteins in bioplastic applications has recently been reviewed by Taylor et al. (2013).

This chapter then, focuses on the physical structure of sorghum grain, the chemistry and composition of major components, and, where relevant, discussion of unique, new, or up and coming uses of sorghum grain related to specific grain components.

Physical Structure of the Grain

Grain Size and Physical Structure

The sorghum grain or seed, technically a caryopsis (Evers and Millar, 2002), can vary widely in physical attributes, including shape, size, color, and hardness. While obviously round, sorghum grains will often have one flattened surface (Reichert et al., 1988). Because of the genetic diversity of sorghum, grain can vary widely in size and shape with thousand-kernel weight for sorghum varying from 30 to 80 g (Rooney and Serna-Saldivar, 2000; Chiremba et al., 2012). A wide range of individual grain weight has been reported for sorghum, ranging from 3 to 80 mg (Serna-Saldivar and Rooney, 1995; Lásztity, 1996; Bean et al., 2006). In a population grown at the same location and grain size measured with the same instrument, Sukumaran et al. (2012) reported grain weight ranging from ~19 to 29 mg and a diameter of ~1.3 to 2 mm across a genetically diverse association mapping panel.

As is typical of cereal grains in general, and as described in the literature, sorghum grain is composed of three main components, the pericarp, endosperm, and germ (Rooney and Miller, 1982, Evers and Millar, 2002) (Fig. 1). Naturally, the amounts of these components will vary, but a general composition of a sorghum grain has been reported to be 3 to 6% pericarp, 84 to 90% endosperm, and 5 to 10% germ (Hubbard et al., 1950; Rooney and Miller, 1982; Rooney and Serna-Saldivar, 2000). The composition of these tissues varies substantially, as is shown in Table 1. The pericarp consists of multiple layers, including the epicarp, mesocarp, and endocarp (Waniska and Rooney, 2000). Sorghum is unique in that it is the only cereal to have starch granules present in the pericarp (Rooney and Miller, 1982; Zeleznak and Varriano-Marston, 1982; Evers and Millar, 2002). Pericarp thickness is variable, is not of uniform thickness within a single grain, and is related to the

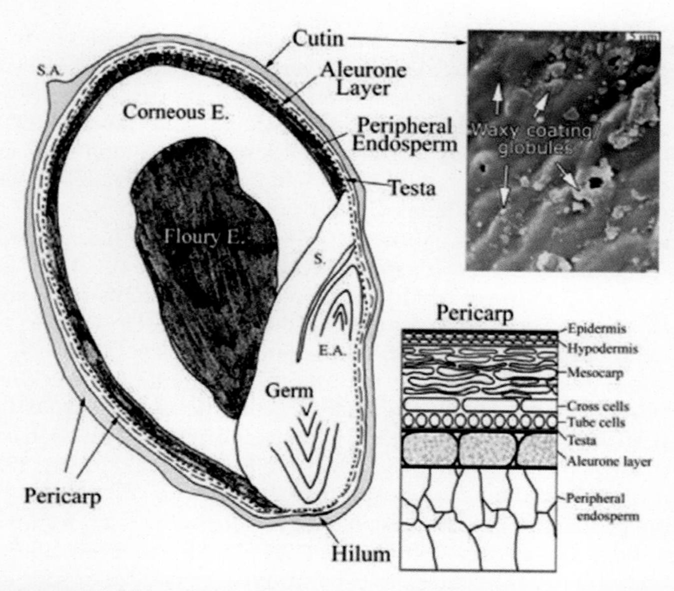

Fig. 1. Structure of the sorghum grain. S.A., stylar area; E.A., embryonic axis; S., scutellum; E., endosperm. From Earp et al. (2004a) with permission.

Table 1. Composition of sorghum grain components. Adapted from Waniska and Rooney (2000) with permission.

	Caryopsis	Endosperm	Germ	Pericarp
Caryopsis	100	84.2	9.4	6.5
Range	–	81.7–86.5	8.0–10.9	4.3–8.7
Protein	11.3	10.5	18.4	6
Range	7.3–15.6	8.7–13.0	17.8–19.2	5.2–7.6
Distribution	100	80.9	14.9	4
Fiber	2.7	–	–	–
Range	1.2–6.6	–	–	–
Distribution	100	–	–	–
Lipid	3.4	0.6	28.1	4.9
Range	0.5–5.2	0.4–0.8	26.9–30.6	3.7–6.0
Distribution	100	13.2	76.2	10.6
Ash	1.7	0.4	10.4	2
Range	1.1–2.5	0.3–0.4	–	–
Distribution	100	20.6	68.6	10.8
Starch	71.8	82.5	13.4	34.6
Range	55.6–75.2	81.3–83.0	–	–
Distribution	100	94.4	1.8	3.8

amount of starch in the mesocarp (Earp et al., 2004a). The outer layer of the pericarp is covered with wax, as is shown in Fig. 1.

The endosperm in cereal grains is composed of the aleurone layer and "starchy endosperm" (Evers and Millar, 2002). In sorghum the starchy endosperm has been divided into the peripheral, vitreous (or corneous), and opaque (or floury) endosperm (Waniska, 2000). The aleurone layer contains both protein and lipid bodies along with inclusion bodies possibly containing phytin (Zeleznak and Varriano-Marston, 1982; Serna-Saldivar and Rooney, 1995). The peripheral endosperm lies below the aleurone layer and is characterized by having dense layers of cells with high protein concentration and small starch granules (Zeleznak and Varriano-Marston, 1982; Serna-Saldivar and Rooney, 1995). Waxy sorghum were found to have less peripheral endosperm than nonwaxy sorghum genotypes, which was speculated to play a role in improved starch digestibility of waxy sorghum (Sullins and Rooney, 1974).

The vitreous endosperm is tightly packed with protein bodies covered with a continuous protein matrix (Seckinger and Wolf, 1973; Hoseney et al., 1974). Starch granules in this area of the sorghum kernel often show indentations where protein bodies were pressed into the surface of the granules (Hoseney et al., 1974). In contrast, the floury endosperm in the center of the kernel is loosely packed with a discontinuous protein matrix and round starch granules (Seckinger and Wolf, 1973; Hoseney et al., 1974). The relative proportions of corneous to floury endosperm can vary widely in sorghum, with overall grain hardness in sorghum correlated to the percent vitreousity of the grain (Hallgren and Murty, 1983). The structure of vitreous and floury endosperm is shown in Fig. 2.

Sorghum proteins appear to play an integral role in the relationship of endosperm type and grain hardness. Shull et al. (1990) investigated the differences in proteins from vitreous and floury endosperm during grain fill of sorghum varieties varying in endosperm texture. They observed the presence of a continuous protein matrix in the vitreous endosperm as well as faster development of a protein matrix correlated with the harder sorghum varieties. Figure 3 shows changes to the physical structure of the sorghum grain during grain fill. Mazhar and Chandrashekar (1993) noted the presence of greater cross-linking within kafirins among predominately hard endosperm cultivars. In a review of "grain strength" in sorghum and maize by Chandrashekar and Mazhar (1999), the role of proteins in sorghum grain strength is summarized.

The germ is composed of the embryonic axis and scutellum and contains lipid, protein, and minerals (Waniska, 2000). The majority of all lipid found in sorghum grain is located in the germ (Waniska and Rooney, 2000). Along with

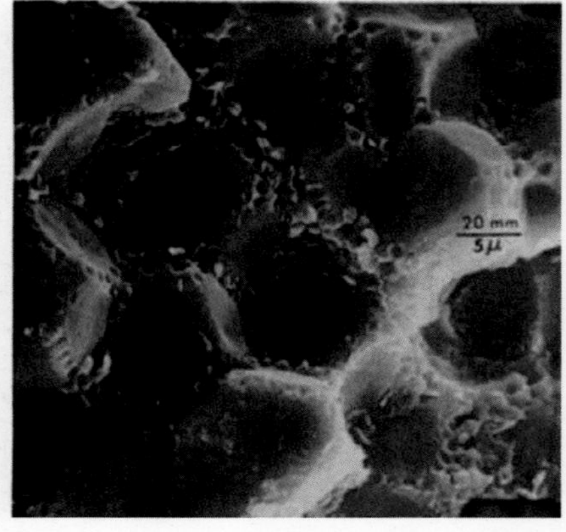

Fig. 2. Scanning electron micrographs of floury endosperm (left) and vitreous endosperm (right). *P*, protein bodies. From Hoseney et al. (1974) with permission.

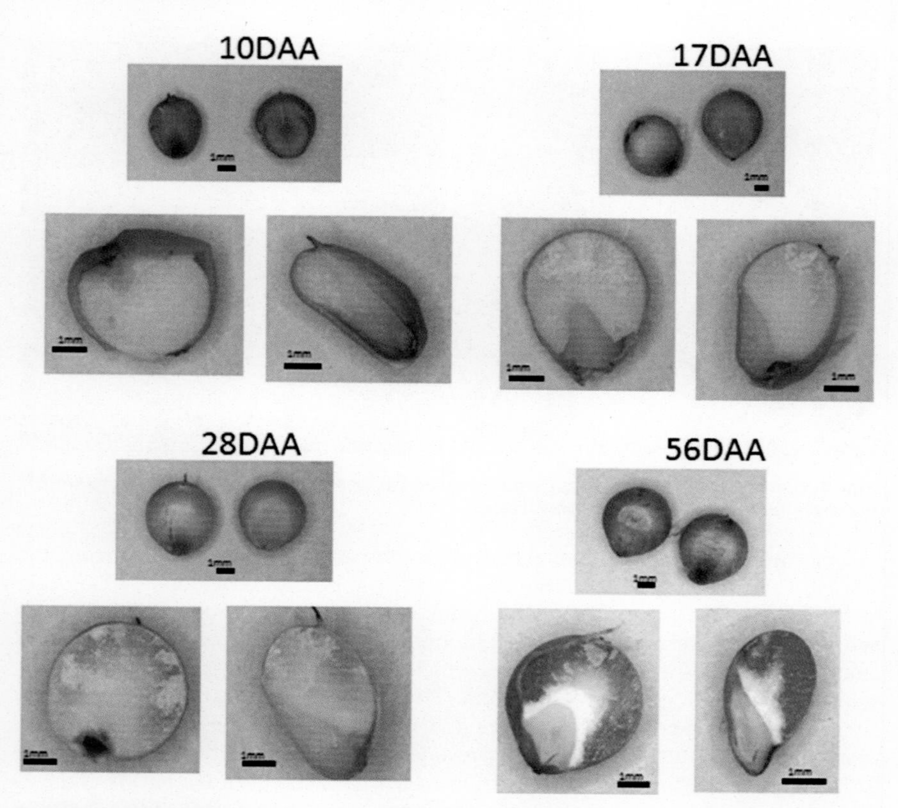

Fig. 3. Changes to sorghum grain physical structure during grain development. DAA, days after anthesis.

the outer layers of the grain, the germ contains most of the vitamins and minerals found in sorghum (Waniska and Rooney, 2000).

Grain Color

The color of sorghum grain varies widely and can be various shades and hues of white, yellow, red, and black. The color is influenced by pericarp color and thickness, the color the endosperm, and the presence of a testa layer (Rooney and Miller, 1982, Rooney, 2000). The genetics of grain color in sorghum have been characterized, and it is known that pericarp color is determined directly by *R* and *Y* genes with *RY* producing red color, *rrY* producing yellow, and *RRyy* or *rryy* producing a white pericarp. Recently there has been increasing interest in sorghum genotypes that have a genetically red pericarp that turns black when exposed to sunlight (Dykes et al., 2005, 2009) (Fig. 4). Pericarp colors are influenced in turn by other genes such as the intensifier, *I*, which increases color in sorghum with red and yellow pericarp (Rooney and Miller, 1982; Rooney, 2000). In addition to genetic factors, biotic and abiotic factors can also impact sorghum grain color; for example, grain infected with molds can be discolored (Waniska, 2000).

Genes which influence the physical properties of the outer layers of the grain can additionally modulate pericarp color. The thickness of the mesocarp

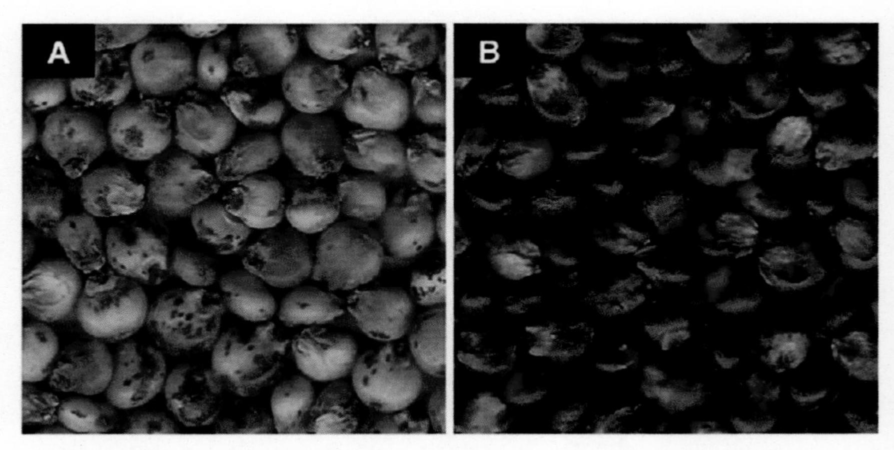

Fig. 4. Black sorghum grain that was covered (a) and uncovered (b) during grain fill. From Dykes et al. (2009) with permission.

is determined by the *Z* gene, and a thick mesocarp (*zz*) masks color from the pericarp and endosperm and induces a "chalky appearance" (Rooney, 2000). Conversely, a thin mesocarp (*ZZ* or *Zz*) produces a "translucent appearance" through which the endosperm color is visible (Rooney, 2000). A sorghum genotype may or may not also have a pigmented testa which is under the control of the B_1 and B_2 genes. When both genes are dominant (B_1B_2), a pigmented testa will be present; when one or both genes are recessive (b_1B_2, B_1b_2, b_1b_2), then there is no pigmented testa (Earp et al., 1983, 2004b). The presence of a pigmented testa can influence kernel color either by being visible in lines with a white pericarp and thin mesocarp or through the action of the spreader gene (*S*) (Rooney and Miller, 1982). A complete list of genes known to influence grain color (and many other traits) in sorghum can be found in Rooney (2000). The development of the testa in sorghum has recently been reexamined by using a combination of microscopy techniques (Earp et al., 2004b).

Endosperm color can also vary in sorghum, being either yellow or white with the yellow color caused by carotenoids in the endosperm (Rooney and Miller, 1982). Development of sorghum with yellow endosperm, and thus increased carotenoid levels, may help to reduce vitamin deficiencies in parts of the world that rely on sorghum as a food staple (Salas Fernandez et al., 2008, 2009). In some cases the color of the endosperm can also influence the outer color of the grain, particularly in the case of genotypes with thin outer layers and no pigmented testa (Serna-Saldivar and Rooney, 1995). The range of sorghum grain and endosperm color and associated genetics responsible for the colors are shown in Fig. 5.

Starch Chemistry and Composition

Sorghum Starch Properties

Starch of cereal grains is synthesized and deposited in the endosperm to function as an energy reserve during reproduction and development. In wild-type grains, starch consists of two distinct glucan polymers, amylose and amylopectin. Amylopectin is a large, highly branched polymer consisting of α-1,4 linked D-glucose

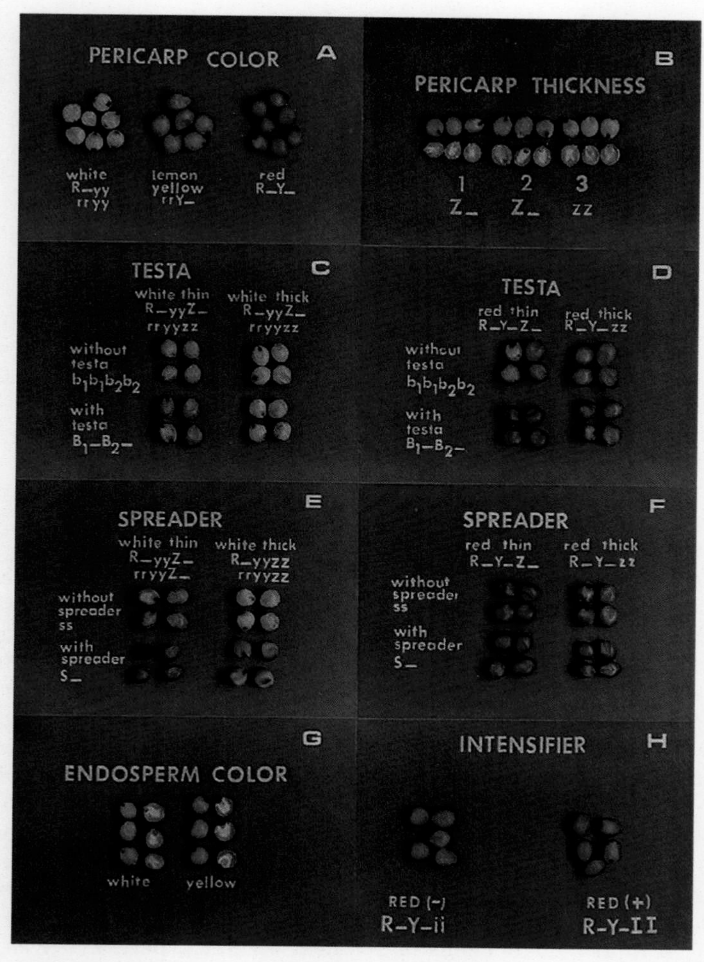

Fig. 5. Pericarp and endosperm color of sorghum along with associated genetics for each color as follows: (a) gene combinations and possible pericarp colors, (b) pericarp thickness examples, (c) gene combinations of white kernels and impact of testa layer, (d) gene combinations of red kernels and impact of testa layer, (e) impact of spreader gene on white kernels with testa layer, (f) impact of spreader gene on red kernels with testa layer, (g) endosperm color variants, and (h) effect of intensifier gene on kernel color. From Rooney and Miller (1982) with permission.

units with branches linked by α-1,6 bonds. Amylose is a mostly linear polymer of α-1,4 linked D-glucose with a few α-1,6 branch points.

The ratio of amylose to amylopectin is important to both the functionality and the nutritional properties of starch and starch based products. Amylose contributes to the thermal characteristics of starch, such as gelatinization and pasting (Jane et al., 1999; Sasaki et al., 2000). The ratio of amylose/amylopectin also influences starch retrogradation, a major issue in the staling of food products (Hug-Iten et al., 2003). Foods with high amylose content have been shown to have a reduction in glycemic index, which promotes many health benefits, such as better control of diabetes and obesity (Behall and Scholfield, 2005). The amylose-to-amylopectin ratio has been reported to vary widely in sorghum, with amylose contents from 0 to 55% (Zhu, 2014). The majority of sorghum has 20 to 30% amylose, which is typical for "normal" cereal grains (Zhu, 2014). However, sorghum, like other cereals, has "waxy" sorghum lines in which little or no amylose is produced (where the term waxy refers only to the amount of amylose present in the starch and is not to be confused with the wax compounds found on the outer layers of the grain as discussed later in the chapter). Several variants of granule bound starch synthase (GBSS) related to the waxy trait have been characterized in sorghum (Pedersen et al., 2005, 2007). Waxy sorghum lines have been reported to have inferior agronomic performance compared with "normal" sorghum lines (Rooney et al., 2005), though selection for specific high-yield waxy sorghum lines may be possible (Jampala et al., 2012). The production of heterowaxy sorghum lines is also possible which phenotypically express the waxy trait in a 1:3 ratio (waxy to nonwaxy) (Lichtenwalner et al., 1978).

In addition to waxy sorghum lines, recently Hill et al. (2012) reported sorghum with an amylose content of 55.8%, which may represent a genetic resource for high amylose sorghum development.

After glucan polymers are synthesized they are bundled into discreet water insoluble granules. The granules are produced in the cellular organelle referred to as the amyloplast; most cereal grains produce one granule per amyloplast. However, in rice and oat multiple small granules are produced in one amyloplast, which results in the formation of compound granules. The entire process of starch granule formation is not fully understood, but some research suggests that the process is more physical than biological. Research using X-ray scattering and nuclear magnetic resonance has demonstrated that amylopectin may be structured as a side-chain liquid crystalline polymer (Waigh et al., 1998; Waigh et al., 2000) which self-assembles into an ordered lamellae. This physical approach to granule formation would only be possible if the amylopectin was synthesized to the proper structure. Thus, the enzymatic processes must be properly controlled so that amylopectin molecules can begin forming a granule. Studies involving mutants with altered functions of synthetic enzymes, particularly starch synthase III and starch synthase IV, have revealed that when the enzymes are not present starch granule formation is altered (Roldán et al., 2007; Szydlowski et al., 2009). Amylose is not necessary for the formation of the starch granule, as waxy mutants exhibit granules with similar physical properties as nonwaxy starches.

The sizes and shapes of the granules are affected by the species as well as the overall architecture of the cereal caryopsis. Starch granule size distribution has been measured by many techniques, including sieving (Evers et al., 1974), image analysis from light microscopy, and laser diffraction sizing (Wilson et al., 2006). An example of sorghum starch granule size distribution as measured by laser diffraction sizing is shown in Fig. 6.

Starch granules may be present in a range of sizes, with size distribution being mainly bimodal or trimodal. The difference in granule size is thought to be the result of production in different phases of endosperm development (Parker, 1985). The small granules are possibly produced by amyloplast stromules (Langeveld et al., 2000; Bechtel and Wilson, 2003). The physiological purpose for the bimodal or trimodal distribution is unclear, but the small granules may be more

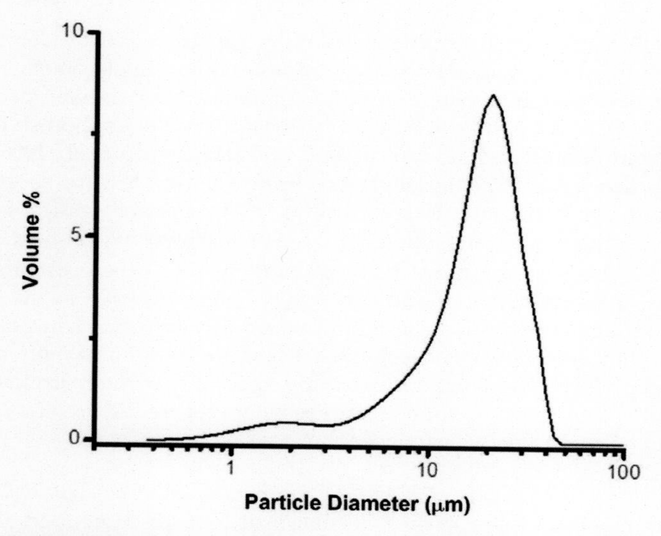

Fig. 6. Example of sorghum starch granule size distribution as measured by laser diffraction sizing.

efficient in carbon and energy storage (Tetlow, 2011). The starch granules in sorghum exhibit a wide distribution of sizes ranging from 4 to 35 μm (summarized in Zhu [2014]). Sorghum starch granules can be either polyhedral or spherical in shape (Benmoussa et al., 2006; Sang et al., 2008) (Fig. 2). Sorghum starch granules also contain channels which penetrate the surface of the granule toward the hilum (Huber and BeMiller, 2000).

The chemical composition of the starch granules consists primarily of a ratio of amylose and amylopectin (~1:4 respectively), but small amounts of protein and lipid are present. In a normal type starch the amylose is the minor component, comprising 15 to 30% by weight of the total starch content. Amylose is a mixture of linear chains of α-(1,4) linked D-glucose units. Some amylose molecules have α-(1,6) linkages occurring at approximately 0.3 to 0.5% of the total linkages. Since these branch points are usually separated by large distances, the molecules tend to act essentially as linear molecules. The molecular weight distribution of amylose in sorghum has been measured between 6.4×10^6 and 1.1×10^7 Da, with the number average degree of polymerization (DP_n) ranging from 1150 to 1390 depending on the method used to determine the DP_n and the type of sorghum (Matalanis et al., 2009; Gaffa et al., 2004).

The remaining 70 to 85% of the starch not comprised of amylose is amylopectin. Amylopectin is a very large and highly branched polymer of D-glucose or even a polymer of amylose chains. Approximately 5% of the total linkages of amylopectin are the branch points or α-(1,6) linkages. The molecular weight distribution of sorghum amylopectin has been reported as ranging from 2.8×10^8 to 3.2×10^8 Da (Matalanis et al., 2009) with a DPn of 8,900 to 12,600 (Gaffa et al., 2004).

In addition to the amylose-to-amylopectin ratio, another important characteristic of the starch is the fine structure of amylopectin. The branch chain-length distribution of the amylopectin is related to the crystalline structure of the starch

(Hizukuri, 1985). The chain-length distribution can be found by debranching the starch with an enzyme, isoamylase, and separating the oligomers by either high-performance anion exchange chromatography with a pulsed amperometric detector (HPAEC-PAD) or fluorophore-assisted capillary electrophoresis by using a laser induced fluorescence detector (FACE-LIF) (Hanashiro et al., 1996; Morell et al., 1998). Amylopectin chain length has been reported to vary substantially in sorghum germplasm (summarized in Zhu [2014]), possibly indicating either strong genetic diversity in this trait (Zhu, 2014) or a strong environmental effect.

There has been a significant body of work in the "other" cereals grains with regard to starch structure and function, whereas research in this area on sorghum has lagged behind. As sorghum usage diversifies and increases it is imperative that cereal starch biochemists examine sorghum with renewed interest to determine fine structure variability, especially with regard to the food industry. Zhu (2014) listed a number of areas where future research on sorghum starch could be most effectively focused.

Starch and Unique Opportunities for Sorghum Utilization

Starch is an important factor in the three major uses of sorghum: food, feed, and fuel. Examples of mainstream uses of sorghum in these three areas are well represented in the literature. However, there is a need to better understand how sorghum and especially sorghum starch fits in with new markets for sorghum utilization.

With regard to animal feed, cattle, broilers, and swine, numerous other animals have been found to be amenable to a diet which includes some form of grain sorghum. A brief review of how sorghum and particularly sorghum starch fits as an animal feed for nontraditional markets follows. For aquaculture, maize and soybean meal have long been the dominant feedstocks because of their widespread availability and ease of storage. However, in India white sorghum grain is significantly cheaper than soybean meal and at parity with maize; accordingly, sorghum is beginning to see use as an ingredient in commercial fish food. In Andhra Pradesh state in the mid-1990s, sorghum usage for aquaculture was reported by 3% of surveyed respondents (de Silva and Anderson, 1995); by 2002 to 2003 approximately 10,000 tons of sorghum was estimated as destined for aquaculture across the country (Food and Agriculture Organization, 2007). Like maize, sorghum can be incorporated into fish feed at up to ~50% of total feed weight (Fagbenro et al., 2003). For tilapia (*O. niloticus* x *O.aureus*), feed and fecal analysis indicates that sorghum carbohydrates are more digestible than those obtained from either maize or soybean meal at an inclusion rate of 25 to 50%. Total digestible energy of steamed, pelleted sorghum grain (12.4 kJ/g) is also higher than that of similarly treated maize (10.7 kJ/g), though lower than that reported for soybean meal (14.9 kJ/g), likely due to the higher protein digestibility of soy (Sklan et al., 2004). Likewise, tilapia (*O. niloticus*) fed a diet comprising 57.9% heated, pelleted sorghum were of similar final weight to those fed a similar diet of wheat or millet, whereas fish fed a maize-based diet were on average 13% heavier (Solomon et al., 2007). While tilapia appears to possess the metabolic processes to digest sorghum starch, other fish are less well suited. Booth et al. (2013) reported that while mulloway are able to digest nearly all of the protein in pelleted 50% sorghum grain, only 24% of the available carbohydrates were digested. For red drum, ~50% of the available carbohydrates in pelleted 30% sorghum grain was utilized, though protein digestibility was relatively reduced at 77% (McGoogan and Reigh, 1996). Future efforts to incorporate sorghum into domestic aquaculture efforts should be cognizant of the species to be cultivated and its individual digestion limitations.

The use of sorghum grain and starch is beginning to make inroads into the pet food market as well. For dogs, it is not clear that sorghum starch differs in digestibility relative to starch from other common sources. One study using mixed-breed dogs found no significant difference in the digestibility of maize, sorghum, or brewer's rice starch, and no significant difference in the digestibility of maize and sorghum protein, while rice was slightly more digestible (Carciofi et al., 2008). Another study, also with mixed-breeds, found maize, sorghum, millet, and rice starch to all be similarly digestible, and no significant difference in protein digestibility among the four grains (Fortes et al., 2010). A third study, using Spitz dogs, found rice starch to be significantly more digestible than starch from millet, maize, or sorghum, but there was no significant difference in protein digestibility among rice, maize, and sorghum with lower digestibility reported for millet (Kore et al., 2009). Ultimately, differences in starch and protein digestibility may come down to differences in hindgut digestion capabilities among individual dogs and breeds; with no difference in metabolizable energy between the various grains (Fortes et al., 2010), price per kilogram grain will likely have the largest impact on dog food formulation trends until further research is completed.

The dog food sector of the pet food market is not the only area that is seeing increased utilization of grain sorghum. Although cats are obligate carnivores and do not require the same starch intake as canines, some cat foods now incorporate ground sorghum grain as a source of carbohydrates and antioxidants, or as a binding agent to hold the rest of the formulation together (Forrester and Kirk, 2009). In a study which measured diet digestibility and insulin response in mixed-breed cats, de Oliveira et al. (2008) reported that carbohydrates from sorghum had lower digestibility than those from maize or brewer's rice, though all were above 90% digestible. Likewise, sorghum protein digestibility was not significantly different from that of maize, but less than that of rice. Of the five diets assessed, maize resulted in the largest change in blood glucose concentration, indicating that other formulations may be more appropriate for cats with limited insulin response.

A discussion of the use of sorghum for novel markets would be incomplete without mention of the use of sorghum for beer production. The use of sorghum in beer is in and of itself not a new phenomenon; in Africa sorghum starch has long been used in the small-scale (and more recently larger-scale) brewing of opaque, fruity-flavored beers with low (1–5%) alcohol content (Daiber and Taylor, 1995; Murty and Kumar, 1995; Rooney and Serna-Saldivar, 2000). In Europe and the United States, on the other hand, barley and wheat have traditionally been the preferred grains used for brewing. Of late this longstanding tradition has changed somewhat, however, due in part to the rapid expansion of the market for gluten-free foods and beverages. Baseline demand for these types of products stems from the small percentage of the population that has celiac disease, an autoimmune disorder triggered by consumption of gliadin, a gluten protein found in wheat, barley, rye, and triticale. In addition to those people with celiac disease, belief that abstaining from gluten leads to weight loss or that the human gastrointestinal system is not capable of digesting gluten has led an estimated 30% of Americans attempting to cut back on its consumption

(NPD Group/Diet Monitor, 2013). Euromonitor, which excludes products that are naturally gluten free (such as rice cakes) from its definition, estimates that the "gluten-free formulation" market was worth $486 million in 2013, but will grow 38% by 2018. While beer makes up only a subset of the potential gluten-free offering, retailers have already noticed the potential value of the market. In 2006 Anheuser-Busch introduced Redbridge, a gluten-free beer brewed from sorghum and maize, and in 2011 Ground Breaker Brewery launched as the first domestic 100% gluten-free brewhouse. Other newcomers to the market include Bard's sorghum malt beer and Lakefront Brewery's New Grist, brewed from sorghum and rice. For more information on the brewing of gluten-free beer and other specialty beers, the recent reviews by Yeo and Liu (2014) and Hager et al. (2014) are recommended.

Sorghum starch is typically utilized as food and as a source of calories by yeast, livestock, and humans; however, sorghum starch has also begun to be utilized in a number of industry-based applications. In the pharmaceutical industry, medication-containing gel capsules are traditionally produced from collagen, bone, or modified cellulose that has been infused with dye. Other polysaccharides, including sorghum starch, have been shown to be satisfactory in this role. Alebiowu and Itiola (2001) have reported that paracetamol tablets formulated with sorghum starch, plantain starch, or maize starch all displayed similar levels of compressibility and tendency to deform when pregelatinized. Sorghum starch also performs comparably to maize starch when utilized as a tablet binding agent and/or disintegrate (Deshpande and Panya, 1987; Garr and Bangudu, 1991). The same gelling and pasting properties that render sorghum starch appropriate for encapsulation also may be used to synthesize biodegradable starch films (Rodríguez-Castellanos et al., 2013).

In a strictly nonconsumable, non-food-related role, sorghum starch has been used in the formulation of packing peanuts, a type of loose-fill packaging designed for shipping delicate items. Unlike polystyrene "peanuts," sorghum-based peanuts have the advantage of being fully biodegradable, water soluble, and edible if accidently consumed by a child or pet. Finally, sorghum starch has been tested as a component of plywood adhesives. Ramos et al. (1984) reported that 20% sorghum flour was an effective plywood adhesive extender, while more recently, Hojilla-Evangelista and Bean (2011) reported a sorghum starch–protein mixture with viscosity and wet-tensile strength equal to those associated with standard plywood glues.

Protein Chemistry and Composition

Protein Content and Distribution

Total protein content of sorghum grain has been reported to range from ~7 to 15%, with the majority of the protein (~80–85%) found in the endosperm (Taylor and Schüssler, 1986; Serna-Saldivar and Rooney, 1995). The germ accounted for 9.4 to 16% of grain protein, and the pericarp contained an additional 3 to 6.5%. Table 2 illustrates a broad range of whole kernel crude protein contents and essential amino acid profiles from a variety of studies to show how these parameters vary across an array of sorghum varieties and types. Of note is the low level of lysine in sorghum, which is typically the lowest in lysine of all cereal grains (Lásztity, 1996).

Table 2. Total protein content and amino acid composition of sorghum summarized from several studies. From Bean and Ioerger (2015) with permission.

Reference†	Protein ‡,§	Amino acid								
		Ile	Leu	Lys	Met	Phe	Thr	Trip	Val	His
	%	g/100 g protein								
Khalil et al. (1984)	15.6	4	13.7	2.6	1.4	5.2	3.2	0.9	4.8	1.8
Pontieri et al. (2010)	11.3	2.06	13.3	2.67	3.62	6.37	3.34	n/a¶	3.61	2.44
Ejeta et al. (1987)	12.0	3.8	13.1	2.5	1.7	5.2	3.3	0.9	5.1	2.2
Ebadi et al. (2005)	11.1	4.01	12.53	1.79	1.43	4.68	3.22	2.24	5.28	2.27
Ahmed et al. (1996)	9.0	4.1	13.5	2.4	1.9	4.7	4.3	n/a	7.6	3.1
Subramanian et al. (1990)	12.1	4.28	14.61	2.44	0.86	4.5	3.32	n/a	5.77	2.07
Sosulski and Imafidon (1990)	11.3	2.26	9	1.32	9	3.02	1.96	0.75	3.64	1.38
da Silva et al. (2011)	10.4	3.75	14.8	2.08	1.65	4.74	2.73	n/a	4.79	1.95
Youssef (1998)	13.3	3.65	13.2	2.11	1.36	5.62	3	n/a	4.55	1.82

† Data adapted from listed references.

‡ Values represent averages when study used multiple samples.

§ Dry basis.

¶ n/a, data not available.

Protein Composition

Cereal proteins have traditionally been divided into fractions or classes based on solubility and/or other chemical properties based on the pioneering work of Osborne (1907, 1924). Over the years various researchers have modified solubility based classification schemes, resulting in a wide range of nomenclature and extraction procedures (e.g., Landry and Moureaux, 1970; Jambunathan et al., 1975; Taylor et al., 1984a; Hamaker et al., 1995; Landry, 1997). Differences in methodology along with significant overlap among the solubility classes can lead to confusion with cereal protein classification, but despite these issues, classification by solubility differences persists as a useful method for the characterization of cereal proteins. In addition, solubility based classification schemes have been improved by also considering protein functionality, structure, and sequence homology (Shewry et al., 1986).

In general, the major cereal protein classes consist of the albumins, globulins, prolamins, and glutelins (Branlard and Bancel, 2006). In sorghum, the albumins (water soluble proteins) and globulins (salt soluble proteins) are found primarily in outer layers of the grain and the germ, and combined have been reported to comprise 10% to more than 30% of total grain protein (Taylor et al., 1984a; Food and Agriculture Organization, 1995; Wall and Paulis, 1978). Molecular mass ranges between 14 and 67 kDa across both albumins and globulins (Taylor and Schüssler, 1986). From a nutritional standpoint, the albumin and globulin classes offer a more desirable balance of amino acids than the sorghum prolamins, particularly with respect to higher lysine contents (Taylor and Schüssler, 1986; Lásztity, 1996). A number of water and salt soluble proteins of sorghum were recently identified and included heat shock proteins, chaperones, enzymes, and metabolic proteins (Cremer et al., 2014).

Prolamins (traditionally classified as alcohol soluble proteins and known to contain high levels of glutamine and proline) form the bulk of sorghum storage protein (Taylor, 1983). In sorghum, the prolamins (known as kafirins) make up from 48% to over 70% of total protein on a whole grain flour basis (Taylor et al., 1984a; Hamaker et al., 1995). As with all cereal prolamins, the kafirins contain high levels of proline and glutamine, which contribute over 30% of the total amino acid residues present in the kafirins (Belton et al., 2006).

The glutelins (traditionally soluble in acid or alkali conditions) are the least studied major group of sorghum proteins, and reports on the amount and composition of sorghum glutelins have varied widely. One of the first studies on sorghum glutelins (Beckwith, 1972) isolated glutelins by enzymatic degradation of starch from flour that had been defatted and the globulin, albumin, and prolamin proteins removed. The resulting glutelin fraction was said to make up more than 50% of the total protein. In contrast, using modifications of the Landry and Moureaux procedure (Landry and Moureaux, 1970), other studies have reported relative average glutelin percentages of 27.7% (Taylor et al., 1985), 31.0% (Subramanian et al., 1990), and 4.5 to 34.3% (Neucere and Sumrell, 1979). While little work has been done to characterize the proteins in the glutelin fraction, recently a number of alkali soluble proteins remaining in the grain after extraction of albumins, globulins, and prolamins were identified and were found to contain heat shock proteins, chaperones, luminal binding proteins, enzymes, vicilin- and legumin-like storage proteins (Cremer et al., 2014).

Kafirin Subclasses

The prolamins of cereals are composed of a homogenous mixture of monomeric, oligomeric, and polymeric protein species (Belton et al., 2006) and have been divided into subclasses based on several criteria. For sorghum, kafirins subclass nomenclature follows that developed from studies of maize prolamins (zeins) with the major kafirin subclass or group being the α-, β-, and γ-kafirins (Esen, 1987; Shull et al., 1991). An additional kafirin category known as δ-kafirin has been proposed by using homology of DNA sequences with δ-zein (Belton et al., 2006). Similarity between kafirin and zein are evidenced based on solubility, molecular mass, and immunological cross-reactivity (Shull et al., 1991; Shull et al., 1992; Mazhar et al., 1993). Zeins and kafirins also show a high degree of sequence homology (DeRose et al., 1989, Belton et al., 2006; Xu and Messing, 2009).

The α-kafirins form the major kafirin subclass and make up approximately 70 to 80% of total grain prolamin protein content (Watterson et al., 1993). Utilizing several techniques, Shull et al. (1991) characterized the α-kafirins as being composed of two bands of molecular mass of 23,000 and 25,000 Da. However, others have reported molecular masses of 22,000 and 28,000 Da (Mazhar et al., 1993), and at the gene level the α-kafirins have been characterized as having molecular masses of 19,000 and 22,000 Da (DeRose et al., 1989; Belton et al., 2006; Holding, 2014). Phylogenetically, α-kafirins belong to the Group I family of prolamins, with the "19 kDa" α-kafirins in the K1α19 subfamily and the "22 kDa" α-kafirins in the K1α22 subfamily (Xu and Messing, 2009). Although sodiumdodecylsulfate polyacrylamide electrophoresis (SDS-PAGE) reveals only two major α-kafirin bands, there are 20 α-kafirin genes and up to 19 of the α-kafirin genes may be expressed (Xu and Messing, 2008). Analytical techniques such as high-performance capillary electrophoresis (HPCE), reverse-phase high-performance liquid chromatography (RP-HPLC), and isoelectric focusing (IEF) that separate proteins by different mechanisms than SDS-PAGE have been reported to resolve a number of α-kafirins proteins, demonstrating that while the α-kafirins may have a narrow molecular weight distribution, they may vary in other chemical properties such as charge density, isoelectric point, and surface hydrophobicity (Taylor and Schüssler, 1986; Sastry et al., 1985; Smith, 1994; Bean et al., 2000; Bean et al., 2011; Blackwell and Bean, 2012).

There has been little research conducted to determine the secondary structure of kafirin in general, but analysis of isolated kafirins and sorghum protein bodies has revealed that α-helix is the predominant structure present (Belton et al., 2006). The structure of individual kafirins, especially the α-kafirins, has been speculated to be very similar to that of α-zein and an excellent discussion of this can be found in the review by Belton et al. (2006).

The γ-kafirins, which belong to the Group II γ1 k2γ27 and k2γ50 phylogenetic group (Xu and Messing, 2009) have been reported to comprise 9 to 12% of vitreous endosperm and 19 to 21% of opaque endosperm (Shull et al., 1991; Watterson et al., 1993). Once reduced, the γ-kafirins are soluble in water and in aqueous organic solvents that span a wide range of polarity such as 10 to 80% tert-butanol (Shull et al., 1991), despite being the most hydrophobic of the kafirins based on free energy of hydration (Belton et al., 2006). γ-Kafirin migrates with an apparent molecular mass of ~28,000 Da on SDS-PAGE, but sequence information places the molecular mass at ~20,000 Da (Belton et al., 2006). The presence of an additional γ-kafirin with molecular mass of ~ 49,000 Da has also been reported (Evans et al., 1987; Belton et al., 2006; Cremer et al., 2014).The cross-linking potential of γ-kafirin is indicated by the cysteine content which is relatively high, having been reported as 7 mol% (Duodu et al., 2003). One γ-kafirin gene is expressed in sorghum (Xu and Messing, 2008), with five allelic variants identified by Laidlaw et al. (2010).

β-Kafirins have also been characterized based on their solubility in 10 to 60% tert-butanol plus reducing agent in addition to cross-reactivity with β-zein antibodies and amino acid sequence, and three different molecular mass components were identified as β-kafirins (16 kDa, 18 kDa, 20 kDa) (Shull et al., 1991, 1992). Later studies by Chamba et al. (2005), using molecular cloning techniques, identified a single gene encoding for a mature molecular mass 18,745 Da β-kafirin species, and Xu and Messing (2008) reported only one β-kafirin gene is expressed. Laidlaw et al. (2010) identified four allelic variants for β-kafirins, including a null mutant. Containing 5.8 mol% cysteine and 10 cysteine residues, β-kafirin may be involved in intra- and interchain disulfide bonding (El Nour et al., 1998).

Possibly the least characterized of the kafirin subclasses, especially at the protein level, are the δ-kafirins. By employing molecular cloning experiments of cDNA encoding for δ-kafirins, Izquierdo and Godwin (2005) reported δ-kafirin as a 147 amino acid polypeptide (M_r 16,000) rich in methionine. Two δ-kafirin DNA sequences (GENPEPT AAK72689 and AAW32936) were described by Belton et al. (2006) that showed extensive homology with molecular mass 14,000 Da δ-zein from maize. δ-Kafirin is a minor protein component, thought to make up less than 1% of total grain storage protein in mature sorghum grain (Laidlaw et al., 2010).

Endosperm Protein Structures

As mentioned above, the endosperm contains the majority of the protein in the sorghum grain. Grain storage proteins evolved as a means of storing nitrogen over potentially long periods of time for later use during plant reproduction and development, and in mature cereal grains may represent from 50% to more than 80% of total protein (Hamaker et al., 1995; Shewry and Halford, 2002). One mechanism for plants to store nitrogen is in the form of specialized organelles known as protein bodies (Shewry et al., 1995). Amino acids needed for growth at germination can be stored for years in the form of proteins, protected in the membrane-bound protein bodies (Müntz, 1998). The protein bodies are formed in the endoplasmic reticulum and consist almost completely of prolamins (Müntz, 1998; Herman and Larkins, 1999; Seckinger and Wolf, 1973; Taylor et al., 1984b). Proteins in the interior of the protein bodies are predominantly α-kafirin, along with smaller quantities of β- and γ-kafirin (Shull et al., 1992). The outer layers of the protein body, on the other hand, have been reported to contain more of the β- and γ-kafirins (Belton et al., 2006). The β- and γ-kafirins contain high levels of cysteine and therefore may form a cross-linked shell around the α-kafirins in the interior of the protein body (Belton et al., 2006). Protein bodies located in the vitreous endosperm were described by Shull et al. (1992) as 0.3 to 1.5 μm spheroids. Likewise, the floury endosperm protein bodies were similar if somewhat smaller in size, but exhibited somewhat irregular shape.

An important development related to the structure of sorghum protein bodies was the discovery of mutant sorghum lines that have improved cooked and uncooked protein digestibility (Weaver et al., 1998; Oria et al., 2000). The improvement in protein digestibility was found to be related to misshapen protein bodies with γ-kafirins concentrated at the bottom of folds in the protein bodies rather than uniformly around the edges as in wild-type sorghum lines (Oria et al., 2000). Recently a mutation in the signal peptide of one α-kafirin gene was reported to be responsible for the formation of misshaped protein bodies in sorghum (Wu et al., 2013).

It is well known that sorghum protein digestibility decreases on cooking (Duodu et al., 2003). This decrease has been linked primarily to the formation of disulfide bonds, and cooking sorghum in the presence of a reducing agent restores protein digestibility to close to that of the raw grain (Hamaker et al., 1986). Interestingly, sorghum proteins form a highly linked web of proteins when cooked under high moisture conditions (Hamaker and Buguso, 2000). In addition to protein bodies, endosperm matrix protein appears to provide a connecting structure within which the protein bodies and starch granules reside (Chandrashekar and Mazhar, 1999). In addition to protein storage, the matrix may also function as an enzyme source for starch and protein hydrolysis (Wu and Wall, 1980, Taylor et al., 1985). On a quantitative basis, the protein matrix was considered the second most important endosperm fraction in a study by Taylor and Schüssler (1986). The primary composition of the protein matrix appears to be influenced by glutelins based on solubility characteristics and amino acid composition (Taylor et al., 1985; Taylor and Schüssler, 1986), though as mentioned above, little work has been done characterizing the proteins in the "glutelin" fraction or direct work characterizing "matrix proteins."

Lipid and Wax Chemistry and Composition

Total Lipid Content (Crude Fat) in Sorghum Kernels and Sorghum Distillers Dried Grains with Solubles

Like maize and other grains, the levels of oil and other lipids (crude fat) in sorghum grains are low, typically 2 to 4 wt% (sorghum is compared with barley, oats, wheat, and maize in Table 3). Rooney (1978) reported that the oil content of

Table 3. Total lipid content (crude fat) in sorghum kernels and sorghum distillers dried grains with solubles.

	Genotype	Oil content	Reference
		% FW	
Sorghum kernels	234 of samples	3.4 (mean)	Rooney (1978)
	Cargill 737	3.22 ± 0.08	Singh et al. (2003)
	Cargill 888Y	3.66 ± 0.30	
	PR6E14	3.77	Liu (2011)
	PR6E6	3.61	
Sorghum germ	Cargill 737	13.67 ± 1.39	Singh et al. (2003)
	Cargill 888Y	18.93 ± 1.67	
Sorghum control "decorticated"	Eight hybrids	3.57 mean	Corredor et al. (2006)
		2.1–2.7 range	
Sorghum DDGS	Mixture of several genotypes	9.32 (hexane)	Wang et al. (2005)
		8.5 (hexane)	Wang et al. (2007)
		15.0 (Supercritical CO_2)	Wang et al. (2008b)
Barley	Baronesse	2.41	Liu, 2011
	Merlin	3.26	
Oat	97AB7761	3.71	Liu, 2011
	99AB12334	6.38	
Wheat	Jefferson	2.18	Liu, 2011
	Brundage	3.26	
Maize	Mean, 1825 samples	4.5%	Rooney, 1978
	Summary of 49 accessions	3.25% (mean)	Moreau et al. (2001)
		2.19–4.83 (range)	

grain sorghum was 3.4% (mean of 234 samples), and the oil content of maize was slightly higher, 4.5% (mean of 1825 samples). However, with all grains, the range of oil content can be broad and is influenced by both genetic and environmental factors. In sorghum, the oil content varies in fractions obtained by wet milling and dry milling, with the germ fraction consistently showing the highest oil content (Singh et al., 2003). During the "dry grind" process for fermenting ground sorghum mash into ethanol, the material remaining after fermentation and distillation of the ethanol are combined and dried and called "distillers dried grains with solubles" (DDGS). The oil content of sorghum DDGS is typically 8 to 10% (Wang et al., 2005).

Triacylglycerols and Fatty Acids

The major lipid class in sorghum seeds is triacylglycerols (comprising about 90% of the total lipids, with linoleic acid the predominant fatty acid). Triacylglycerols are a storage lipid that is used for energy and a source of carbons during seed germination (Table 4). In addition to linoleic acid (18 carbons and 2 C-C double bonds), the second most abundant fatty acid is oleic acid (18 carbons and 1 C-C double bond), followed by palmitic acids (16 carbons and no C-C double bonds) and approximately 1% each of stearic acid (18 carbons and no C-C double bonds) and linolenic acid (18 carbons and 3 C-C double bonds). Although linoleic acid is the most abundant fatty acid in both sorghum and maize, the levels of linoleic acid are generally slightly higher in maize and the levels of oleic acid are slightly higher in sorghum (Table 4).

Sterols and Squalene

The second most abundant lipid class in sorghum and most seeds after triacylglycerols (which include most of the esterified fatty acids) are plant sterols (phytosterols).

Table 4. Fatty acid composition of sorghum and maize.†

Sample	Oil	C16:0	C18:0	C18:1	C18:2	C18:3	Others	Reference
	%	% (w/w)						
Sorghum, mean	3.16	12	1	34	50	1		Rooney (1978)
Sorghum	NR‡	11.34	1.25	36.39	46.43	2.45		Zhang and Hamaker (2005)
10 varieties Mean	NR	11.73–20.18	1.09–2.59	31.12–48.99	29.67–50.72	1.71–3.89	NR	Mehmood et al. (2008)
PR6E14	3.77	16.83	2.44	36.64	40.10	2.27	1.72	Liu (2011)
PR6E6	3.61	17.21	1.85	31.60	45.44	2.16	1.74	Liu (2011)
Sorghum, mean	NR	12.25	5.58	11.73	65.17	3.38	1.89	Bhandari and Lee (2013)
Maize (germ oil)	NR	10.72 ± 0.02	1.85 ± 0.01	27.65 ± 0.02	57.26 ± 0.04	1.22 ± 0.01	1.30 ± 0.002	Moreau et al. (2009)

† C16:0, palmitic acid; C18:0, stearic acid; C18:1, oleic acid; C18:2, linoleic acid; C18:3, linolenic acid.

‡ NR, not reported.

Sterols in sorghum comprise about 1.2 to 1.6% of the extractable lipids (Singh et al., 2003). In sorghum approximately half of the plant sterols are fatty acid esters and half occur in the "free OH" form (Singh et al., 2003). The most abundant plant sterol in sorghum is sitosterol, followed by stigmasterol or campesterol (Table 5). Squalene is a hydrocarbon precursor to sterols, and it has been reported to occur in concentrations similar to plant sterols in some sorghum seeds (Bhandari and Lee, 2013). Squalene has been reported to possess serum cholesterol-lowering and other health-promoting properties (Bhandari and Lee, 2013).

Tocopherols and Tocotrienols

Sorghum, like most other grains, contains two forms of vitamin E, which include four tocoperhols (α, β, γ, and δ) and two tocotrienols (α and γ). Like maize the most abundant tocopherols in sorghum kernels is γ-tocopherol, followed by α-tocopherol (Table 6). In addition to α- and γ-tocopherols, Chung et al. (2013) also reported β- and δ-tocopherols in sorghum seeds (Table 6). Bhandari and Lee (2013) also reported α-, γ-, and δ-tocopherols and α- and γ-tocotrienols in sorghum kernels (Table 6).

Carotenoids

Carotenoids are the normal yellow or orange pigments found in many seeds, leaves, and other plant materials. There are two major classes of carotenoids: carotenes and xanthophylls. Carotenes are hydrocarbons and include mainly α- and β-carotenes and lycopene. Xanthophylls contain one or more oxygen and include lutein and zeaxanthin. Grain sorghums are mainly classified as having a white endosperm which has low levels of carotenoids, or a yellow endosperm which has higher levels of carotenoids, mainly lutein and zeaxanthin. Kean et al. (2007, 2011) reported that lutein and zeaxanthin were the most abundant carotenoids in grain sorghum, and the concentrations of both were lower than the concentrations of the same carotenoids in yellow maize (Table 7). Lipkie et al. (2013)

Table 5. Sterols and squalene in sorghum and maize (note that different units were reported by the different references).

Sample	Campesterol	Stigmasterol	Sitosterol	Other sterols	Squalene	Reference
Sorghum, mg/g of lipid	2.69–2.86	1.73–1.74	5.07–5.66	NR†	NR	Leguizamon et al. (2009)
Sorghum, average of 10 landraces, mg/kg of seeds	75.5	96.5	345.9	NR	84.7	Bhandari and Lee (2013)
Maize, mg/g of lipid	1.54–2.46	0.45- 0.71	4.60–6.97	NR	NR	Leguizamon et al. (2009)
Maize, mg/100 g of seeds	9.1 ± 0.5	0.4 ± 0.0	34.1 ± 1.1	NR	1.6 ± 0.6	Ryan et al. (2007)

† NR, not reported.

Table 6. Tocoperols and tocotrienols in sorghum and maize.

	α-tocopherol	β-tocopherol	γ-tocopherol	δ-tocopherol	α-tocotrienol	γ-tocotrienol	Reference
	mg/kg of seeds						
Sorghum flour	0.85 ± 0.21	NR†	2.01 ± 0.45	NR	NR	NR	Pinheiro-Sant'Ana et al. (2011)
Sorghum seed (mg/100 g of seeds)	1.25 ± 0.009	NR	2.24 ± 0.48	ND	NR	NR	Pinheiro-Sant'Ana et al. (2011)
Sorghum Seeds (mean of 5 varieties)	43.17 ± 1.51	70.66 ± 4.82	39.58 ± 4.27	34.94 ± 1.70	NR	NR	Chung et al. (2013)
Sorghum seeds (mean of 10 varieties)	4.4	NR	37.5	NR	1.8	NR	Bhandari and Lee (2013)
Maize kernels‡	17.02 ± 0.44	NR	41.38 ± 0.21	4.50 ± 0.00	6.94 ± 0.02	13.69 ± 0.02	Moreau and Hicks (2005)

† NR, not reported.

‡ Concentration of tocols in maize kernels extrapolated from reported values of tocols in extracted oils assuming crude fat value of 4.0%

Table 7. Carotenoids in sorghum and maize seeds.

	Lutein	Zeaxanthin	β- carotene	Total	Reference
	mg/kg of seeds				
Sorghum, P88–50DAHB	0.217 ± 0.104	0.157 ± 0.059	0.067 ± 0.010	0.604 ± 0.222	Kean et al. (2011)
Sorghum, P1222–50DAHB	0.301 ± 0.042	0.362 ± 0.027	0.063 ± 0.007	0.980 ± 0.102	Kean et al. (2011)
Sorghum, range, of 7 cultivars	0.003–0.173	0.007–0.142	0.005–0.010	0.010–0.315	Kean et al. (2007)
Maize, Becks-5538	0.511 ± 0.016	0.286 ± 0.011	0.027 ± 0.006	1.152 ± 0.023	Kean et al. (2007)
Maize, Becks 5856	1.101 ± 0.311	0.983 ± 0.254	0.095 ± 0.007	3.423 ± 0.756	Kean et al. (2011)

reported on the development and increased bioavailability of provitamin A in some transgenic lines of sorghum.

Waxes: Alkanes, Fatty Alcohols, Fatty Aldehydes, and Wax Esters

Kummerow (1946a) reported that sorghum grain oil contained about 0.5% wax and about 2.5% oil. This was about 50 times more wax and two thirds less oil compared with maize. He also compared grain sorghum varieties and reported an average yield of 0.32% wax and 2.76% oil (Kummerow, 1946b). Bunger and Kummerow (1951) compared wax extracted from four varieties of grain sorghum with carnauba wax. They reported that grain sorghum wax contained 6 to 28% esters and carnauba wax contained 71% esters. Table 8 shows the composition of sorghum wax compared with that of carnuba wax.

Dalton and Mitchell (1959) reported that the sorghum wax was composed of 49% esters, 46% free alcohols, and 5% paraffin, as compared with carnauba wax which was thought to be composed of 80% esters, 12% free alcohols, 1% paraffin, 3% lactone, and 4% resins. Seitz (1977) reported that wax from sorghum grain contains 4 to 5% alkanes, 46 to 50% esters, 40 to 46% free alcohols, and 8% other lipids.

Bianchi et al. (1979) reported that sorghum wax was composed of 1.3% alkanes (C25–C31, with C29 the most abundant), 31.9% fatty aldehydes (C26–C32, with C28 and C30 the most abundant), 33.7% fatty alcohols (C22–C30, with C28 the most abundant), 4.0% esters, 24.2% acids (C16–C30, with C28 and C30 the most abundant), and 4.7% unidentified components.

Weller et al. (1998) extracted sorghum wax from intact sorghum kernels by refluxing the intact kernels in petroleum ether at 43°C for 1.5 h and reported a mean yield of 0.195 ± 0.004%. They found that when gelatin-based candies were prepared with carnauba wax or sorghum wax, both had similar physical and sensory properties. Unlike sorghum wax, which is extracted from the outer surface of sorghum seeds, carnauba wax is obtained by extracting the leaves of the palm tree, *Coernica cerivera*, and is mainly manufactured in Brazil.

Lochte-Watson et al. (2000) attempted to use abrasive decortication to fractionate grain sorghum into a starch-enriched fraction. They concluded, "Overall the use of abrasive decortication, as facilitated by the tangential abrasive dedhulling devise (TADD), to obtain bran fractions with high levels of coverable wax (>85%) with limited starch contamination (<10%) remains elusive."

Hwang et al. (2002a) used silica gel column chromatography to fractionate sorghum wax. Their major fraction which comprised about 40% by thin layer chromatography was found to contain aldehydes. This group also used normal phase high-performance liquid chromatography (HPLC) with an evaporative light-scattering detector (Hwang et al., 2002b) and found that grain sorghum wax was composed of 46.3% fatty aldehydes, 7.5% fatty acids, 41.0% fatty alcohols, 0.7% hydrocarbons, 1.4% wax esters, and 0.9% triglycerides. Using the same HPLC system, they reported that carnauba wax contained 34.4% wax esters, 5.1% fatty acids, and 3.0% triglycerides. They also compared some of the physical properties of sorghum wax and carnauba wax and reported that the melting points were 84.94 ± 0.12 and 82.19 ± 0.06°C, respectively (Hwang et al., 2002c). Hwang et al. (2004) used gas chromatography to quantify the policosanols (fatty alcohols) in sorghum kernel wax and in lipids extracted from sorghum DDGS. They reported that the policosanols ranged from C22 to C32, with C28 and C30 being the most abundant of the aldehydes. In another paper they reported that the level of policosanols in grain

Table 8. Waxes in sorghum.

Sample	Hydrocarbons	Alcohols	Aldehydes	Acids	Esters	Other	Reference
	% (w/w) of total wax fraction						
Sorghum	5	46	NR†	NR	49	NR	Dalton and Mitchell (1959)
Carnauba	1	12			80	7	
Sorghum	1.3	33.7	31.9	24.2	4.0	4.7	Bianchi et al. (1979)
Sorghum	0.7	41.0	46.3	7.5	1.4	0.9	Hwang et al. (2002b)
Carnauba	NR	NR	NR	5.1	34.4	3.0	

† NR, not reported.

sorghum from Korea was 74.5 mg/100 g grain. The alcohols ranged from C20 to C30, with the C28 aldehyde being the most abundant (Hwang et al., 2005). Several clinical studies reported that policosanols (from sugarcane) possessed valuable health-promoting properties (Castano et al., 2001). However, more recent studies, with both laboratory animals and human subjects, have not been able to confirm the previous results, thus raising doubt about the health-promoting properties of policosanols from sugarcane and from sorghum (Marinangeli et al., 2010).

Singh et al. (2003), using an alumina normal phase HPLC method with detection via evaporative light scattering (Fig. 7), reported that the levels of fatty aldehyde in oil extracted from sorghum kernels was 1.34 to 1.47 wt %. They

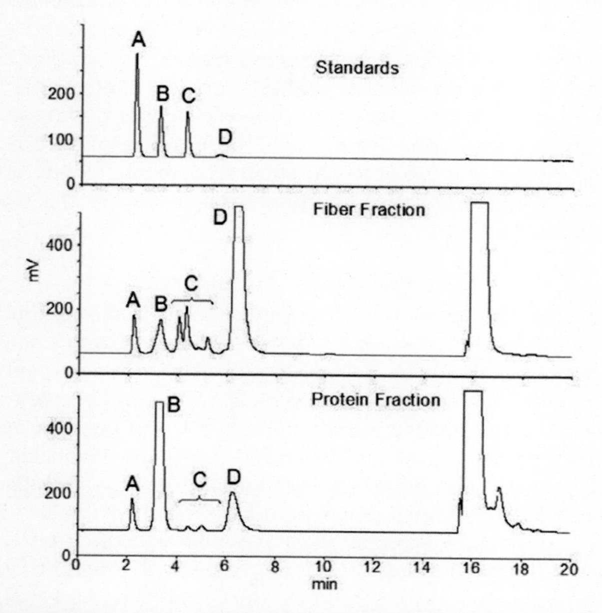

Fig. 7. high performance liquid chromatography (HPLC) chromatogram of very nonpolar lipids in extracts of wet milled fractions of sorghum kernels. A, hydrocarbons; B, wax esters (stearyl-stearate as standard); C, sterol esters (cholesterol stearate as standard); D, fatty aldehyde (C28 standard). Reproduced with permission from Singh et al. (2003).

reported that the concentration of fatty aldehydes was higher in the wet milled fiber fraction (2.54–2.93%) and lower in the wet milled germ fraction (0.08–0.13%).

The composition and properties of sorghum oil are similar to those of commercial maize oil, including compositions of fatty acids, sterols, tocols, and carotenoids. In one recent report, the levels of squalene were revealed to be about 40 times higher in sorghum oil compared with maize oil (Bhandari and Lee, 2013). The levels of waxes are also considerably higher in sorghum oil than in maize oil, being 50 times higher according to Kummerow (1946a). The physical properties of sorghum oil also appear to be similar to the physical properties of carnauba wax, but additional research is necessary to more rigorously compare the two.

Phenolic Chemistry and Composition

Phenolic compounds exist in plants as structural components, intermediates of various metabolic pathways, natural defense molecules, and signaling molecules, among other important functions. Such compounds have been extensively investigated for their health benefits, and a significant body of evidence demonstrates their important role in preventing chronic disease in humans. In sorghum, like most cereal grains, the major polyphenol groups include phenolic acids and flavonoids (Awika and Rooney, 2004; Dykes and Rooney, 2006). Sorghum is unique among cereal grains in that some varieties contain high levels of phenolic acid esters and flavonoids with important implications to food quality and human health.

In sorghum, as with other cereals, the phenolic compounds are mostly concentrated in the bran layers. The type and level of flavonoids in sorghum are controlled by a set of well documented genes (Rooney, 2000). For example, the *yellow seed*1 (*ys*1) gene found in most sorghum varieties controls the biosynthetic pathway that leads to accumulation of 3-deoxyflavanoid compounds (Boddu et al., 2005). These compounds are not usually found in other cereal grains in meaningful quantities and are believed to be responsible for the unique beneficial properties of sorghum relative to other grains (Awika et al., 2003; Shih et al., 2007; Yang et al., 2009).

Phenolic Acids in Sorghum

Sorghum contains the same major phenolic acids that are found in other cereal grains with ferulic acid and its derivatives being dominant. The classic composition of phenolic acids in diverse sorghum varieties have been extensively reviewed (Subba Rao and Muralikrishna, 2002; Awika and Rooney, 2004; Dykes and Rooney, 2006; Chandrasekara and Shahidi, 2011). In general, most phenolic acids in grains exist in bound and esterified forms and are thus not usually characterized or quantified based on standard laboratory analytical protocols. In fact, new evidence indicates that sorghum phenolic acids are more readily extractable than previously thought; these phenolic acids are mostly released in solvent as glycerol mono or di-esters that were previously unidentified because of the limitations of the analytical techniques employed (Svensson et al., 2010; Yang et al., 2012). Red sorghums appear to accumulate the highest levels of extractable (nonbound) phenolic acid esters, with values above 2000 μg/g observed in some varieties (Yang et al., 2012). The implication is that the overall contribution of phenolic acids to antioxidant activity and other bioactive properties of extracts from sorghum may be much higher than previously assumed.

Flavonoids in Sorghum

The flavonoids in sorghum have been the subject of long-term investigations, primarily because of the negative impact they have on feed conversion efficiency in monogastric animals as well as the reduction of micronutrient bioavailability. The most important group of flavonoids in this regard is the proanthocyanidins (condensed tannins) that can exist in high concentrations in specific varieties of sorghum. The proanthocyanidins bind principally with proteins to form non-digestible complexes. However, recent evidence has shown that tannins have important health benefits (Awika and Rooney, 2004). In general, most cultivated sorghum varieties have inconsequential quantities of the tannins because their accumulation is genetically controlled. Thus, when discussing the impact of sorghum tannins on diet, it is important to define the genetic background of the material. Besides the tannins, sorghum has an array of flavonoids not commonly found in other grains. The major ones are summarized below.

3-Deoxyanthocyanins

The 3-deoxyanthocyanins (Fig. 8a) are unique flavonoids mostly found in sorghum and rarely in other food plants. These compounds are responsible for most of the red to black pigmentation in sorghum grain, glumes, sheath, stem, and leaves. The 3-deoxyanthocyanins are analogous to the anthocyanins found in most other plants (Fig. 8b), but without a substitution at the C-3 position. Black sorghum grains contains high levels of these compounds in the bran (4–16 mg/g) (Awika et al., 2004, 2005). The 3-deoxyanthocyanins are of growing interest because of their potential as natural food colorants with superior stability compared with the anthocyanin pigments (Ojwang and Awika, 2008; Geera et al., 2012), as well as specific bioactive properties (Shih et al., 2007; Yang et al., 2009).

Flavones

Flavones are a group of flavonoids that contain a 2-phenyl-1-benzopyran-4-one skeleton (Fig. 9a). Flavones are mainly reported in herbs such as parsley and celery (Yao et al., 2004). However, specific sorghum varieties accumulate flavones and their derivatives at nutritionally significant levels, primarily as derivatives of luteolin and apigenin (Salunkhe et al., 1983; Awika, 2011). The high accumulation

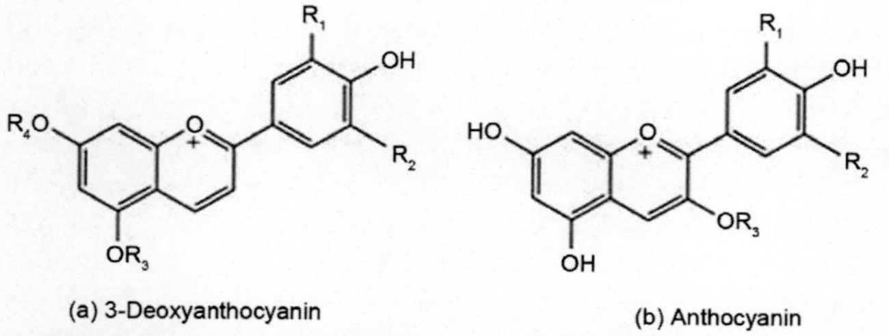

(a) 3-Deoxyanthocyanin (b) Anthocyanin

Fig. 8. Skeletal structure of 3-deoxyanthocyanin compounds found in sorghum (a) and their anthocyanins analogs (b). The R_x are commonly OH, OCH_3, sugars, or phenolic acid esters.

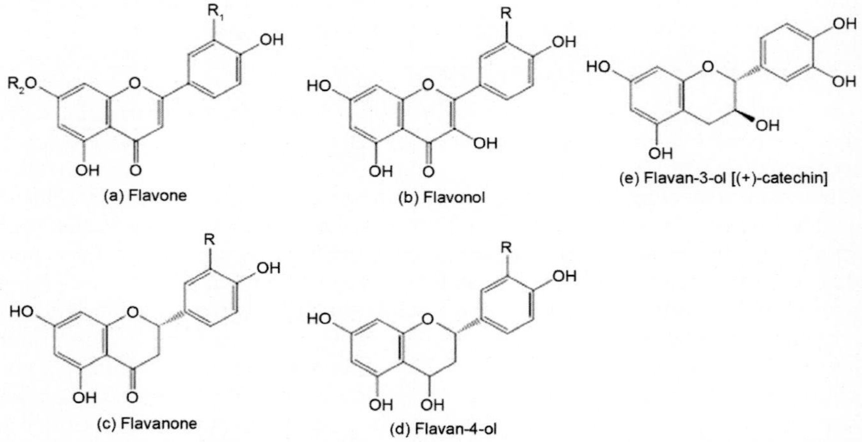

(a) Flavone

(b) Flavonol

(e) Flavan-3-ol [(+)-catechin]

(c) Flavanone

(d) Flavan-4-ol

Fig. 9. Major classes of monomeric flavonoids found in sorghum. Figure 9. Major classes of monomeric flavonoids found in sorghum showing the structures for the following types: (a) flavone, (b) flavonol, (c) flavanone, (d) favan-4-0l, and (e) flavan-3-ol [(+)-catechin].

of flavones in cereals is unique to sorghum and some millets (fonio and pearl). For example, tan plant sorghums with a pigmented pericarp contain 60 to 386 μg/g flavones (Awika, 2011). These levels are much higher than reported for most plant foods and thus may have important implications for human health.

Flavonols

Flavonols share the same backbone structure with flavones, except for the hydroxyl group at C-3 of the flavonols (Fig. 9b). Various flavonols have been identified in cereal grains. However, these compounds are relatively rare in sorghum and have only been sparsely reported. The original report of flavanol in sorghum was by Nip and Burns (1969) who identified kaempferol-3-rutinoside-7-glucuronide in a red pericarp sorghum. More recently, quercetin 3, 4′-dimethyl ether was identified in nongrain sorghum tissue (Yong and Chang, 2003).

Flavanones

Flavanones have the basic 2,3-dihydroflavone structure, that is, they differ from flavones by the lack of a double bond between C-2 and C-3 (Fig. 9c), and thus have a chiral center at the C-2 position. Flavanones are widely distributed in nature since they are key intermediates in flavonoid biosynthetic pathway. In food plants, they are most readily associated with citrus fruits; for example, naringenin in grapefruit has been widely studied for its health benefits. Among grains, sorghum accumulates relatively high levels of flavanones; up to 1800 μg/g has been reported (Dykes et al., 2011; USDA, 2007). The major flavanones identified in sorghum are primarily eriodictyol and naringenin as well as their O-methyl derivatives and glycosides (Kambal and Bate-Smith, 1976; Yasumatsu et al., 1965; Gujer et al., 1986).

Flavan-3-ols

Flavan-3-ols (sometimes simply referred to as flavanols) are a subclass of flavonoids that contain a 2-phenyl-3,4-dihydro-2H-chromen-3-ol skeleton (Fig. 9e). Flavan-3-ols are widely distributed in fruits, vegetables, and other food plants like tea and cocoa beans, and are the main building blocks for procyanidins (a form of condensed tannins). The most common flavan-3-ol monomers found in food plants include catechin, epicatechin, and epigallocatechin. In cereal grains, barley, sorghum, and finger millet are the major commodities that contain catechin and epicatechin, along with their polymers (condensed tannins).

Phenolics and Unique Opportunities for Sorghum Utilization

The phenolic content and composition of sorghum reviewed above provides several opportunities for unique utilization of sorghum in human foods which could impact human health on several fronts, including oxidative stress, inflammation, and cancer. Evidence demonstrates that the unusual compounds in some cell lines of sorghum and the high levels at which they are present in the grain may produce specific health benefits that are not observed for other grains like maize, rice, or wheat.

Oxidative stress plays a central role in the development of many chronic diseases, thus most investigations of health benefits of various food commodities invariably begin with antioxidant assays. Antioxidant activity is directly related to the polyphenol content of grains (Awika et al., 2003); thus it is not surprising that some cell lines of sorghum have much higher free radical scavenging activity when compared with other cereals grains, or even fruits and vegetables (Awika and Rooney, 2004). For example, pigmented sorghum varieties show 10 to >20 times the antioxidant activity of red wheat (Awika et al., 2005). Among sorghum, tannin containing varieties have the strongest antioxidant capacity; this is attributed to the generally higher free radical scavenging power of tannins relative to simple flavonoids (Awika et al., 2005; Hagerman et al., 1998). Tannins were recently reported to exert powerful antioxidant activity in vivo, particularly in the gastrointestinal tract (Tian et al., 2012), which could mean they are more important to health than previously assumed.

In addition to their role in direct free radical scavenging, various phenolic compounds are capable of stimulating synthesis of various endogenous antioxidant and detoxifying enzymes, for example, glutathione reductase, quinone oxidoreductase, among others. For example, it was recently reported that O-methyl substituted 3-deoxyanthocyanidins in black sorghum induce the phase II enzymes, quinone oxidoreductase in the murine hepatoma cell model in vitro, while nonmethoxylated samples did not (Awika et al., 2009; Yang et al., 2009).

Another common effect of many flavonoid antioxidants is their anti-inflammatory activity. Chronic inflammation, which is directly related to oxidative stress, is a common pathway to various chronic diseases. Thus the ability of bioactive compounds to reduce inflammation is considered one of the most important predictors of health-promoting potential. Emerging data confirms that sorghum flavonoids are especially potent anti-inflammatory agents (Bralley et al., 2008; Burdette et al., 2010; Sugahara et al., 2009; Oboh et al., 2010).

The role of whole grains in cancer prevention is well documented. The evidence for cancer prevention is strongest for gastrointestinal cancers (Levi et al.,

2000; Larsson et al., 2005). Among cereal grains, sorghum has had the most striking evidence for its potential benefit as a chemopreventive agent. Controlled experiments using cell culture and animal models confirm the unique benefits of sorghum polyphenols in chemoprevention. Shih et al. (2007) found that the sorghum 3-deoxyanthocyanidins luteolinidin and apigeninidin were much more effective at reducing HL-60 leukemia and HepG2 cancer cell proliferation than their analogs cyanidin and pelargonidin at all the concentrations tested. More recently, it was found that activation of estrogen receptor β in the colon is an additional mechanism by which sorghum flavonoids may contribute to chemoprevention, particularly those high in flavones (Yang et al., 2012).

Antinutrient–Nutraceutical Factors

Naturally occurring antinutrients in sorghum vary and can cause reduced nutritional value compared with wheat or maize as a human food or animal feed source. The nutrient value of any food depends on a number of factors. In addition to the total nutritional composition of the food, the consumer's accessibility to the nutrient and caloric components must be considered. Antinutrients are ordinary components of a food that inhibit, interfere, or antagonize the digestion and absorption of proteins, minerals, carbohydrates, and micronutrients of the food source reducing their bioavailability and, thus, reducing the net nutritional value realized on consumption. Because of various antinutrient interactions, traditional cereal crude protein content quantitation is insufficient to determine the nutritional availability of sorghum because of the limited bioavailability of sorghum compared with other grains (Soetan and Oyewole, 2009).

Antinutrients are not unique to sorghum, and most plants produce antinutrients as naturally occurring components of secondary metabolism. Some sorghum lines produce antinutrient compounds more robustly than others. The negative connotation associated with the name "antinutrient" is not entirely appropriate for these compounds because many of these factors have been reported to be beneficial in varying degrees for human or animal health (Thompson, 1993). Antinutrients are generated through the metabolic process of plants, and some serve to protect the plant or facilitate germination or pollination, or can function as colorants, attractants, or repellants, as a few examples. Antinutrients are also called nonnutrients, phytochemicals, nutraceuticals, phytoceuticals, or phytopharmaceuticals depending on context. Harnessing the potential of sorghum through decreasing antinutrients should be an important goal for nutritionally poor peoples of developing arid lands. On the other hand, sorghum "antinutrients" can potentially be utilized to combat cancer, Alzheimer's, inflammation, cardiovascular disease or obesity, thus suggesting that these sorghum developments are on the cutting edge of biotechnology research and biopharmaceutical development importance.

Antinutritional–nutraceutical factors found in sorghum include phytic acid, phenolic compounds, enzyme inhibitors, saponins, and, in a larger sense, molecular structure of proteins and starches that reduce digestibility of the grain. This section touches on research of the past which focused on the negative nutritional effects of sorghum antinutrients and also on the biopharmaceutical potential of sorghum as a candidate to benefit human health and fight disease.

Phytic acid (phytate, myo-inositol hexaphosphoric acid, IP6) is the primary storage form of phosphate in many seeds. Phytates are indigestible to humans but bioavailable to ruminant animals via digestion by rumen microbes. Phytic acid chelates positively charged minerals like zinc, iron, potassium, calcium, magnesium, and manganese, forming insoluble complexes and causing decreased mineral absorption, in addition to binding starch and protein (Reddy et al., 1989). Variations in phytate quantity are due, in part, to growing conditions, fertilization, harvesting techniques, processing methods, testing methods, and age of sample. Complexes of phytic acid, protein, and starch may form and slow the digestion rate (Thompson, 1986). Situational attention to dietary attributes of phytic acid needs to be considered when sorghum is a primary food source. For example, in developing poor nations where dietary iron absorption is inefficient, phytates can inhibit iron absorption especially critical to pregnant women and their developing fetus. Conversely, phytic acid is one of the few ways, besides therapeutic phlebotomy, that excess body iron levels can be removed (Coulibaly et al., 2011). Modest phytic acid effects on proteins associated with Alzheimer's suggest phytic acid may play a role in a viable treatment option for Alzheimer's disease (Anekonda et al., 2011).

As discussed above sorghum grain is a rich source of phenolic compounds that have been reported to have many beneficial attributes. Phenolic compounds also have negative implications, especially with regard to animal feed (Awika and Rooney, 2004) and ethanol production (Yan et al., 2012). In addition to impacting utilization of sorghum both positively and negatively, phenolic compounds have important agronomic benefits such as reducing predation of sorghum by birds and insects (Waniska et al., 1989; Waniska, 2000).

Saponins of sorghum are glycosides that form stable foam. Crude sorghum saponin extracts revealed antimicrobial properties also reported in other plants (Soetan et al., 2006). A recent review of saponins in general discusses their different amphiphilic membrane interactions and the potential for new saponin cancer treatments (Lorent et al., 2014).

Several enzyme inhibitors have been identified in sorghum. Alpha amylase inhibitors reduce carbohydrate digestibility, reducing the caloric potential (Daiber, 1975), and in cases of obesity or diabetes can reduce the glycemic effect of starch (Hargrove et al., 2011). Protease inhibitors are associated with growth inhibition in animals (Hathcock, 1991). A number of enzyme inhibitors have been identified in sorghum, including protease inhibitors (Kumar et al., 1978) and amylase inhibitors (Cremer et al., 2014). Various forms of enzyme inhibitors are found in all cereal grains (Boisen, 1983; Cordain, 1999; Piasecka-Kwiatkowska et al., 2012).

To overcome nutritional deficits generated by antinutrients, an increase in net nutritional value is achieved by various cereal processing methods. Thermal or mechanical processing, soaking, fermentation, germination, and malting enhance the bioavailability of micronutrients in plant-based diets, while combining processing methods can increase desired results (Hotz and Gibson, 2007). Wet cooking sorghum decreases protein digestibility, while popping had no effect on protein digestibility; however, dry-heated samples had the same digestibility values as unprocessed samples, indicating a deleterious effect of water in cooking sorghum (Correia et al., 2010). Wet heat processing increased antioxidant activity, while dry heat did not affect the phenolic compounds and 3-deoxyanthocyanidins of sorghum (Cardoso et al., 2014). Twenty-four hour fermentation of sorghum in the production of Khamir bread was found to significantly reduce enzyme inhibition, phytic acid, and tannin content (Osman, 2004). Proteases activated by germination of sorghum led to an increase in in vitro protein digestibility (Afify et al., 2012). During processing of the injera (an African sourdough–risen flatbread), polyphenols, phytates, and tannins were lower than raw and fermented sorghum flour (Mohammed et al., 2011). Malt pretreatment of fermentation was found to further reduce phytate and tannin antinutritional factors compared with raw or fermentation alone (Wedad et al., 2008). Cooking sorghum in alkaline lime solution resulted in flour of higher protein, water and oil absorption, pH, phytates, and trypsin inhibitor producing lower ash, tannin and cyanide content than samples untreated or water treated (Boniface and Gladys, 2011). Gamma irradiation of sorghum flour, in combination with traditional African sorghum food processing, resulted in improved protein and mineral bioavailability (Ismat et al., 2013). Compared with untreated samples, defatting sorghum flour results in lower amounts of phenolic compounds with higher antioxidant activities (Buitimea-Cantua et al., 2013).

In addition to various processing methods, environmental conditions and natural genetic variation of sorghum has to be considered. Landrace selection is important when considering nutritional availability due to variations in antinutrients and bioavailability between different sorghum lines (Proietti et al., 2014). Micronutrient fertilization was found to increase sorghum digestibility while phytate and tannin content dropped significantly (Ahmed et al., 2014). Understanding the net nutritional gains achieved by processing, environmental supplementation, varietal selection, or a combination of these is important to the millions of people who depend on sorghum cereals as a primary food source.

Antinutrient–Nutraceutical Factors and Sorghum Utilization

Research continues to elucidate the human health benefit potential of sorghum components. A small 2-kDa sorghum peptide was found to inhibit the initiation and spread of the herpes simplex virus and had an in vitro prophylactic effect against infection (Camargo Filho et al., 2008). Phenolic compounds obviously may have a major role to play in influencing future utilization of sorghum.

The knowledge of the many different antinutrient components of sorghum and their characterization correlated to their potential roles and specific function in human health is lacking. As the knowledge base for sorghum continues to develop, especially involving advances in genetics and technology, a clearer picture may help to elevate the importance of sorghum in human health. Lack of research pertaining to the effects on animal and human health and disease prevention may suppress the acceptance of sorghum as a major food staple in developed countries (Farrar et al., 2008). In addition to emphasizing the need for fully characterizing sorghum phytochemical compositions, a recent review suggested sorghum food products should be studied in animals to potentially replace raw flours and extracts in well-controlled human studies and intervention trials (Taylor and Duodu, 2014). In vivo studies verifying the human health benefits and disease prevention of sorghum may provide evidence for the value of increased incorporation of sorghum into staple foods.

For further reading there are several reviews discussing the potential health benefits and antinutrient effects of sorghum (Thompson, 1993; Awika and Rooney,

2004; Taylor et. al., 2006; Dykes and Rooney, 2006; Taylor et al., 2014). In addition to sorghum food or feed products, antinutrients can be manipulated to modulate mechanical properties of novel biomaterials produced from sorghum films (Taylor et al., 2006), increasing the biodiversity of sorghum cereal products. Regardless, standardized methods are needed for sorghum data comparison across published research (Awika et al., 2003).

Vitamin and Mineral Content

Whole sorghum has been reported to be a good source of several vitamins and minerals, including B vitamins (with the exception of B-12), phosphorous, potassium, iron, zinc, copper, and magnesium (Serna-Saldivar and Rooney, 1995). Of the vitamins and minerals found in cereal grains, vitamin A, iron, and zinc are especially important micronutrients for populations around the world where deficiencies exist (Zimmermann and Hurrell, 2007; Prasad, 2012). Kayodé et al. (2006a) evaluated Fe and Zn levels in farmers' samples grown in Benin and reported ranges of 30 to 113 mg/kg of Fe and 11 to 44 mg/kg of Zn. Kumar et al. (2012) evaluated Zn and Fe content of 29 accessions identified as having high levels of these minerals in screening of over 2200 lines and compared values to commercial sorghum lines. Fe values varied from 26 to 60 mg/kg and Zn ranged from 21 to 57 mg/kg in this study. It is important to note that the presence of other components such as phytate can influence the bioavailability of minerals. Likewise, grain processing can alter composition and bioavailability as well. Milling, cooking, germinating, and fermenting can all impact vitamin and mineral content of sorghum (Pedersen and Eggum, 1983; Kayodé et al., 2006b; Kayodé et al., 2007; Hotz and Gibson, 2007; Omary et al., 2012). Biotechnology has also been identified as a potential mechanism for improving Zn and Fe in sorghum (O'Kennedy et al., 2006).

Acknowledgments

Names are necessary to report factually on available data; however, the USDA neither guarantees nor warrants the standard of the product, and use of the name by the USDA implies no approval of the product to the exclusion of others that may also be suitable.

References

Afify, A.E.-M.M.R., H.S. El-Beltagi, S.M.A. El-Salam, and A.A. Omran. 2012. Protein solubility, digestibility and fractionation after germination of sorghum varieties. PLoS ONE 7:e31154. doi:10.1371/journal.pone.0031154

Ahmed, S.O., A.W. Abdalla, T. Inoue, A. Ping, and E.E. Babiker. 2014. Nutritional quality of grains of sorghum cultivar grown under different levels of micronutrients fertilization. Food Chem. 159:374–380. doi:10.1016/j.foodchem.2014.03.033

Ahmed, Z.S., G.M.A. El-Moniem, and A.A.E. Yassen. 1996. Comparative studies on protein fractions and amino acid composition from sorghum and pearl millet. Nahrung 40:305–309. doi:10.1002/food.19960400603

Alebiowu, G., and Itiola, O.A. 2001. Effects of natural and pregelatinized sorghum, plantain, and corn starch binders on the compressional characteristics of a paracetamol tablet formulation. Pharm. Tech. Drug Delivery 2001:26–30.

Anekonda, T.S., T.L. Wadsworth, R. Sabin, K. Frahler, C. Harris, B. Petriko, M. Ralle, R. Woltjer, and J.F. Quinn. 2011. Phytic acid as a potential treatment for Alzheimer's pathology: Evidence from animal and in vitro models. J. Alzheimers Dis. 23:21–35.

Awika, J.M. 2011. Sorghum flavonoids: Unusual compounds with promising implications for health. In: J.M. Awika, V. Piironen, and S.R. Bean, editors, Advances in cereal science: Implications to food processing and health promotion. ACS Symp. Ser. 1089. Am. Chem. Soc., Washington, DC. p. 171–200.

Awika, J.M., C.M. McDonough, and L.W. Rooney. 2005. Decorticating sorghum to concentrate healthy phytochemicals. J. Agric. Food Chem. 53:6230–6234. doi:10.1021/jf0510384

Awika, J.M., and L.W. Rooney. 2004. Sorghum phytochemicals and their potential impact on human health. Phytochemistry 65:1199–1221. doi:10.1016/j.phytochem.2004.04.001

Awika, J.M., L.W. Rooney, and R.D. Waniska. 2004. Properties of 3-deoxyanthocyanins from sorghum. J. Agric. Food Chem. 52:4388–4394. doi:10.1021/jf049653f

Awika, J.M., L.W. Rooney, X. Wu, R.L. Prior, and L. Cisneros-Zevallos. 2003. Screening methods to measure antioxidant activity of sorghum (*Sorghum bicolor*) and sorghum products. J. Agric. Food Chem. 51:6657–6662. doi:10.1021/jf034790i

Awika, J.M., L. Yang, J.D. Browning, and A. Faraj. 2009. Comparative antioxidant, antiproliferative and phase II enzyme inducing potential of sorghum (*Sorghum bicolor*) varieties. LWT - Food Sci. Technol. 42:1041–1046.

Bean, S.R., O.K. Chung, J.F. Tuinstra, and J. Erpelding. 2006. Evaluation of the single kernel characterization system (SKCS) for measurement of sorghum grain attributes. Cereal Chem. 83:108–113. doi:10.1094/CC-83-0108

Bean, S.R., and B.P. Ioerger. 2015. Sorghum and millet proteins. In: Z. Ustunol, editor, Applied food protein chemistry. Wiley, London. p. 323–359.

Bean, S.R., B.P. Ioerger, and D.L. Blackwell. 2011. Separation of kafirins on surface porous reversed phase-high performance liquid chromatography columns. J. Agric. Food Chem. 59:85–91. doi:10.1021/jf1036195

Bean, S.R., G.L. Lookhart, and J.A. Bietz. 2000. Acetonitrile as a buffer additive for the separation of maize (*Zea mays* L.) and sorghum (*Sorghum bicolor* L.) storage proteins by HPCE. J. Agric. Food Chem. 48:318–327. doi:10.1021/jf990786o

Bechtel, D.B., and J.D. Wilson. 2003. Amyloplast formation and starch granule development in hard red winter wheat. Cereal Chem. 67:59–63.

Beckwith, A.C. 1972. Grain sorghum glutellin: Isolation and characterization. J. Agric. Food Chem. 20:761–764. doi:10.1021/jf60182a020

Behall, K.M., and D.J. Scholfield. 2005. Food amylose content affects postprandial glucose and insulin responses. Cereal Chem. 82:654–659. doi:10.1094/CC-82-0654

Belton, P.S., and J.R.N. Taylor. 2004. Sorghum and millets: Protein sources for Africa. Trends Food Sci. Technol. 15:94–98. doi:10.1016/j.tifs.2003.09.002

Belton, P.S., I. Delgadillo, N.G. Halford, and P.R. Shewry. 2006. Kafirin structure and functionality. J. Cereal Sci. 44:272–286. doi:10.1016/j.jcs.2006.05.004

Benmoussa, M., B. Suhendra, A. Aboubacar, and B.R. Hamaker. 2006. Distinctive sorghum starch granule morphologies appear to improve raw starch digestibility. Starke 58:92–99. doi:10.1002/star.200400344

Bhandari, S.R., and Y.S. Lee. 2013. The contents of phytosterols, squalene, and vitamin E and the composition of fatty acids of Korean landrace *Setaria itlica* and *Sorghum bicolar* seeds. Korean J. Plant Resour. 26:663–672. doi:10.7732/kjpr.2013.26.6.663

Bianchi, G., P. Avato, and G. Mariani. 1979. Composition of surface wax from sorghum grain. Cereal Chem. 56:491–492.

Blackwell, D.L., and S.R. Bean. 2012. Separation of alcohol soluble sorghum proteins using non-porous cation-exchange columns. J. Chromatogr. A 1230:48–53. doi:10.1016/j.chroma.2012.01.063

Boddu, J., C. Svabek, F. Ibraheem, A.D. Jones, and S. Chopra. 2005. Characterization of a deletion allele of a sorghum Myb gene yellow seed1 showing loss of 3-deoxyflavonoids. Plant Sci. 169:542–552. doi:10.1016/j.plantsci.2005.05.007

Boisen, S. 1983. Protease inhibitors in cereals. Acta Agric. Scand. 33:369–381. doi:10.1080/00015128309435377

Boniface, O.O., and M.E. Gladys. 2011. Effect of alkaline soaking and cooking on the proximate, functional and some anti-nutritional properties of sorghum flour. Au J.T. 14:210–216.

Booth, M.A., G.L. Allan, and R.P. Smullen. 2013. Digestibility of common feed ingredients by juvenile mulloway *Argyrosomus japonicas*. Aquaculture 414-415:140–148. doi:10.1016/j.aquaculture.2013.07.045

Bralley, E., P. Greenspan, J.L. Hargrove, and D.K. Hartle. 2008. Inhibition of hyaluronidase activity by select sorghum brans. J. Med. Food 11:307–312. doi:10.1089/jmf.2007.547

Branlard, G., and E. Bancel. 2006. Protein extraction from cereal seeds. In: H. Thiellment, M. Zivy, C. Damerval, and V. M echin, editors, Methods in molecular biology. Vol. 335: Plant proteomics: Methods and protocols. Humana Press, Totowa, NJ. p. 15–35.

Buitimea-Cantua, N.E., P.I. Torrez-Chavez, A.I. Ledesma-Osuna, B. Ramirez-Wong, R.M. Robles-Sanchez, and S.O. Serna-Saldivar. 2013. Effect of defatting and decortication on distribution of fatty acids, phenolic and antioxidant compounds in sorghum (*Sorghum bicolor*) bran fractions. Int. J. Food Sci. Technol. 48:2166–2175.

Bunger, W.B., and F.A. Kummerow. 1951. A comparison of several methods for the separation of unsaponifiable material from carnauba and sorghum grain waxes. J. Am. Oil Chem. Soc. 28:121–123. doi:10.1007/BF02612208

Burdette, A., P.L. Garner, E.P. Mayer, J.L. Hargrove, D.K. Hartle, and P. Greenspan. 2010. Anti-inflammatory activity of select sorghum (*Sorghum bicolor*) brans. J. Med. Food 13:879–887. doi:10.1089/jmf.2009.0147

Camargo Filho, I., D.A.G. Cortez, T. Ueda-Nakamura, C.V. Nakamura, and B.P. Dias Filho. 2008. Antiviral activity and mode of action of a peptide isolated from *Sorghum bicolor*. Phytomedicine 15:202–208. doi:10.1016/j.phymed.2007.07.059

Carciofi, A.C., F.S. Takakura, L.D. de-Oliveira, E. Teshima, J.T. Jeremias, M.A. Brunetto, and F. Prada. 2008. Effects of six carbohydrate sources on dog diet digestibility and post-prandial glucose and insulin response. J. Anim. Physiol. Anim. Nutr. 92:326–336. doi:10.1111/j.1439-0396.2007.00794.x

Cardoso, L.M., T.A. Montini, S.S. Pinheiro, H.M. Pinheiro-Santana, H.S.D. Martino, and A.V.B. Moreira. 2014. Effects of processing with dry heat and wet heat on the antioxidant profile of sorghum. Food Chem. 152:210–217. doi:10.1016/j.foodchem.2013.11.106

Castano, G., R. Mas, L. Fernadez, J. Illnait, R. Gamez, and E. Alvarez. 2001. Effects of policosanol 20 versus 40 mg/day in the treatment of patients with type II hypercholesterolemia: A 6-month double–blind study. Int. J. Clin. Pharmacol. Res. 21:43–57.

Chamba, E.B., N.G. Halford, J. Forsyth, M. Wilkinson, and P.R. Shewry. 2005. Molecular cloning of b-kafirin, a methionine-rich protein of sorghum grain. J. Cereal Sci. 41:381–383. doi:10.1016/j.jcs.2004.09.004

Chandrasekara, A., and F. Shahidi. 2011. Determination of antioxidant activity in free and hydrolyzed fractions of millet grains and characterization of their phenolic profiles by HPLC-DAD-ESI-MSn. J. Funct. Foods 3:144–158. doi:10.1016/j.jff.2011.03.007

Chandrashekar, A., and H. Mazhar. 1999. The biochemical basis and implications of grain strength in sorghum and maize. J. Cereal Sci. 30:193–207. doi:10.1006/jcrs.1999.0264

Chiremba, C., J.R.N. Taylor, L.W. Rooney, and T. Beta. 2012. Phenolic acid content of sorghum and maize cultivars varying in hardness. Food Chem. 134:81–88. doi:10.1016/j.foodchem.2012.02.067

Chung, I.M., S.J. Yong, J. Lee, and S.H. Kim. 2013. Effect of genotype and cultivation location on b sitosterols and a, b, g, d, tocopherols. Food Res. Int. 51:971–976. doi:10.1016/j.foodres.2013.02.027

Cordain, L. 1999. Cereal grains: Humanity's double-edged sword. World Rev. Nutr. Diet. 84:19–73. doi:10.1159/000059677

Corredor, D.Y., S.R. Bean, T. Schober, and D. Wang. 2006. Effect of decorticating sorghum on ethanol production and composition of DDGS. Cereal Chem. 83:17–21. doi:10.1094/CC-83-0017

Correia, I., A. Nunes, A. Barros, and I. Delgadillo. 2010. Comparison of the effects induced by different processing methods on sorghum proteins. J. Cereal Sci. 51:146–151. doi:10.1016/j.jcs.2009.11.005

Coulibaly, A., B. Kouakou, and J. Chen. 2011. Phytic acid in cereal grains: Structure, healthy or harmful ways to reduce phytic acid in cereal grains and their effects on nutritional quality. Am. J. Plant Nutr. Fert. 1:1–22. doi:10.3923/ajpnft.2011.1.22

Cremer, J.E., S.R. Bean, M. Tilley, B.P. Ioerger, J.-B. Ohm, R.C. Kaufmann, J.D. Wilson, D.D. Innes, E.K. Gilding, and I.D. Godwin. 2014. Grain sorghum proteomics: An integrated approach towards characterization of seed storage proteins in kafirin allelic variants. J. Agric. Food Chem. 62:9819–9831. doi:10.1021/jf5022847

da Silva, L.S., J. Taylor, and J.R.N. Taylor. 2011. Transgenic sorghum with altered kafirin synthesis: Kafirin solubility, polymerization, and protein digestion. J. Agric. Food Chem. 59:9265–9270. doi:10.1021/jf201878p

Daiber, K.H. 1975. Enzyme inhibition by polyphenols of sorghum grain and malt. J. Sci. Food Agric. 26:1399–1411. doi:10.1002/jsfa.2740260920

Daiber, K.H., and J.R.N. Taylor. 1995. Opaque beers. In: D.A.V. Dendy, editor, Sorghum and millets: Chemistry and technology. AACC, St. Paul, MN. p. 299–323.

Dalton, J.L., and H.L. Mitchell. 1959. Fractionation of sorghum grain wax. J. Agric. Food Chem. 7:570–573. doi:10.1021/jf60102a009

de Oliveira, L.D., A.C. Carciofi, M.C.C. Oliveira, R.S. Vasconcellos, R.S. Bazolli, G.T. Pereira, and F. Prada. 2008. Effects of six carbohydrate sources on the diet digestibility and postprandial glucose and insulin responses in cats. J. Anim. Sci. 86:2237–2246. doi:10.2527/jas.2007-0354

de Silva, S.S., and T.A. Anderson. 1995. Fish nutrition in aquaculture. Springer, New York.

DeRose, R.T., D.-P. Ma, I.-S. Kwon, S.E. Hasnain, R.C. Klassy, and T.C. Hall. 1989. Characterization of the kafirin gene family from sorghum reveals extensive homology with zein from maize. Plant Mol. Biol. 12:245–256. doi:10.1007/BF00043202

Deshpande, A.V., and L.B. Panya. 1987. Evaluation of sorghum starch as a tablet disintegrant and binder. J. Pharm. Pharmacol. 39:495–496. doi:10.1111/j.2042-7158.1987.tb03431.x

Duodu, K.G., J.R.N. Taylor, P.S. Belton, and B.R. Hamaker. 2003. Factors affecting sorghum protein digestibility. J. Cereal Sci. 38:117–131. doi:10.1016/S0733-5210(03)00016-X

Dykes, L., G.C. Peterson, W.L. Rooney, and L.W. Rooney. 2011. Flavonoid composition of lemon-yellow sorghum genotypes. Food Chem. 128:173–179. doi:10.1016/j.foodchem.2011.03.020

Dykes, L., and L.W. Rooney. 2006. Sorghum and millet phenols and antioxidants. J. Cereal Sci. 44:236–251. doi:10.1016/j.jcs.2006.06.007

Dykes, L., L.W. Rooney, R.D. Waniksa, and W.L. Rooney. 2005. Phenolic compounds and antioxidant activity of sorghum grains of varying genotypes. J. Agric. Food Chem. 53:6813–6818. doi:10.1021/jf050419e

Dykes, L., L.M. Seitz, W.L. Rooney, and L.W. Rooney. 2009. Flavonoid composition of red sorghum genotypes. Food Chem. 116:313–317. doi:10.1016/j.foodchem.2009.02.052

Earp, C.F., C.A. Doherty, and L.W. Rooney. 1983. Fluorescence microscopy of the pericarp, aleurone layer and endosperm cell walls of three sorghum cultivars. Cereal Chem. 60:408–410.

Earp, C.F., C.M. McDounough, and L.W. Rooney. 2004a. Microscopy of pericarp development in the caryopsis of *Sorghum bicolor* (L.) *Moench*. J. Cereal Sci. 39:21–27. doi:10.1016/S0733-5210(03)00060-2

Earp, C.F., C.M. McDonough, J. Awika, and L.W. Rooney. 2004b. Testa development in the caryopsis of *Sorghum bicolor* (L.) *Moench*. J. Cereal Sci. 39:303–311. doi:10.1016/j.jcs.2003.11.005

Ebadi, M.R., J. Pourreza, J. Jamalian, M.A. Edriss, A.H. Samie, and S.A. Mirhadi. 2005. Amino acid content and availability in low, medium and high tannin sorghum grain for poultry. Int. J. Poult. Sci. 4:27–31. doi:10.3923/ijps.2005.27.31

Ejeta, G., M.M. Hassen, and E.T. Mertz. 1987. In vitro digestibility and amino acid composition of pearl millet (*Pennisetum typhoides*) and other cereals. Proc. Natl. Acad. Sci. USA 84:6016–6019. doi:10.1073/pnas.84.17.6016

Elkhalifa, A.E.O., D.M.R. Georget, S.A. Barker, and P.S. Belton. 2009. Study of the physical properties of kafirin during the fabrication of tablets for pharmaceutical applications. J. Cereal Sci. 50:159–165. doi:10.1016/j.jcs.2009.03.010

El-Nour, I.N.A., A.D.B. Peruffo, and A. Curioni. 1998. Characterization of sorghum kafirins in relation to their cross-linking behavior. J. Cereal Sci. 28:197–207. doi:10.1006/jcrs.1998.0185

Esen, A. 1987. A proposed nomenclature for the alcohol-soluble proteins (zeins) of maize (*Zea mays L.*). J. Cereal Sci. 5:117–128. doi:10.1016/S0733-5210(87)80015-2

Evans, D.J., L. Schuessler, and J.R.N. Taylor. 1987. Isolation of reduced soluble protein from sorghum starchy endosperm. J. Cereal Sci. 5:61–65. doi:10.1016/S0733-5210(87)80010-3

Evers, A.D., C.T. Greenwood, D.D. Muir, and C. Venables. 1974. Studies on the biosynthesis of starch granules. Starke 26:42–46. doi:10.1002/star.19740260203

Evers, T., and S. Millar. 2002. Cereal grain structure and development: Some implications for quality. J. Cereal Sci. 36:261–284. doi:10.1006/jcrs.2002.0435

Fagbenro, O.A., E. Adeparusi, and O. Fapohunda. 2003. Feed stuffs and dietary substitutions for farmed fish in Nigeria. Proceedings of the National Workshop on Fish Feed Development and Feeding Practices in Aquaculture, (NWFFDFPA'03), New Bussa, Nigeria. 15–19 Sept. Natl. Fresh Water Fisheries Res. Inst., New Bussa, Nigeria. p. 60–65.

Farrar, J.L., D.K. Hartle, J.L. Hargrove, and P. Greenspan. 2008. A novel nutraceutical property of select sorghum (*Sorghum bicolor*) brans: Inhibition of protein glycation. Phytother. Res. 22:1052–1056. doi:10.1002/ptr.2431

Food and Agriculture Organization. 1995. Protein content and quality. In: Sorghum and millets in human nutrition. FAO Food and Nutr. Ser. 27. FAO, Rome.

Food and Agriculture Organization. 2007. Study and analysis of feeds and fertilizers for sustainable aquaculture development. FAO Fisheries Technical Paper 497. Edited by M.R. Hasan et al. FAO, Rome.

Food and Agriculture Organization. 2014. FAOSTAT database. http://faostat3.fao.org/home/E (accessed 30 Dec. 2014).

Forrester, S.D., and C.A. Kirk. 2009. Cats and Carbohydrates—What are the Concerns? Hill's Pet Nutrition. Nutrition Myths and Truths, Facts and Fallacies, Topeka, KA.

Fortes, C.M.L.S., A.C. Carciofi, N.K. Sakomura, I.M. Kawauchi, and R.S. Vasconcellos. 2010. Digestibility and metabolizable energy of some carbohydrate sources for dogs. Anim. Feed Sci. Technol. 156:121–125. doi:10.1016/j.anifeedsci.2010.01.009

Gaffa, T., Y. Yoshimoto, I. Hanashiro, O. Honda, S. Kawaski, and Y. Takeda. 2004. Physicochemical properties and molecular structures of starches from millet (*Pennisetum typhoides*) and sorghum (*Sorghum bicolor* L. Moench) cultivars in Nigeria. Cereal Chem. 81:255–260. doi:10.1094/CCHEM.2004.81.2.255

Garr, J.S.M., and A.B. Bangudu. 1991. Evaluation of sorghum starch as a tablet excipient. Drug Dev. Ind. Pharm. 17:1–6. doi:10.3109/03639049109043805

Geera, B., L.O. Ojwang, and J.M. Awika. 2012. New highly stable dimeric 3-deoxyanthocyanidin pigments from *Sorghum bicolor* leaf sheath. J. Food Sci. 77:C566–C572. doi:10.1111/j.1750-3841.2012.02668.x

Gujer, R., D. Magnolato, and R. Self. 1986. Glucosylated flavonoids and other phenolic compounds from sorghum. Phytochemistry 25:1431–1436. doi:10.1016/S0031-9422(00)81304-7

Hager, A.-S., J.P. Taylor, D.M. Waters, and E.K. Arendt. 2014. Gluten free beer—A review. Trends Food Sci. Technol. 36:44–54. doi:10.1016/j.tifs.2014.01.001

Hagerman, A.E., K.M. Riedl, G.A. Jones, K.N. Sovik, N.T. Ritchard, P.W. Hartzfeld, and T.L. Riechel. 1998. High molecular weight plant polyphenolics (tannins) as biological antioxidants. J. Agric. Food Chem. 46:1887–1892. doi:10.1021/jf970975b

Hallgren, L., and D.S. Murty. 1983. A screening test for grain hardness in sorghum employing density grading in sodium nitrate solution. J. Cereal Sci. 1:265–274. doi:10.1016/S0733-5210(83)80014-9

Hamaker, B.R., and B.A. Bugusu. 2000. IAFRIPRO: Workshop on the proteins of sorghum and millets: Enhancing nutritional and functional properties for Africa. http://www.afripro.org.uk/papers/Paper08Hamaker.pdf (accessed 7 Jan. 2015).

Hamaker, B.R., A.W. Kirleis, E.D. Mertz, and J.D. Axtell. 1986. Effect of cooking on the protein profiles and *in-vitro* digestibility of sorghum and maize. J. Agric. Food Chem. 34:647–649. doi:10.1021/jf00070a014

Hamaker, B.R., A.A. Mohamed, J.E. Habben, C.P. Huang, and B.A. Larkins. 1995. Efficient procedure for extracting maize and sorghum kernel proteins reveals higher prolamin contents than the conventional method. Cereal Chem. 72:583–588.

Hanashiro, I., J. Abe, and S. Hizukuri. 1996. A periodic distribution of the chain length of amylopectin as revealed by high-performance anion-exchange chromatography. Carbohydr. Res. 283:151–159. doi:10.1016/0008-6215(95)00408-4

Hargrove, J.L., P. Greenspan, D.K. Hartle, and C. Dowd. 2011. Inhibition of aromatase and a-amylase by flavonoids and proanthocyanidins from Sorghum bicolor bran extracts. J. Med. Food 14:799–807. doi:10.1089/jmf.2010.0143

Hathcock, J.N. 1991. Residue trypsin inhibitor: Data needs for risk assessment. In: M. Friedman, editor, Nutritional and toxicological consequences of food processing. Plenum Press, New York. p. 273–279.

Herman, E.M., and B.A. Larkins. 1999. Protein storage bodies and vacuoles. Plant Cell 11:601–613. doi:10.1105/tpc.11.4.601

Hill, H., L.S. Lee, and R.J. Henry. 2012. Variation in sorghum starch synthesis genes associated with differences in starch phenotype. Food Chem. 131:175–183. doi:10.1016/j.foodchem.2011.08.057

Hizukuri, S. 1985. Relationship between the distribution of the chain length of amylopectin and the crystalline structure of starch granules. Carbohydr. Res. 141:295–306. doi:10.1016/S0008-6215(00)90461-0

Hojilla-Evangelista, M.P., and S.R. Bean. 2011. Evaluation of sorghum flour as extender in plywood adhesives for sprayline coaters or foam extrusion. Ind. Crops Prod. 34:1168–1172. doi:10.1016/j.indcrop.2011.04.005

Holding, D.R. 2014. Recent advances in the study of prolamin storage protein organization and function. Front. Plant Sci. 5:276. doi:10.3389/fpls.2014.00276

Hoseney, R.C., A.B. Davis, and L.H. Herbers. 1974. Pericarp and endosperm structure of sorghum grain shown by scanning electron microscopy. Cereal Chem. 51:552–558.

Hotz, C., and R. Gibson. 2007. Traditional food-processing and preparation practices to enhance the bioavailability of micronutrients in plant-based diets. J. Nutr. 137:1097–1100.

Hubbard, J.E., H.H. Hall, and F.R. Earle. 1950. Composition of the component parts of the sorghum kernel. Cereal Chem. 51:825–829.

Huber, K.C., and J.N. BeMiller. 2000. Channels of maize and sorghum starch granules. Carbohydr. Polym. 41:269–276. doi:10.1016/S0144-8617(99)00145-9

Hug-Iten, S., F. Escher, and B. Conde-Petit. 2003. Staling of bread: Role of amylose and amylopectin and influence of starch-degrading enzymes. Cereal Chem. 80:654–661. doi:10.1094/CCHEM.2003.80.6.654

Hwang, T., S.L. Cuppett, C.L. Weller, M.A. Hanna, and R.K. Shoemaker. 2002a. Aldehydes in grain sorghum wax. J. Am. Oil Chem. Soc. 79:529–533. doi:10.1007/s11746-002-0516-4

Hwang, K.T., C.L. Weller, S.L. Cuppett, and M.A. Hanna. 2002b. HPLC of grain sorghum wax classes highlighting separation of aldehydes from wax esters and steryl esters. J. Sep. Sci. 25:619–623. doi:10.1002/1615-9314(20020601)25:9<619::AID-JSSC619>3.0.CO;2-J

Hwang, T., S.L. Cuppett, C.L. Weller, M.A. Hanna, and R.K. Shoemaker. 2002c. Properties, composition, and analysis of grain sorghum wax. J. Am. Oil Chem. Soc. 79:521–527. doi:10.1007/s11746-002-0515-5

Hwang, K.T., J.E. Kim, and C.L. Weller. 2005. Policosanol contents and compositions in wax-like materials extracted from selected cereals of Korean origin. Cereal Chem. 82:242–245. doi:10.1094/CC-82-0242

Hwang, K.T., C.L. Weller, S.L. Cuppet, and M.A. Hanna. 2004. Policosanol content and composition of grain sorghum kernels and dried distillers grains. Cereal Chem. 81:345–349. doi:10.1094/CCHEM.2004.81.3.345

Ismat, G.A., E.A. Khogali, O.A. Azhari, and E.E. Babiker. 2013. Effect of radiation process followed by traditional treatments on nutritional and antinutritional attributes of sorghum cultivar. Int. Food Res. J. 20:3221–3228.

Izquierdo, L., and I.D. Godwin. 2005. Molecular characterization of a novel methionine-rich d-kafirin seed storage protein gene in sorghum (*Sorghum bicolor* L.). Cereal Chem. 82:706–710. doi:10.1094/CC-82-0706

Jambunathan, R., E.T. Mertz, and J.D. Axtell. 1975. Fractionation of soluble proteins of high lysine and normal sorghum grain. Cereal Chem. 52:119–121.

Jampala, B., W.L. Rooney, G.C. Peterson, S. Bean, and D.B. Hays. 2012. Estimating the relative effects of the endosperm traits of waxy and high protein digestibility on yield in grain sorghum. Field Crops Res. 139:57–62. doi:10.1016/j.fcr.2012.09.021

Jane, J., Y.Y. Chen, L.F. Lee, A.E. McPherson, K.S. Wong, M. Radosavljevic, and T. Kasemsuwan. 1999. Effects of amylopectin branch chain length and amylose content on the gelatinization and pasting properties of starch. Cereal Chem. 76:629–637. doi:10.1094/CCHEM.1999.76.5.629

Jessen, H. 2010. Sorghum surges. Ethanol Producer Magazine. http://www.ethanolproducer.com/articles/6338/sorghum-surges (accessed 27 Oct. 2014).

Kambal, A.E., and E.C. Bate-Smith. 1976. A genetic and biochemical study on pericarp pigments in a cross between two cultivars of grain sorghum, Sorghum bicolor. Heredity 37:413–416. doi:10.1038/hdy.1976.106

Kayodé, A.P.P., J.D. Hounhouigan, and M.A.J.S. Van Boekel. 2007. Impact of brewing process operations on phytate, phenolic compounds and in vitro solubility of iron and zinc in opaque sorghum beer. LWT - Food Sci. Technol. 40:834–841.

Kayodé, A.P.P., A.R. Linnemann, J.D. Hounhouigan, M.J.R. Nout, and M.A.J.S. Van Boekel. 2006a. Genetic and environmental impact on iron, zinc, and phytate in food sorghum grown in Benin. J. Agric. Food Chem. 54:256–262. doi:10.1021/jf0521404

Kayodé, A.P.P., M.J.R. Nout, E.J. Bakker, and M.A.J.S. Van Boekel. 2006b. Evaluation of the simultaneous effects of processing parameters on the iron and zinc solubility of infant sorghum porridge by response surface methodology. J. Agric. Food Chem. 54:4253–4259. doi:10.1021/jf0530493

Kean, E.G., N. Bordenave, G. Ejeta, B.R. Hamaker, and M.G. Ferruzzi. 2011. Carotenoid biaccessibility from whole grain and decorticated yellow endosperm sorghum porridge. J. Cereal Sci. 54:450–459. doi:10.1016/j.jcs.2011.08.010

Kean, E.G., G. Ejeta, B.R. Hamaker, and M.G. Ferruzzi. 2007. Characterization of carotenoid pigments in mature and developing kernels of selected yellow-endosperm sorghum varieties. J. Agric. Food Chem. 55:2619–2626. doi:10.1021/jf062939v

Khalil, J.K., W.N. Sawaya, W.J. Safi, and H.M. Al-Mohammad. 1984. Chemical composition and nutritional quality of sorghum flour and bread. Plant Foods Hum. Nutr. 34:141–150. doi:10.1007/BF01094842

Kore, K.B., A.K. Pattanaik, A. Das, and K. Sharma. 2009. Evaluation of alternative cereal sources in dog diets: Effect on nutrient utilization and hindgut fermentation characteristics. J. Sci. Food Agric. 89:2174–2180. doi:10.1002/jsfa.3698

Kumar, A., B.V.S. Reddy, B. Ramaiah, K.L. Sahrawat, and W.H. Pfeiffer. 2012. Genetic variability and character association for grain iron and zinc contents in sorghum germplasm accessions and commercial cultivars. Eur. J. Plant Sci. Biotechnol. 6:1–5.

Kumar, P.M., T.K. Virupaksha, and P.J. Vithayathil. 1978. Sorghum proteinase inhibitors: Purification and some biochemical properties. Int. J. Pept. Protein Res. 12:185–196. doi:10.1111/j.1399-3011.1978.tb02886.x

Kummerow, F.A. 1946a. The composition of sorghum grain oil *Andropogon Sorghum* var vulgaris. Oil Soap 23:167–170. doi:10.1007/BF02640963

Kummerow, F.A. 1946b. The composition of oil extracted from 14 different varieties of *Andropogon Sorghum* var vulgaris. Oil Soap 23:273–275. doi:10.1007/BF02545394

Laidlaw, H.K.C., E.S. Mace, S.B. Williams, K. Sakrewski, A.M. Mudge, P.J. Prentis, D.R. Jordan, and I.D. Godwin. 2010. Allelic variation of the b-, g- and d-kafirin genes in diverse *Sorghum* genotypes. Theor. Appl. Genet. 121:1227–1237. doi:10.1007/s00122-010-1383-9

Landry, J., and T. Moureaux. 1970. Heterogeneity of corn seed glutelin: Selective extraction and amino acid composition of the 3 isolated fractions. Bull. Soc. Chim. Biol. (Paris) 52:1021–1037.

Landry, J. 1997. Comparison of extraction methods for evaluating zein content of maize grain. Cereal Chem. 74:188–189. doi:10.1094/CCHEM.1997.74.2.188

Langeveld, S.M.J., R. van Wijk, N. Stuurman, J.W. Kijne, and S. de Pater. 2000. B-type granule containing protrusions and interconnections between amyloplasts in developing wheat endosperm revealed by transmission electron microscopy and GFP expression. J. Exp. Bot. 51:1357–1361. doi:10.1093/jexbot/51.349.1357

Larsson, S.C., E. Giovannucci, L. Bergkvist, and A. Wolk. 2005. Whole grain consumption and risk of colorectal cancer: A population-based cohort of 60,000 women. Br. J. Cancer 92:1803–1807. doi:10.1038/sj.bjc.6602543

Lásztity, R. 1996. The chemistry of cereal proteins. CRC Press, Boca Raton, FL.

Leguizamon, C., C.L. Weller, V.L. Schlegel, and T.P. Carr. 2009. Plant sterol and policosanol characterization of hexane extracts from grain sorghum, corn, and their DDGS. J. Am. Oil Chem. Soc. 86:707–716. doi:10.1007/s11746-009-1398-z

Levi, F., C. Pasche, F. Lucchini, L. Chatenoud, D.R. Jacobs, and C. La Vecchia. 2000. Refined and whole grain cereals and the risk of oral, oesophageal and laryngeal cancer. Eur. J. Clin. Nutr. 54:487–489. doi:10.1038/sj.ejcn.1601043

Li, N., Y. Wang, M. Tilley, S.R. Bean, X. Wu, X.S. Sun, and D. Wang. 2011. Adhesive performance of sorghum protein extracted from sorghum DDGS and flour. J. Polym. Environ. 19:755–765. doi:10.1007/s10924-011-0305-5

Lichtenwalner, R.E., E.B. Ellis, and L.W. Rooney. 1978. Effect of incremental dosages of the waxy gene of sorghum on digestibility. J. Anim. Sci. 46:1113–1119.

Lipkie, T.E., F.F. DeMoura, Z.Y. Zhao, M.C. Albertsen, P. Che, K. Glassman, and M.G. Ferruzzi. 2013. Bioaccessibility of carotenoids from transgenic provitamin A biofortified sorghum. J. Agric. Food Chem. 61:5764–5771. doi:10.1021/jf305361s

Liu, K.S. 2011. Comparison of lipid content and fatty acid composition and their distribution within seeds of 5 small grain species. J. Food Sci. 76:C334–C342. doi:10.1111/j.1750-3841.2010.02038.x

Lochte-Watson, K.R., C.L. Weller, and D.S. Jackson. 2000. Fractionation of grain sorghum using abrasive decortication. J. Agric. Eng. Res. 77:203–208. doi:10.1006/jaer.2000.0583

Lorent, J.H., J. Quetin-Leclercq, and M. Mingeot-Leclercq. 2014. The amphiphilic nature of saponins and their effects on artificial and biological membranes and potential consequences for red blood and cancer cells. Org. Biomol. Chem. 12:8803–8822. doi:10.1039/C4OB01652A

Marinangeli, C.P.F., P.J.H. Jones, A.N. Kassis, and M.N.A. Eskin. 2010. Policosanols as nutraceuticals: Fact or fiction. Crit. Rev. Food Sci. Nutr. 50:259–267. doi:10.1080/10408391003626249

Matalanis, A.M., O.H. Campanella, and B.R. Hamaker. 2009. Storage retrogradation behavior of sorghum, maize and rice starch pastes related to amylopectin fine structure. J. Cereal Sci. 50:74–81. doi:10.1016/j.jcs.2009.02.007

Mazhar, H., and A. Chandrashekar. 1993. Differences in kafirin composition during endosperm development and germination in sorghum cultivars of varying hardness. Cereal Chem. 70:667–671.

Mazhar, H., A. Chandrashekar, and H.S. Shetty. 1993. Isolation and immunochemical characterization of the alcohol-extractable proteins (kafirins) of *Sorghum bicolor* (L.) Moench. J. Cereal Sci. 17:83–93. doi:10.1006/jcrs.1993.1009

McGoogan, B.B., and R.C. Reigh. 1996. Apparent digestibility of selected ingredients in red drum (*Sciaenops ocellatus*) diets. Aquaculture 141:233–244. doi:10.1016/0044-8486(95)01217-6

Mehmood, S., I. Orhan, Z. Ahsan, S. Aslan, and M. Gulfraz. 2008. Fatty acid composition of seed oil of different *Sorghum bicolor* varieties. Food Chem. 109:855–859. doi:10.1016/j.foodchem.2008.01.014

Mohammed, N.A., E.E. Babiker, and I.A. Ahmed. 2011. Nutritional evaluation of sorghum flour (*Sorghum bicolor L. Moench*) during processing of injera. World Acad. Sci. Eng. Technol. 5:58–62.

Moreau, R.A., and K.B. Hicks. 2005. The composition of corn oil obtained by alcohol extraction of ground corn. J. Am. Oil Chem. Soc. 82:809–815. doi:10.1007/s11746-005-1148-4

Moreau, R.A., A.M. Lampi, and K.B. Hicks. 2009. Fatty acid, phytosterol, and polyamine conjugate profiles of edible oils extracted from corn germ, corn fiber, and corn kernels. J. Am. Oil Chem. Soc. 86:1209–1214. doi:10.1007/s11746-009-1456-6

Moreau, R.A., V. Singh, and K.B. Hicks. 2001. Comparison of oil and phytosterol levels in germplasm accessions of corn, teosinte, and Job's Tears. J. Agric. Food Chem. 49:3793–3795. doi:10.1021/jf010280h

Morell, M.K., M.S. Samuel, and M.G. O'Shea. 1998. Analysis of starch structure using fluorophore-assisted capillary electrophoresis. Electrophoresis 19:2603–2611. doi:10.1002/elps.1150191507

Müntz, K. 1998. Deposition of storage proteins. Plant Mol. Biol. 38:77–99. doi:10.1023/A:1006020208380

Murty, D.S., and K.A. Kumar. 1995. Traditional uses of sorghum. In: D.A.V. Dendy, editor, Sorghum and millets: Chemistry and technology. AACC, St. Paul, MN. p. 185–217.

Nip, W.K., and E.E. Burns. 1969. Pigment characterization in grain sorghum. I. Red varieties. Cereal Chem. 46:490–495.

NPD Group/Dieting Monitor. 2013. 52 week data year ending January 30, 2013. https://www.npd.com/wps/portal/npd/us/news/press-releases/percentage-of-us-adults-trying-to-cut-down-or-avoid-gluten-in-their-diets-reaches-new-high-in-2013-reports-npd/ (accessed 27 Oct. 2014).

Neucere, N.J., and G. Sumrell. 1979. Protein fractions from five varieties of grain sorghum: Amino acid composition and solubility properties. J. Agric. Food Chem. 27:809–812. doi:10.1021/jf60224a059

Oboh, G., T.L. Akomolafe, and A. O. Adetuyi. 2010. Inhibition of cyclophosphamide-induced oxidative stress in brain by dietary inclusion of red dye extracts from sorghum (Sorghum bicolor) stem. J. Med. Food 13:1075–1080. doi:10.1089/jmf.2009.0226

Ojwang, L., and J.M. Awika. 2008. Effect of pyruvic acid and ascorbic acid on stability of 3-deoxyanthocyanidins. J. Sci. Food Agric. 88:1987–1996. doi:10.1002/jsfa.3308

O'Kennedy, M.M., A. Grootboom, and P.R. Shewry. 2006. Harnessing sorghum and millet biotechnology for food and health. J. Cereal Sci. 44:224–235. doi:10.1016/j.jcs.2006.08.001

Omary, M.B., C. Fong, J. Rothschild, and P. Finney. 2012. Effects of germination on the nutritional profile of gluten-free cereals and pseudocereals: A review. Cereal Chem. 89:1–14. doi:10.1094/CCHEM-01-11-0008

Oria, M.P., B.R. Hamaker, J.D. Axtell, and C.-P. Huang. 2000. A highly digestible sorghum mutant cultivar exhibits a unique folded structure of endosperm protein bodies. Proc. Natl. Acad. Sci. USA 97:5065–5070. doi:10.1073/pnas.080076297

Osborne, T.B. 1907. The proteins of the wheat kernel. Carnegie Inst., Washington, DC.

Osborne, T.B. 1924. The vegetable proteins. 2nd ed. Longmans Green, London.

Osman, M.A. 2004. Changes in sorghum enzyme inhibitors, phytic acid, tannins and in vitro protein digestibility occurring during Khamir (local bread) fermentation. Food Chem. 88:129–134. doi:10.1016/j.foodchem.2003.12.038

Parker, M.L. 1985. The relationship between A-type and B-type starch granules in the developing endosperm of wheat. J. Cereal Sci. 3:271–278. doi:10.1016/S0733-5210(85)80001-1

Pedersen, B., and B.O. Eggum. 1983. The influence of milling on the nutritive value of flour from cereal grains. 6. Sorghum. Plant Foods Hum. Nutr. 33:313–326. doi:10.1007/BF01094756

Pedersen, J.F., S.R. Bean, R.A. Graybosch, S.H. Park, and M. Tilley. 2005. Characterization of waxy grain sorghum lines in relation to granule-bound starch synthase. Euphytica 144:151–156. doi:10.1007/s10681-005-5298-5

Pedersen, J.F., R.A. Graybosch, and D.L. Funnell. 2007. Occurrence of the waxy alleles wx^a and wx^b in waxy sorghum plant introductions and their effect on starch thermal properties. Crop Sci. 47:1927–1933. doi:10.2135/cropsci2006.10.0652

Piasecka-Kwiatkowska, D., J.R. Warchalewski, M. Zeilińska-Dawidziak, and M. Michalak. 2012. Digestive enzyme inhibitors from grains as potential components of nutraceuticals. J. Nutr. Sci. Vitaminol. (Tokyo) 58:217–220. doi:10.3177/jnsv.58.217

Pinheiro-Sant'Ana, H.M., M. Guinazi, D. Oliverira, C.M.D.Lucia, B.D.L. Reis, and S.C.C Brandao. 2011. Method for simultaneous analysis of eight vitamin E isomers in various foods by high performance liquid chromatography and fluorescence detection. J. Chromatogr. A 1218:8496–8502. doi:10.1016/j.chroma.2011.09.067

Pontieri, P., A. Di Maro, R. Tamburino, M. De Stefano, M. Tilley, S.R. Bean, E. Roemer, P. De Vita, P. Alifano, L. Del Giudice, and D.R. Massardo. 2010. Chemical composition of selected food-grade sorghum varieties grown under typical Mediterranean conditions. Maydica 55:139–143.

Prasad, R. 2012. Micro mineral nutrient deficiencies in humans, animals and plants and their amelioration. Proc. Natl. Acad. Sci. India 82:225–233.

Proietti, I., S. Tait, F. Aureli, and A. Mantovani. 2014. Modulation of sorghum biological activities by varieties and two traditional processing methods: An integrated in vitro/ modelling approach. Int. J. Food Sci. Technol. 49:1593–1599. doi:10.1111/ijfs.12460

Ramos, J.R., R.R. Cabral, and F.D. Chan. 1984. Utilization of sorghum flour as extender for plywood adhesive (Philippines). NSDB Technol. J. 9:29–48.

Reddy, N.R., M.D. Pierson, S.K. Sathe, and D.K. Salunkhe. 1989. Phytates in cereals and legumes. CRC Press. Boca Raton, FL.

Reichert, R., M. Mwararu, and S. Mukuru. 1988. Characterization of colored grain sorghum lines and identification of high tannin lines with good dehulling characteristics. Cereal Chem. 65:165–170.

Rodríguez-Castellanos, W., F. Martínez-Bustos, O. Jiménez-Arévalo, R. González-Núñez, and T. Galicia-García. 2013. Functional properties of extruded and tubular films of sorghum starch-based glycerol and *Yucca schidigera* extract. Ind. Crops Prod. 44:405–412. doi:10.1016/j.indcrop.2012.11.027

Roldán, I., M.M. Lucas, D. Devallé, V. Planchot, S. Jimenez, R. Perez, S. Ball, C. D'Hulst, and A. Mérida. 2007. The phenotype of soluble starch synthase IV defective mutants of *Arabidopsis thaliana* suggests a novel function of elongation enzymes in the control of starch granule formation. Plant J. 49:492–504. doi:10.1111/j.1365-313X.2006.02968.x

Rooney, L.W. 1978. Sorghum and pearl millet lipids. Cereal Chem. 55:584–590.

Rooney, L.W., and F.R. Miller. 1982. Variation in the structure of the kernel characteristics of sorghum. In: J.V. Mertin, editor, Proceedings of the international symposium on sorghum grain quality. Patancheru, India. 28–31 Oct. 1981. ICRISAT, Patancheru, India. p. 143–162.

Rooney, W.L., and S.O. Serna-Saldivar. 2000. Sorghum. In: K. Kulp, editor, Handbook of cereal science and technology. CRC Press, Boca Raton, FL. p. 149–176.

Rooney, W.L., and R.D. Waniska. 2000. Sorghum food and industrial utilization. In: C.W. Smith and R.A. Frederiksen, editors, Sorghum: Origin, history, technology, and production. John Wiley, New York. p. 689–729.

Rooney, W.L. 2000. Genetics and cytogenetics. In: C.W. Smith and R.A. Frederiksen, editors, Sorghum: Origin, history, technology, and production. John Wiley, New York. p. 261–308.

Rooney, W.L., S. Aydin, and L.C. Kuhlman. 2005. Assessing the relationship between endosperm type and grain yield potential in sorghum (*Sorghum bicolor* L. *Moench*). Field Crops Res. 91:199–205. doi:10.1016/j.fcr.2004.07.011

Ryan, E., K. Galvin, T.P. O'Connor, A.R. Maguire, and N.M. O'Brien. 2007. Phytosterol, squalene, tocopherol content and fatty acid profile of selected seeds, grains, and legumes. Plant Foods Hum. Nutr. 62:85–91. doi:10.1007/s11130-007-0046-8

Salas Fernandez, M.G., M. Hamblin, L. Li, W.L. Rooney, M.R. Tuinstra, and S. Kresovich. 2008. Quantitative trait loci analysis of endosperm color and carotenoid content in sorghum grain. Crop Sci. 48:1732–1743. doi:10.2135/cropsci2007.12.0684

Salas Fernandez, M.G., I. Kapran, S. Souley, M. Abdou, I.H. Maiga, C.B. Acharya, M.T. Hamblin, and S. Kresovich. 2009. Collection and characterization of yellow endosperm sorghums from West Africa for biofortification. Genet. Resour. Crop Evol. 56:991–1000. doi:10.1007/s10722-009-9417-3

Salunkhe, D.K., S.J. Jadhav, S.S. Kadam, J.K. Chavan, and B.S. Luh. 1983. Chemical, biochemical, and biological significance of polyphenols in cereals and legumes. CRC Crit. Rev. Food Sci. Nutr. 17:277–305. doi:10.1080/10408398209527350

Sang, Y., S.R. Bean, P.A. Seib, J.F. Pedersen, and Y.-C. Shi. 2008. Structure and functional properties of sorghum starches differing in amylose content. J. Agric. Food Chem. 56:6680–6685. doi:10.1021/jf800577x

Sasaki, T., T. Yasui, and J. Matsuki. 2000. Effect of amylose content on gelatinization, retrogradation, and pasting properties of starches from waxy and nonwaxy wheat and their F1 seeds. Cereal Chem. 77:58–63. doi:10.1094/CCHEM.2000.77.1.58

Sastry, L.V.S., J.W. Paulis, J.A. Bietz, and J.S. Wall. 1985. Genetic variation of storage proteins in sorghum grain: Studies by isoelectric focusing and high-performance liquid chromatography. Cereal Chem. 63:420–427.

Seckinger, H.L., and M.J. Wolf. 1973. Sorghum protein ultrastructure as it relates to composition. Cereal Chem. 50:455–465.

Seitz, L.M. 1977. Composition of sorghum wax. Cereal Food Word 22:170.

Serna-Saldivar, S., and L.W. Rooney. 1995. Structure and chemistry of sorghum and millets. In: D.V. Dendy, editor, Structure and chemistry of sorghum and millets. AACC, St. Paul, MN. p. 69–124.

Shewry, P.R., and N.G. Halford. 2002. Cereal seed storage proteins: Structures, properties and role in grain utilization. J. Exp. Bot. 53:947–958. doi:10.1093/jexbot/53.370.947

Shewry, P.R., J.A. Napier, and A.S. Tatham. 1995. Seed storage proteins: Structures and biosynthesis. Plant Cell 7:945–956. doi:10.1105/tpc.7.7.945

Shewry, P.R., A.S. Tatham, J. Forde, M. Kreis, and B.J. Miflin. 1986. The classification and nomenclature of wheat gluten proteins: A reassessment. J. Cereal Sci. 4:97–106. doi:10.1016/S0733-5210(86)80012-1

Shih, C.H., S.O. Siu, R. Ng, E. Wong, L.C.M. Chiu, I.K. Chu, and C. Lo. 2007. Quantitative analysis of anticancer 3-deoxyanthocyanidins in infected sorghum seedlings. J. Agric. Food Chem. 55:254–259. doi:10.1021/jf062516t

Shull, J.M., A. Chandrashekar, A.W. Kirleis, and G. Ejeta. 1990. Development of sorghum (*Sorghum bicolor* (L.) Moench) endosperm in varieties of varying hardness. Food Struct. 9:253–267.

Shull, J.M., J.J. Watterson, and A.W. Kirleis. 1991. Proposed nomenclature for the alcohol-soluble proteins (kafirins) of Sorghum bicolor (L. Moench) based on molecular weight, solubility, and structure. J. Agric. Food Chem. 39:83–87. doi:10.1021/jf00001a015

Shull, J.M., J.J. Watterson, and A.W. Kirleis. 1992. Purification and immunocytochemical localization of kafirins in *Sorghum bicolor* (L. Moench) endosperm. Protoplasma 171:64–74. doi:10.1007/BF01379281

Singh, V., R.A. Moreau, and K.B. Hicks. 2003. Yield and phytosterol composition of oil extracted from grain sorghum and its wet milled fractions. Cereal Chem. 80:126–129. doi:10.1094/CCHEM.2003.80.2.126

Sklan, D., T. Prag, and I. Lupatsch. 2004. Apparent digestibility coefficients of feed ingredients and their prediction in diets for tilapia *Oreochromis niloticus x Oreochromis aureus* (Teleostei, Cichlidae). Aquacult. Res. 35:358–364. doi:10.1111/j.1365-2109.2004.01021.x

Smith, J.S.C. 1994. RP-HPLC for varietal identification in cereals and legumes: Sorghum. In: J.E. Kruger and J.A. Bietz, editors, High performance liquid chromatography of cereal and legume proteins. AACC, St. Paul, MN. p. 201–205.

Soetan, K.O., M.A. Oyewole, O.O. Aiyelaagve, and M.A. Fafunso. 2006. Evaluation of the antimicrobial activity of saponins extract of *Sorghum bicolor* L. Moench. Afr. J. Biotechnol. 5:2405–2407.

Soetan, K.O., and O.E. Oyewole. 2009. The need for adequate processing to reduce the antinutritional factors in plants used as human foods and animal feeds: A review. Afr. J. Food Sci. 3:223–232.

Solomon, S.G., L.O. Tiamiyu, and U.J. Agaba. 2007. Effect of feeding different grain sources on the growth performance and body composition of tilapia (*Oreochromis niloticus*) fingerlings fed in outdoor hapas. Pakistan J. Nutr. 6:271–275. doi:10.3923/pjn.2007.271.275

Sosulski, F.W., and G.I. Imafidon. 1990. Amino acid composition and nitrogen-to-protein conversion for animal and plant foods. J. Agric. Food Chem. 38:1351–1356. doi:10.1021/jf00096a011

Subba Rao, M., and G. Muralikrishna. 2002. Evaluation of the antioxidant properties of free and bound phenolic acids from native and malted finger millet (Ragi, *Eleusine coracana* Indaf-15). J. Agric. Food Chem. 50:889–892. doi:10.1021/jf011210d

Subramanian, V., N. Seetharama, R. Jambunathan, and P.V. Rao. 1990. Evaluation of protein quality of sorghum [*Sorghum bicolor* (L.) *Moench*]. J. Agric. Food Chem. 38:1344–1347. doi:10.1021/jf00096a009

Sugahara, T., S. Nishimoto, Y. Morioka, and K. Nakano. 2009. White sorghum (Sorghum bicolor (L.) moench) bran extracts suppressed IgE production by U266 cells. Biosci. Biotechnol. Biochem. 73:2043–2047. doi:10.1271/bbb.90245

Sukumaran, S., W. Xiang, S.R. Bean, J.F. Pedersen, M.R. Tuinstra, T.T. Tesso, M.T. Hamblin, and J. Yu. 2012. Genetic structure of a diverse sorghum collection and association mapping for grain quality. Plant Gen. 5:126–135. doi:10.3835/plantgenome2012.07.0016

Sullins, R.D., and L.W. Rooney. 1974. Microscopic evaluation of the digestibility of sorghum lines that differ in endosperm characteristics. Cereal Chem. 51:134–142.

Svensson, L., B. Sekwati-Monang, D.L. Lutz, A. Schieber, and M.G. Ganzle. 2010. Phenolic acids and flavonoids in nonfermented and fermented red sorghum (*Sorghum bicolor* (L.) Moench). J. Agric. Food Chem. 58:9214–9220. doi:10.1021/jf101504v

Szydlowski, N., P. Ragel, S. Raynaud, M.M. Lucas, I. Roldan, M. Montero, F.J. Munoz, M. Ovecka, A. Bahaji, V. Planchot, J. Pozueta-Romero, C. D'Hulst, and A. Merida. 2009. Starch granule initiation in *Arabidopsis* requires the presence of either class IV or class III starch synthases. Plant Cell 21:2443–2457. doi:10.1105/tpc.109.066522

Taylor, J., J.O. Anyango, and J.R.N. Taylor. 2013. Developments in the science of zein, kafirin, and gluten protein bioplastic materials. Cereal Chem. 90:344–357. doi:10.1094/CCHEM-12-12-0165-IA

Taylor, J., T. Schober, and S.R. Bean. 2006. Non-traditional uses of sorghum and pearl millet. J. Cereal Sci. 44:252–271. doi:10.1016/j.jcs.2006.06.009

Taylor, J., J.R.N. Taylor, P.S. Belton, and A. Minnaar. 2009. Kafirin microparticle encapsulation of catechin and sorghum condensed tannins. J. Agric. Food Chem. 57:7523–7528. doi:10.1021/jf901592q

Taylor, J., J.R.N. Taylor, M.F. Dutton, and S. de Kock. 2005. Identification of kafirin film casting solvents. Food Chem. 90:401–408. doi:10.1016/j.foodchem.2004.03.055

Taylor, J.R.N. 1983. Effect of malting on the protein and free amino nitrogen composition of sorghum. J. Sci. Food Agric. 34:885–892. doi:10.1002/jsfa.2740340817

Taylor, J.R.N., P.S. Belton, T. Beta, and K.G. Duodu. 2014. Increasing the utilization of sorghum, millets, and pseudocereals: Developments in the science of their phenolic phytochemicals, biofortification and protein functionality. J. Cereal Sci. 59:257–275.

Taylor, J.R.N., and K.G. Duodu. 2014. Effects of processing sorghum and millets on their phenolic phytochemicals and the implications of this to the health-enhancing properties of sorghum and millet food and beverage products. J. Cereal Sci. 59:257–275. doi:10.1016/j.jcs.2013.10.009

Taylor, J.R.N., L. Novellie, and N.V.D.W. Liebenberg. 1985. Protein body degradation in the starchy endosperm of germinating sorghum. J. Exp. Bot. 36:1287–1295. doi:10.1093/jxb/36.8.1287

Taylor, J.R.N., L. Novellie, and N.V.D.W. Liebenberg,. 1984b. Sorghum protein body composition and ultrastructure. Cereal Chem. 61:69–73.

Taylor, J.R.N., and L. Schüssler. 1986. The protein compositions of the different anatomical parts of sorghum grain. J. Cereal Sci. 4:361–369. doi:10.1016/S0733-5210(86)80040-6

Taylor, J.R.N., L. Schüssler, and W.H. van der Walt. 1984a. Fractionation of proteins from low-tannin sorghum grain. J. Agric. Food Chem. 32:149–154. doi:10.1021/jf00121a036

Tetlow, I.J. 2011. Starch biosynthesis in developing seeds. Seed Sci. Res. 21:5–32. doi:10.1017/S0960258510000292

Thompson, L.U. 1993. Potential health benefits and problems associated with antinutrients in foods. Food Res. Int. 26:131–149. doi:10.1016/0963-9969(93)90069-U

Thompson, L.V. 1986. Phytic acid: Factor influencing starch digestibility and blood glucose response. In: E. Graf, editor, Phytic acid: Chemistry and applications. Pilatus Press, Minneapolis, MN. p. 179–194.

Tian, Y., B. Zou, C.M. Li, J. Yang, S.F. Xu, and A.E. Hagerman. 2012. High molecular weight persimmon tannin is a potent antioxidant both ex vivo and in vivo. Food Res. Int. 45:26–30. doi:10.1016/j.foodres.2011.10.005

USDA. 2007. USDA Database for the Flavonoid Content of Selected Foods. Release 2.1. USDA, Washington, DC.

Waigh, T.A., K.L. Kato, A.M. Donald, M.J. Gidley, C.J. Clarke, and C. Riekel. 2000. Side-chain liquid-crystalline model for starch. Starke 52:450–460. doi:10.1002/1521-379X(200012)52:12<450::AID-STAR450>3.0.CO;2-5

Waigh, T.A., P. Perry, C. Reikel, M.J. Gidley, and A.M. Donald. 1998. Chiral side-chain liquid crystalline polymeric properties of starch. Macromolecules 31:7980–7984. doi:10.1021/ma971859c

Wall, J.S., and J.W. Paulis. 1978. Corn and sorghum grain proteins. In: J. Pomeranz, editor, Advances in cereal science and technology. Vol. II. AACC, St. Paul, MN. p. 135–219.

Wang, D., S. Bean, J. McClaren, P. Seib, M. Tuinstra, M. Lenz, X. Wu, and R. Zhao. 2008a. Grain sorghum is a viable feedstock for ethanol production. J. Ind. Microbiol. Biotechnol. 35:313–320. doi:10.1007/s10295-008-0313-1

Wang, L., C.L. Weller, and K.T. Hwang. 2005. Extraction of lipids from grain sorghum DDG. Trans. ASAE 48:1883–1888. doi:10.13031/2013.19986

Wang, L., C.L. Weller, V.L. Schlegel, T.P. Carr, and S.L. Cuppett. 2007. Comparison of supercritical CO_2 and hexane extraction of lipids from sorghum distillers grains. Eur. J. Lipid Sci. Technol. 109:567–574. doi:10.1002/ejlt.200700018

Wang, L., C.L. Weller, V.L. Schlegel, T.P. Carr, and S.L. Cuppett. 2008b. Supercritical CO_2 extraction of lipids from sorghum distillers grains with solubles. Bioresour. Technol. 99:1373–1382. doi:10.1016/j.biortech.2007.01.055

Waniska, R.D., J.H. Poe, and R. Bandyopadhyay. 1989. Effects of growth conditions on grain molding and phenols in sorghum caryopsis. J. Cereal Sci. 10:217–225. doi:10.1016/S0733-5210(89)80051-7

Waniska, R.D. 2000. Structure, phenolic compounds, and antifungal proteins of sorghum caryopses. p. 72-106 in Technical and institutional options for sorghum grain mold management: Proceedings of an international consultation, Patancheru, India. 18–19 May 2000. ICRISAT, Patancheru, India.

Waniska, R.D., and L.W. Rooney. 2000. Structure and chemistry of the sorghum caryopsis. In: C.W. Smith and R.A. Frederiksen, editors, Sorghum: Origin, history, technology, and production. John Wiley, New York. p. 649–688.

Watterson, J.J., J.M. Shull, and A.W. Kirleis. 1993. Quantitation of a-, b-, and g-kafirins in vitreous and opaque endosperm of *Sorghum bicolor.* Cereal Chem. 70:452–457.

Weaver, C.A., B.R. Hamakar, and J.D. Axtell. 1998. Discovery of grain sorghum germplasm with high uncooked and cooked in vitro protein digestibilities. Cereal Chem. 75:665–670. doi:10.1094/CCHEM.1998.75.5.665

Wedad, W., H. Abdelhaleem, A.H. El Tinay, A.I. Mustafa, and E.E. Babiker. 2008. Effect of fermentation, malt-pretreatment and cooking on antinutritional factors and protein digestibility of sorghum cultivars. Pakistan J. Nutr. 7:335–341. doi:10.3923/pjn.2008.335.341

Weller, C.L., A. Gennadios, R.A. Saraiva, and S.L. Cuppett. 1998. Grain sorghum wax as an edible coating for gelatin-based candies. J. Food Qual. 21:117–128. doi:10.1111/j.1745-4557.1998.tb00509.x

Wilson, J.D., D.B. Bechtel, T.C. Todd, and P.A. Seib. 2006. Measurement of wheat starch granule size distribution using image analysis and laser diffraction technology. Cereal Chem. 83:259–268. doi:10.1094/CC-83-0259

Wu, Y., L. Yuan, X. Guo, D.R. Holding, and J. Messing. 2013. Mutation in the seed storage protein kafirin creates a high-value food trait in sorghum. Nat. Commun. 4:2217. doi:10.1038/ncomms3217.

Wu, Y.V., and J.S. Wall. 1980. Lysine content of protein increased by germination of normal and high-lysine sorghums. J. Agric. Food Chem. 28:455–458. doi:10.1021/jf60228a046

Xu, J.-H., and J. Messing. 2008. Organization of the prolamins gene family provides insight into the evolution of the maize genome and gene duplication in grass species. Proc. Natl. Acad. Sci. USA 105:14330–14335. doi:10.1073/pnas.0807026105

Xu, J.-H., and J. Messing. 2009. Amplification of prolamin storage protein genes in different subfamilies of the Poaceae. Theor. Appl. Genet. 119:1397–1412. doi:10.1007/s00122-009-1143-x

Yan, S., X. Wu, J. Faubion, S. Bean, L. Cai, Y.C. Shi, X.S. Sun, and D. Wang. 2012. Ozone treatment on high-tannin grain sorghum flour and its ethanol production performance. Cereal Chem. 89:30–37. doi:10.1094/CCHEM-06-11-0075

Yang, L., K.F. Allred, B. Geera, C.D. Allred, and J.M. Awika. 2012. Sorghum phenolics demonstrate estrogenic action and induce apoptosis in nonmalignant colonocytes. Nutr. Cancer 64:419–427. doi:10.1080/01635581.2012.657333

Yang, L.Y., J.D. Browning, and J.M. Awika. 2009. Sorghum 3-deoxyanthocyanins possess strong phase II enzyme inducer activity and cancer cell growth inhibition properties. J. Agric. Food Chem. 57:1797–1804. doi:10.1021/jf8035066

Yao, L.H., Y.M. Jiang, J. Shi, F.A. Tomas-Barberin, N. Datta, R. Singanusong, and S.S. Chen. 2004. Flavonoids in food and their health benefits. Plant Foods Hum. Nutr. 59:113–122. doi:10.1007/s11130-004-0049-7

Yasumatsu, K., T.O.M. Nakayama, and C.O. Chichester. 1965. Flavonoids of sorghum. J. Food Sci. 30:663–667. doi:10.1111/j.1365-2621.1965.tb01821.x

Yeo, H.Q., and S.-Q. Liu. 2014. An overview of selected specialty beers: Developments, challenges and prospects. Int. J. Food Sci. Technol. 49:1607–1618. doi:10.1111/ijfs.12488

Yong, S.K., and M.K. Chang. 2003. Antioxidant constituents from the stem of Sorghum bicolor. Arch. Pharm. Res. 26:535–539. doi:10.1007/BF02976877

Youssef, A.M. 1998. Extractability, fractionation and nutritional value of low and high tannin sorghum proteins. Food Chem. 63:325–329. doi:10.1016/S0308-8146(98)00028-4

Zeleznak, K., and E. Varriano-Marston. 1982. Pearl millet (*Pennisetum americanum* (L.) Leeke) and grain sorghum (*Sorghum bicolor* (L.) Moench) ultrastructure. Am. J. Bot. 69:1306–1313. doi:10.2307/2442755

Zhang, G., and B.R. Hamaker. 2005. Sorghum (*Sorghum bicolor* L., Moench) flour pasting properties influenced by free fatty acids and proteins. Cereal Chem. 82:534–540.

Zhu, F. 2014. Structure, physicochemical properties, modifications, and uses of sorghum starch. Compr. Rev. Food Sci. Food Safety 13:597–610. doi:10.1111/1541-4337.12070

Zimmermann, M.B., and R.F. Hurrell. 2007. Nutritional iron deficiency. Lancet 370:511–520. doi:10.1016/S0140-6736(07)61235-5

Sorghum Crop Modeling and Its Utility in Agronomy and Breeding

Graeme Hammer,* Greg McLean, Al Doherty, Erik van Oosterom, and Scott Chapman

Abstract

Crop models are simplified mathematical representations of the interacting biological and environmental components of the dynamic soil–plant–environment system. Sorghum crop modeling has evolved in parallel with crop modeling capability in general, since its origins in the 1960s and 1970s. Here we briefly review the trajectory in sorghum crop modeling leading to the development of advanced models. We then (i) overview the structure and function of the sorghum model in the Agricultural Production System sIMulator (APSIM) to exemplify advanced modeling concepts that suit both agronomic and breeding applications, (ii) review an example of use of sorghum modeling in supporting agronomic management decisions, (iii) review an example of the use of sorghum modeling in plant breeding, and (iv) consider implications for future roles of sorghum crop modeling. Modeling and simulation provide an avenue to explore consequences of crop management decision options in situations confronted with risks associated with seasonal climate uncertainties. Here we consider the possibility of manipulating planting configuration and density in sorghum as a means to manipulate the productivity–risk trade-off. A simulation analysis of decision options is presented and avenues for its use with decision-makers discussed. Modeling and simulation also provide opportunities to improve breeding efficiency by either dissecting complex traits to more amenable targets for genetics and breeding, or by trait evaluation via phenotypic prediction in target production regions to help prioritize effort and assess breeding strategies. Here we consider studies on the stay-green trait in sorghum, which confers yield advantage in water-limited situations, to exemplify both aspects. The possible future roles of sorghum modeling in agronomy and breeding are discussed as are opportunities related to their synergistic interaction. The potential to add significant value to the revolution in plant breeding associated with genomic technologies is identified as the new modeling frontier.

Abbreviations: APSIM, Agricultural Production System sIMulator; ENSO, El Niño–Southern Oscillation; G × M × E, genotype by environment interaction; IR, intercepted radiation; LAI, leaf area index; QTLs, quantitative trait loci; RUE radiation use efficiency; SLN, specific leaf nitrogen; TE, transpiration efficiency; TPLA, total plant leaf area; vpd, vapor pressure deficit.

Graeme Hammer and Erik van Oosterom, The Univ. of Queensland, Centre for Plant Science, Queensland Alliance for Agriculture and Food Innovation, Brisbane, QLD 4072, Australia; Greg McLean, Agri-Science Queensland, Dep. of Agriculture and Fisheries, 203 Tor St., Toowoomba, QLD 4350, Australia; Al Doherty, The Univ. of Queensland, Centre for Plant Science, Queensland Alliance for Agriculture and Food Innovation, 203 Tor St., Toowoomba, QLD 4350, Australia; and Scott Chapman, CSIRO Agriculture Flagship, Queensland Biosciences Precinct, 306 Carmody Rd., St. Lucia, QLD 4067, Australia. *Corresponding author (g.hammer@uq.edu.au).

doi:10.2134/agronmonogr58.2014.0064

 Sorghum: State of the Art and Future Perspectives, Ignacio Ciampitti and Vara Prasad, editors. Agronomy Monograph 58.

Crop models are simplified mathematical representations of the interacting biological and environmental components of the dynamic soil–plant–environment system. Such models capture the dynamics of major plant growth and development processes as they predict trajectories of crop attributes through the crop life cycle. Environmental (E), genetic (G), and management (M) influences can be incorporated via the nature and coefficients of the response and control equations in the model and aspects of its initialization. Crop models provide an interpretive and predictive potential that moves beyond the capability of traditional statistics via its enhanced explanatory and extrapolation abilities.

While crop modeling had its origins as far back as the 1960s and 1970s (e.g., de Wit, 1970), it took a long time to become accepted as a legitimate scientific endeavor in crop and other sciences. Hammer et al. (2002) reviewed the evolution of crop models in relation to their approaches and motivations in scientific investigation, crop management decisions, and education. At that time, they also suggested a major opportunity for crop modeling to play a significant role in genetic improvement and plant breeding, given the rapid developments occurring in genotyping technologies. This has subsequently become a major field of endeavor with some realization of effective application (e.g., Messina et al., 2009; Cooper et al., 2014a,b).

Crop modeling capability evolved along with rapid increases in computational capacity, maturation of concepts related to agricultural decision/discussion support, and the rapid advances in plant genotyping capacity. Advanced platforms, such as APSIM (McCown et al., 1996; Keating et al., 2003; Holzworth et al., 2014), evolved with this capability and the broadened relevance it opened for modeling technology and its application. The APSIM initiative (www.apsim.com) has provided a collaborative environment for scientists and software engineers to pursue the ongoing evolution in modeling in a manner that can capture advances generated by simultaneous pursuit of explanation, prediction, robust software practice, and effective application. The APSIM platform incorporates structured templates for crop models that facilitate ease of crop model development, testing, and process-level comparison (Wang et al., 2002; Brown et al., 2014). The Decision Support System for Agrotechnology Transfer platform (DSSAT) (Jones et al., 2003) is based around a collaborative effort that developed along similar lines. While there are numerous other active crop modeling efforts around the world, and developments to introduce architectural capability into crop models are worthy of note (e.g., Dingkuhn et al., 2005; Drouet and Pages, 2003), the intent in this paper is to focus on crop modeling in sorghum.

Sorghum modeling has shared a similar evolutionary path to crop modeling in general. Its roots reside in the innovative activities of the research groups in Texas and Kansas in the 1970s (Arkin et al., 1976; Vanderlip and Arkin, 1977) that generated the SORGF and SORKAM models (Rosenthal et al., 1989). It was later that simpler mechanistic sorghum models emerged (Hammer and Muchow, 1991; Hammer and Muchow, 1994; Sinclair et al., 1997) in Australia and the United States, and this coincided with other attempts to modify the Ceres maize model for sorghum in Australia (Birch et al., 1990) and elsewhere (Ritchie and Alagarswamy, 1989; White et al., 2015). The novel modeling approach based on concepts of (light and water) resource capture was also developed for sorghum at ICRISAT in India at about this time (Monteith et al., 1989). These early efforts, along with advances in understanding of various growth and development processes, such as plant nitrogen dynamics (van Oosterom et al. (2010a, 2010b), informed development of the advanced sorghum model that now resides in the APSIM platform (Hammer et al., 2010), which we will use as the focus of this paper.

Early sorghum modeling was motivated by the need to move inference from agronomic experimentation beyond the traditional experimental confines associated with specific seasons and locations. Results could be highly variable, depending on the experimental circumstances. Hammer et al. (1996b) noted this issue in considering the effect of sowing date on sorghum yield in key locations in NE Australia. While, when averaged across many years, there was little simulated response to sowing date, individual years (or even decades) demonstrated advantage to either early or late sowing. It was clear that an understanding of climatic risks and their interaction with crop management decisions could be considerably enhanced via modeling and simulation that could explore the likely risk consequences of management decisions in a more comprehensive manner than possible with just shorter term conventional agronomic experimentation. Heiniger et al. (1997a) were making similar use of sorghum models to explore replanting decisions for sorghum in the Midwest United States. This modeling capability underpinned interactions with decision-makers via discussions of decision options and risks (McCown, 2001; Nelson et al., 2002; Carberry et al., 2002), ultimately leading to development of effective decision support for farmers (Carberry et al., 2009; Hochman et al., 2009).

The potential to utilize crop modeling capability in a plant-breeding context was first explored seriously in the 1990s by thinking about prediction of genotype-by-management-by-environment (G × M × E) interactions in the context of breeding strategies (Cooper and Hammer, 1996). This led to the notion of using crop simulation to characterize breeding trial and target environments by quantifying trajectories of water stress through the crop cycle (Muchow et al., 1996) as a means to unravel the complexity of G × E. This approach to environmental characterization has since gained considerable application in supporting plant breeding (Chapman et al., 2000a, 2000b; Podlich et al., 1999). Further, the idea to capture the physiological understanding of the dynamics of crop growth and development inherent in crop models to simulate consequences of trait manipulation (Bidinger et al., 1996; Hammer et al., 1996a) led to the notion of simulating the comprehensive G × M × E adaptation landscape that could be subjected to evaluating plant-breeding search strategies in silico (Cooper et al., 1999, 2002). With the rapid development of genotyping capacity the potential to develop closer linkages between crop modeling and genetic regulation became evident (Hammer et al., 2002; Yin et al., 2004). Studies using crop models to link genetic regulation to phenotypic complexity emerged (Hammer et al., 2005), and the potential for using this approach in breeding was reviewed (Hammer et al., 2006). For control of flowering in sorghum, there were attempts to link known key loci to existing coefficients for prediction of phenology, as well as attempts to develop and interface simple models of controlling gene networks (van Oosterom et al., 2006). The development of effective use of crop modeling in plant breeding is exemplified by the studies of Messina et al. (2009, 2011) in maize. With the recent developments in genomic prediction capabilities and further advances in genotyping and phenotyping (Morrell et al., 2012; Furbank and Tester, 2011), the potential for models to help unravel complexity is now more compelling (Cooper et al. (2014a,b).

We will (i) overview the structure and function of the sorghum model in APSIM to exemplify advanced modeling concepts that suit both agronomic and breeding applications, (ii) review an example of use of sorghum modeling in supporting agronomic management decisions, (iii) review an example of the use of sorghum modeling in plant breeding, and (iv) consider implications for future roles of sorghum crop modeling.

Structure and Function of the Sorghum Model in APSIM

The structure and function of the sorghum model in APSIM (APSIM-sorghum) has been detailed recently by Hammer et al. (2010). An extended summary of key features is reproduced here.

The APSIM-sorghum model is based on a framework of the physiological determinants of crop growth and development (Charles-Edwards, 1982) and is focused at organ scale. It generates the phenotype of a crop as a consequence of underlying physiological processes (Fig. 1) by using the concept of supply and demand balances for light, carbon, water, and nitrogen (Hammer et al., 2001b). The approach is focused around quantifying capture and use of radiation, water, and nitrogen within a framework that predicts the dynamics of crop development and the realized growth of major organs based on their potential growth and whether the supply of carbohydrate and nitrogen can satisfy this potential. Demand for resources is defined by potential organ growth and potential supply by resource capture (Monteith, 1977; Passioura, 1983; Monteith, 1986) (Fig. 1). Arbitration rules and organ level responses are invoked when resource capture cannot satisfy demand. The APSIM-sorghum model retains some features and concepts of earlier models (Sinclair, 1986; Rosenthal et al., 1989; Birch et al., 1990; Sinclair and Amir, 1992; Chapman et al., 1993; Hammer and Muchow, 1994), but has been adapted and redesigned to generate a more explanatory approach to the modeling of the underlying physiology (Hammer et al., 2006).

APSIM-sorghum operates via the dynamic interaction of crop development, crop growth, and crop nitrogen with soil and weather attributes (Fig. 1). Predictive schemas can be separated into crop growth and development dynamics (Fig. 1a) and crop nitrogen dynamics (Fig. 1b) for purposes of description, but the interactions between these major components are critical.

Crop Growth and Development Dynamics

Phenology is simulated through a number of development stages by using a thermal time approach (Muchow and Carberry, 1990; Hammer and Muchow, 1994), with the temperature response characterized by a base (T_b), optimum (T_{opt}), and maximum (T_m) temperature. Hammer et al. (1993) and Carberry et al. (1993) reported values of T_b, T_{opt}, and T_m for sorghum of 11, 32, and 42°C, respectively. The thermal time target for the phase between emergence and panicle initiation is also a function of day length (Hammer et al., 1989; Ravi Kumar et al., 2009), and its duration, when divided by the plastochron (°Cd per leaf), determines total leaf number once an allowance for leaf initials in the embryo has been included. Total leaf number multiplied by the phyllochron (°Cd per leaf) determines the thermal time to reach flag leaf stage, which is thus an emergent property of the model. The duration of the phases between the stages of flag leaf, anthesis, and start and end of grain filling are also simulated through thermal time targets

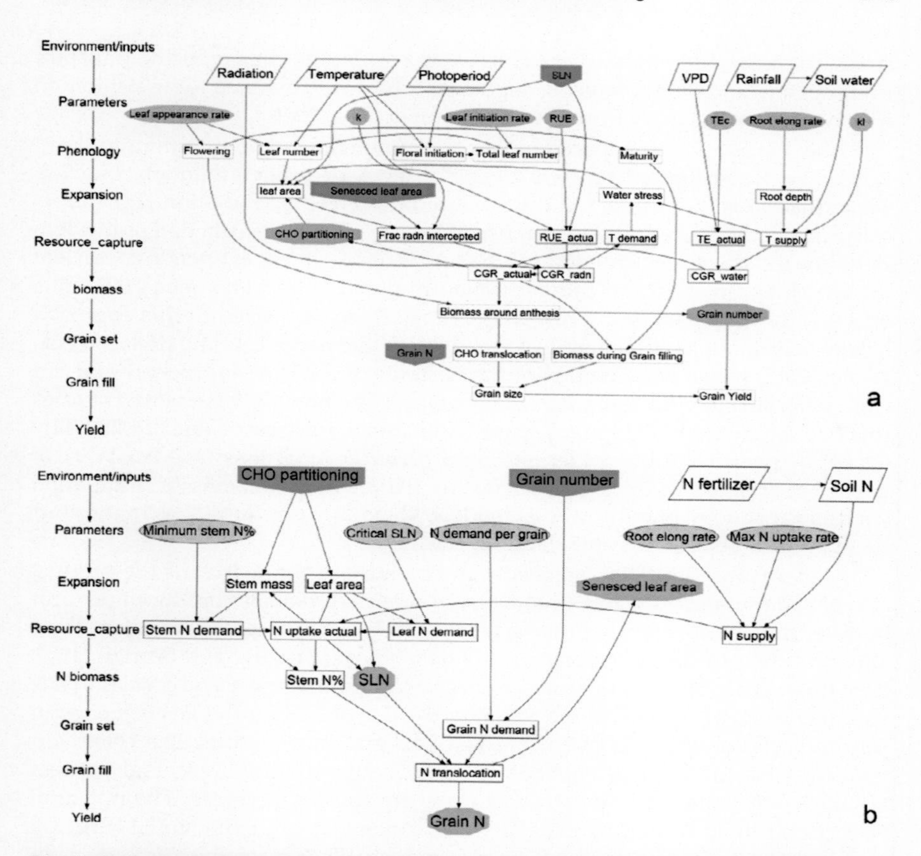

Fig. 1. Schematic representation of crop growth and development dynamics (a) and crop nitrogen dynamics (b) in Agricultural Production System sIMulator (APSIM)-sorghum model. Connection points between the two schematics are shown by the shaded boxes. After Hammer et al. (2010). SLN, specific leaf nitrogen.

(Muchow and Carberry, 1990; Hammer and Muchow, 1994; Ravi Kumar et al., 2009). Drought stress and N stress can both reduce the leaf appearance rate and hence delay phenology during the vegetative stages (Craufurd et al., 1993; van Oosterom et al., 2010a).

Canopy development is simulated on a whole plant basis through a relationship between total plant leaf area (TPLA) and thermal time. The TPLA integrates the number of fully expanded leaves, their individual size, and tiller number, and includes an adjustment for the area of expanding leaves (Hammer et al., 1993). The object-oriented design of the software provides flexibility to readily model canopy development by using other options, such as via leaf size distribution (Carberry et al., 1993; van Oosterom et al., 2001), or from extension rate of each leaf (Chenu et al., 2008). The number of fully expanded leaves is the product of thermal time elapsed since emergence and the leaf appearance rate. Actual crop leaf area is the product of plant density and leaf area per plant. Green leaf area index (LAI) is the difference between the total plant leaf area and the senesced leaf area. Under drought stress, the crop will initially cease expanding new leaves, thus reducing transpiration demand, and then commence senescing leaves until demand for transpiration no longer exceeds supply from uptake (Hammer et al., 2001b).

Aboveground biomass accumulation is simulated as the minimum of light-limited or water-limited growth. In the absence of water limitation, biomass accumulation is the product of the amount of intercepted radiation (IR) and its conversion efficiency, the radiation use efficiency (RUE). The fraction of incident radiation intercepted is a function of the LAI and the canopy extinction coefficient (k), which is a measure of canopy structure (Lafarge and Hammer, 2002). The effects of N supply on crop growth are implicitly incorporated in this approach. Nitrogen limitation will reduce leaf area growth and hence LAI and IR. It can also reduce RUE, which is a function of the N status of the leaves (Muchow and Sinclair, 1994; Sinclair and Amir, 1992). Sinclair and Muchow (1998) reviewed studies that had measured RUE in many crops and noted a consistent value of 1.25 g MJ^{-1} for triple-dwarf sorghum under optimum growing conditions. The flexibility of the object-oriented template also allows simulation of crop biomass accumulation via diurnal canopy photosynthesis models where this is required, as in the studies of Sinclair et al. (2005) and Hammer et al. (2009).

Under water limitation, aboveground biomass accumulation is the product of realized transpiration and its conversion efficiency, biomass produced per unit of water transpired, or transpiration efficiency (TE). It is necessary to adjust TE to allow for the prevailing vapor pressure deficit (vpd) (Tanner and Sinclair, 1983; Kemanian et al., 2005). Numerous studies in sorghum (Tanner and Sinclair, 1983; Hammer et al., 1997) have found a standard value of 9 Pa for the TE coefficient in sorghum, so that at a vpd of 2 kPa a TE of 4.5 gm^{-2} mm^{-1} results. The water supply accessible to the plant depends on the effective rooting depth and the rate at which soil water can be extracted from the soil by the roots. The potential extraction rate is related to the soil water content via an exponential function, parameterized via an extraction decay constant (kl) that incorporates effects of both soil hydraulic conductivity and root length density on water uptake (Passioura, 1983; Monteith, 1986; Robertson et al., 1993; Hammer et al., 2001b). Water extraction occurs from multiple layers, and the total extraction is the sum of that calculated for individual layers. As RUE and TE are based on aboveground biomass only, root mass is not explicitly modeled, but is added to the aboveground biomass accumulation according to a root/shoot ratio that declines with successive growth stages of the crop.

Daily aboveground biomass accumulation is partitioned to plant parts in ratios that depend on the growth stage of the crop via functions that have been found to describe these ratios well (Jones and Kiniry, 1986). Before the flag leaf stage, new biomass is allocated to stem and leaves. Leaves are partitioned a fraction that decreases with increasing node number up to a maximum absolute allocation to leaf that is set by the ratio of the new leaf area to be grown (described above) and a minimum specific leaf area (cm^2 g^{-1}). The remaining biomass is partitioned to stem and rachis. The stem fraction incorporates leaf sheaths, but a distinct allocation to rachis commences after panicle initiation. Between flag leaf and anthesis, accumulated biomass is allocated to the stem and rachis in a fixed ratio.

Grain yield is simulated as the product of grain number and grain size. Maximum grain number is a function of the change in plant biomass between panicle initiation and start grain filling (Rosenthal et al., 1989), while grain size is determined by grain growth rate, effective grain filling period, and redistribution of assimilates postanthesis (Heiniger et al., 1997b). If grain mass demand for a day exceeds the daily increase in biomass, the shortfall will first be met through translocation from stem and, if that is insufficient to meet demand of the grain, through translocation from leaves, accelerating their senescence. Conversely, if the daily increase in biomass exceeds the grain mass demand, the excess biomass production is allocated to the stem.

Crop Nitrogen Dynamics

Crop N dynamics are modeled based on a physiological approach that accounts for the fact that the bulk of reduced N present in leaves is associated with photosynthesis structures and enzymes (Grindlay, 1997) (Fig. 1b). The rate of light-saturated net photosynthesis has been shown to be a linear function of the amount of leaf N per unit leaf area (specific leaf nitrogen [SLN]), until a species-specific maximum rate of photosynthesis has been reached (Sinclair and Horie, 1989; Anten et al., 1995; Grindlay, 1997). Expressing crop N demand relative to canopy expansion thus provides a physiological link between crop N status, light interception, and dry matter accumulation. In addition, the cardinal SLN values for new leaf growth and for leaf death in response to N deficiency are independent of growth stage (van Oosterom et al., 2010a).

During the preanthesis period, only stems (including rachis) and leaves are expanding, and their N demand is met in a hierarchical fashion (van Oosterom et al., 2010a). First, structural N demand of the stem (and rachis) is met, as structural stem mass is required to support leaf growth. Structural stem N demand is represented by the minimum stem N concentration. If insufficient N has been taken up to meet structural stem N requirement, N can be translocated from leaves by dilution or, in extreme cases of early season N deficiency, by leaf senescence. Second, the N demand of expanding new leaves will be met, and this is determined from their critical SLN. Any additional N uptake will first be allocated to leaves to meet their target SLN and then to stem. For leaves, this N uptake represents "luxury" uptake that can occur after full expansion of a leaf, and which does not affect growth and development (van Oosterom et al., 2010a). This hierarchical allocation of N is consistent with observations that under N stress a relatively larger proportion of N is allocated to the leaves (van Oosterom et al., 2010a). Hence, preanthesis N allocation ratios are a consequence of model dynamics, rather than a model input.

After anthesis, grain becomes the major sink for N, and grain N demand is determined as the product of grain number and N demand per grain. During the first part of grain filling, N demand per grain is constant and independent of grain growth rate and N status of the crop (van Oosterom et al., 2010b). At this time endosperm cells are dividing, so that the accumulation of structural (metabolic) proteins in the grain is the key driver. During the second half of grain filling, grain N demand is linked with grain growth rate as cell division and simultaneous storage of carbohydrate and proteins assumes a greater role (Martre et al., 2006). Grain protein content can thus vary depending on the N supply–demand balance and the carbohydrate supply to the grain. Grain N demand is initially met through stem (plus rachis) N translocation, and if this becomes insufficient, then N translocation from leaf can occur. Maximum N translocation rates from stem and per unit leaf area are a function of the N status of these organs, so that sink demand determines the amount of leaf area that is senescing at any one time (van Oosterom et al., 2010b). The source regulation of N translocation follows a first-order kinetic relationship that is representative of enzyme activity. Leaf SLN thus declines to its structural (minimum) level, and the amount of leaf area senesced, in the absence of other factors that can affect senescence, such as water limitation and shading, depends on the N supply–demand balance.

The daily rate of crop N uptake is the minimum of demand for N by the crop and potential supply of N from the soil and senescing leaves, capped at a maximum N uptake rate (van Oosterom et al., 2010a). Potential N supply from the soil depends on the available soil N through the profile and on the extent to which roots have explored the soil. N supply from the soil is calculated from the combination of passive uptake, through mass flow of N taken up with the transpiration stream, and active uptake (van Keulen and Seligman, 1987). Soil N transformations and their modeling in APSIM have been detailed by Probert et al. (1998).

Model Testing

Hammer et al. (2010) presented comprehensive tests of the predictive capability of APSIM-sorghum and found that (i) simulated crop growth and development trajectories throughout the crop life cycle accurately reflected observed effects and responses to water and N availability measured in detailed experiments, and (ii) simulated values of crop attributes reflected observations well across a diverse range of experiments. They noted that the general adequacy of relationships between predicted and observed values for phenology, biomass, yield, LAI, and organ N across the diverse range of genotypes and environments found in these experiments indicated a robust predictive capability of the model.

Use of Modeling in Crop Management

Modeling and simulation provide an avenue to explore consequences of crop management decision options in situations confronted with risks associated with seasonal climate uncertainties. Rain-fed sorghum production in the subtropics and tropics is a risky enterprise because of high rainfall variability. When planting opportunities occur, farmers face risky choices because the consequences of decisions made at planting are uncertain. Crop failures due to water stress can generate food insecurity and income variability. Via simulation it is feasible to look at risk a priori by examining outcomes over many years. The best way to do this is by using a reliable crop model with long-term historical climate data to simulate what might happen when agronomic practices are changed. Muchow et al. (1994) present an analysis to support planting decisions for sorghum in NE Australia. They used simulation with long-term sequences of climatic data to provide probabilistic estimates of yield for a range of decision options, such as planting time, antecedent soil moisture status, and cultivar maturity for a range of soil types.

Adapting sowing date, planting configuration, plant density, and cultivar maturity are options often available to farmers that will influence the dynamic of the crop water balance, and hence, production level and risk. Here we consider the possibility of manipulating planting configuration and density as a means to manipulate productivity–risk trade-off. Moisture conserving agronomic practices, like single- and double-skip row systems (Fig. 2) or clump planting, have been considered and implemented in risky environments across the world (Routley et al., 2003; McLean et al., 2003; Whish et al., 2005; Bandaru et al., 2006; Abunyewa et al., 2010, 2011). But it is not clear how much is sacrificed in good years to get some production/income in the bad years. Can modeling and simulation help in deciding which system to adopt?

Skip row configurations are thought to improve yield reliability by slowing water use due to reduced canopy cover as well as by delaying utilization of soil moisture in the center of the skip area until late in the growing season when roots extend into this area. As a result, soil moisture in the center of the skip is more likely to be available during the grain filling stage, allowing higher yield and increased harvest index in moisture-limited growing conditions. However, in growing conditions with more favorable moisture, skip row yields are likely to yield less than solid row planted systems because of their reduced crop leaf area and associated light interception. Routley et al. (2003), McLean et al. (2003), and Whish et al. (2005) report details of an experimental and modeling study in NE Australia to quantify production levels and risks associated with various skip row systems. By incorporating effects of row configuration on canopy and root system architecture in the APSIM-sorghum model they were able to simulate the reduced leaf area and slower water use of the skip row configurations (Fig. 3) and reliably predict yield levels of their experiments (Whish et al., 2005).

To quantify the production–risk trade-off for double-skip vs. solid-planted row configuration systems in NE Australia, long-term simulations were conducted for locations in the dryland sorghum-growing region and results are presented here for Dulacca (McLean et al., 2003). Long-term meteorological data was obtained from the Dulacca weather station. The common soil in the region is a uniform medium clay with a plant available water holding capacity of 160 mm. The sowing date was set to 15 November each year and the soil water available

Fig. 2. Sorghum field grown under a double-skip row configuration near Goondiwindi in NE Australia.

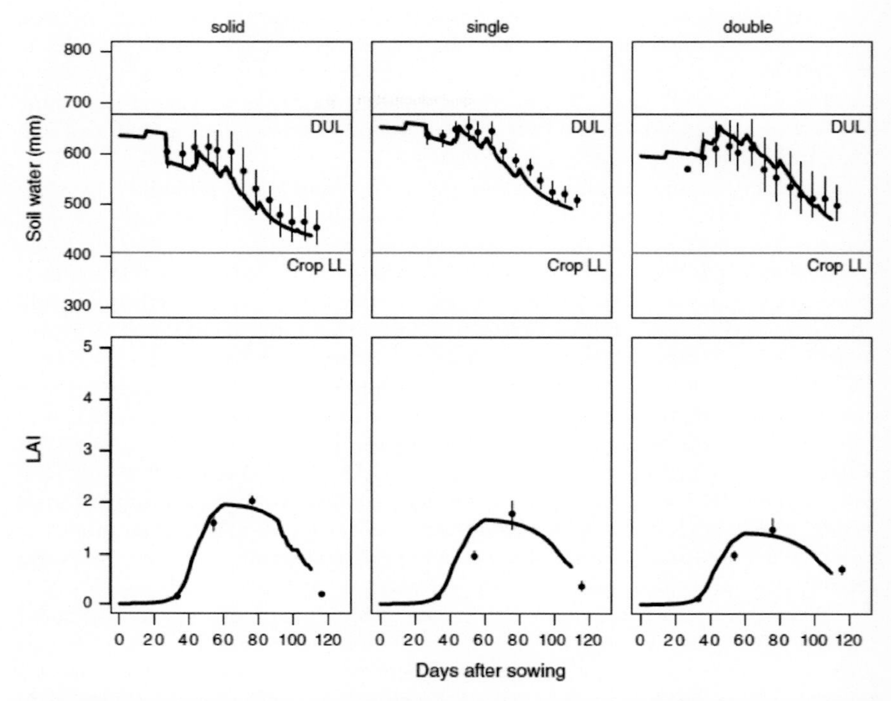

Fig. 3. Observed (points) and simulated (line) soil water and leaf area index (LAI) for solid, single, and double-skip row configurations for a field experiment in NE Australia. Horizontal lines indicate the crop lower limit (Crop LL) and drained upper limit (DUL) for soil water. Vertical bars indicate the range of observed data. After Whish et al. (2005).

at planting was set at two-thirds full profile. Nitrogen was assumed to be non-limiting. The simulation results thus reflect only the effects of seasonal climate variability from year-to-year.

The long-term simulations for this specific situation indicated that the median grain yield from the double-skip configuration (2.33 t/ha) was slightly lower than that obtained from the solid configuration (2.47 t/ha) (Fig. 4). However, there was a considerably reduced risk associated with adopting a double-skip row configuration. While the double-skip configuration yielded well below that of the solid configuration in high-yielding years, the converse occurred in low-yielding years. This is more evident if the yield ratio of the two row configuration options is considered for each year of the long-term simulation (Fig. 5). The trend line indicates yield advantage of the double-skip system for years with yield levels below about 2.2 t/ha under solid planting, but yield sacrifice for seasons resulting in higher yield levels. This result is consistent with available experimental evidence (Routley et al., 2003; Whish et al., 2005).

The quantification of the production–risk trade-off by modeling and simulation provides useful information for discussion with the decision-maker about the choice of system, but the risk preference of the decision-maker and other influencing factors will interact with this information. Each scenario will differ in risk

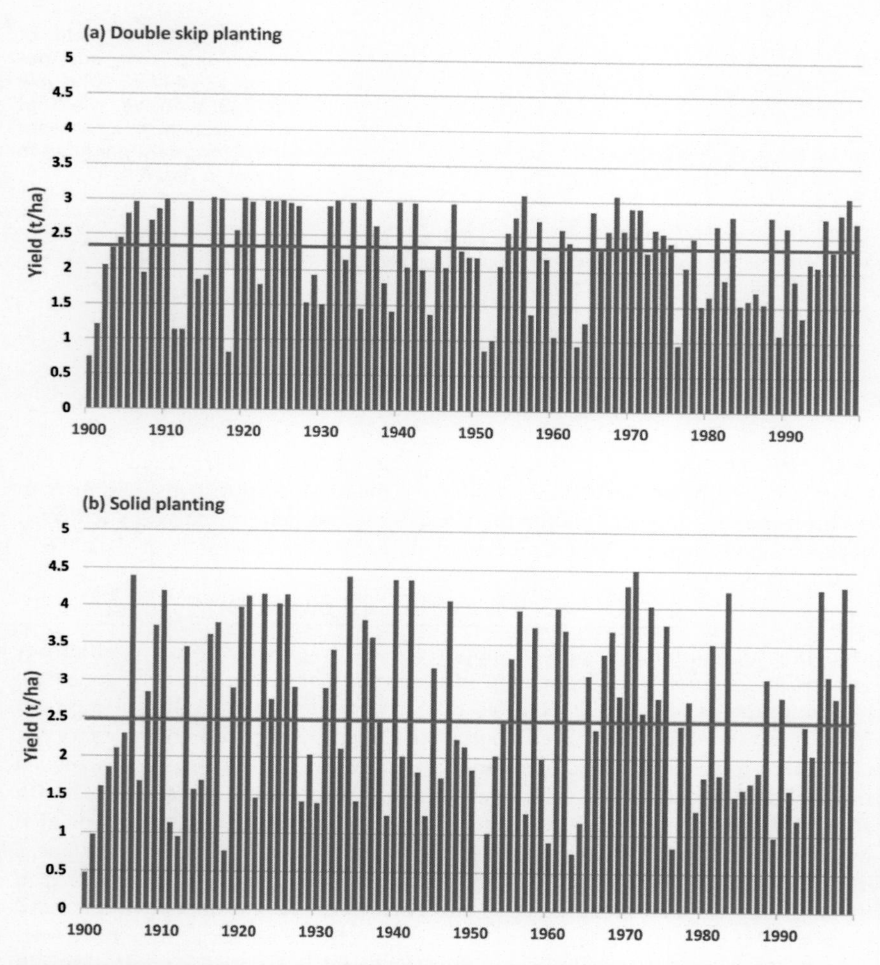

Fig. 4. Simulated grain yield for double-skip (a) and solid-planted (b) row configurations for a long-term simulation at Dulacca in southern Queensland (NE Australia), assuming a 15 November planting each year with the same available stored soil water (see text). The horizontal line is the median yield over the long-term series.

profile with other factors, such as level of stored soil water at planting (Dalgliesh et al., 2009) or skill in seasonal climate forecasts based on the El Niño–Southern Oscillation (ENSO) status before sowing (Hammer, 2000; Meinke and Hochman, 2000), which can moderate the risk situation. For example, repeating the simulations for Dulacca with a full profile at planting resulted in higher median yield and very few years in which the double-skip row system was advantageous. This highlights the value of knowing stored soil water status in this situation and is a factor that would be readily clear from discussion of simulation outputs. While the value of ENSO-based seasonal climate forecasts is not as clear cut in this situation, it has the potential to moderate risk profiles in this cropping region (Stone

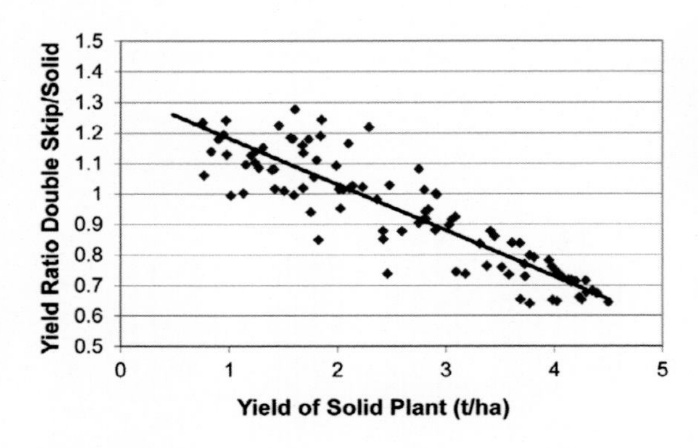

Fig. 5. Ratio of yield of double-skip row configuration to yield of solid row configuration vs. yield of solid system for each year of the long-term simulation shown in Fig. 4.

et al., 1996; Hammer et al., 2001a). Information systems underpinned by model-based analysis have evolved as an effective means to support discussion with decision-makers (farmers and/or their advisers) and help in making such risky decisions (Nelson et al., 2002; Carberry et al., 2002, 2009; Hochman et al., 2009).

Use of Modeling in Plant Breeding

Modeling and simulation provide opportunities to improve breeding efficiency by introducing biological knowledge in a manner that can enhance the genotype-to-phenotype prediction capabilities that underpin plant breeding (Chapman, 2008; Hammer et al., 2016). Prediction of the phenotype based on the genotype is required for yield advance in breeding (Cooper et al., 2002). This was achieved traditionally by extensive field phenotyping of many genotypes combined with advanced statistical procedures to identify components of variance associated with observations from these multi-environment trials. But recently, advances in genotyping and whole genome prediction methodologies have revolutionized the approach to plant breeding (Cooper et al., 2014b). While such advances have been effective in enhancing rate of genetic gain from breeding, they have only heightened the potential role of crop modeling and associated biological insight to add value in plant breeding (Technow et al., 2015).

There are two main avenues by which crop ecophysiology and modeling can enhance breeding efficiency (Fig. 5). The first involves use of ecophysiological insight from dynamic models to enhance phenotyping strategies by dissecting complex traits into more robust targets that help to deal with G × E interactions, thus aiding connections to genetic regulation and predictive capability. The second involves using crop growth and development models for trait evaluation and phenotypic prediction in target production regions to help prioritize effort and assess breeding strategies. We consider studies on the stay-green trait in sorghum, which confers yield advantage in water-limited situations (Jordan et al., 2012), to exemplify both aspects.

Stay-green is an integrated drought-adaptation trait in sorghum. Delayed leaf senescence during grain filling is now largely considered as an emergent consequence of dynamics occurring earlier in crop growth and is largely due to enhanced water availability postanthesis. van Oosterom et al. (2011) set out the crop physiological processes that can affect grain yield under drought stress by increasing postanthesis water availability. Borrell et al. (2014a,b) quantify the connections of these component processes with expression of the stay-green trait and their association with quantitative trait loci (QTLs) known to underpin stay-green. They indicate how the positive effect of stay-green QTLs on grain yield under drought can be explained as emergent consequences of their effects on temporal and spatial water use patterns that result from changes in canopy development and water use efficiency. Crop water use during grain filling can be enhanced by increasing water availability at anthesis because of reduced canopy development and water use earlier in the crop cycle. For example, studies on the genetic regulation of tillering in sorghum (Kim et al., 2010; Alam et al. (2014a,b) expose QTLs that colocate with those for stay-green. Transpiration efficiency is known to vary genetically in sorghum (Henderson et al., 1998) as is the maximum transpiration rate (Gholipoor et al.,2010), which can also generate water saving by limiting transpiration during the time of day when crop water use is least efficient (Sinclair et al., 2005). Other factors capable of enhancing crop water use during grain filling and known to vary genetically in sorghum are root system architecture (Singh et al., 2011, 2012; Mace et al., 2012) and leaf attributes (size and appearance rate) (van Oosterom et al., 2011; Borrell et al. (2014a,b).

The capacity of a crop model to simulate complex traits as emergent consequences of the internal dynamics of processes incorporated in the model is critical to effective dissection of complex traits and hence to their more robust connection to underpinning genetic control. An intrinsic ability to unravel G × E interactions in a manner that enhances heritability of component traits, and thus highlights phenotyping targets (Fig. 6), requires enhanced biological realism in the structure of the model (Hammer et al., 2006, 2010). A simulation study was undertaken with the APSIM-sorghum model for two component traits affecting stay-green—tillering and TE. The tillering routine was adapted to generate a

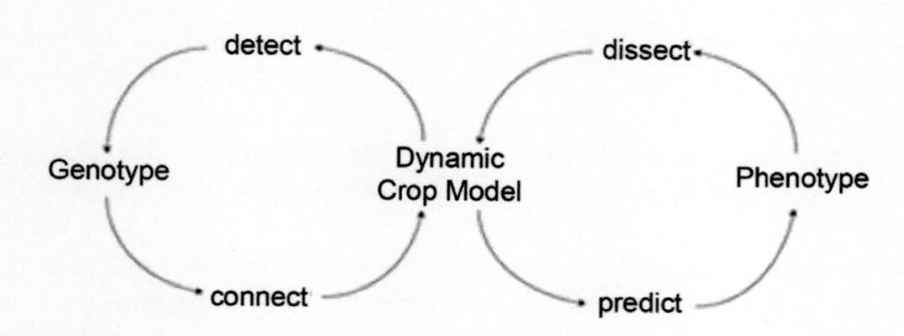

Fig. 6. Schematic of transdisciplinary approach to breeding systems highlighting integration and roles of physiology and modeling with genetics. After Messina et al. (2009) and Hammer et al. (2014). TPE, target population of environments; EC, environment characterization; G × X, genotype by environment interaction; QTL, quantitative trait loci.

reduced tillering type when compared with a standard hybrid, and a maximum transpiration rate was introduced to restrict water use in the middle of the day if potential transpiration exceeded that maximum. The latter required invoking an hourly calculation and used the routine developed by Hammer et al. (2009), with transpiration limits imposed in the same manner set out by Sinclair et al. (2005). The simulations were conducted for Emerald, a location in the sorghum-growing region of NE Australia, by using long-term historical climate data and local soil information. The 1-m-deep uniform clay held 160 mm of plant available water and was assumed to hold 120 mm at the time of sowing, which was set to 15 December each year. A standard agronomy of 1-m rows and 50,000 plants ha^{-1} was used in all cases, and nitrogen was assumed nonlimiting. This reflects a common situation at this site. Kholová et al. (2014) report a similar study using APSIM-sorghum to model the effect of plant water use traits on yield and stay-green expression in post–rainy season sorghum in India.

The simulated results for a single year, chosen because of its terminal moisture stress pattern, indicate the capacity of the model to generate stay-green as an emergent consequence in this type of situation (Fig. 7). The LAI of either the reduced tillering type or the type with limited maximum transpiration remained green for longer into the postanthesis period of the crop cycle. This was associated with reduced canopy size (LAI) causing reduced preanthesis water use in the case of reduced tillering, or enhanced TE causing reduced preanthesis water use in the case of limited maximum transpiration, despite similar canopy size. In both of these situations there was increased simulated yield associated with the modified type (data not shown). In the case of reduced tillering, this was not linked to any increase in total biomass at maturity, but rather to increased harvest index as a consequence of the increased water availability and crop growth postanthesis. In the case of limited maximum transpiration rate, both total and grain biomass were simulated to increase as a result of the enhanced TE this effect generated.

When viewed over the long-term climate for this location, there is considerable yield advantage to both reduced tillering and limited maximum transpiration for seasons where yield of the standard hybrid was below about 4.5 t ha^{-1} (Fig. 8). This is consistent with observations from breeding trials for the stay-green trait (Jordan et al., 2012) and reflects the majority of farmer field situations in NE Australia, where the long-term average yield level is around 3 t ha^{-1}. However, for the situation and agronomy of this simulation scenario, there was yield reduction in high-yielding seasons. This was associated with the conservative growth (and resource use) of the adaptations simulated. This result highlights the likely value of these component traits and suggests phenotyping strategies to target them directly could be advantageous. This exemplifies the concept of dissecting a complex trait to component targets that might be more tightly linked to genetic control (higher heritability) and thus offer potential to enhance genetic gain if effective phenotyping systems can be designed.

As in this case for stay-green (Fig. 7 and 8), the value of any specific trait is often dependent on the growing environment (Tardieu, 2012), and this G × E interaction complicates breeding programs by confounding predictive capability (Cooper et al., 2002; Hammer et al., 2005). Crop models can help in unraveling this complexity in a number of ways. First, by using crop simulation to characterize and classify production environments in a biologically meaningful way,

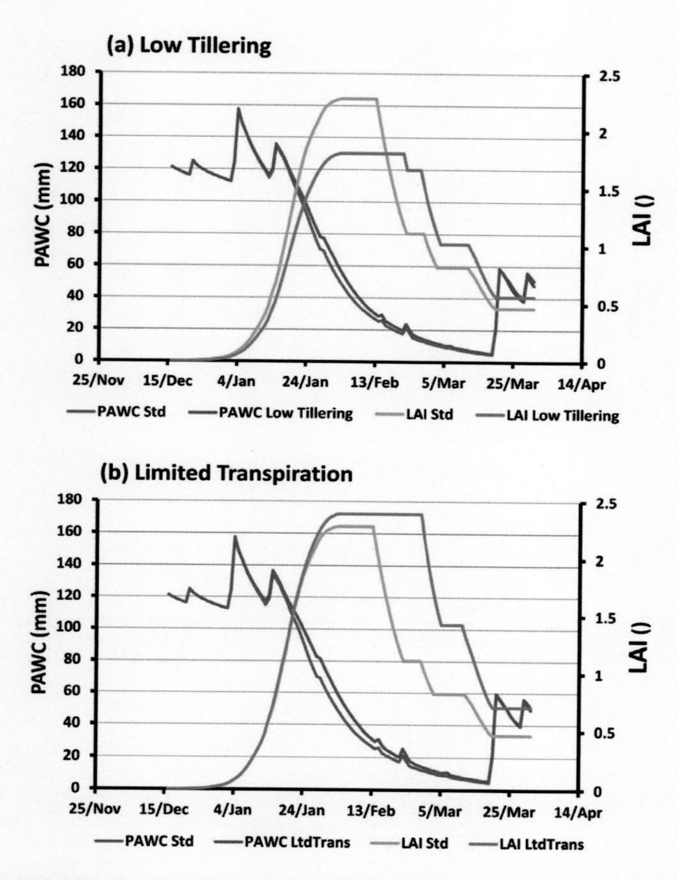

Fig. 7. Simulated soil water content and crop leaf area index through the crop life cycle for a terminal drought year from the long-term sorghum simulation at Emerald (NE Australia) for a contrast with the standard hybrid showing either reduced tillering (a) or a limited maximum transpiration rate (b). PAWC, plant available water content.

some G × E can be explained (Chapman et al., 2000a). It is possible to make use of this enhanced explanatory power to improve rate of genetic gain by weighting selection decisions depending on the representativeness of each particular test environment relative to its frequency in the target population of environments (Podlich et al., 1999; Hammer et al., 2005). Second, capturing trait understanding in biologically well-structured crop models provides an avenue to predict the nature of the genotype-to-phenotype adaptation landscape in the target environments. This provides a basis to explore the potential value of breeding strategies (Chapman et al., 2003). Hammer et al. (2014) used a simulation-based evaluation of a specific adaptation breeding strategy for sorghum in NE Australia to quantify the level of benefits that would likely arise. Messina et al. (2009, 2011) linked such trait-yield performance landscapes with breeding system simulation algorithms to explore opportunities in breeding maize for drought tolerance.

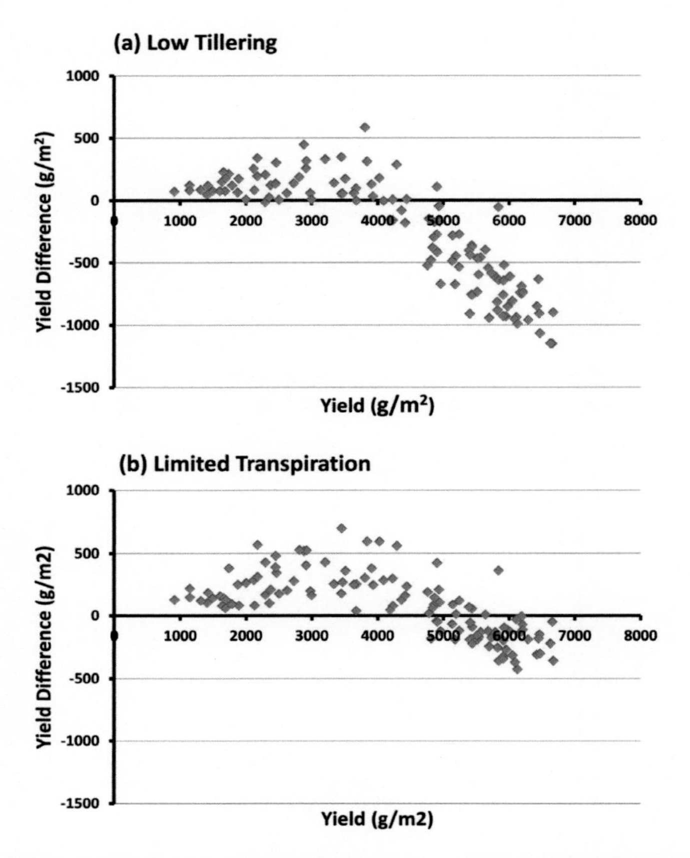

Fig. 8. Simulated difference in grain yield relative to a standard sorghum hybrid vs. simulated yield of the standard hybrid for a low tillering (a) and a limited maximum transpiration (b) sorghum hybrid for a long-term simulation at Emerald in central Queensland (NE Australia) assuming a 15 December planting each year with the same available stored soil water and standard agronomy (see text).

Implications for Future Roles of Sorghum Crop Modeling

Sorghum crop modeling and simulation has reached a degree of maturation where its utility in agronomy and breeding has been established, if not comprehensively implemented. In agronomy, the understanding and modeling of key factors related to water and nitrogen dynamics has enabled its application in exploring production–risk trade-offs associated with management options in developed-world agricultural systems. Effective application at this level is advanced but can only be enhanced by development of improved tools and procedures for connectivity to, and interaction with, decision-makers and advisers (Hochman et al., 2009). In developing-world agricultural systems, there remain issues of model calibration for development and growth aspects of local varieties/landraces, as well as attention to issues, such as cultivation practices, crop nutrition, and soil factors. While none of these prohibit progress, and a number of major projects are underway to address them, it requires good connection between experimentation and modeling effort to underpin model testing and any enhancements required for application in these systems. Beyond these crop level issues, it is important to consider the cropping system perspective as this will interact with crop agronomy. While this has not been addressed here, modeling platforms, such as ASPIM, have the capacity to simulate cropping systems and their implications on individual crops (Holzworth et al., 2014). However, some issues, such as those associated with management of system level weeds and/or pests, remain as constraints around choice of management options, rather than aspects incorporated directly into modeling approaches. Nonetheless, by placing the crop modeling capability into the cropping–farming system context, it provides significant opportunity for contribution to thinking and practice.

Impacts of climate change and possible adaptations to it (Lobell et al., 2011) will be another area where sorghum crop modeling will contribute significantly in the future. It is clear that future and current climates will differ so that simulation analyses based on historical climate data have limited applicability moving forward. Analyses based on predictions of climate in the future rely on the extrapolation capability of crop models. Lobell et al. (2015) report on a modeling study quantifying the shifting influence of drought and heat stress on sorghum in NE Australia based on climate model projections (Taylor et al., 2012). Although climate models diverged considerably in predictions of likely rainfall scenarios, there was convergence on temperature increase. In the modeling study, they found that even in the absence of rainfall effects, warming exacerbated drought stresses by raising the atmospheric vpd and thus reducing TE. However, a relative increase in TE due to elevated CO_2 more than offset these effects. The influence of increased temperature on crop water balance via effects on vpd was also a critical issue in explaining potential effects of climate change on maize in the United States (Lobell et al., 2013). In the sorghum study in NE Australia, in addition to indirect effects of extreme heat via vpd, it was noted that direct effects of heat stress on grain set will grow in importance, indicating that an emphasis on heat tolerance is warranted in breeding programs.

In breeding, the putative role of crop modeling as a support technology has been tested intensively over the past decade or two. Crop modeling, utilized in an appropriate manner, has emerged from this process as a useful contributing component of comprehensive plant breeding programs (Cooper et al., 2014a,b; Hammer et al., 2016). Again, as with agronomic application, this requires connectivity with practitioners, and the key driver must be the ability to generate innovation that delivers added value to decisions in practice. Advance in the effective application of crop modeling in breeding will undoubtedly occur. As the technology of genomic prediction gains impetus, so will the awareness of the significant value-adding role crop modeling can play in adding biological knowledge to these advanced statistical methods. The recent study of Technow et al. (2015) is a portal to this future. This advance will require attention to underpinning biology in crop models while limiting their complexity, as they strive to make more effective connections between genotype and phenotype than could otherwise occur. The importance of the modeling adage "the right answer for the right reason" and Einstein's remark of "as simple as possible but no simpler," will become evident!

Beyond separate futures in agronomy and breeding, there remains a significant role for crop modeling in crop design via their integration—the G × M × E concept of Cooper and Hammer (1996). Simulation remains the only feasible avenue to explore the myriad possibilities associated with such interactions, but this also requires a crop model of sufficient capability to simulate consequences realistically. The simulation study on specific adaptation of sorghum in NE Australia (Hammer et al., 2014) explored the productivity–risk trade-offs of combinations of G × M in different E to identify those that might be options for advance at industry scale. This provides targets for testing and evaluation by appropriate experimentation and consideration of associated practical issues, such as the economics of seed supply.

Sorghum crop modeling has evolved in capability and reached a credible level of acceptance in agronomy. This has been associated with strong connections between crop physiological experimentation, model development, and agronomists. While advances in knowledge, model improvements, and enhanced interactions with decision-makers will undoubtedly further advance the utility of modeling in agronomy, it is the potential to add significant value to the revolution in plant breeding associated with genomic technologies that is the new modeling frontier. This will require models where capturing biological understanding in a crop growth and development context is as important as the predictive capability of the model—the right answer for the right reason. Models developed for agronomic application will likely not be sufficient. Models with more robust biological underpinning and the ability to link parameters with the genetic architecture of adaptive traits in a stable manner will come to the fore. Above all, modeling cannot advance in isolation. It will remain an integrative activity dependent on transdisciplinary effort with experimental and theoretical scientists and practitioners to generate the model development ideas and provide the data for their testing via serious play.

Acknowledgments

This paper summarizes the research of a team of people, which would not have been possible without the long-term and ongoing support of their respective organizations. Additional financial support from a number of research funding agencies/collaborators, including Australian Research Council, Grains Research and Development Corporation, DuPont Pioneer, European Union FP7, and Australian Centre for International Agricultural Research, is gratefully acknowledged. The authors acknowledge *Crop & Pasture Science* (CSIRO Publishing) and *Journal of Experimental Botany* (Oxford Journals) for permissions to reuse figures.

References

Abunyewa, A.A., R.B. Ferguson, C.S. Wortmann, D.J. Lyon, S.C. Mason, S. Irmak, and R.N. Klein. 2011. Grain sorghum water use with skip-row configuration in the Central Great Plains of the USA. African J. Agric. Res. 6:5328–5338.

Abunyewa, A.A., R.B. Ferguson, C.S. Wortmann, D.J. Lyon, S.C. Mason, and R.N. Klein. 2010. Skip-row and plant population effects on sorghum grain yield. Agron. J. 102:296–302. doi:10.2134/agronj2009.0040

Alam, M.M., G.L. Hammer, E.J. van Oosterom, A. Cruickshank, C. Hunt, and D.R. Jordan. 2014a. A physiological framework to explain genetic and environmental regulation of tillering in sorghum. New Phytol. 203:155–167. doi:10.1111/nph.12767

Alam, M.M., E.S. Mace, E.J. van Oosterom, A. Cruickshank, C.H. Hunt, G.L. Hammer, and D.R. Jordan. 2014b. QTL analysis in multiple sorghum populations facilitates the dissection of the genetic and physiological control of tillering. Theor. Appl. Genet. 127:2253–2266. doi:10.1007/s00122-014-2377-9

Anten, N.P.R., F. Schieving, and M.J.A. Werger. 1995. Patterns of light and nitrogen distribution in relation to whole canopy carbon gain in C3 and C4 mono- and dicotyledonous species. Oecologia 101:504–513. doi:10.1007/BF00329431

Arkin, G.F., R.L. Vanderlip, and J.T. Ritchie. 1976. A dynamic grain sorghum growth model. Trans. ASAE 19:0622–0626. doi:10.13031/2013.36082

Bandaru, V., B.A. Stewart, R.L. Baumhardt, S. Ambati, C.A. Robinson, and A. Schlegel. 2006. Growing dryland grain sorghum in clumps to reduce vegetative growth and increase yield. Agron. J. 98:1109–1120. doi:10.2134/agronj2005.0166

Bidinger, F.R., G.L. Hammer, and R.C. Muchow. 1996. The physiological basis of genotype x environment Interaction in crop adaptation. In: M. Cooper and G.L. Hammer, editors, Plant adaptation and crop improvement. CAB Int., Wallingford, UK. p. 329–347.

Birch, C.J., P.S. Carberry, R.C. Muchow, R.L. McCown, and J.N.G. Hargreaves. 1990. Development and evaluation of a sorghum model based on CERES-Maize in a semi-arid tropical environment. Field Crops Res. 24:87–104. doi:10.1016/0378-4290(90)90023-5

Borrell, A.K., J.E. Mullet, B. George-Jaeggli, E.J. van Oosterom, G.L. Hammer, P.E. Klein, and D.R. Jordan. 2014a. Drought adaptation of stay-green cereals is associated with canopy development, leaf anatomy, root growth and water uptake. J. Exp. Bot. 65:6251–6263. doi:10.1093/jxb/eru232

Borrell, A.K., E.J. van Oosterom, J.E. Mullet, B. George-Jaeggli, D.R. Jordan, P.E. Klein, and G.L. Hammer. 2014b. Stay-green alleles individually enhance grain yield in sorghum under drought by modifying canopy development and water uptake patterns. New Phytol. 203:817–830. doi:10.1111/nph.12869

Brown, H., N.I. Huth, D.P. Holzworth, R.F. Zyskowski, E.L. Teixera, J.N.L. Hargreaves, and D. Moot. 2014. Plant modelling framework: Software for building and running crop models on the APSIM platform. Environ. Model. Softw. 62:385–398. doi:10.1016/j.envsoft.2014.09.005

Carberry, P.S., Z. Hochman, R.L. McCown, N.P. Dalgliesh, M.A. Foale, P.L. Poulton, J.N.G. Hargreaves, D.M.G. Hargreaves, S. Cawthray, N.S. Hillcoat, and M.J. Robertson. 2002. The FARMSCAPE approach to decision support: Farmers, advisers, researchers, monitoring, simulation, communication and performance evaluation. Agric. Syst. 74:141–177. doi:10.1016/S0308-521X(02)00025-2

Carberry, P.S., Z. Hochman, J.R. Hunt, N.P. Dalgliesh, R.L. McCown, J.P.M. Whish, M.J. Robertson, M.A. Foale, P.L. Poulton, and H. van Rees. 2009. Re-inventing model-based decision support with Australian dryland farmers. 3. Relevance of APSIM to commercial crops. Crop Pasture Sci. 60:1044–1056. doi:10.1071/CP09052

Carberry, P.S., R.C. Muchow, and G.L. Hammer. 1993. Modelling genotypic and environmental control of leaf area dynamics in grain sorghum. II. Individual leaf level. Field Crops Res. 33:311–328. doi:10.1016/0378-4290(93)90088-5

Chapman, S.C. 2008. Use of crop models to understand genotype by environment interactions for drought in real-world and simulated plant breeding trials. Euphytica 161:195–208. doi:10.1007/s10681-007-9623-z

Chapman, S.C., M.C. Cooper, G.L. Hammer, and D. Butler. 2000a. Genotype by environment interactions affecting grain sorghum. II. Frequencies of different seasonal patterns of drought stress are related to location effects on hybrid yields. Aust. J. Agric. Res. 51:209–222. doi:10.1071/AR99021

Chapman, S.C., M. Cooper, D. Podlich, and G.L. Hammer. 2003. Evaluating plant breeding strategies by simulating gene action and dryland environment effects. Agron. J. 95:99–113. doi:10.2134/agronj2003.0099

Chapman, S.C., G.L. Hammer, D. Butler, and M. Cooper. 2000b. Genotype by environment interactions affecting grain sorghum. III. Temporal sequences and spatial patterns in the target population of environments. Aust. J. Agric. Res. 51:223–234. doi:10.1071/AR99022

Chapman, S.C., G.L. Hammer, and H. Meinke. 1993. A sunflower simulation model: I. Model development. Agron. J. 85:725–735. doi:10.2134/agronj1993.00021962008500030038x

Charles-Edwards, D.A. 1982. Physiological determinants of crop growth. Academic Press, Sydney, Australia.

Chenu, K., S.C. Chapman, G.L. Hammer, G. McLean, and F. Tardieu. 2008. Short term responses of leaf growth rate to water deficit scale up to whole plant and crop levels. An integrated modelling approach in maize. Plant Cell Environ. 31:378–391. doi:10.1111/j.1365-3040.2007.01772.x

Cooper, M., and G.L. Hammer. 1996. Synthesis of strategies for crop improvement. In: M. Cooper and G.L. Hammer, editors, Plant adaptation and crop improvement. CAB Int., Wallingford, UK. p. 591–623.

Cooper, M., S.C. Chapman, D.W. Podlich, and G.L. Hammer. 2002. The GP problem: Quantifying gene-to-phenotype relationships. In Silico Biol. 2:151–164.

Cooper, M., C. Gho, R. Leafgren, T. Tang, and C. Messina. 2014a. Breeding drought-tolerant maize hybrids for the US corn-belt: Discovery to product. J. Exp. Bot. 65:6191–6204. doi:10.1093/jxb/eru064

Cooper, M., C.D. Messina, D. Podlich, L.R. Totir, A. Baumgarten, N.J. Hausmann, D. Wright, and G. Graham. 2014b. Predicting the future of plant breeding: Complementing empirical evaluation with genetic prediction. Crop Pasture Sci. 65:311–336. doi:10.1071/CP14007

Cooper, M., D.W. Podlich, N.M. Jensen, S.C. Chapman, and G.L. Hammer. 1999. Modelling plant breeding programs. Trends Agron. 2:33–64.

Craufurd, P.Q., D.J. Flower, and J.M. Peacock. 1993. Effect of heat and drought stress on sorghum (Sorghum bicolor). I. Panicle development and leaf appearance. Exp. Agric. 29:61–76. doi:10.1017/S001447970002041X

Dalgliesh, N.P., M.A. Foale, and R.L. McCown. 2009. Re-inventing model-based decision support with Australian dryland farmers. 2. Pragmatic provision of soil information for paddock-specific simulation and farmer decision making. Crop Pasture Sci. 60:1031–1043. doi:10.1071/CP08459

de Wit, C.T. 1970. Dynamic concepts in biology. In: I. Setlik, editor, Prediction and measurement of photosynthetic activity. Pudoc, Wageningen, The Netherlands. p. 17–23.

Dingkuhn, M., D. Luquet, B. Quilot, and P. de Reffye. 2005. Environmental and genetic control of morphogenesis in crops: Towards models simulating phenotypic plasticity. Aust. J. Agric. Res. 56:1289–1302. doi:10.1071/AR05063

Drouet, J.L., and L. Pages. 2003. GRAAL: A model of growth, architecture and carbon allocation during vegetative phase of the whole plant: Model description and parameterisation. Ecol. Modell. 165:147–173. doi:10.1016/S0304-3800(03)00072-3

Furbank, R.T., and M. Tester. 2011. Phenomics—Technologies to relieve the phenotyping bottleneck. Trends Plant Sci. 16:635–644. doi:10.1016/j.tplants.2011.09.005

Gholipoor, M., P.V.V. Prasad, R.N. Mutava, and T.R. Sinclair. 2010. Genetic variability of transpiration response to vapor pressure deficit among sorghum genotypes. Field Crops Res. 119:85–90. doi:10.1016/j.fcr.2010.06.018

Grindlay, D.J.C. 1997. Towards an explanation of crop nitrogen demand based on the optimization of leaf nitrogen per unit leaf area. J. Agric. Sci. 128:377–396. doi:10.1017/S0021859697004310

Hammer, G.L. 2000. A general approach to applying seasonal climate forecasts. In: G.L. Hammer, N. Nicholls, and C. Mitchell, editors, Applications of seasonal climate forecasting in agricultural and natural ecosystems: The Australian experience. Kluwer Acad., Dordrecht, The Netherlands. p. 51–65.

Hammer, G.L., and R.C. Muchow. 1991. Quantifying climatic risk to sorghum in Australia's semiarid tropics and subtropics: Model development and simulation. In: R.C. Muchow and J.A. Bellamy, editors, Climatic risk in crop production: Models and management for the semiarid tropics and subtropics. CAB Int., Wallingford, UK. p. 205–232.

Hammer, G.L., and R.C. Muchow. 1994. Assessing climatic risk to sorghum production in water-limited subtropical environments. I. Development and testing of a simulation model. Field Crops Res. 36:221–234. doi:10.1016/0378-4290(94)90114-7

Hammer, G.L., D. Butler, R.C. Muchow, and H. Meinke. 1996a. Integrating physiological understanding and plant breeding via crop modelling and optimisation. In: M.

Cooper and G.L. Hammer, editors, Plant adaptation and crop improvement. CAB Int. Wallingford, UK. p. 419–441.

Hammer, G.L., P.S. Carberry, and R.C. Muchow. 1993. Modelling genotypic and environmental control of leaf area dynamics in grain sorghum. I. Whole plant level. Field Crops Res. 33:293–310. doi:10.1016/0378-4290(93)90087-4

Hammer, G., S. Chapman, E. van Oosterom, and D. Podlich. 2005. Trait physiology and crop modelling as a framework to link phenotypic complexity to underlying genetic systems. Aust. J. Agric. Res. 56:947–960. doi:10.1071/AR05157

Hammer, G.L., S.C. Chapman, and R.C. Muchow. 1996b. Modelling sorghum in Australia: The state of the science and its role in the pursuit of improved practices. In: M.A. Foale, R.G. Henzell, and J.F. Kneipp, editors, Proceedings Third Australian Sorghum Conference, Tamworth, 20–22 Feb. 1996, Australian Institute of Agricultural Science, Melbourne, Occasional Publication No. 93. p. 43–61

Hammer, G., M. Cooper, F. Tardieu, S. Welch, B. Walsh, F. van Eeuwijk, S. Chapman, and D. Podlich. 2006. Models for navigating biological complexity in breeding improved crop plants. Trends Plant Sci. 11:587–593. doi:10.1016/j.tplants.2006.10.006

Hammer, G.L., Z. Dong, G. McLean, A. Doherty, C. Messina, J. Schussler, C. Zinselmeier, S. Paszkiewicz, and M. Cooper. 2009. Can changes in canopy and/or root system architecture explain historical maize yield trends in the US corn belt? Crop Sci. 49:299–312. doi:10.2135/cropsci2008.03.0152

Hammer, G.L., G.D. Farquhar, and I.J. Broad. 1997. On the extent of genetic variation for transpiration efficiency in sorghum. Aust. J. Agric. Res. 48:649–655. doi:10.1071/A96111

Hammer, G.L., J.W. Hansen, J.G. Phillips, J.W. Mjelde, H. Hill, A. Love, and A. Potgieter. 2001a. Advances in application of climate prediction in agriculture. Agric. Syst. 70:515–553. doi:10.1016/S0308-521X(01)00058-0

Hammer, G.L., M.J. Kropff, T.R. Sinclair, and J.R. Porter. 2002. Future contributions of crop modelling– from heuristics and supporting decision-making to understanding genetic regulation and aiding crop improvement. Eur. J. Agron. 18:15–31. doi:10.1016/S1161-0301(02)00093-X

Hammer, G.L., G. McLean, S. Chapman, B. Zheng, A. Doherty, M.T. Harrison, E. van Oosterom, and D. Jordan. 2014. Crop design for specific adaptation in variable dryland production environments. Crop Pasture Sci. 65:614–626.

Hammer, G., C. Messina, E. van Oosterom, C. Chapman, V. Singh, A. Borrell, D. Jordan, and M. Cooper. 2016. Molecular breeding for complex adaptive traits: How integrating crop ecophysiology and modelling can enhance efficiency. In: X. Yin and P. Struik, editors, Crop systems biology: Narrowing the gap between genotype and phenotype. Springer, Cham, Switzerland. p. 147–162.

Hammer, G.L., R.L. Vanderlip, G. Gibson, L.J. Wade, R.G. Henzell, D.R. Younger, J. Warren, and A.B. Dale. 1989. Genotype-by-environment interaction in grain sorghum. II. Effects of temperature and photoperiod on ontogeny. Crop Sci. 29:376–384. doi:10.2135/cropsci1989.0011183X002900020029x

Hammer, G.L., E.J. van Oosterom, S.C. Chapman, and G. McLean. 2001b. The economic theory of water and nitrogen dynamics and management in field crops. In: A.K. Borrell and R.G. Henzell, editors, Proceedings of the Fourth Australian Sorghum Conference, Kooralbyn, QLD, Australia. 5–8 Feb. 2001. Range Media, Toowoomba, QLD, Australia.

Hammer, G.L., E. van Oosterom, G. McLean, S.C. Chapman, I. Broad, P. Harland, and R.C. Muchow. 2010. Adapting APSIM to model the physiology and genetics of complex adaptive traits in field crops. J. Exp. Bot. 61:2185–2202. doi:10.1093/jxb/erq095

Heiniger, R.W., R.L. Vanderlip, and S.M. Welch. 1997a. Developing guidelines for replanting grain sorghum: I. Validation and sensitivity analysis of the SORKAM sorghum growth model. Agron. J. 89:75–83. doi:10.2134/agronj1997.00021962008900010012x

Heiniger, R.W., R.L. Vanderlip, S.M. Welch, and R.C. Muchow. 1997b. Developing guidelines for replanting grain sorghum: II. Improved methods of simulating caryopsis weight and tiller number. Agron. J. 89:84–92. doi:10.2134/agronj1997.00021962008900010013x

Henderson, S.A., S. von Caemmerer, G.D. Farquhar, L.J. Wade, and G.L. Hammer. 1998. Correlation between carbon isotope discrimination and transpiration efficiency in lines of the C4 species *Sorghum bicolor* in the glasshouse and the field. Aust. J. Plant Physiol. 25:111–123. doi:10.1071/PP95033

Hochman, Z., H. van Rees, P.S. Carberry, J.R. Hunt, R.L. McCown, A. Gatmann, D. Holzworth, S. van Rees, N.P. Dalgliesh, W. Long, A.S. Peake, P.L. Poulton, and T. McClelland. 2009. Re-inventing model-based decision support with Australian dryland farmers. 4. Yield Prophet® helps farmers monitor and manage crops in a variable climate. Crop Pasture Sci. 60:1057–1070. doi:10.1071/CP09020

Holzworth, D.P., N.I. Huth, P.G. deVoil, E.J. Zurcher, N.I. Herrmann, G. McLean, K. Chenu, E. van Oosterom, V. Snow, C. Murphy, A.D. Moore, H.E. Brown, J.P. Whish, S. Verrall, J. Fainges, L. Bell, A. Peake, P. Poulton, Z. Hochman, P.J. Thorburn, D.S. Gaydon, N. Dalgliesh, D. Rodriguez, H. Cox, S. Chapman, A. Doherty, E. Teixeira, J. Sharp, R. Cichota, I. Vogeler, F. Li, E. Wang, G.L. Hammer, M.J. Robertson, J. Dimes, P. Carberry, J.N.G. Hargreaves, N. MacLeod, C. McConald, J. Harsdorf, S. Wedgewood, and B.A. Keating. 2014. APSIM—Evolution towards a new generation of agricultural systems simulation. Environ. Modell. Software. 62:327–350.

Jones, C.A., and J.R. Kiniry, editors. 1986. Ceres-maize. A simulation model of maize growth and development. Texas A&M Univ. Press, College Station, TX.

Jones, J.W., G. Hoogenboom, C.H. Porter, K.J. Boote, W.D. Batchelor, L.A. Hunt, P.W. Wilkens, U. Singh, A.J. Gijsman, and J.T. Ritchie. 2003. The DSSAT cropping system model. Eur. J. Agron. 18:235–265. doi:10.1016/S1161-0301(02)00107-7

Jordan, D.R., C.H. Hunt, A.W. Cruickshank, A.K. Borrell, and R.G. Henzell. 2012. The relationship between the stay-green trait and grain yield in elite sorghum hybrids grown in a range of environments. Crop Sci. 52:1153–1161. doi:10.2135/cropsci2011.06.0326

Keating, B.A., P.S. Carberry, G.L. Hammer, M.E. Probert, M.J. Robertson, D. Holzworth, N.I. Huth, J.N.G. Hargreaves, H. Meinke, Z. Hochman, G. McLean, K. Verburg, V. Snow, J.P. Dimes, M. Silburn, E. Wang, S. Brown, K.L. Bristow, S. Asseng, S. Chapman, R.L. McCown, D.M. Freebairn, and C.J. Smith. 2003. An overview of APSIM, a model designed for farming systems simulation. Eur. J. Agron. 18:267–288. doi:10.1016/S1161-0301(02)00108-9

Kemanian, A.R., C.O. Stockle, and D.R. Huggins. 2005. Transpiration-use efficiency of barley. Agric. For. Meteorol. 130:1–11. doi:10.1016/j.agrformet.2005.01.003

Kholová, J., T. Murugesan, S. Kaliamoorthy, S. Malayee, R. Baddam, G.L. Hammer, G. McLean, S. Deshpande, C.T. Hash, P.Q. Craufurd, and V. Vadez. 2014. Modelling the effect of plant water use traits on yield and stay-green expression in sorghum. Funct. Plant Biol. 41:1019–1034. doi:10.1071/FP13355

Kim, H.K., D. Luquet, E. van Oosterom, M. Dingkuhn, and G. Hammer. 2010. Regulation of tillering in sorghum: Genotypic effects. Ann. Bot. (Lond.) 106:69–78. doi:10.1093/aob/mcq080

Lafarge, T.A., and G.L. Hammer. 2002. Predicting plant leaf area production: Shoot assimilate accumulation and partitioning, and leaf area ratio, are stable for a wide range of sorghum population densities. Field Crops Res. 77:137–151. doi:10.1016/S0378-4290(02)00085-0

Lobell, D.B., G.L. Hammer, K. Chenu, B. Zheng, G. McLean, and S.C. Chapman. 2015. The shifting influence of drought and heat stress for crops in Northeast Australia. Glob. Change Biol.

Lobell, D.B., G.L. Hammer, G. McLean, C. Messina, M.J. Roberts, and W. Schlenker. 2013. The critical role of extreme heat for maize production in the United States. Nat. Clim. Change 3:497–501. doi:10.1038/nclimate1832

Lobell, D.B., W.S. Schlenker, and J. Costa-Roberts. 2011. Climate trends and global crop production since 1980. Science 333:616–620. doi:10.1126/science.1204531

Mace, E.S., V. Singh, E.J. van Oosterom, G.L. Hammer, C.H. Hunt, and D.R. Jordan. 2012. QTL for nodal root angle in sorghum (*Sorghum bicolor* L. Moench) co-locate with QTL for traits associated with drought adaptation. Theor. Appl. Genet. 124:97–109. doi:10.1007/s00122-011-1690-9

Martre, P., P.D. Jamieson, M.A. Semenov, R.F. Zyskowski, J.R. Porter, and E. Triboï. 2006. Modelling protein content and composition in relation to crop nitrogen dynamics for wheat. Eur. J. Agron. 25:138–154. doi:10.1016/j.eja.2006.04.007

Meinke, H., and Z. Hochman. 2000. Using seasonal climate forecasts to manage dryland crops in northern Australia—Experiences from the 1997/98 seasons. In: G.L. Hammer, N. Nicholls, and C. Mitchell, editors, Applications of seasonal climate forecasting in agricultural and natural ecosystems: The Australian Experience. Kluwer Acad., Dordrecht, The Netherlands. p. 149–165.

Messina, C., G. Hammer, Z. Dong, D. Podlich, and M. Cooper. 2009. Modelling crop improvement in a G*E*M framework via gene-trait-phenotype relationships. In: V.O. Sadras and D. Calderini, editors, Crop physiology: Applications for genetic improvement and agronomy. Acad. Press, Amsterdam, The Netherlands. p. 235–265.

Messina, C.D., D. Podlich, Z. Dong, M. Samples, and M. Cooper. 2011. Yield-trait performance landscapes: From theory to application in breeding maize for drought tolerance. J. Exp. Bot. 62:855–868. doi:10.1093/jxb/erq329

McCown, R.L., G.L. Hammer, J.N.G. Hargreaves, D.P. Holzworth, and D.M. Freebairn. 1996. APSIM: A novel software system for model development, model testing, and simulation in agricultural systems research. Agric. Syst. 50:255–271. doi:10.1016/0308-521X(94)00055-V

McCown, R.L. 2001. Learning to bridge the gap between science-based decision support and the practice of farming: Evolution in paradigms of model-based research and intervention from design to dialogue. Aust. J. Agric. Res. 52:549–571. doi:10.1071/AR00119

McLean, G., J. Whish, R. Routley, I. Broad, and G. Hammer. 2003. The effect of row configuration on yield reliability in grain sorghum: II. Modelling the effects of row configuration. Proceedings of the Eleventh Australian Agronomy Conference, Geelong, Jan 2003. http://www.regional.org.au/au/asa/2003/c/9/mclean.htm (accessed 29 Feb. 2016).

Monteith, J.L. 1977. Climate and the efficiency of crop production in Britain. Philos. Trans. R. Soc. Lond. B Biol. Sci. 281:277–294. doi:10.1098/rstb.1977.0140

Monteith, J.L. 1986. How do crops manipulate water supply and demand? Philos. Trans. R. Soc. Lond. A 316:245–259. doi:10.1098/rsta.1986.0007

Monteith, J.L., A.K.S. Huda, and D. Midya. 1989. RESCAP: A resource capture model for sorghum and pearl millet. Res. Bulletin No. 12, ICRISAT, Patancheru, India. p. 30–34.

Morrell, P.L., E.S. Buckler, and J. Ross-Ibarra. 2012. Crop genomics: Advances and applications. Nat. Rev. Genet. 13:85–96.

Muchow, R.C., and P.S. Carberry. 1990. Phenology and leaf area development in a tropical grain sorghum. Field Crops Res. 23:221–237. doi:10.1016/0378-4290(90)90056-H

Muchow, R.C., and T.R. Sinclair. 1994. Nitrogen response of leaf photosynthesis and canopy radiation use efficiency in field-grown maize and sorghum. Crop Sci. 34:721–727. doi:10.2135/cropsci1994.0011183X003400030022x

Muchow, R.C., M. Cooper, and G.L. Hammer. 1996. Characterising environmental challenges using models. In: M. Cooper and G.L. Hammer, editors, Plant adaptation and crop improvement. CAB Int., Wallingford, UK. p. 349–364.

Muchow, R.C., G.L. Hammer, and R.L. Vanderlip. 1994. Assessing climatic risk to sorghum production in water-limited subtropical environments. II. Effects of planting date, soil water at planting, and cultivar phenology. Field Crops Res. 36:235–246. doi:10.1016/0378-4290(94)90115-5

Nelson, R.A., D.P. Holzworth, G.L. Hammer, and P.T. Hayman. 2002. Infusing the use of seasonal climate forecasting into crop management practice in north east Australia using discussion support software. Agric. Syst. 74:393–414. doi:10.1016/S0308-521X(02)00047-1

Passioura, J.B. 1983. Roots and drought resistance. Agric. Water Manage. 7:265–280. doi:10.1016/0378-3774(83)90089-6

Podlich, D.W., M. Cooper, and K.E. Basford. 1999. Computer simulation of a selection strategy to accommodate genotype-by-environment interaction in a wheat recurrent selection program. Plant Breed. 118:17–28. doi:10.1046/j.1439-0523.1999.118001017.x

Probert, M.E., J.P. Dimes, B.A. Keating, R.C. Dalal, and W.M. Strong. 1998. APSIM's water and nitrogen modules and simulation of the dynamics of water and nitrogen in fallow systems. Agric. Syst. 56:1–28. doi:10.1016/S0308-521X(97)00028-0

Ravi Kumar, S., G.L. Hammer, I. Broad, P. Harland, and G. McLean. 2009. Modelling environmental effects on phenology and canopy development of diverse sorghum genotypes. Field Crops Res. 111:157–165. doi:10.1016/j.fcr.2008.11.010

Ritchie, J.T., and G. Alagarswamy. 1989. Simulation of sorghum growth and development in CERES models. Research Bull. 12, ICRISAT, Patancheru, India. p. 34–38.

Robertson, M.J., S. Fukai, M.M. Ludlow, and G.L. Hammer. 1993. Water extraction by grain sorghum in a sub-humid environment. I. Analysis of the water extraction pattern. Field Crops Res. 33:81–97. doi:10.1016/0378-4290(93)90095-5

Rosenthal, W.D., R.L. Vanderlip, B.S. Jackson, and G.F. Arkin. 1989. SORKAM: A grain sorghum growth model. TAES Computer Software Documentation Series No. MP-1669. Texas Agric. Exp. Stn., College Station, TX.

Routley, R., I. Broad, G. McLean, J. Whish, and G. Hammer. 2003. The effect of row configuration on yield reliability in grain sorghum: I. Yield, water use efficiency and soil water extraction. Proceedings of the 11th Australian Agronomy Conference, Geelong, Australia. January 2003. Regional Institute, Erina, NSW, Australia.

Sinclair, T.R. 1986. Water and nitrogen limitations in soybean grain production. I. Model development. Field Crops Res. 15:125–141. doi:10.1016/0378-4290(86)90082-1

Sinclair, T.R., and J. Amir. 1992. A model to assess nitrogen limitations on the growth and yield of spring wheat. Field Crops Res. 30:63–78. doi:10.1016/0378-4290(92)90057-G

Sinclair, T.R., and T. Horie. 1989. Leaf nitrogen, photosynthesis, and crop radiation use efficiency: A review. Crop Sci. 29:90–98. doi:10.2135/cropsci1989.0011183X002900010023x

Sinclair, T.R., G.L. Hammer, and E.J. van Oosterom. 2005. Potential yield and water-use efficiency benefits in sorghum from limited maximum transpiration rate. Funct. Plant Biol. 32:945–952. doi:10.1071/FP05047

Sinclair, T.R., and R.C. Muchow. 1998. Radiation use efficiency. Adv. Agron. 65:215–265. doi:10.1016/S0065-2113(08)60914-1

Sinclair, T.R., R.C. Muchow, and J.L. Monteith. 1997. Model analysis of sorghum response to nitrogen in subtropical and tropical environments. Agron. J. 89:201–207. doi:10.2134/agronj1997.00021962008900020009x

Singh, V., E.J. van Oosterom, D.R. Jordan, and G.L. Hammer. 2011. Genetic variability and control of root architecture in sorghum. Crop Sci. 51:2011–2020. doi:10.2135/cropsci2011.01.0038

Singh, V., E.J. van Oosterom, D.R. Jordan, and G.L. Hammer. 2012. Genetic control of nodal root angle in sorghum and its implications on water extraction. Eur. J. Agron. 42:3–10. doi:10.1016/j.eja.2012.04.006

Stone, R.C., G.L. Hammer, and T. Marcussen. 1996. Prediction of global rainfall probabilities using phases of the Southern Oscillation Index. Nature 384:252–255.

Tanner, C.B., and T.R. Sinclair. 1983. Efficient water use in crop production: Research or research? In: H.M. Taylor, W.R. Jordan, and T.R. Sinclair, editors, Limitations to efficient water use in crop production. Am. Soc. of Agron., Madison, WI. p. 1–27.

Tardieu, F. 2012. Any trait or trait-related allele can confer drought tolerance: Just design the right drought scenario. J. Exp. Bot. 63:25–31. doi:10.1093/jxb/err269

Taylor, K.E., R.J. Stouffer, and G.A. Meehl. 2012. An overview of CMIP5 and the experiment design. Bull. Am. Meteorol. Soc. 93:485–498. doi:10.1175/BAMS-D-11-00094.1

Technow, F., C.D. Messina, L.R. Totir, and M. Cooper. 2015. Integrating crop growth models with whole genome prediction through approximate Bayesian computation. PLoS ONE 10(6):e0130855. doi:10.1371/journalpone.0130855

van Keulen, H., and N.G. Seligman. 1987. Simulation of water use, nitrogen nutrition and growth of a spring wheat crop. PUDOC, Wageningen, The Netherlands.

van Oosterom, E.J., A.K. Borrell, S.C. Chapman, I.J. Broad, and G.L. Hammer. 2010a. Functional dynamics of the nitrogen balance of sorghum. I. Nitrogen demand of vegetative plant parts. Field Crops Res. 115:19–28. doi:10.1016/j.fcr.2009.09.018

van Oosterom, E.J., A.K. Borrell, K. Deifel, and G.L. Hammer. 2011. Does increased leaf appearance rate enhance adaptation to postanthesis drought stress in sorghum? Crop Sci. 51:2728–2740. doi:10.2135/cropsci2011.01.0031 van Oosterom, E.J., P.S. Carberry, and G.J. O'Leary. 2001. Simulating growth, development, and yield of tillering pearl millet. I. Leaf area profiles on main shoots and tillers. Field Crops Res. 72:51–66. doi:10.1016/S0378-4290(01)00164-2

van Oosterom, E.J., S.C. Chapman, A.K. Borrell, I.J. Broad, and G.L. Hammer. 2010b. Functional dynamics of the nitrogen balance of sorghum. II. Grain filling period. Field Crops Res. 115:29–38. doi:10.1016/j.fcr.2009.09.019

van Oosterom, E.J., G.L. Hammer, S.C. Chapman, A. Doherty, E. Mace, and D.R. Jordan. 2006. Predicting flowering time in sorghum using a simple gene network: Functional physiology or fictional functionality? In: A.K. Borrell, R.G. Henzell, and D.R. Jordan, editors, Proceedings from the 5th Australian Sorghum Conference, Gold Coast, Australia. 30 Jan.–2 Feb. 2006. Range Media, Toowoomba, QLD, Australia.

Vanderlip, R.L., and G.F. Arkin. 1977. Simulating accumulation and distribution of dry matter in grain sorghum. Agron. J. 69:917–923. doi:10.2134/agronj1977.00021962006900060007x

Wang, E., M.J. Robertson, G.L. Hammer, P.S. Carberry, D. Holzworth, H. Meinke, S.C. Chapman, J.N.G. Hargreaves, N.I. Huth, and G. McLean. 2002. Development of a generic crop model template in the cropping system model APSIM. Eur. J. Agron. 18:121–140. doi:10.1016/S1161-0301(02)00100-4

Whish, J., G. Butler, M. Castor, S. Cawthray, I. Broad, P. Carberry, G. Hammer, G. McLean, R. Routley, and S. Yeates. 2005. Modelling the effects of row configuration on sorghum yield in north-eastern Australia. Aust. J. Agric. Res. 56:11–23. doi:10.1071/AR04128

White, J.W., G. Alagarswamy, M.J. Ottman, C.H. Porter, U. Singh, and G. Hoogenboom. 2015. An overview of CERES-sorghum as implemented in the Cropping System Model version 4.5. Agron. J. 107:1987–2002.

Yin, X., P.C. Struik, and M.J. Kropff. 2004. Role of crop physiology in predicting gene-to-phenotype relationships. Trends Plant Sci. 9:426–432. doi:10.1016/j.tplants.2004.07.007

Drought and High Temperature Stress and Traits Associated with Tolerance

P.V.V. Prasad,* M. Djanaguiraman, S.V.K. Jagadish, I.A. Ciampitti

Abstract

Sorghum is an important food crop in arid and semiarid regions of the world. The resilience of sorghum production to changes in climate can be improved via better understanding of physiological bases of abiotic stress tolerance or susceptibility. Among the various abiotic stresses that limits sorghum production, drought and high temperature (HT) are of foremost importance. Physiological mechanisms of drought resilience are (i) drought avoidance, (ii) drought escape, and (iii) drought tolerance. Primary effects of drought stress are cellular dehydration and increase in hydraulic resistance. The secondary effects include reduced cellular and metabolic activities, stomatal closure, and production of reactive oxygen species. The primary and secondary effects finally leads to wilting by loss of turgor. In sorghum, the most sensitive stage to drought is during the panicle development, flowering, and grain-filling stages. The traits like stay green, limited transpiration under high vapor pressure deficit, canopy temperature depression, and root architecture can be used to identify drought stress tolerant genotypes. High temperature stress during reproductive stage (sporogenesis and anthesis) is most sensitive stage in sorghum. In sorghum, HT causes decreases the CO_2 assimilation by causing damage to thylakoid membrane, stomatal conductance, and rubisco activase enzyme activity. Apart from this, HT stress decreases pollen viability leading to decreased seed numbers per panicle. The decreased seed number is not compensated with increased individual seed mass. The traits like canopy temperature depression, favorable respiration, time of day of anthesis, pollen germination percentage, seed-set percentage and seed size were the traits that has significant in identification of HT stress tolerant sorghum genotypes in the breeding program.

Introduction

In arid and semiarid regions of the world, sorghum [*Sorghum bicolor* (L.) Moench] is an important crop that addresses food and nutritional security. Sorghum is a crop with more efficient photosynthetic machinery (C_4 plant) and increased water use efficiency. It has multifaceted uses like food, feed, fuel, and fiber. Despite

---

Abbreviations: CTD, canopy temperature depression; CMT, cellular membrane thermostability; HT, high temperature; H_2O_2, hydrogen peroxide; OH-, hydroxyl radical; MDA, malondialdehyde; T_{max}, maximum temperature; T_{min}, minimum temperature; T_{opt}, optimum temperature; OAC, osmotic adjustment capacity; QTL, quantitative trait loci; ROS, Reactive oxygen species; RWC, Relative water content; O_2^-, superoxide radical; VPD, vapour pressure deficit.

Department of Agronomy, Kansas State University, Manhattan, Kansas, 66506. *Corresponding author (vara@ksu.edu)

doi:10.2134/agronmonogr58.2014.0065

greater climate resilience of sorghum to environmental stress, both drought and high temperature (HT) stress limits sorghum productivity in several regions of the world including the United States (Tack et al., 2017). Drought is one of the major constraints limiting crop production worldwide. A continuous reduction or erratic distribution in precipitation coupled with higher evaporation leads to drought. Drought stress severely affects plant growth and development, leading to reduced grain yield. Climate models have predicted that due to climate change, the severity and frequency of drought will be increased (Metz et al., 2007; Field et al., 2012). Apart from this, global climate change will cause significant alteration in precipitation patterns and temperature regimes as well as an increase in both the frequency and intensity of extreme events. The occurrence and intensity of HT episodes is likely to increase in future climates (Field et al., 2012). Increased frequency of HT stress on crops can cause significant yield losses depending on timing, intensity and duration of stress. It is quantified that crops attain only 25% of their potential yield, due to the detrimental effects of abiotic stresses (Boyer, 1982). Understanding the response of sorghum plants to increasing drought and HT would be desirable to adjust crop management practices to improve grain yield under stress. In this chapter we summarize the impacts of drought and high temperature stress on various physiological, growth and yield traits of sorghum, and elicit current knowledge and future opportunities.

Drought Stress

Sorghum growth and yield are impacted by drought stress, which is caused by intermittent to continuous periods without precipitation. Plants experience drought when the water supply from the soil is not sufficient to meet the crop transpiration demand.

Mechanisms of Drought Stress Tolerance

Physiological mechanisms for drought tolerance can be categorized into three components (i) drought avoidance, (ii) drought escape, and (iii) drought tolerance. Drought avoidance is maintenance of adequate cell water content and/or water potential, even though the soil or atmosphere has lower water potential than the plant tissue. At ample soil moisture availability, the environment largely drives transpiration. Atmospheric vapor pressure deficit (VPD) plays an important role in transpiration and leaf water potential. The water use depends on developmental stage and leaf area of the plant. The maximum water use in sorghum is known to coincide with leaf area index of 3.5 m^{-2} m^{-2}, after that it starts to decrease. Drought avoidance mechanism will allow the plant to maintain the turgor and cell water content by maintaining water uptake by the roots and decreased transpiration rate. Most of the sorghum genotypes have a thick waxy cuticle, which reflects the incoming solar radiation and also decreases transpiration. Sorghum plants also have a deep root system compared to other cereals, exploring and absorbing moisture from deeper soil layer. Sorghum plant transpiration decreases significantly at a soil moisture level of 20 to 40% of the total available or extractable moisture (Blum and Arkin, 1984), and this occurs by reduction in leaf area per plant, causing death of the lower leaves, with the actively growing top young leaves maintaining near or normal transpiration (Stout and Simpson, 1978). Evidence also indicates that the leaf water potential decreases due to osmoregulation a drought avoidance mechanism. The osmoregulation mechanism occurs by the accumulation of osmotically active cellular solutes. In general, sorghum displays a sigmoid relationship between leaf water potential and relative water content (Acevedo et al., 1979). This relationship is highly dependent on osmotic adjustment capacity (OAC). If the leaf OAC is high, by accumulation of osmotically-active substances, the relative water content is maintained at a higher level for a given leaf water potential. For short-term drought stress, the drought avoidance will work; however, over dependence of drought avoidance will decrease the photosynthetic rate and activity.

In sorghum (C_4 plant), the stomatal conductance decreases with declining leaf water potential [mild drought stress; relative water content (RWC) ~70%] leading to a lower photosynthetic rate. This is based on the following evidence: (i) reduced intercellular CO_2 concentration (C_i), (ii) recovery of photosynthetic rate at elevated CO_2 level, (iii) occurrence of photorespiration, and (iv) recovery of photosynthetic rate following rehydration. Decreased C_i indicates the stomatal closure and CO_2 limitation for C_4 photosynthesis. In sorghum, the C_i decreases during the early phases of drought stress and increases at later stages as net photosynthetic rate continues to decline. The increase in intercellular CO_2 concentration also leads to decrease the photorespiration.

Drought escape is a phenomenon by which the plants complete their life cycle early under drought stress by decreasing the duration of the vegetative or reproductive growth. Hall (2000) defined drought escape a phenomenon wherein drought sensitive stages of plant development are completed during part of the season when drought is not present. If the rainy season is short, then the plants complete their reproductive stage before the onset of the severe drought stress. For example, if the rain ends a week before, the cultivar having the drought escape mechanism will flower early and completes its life cycle with the residual moisture compared to genotypes that have limited plasticity. This one-week difference will be equivalent to about 20 to 25% of grain filling period, which is significant enough to cause yield losses through seed filling duration, seed filling rate, or both, leading to smaller seed size. Remobilization of preanthesis assimilates from the leaf and stem is one of the drought escape mechanism. One major problem anticipated with this mechanism is the low yield potential and the inadequate plasticity to cope with mild or intermittent drought. Photoperiod sensitivity is another method of drought escape. Earlier studies have shown that in sorghum homeostasis to heading date was observed in local varieties. In Nigeria, a local sorghum variety from Samaru was planted from 9 May to 15 July (difference of 67 d), and all the plants headed between 6 and 17 October (difference of 11 d). These heading dates homeostasis is to avoid drought stress during seed filling (Kassam and Andrew, 1975; Craufurd and Qi, 2001). The sensitive reproductive stages are photoperiodically controlled to coincide with the favorable environment, making the plant to complete the grain filling before the onset of drought stress. This type adaptation will be suitable to the regions where the rainfall follows bimodal distribution or if the monsoon ends abruptly.

Sorghum drought tolerance mechanisms function is expressed at the tissue or cellular or plant levels. Drought tolerance allows plants to maintain turgor and cell volume at low leaf water potential, thus helping to continue metabolic activity longer under water stress through osmotic adjustment, antioxidant capacity, and cell membrane stability. Osmotic adjustment is defined as a lowering of osmotic potential of the plant cell resulting from a net accumulation of osmotic activity (compatible inorganic and organic solutes) in the cytosol in response to drought stress. It allows plants to maintain higher turgor to sustain normal physiological functions under drought stress conditions. Genetic variability for osmotic adjustment exists in sorghum (Blum and Sullivan, 1986). Although osmotic adjustment is considered as an important mechanism for drought tolerance, it has not been used in sorghum breeding programs because it provides a short-term adaptation to drought stress tolerance. Breeding programs typically aim at identification and developing true drought tolerant varieties. The expression of drought tolerance is dependent on the stage of the crop when drought stress occurs. Susceptibility to drought can occur during the early vegetative seedling stage, period of panicle development and during the post-flowering stage. An important criterion for drought tolerance is a high water-use–efficiency which is determined by (i) the amount of water taken up by the roots during the growing season, and (ii) the efficiency of using the absorbed water for dry matter production. The crops will yield more if it has deeper root system, and if it is dependent on stored soil moisture. Compared with maize, sorghum penetrates the soil faster, more intensively, and to a greater depth. If the soil is shallow, then the advantage of deep rooting system on yield benefit will be less (Prasad et al., 2008a). Sorghum survives severe drought stress at vegetative stage probably due to small plant size (small leaf area consequently a slow rate of water loss). However, drought stress during peak vegetative stage causes delay in panicle initiation and flowering. Drought stress during flowering and seed-set stage causes decreased grain number by affecting the gametes viability. Similarly, drought stress during seed filling stage

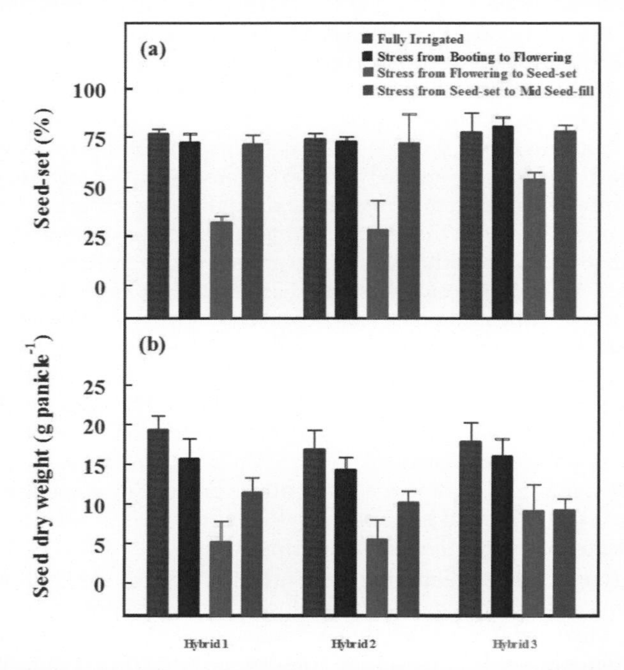

Fig. 1. Impact of timing of drought stress on seed-set percentage and seed dry weight of three different hybrids under controlled environment conditions.

have resulted in decreased grain yield by affecting leaf senescence and seed-size (seed filling duration and rate).

Sensitive Stages to Drought Stress

The impacts of drought stress on sorghum growth and yield depend on the plant developmental stage (timing), duration and severity at which drought stress occurs. In general, drought stress tolerance is relatively higher during early seedling stages and decreases through later stages until flowering and early grain filling stage. The most sensitive stages for drought stress are during panicle development, flowering, and grain filling just like other cereals (Prasad et al., 2008a). However, until recently there has been little information available on sensitive stages to drought stress. Most of the investigations have been focused on independent stages (vegetative, flowering and post flowering), rather than comparing multiple stages to drought stress. Sorghum is most vulnerable to drought at preanthesis and postanthesis stages (Rosenow et al., 1983). Most of the sorghum cultivars are tolerant to drought at preanthesis, but not many during postanthesis stage (Sanchez et al., 2002). Our experiment with sorghum hybrids indicated that drought stress during flowering to seed-set stages decreases seed-set percentage more than the drought stress from booting to flowering stages and seed-set to mid seed fill stages. Similarly, the seed dry weight also decreased more when the drought stress occurred during flowering to seed-set stage, than seed-set to mid-seed fill and booting to flowering (Fig. 1.; Prasad, personal communication).

Important Traits for Drought Tolerance

Drought tolerance is a function of various physiological functions. In the breeding program selection for physiological traits should be regarded as complementing and not replacing empirical approaches for screening for yield and quality. Some of the important physiological traits that are significant for improving drought tolerance include: stay green, limited transpiration, canopy temperature depression, and root architecture.

Stay-Green

Sorghum is susceptible to postanthesis drought stress, as it suffers from premature leaf and plant senescence, and genotypes tolerant to postanthesis drought condition are called stay-green because they retain higher levels of chlorophyll in their leaves during terminal drought stress conditions. Stay-green phenotype can be classified into five types (Thomas and Howarth, 2000): (i) Type A, will show delayed onset of leaf senescence, (ii) Type B, the rate of senescence is reduced, (iii) Type C, chlorophyll is retained but photosynthesis declines, (iv) Type D, greenness is retained due to rapid death at harvest, and (v) Type E, the phenotype is greener to begin with. This classification indicates that stay-green may be functional or superficial. Functional stay-green is characterized by the maintenance of leaf photosynthesis during grain filling (Types A, B, and E), while, superficial stay-green occurs when photosynthetic capacity is disconnected from leaf greenness (Types C and D). However, not all functional stay-green is necessarily productive. Greenness as such does not indicate continued photosynthesis, and it will be detrimental if it has weak sink and/or an inability to remobilize the stem reserves. However, in sorghum stay-green genotype, the photosynthesis is maintained for a longer period of time to support the developing grain for carbohydrates.

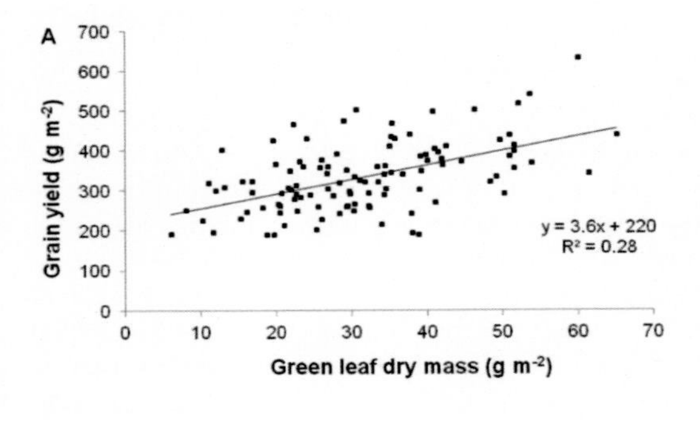

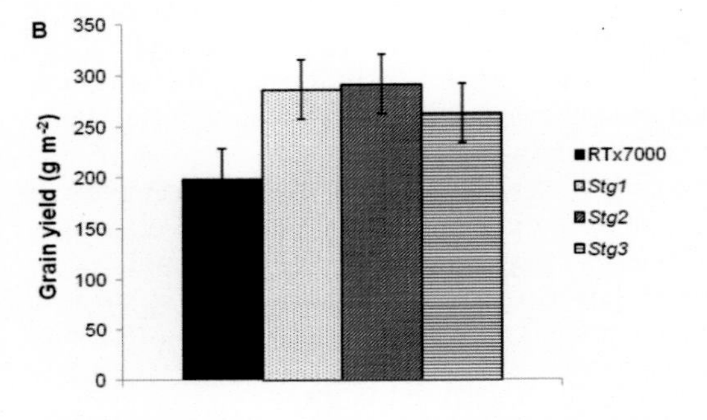

Fig. 2. (a) Relationship between grain yield and green leaf dry mass of a set of recombinant inbred lines from the cross between BQL39 (senescent) and BQL41 (stay-green) (Borrell et al., 2014b); and (b). Grain yield of Stg NILs and RTx7000 under postanthesis drought stress (Borrell et al., 2014b).

Therefore, selection for both stay-green and grain yield should be considered together in breeding programs to ensure that delayed senescence is not due to low sink demand. Stay-green also provides resistance to premature leaf and stalk death induced by post-flowering drought. The stay green trait has been recognized as a major mechanism of post-flowering drought resistance in grain sorghum. The genotype B35 is resistant to post-flowering drought stress, but susceptible to pre-flowering drought stress. A stay green line QL41 was developed from a cross between B35 and QL33. The genotype SC56 is a stay green line, selected from Sudan. The genotype TX7000 is resistant to pre-flowering stress, but susceptible to post-flowering drought stress. A number of stay green QTLs (quantitative trait loci) have been detected by several authors using populations developed from the different genetic backgrounds (Subudhi et al., 2000). A positive correlation between stay green and yield has been demonstrated in many studies. The most convincing evidence showing a strong positive relationship between green leaf dry mass at 25 d after anthesis and grain yield in a set of 160 RILs [BQL39 (senescent) × BQL41 (stay-green)] was observed by Borrell et al. (1999; Fig. 2a). In a controlled environment experiment the performance of stay green QTLs was evaluated under severe terminal drought stress, the result indicated that *Stg1* (31%), *Stg2* (32%), *and Stg3* (31%) yielded significantly more than RTx7000 (Fig. 2b). The increased yield was attributed to individual grain mass (Borrell et al., 2014b). The contribution of *Stg* loci to higher grain yield may be due to the continued higher photosynthetic rate of leaf. The increased functional leaf area duration during grain filling is the key mechanism (Borrell et al., 2014a).

Delayed leaf senescence during postanthesis stage is a consequence of changes occurring during early crop growth stages, particularly the supply and demand of water and water use efficiency. Increased water availability at grain filling stage can be achieved by increased water accessibility; and/or conserving water by decreasing leaf area (decreasing transpiration per unit leaf area). Reduced canopy size has been linked to increased grain yield under postanthesis drought stress (Borrell et al., 2014a). The biomass accumulation during the grain filling period was significantly negatively correlated with green leaf area at anthesis. Under drought stress, where biomass production is a function of water availability, this negative relationship likely represents increased water availability during grain filling. Thus, the increased grain yield of stay-green QTLs under postanthesis drought stress may be due to preanthesis leaf area dynamics. The reduced canopy size in the stay green QTLs was achieved by reduced tillering and small upper leaves, and this decrease in canopy architecture can shift crop water use. Conservation of soil water before anthesis for utilization during grain filling is the key physiological mechanism by which stay-green confers drought adaptation under terminal water deficit. Small increases in water use during grain filling can significantly impact grain yield. Apart from this, it is also observed that, increased postanthesis water use by introducing stay green QTLs could be due to increased access to water by either better water extraction from the soil by deeper roots or greater lateral spread. Genetic studies of stay-green have generally indicated a complex pattern of inheritance and expression of stay-green is strongly influenced by the environment.

Limited Transpiration in Response to Higher Vapour Pressure Deficit

Conservation of soil moisture is one of the most important strategies under drought conditions. Passioura (1972) proposed that decreased diameters of the xylem elements in the seminal roots of wheat might result in decreased hydraulic conductance. Subsequently, Richards and Passioura (1989) reported that genetic variation within wheat existed for the trait. Similarly, identification of sorghum genotypes that has limited transpiration rate at high evaporative demand will conserve moisture. This trait could be useful for early season water conservation so that more water is available later in the season, especially during post-flowering drought stress. The early season water conservation is an expression of limited transpiration trait under high VPD. The hypothesized mechanism for the maximum-transpiration-rate trait is a leaf level hydraulic limitation that results in limited water transport and decreased stomatal conductance at high VPD. Quick changes in transpiration occur after certain VPD thresholds are crossed. This indicates that hydraulic signals are needed to induce fairly rapid stomatal closure and avoid loss of turgor. Atmospheric VPD and transpiration rate follow a diurnal pattern, however if the VPD is very high and plants are unable to transport water to match this transpiration demand, this diurnal pattern breaks. The decreased hydraulic conductance in a plant renders it unable to support continuing increases in transpiration to keep pace with increasing atmospheric VPD. Hence, at the VPD where increases in transpiration rate cannot be sustained, transpiration rate will exhibit a breakpoint. After which, further increases in VPD result in little or no increase in transpiration rate. This phenomenon is mostly expressed during mid-day, so called as midday stomatal closure. This trait has two promising benefits, (i) decreased stomatal conductance earlier in the soil drying cycle; and (ii) limited transpiration rate under high atmospheric VPD. In the case of soil drying, a lower hydraulic conductance in the plant means that decreases in transpiration rate will occur at higher soil water content in the soil prior to drying cycle. Evidence has shown that there are differences among sorghum genotypes in their VPD response, such that at high VPD there was little or no further increases in transpiration with increasing VPD (Gholipoor et al., 2010). Genotypes BTx2752, SC599, SC982, and B35 exhibited the limited transpiration trait (Shekoofa et

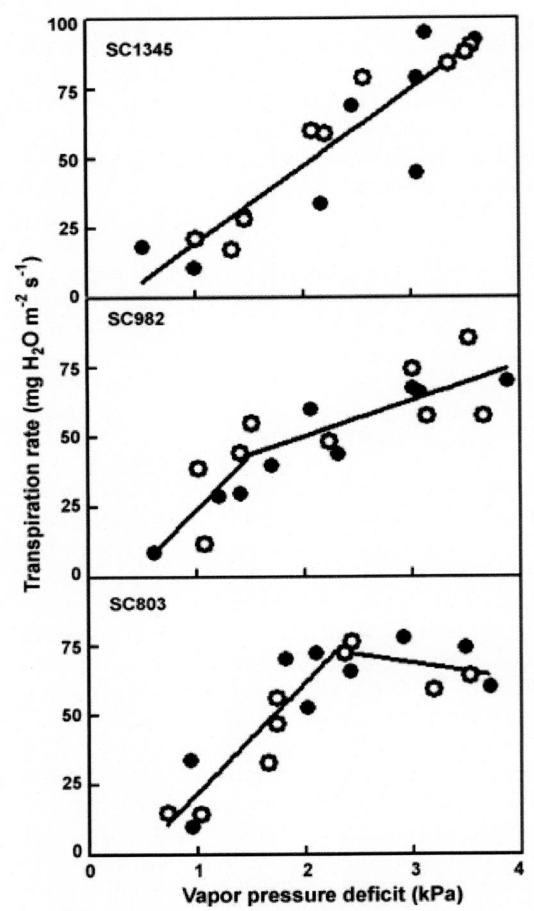

Fig. 3. Transpiration rate response of three genotypes to increasing vapor pressure deficit. The two symbols indicate data recorded on two consecutive days of observations (Gholipoor et al., 2010).

al., 2014). This trait would be desirable in less humid environments for increasing yields in water-deficit seasons. Our earlier experiment in sorghum showed genetic variability for transpiration rate under drying soil (Gholipoor et al., 2010). Transpiration rate response of three sorghum genotypes to increasing VPD was documented as three general patterns of response (Fig. 3).

The transpiration response to VPD clearly diverged among the sorghum genotypes. A low breakpoint value would provide the greatest water conservation when soil water is still available, while a high breakpoint value imposes a less-restrictive water conservation strategy. The genotypes with the high fraction of transpirable soil water threshold, low breakpoint, are classified as "water conserver", restricting transpiration rate as the soil dries down. The response of transpiration to a high VPD appears to have a hydraulic basis, in which aquaporin might play a role. However, the conditions at which the stomatal closure happens due to altered hydraulic conductance and increases in transpiration rate have to be studied in detail to explain the role of aquaporin on restricted transpiration rate trait.

Root Architecture

The rooting system of a plant is comprised of a taproot and/or seminal roots, and lateral roots, which are produced from the taproot and seminal roots during the lifetime of a plant. The rooting system of the plant has the ability to exhibit morphological, structural, and physiological responses to changes in environment. This is called as root developmental plasticity. The root developmental plasticity includes changes in tap and/or seminal root elongation, lateral root and root hair formation, elongation and distribution, and the ability to absorb water and nutrients. In general, the plants growing under drought stress condition will have increased root to shoot ratio, as a mechanism of exploring more soil volume for moisture than plants growing under well-watered conditions. The distribution of roots in soil can influence the extent of access to water to support crop growth. Two different drought-tolerance mechanisms were proposed considering rooting depth. First, a deep, wide-spreading and more branched root system and a second mechanism related to a small and shallow root system. The former is called as "water spender" and later as "water savers". Anatomical studies showed that under drought stress, the suberization of cell wall increases during root development to impose limitation to root radial hydraulic conductivity (North and Nobel, 1995). Similarly, the number of cortical layers is an adaptive mechanism under drought stress; the decreased number shortens the way between the soil and stele, favoring quick radial water transport (Fahn, 1964). However, under drought conditions, rupture of cortical cells is a strategy operated by the plant to avoid inverse flux of water to soil. Another mechanism is creating xylem vessel cavitation during drought stress, which could optimize water flow according to water availability. Root thickness and rooting depth are important for water extraction from deeper soil horizon. A positive correlation between root elongation rate and root diameter has been observed. Under drought stress, the small diameter roots are considered as a strategy aimed to maximize absorptive surfaces, thus increasing rates of water and nutrient uptake. Differences in root angle and direction of growth of major root axes is related to degree of horizontal and vertical soil exploration; this decides the amount of water extracted from the soil. A deeper root system will be advantageous during terminal or post-flowering drought stress.

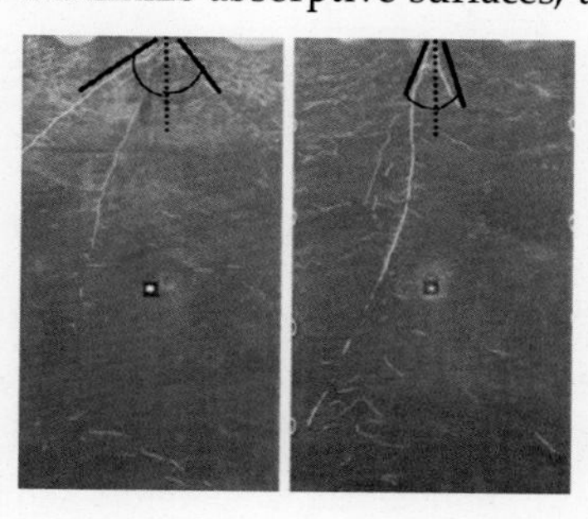

Fig. 4. Nodal roots, visible on the glass surface of root chambers, wide angle, left panel and narrow angle, right panel. Thick solid lines indicate first flush of nodal roots, dotted lines indicate the vertical plane, and arcs indicate the estimated root angle (Mace et al., 2012).

Genotypic variation in the nodal root flush angle and mean diameter of nodal roots (Fig. 4) was observed and could be trait for large scale screening for root architecture (Singh et al., 2010). Genetic variation in nodal root angle (narrow and wide) was also observed among sorghum parental inbred lines with moderate heritability for this trait. The narrow root angle had relatively more root length than wide root angle, which may increase water accessibility in deep soils. These results support the hypothesis of inclusion of root angle as a selection criterion in sorghum breeding programs. The root angle of parental inbred lines would be a poor predictor of the root angle of their hybrids. The root angle was mainly independent of plant size and these two traits represented contrasting drought adaptation mechanisms; thus, these two

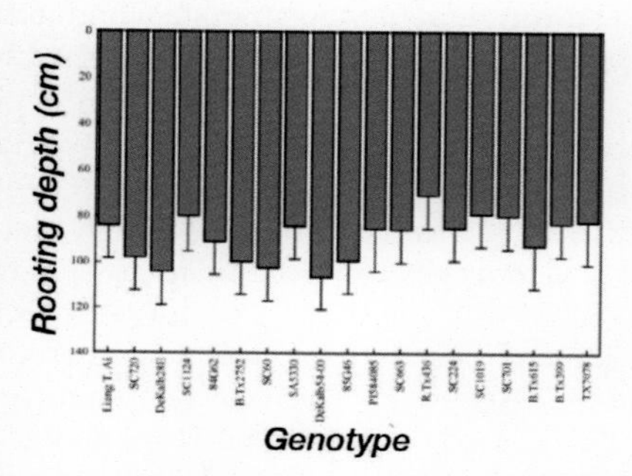

Fig. 5. Variation in rooting depth among sorghum genotypes grown in controlled environment (Mutava, 2012).

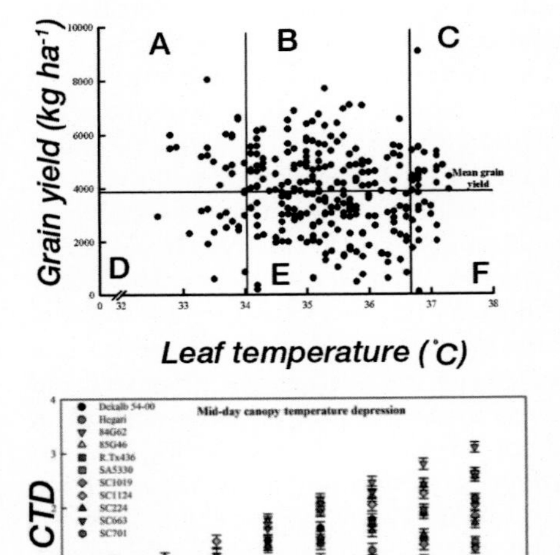

Fig. 6. (a) Classification of 300 sorghum genotypes based on leaf temperature and grain yield. Quadrants identifies genotypes with (A) low leaf temperature and high yield, (B) normal leaf temperature and high yield, (C) high leaf temperature and high yield, (D) low leaf temperature and low yield, (E) normal leaf temperature and low yield and (F) high leaf temperature and low yield (Mutava et al., 2011); and (b) variation in midday canopy temperature depression (CTD) among sorghum genotypes grown under rainfed conditions (Mutava, 2012).

mechanisms could be stacked in a single hybrid (Singh et al., 2011).

Our earlier studies on sorghum indicated that drought stress significantly increased the rooting depth, total root length and root surface area, and also showed genetic variability among the sorghum lines for rooting depth (Mutava, 2012; Fig. 5). Under drought stress conditions an increased rooting depth would contribute to better drought tolerance. Genotypic variation in the nodal root flush angle and mean diameter of nodal roots was observed and could be trait for large scale screening for root architecture (Singh et al., 2010).

Canopy Temperature Depression

Canopy temperature depression (CTD), the difference between air temperature (T_a) and canopy temperature (T_c), is positive when the canopy is cooler than the air (CTD = T_a–T_c). The key role of transpiration in plants is leaf cooling, and therefore changes in canopy temperature relative to air temperature are an indication of the capability of transpiration in cooling plant leaves under a demanding environmental load. Plants that restrict transpiration rate will maintain a warmer canopy, leading to conservation of soil available moisture for later stages. On the other hand, plants that use more water for transpiration, cooling under higher VPD through increased stomatal conductance, will have higher CTD. The earlier is "true drought tolerant" and later is a "short-term drought escaper".

Our study with 300 sorghum genotypes showed presence of these two mechanisms in sorghum diversity panel. We have identified sorghum genotypes that had high leaf temperature with high yield (tolerance mechanism) and genotypes with low leaf temperature and high yields (escape mechanism; Mutava et al., 2011). Genotypes were classified as (i) high leaf temperature and high yield (e.g., SC782, SC284, and SC987), (ii) low leaf temperature and high yield (e.g., RTx2783, SC332 and SC60), (iii) low leaf temperature and low yield (e.g., SC25, RTx2917, and SC224) and (iv) high leaf temperature and low yield (e.g., SC949, SC970, and SC1277) based on leaf temperature and grain yield factors (Fig. 6 – Panel 1).

A further analysis of these genotypes revealed that those with high leaf temperature had a breakpoint in their transpiration rate with increased VPD, but this was absent in genotypes with low leaf temperature (Fig. 3; Gholipoor et al., 2010).

The CTD represents an overall, integrated physiological response to drought and has therefore been used generally to assess plant water status. Canopy temperature measurements have been widely used to study the response of crops to drought. This approach is based on inverse relationship between transpiration and leaf temperature. Blum et al. (1989) used canopy temperatures of drought stressed wheat genotypes to characterize yield stability under various moisture conditions and found a positive correlation between drought susceptibility and canopy temperature in stressed environments.

Our experiments on sorghum showed significant variation in mean CTD diurnal patterns under both rainfed and irrigated conditions with most negative CTD occurring at 8:00 to 11:00 a.m., most positive at 4:00 to 7:00 p.m. There were also genotypic differences in midday and predawn CTD (Fig. 6, Panel 2). Nighttime measurements may provide more stable conditions for CTD comparison among genotypes because of consistent differences among genotypes between 2:00 am and 8:00 a.m. There was a positive relationship between midday CTD and yield and harvest index (Mutava, 2012). The CTD measurements under drought stress will help identifying cooler canopies with higher yield because it is associated with transpiration rates, while CTD measurement under well-watered conditions is to identify warmer canopies with high yield because it is associated with stomatal conductance. However, the slow wilting lines that have warmer canopies during drought should not be avoided in the drought selection process.

High Temperature Stress

High temperature (HT) stress is another important abiotic stress that limits sorghum productivity. Sorghum is generally grown in arid and semiarid regions of the world; the mean temperatures of those regions already above optimum for sorghum growth and development. Further increase in temperatures will severely impact sorghum productivity (Prasad et al., 2006). The increase in temperature is associated with extreme change in weather patterns (uneven rainfall pattern, severe droughts, short or long episodes of HT); as well, the frequency of warm nights and days is increasing (Intergovernmental Panel on Climate Change, 2013). Temperature extremes can negatively impact different crop species and genotypes within a species (Hatfield et al., 2011). Thus understanding the physiological and biochemical basis of HT tolerance or susceptibility will help in sorghum breeding and selection strategy. The optimum temperature for sorghum vegetative stage is 26 to 34°C (Maiti, 1996). Temperature above and below the optimum temperature will have a significant negative impact on sorghum growth and yield.

Temperature Thresholds

Crop species differ for response to temperature throughout their life cycle. In general, for each species, there is a clear defined minimum, optimum, and maximum temperatures for each growth or developmental stages. For each crop there is a base temperature (minimum temperature; T_{min}) and maximum temperature (maximum threshold; T_{max}) below and above which growth stops, respectively. The optimum temperature is at which the growth will at maximum (T_{opt}). Above the T_{min}, the growth commences, and at T_{opt} the plant grows at a rapid pace and beyond this the growth slows and stops at T_{max}. For most of the crops, the

Table 1. Summary of cardinal temperatures for different growth stages, phenology, canopy development, root growth, resource use efficiency, and physiological processes in sorghum.†

Variable	T_{min}	T_{opt}- I	T_{opt}- II	T_{max}	Reference
	°C; daytime maximum temperature				
		Growth stages			
Germination	10	21	–	> 45	Gerik et al. (2003), Soman and Peacock (1985)
Vegetative	8	34	–	44	Alagarswamy and Ritchie, (1991)
Reproductive	–	26–34	–	40	Hatfield et al. (2008), Prasad et al. (2006)
		Biomass			
Total biomass at vegetative stage	14	27	29	> 40	White et al. (2005)
Total biomass at anthesis	< 14	16	19	42	White et al. (2005)
Total biomass at harvest	< 14	16	19	42	White et al. (2005)
		Phenology			
Emergence rate	< 14	30	35	> 40	White et al. (2005)
Vegetative development rate	< 14	36	–	–	White et al. (2005)
Reproductive development rate	< 14	34	–	–	White et al. (2005)
		Canopy development			
Leaf area index	12	22	25	42	White et al. (2005)
Crop growth rate	12	29	33	42	White et al. (2005)
		Root growth			
Root biomass	12	19	21	42	White et al. (2005)
Maximum rooting depth	< 14	34	–	–	White et al. (2005)
		Resource use efficiency			
Water use efficiency-evapotranspiration basis	< 14	23	30	–	White et al. (2005)
Water use efficiency–transpiration basis	< 14	24	31	–	White et al. (2005)
Radiation use efficiency	< 14	29	33	–	White et al. (2005)
Nitrogen use efficiency	< 14	34	36	–	White et al. (2005)
		Physiological processes			
Photosynthesis	–	36	–	–	Prasad et al. (2006),
Panicle exsertion	–	36	–	44	Prasad et al. (2006)
Pollen germination	15	29	–	43	Djanaguiraman et al. (2014)
Seed-set	–	32	–	40	Prasad et al. (2006)
Seed numbers	10	26	29	42	White et al. (2005)
Seed size	–	32	–	44	Prasad et al. (2006)
Harvest index	–	32	–	40	Prasad et al. (2006)
Grain yield	–	32	–	40	Prasad et al. (2006)

† Data not available, the duration, intensity and stage of imposition of high temperature stress varies with study. The data from Prasad et al. (2006) are from temperate short season, photoperiod insensitive grain sorghum hybrid. Similarly, the data on pollen germination percentage (Djanaguiraman et al., 2014) are from eight sorghum genotypes representing temperate short season, photoperiod insensitive grain sorghum hybrids, sorghum conversion lines, and elite lines. The data on chlorophyll fluorescence trait (Yan et al., 2013) are from sweet sorghum genotype. The data from White et al. (2005) are from temperate short season, photoperiod insensitive grain sorghum hybrid and temperate, indeterminate, day neutral and small seeded genotype.

vegetative development has higher optimum temperatures and thresholds than reproductive stages. For sorghum, the base and optimum temperature for vegetative stage are 8 and 34°C, respectively. Similarly, for reproductive growth the base and optimum temperatures are 8 and 31°C, respectively (Hatfield et al., 2008). In general, time from seedling emergence to initiation of reproductive phase will be shorter, if the temperature increase occurs within the optimal range. However, at extreme HT, the duration to panicle initiation and anthesis will be delayed. For example in grain sorghum, severe HT (> 40°C) at vegetative stage inhibited the reproductive stage and the plants remained vegetative till the stress is relieved (Prasad et al., 2006). In sorghum, panicle emergence was delayed by 20 d at day/night temperature > 36/26 to 40/30°C, and no panicles were formed at 44/34°C (Prasad et al., 2006). High temperature stress during later stages of panicle or flower development can increase the time from flowering to initiation of rapid seed filling in peanut (Prasad et al., 1999; Table 1).

Mechanisms of High Temperature Stress

Unlike drought, for HT stress there is no specific classification associated with tolerance (avoidance, escape, tolerance). However, these principles can be explored. In this section, the responses of physiological process to HT will be summarized. The maintenance of cellular membrane integrity and function under HT stress is essential for a continued photosynthetic and respiratory activity. High temperature stress induces the oxidants or reactive oxygen species (ROS) [superoxide radical; O_2^-, hydrogen peroxide; H_2O_2 and hydroxyl radical; OH^-] production. In sorghum, HT increased the ROS in leaves (Djanaguiraman et al., 2010) and pollen (Prasad and Djanaguiraman, 2011). The generation of ROS under HT stress is a symptom of cellular damage, where membrane lipid peroxidation compromises membrane permeability and function. The reaction centers of PS I and PS II in chloroplasts, peroxisomes and mitochondria are the major sites of ROS. Hydroxyl radical is formed due to the reaction of H_2O_2 with O_2^- (Haber-Weiss reaction) and reactions of H_2O_2 with Fe^{2+} (Fenton reaction). High temperature stress increases leaf temperature, which reduces antioxidant enzyme activities increasing malondialdehyde (MDA) content in leaves of sorghum plant (Djanaguiraman et al., 2010).

Moderate HT stress causes a reversible decrease in photosynthesis; increased HT stress causes irreversible damage to the photosynthetic apparatus, resulting in greater inhibition of plant growth. Sorghum plants have higher optimum temperature for photosynthesis (30–42°C), and also higher photosynthetic capacity, particularly at HT (Matsuoka et al., 2001). Many studies have examined the effects of HT stress on the pigments, chlorophyll fluorescence kinetics, electron transport system, photosynthesis related enzyme activities, and gas exchange in sorghum plants (Al-Khatib and Paulsen, 1984; Jagtap et al., 1998; Sharkey 2005; Prasad et al., 2006; Yan et al., 2011; 2013; Djanaguiraman and Prasad, 2014). Outcomes from this research includes (i) decreased chlorophyll content in sensitive genotypes, thereby changing chlorophyll *a:b* ratio causing premature leaf senescence, (ii) more damage to thylakoid membranes, specifically to photosystem (PS) II than PS I, (iii) PS II reaction center, oxygen evolving complex and electron transport chain of PS II acceptor side is more susceptible to HT stress compared with PS II donor side, (iv) the enzyme ribulose 1,5-bisphosphate carboxylase/oxygenase (Rubisco) does not appear to limit photosynthesis at HT because the in vitro capacity of Rubisco is well in excess of the net photosynthetic rate, (v) Rubisco

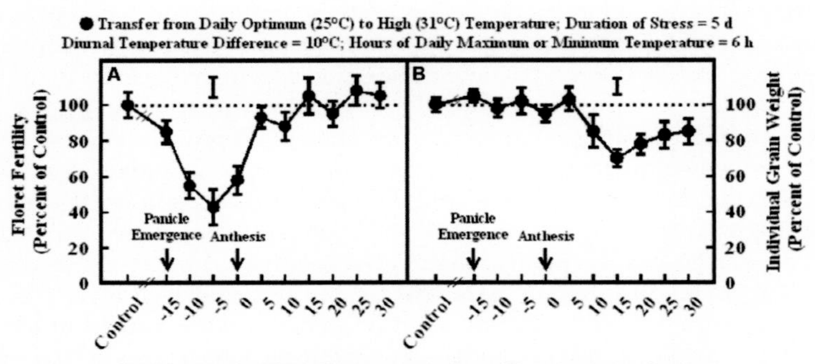

Fig. 7. Impact of high temperature stress at different times relative to anthesis on (a) floret fertility, and (b) individual grain weight (Prasad et al., 2015).

regeneration capacity, content and activity were decreased, (vi) increased PEP-case activity, (vii) decreased net photosynthetic rate, and (viii) increased stomatal conductance and transpiration rate.

Sensitive Stages and Thresholds During Reproductive Stages of Development

The impacts of HT stress on sorghum growth and yield depend on (i) plant developmental stage ("timing"), (ii) duration of stress, and (iii) severity of the stress. Sorghum reproductive stages are highly sensitive to HT stress compared to vegetative stages. The major yield deciding factor namely seed number and seed weight are sensitive to HT. The formation of seed (grain) is a result of successful fertilization, which in turn depends on pollen and ovule functionality. In sorghum, periods between 10 and 5 d before anthesis [coinciding with gametogenesis– both microsporogenesis (pollen development) and megasporogenesis (ovule development)]; and at anthesis (0 d before anthesis) were most sensitive to HT causing maximum decreases in floret fertility (Fig. 7. Prasad et al., 2015, 2017).

Genetic variability for pollen germination and seed numbers was observed in sorghum under HT stress (Nguyen et al., 2013). High temperature during day or night or combined day and night time decreased pollen viability and germination (Nguyen et al., 2013; Prasad and Djanaguiraman, 2011; Djanaguiraman et al., 2014). The decrease in the pollen germination is possibly due to increased oxidative damage and decreased unsaturation of phospholipids (Prasad and Djanaguiraman, 2011). Membrane lipid saturation is therefore considered an important element in HT tolerance. In another study, it was observed that impaired carbohydrate metabolism in the anthers is linked with pollen sterility (Jain et al., 2010). It is also speculated that under HT stress the pollen production per anther and pollen reception per stigma will also be reduced like previously reported in flax (*Linum usitatissimum* L., Cross et al., 2003). Similarly, HT during the grain filling period decreases individual grain size due to shorter grain filling duration (Prasad et al., 2008b) and/or grain filling rate (Prasad et al., 2006; 2008b). Decreases in grain number and individual grain weight leads to lower grain yields.

Important Traits for High Temperature Tolerance

The success of the crop improvement program is dependent on efficiency of phenotyping and its robustness. High temperature stress leads to morphological, physiological, biochemical and molecular changes that adversely affect plant growth and productivity. To reduce the negative effects of HT stress, there is a need to develop cultivars or hybrids that have tolerance to HT. Breeding for stress tolerance requires: (i) identification of key traits, (ii) robust phenotyping procedures, (iii) identification of suitable and genetically diverse donor lines, and (iv) a proper understanding of the genetics and inheritance of the traits of interest. Yield cannot always be an indicator of stress tolerance; basic physiological process such as photosynthesis and reproductive ability are also relevant and need to be considered during breeding for increased stress resilience. Several traits are associated with yield under HT tolerance. Most of the traits listed in the table are good for cereals and specific traits for sorghum are also included. Availability of genetic variability for these traits, knowledge about their heritability and inheritance, and availability of effective phenotyping methodologies are essential for a successful breeding program.

Membrane Stability, Chlorophyll Fluorescence and Index

The photosynthetic efficiency of a plant is directly proportional to the quantity of chlorophyll *a* fluorescence and chlorophyll content in the leaf tissue. Chlorophyll *a* fluorescence, typically the maximum potential quantum efficiency of PS II (F_v/F_m ratio) and basal fluorescence (F_o) are the most important physiological traits that have high correlation with HT stress tolerance. These fluorescence traits are directly associated with thylakoid membrane damage. Membrane damage and chlorophyll content are very closely correlated when plants are exposed to a long period of HT stress, and are most likely under similar genetic control (Talukder et al., 2014). Images of the fluorescence parameter can also be used to detect stress in plants (Maxwell and Johnson, 2000). Chlorophyll fluorescence is indicative of thylakoid membrane damage, which can decrease the photosynthesis. Studies in wheat suggested that there is a strong relationship between chlorophyll content and chlorophyll *a* fluorescence under HT stress (Talukder et al., 2014).

The differences in chlorophyll index can be captured by using chlorophyll meter (SPAD meter), which is very robust in nature. The genetic variability for photosynthesis under HT stress have been shown to be associated with a loss of chlorophyll and a change in chlorophyll *a*/*b* ratio. Reynolds et al. (1994) demonstrated that loss of chlorophyll during grain fill is associated with reduced wheat yield. Spectral reflectance measurements hold promise for the assessment of some physiological parameters at the leaf level.

Sullivan (1972) developed a HT tolerance test based on cellular membrane thermostability (CMT) by measuring the amount of electrolyte leakage from leaf discs bathed in deionized water after exposure to HT treatment. Increased CMT indicates less plasma membrane damage and intact membrane and well associated with photosynthesis. Blum et al. (2001) reported that heritability for CMT for HT tolerance was 71% in winter wheat and 67% in spring wheat. However, the relationship between CMT and yield under HT may vary from plant to plant; hence, it has to be verified in sorghum before using as a screening tool. Studies have indicated that vegetative tissue and reproductive tissue vary in their tolerance level to HT stress (Young et al., 2004). Mostly, vegetative tissue tolerance does not have a correlation with reproductive tissue tolerance. Hence, sufficient caution needs to be taken during selection process, identification of HT stress tolerant lines based on vegetative tissue will confer vegetative stage HT stress, similarly, reproductive tissue based screening will confer reproductive stage HT stress tolerance.

Canopy Temperature Depression

As previously introduced in this chapter, the deviation of temperature of plant canopies in comparison to ambient temperature is known as canopy temperature depression (CTD). The ability of the plant to decrease temperature through transpiration cooling will keep the plant cool and benefits plants at above optimal stress conditions. The CTD has been a good criterion for screening HT stress tolerance (as for drought also). The CTD is a function of soil water status, air temperature, relative humidity, and incident radiation and it is best expressed at high VPD conditions associated with low relative humidity and warm or hot air temperature. The CTD measurement using infrared thermometer have some common genetic bases under HT and drought stress (Pinto et al., 2010). Under HT stress, well-watered plants increase their transpiration rate due to high vapor pressure deficit, which permits evaporative canopy cooling. Cool canopy temperatures have been associated with increased stomatal conductance and deeper roots leading to more transpiration cooling. This type of genotype is classified as "HT escaper". In the absence of subsoil moisture for transpiration cooling, these genotypes will exhibit yield penalty due to stomatal closure. True HT tolerant genotypes will have high canopy temperature and grain yield, indicating its endurance to HT tolerance stress by different mechanisms (still unknown, but may be by alteration in membrane lipid unsaturation index and lipid species content). Hence, by selecting the CTD in the parental lines, indirectly the assimilation capacity of the plant is also improved. Mutava et al. (2011) have identified sorghum genotypes having high leaf temperature and high yield in the sorghum diversity panel (Fig. 6, Panel 1), which can be exploited for improving HT stress tolerance in sorghum.

Optimum Respiration

The regulation of respiration under HT stress conditions is relatively less understood. It is important to understand these responses, as photosynthesis is temporally and spatially restricted, while respiration occurs continuously and in all organs. Respiration exponentially increases with increasing temperatures from 0 to 35 or 40°C, reaching a plateau at 40 to 50°C. At temperature above 50°C, respiration decreases because of damage to respiratory mechanism. In general, 30 to 80% of carbohydrates fixed by the plants are used for respiration per day and it depends on stage of the crop and growth temperature. High day or nighttime temperatures increased the night respiration rate (Djanaguiraman et al., 2013). The increase in respiration rate indicates increased consumption of assimilates for maintenance or growth respiration. Tjoelker et al. (1999) reported foliar respiration was linearly related to total non-structural carbohydrate concentration. Field trails across hottest wheat growing environment showed that yield is negatively associated with average daily temperature and nighttime temperature explained the most of the variation, indicating respiration is a major trait determining yield (Reynolds et al., 1994). Hence, selection of genotypes with low respiration rate and high biomass under HT stress will yield better. High temperature stress induced changes in respiration leads to a shortened life cycle and decreased crop productivity. High efficiency in respiration may increase growth rate, yield, or the tolerance of plants to HT stress. Mitochondria are very stable; however, HT stress is more destructive to mitochondrial activity. The research on *Proteus vulgaris* indicated that HT stress decreased electron transport by four-to five fold and decreased oxidative phosphorylation by twofold. Differences in the sensitivity of oxidative phosphorylation may explain the genotypic differences under HT stress tolerance (Lin and Markhart, 1990). Thus, increasing the efficiency of respiration and its resistance to HT stress could improve tolerance to growth and yield. In rice, Bahuguna et al. (2017) observed that high nighttime temperature during flowering stage increased night respiration leading to decreased grain size and grain yield in temperature sensitive variety and in tolerant variety no significant differences were observed.

Increased Seed-Set Percentage

Seed number is function of successful pollination, fertilization and seed-set. In sorghum, there is a strong positive correlation between number of seeds and seed yield. Lower seed-set at HT was due to lower pollen production and lower pollen germination; however, the role of stigma receptivity cannot be completely overruled (Prasad et al., 2006). Decreased pollen production at HT may be related to poor anther dehiscence (Porch and Jahn, 2001). Lower pollen viability at HT is due

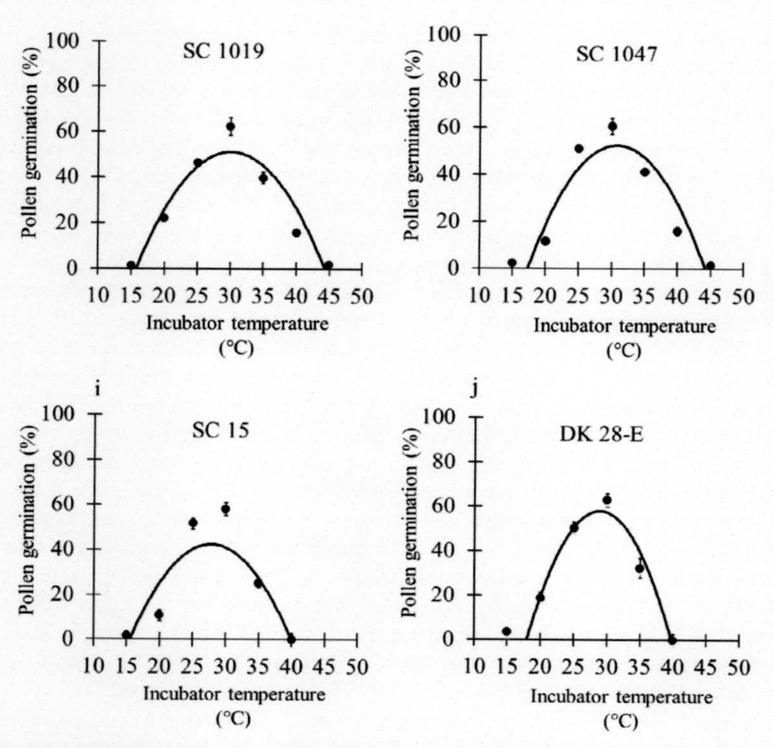

Fig. 8. Variation in cardinal temperature for pollen germination percentage among sorghum genotypes (Djanaguiraman et al., 2014).

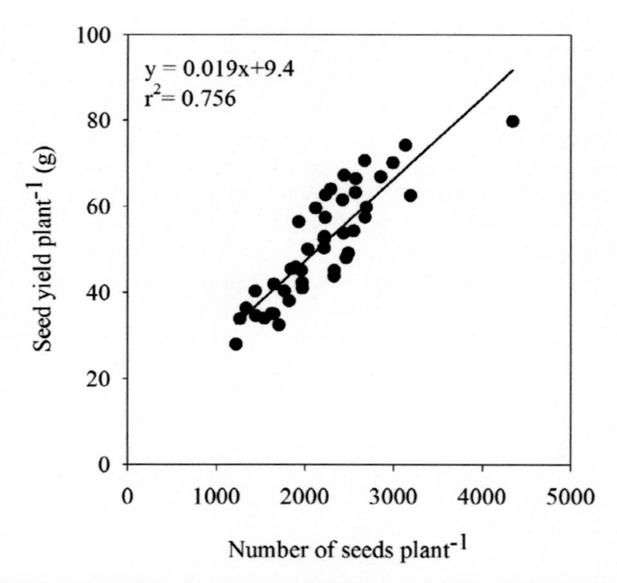

Fig. 9. Number of seeds per plant versus seed yield per plant for various sorghum genotypes (Prasad, personal communication).

to degeneration of tapetum layer, and/or decreased carbohydrate metabolism. The degradation of tapetum cells under HT stress could negatively influence nourishment of pollen mother cells that could lead to sterile pollen. Jain et al. (2007; 2010) observed a decrease in carbohydrates in the anther wall and pollen of sorghum along with reduced pollen germination under HT. In addition, there was a decrease in the expression of transporters associated with hexokinase and sugar transporters in pollen grains. Prasad and Djanaguiraman (2011) observed loss of pollen viability under high nighttime temperature, associated with decreased phospholipids and increased ROS in the pollen grains. This indicates that pollen growth and development are occurring throughout the day and do not depend on photoperiod. Identifying the processes that control pollen production and pollen fertility at HT would be essential to improve current cultivars and/or to develop new HT tolerant grain sorghum cultivars. Nguyen et al. (2013) reported that under HT stress, sorghum genotypes differed significantly for pollen viability and seed-set percentage, and these two traits were strongly and positively associated. The genetic variation observed for pollen and seed-set traits among the genotypes can be exploited through breeding efforts to develop HT tolerant varieties. Djanaguiraman et al. (2014) observed genetic variability for pollen germination and seed-set percentage in sorghum under HT stress. Genetic variability for cardinal temperature for pollen germination was documented by Prasad et al. (2006) and Djanaguiraman et al. (2014). The genotypes having higher optimum temperature (T_{opt}) and maximum temperature (T_{max}) have greater HT stress tolerance (Fig. 8). However, field level confirmation and multilocation trials are required before using this trait- because apart from pollen, the pistils are also sensitive to HT stress.

Number of seeds is a good indicator of reproductive success and yield formation under HT stress (Fig. 9). Selection for improved seed set and seed numbers under HT will be useful if increased seed numbers is not offset by a reduction in individual seed size. Under optimum growing conditions, a reduction in seed number can increase individual seed mass as a consequence of increased assimilate availability per grain (Yang et al., 2009). Singh et al. (2015) observed that the effect of temperature on seed-set percentage was much greater than the effect on individual seed mass, indicating that reduced seed-set in temperature susceptible genotypes was not compensated by increased seed mass. The HT treatment differences in grain yield was mainly associated with differences in seed-set percentage. Similarly, Prasad et al. (2008b) found that an increase in day to night temperatures from 32/22 to 36/26°C had no effect on individual seed mass of sorghum. A possible reason for the lack of compensation under HT stress may be due to restriction in the supply of assimilates to the grains by reduction in expression of genes involved in sucrose transport and starch synthesis (Phan et al., 2013). In addition, the maximum seed-size of sorghum is already determined before anthesis, possibly by the size of the meristem that affects the cell number in the pericarp (Yang et al., 2009). Nguyen et al. (2013) reported that under HT stress sorghum genotypes differed significantly in seed-set percentage, and pollen germination and seed-set were strongly and positively associated.

Early Morning Flowering (Time-of-Day of Anthesis)

Early morning flowering (time of day of anthesis) is expressed as hours relative to sunrise (as daylength, as the time of solar noon varies among environments). In a typical day, temperature starts to rise from sunrise and reaches peak during solar noon and decreases thereafter. Early morning flowering is an important trait that can be exploited to avoid or escape from HT stress. If plants can flower (anthesis) early in the morning and complete fertilization, then they are avoiding the high temperatures of the day. In sorghum the time from anthesis to fertilization can range from 4 to 12 h. Once pollination occurs and pollen germinates in the style, it is protected from the high temperatures. However, the rate of pollen tube can be lower under HT stress, thus taking more time to fertilize. In sorghum, the most sensitive stages to HT stress are sporogenesis and anthesis. Time of day of flowering is regulated at the level of the individual spikelet because anthesis events on a panicle are spread over a period. It is due to different physiological age and position of spikelet in a panicle.

In rice, the average environmental conditions during the seven-day preceding anthesis event explained time of flowering better than those observed on the day of anthesis, suggesting the involvement of regulatory processes happening prior to the anthesis event itself (Julia and Dingkuhn, 2012). Ishimaru et al. (2010) demonstrated that introgression of an early-anthesis trait effectively reduces sterility under HT conditions in rice, which has been documented to be effective under tropical field conditions (Bheemanahalli et al., 2016). Similarly, identification of sorghum genotypes that flower very early in the morning, before the daytime temperature rises, so they can escape from HT during anther dehiscence and pollen germination, pollen tube growth, and fertilization process; leading to successful seed-set. The adaptive value of anthesis occurring early in the day is thus an established fact in rice, however, it is not exploited in sorghum and further research is needed on this trait.

Rate and Duration Seed Filling

Yield is mainly a function of various components namely number of plants, panicles, and seed numbers per unit area and seed size. The seed size is a product of seed-filling rate and seed-filling duration. Seed-filling duration is the time from seed-set to physiological maturity. High temperature stress decreases seed-filling duration and thus leads to smaller seeds (Prasad et al., 2006). There might be compensation mechanisms for increased seed size, which may compensate for lower seed numbers; however, these responses may vary depending on crop species or varieties within species. For sorghum, there may be a slight increase in the seed-filling rate, but a large decrease in seed-filling duration under HT stress (Prasad et al., 2006). Although increasing temperatures sometimes stimulate seed-filling rate, this effect often does not fully compensate for shortening seed-filling duration. This results in smaller seed size and seed yields. The genetic variability for rate of seed-filling need detailed exploitation to enhance stress tolerance and yield improvements in sorghum.

Conclusions

Temperature increases can be a major factor in driving drought stress because VPD increases nonlinearly with higher air temperatures. Studies have indicated that the main reason for increased drought frequency was associated with higher VPD that results from the projected warming. Although HT and drought stresses often occur together, it is possible to get one without the other. Therefore, breeding strategies to target HT stress tolerance may differ from those targeting drought stress tolerance. This includes screening environment and approach, and the selection criteria. Better understanding of the possible impacts of drought and HT stress on sorghum would help in mitigating the adverse effects of these stresses. Sorghum is sensitive to drought and HT stress during critical stages of reproductive development, particularly during gametogenesis, flowering and seed filling. Respiration rates increase at HT stress and result in loss of biomass and seed yield. Both drought and HT stresses results in decreased photosynthesis and grain yield. Physiological trait based breeding strategy offers the benefit of maximizing the probability to harness more relevant additive gene actions. Sorghum breeding should be aimed at physiological traits that are related to canopy structure, delayed senescence, photosynthetic and water use efficiency, CTD, EMF, lower respiration rates, higher reproductive success, and harvest index for development of stress escapers and true tolerant genotypes. Genetic variability for the above mentioned traits under drought and HT stress is available, hence breeding strategies should aim at targeted exploration of diverse germplasm (landraces, diversity panel, sorghum conversion program) and harnessing of novel alleles from wild gene pool. The whole sorghum genome sequence, genetic and physical maps are available to draw better link between phenome to the genome to identify effective and stable markers. In parallel, through molecular breeding discovery of candidate genes conferring drought and HT stress tolerance in sorghum can be expedited through genome-wide expression profiling. Transgenic approaches hold promise for transferring drought and HT tolerant genes across the species. The hurdle of phenotyping of complex traits can be overcome by the use of high-throughput ground and aerial phenotyping platforms. It is expected that in the future, the semiarid and arid regions of the world will be more affected by climate change and sorghum is the major crop grown in those areas. Hence, improvement of drought and HT stress tolerance in sorghum can improve the livelihood of the farmers in those regions of the world.

Acknowledgments

We thank Kansas Grain Sorghum Commission, Center for Sorghum Improvement, and United States Agency for International Development. Contribution no. 18-135-B from the Kansas Agricultural Experiment Station.

References

Acevedo, E., E. Fereres, T.C. Hsiao, and D.W. Henderson. 1979. Diurnal growth trends water potential and osmotic adjustment of maize and sorghum leaves in the field. Plant Physiol. 64:476–480. doi:10.1104/pp.64.3.476

Alagarswamy, G., and J.T. Ritchie. 1991. Phasic development in CERES-sorghum model. In: T. Hodges, editor, Predicting crop phenology. chap 13. CRC, Boca Raton. p. 143–152.

Al-Khatib, K., and G.M. Paulsen. 1984. Mode of high temperature injury to wheat during grain development. Physiol. Plant. 61:363–368.

Bahuguna, R.N., C.A. Solis, W. Shi, and K.S.V. Jagadish. 2017. Post-flowering night respiration and altered sink activity account for high night temperature-induced grain yield and quality loss in rice (Oryza sativa L.). Physiol. Plant. 159:59–73. doi:10.1111/ppl.12485

Bheemanahalli, R., R. Sathishraj, M. Manoharan, H.N. Sumanth, R. Muthurajan, T. Ishimaru, and S.V.K. Jagadish. 2016. Is early morning flowering an effective trait to minimize heat stress damage during flowering in rice? Field Crops Res. 203:1–5.Blum, A. 1997. Constitutive traits affecting plant performance under stress. In: G.O. Edmeades, M. Banziger, H.R. Mickelson, C.B. Bena-Valdivia, Developing drought and low N tolerant maize. Proceedings of a Symposium, El-Batan, Mexico. March 25-29, 1996. CIMMYT, El-Batan, Mexico, p. 131-141.

Blum, A., and G.F. Arkin. 1984. Sorghum root growth and water-use as affected by water supply and growth duration. Field Crops Res. 9:131–142. doi:10.1016/0378-4290(84)90019-4

Blum, A., and C.Y. Sullivan. 1986. The comparative drought resistance of landraces of sorghum and millet from dry and humid regions. Ann. Bot. (Lond.) 57:835–846. doi:10.1093/oxfordjournals.aob.a087168

Blum, A., N. Klueva, and H.T. Nguyen. 2001. Wheat cellular thermotolerance is related to yield under heat stress. Euphytica 117:117–123. doi:10.1023/A:1004083305905

Blum, A., L. Shipiler, G. Golan, and J. Mayer. 1989. Yield stability and canopy temperature of wheat genotypes under drought stress. Field Crops Res. 22:289–296. doi:10.1016/0378-4290(89)90028-2

Borrell, A.K., F.R. Bidinger, and K. Sunitha. 1999. Stay-green associated with yield in recombinant inbred sorghum lines varying in rate of leaf senescence. International Sorghum and Millets Newsletter 40:31–34.

Borrell, A.K., J.E. Mullet, B.J. Jaeggli, E.J. van Oosterom, G.L. Hammer, P.E. Klein, and D.R. Jordan. 2014b. Drought adaptation of stay-green sorghum is associated with canopy development, leaf anatomy, root growth, and water uptake. J. Exp. Bot. doi:10.1093/jxb/eru232

Borrell, A.K., E.J. van Oosterom, J.E. Mullet, B. George-Jaeggli, D.R. Jordan, P.E. Klein, and G.L. Hammer. 2014a. Stay-green alleles individually enhance grain yield in sorghum under drought by modifying canopy development and water uptake patterns. New Phytol. doi:10.1111/nph.12869

Boyer, J.S. 1982. Plant productivity and environment. Science 218:443–448. doi:10.1126/science.218.4571.443

Craufurd, P.Q., and A. Qi. 2001. Photothermal adaptation of sorghum (Sorghum bicolor) in Nigeria. Agric. For. Meteorol. 108:199–211. doi:10.1016/S0168-1923(01)00241-6

Cross, R.H., S.A.B. McKay, A.G. McHughen, and P.C. Bonham-Smith. 2003. Heat stress effects on reproduction and seed set in Linum usitatissimum L. (flax). Plant Cell Environ. 26:1013–1020. doi:10.1046/j.1365-3040.2003.01006.x

Djanaguiraman, M., and P.V.V. Prasad. 2014. High temperature stress. In: M. Jackson, B. Ford-Lloyd, and M. Parry, editors, Plant genetic resources and climate change. CABI publisher, Oxfordshire, UK. p. 201–220.

Djanaguiraman, M., P.V.V. Prasad, and W.T. Schapaugh. 2013. [Glycine max (L.) Merr.] High day- or nighttime temperature alters leaf assimilation, reproductive success, and phosphotidic acid of pollen grain in soybean. Crop Sci. 53:1594–1604. doi:10.2135/cropsci2012.07.0441

Djanaguiraman, M., P.V.V. Prasad, and M. Seppanen. 2010. Selenium protects sorghum leaves from oxidative damage under high temperature stress by enhancing antioxidant defence system. Plant Physiol. Biochem. 48:999–1007. doi:10.1016/j.plaphy.2010.09.009

Djanaguiraman, M., P.V.V. Prasad, M. Murugan, M. Perumal, and U.K. Reddy. 2014. Physiological differences among sorghum (Sorghum bicolor L. Moench) genotypes under high temperature stress. Environ. Exp. Bot. 100:43–54. doi:10.1016/j.envexpbot.2013.11.013

Fahn, A. 1964. Some anatomical adaptations in desert plants. Phytomorphology 14:93–102.

Field, C.B., V. Barros, T.F. Stocker, D. Qin, D.J. Dokken, and K.L.M.D. Ebi, editors. 2012. Managing the risks of extreme events and disasters to advance climate change adaptation. A special report of working groups I and II of the Intergovernmental Panel on Climate Change. Intergovernmental Panel on Climate Change. Cambridge Univ. Press, Cambridge, UK. p. 582.

Gerik, T., B.W. Bean, and R. Vanderlip. 2003. Sorghum growth and development. Agric. Communications. The Texas A&M University, College Station, TX.

Gholipoor, M., P.V.V. Prasad, R.N. Mutava, and T.R. Sinclair. 2010. Genetic variability of transpiration response to vapor pressure deficit among sorghum genotypes. Field Crops Res. 119:85–90. doi:10.1016/j.fcr.2010.06.018

Hall, A.E. 2000. Crop response to environment. CRC Press LLC, New York. doi:10.1201/9781420041088

Hatfield, J.L., K. Boote, P. Fay, L. Hahn, C. Izaurralde, B.A. Kimball, T. Mader, J. Morgan, D. Ort, W. Polley, A. Thomson, and D. Wolfe. 2008. Agriculture In: P. Backlund, A. Janetos, D. Schimel, J. Hatfield, K. Boote, P. Fay, L. Hahn, C. Izaurralde, B.A. Kimball, T. Mader, J. Morgan, D. Ort, W. Polley, A. Thomson, D. Wolfe, M. Ryan, S. Archer, R. Birdsey, C. Dahm, L. Heath, J. Hicke, D. Hollinger, T. Huxman, G. Okin, R. Oren, J. Randerson, W. Schlesinger, D. Lettenmaier, D. Major, L. Poff, S. Running, L. Hansen, D. Inouye, B.P. Kelly, L Meyerson, B. Peterson, R. Shaw. The effects of climate change on agriculture, land resources, water resources, and biodiversity. A report by the U.S. Climate Change Science Program and the Subcommittee on Global Change Research, Washington, D.C. p. 21–74.

Hatfield, J.L., K.J. Boote, B.A. Kimball, L.H. Ziska, R.C. Izaurralde, D. Ort, A.M. Thomson, and D. Wolfe. 2011. Climate impacts on agriculture: Implications for crop production. Agron. J. 103:351–370. doi:10.2134/agronj2010.0303

Intergovernmental Panel on Climate Change. 2013. Summary for policymakers. In: T.F. Stocker, G.K. Qin, M. Plattner, S.K. Tignor, J. Allen, A. Boschung, A. Nauels, Y. Xia, V. Bex, and P.M. Midgley, editors, Climate change 2013: The physical science basis. Contribution of working group I to V assessment report of the intergovernmental panel on climate change. Cambridge Univ. Press, Cambridge, UK.

Ishimaru, T., H. Hirabayashi, M. Ida, T. Takai, Y.A. San-Oh, S. Yoshinaga, I. Ando, T. Ogawa, and M. Kondo. 2010. A genetic resource for early-morning flowering trait of wild rice Oryza officinalis to mitigate high temperature induced spikelet sterility at anthesis. Ann. Bot. (Lond.) 106:515–520. doi:10.1093/aob/mcq124

Jagtap, V., S. Bhargava, P. Streb, and J. Feierabend. 1998. Comparative effect of water, heat and light stresses on photosynthetic reactions in Sorghum bicolor (L.) Moench. J. Exp. Bot. 49:1715–1721.

Jain, M., P.S. Chourey, K.J. Boote, and L.H. Allen, Jr. 2010. Short-term high temperature growth conditions during vegetative-to-reproductive phase transition irreversibly compromise cell wall invertase-mediated sucrose catalysis and microspore meiosis in grain sorghum (Sorghum bicolor). J. Plant Physiol. 167:578–582. doi:10.1016/j.jplph.2009.11.007

Jain, M., P.V.V. Prasad, K.J. Boote, A.L. Hartwell, and P.S. Chourey. 2007. Effects of season-long high temperature growth conditions on sugar-to-starch metabolism in developing microspores of grain sorghum (Sorghum bicolor L. Moench). Planta 227:67–79. doi:10.1007/s00425-007-0595-y

Julia, C., and M. Dingkuhn. 2012. Variation in time of day of anthesis in rice in different climatic environments. Eur. J. Agron. 43:166–174. doi:10.1016/j.eja.2012.06.007

Kassam, A.H., and D.J. Andrew. 1975. Effects of sowing dates on growth, development and yield of photosensitive sorghum at Samaru, Northern Nigeria. Exp. Agric. 11:227–240. doi:10.1017/S0014479700006761

Lin, T., and A.H. Markhart, III. 1990. Temperature effects on mitochondrial respiration in Phaseolus acutffolius A. Gray and Phaseolus vulgaris L. Plant Physiol. 94:54–58. doi:10.1104/pp.94.1.54

Mace, E., V. Singh, E. van Oosterom, G. Hammer, C. Hunt, and D. Jordan. 2012. QTL for nodal root angle in sorghum (Sorghum bicolor L. Moench) co-locate with QTL for traits associated with drought adaptation. Theor. Appl. Genet. 124:97–109. doi:10.1007/s00122-011-1690-9

Maiti, R.K. 1996. Sorghum science. Science Publication, Lebanon, NH.

Matsuoka, M., R.T. Furbank, H. Fukayama, and M. Miyao. 2001. Molecular engineering of C4 photosynthesis. Annu. Rev. Plant Physiol. Plant Mol. Biol. 52:297–314. doi:10.1146/annurev.arplant.52.1.297

Maxwell, K., and G.N. Johnson. 2000. Chlorophyll fluorescence: A practical guide. J. Exp. Bot. 51:659–668. doi:10.1093/jexbot/51.345.659

Metz, B., O. Davidson, P. Bosch, R. Dave, and L. Meyer, editors. 2007. Climate change 2007: Mitigation. Contribution of working group III to the fourth assessment report of the intergovernmental panel on climate change. Intergovernmental Panel on Climate Change. Cambridge Univ. Press, Cambridge, UK.

Mutava, R.N. 2012. Evaluation of sorghum genotypes for variation in canopy temperature and drought tolerance. Ph. D thesis. Kansas State University, Manhattan, KS.

Mutava, R.N., P.V.V. Prasad, M.R. Tuinstra, K.D. Kofoid, and J. Yu. 2011. Characterization of sorghum genotypes for traits related to drought tolerance. Field Crops Res. 123:10–18. doi:10.1016/j.fcr.2011.04.006

Nguyen, C.T., V. Singh, E.J. van Oosterom, S.C. Chapman, D.R. Jordan, and G.L. Hammer. 2013. Genetic variability in high temperature effects on seed-set in sorghum. Funct. Plant Biol. 40:439–448. doi:10.1071/FP12264

North, G.B., and P.S. Nobel. 1995. Hydraulic conductivity of concentric root tissues of Agave deserti Engelm. under wet and drying conditions. New Phytol. 130:47–57. doi:10.1111/j.1469-8137.1995.tb01813.x

Passioura, J.B. 1972. The effect of root geometry on the yield of wheat growing on stored water. Aust. J. Agric. Res. 23:745–752. doi:10.1071/AR9720745

Phan, T.T.T., Y. Ishibashi, M. Miyazaki, H.T. Tran, K. Okamura, S. Tanaka, J. Nakamura, T. Yuasa, and M. Iwaya-Inoue. 2013. High temperature-induced repression of the rice sucrose transporter (OsSUT1) and starch synthesis-related genes in sink and source organs at milky ripening stage causes chalky grains. J. Agron. Crop Sci. 199:178–188. doi:10.1111/jac.12006

Pinto, R.S., M.P. Reynolds, K.L. Mathews, C.L. McIntyre, J.-J. Olivares-Villegas, and S.C. Chapman. 2010. Heat and drought adaptive QTL in a wheat population designed to minimize confounding agronomic effects. Theor. Appl. Genet. 121:1001–1021. doi:10.1007/s00122-010-1351-4

Porch, T.G., and M. Jahn. 2001. Effects of high-temperature stress on microsporogenesis in heat-sensitive and heat-tolerant genotypes of Phaseolus vulgaris. Plant Cell Environ. 24:723–731. doi:10.1046/j.1365-3040.2001.00716.x

Prasad, P.V.V., and M. Djanaguiraman. 2011. High night temperature decreases leaf photosynthesis and pollen function in grain sorghum. Funct. Plant Biol. 38:993–1003. doi:10.1071/FP11035

Prasad, P.V.V., K.J. Boote, and L.H. Allen, Jr. 2006. Adverse high temperature effects on pollen viability, seed-set, seed yield and harvest index of grain-sorghum [Sorghum bicolor (L.) Moench] are more severe at elevated carbon dioxide due to high tissue temperature. Agric. For. Meteorol. 139:237–251. doi:10.1016/j.agrformet.2006.07.003

Prasad, P.V.V., P.Q. Craufurd, and R.J. Summerfield. 1999. Sensitivity of peanut to timing of heat stress during reproductive development. Crop Sci. 39:1352–1357. doi:10.2135/cropsci1999.3951352x

Prasad, P.V.V., M. Djanaguiraman, R. Perumal, and I.A. Ciampitti. 2015. Impact of high temperature stress on floret fertility and individual grain weight of grain sorghum: Sensitive stages and thresholds for temperature and duration. Front. Plant Sci. 6:1–11. doi:10.3389/fpls.2015.00820

Prasad, P.V.V., R. Bheemanahalli, and S.V.K. Jagadish. 2017. Field crops and the fear of heat stress - Opportunities, challenges, and future directions. Field Crops Res. 200:114–121.

Prasad, P.V.V., S.A. Staggenborg, and Z. Ristic. 2008a. Impacts of drought and/or heat stress on physiological, developmental, growth, and yield processes of crop plants In: L.R. Ahuja, V.R. Reddy, S.A. Saseendran, and Q. Yu, editors, Response of crops to limited water: Understanding and modelling water stress effects on plant growth processes. Advances in Agricultural Systems Modelling Series 1. ASA, CSSA, SSSA, Madison, WI. p. 301–355. doi:10.2134/advagricsystmodel1.c11

Prasad, P.V.V., S.R. Pisipati, R.N. Mutava, and M.R. Tuinstra. 2008b. Sensitivity of grain sorghum to high temperature stress during reproductive development. Crop Sci. 48:1911–1917. doi:10.2135/cropsci2008.01.0036

Reynolds, M.P., M. Balota, M.I.B. Delgado, I. Amani, and R.A. Fisher. 1994. Physiological and morphological traits associated with spring wheat yield under hot irrigated conditions. Aust. J. Plant Physiol. 21:717–730. doi:10.1071/PP9940717

Richards, R.A., and J.B. Passioura. 1989. A breeding program to reduce the diameter of the major xylem vessel in the seminal roots of wheat and its effect on grain yield in rainfed environments. Aust. J. Agric. Res. 40:943–950. doi:10.1071/AR9890943

Rosenow, D.T., J.E. Quisenberry, C.W. Wendt, and L.E. Clark. 1983. Drought tolerant sorghum and cotton germplasm. Agric. Water Manage. 7:207–222. doi:10.1016/0378-3774(83)90084-7

Sanchez, A.C., P.K. Subudhi, D.T. Rosenow, and H.T. Nguyen. 2002. Mapping QTLs associated with drought resistance in sorghum (Sorghum bicolor L. Moench). Plant Mol. Biol. 48:713–726. doi:10.1023/A:1014894130270

Sharkey, T.D. 2005. Effects of moderate heat stress on photosynthesis: Importance of thylakoid reactions, rubisco deactivation, reactive oxygen species, and thermotolerance provided by isoprene. Plant Cell Environ. 28:269–277.

Shekoofa, A., M. Balota, and T.R. Sinclair. 2014. Limited-transpiration trait evaluated in growth chamber and field for sorghum genotypes. Environ. Exp. Bot. 99:175–179. doi:10.1016/j.envexpbot.2013.11.018

Singh, V., C.T. Nguyen, E.J. van Oosterom, S.C. Chapman, D.R. Jordan, and G.L. Hammer. 2015. Sorghum genotypes differ in high temperature responses for seed set. Field Crops Res. 171:32–40. doi:10.1016/j.fcr.2014.11.003

Singh, V., E.J. van Oosterom, D.R. Jordan, C.H. Hunt, and G.L. Hammer. 2011. Genetic variability and control of nodal root angle in sorghum. Crop Sci. 51:2011–2020. doi:10.2135/cropsci2011.01.0038

Singh, V., E.J. van Oosterom, D.R. Jordan, C.D. Messina, M. Cooper, and G.L. Hammer. 2010. Morphological and architectural development of root systems in sorghum and maize. Plant Soil 333:287–299. doi:10.1007/s11104-010-0343-0

Soman, P., and J.P. Peacock. 1985. A laboratory technique to screen seedling emergence of sorghum and pearl millet at high soil temperature. Exp. Agric. 21:335–341. doi:10.1017/S0014479700013168

Stout, D., and G.M. Simpson. 1978. Drought resistance of Sorghum bicolor. I. Drought avoidance mechanisms related to leaf water status. Can. J. Plant Sci. 58:213–224. doi:10.4141/cjps78-031

Subudhi, P.K., D.T. Rosenow, and H.T. Nguyen. 2000. Quantitative trait loci for the stay-green trait in sorghum (Sorghum bicolor L. Moench): Consistency across genetic backgrounds and environments. Theor. Appl. Genet. 101:733–741. doi:10.1007/s001220051538

Sullivan, C.Y. 1972. Mechanism of heat and drought resistance in grain sorghum and methods of measurement. In: Sorghum in the Seventies. Eds. Rao, N.G.P. and House L.R., Oxford and IBH Publishing Co., New Delhi, India.

Tack, J., J. Lingenfelser, and S.V.K. Jagadish. 2017. Disaggregating sorghum yield reductions under warming scenarios exposes narrow genetic diversity in US breeding programs. Proc. Natl. Acad. Sci. USA 114:9296–9301.

Talukder, S.K., M.A. Babar, K. Vijayalakshmi, J. Poland, P.V.V. Prasad, R. Bowden, and A. Fritz. 2014. Mapping QTL for the traits associated with heat tolerance in wheat (Triticum aestivum L.). BMC Genet. 15:97. doi:10.1186/s12863-014-0097-4

Thomas, H., and C.J. Howarth. 2000. Five ways to stay green. J. Exp. Bot. 51:329–337. doi:10.1093/jexbot/51.suppl_1.329

Tjoelker, M.G., P.B. Reich, and J. Oleksyn. 1999. Changes in leaf nitrogen and carbohydrates underlie temperature and CO2 acclimation of dark respiration in five boreal tree species. Plant Cell Environ. 22:767–778. doi:10.1046/j.1365-3040.1999.00435.x

White, J.W., G. Hoogenboom, and L.A. Hunt. 2005. A structured procedure for assessing how crop models respond to temperature. Agron. J. 97(2):426–439. doi:10.2134/agronj2005.0426

Yan, K., P. Chen, H. Shao, C. Shao, S. Zhao, and M. Brestic. 2013. Dissection of photosynthetic electron transport process in sweet sorghum under heat stress. PLoS One 8(5):E62100. doi:10.1371/journal.pone.0062100

Yan, K., P. Chen, H. Shao, L. Zhang, and G. Xu. 2011. Effects of short-term high temperature on photosynthesis and photosystem ii performance in sorghum. J. Agron. Crop Sci. 197:400–408. doi:10.1111/j.1439-037X.2011.00469.x

Yang, Z., E.J. van Oosterom, D.R. Jordan, and G.L. Hammer. 2009. Preanthesis ovary development determines genotypic differences in potential kernel weight in sorghum. J. Exp. Bot. 60:1399–1408. doi:10.1093/jxb/erp019

Young, L.W., R.W. Wilen, and P.C. Bonham-Smith. 2004. High temperature stress of Brassica napus during flowering reduces micro- and megagametophyte fertility, induces fruit abortion, and disrupts seed production. J. Exp. Bot. 55:485–495. doi:10.1093/jxb/erh038

Water-Use Efficiency

Vincent Vadez

Abstract

Much has been told about transpiration efficiency (TE) and much more is still needed in an agriculture era marked by water scarcity, not only from the standpoint of dry climate but also from competition for water sources in agriculture. Therefore, research is needed to improve the water-use efficiency (WUE) of drought-tolerant sorghum [*Sorghum bicolor* (L.) Moench]. This paper presents the theory around WUE and TE. It covers the methods used to measure WUE that have been used in sorghum and reports the range of genetic variation found including in genetic material introgressed with stay-green quantitative trait loci (QTL). The last part explores the main avenues currently being investigated to increase WUE in sorghum.

As a C4 crop equipped with a specialized means of concentrating carbon dioxide (CO_2) in the mesophyll cells, sorghum physiologically has a higher efficiency of trading C for water at the stomata level. However, sorghum grows in environments where water is limited, and improving WUE much further than its specie baseline is an important avenue of research.

Water-use efficiency can be defined at different levels. At a large scale, community to watershed level, it can be defined as the economic yield (often grain) divided by the water applied to the crop in form of rain or irrigation (in kg mm^{-1}). In this case, runoff and deep drainage are also accounted in the water used by the crops so that this definition provides only a crude assessment of WUE. At the field level, a finer definition is the crop yield (either grain or grain plus stover yield) divided by the crop evapotranspiration. The units are also in kilogram per millimeter (or kg m^{-3}), but in this case only the water evaporated from the soil and that transpired by the crop are accounted for in the calculation of the water input. At the plant level, the WUE definition gives way to TE, that is, the ratio of biomass produced (vegetative or grain plus stover yield) divided by the water transpired (in g biomass kg^{-1} water transpired). Transpiration efficiency is the component of WUE that encapsulates much of the genetic determinacy of TE. Finally, at the leaf level, TE is defined by the ratio of the photosynthetic rate divided by the leaf conductance (A/E), and this is a transient measurement often called intrinsic TE. In this review, we will explore only TE (biomass produced per unit of water

---

Abbreviations: HI, harvest index; QTL, quantitative trait loci; RUE, radiation use efficiency; TE, transpiration efficiency; VPD, vapor pressure deficit; WUE, water-use efficiency.

V. Vadez, ICRISAT, Crop Physiology Laboratory, Patancheru 502324, Greater Hyderabad, Telangana, India (v.vadez@cgiar.org).

doi:10.2134/agronmonogr58.2014.0066

 Sorghum: State of the Art and Future Perspectives, Ignacio Ciampitti and Vara Prasad, editors. Agronomy Monograph 58.

transpired), which can be referred to either as WUE or TE, and then explore how TE or WUE can be improved genetically in a crop like sorghum.

In a second section, previous research on TE in sorghum and methods to assess TE will be discussed, looking at reports on the range of variation for TE or WUE and analyzing possible links with important phenotypes like staygreen. It is indeed important to assess genetic variation for TE in sorghum, but it is also quite critical to do that with an adequate method. his section will present one method that allows a robust assessment of TE and a comparison to agronomic characteristics. Genotypic differences for TE have been reported in sorghum under well-watered conditions (Hammer et al. 1997; Xin et al. 2009). Other studies have looked at TE under both fully irrigated and water stress conditions (Donatelli et al. 1992; Balota et al. 2008). Also, except Balota et al. (2008), TE has been measured over relatively limited periods of time.

Of course, TE is an integrated measurement that depends not only on photosynthetic efficiency and control of stomata opening (Condon et al., 2002) but also on a component of TE that was until now considered as an environmental factor, that is, vapor pressure deficit (VPD) at the leaf level, and for which we have shown evidence of genetic factors potentially affecting it (see a review in Vadez et al., 2014). This third section will then review the research at stake in these different aspects. In a recent review, we dealt with the stomatal control and have indicated that there were two ways of improving TE in crops by means of stomata control (Vadez et al., 2014). The first one was about restricting transpiration under high VPD conditions, as this would decrease the effective VPD at the leaf level, integrated during the course of an entire day, if there was a restriction in transpiration at the times of the day with highest VPD conditions as shown from a modeling study (Sinclair et al., 2005). The second way would be to have plants restricting stomatal conductance at high level of soil moisture. The indirect consequence of this trait would be that plants equipped with this trait would likely close stomata, at least partially, during the times of highest evaporative demand, which would also alter the effective VPD at the leaf level. The last part of that section will then explore the possibility to enhance the photosynthetic efficiency of sorghum. In this part, we will compare photosynthetic efficiency of sorghum vs. that of related cereals {pearl millet [*Pennisetum glaucum* (L.) R. Br.] and maize (*Zea mays* L.)}. Our hypothesis is that there is a great scope for improvement of radiation use efficiency (RUE) in sorghum, which would likely also lead to increases in TE.

Theoretical Basis for Transpiration Efficiency: Transpiration Efficiency Versus Water-Use Efficiency

The definition of intrinsic TE from (Condon et al., 2002) is as follows:

$$TE = 0.6C_a(1 - C_i/C_a)/(W_i - W_a)$$

where C_i and C_a are the CO_2 concentrations in the stomatal chamber and the ambient atmosphere, respectively, and W_i and W_a are the vapor pressures in the stomatal chamber and ambient atmosphere, respectively. The first means of increasing TE is to reduce the C_i/C_a ratio, which can be achieved by (i) a high photosynthetic efficiency or (ii) a low stomatal conductance, since C_a can be considered constant for a given crop. The numerator of this equation then represents a term that reflects the genetics of the crop. On the contrary, the denominator is a representation of the gradients of vapor pressure prevailing between the atmosphere and the stomata chamber and is then a factor that represents the environment. Therefore, the TE ratio depends on both genetic and an environmental components. Based on earlier works (Bierhuizen and Slatyer, 1965; Tanner and Sinclair, 1983), Sinclair and colleagues have reformulated it in the following ratio:

$$TE = k_d/(e^*_a - e)_d$$

where the denominator term in parentheses reflects the VPD, and k_d is a coefficient that reflects the crop and the CO_2 concentration in the stomatal chamber (Sinclair et al., 1984; Sinclair, 2012). For C4 crops like sorghum, the C_i value can be kept at lower values than in C3 species thanks to an active process of CO_2 transport and fixation onto organic acids. Recent research shows the involvement of aquaporins in the transport of CO_2 in the leaves, and there could be an avenue of research trying to explore possible relationships between a low C_i content and activity of CO_2–specific aquaporins (Kaldenhoff, 2012). In any case, this definition shows clearly that both genetics and environmental effects can alter TE. However, much of the research on the genetics of TE over the last three decades has disregarded the aspects of TE that concern this environmental factor, instead concentrating on aspects related to CO_2 in the leaf. This was despite the fact that model prediction on the effect of maximizing transpiration rate at the time of the day with highest VPD would have a positive effect on yield of sorghum via an improvement in TE (Sinclair et al., 2005). We will analyze, below, which genetic traits could alter this environmental factor.

Methods to Assess Transpiration Efficiency and Genetic Variation Identified

Methods

A lysimetric system is used on a routine basis for assessing TE. In the case of sorghum, plants are grown in PVC tubes of 25 cm diam. and 2.0 m length filled with Alfisol. A PVC end plate is placed on top of four screws at the bottom of the cylinders 3 cm from the very bottom to prevent the soil from seeping through and allowing water drainage. The lysimeters were initially filled with Alfisol collected from the ICRISAT farm and sieved to particles <1 cm. After each crop, the soil is tilled superficially with sickles and soil topped up if needed. This allowed us to create a soil profile that is undisturbed from previous cropping, except for minimum tillage of the surface, and which represents a real soil profile. Transpiration from these lysimeters is obtained from consecutive weightings. The top of the cylinders is tied with a metal collar. Weighing of the cylinders is done by lifting them with a block–chain pulley and an S-type load cell (Mettler-Toledo International Inc). The scale (200 kg capacity) allows repeated measurements with an accuracy of 20 g on each weighing. The lysimeters are separated from one another by a distance of 3 to 5 cm. Thus, the sorghum crop was planted at a density of ~10 plants m^{-2}, a plant population similar to typical field plantings at ICRISAT (row-to-row distance of 60 cm and plant-to-plant spacing of 15 cm). This allows us to accurately assess the water extraction pattern of a crop cultivated in conditions similar to the field and then to have a complete profile of the water extraction along with a highly relevant agronomic assessment.

Another method has been developed (Hammer et al., 1997) in which the plants were grown in a system of double pots with one pot serving as a water tank to constantly keep the plants in a soil at field capacity. The simplicity of the system was then in counting how much water was used to replenish the storage tank. Other gravimetric methods that have been used for pearl millet both in small pots (Donatelli et al., 1992; Xin et al., 2009) and in larger containers (Payne et al., 1992). In sorghum, except for one study (Balota et al., 2008), these gravimetric measurements were relatively short term and used a limited number of genotypes (except Xin et al., 2009), compared to long-term assessment in lysimeters and natural conditions (Vadez et al., 2011a,b). Transpiration efficiency based on leaf gas exchange measurements has also been performed (Peng and Krieg 1992; Kapanigowda et al., 2014). These measurements provide an instantaneous assessment of WUE based on the measurement of photosynthetic activity together with the transpiration assessment. While these measurements allow us to carefully look into both factors potentially affecting the CO_2 concentration in the stomata chamber (C_i), these methods have several issues to be aware of: (i) they are transient, and the measurement depends much on the environmental conditions prevailing at the time of measurement; (ii) they can vary from spot to spot sampled on the leaf, which raises the question as how to extrapolate the measurement to the entire crop canopy (scaling-up issue); (iii) they are extremely time consuming. Therefore, this latter method can only be used for detailed studies on TE (e.g., Masle et al., 2005).

Genetic Variation in Transpiration Efficiency and Water-use Efficiency

The evaluation of a portion (~40%) of the reference collection of sorghum germplasm showed a two-fold range of variation for TE (~3 to 6 g kg^{-1} water transpired with an average VPD during the season of ~2.0 kPa) under water stress, and values were ~10% higher than under fully irrigated conditions (Vadez et al., 2011a). This range of variation, ~100%, was higher than the 20% differences (Donatelli et al. 1992), 25% differences (Hammer et al. 1997), or 50% differences (Balota et al., 2008) reported earlier. In addition, the low CV percentage of the lysimetric measurements (13.6%) illustrated the high quality of data from the lysimetric method (Vadez et al., 2011a). In these trials, the effect of TE on yield was assessed. After removing the proportion of yield differences explained by the harvest index (HI), there was still a substantial yield variation unexplained, and this was highly significantly related to TE. Also, the absence of relationship between TE and total water use, tested in a large and representative set of sorghum germplasm, showed it was possible to find germplasm capable of exhibiting both high water extraction and high TE in contrast to previous speculation that TE and water use ought to be negatively related (Blum, 2005). This is in agreement with an earlier report (Peng and Krieg, 1992). An interesting finding in the assessment of the sorghum reference collection was the significantly lower TE in the germplasm of Guinea race than in the other races (~1 g kg^{-1}). By contrast, germplasm from the Caudatum and Durra had overall higher values of TE than other landraces. Some of these variants are currently being used in the development of backcross-nested association mapping populations, some of them targeting Guinea recurrent parents. These results also open the prospect of screening more entries from the Durra and Caudatum races for potential additional variants.

In relation to possible relationship between TE and the staygreen trait, there has been limited investigation on this. Tx7078, a genotype assumed to be tolerant to preflowering drought (Tuinstra et al., 1998), had low TE. Borrell and Hammer (2000) showed higher TE in one line having B35 as its source of staygreen. They also reported that differences among sorghum genotypes in biomass production under terminal water deficit were associated with variation in transpiration and TE for both the A35 and RQL12 sources of staygreen. In that study, A35 (the male-sterile line counterpart of staygreen donor B35) increased TE relative to AQL39 in two out of three genetic backgrounds. In our lysimetric assessments, StgB QTL introgression lines also increased TE in the R16 background but not in S35 background, and this was likely related to the fact that R16 had relatively low TE, whereas TE was relatively high in S35 compared with germplasm from the reference collection. There was also a trend for higher TE in three out of four StgB QTL introgression lines in R16 background. In the staygreen introgression lines, we also had a significant positive relationship between TE and the residual yields that were not explained by the HI (Vadez et al., 2011b). The slope of that regression provided an estimate of the yield gains per unit of TE (g kg^{-1}), and these amounted to 3 g grain $plant^{-1}$, which could be extrapolated to an additional 330 kg ha^{-1} of grain for each unit of TE or ~16% of the existing mean yield under terminal stress. Therefore, staygreen expression has some relationship with an improvement in TE. In any case, there are now clear evidences that high TE is not inversely related to low total water use (which would lead to low yield potential) (Vadez et al., 2014; Peng and Krieg 1992), which goes against what was earlier asserted (Blum, 2005).

Efficiency of Water Use versus Water-Use Efficiency or Transpiration Efficiency

At the canopy level, plants can curb water use by reducing the leaf area being developed or by reducing the conductance of the canopy (Kholová et al., 2010a,b). The latter has been shown to be one of the reasons for the expression of the staygreen phenotype in *Miscanthus* (Clifton-Brown et al., 2002). Hammer (2006) used crop simulation modeling to evaluate the effect of a reduced leaf area on the yield across locations receiving different amounts of rainfall and showed that a leaf area index of 1 would bring a substantial yield benefit in situations where rainfall is <75 mm. A reduction in the leaf area at anthesis in staygreen genotypes could be a way of reducing preanthesis water losses. It has been shown also that higher tolerance could be related to a slower rate of soil water use (Kirkham, 1988). It is exemplified that the success of the sorghum crop under limited water supply is to have sufficient water available during the grain-filling period, which comes from different genetic or agronomic alterations that reduce the amount of preanthesis water use. This has also been reported by others (Araus et al., 2003; Siddique et al., 2001).

We have termed this *efficiency of water use*, and this bring a dimensions of time in the traditional thinking around WUE. Thinking back of the Passioura equation (Yield = $T \times$ TE $\times$ HI; Passioura, 1977), this would then indicate that the T component would not be linear all along the cropping cycle but would have time-bound subcomponents having a higher weight with higher importance of the water used during the grain filling (Wasson et al., 2012). For instance, we have shown that the WUE of the water transpired during the grain-filling period under water stress was between 37 and 45 kg ha^{-1} mm^{-1} in pearl millet (Vadez et al., 2013) or ~40 kg ha^{-1} mm^{-1} in chickpea (*Cicer arietinum* L.) (Zaman-Allah et al., 2011a). These results are in full agreement with a similar range published earlier in wheat (*Triticum aestivum* L.) (55 kg ha^{-1} mm^{-1} [Manschadi et al., 2006] and 59 kg ha^{-1} mm^{-1} [Kirkegaard et al., 2007]). The reasons for these high WUE values of water used in certain periods of time are not related to any physiological difference in how water is traded off for C. It is simply related to the fact that (i) water extracted during the grain-filling period comes from deep layers, from which the component of evaporation is virtually nil, and (ii) water is used at a stage when vegetative growth has stopped and then supports photosynthesis purely for the sake of grain growth. In other words, gain weight during that stage is not discounted by any HI factor.

Prospects to Further Improve Transpiration Efficiency in Sorghum

Control of Stomata Opening

Here, we come back to the equation of Sinclair et al. (1984), where $G_d/T = k_d/(e^*_a - e)_d$, with G_d representing the daily increase in biomass and TE showing an inverse relationship with the VPD faced by the plants. The term $(e^*_a - e)_d$ represents the difference between the saturated vapor pressure (e^*_a) and the ambient one (e). This term is calculated as a mean over an entire day and it assumes that all genotypes that would be tested would be following the same type of integration. This leads then to the possibility of obtaining altered values if there are genotypic variations in the transpiration profile throughout the day (Sinclair, 2012). Indeed, if there are high VPD periods of the day when transpiration is restricted, it means also there is less weight of those VPD values on the mean daily VPD. The search for possible genetic variation in this came from a simulation study where restricting the maximum transpiration rate, putatively under high VPD conditions, increased TE and yield of sorghum (Sinclair et al., 2005). The sensitivity of the stomatal aperture has been known for a long time (Turner et al., 1984), but genotypic variation has shown up only recently in different species such as soybean [*Glycine max* (L.) Merr.] (Fletcher et al., 2007), chickpea (Zaman-Allah et al., 2011b), cowpea [*Vigna unguiculata* (L.) Walp.] (Belko et al., 2012), peanut (*Arachis hypogaea* L.) (Devi et al., 2010), sorghum (Gholipoor et al., 2010), pearl millet (Kholová et al., 2010b), and wheat (Schoppach and Sadok, 2012). The discovery of these large genetic differences therefore reoriented research on TE, providing new and exciting insights into ways to alter the negative influence of VPD on TE. This is probably the main avenue to increasing TE from the angle of controlling stomata aperture. Of course, limiting stomata opening can also lead to reductions in C fixation, although recent evidence showed that within certain limits there would be only a limited or even no trade off in soybean (Gilbert et al., 2011a, 2011b). Another means of altering that integrated VPD value may come from stomata closure at fairly high soil moisture content. Indeed this would lead to plants having to restrict transpiration at the time of highest transpiration demand, that is, when VPD is high, therefore having a similar effect than restricting transpiration under high VPD conditions.

Increasing Radiation Use Efficiency?

In the section above, we dealt with genetic means of altering a component of the TE definition that has so far been seen as a purely environmental characteristic. On the other hand, the k_d coefficient reflects the C_i/C_a ratio, and as we have seen above, increasing TE implies minimizing this ratio, which is akin to reducing the C_i component, that is, the CO_2 concentration in the stomata chamber. Traits restricting stomata opening, like those described in the previous section, would also bring this benefit. Much research has also been undertaken so far to decrease C_i by means of improving the photosynthetic efficiency (e.g., Hubick et al., 1986; Wright et al., 1994). Therefore, the question is whether in a crop like sorghum, improving RUE would indirectly lead to increasing TE by means of increasing the CO_2 assimilation rate per unit surface area. For instance, it was shown in sorghum that single dwarfing gene line (taller) had higher RUE than triple dwarfing line (shorter) and, overall, that taller genotypes have higher yield (Hammer et al., 2010). The research on RUE will need to address possible differences in the efficiency of the photosynthesis, that is, interaction between height and light interception in the canopy where differences in leaf size and angle may have a major role to play, and possible difference in the root to shoot allometric coefficients.

Conclusions

Despite the intrinsic adaptation of sorghum to dry environments, further improvements in its productivity under such environments remain possible because large variation for WUE has been reported in germplasm and can be used in breeding. A large number of methods are now available to measure WUE accurately and at a large scale, especially those using lysimeters for a crop lifelong assessment of water use and biomass production. High WUE relates directly to yield, especially under water-limited conditions, and the physiological mechanisms responsible for large WUE difference are now getting deciphered. One of those is the capacity of certain genotypes to control stomata aperture under high VPD, which both increases WUE and saves water to support grain filling, a key crop stage. Genetic variation for RUE in sorghum remains an unexplored avenue to increase WUE in sorghum that merits attention.

References

Araus, J.L., D. Villegas, N. Aparicio, L.F.G. del Moral, S. El Hani, Y. Rharrabti, J.P. Ferrio, and C. Royo. 2003. Environmental factors determining carbon isotope discrimination and yield in durum wheat under Mediterranean conditions. Crop Sci. 43:170–180. doi:10.2135/cropsci2003.1700

Balota, M., W.A. Payne, W. Rooney, and D. Rosenow. 2008. Gas exchange and transpiration ratio in sorghum. Crop Sci. 48:2361–2371. doi:10.2135/cropsci2008.01.0051

Belko, N., M. Zaman-Allah, N. Cisse, N.N. Diop, G. Zombre, J.D. Ehlers, and V. Vadez. 2012. Lower soil moisture threshold for transpiration decline under water deficit correlates with lower canopy conductance and higher transpiration efficiency in drought-tolerant cowpea. Funct. Plant Biol. 39:306–322. doi:10.1071/FP11282

Bierhuizen, J.F., and R.O. Slatyer. 1965. Effect of atmospheric concentration of water vapor and CO_2 in determining transpiration-photosynthesis relationships of cotton leaves. Agric. Meteorol. 2:259–270. doi:10.1016/0002-1571(65)90012-9

Blum, A. 2005. Drought resistance, water use efficiency, and yield potential: Are they compatible, dissonant, or mutually exclusive. Aust. J. Agric. Res. 56:1159–1168. doi:10.1071/AR05069

Borrell, A.K., and G.L. Hammer. 2000. Nitrogen dynamics and the physiological basis of stay-green in sorghum. Crop Sci. 40:1295–1307. doi:10.2135/cropsci2000.4051295x

Clifton-Brown, J.C., I. Lewandowski, F. Bangerth, and M.B. Jones. 2002. Comparative response to water stress in staygreen, rapid- and slow senescing genotypes of the biomass crop Miscanthus. New Phytol. 154:335–345. doi:10.1046/j.1469-8137.2002.00381.x

Condon, A.G., R.A. Richards, G.J. Rebetzke, and G.D. Farquhar. 2002. Improving intrinsic water-use efficiency and crop yield. Crop Sci. 42:122–131. doi:10.2135/cropsci2002.0122

Devi, M.J., T.R. Sinclair, and V. Vadez. 2010. Genotypic variation in peanut for transpiration response to vapor pressure deficit. Crop Sci. 50:191–196. doi:10.2135/cropsci2009.04.0220

Donatelli, M., G.L. Hammer, and R.L. Vanderlip. 1992. Genotype and water limitation effects on phenology, growth, and transpiration efficiency in grain sorghum. Crop Sci. 32:781–786. doi:10.2135/cropsci1992.0011183X003200030041x

Fletcher, A.L., T.R. Sinclair, and L.H. Allen. 2007. Transpiration responses to vapor pressure deficit in well watered 'slow-wilting' and commercial soybean. Environ. Exp. Bot. 61:145–151. doi:10.1016/j.envexpbot.2007.05.004

Gholipoor, M., P.V.V. Prasad, R.N. Mutava, and T.R. Sinclair. 2010. Genetic variability of transpiration response to vapor pressure deficit among sorghum genotypes. Field Crops Res. 119:85–90. doi:10.1016/j.fcr.2010.06.018

Gilbert, M.E., N.M. Holbrook, M.A. Zwieniecki, W. Sadok, and T.R. Sinclair. 2011a. Field confirmation of genetic variation in soybean transpiration response to vapor pressure deficit and photosynthetic compensation. Field Crops Res. 124:85–92. doi:10.1016/j.fcr.2011.06.011

Gilbert, M.E., M.A. Zwieniecki, and N.M. Holbrook. 2011b. Independent variation in photosynthetic capacity and stomatal conductance leads to differences in intrinsic water use efficiency in 11 soybean genotypes before and during mild drought. J. Exp. Bot. 62:2875–2887. doi:10.1093/jxb/erq461

Hammer, G.L. 2006. Pathways to prosperity: Breaking the yield barrier in sorghum. In: A.K. Borrell and D.R. Jordan, editors, 5th Australian Sorghum Conference. Gold Coast, QLD, Australia, 30 Jan–2 Feb 2006. Range Media Pty Ltd, Harlaxton QLD, Australia. p. 1–19.

Hammer, G.L., G.D. Farquhar, and I.J. Broad. 1997. On the extent of genetic variation for transpiration efficiency in sorghum. Aust. J. Agric. Res. 48:649–655. doi:10.1071/A96111

Hammer, G.L., E. van Oosterom, G. McLean, S.C. Chapman, I. Broad, P. Harland, and R.C. Muchow. 2010. Adapting APSIM to model the physiology and genetics of complex adaptive traits in field crops. J. Exp. Bot. 61:2185–2202. doi:10.1093/jxb/erq095

Hubick, K.T., G.D. Farquhar, and R. Shorter. 1986. Correlation between water-use efficiency and carbon isotope discrimination in diverse peanut (*Arachis*) germplasm. Aust. J. Plant Physiol. 13:803–816. doi:10.1071/PP9860803

Kaldenhoff, R. 2012. Mechanisms underlying CO2 diffusion in leaves. Curr. Opin. Plant Biol. 15:276–281. doi:10.1016/j.pbi.2012.01.011

Kapanigowda, M.H., W.A. Payne, W.L. Rooney, J.E. Mullet, and M. Balota. 2014. Quantitative trait locus mapping of the transpiration ratio related to preflowering drought tolerance in sorghum (*Sorghum bicolor*). Funct. Plant Biol. 41:1049–1065. doi:10.1071/FP13363

Kholová, J., C.T. Hash, A. Kakkera, M. Kocova, V. Vadez. 2010a. Constitutive water-conserving mechanisms are correlated with the terminal drought tolerance of pearl millet [*Pennisetum glaucum* (L.) R. Br.]. J. Exp. Bot. 61:369–377. doi:10.1093/jxb/erp314

Kholová, J., C.T. Hash, L.K. Kumar, R.S. Yadav, M. Kocova, V. Vadez. 2010b. Terminal drought tolerant pearl millet [*Pennisetum glaucum* (L.) R. Br.] have high leaf ABA and limit transpiration at high vapor pressure deficit. J. Exp. Bot. 61:1431–1440. doi:10.1093/jxb/erq013

Kirkegaard, J.A., J.M. Lilley, G.N. Howe, and J.M. Graham. 2007. Impact of subsoil water use on wheat yield. Aust. J. Agric. Res. 58:303–315. doi:10.1071/AR06285

Kirkham, M.B. 1988. Hydraulic resistance of two sorghums varying in drought resistance. Plant Soil 105:19–24. doi:10.1007/BF02371138

Manschadi, A.M., J.T. Christopher, P. Peter deVoil, and G.L. Hammer. 2006. The role of root architectural traits in adaptation of wheat to water-limited environments. Funct. Plant Biol. 33:823–837. doi:10.1071/FP06055

Masle, J., S.R. Gilmore, and G.D. Farquhar. 2005. The *ERECTA* gene regulates plant transpiration efficiency in *Arabidopsis*. Nature 436:866–870, doi:10.1038/nature03835

Passioura, J.B. 1977. Grain yield, harvest index and water use of wheat. J. Aust. Inst. Agric. Sci. 43:117–121.

Payne, W.A., M.C. Drew, L.R. Hossner, R.J. Lascano, A.B. Onken, and C.W. Wendt. 1992. Soil phosphorus availability and pearl millet water use efficiency. Crop Sci. 32:1010–1015. doi:10.2135/cropsci1992.0011183X003200040035x

Peng, S., and D.R. Krieg. 1992. Gas exchange traits and their relationship to water use efficiency. Crop Sci. 32:386–391. doi:10.2135/cropsci1992.0011183X003200020022x

Schoppach, R., and W. Sadok. 2012. Differential sensitivities of transpiration to evaporative demand and soil water deficit among wheat elite cultivars indicate different strategies for drought tolerance. Environ. Exp. Bot. 84:1–10. doi:10.1016/j.envexpbot.2012.04.016

Siddique, K.H.M., K.L. Regan, D. Tennant, and B.D. Thomson. 2001. Water use and water use efficiency of cool season grain legumes in low rainfall Mediterranean-type environments. Eur. J. Agron. 15:267–280. doi:10.1016/S1161-0301(01)00106-X

Sinclair, T.R. 2012. Is transpiration efficiency a viable plant trait in breeding for crop improvement? Funct. Plant Biol. 39:359–365. doi:10.1071/FP11198

Sinclair, T.R., G.L. Hammer, and E.J. van Oosterom. 2005. Potential yield and water-use efficiency benefits in sorghum from limited maximum transpiration rate. Funct. Plant Biol. 32:945–952. doi:10.1071/FP05047

Sinclair, T.R., C.B. Tanner, and J.M. Bennett. 1984. Water-use efficiency in crop production. Bioscience 34:36–40. doi:10.2307/1309424

Tanner, C.B., and T.R. Sinclair. 1983. Efficient water use in crop production: Research or research? In: H.M. Taylor, W.R. Jordan, and T.R. Sinclair, editors, Limitations to efficient Water Use in Crop Production. ASA, CSSA, and SSSA, Madison, WI. p. 1–27.

Tuinstra, M.R., G. Ejeta, and P. Goldsbrough. 1998. Evaluation of near isogenic sorghum lines contrasting for QTL markers associated with drought tolerance. Crop Sci. 38:835–842. doi:10.2135/cropsci1998.0011183X003800030036x

Turner, N.C., E.D. Schulze, and T. Gollan. 1984. The response of stomata and leaf gas exchange to vapour pressure deficits and soil water content. Oecologia 63:338–342. doi:10.1007/BF00390662

Vadez, V., S.P. Deshpande, J. Kholova, G.L. Hammer, A.K. Borrell, H.S. Talwar, and C.T. Hash. 2011a. Stay-green quantitative trait loci's effects on water extraction, transpiration efficiency and seed yield depend on recipient parent background. Funct. Plant Biol. 38:553–566. doi:10.1071/FP11073

Vadez, V., J. Kholova, S. Medina, K. Aparna, and H. Anderberg. 2014. Transpiration efficiency: New insights into an old story. J. Exp. Bot. doi:10.1093/jxb/eru040

Vadez, V., J. Kholova, R.S. Yadav, and C.T. Hash. 2013. Small temporal differences in water uptake among varieties of pearl millet (*Pennisetum glaucum* (L.) R. Br.) are critical for grain yield under terminal drought. Plant Soil 371:447–462. doi:10.1007/s11104-013-1706-0

Vadez, V., L. Krishnamurthy, C.T. Hash, H.D. Upadhyaya, and A.K. Borrell. 2011b. Yield, transpiration efficiency, and water-use variations and their interrelationships in the sorghum reference collection. Crop Pasture Sci. 62:645–655. doi:10.1071/CP11007

Wasson, A.P., R.A. Richards, R. Chatrath, S.C. Misra, S.V.S. Prasad, G.J. Rebetzke, J.A. Kirkegaard, J. Christopher, and M. Watt. 2012. Traits and selection strategies to improve root systems and water uptake in water-limited wheat crops. J. Exp. Bot. 63:3485–3498. doi:10.1093/jxb/ers111

Wright, G.C., R.C. Nageswara Rao, and G.D. Farquhar. 1994. Water-use efficiency and carbon isotope discrimination in peanut under water deficit conditions. Crop Sci. 34:92–97. doi:10.2135/cropsci1994.0011183X003400010016x

Xin, Z., R. Aiken, and J.J. Burke. 2009. Genetic diversity of transpiration efficiency in sorghum. Field Crops Res. 111:74–80. doi:10.1016/j.fcr.2008.10.010

Zaman-Allah, M., D.M. Jenkinson, and V. Vadez. 2011a. Chickpea genotypes contrasting for seed yield under terminal drought stress in the field differ for traits related to the control of water use. Funct. Plant Biol. 38:270–281. doi:10.1071/FP10244

Zaman-Allah, M., D.M. Jenkinson, and V. Vadez. 2011b. A conservative pattern of water use, rather than deep or profuse rooting, is critical for the terminal drought tolerance of chickpea. J. Exp. Bot. 62:4239–4252. doi:10.1093/jxb/err139

Genotype × Environment × Management Interactions: US Sorghum Cropping Systems

Ignacio A. Ciampitti,* P.V.V. Prasad, Alan J. Schlegel, Lucas Haag, Ronnie W. Schnell, Brian Arnall, and Josh Lofton

Abstract

The United States accounts for 24% of the global sorghum production, producing 15 Mt annually. At the global scale, the United States is one of the top five worldwide sorghum producers, which also include Nigeria, India, Ethiopia, and Argentina. The Great Plains region is the most important US sorghum-producing area, accounting for 75% of the production. In the last century, a continuous positive sorghum yield improvement was documented but not at same rate as that for other cereals such as maize. Therefore, efforts dedicated to germplasm improvement, hybrid selection, and agronomic practices are needed for reducing yield barriers. Closing sorghum yield gaps (potential minus actual yields) can be approached by improving the understanding of the complexity of the genotype × environment × management (G×E×M) interactions. Information on the best crop production practices for improving yields in modern hybrids is limited. A summary is needed of regional data on sorghum yield interaction with crop production practices such as planting date, seeding depth, hybrid selection, row spacing, plant density, N dynamics, and crop rotation, among other factors. This chapter describes the best management practices for improving grain sorghum yields under diverse environments across the US sorghum-producing areas. Major factors affecting grain sorghum yields and a review-analysis on nitrogen use efficiency (NUE) is also discussed.

Together with Nigeria, India, Ethiopia, and Argentina, the United States is one of the top five sorghum (*Sorghum bicolor* L. Moench) producers in the world (FAOSTAT, 2012). Within the USA, more than 75% of sorghum production is in the central and south-central region known as the Great Plains, with a majority of this production located in the states of Kansas and Texas (Fig. 1; USDA-NASS, 2013). Other states that lead sorghum production in the United States are Oklahoma, South Dakota, Arkansas, Louisiana, and Nebraska (USDA-NASS, 2013). Management of

---

Abbreviations: CT, conventional tillage; GDD, growing degree day; G×E×M, genotype × environment × management; NT, no tillage; NUE, nitrogen use efficiency; PAR, photosynthetically active radiation; RT, reduced tillage; WCF, wheat-corn-fallow; WF, wheat-fallow; WSF, wheat-sorghum-fallow; WSSF, wheat-sorghum-sorghum-fallow; WW, winter wheat; WWSF, wheat-wheat-sorghum-fallow.

I.A. Ciampitti, P.V.V. Prasad (vara@ksu.edu), A.J. Schlegel (schlegel@ksu.edu), L. Haag (haag@ksu.edu), Dep. of Agronomy, Kansas State Univ., Manhattan, KS; R.W. Schnell, Soil and Crop Science Dep., Texas A&M Univ., College Station, TX (Ronnie.schnell@ag.tamu.edu); B. Arnall (b.arnall@okstate.edu) and J. Lofton (josh.lofton@okstate.edu), Dep. of Plant and Soil Sciences, Oklahoma State Univ., Stillwater, OK. *Corresponding author (ciampitti@ksu.edu).

doi:10.2134/agronmonogr58.2014.0067

 Sorghum: State of the Art and Future Perspectives, Ignacio Ciampitti and Vara Prasad, editors. Agronomy Monograph 58.

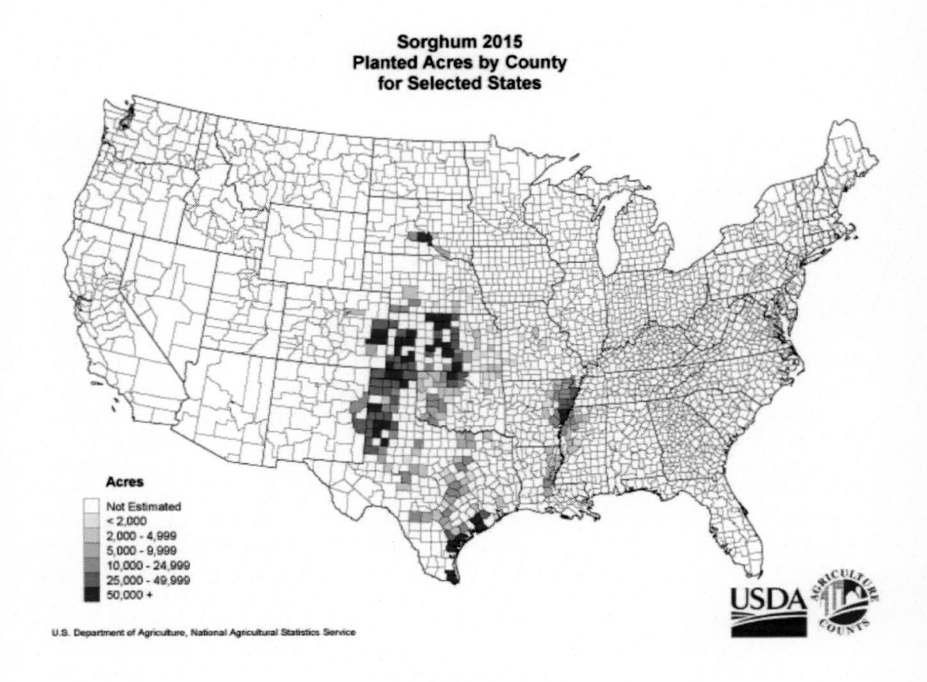

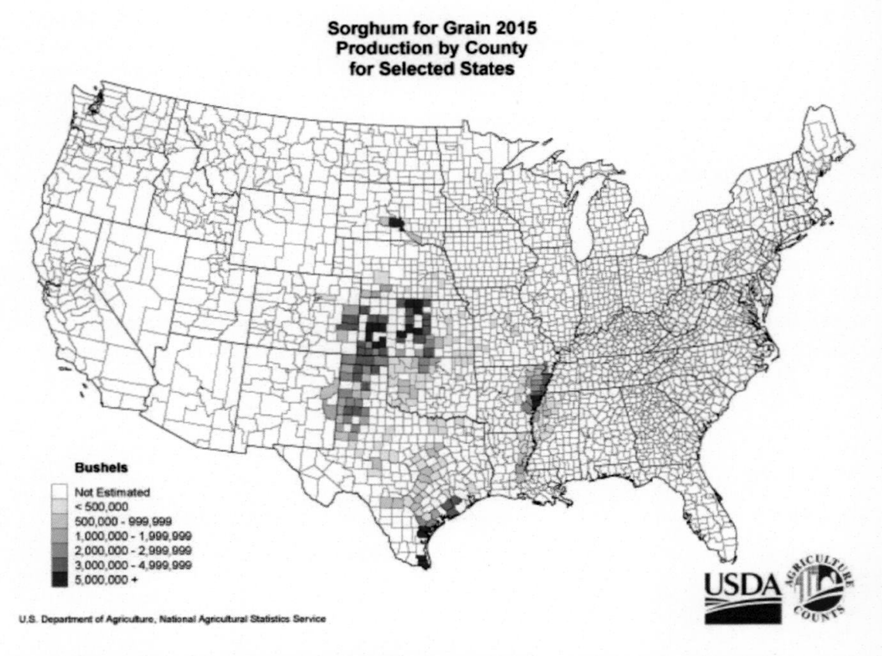

Fig. 1. Map of the United States showing the sorghum (upper panel) planted acreage and (lower panel) total production by county. From USDA National Agricultural Statistics Service.

the crop across these regions can be drastically different. Less than 50% of the sorghum acreage is irrigated, much of it in the state of Texas (Census of Agriculture, 2007). The present chapter describes from an agronomic viewpoint the main production factors impacting sorghum yield and yield potential.

Sorghum Improvement

Sorghum improvement during the last decades has been associated with marked changes in genotype (G component) and management practices (M component), specifically fertilization rates, irrigation, and tillage practices (Duvick, 1999; Assefa and Staggenborg, 2010). Compared with corn, yield improvement in sorghum has evolved at a slower rate (Mason et al., 2008). A similar rate of genetic gain in sorghum was documented in the last 50–60 yr (Miller and Kebede, 1984; Unger and Baumhardt, 1999). Unger and Baumhardt (1999) reported a 50 kg ha^{-1} annual increase in grain sorghum yields across 502 treatment-years, which represented a 139% increase. Of these yield increases, one-third was attributed to the G component and two-thirds to M factor in interaction with the environment (E component). Genetic improvements during the last several decades have focused on improving sorghum drought avoidance (Assefa and Staggenborg, 2011), which has ultimately increased crop production potential, especially under rainfed conditions. However, Assefa and Staggenborg (2011) documented changes in physiological characteristics of new and older sorghum hybrids under sufficient and water-deficit environments. These observations emphasize the overall importance of the environment; thus, endpoint sorghum productivity may be considered the outcome of the complex interaction of G and E and M.

Little is understood, however, about the relative contribution of each component (G, E, and M) and their interactions that influence sorghum yield. A better understanding of sorghum response under diverse management crop production practices would allow optimizing the use of all soil-plant resources and then closing yield gaps (i.e., maximizing sorghum yield at each specific environment, depending on the soil × weather interaction).

Crop Management

Management practices such as planting date are primarily governed by crop rotation and are adjusted according to thermal conditions (soil temperature). Average planting dates for the Great Plains region are dependent on the soil temperature. The recommended temperature for optimal germination ranges from 15 to 23°C. Cold soils can significantly affect sorghum germination and emergence. Temperatures below 10°C can impede sorghum germination (Anda and Pinter, 1994). While germination and stand establishment are controlled by soil temperature and environmental conditions, crop rotations determine the planting date. For rotations in which corn and soybean are the primary focus, those crops will be planted on better ground and earlier, leaving sorghum to be planted later, frequently outside the optimal window.

In the last 5-yr period, the overall 50%-planting date in Texas (state average) was approximately 10 April, whereas in Kansas, the 50%-planting date was reached on approximately 4 June (Fig. 2). In Texas, a consistent trend from 1980 to 1999 and from 2008 to 2014 depicted a historical planting date from 10 to 20 d

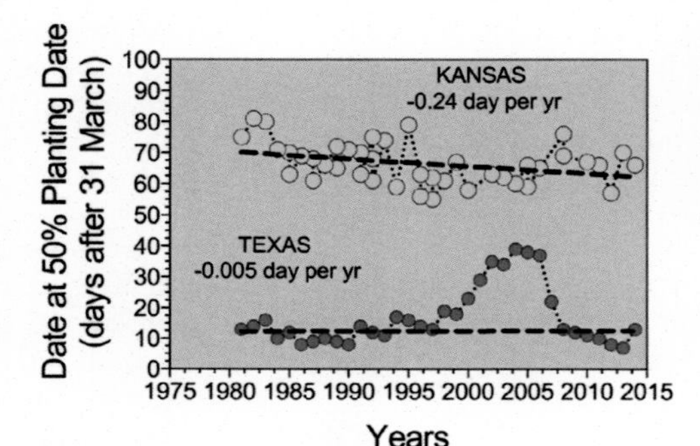

Fig. 2. Historical trend of the date at which 50% of planting progress was achieved for total planted area of sorghum between 1980 and 2014 in Kansas and Texas. From USDA National Agricultural Statistics Service.

after 31 March, with a negligible variation when the period from 2000 to 2007 is excluded (Fig. 2). It is worth clarifying that the optimal planting date for sorghum presented an ample range of variation for Texas from south to north. In Oklahoma, the overall 50%-planting date was achieved on approximately 1 June in the last 5-yr period, with the exception of 2012 and 2013, when the 50%-planting date was attained about 1 wk after 1 June (USDA-NASS). In Kansas, the historical trend portrayed a change to earlier planting dates at a rate of about 0.24 d per year (Fig. 2). This change can be attributed to warmer springs, changes in agronomic technologies related to machinery, and improvements in seed treatment and genetics. If sorghum is planted too early during the growing season, delays in emergence can be reflected in poor plant-to-plant uniformity and reductions in stands. Late planting dates may jeopardize the ability to reach full maturity before a damaging fall freeze, particularly in the northern Great Plains region. The probability for the crop to reach maturity can be calculated with more accuracy when it is based on the blooming time. The length of the growing season for sorghum can be estimated via the calculation of the growing degree days (GDDs) for each state producing this crop. Seasonal 30-yr GDD information (base temperature = 10°C; if daily_min. < 10°C, daily_min. = 10°C; if daily_max. > 37.8°C, daily_max. = 37.8°C; GDD = [(daily_max. + daily_min. air temp.)/2] − base temp.) was obtained to estimate the length of the crop season (Fig. 3). For this purpose the sorghum growing region was divided into two areas: (i) the Southern Great Plains/Early Sorghum Production region (Texas, Louisiana, New Mexico, and Mississippi) and (ii) the Northern Great Plains/Late Sorghum Production region (Oklahoma, Kansas, Colorado, Arkansas, and Missouri). As expected, the cumulative GDDs increased from north to south, increasing the potential length of the growing season for the sorghum cropping system.

Planting Date

The planting window for maximizing sorghum productivity under diverse environments (e.g., moisture-limiting) can assist in closing the yield gap between the maximum potential yield and the yield attainable in each region. Heat and drought can severely impact yield formation in sorghum. A recent study investigating the effect of heat on sorghum (Prasad et al., 2015) portrayed the critical period for yield formation as 10 d before and 5 d after flowering. Planting date is a critical management tool for determining flowering time.

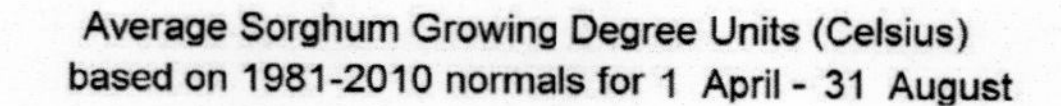

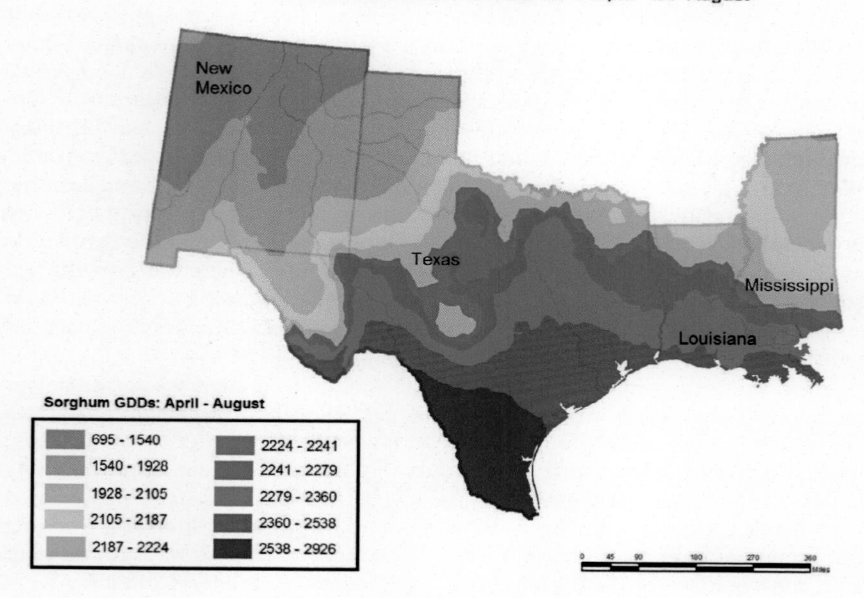

Average Sorghum Growing Degree Units (Celsius)
based on 1981-2010 normals for 1 June - 30 September

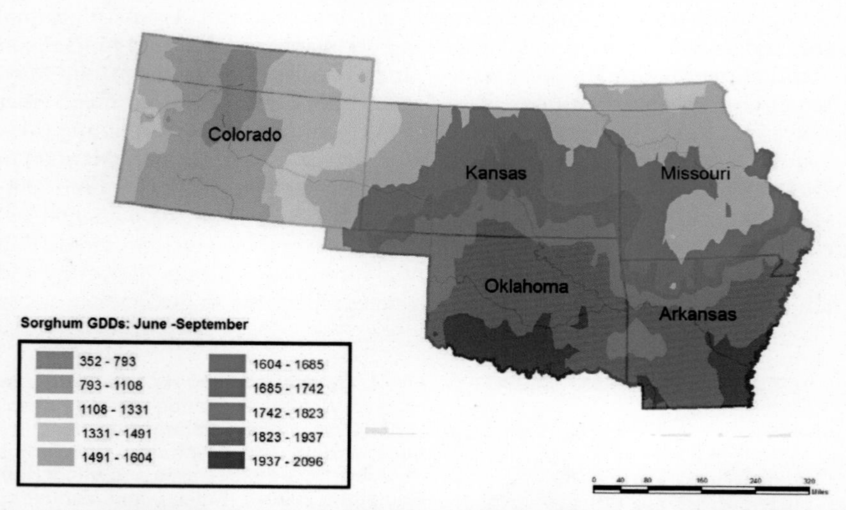

Fig. 3. Cumulative growing degree days from April to August for the Southern Great Plains/Early Production region (Texas, Louisiana, New Mexico, and Mississippi) (upper panel) and from June to September for the Northern Great Plains/Late Production region (Oklahoma, Kansas, Colorado, Arkansas, and Missouri) (lower panel). Produced by the K-State Research and Extension Weather Data Library, Dep. of Agronomy, Kansas State Univ.

At the state level, the agronomic optimal planting date for grain sorghum in Texas is 2 wk after the average final freeze date. For south and central Texas, the goal is to plant early to avoid the extended period of heat and drought stress that is common in Texas summers. In addition, early planting is advised so that plants reach flowering before the potential for sorghum midge (*Stenodiplosis sorghicola*) damage increases (Cronholm et al., 2007). Preferred planting dates are from late January to late February in the Lower Rio Grande Valley, from late February through mid-March in the Coastal Bend and Upper Gulf Coast, and from mid-March through mid-April in central and north-central Texas (Trostle and Fromme, 2011). With the long growing season in south and central Texas, hybrid maturity is generally selected based on available soil moisture. Planting in the Panhandle of Texas generally occurs from mid-April through June, where the growing season is much shorter. Therefore, the last recommended planting date is based on sorghum maturity, with 30 June for mid-maturity and 15 July for early-maturity hybrids (Barber et al., 2007).

For Kansas, a summary of research information from diverse site-years demonstrated that an early June planting date provided the highest-yield sorghum environment (and more predictable, with lower yield variation) than planting dates ranging from late April to early July (Fig. 4). The increase in productivity experienced with an early June (late spring) planting date was strictly associated with better conditions during blooming time (late-summer rains), minimizing the impact of stress conditions (primarily drought and heat) on grain number and, consequently, on final grain yield. Interactions with hybrid maturity were not clearly observed, presenting similar trends for medium- and late-maturing hybrids (Fig. 4), with early June planting dates maximizing yield. Martin and Vanderlip (1997) documented the widest planting window with early hybrids, the planting window decreasing with increased hybrid maturity. A similar optimal planting date was reported, with a shift to later planting dates (mid-June) if conditions became abnormally wet and cool, and to earlier planting times (mid-May) when environmental conditions were dry and hot during mid-summer. Wider positive yield-response ranges (from 7 to 24%) resulted with early planting (May vs. June) in different locations around Kansas (Belleville, Ottawa, Manhattan, and Hutchinson; Maiga, 2012). As documented by several authors (Khan, 2000; Naeem, 2001; Maiga, 2012), early planting times produced more biomass and leaf area, which increased early interception of photosynthetically active radiation (PAR), and presented a higher yield potential if conditions around flowering and grain-fill were favorable. Delayed planting dates reduce the length of the growing

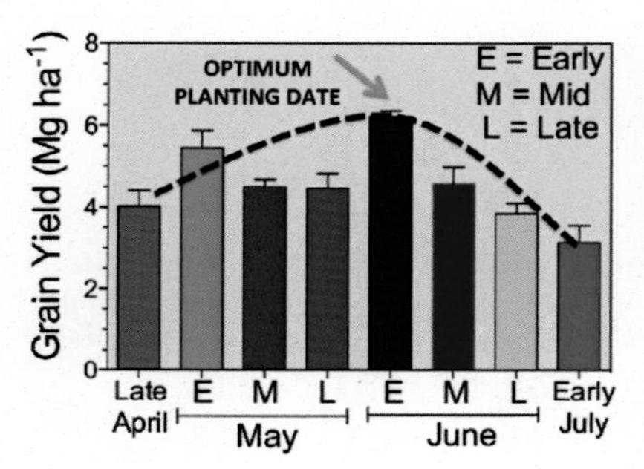

Fig. 4. Optimal planting date for sorghum. Summary of information from Kansas (1993–2014), incorporating data from Tribune, Hutchinson, Manhattan (Vanderlip), St. John 1993–1995 (Martin and Vanderlip), Columbus 2000–3 (Kelley), and Manhattan 2014 (Ciampitti).

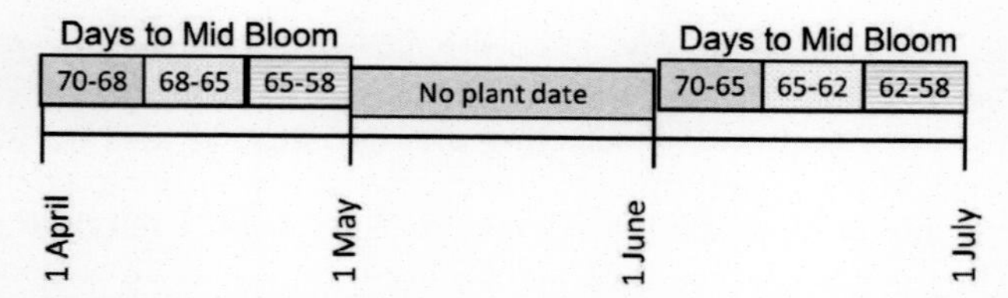

Fig. 5. Optimal maturity group based on planting date for sorghum in Oklahoma.

season, which affects PAR interception (effective duration), resulting in a reduced canopy size, and consequently, an impaired carbon supply to the reproductive organs (grains).

In Oklahoma the first planting date is recommended based on the soil temperature and Risk Management Agency first insurable planting date. Planting can occur from April through August depending on the available moisture. It is generally recommended not to plant sorghum during the month of May, to avoid anthesis from occurring during mid-July to mid-August, which is historically the hottest period in the state. For consideration of maturity-group selection, the days to mid-bloom for both early- and late-season-planted sorghum are broken into thirds (Fig. 5). Both planting windows begin with the longest maturity groups, and as the planting window shortens, the days to mid-bloom should be shortened to avoid either the heat or first freeze.

Seeding Depth

One of the most important practices for planting sorghum is planting depth. An optimal planting depth is not uniform across soil conditions and environments. It differs primarily with soil type, residue cover (effect on temperature), and moisture conditions. Optimal seeding depth varies from 2.5 to 5 cm. Early planting can produce adequate emergence with a seeding depth of 2.5 cm in high-clay soils and 5 cm in sandier soil textures. Deeper (>5-cm seeding depth) placement can reduce emergence, potentially affecting the final stand count and/or uniformity. For late planting (soil temperatures > 23°C), sorghum seed can be placed deeper if beneficial soil moisture is present at those soil depths (Fig. 6).

Fig. 6. Sorghum planting depth effects on plant growth and development. Shallower placement (0.5 inches, or 1.2 cm) shows lower plant growth compared with both normal (1.5 inches, or 3.8 cm) and deeper (3 inches, or 7.7 cm) seed placement with better moisture conditions.

Hybrid Selection

Genotype selection is one of the most crucial factors for improving sorghum productivity. The complexity of selecting the right genotypic fit for each E (soil and weather) × M (e.g., planting date, plant density, row spacing, among others) combination should be properly understood. Thus, hybrid selection should be based on key traits such as (i) maturity, (ii) head exertion (full vs. poor exertion, less susceptible to biotic stress, e.g., head mold), (iii) seedling vigor (early planting with no-till stubble or deep seed placement), (iv) resistance to pests (insects and diseases), (v) plant standability (more important in adverse environments), (vi) grain yield and quality, and (vii) probability of maturing before the first fall freeze. In regard to the last point, the selection of a hybrid that needs to be fully mature before the first fall freeze will also be related to the planting date, which is emphasized by planting shorter-season hybrids when planting late (e.g., mid- to late June in Kansas). Physiological maturity is reached when a black layer is formed at the base of the grain (Fig. 7), coinciding with the cessation of dry-matter and nutrient accumulation.

A compilation of diverse research studies performed in Kansas (with planting dates ranging from the end of May to the first week of June) showed the complexity of the hybrid-selection component (Fig. 8). Early-maturing hybrids produced better in environments with lower yield potential (<3 Mg ha^{-1}). The benefits of using medium- to late-maturing sorghum hybrids were observed under higher-yield environments (>4 Mg ha^{-1}). Martin and Vanderlip (1997) documented a lack of hybrid effect on final yield when sorghum hybrids with a wide range of maturity were planted at early, mid-, and late dates in Kansas (mid-May, early June, and mid-June).

Hybrid selection and performance should be critically considered when planting sorghum. Well-adapted sorghum hybrids can produce very stable yields across low- to high-yield environments. Still, local information on hybrid performance should be pursued for hybrid selection. In a 2-yr research study (I.A.

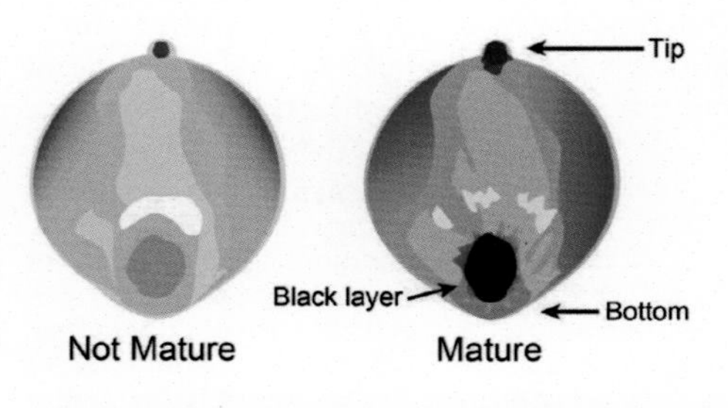

Fig. 7. Immature and physiologically mature (black layer formation) grain sorghum.

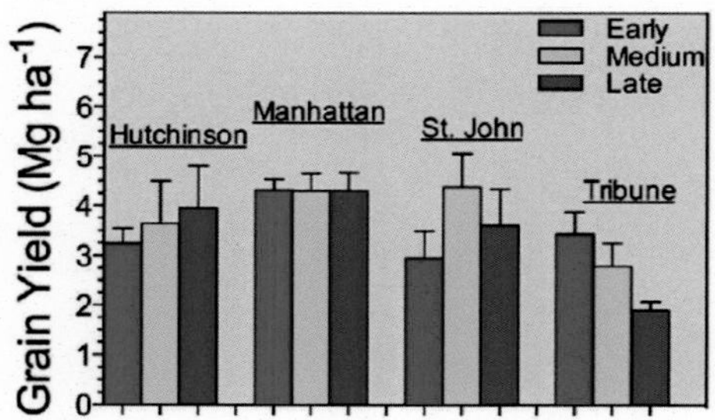

Fig. 8. Sorghum hybrid selection based on the plant maturity trait (early-, medium-, and late-maturing sorghum hybrids). Summary of information from Kansas (1993–2014), incorporating data from Tribune, Hutchinson, Manhattan (Vanderlip), St. John 1993–1995 (Martin and Vanderlip), Columbus 2000–3 (Kelley), and Manhattan 2014 (Ciampitti).

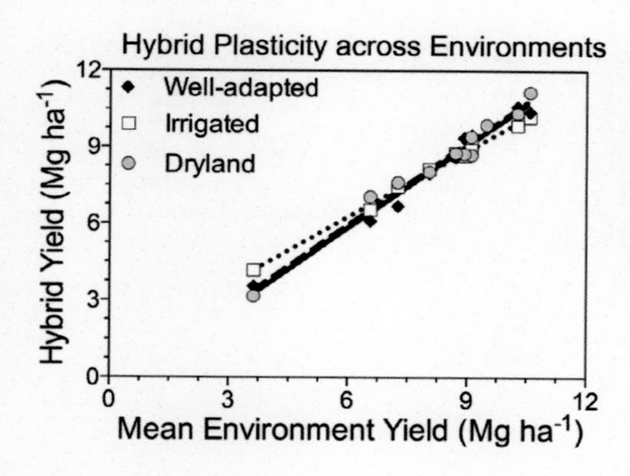

Fig. 9. Sorghum hybrid yield plasticity (yield treatment mean to environmental yield mean ratio) across 11 site-years for three hybrids (dryland suited, irrigated suited, and well-adapted) during the 2014–15 season in Kansas (I.A. Ciampitti and J. Broeckelman, unpublished data, 2014–2015).

Ciampitti and J. Broeckelman, unpublished data, 2014–2015) evaluating three contrasting sorghum hybrids (dryland-suited, irrigation-suited, and well-adapted) under different water regimes, a yield advantage was documented in a hybrid recommended for irrigated environments under rainfed conditions (yield environment < 4 Mg ha^{-1}), while a similar yield was observed when hybrids attained higher yields (yield > 10 Mg ha^{-1}) under full irrigation (Fig. 9). Hybrid selection under rainfed conditions portrayed a yield difference of from 0.5 to 1 Mg ha^{-1}, emphasizing the importance of site-specific information on hybrid performance as a critical management practice for improving sorghum yields.

Row Spacing

Reducing the row spacing under favorable conditions can promote fast canopy closure, decrease evaporation, or increase evapotranspiration efficiency (Steiner, 1986, 1987; Sanabria et al., 1995). Under non-stress conditions, yield response to narrow rows is directly associated with an improvement in light interception early in the season, which can be translated into greater yields. In Kansas, Staggenborg et al. (1999) documented superior grain sorghum yields (10% higher) when the row spacing decreased from 75 cm to 25 cm under high-yield environments. Similar yield improvement with narrow row spacing, when compared with wide spacing (25 cm vs. 75 cm), was documented by Maiga (2012), presenting a benefit in yield ranging from 3 to 14% when tested in different environments during the same season (Belleville, Ottawa, Manhattan, and Hutchinson, KS). In Missouri, Conley et al. (2005) documented a positive yield response with narrower row spacings in 1 yr of a 2-yr study, with a yield benefit close to 1 Mg ha^{-1} and highest yield levels greater than 7 Mg ha^{-1}. In Texas, Steiner (1986) reported that narrow geometry increased evapotranspiration partitioning to the transpiration factor; however, this failed to result in higher yields. Similar results were documented by Sanabria et al. (1995) under conditions of better water conservation with narrow row spacing and high evaporative demand. A similar outcome was previously documented by Welch et al. (1966): when planting sorghum at 98,000 plants ha^{-1}, narrow (50-cm) row spacing showed a minor yield benefit under adequate N application (112 kg ha^{-1}) than wider (100-cm) row spacing. For the Coastal Bend Region of Texas, Fernandez et al. (2012) documented a lack of response to narrower rows in sorghum (38 cm vs. 76 cm) even when favorable conditions were present at critical crop

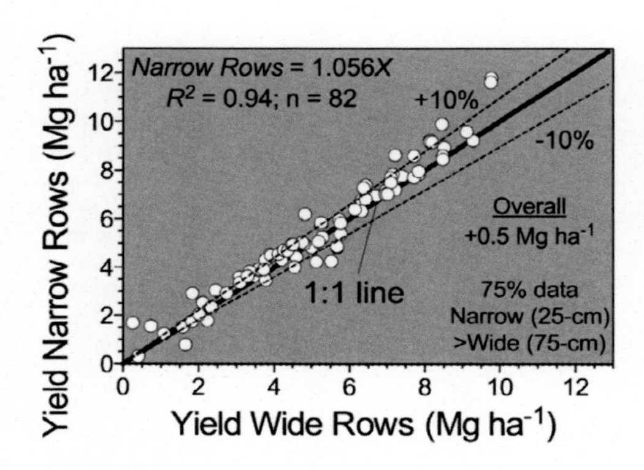

Fig. 10. Narrow (25-cm) versus wide (75-cm) row spacing for sorghum hybrids under diverse yield environments. Summary of information from Kansas (1993–2014), incorporating data from Tribune and Hutchinson (Vanderlip) and Maiga, 2012.

stages such as panicle formation, flowering, or early boot. In addition, when water limited biomass production and yield formation, wide rows outyielded narrow rows. A summary of research studies in Kansas from the 1980s to 2011 documented an increase in sorghum yields when narrow row spacing was employed (+0.5 Mg ha^{-1} in >75% of all observations). In addition, sorghum yield increases were consistent when grain yields were above 6 Mg ha^{-1} (Fig. 10). In a study from Texas, Fromme et al. (2012) documented that narrow rows (51 cm) slightly improved sorghum yields compared with wide rows in lower-yield environments (below 7 Mg ha^{-1}). However, they documented that this was not consistently found across plant densities and hybrids. This observation emphasizes the impact of the G and other M factors on management decisions.

Plant Density

For sorghum, the yield response to plant density is not as consistent as with other crops, such as corn. The ability of sorghum under low plant density to compensate with the production of tillers provides a unique feature that minimizes the impact of plant density on grain yield. Recent studies performed in Kansas documented a positive yield trend at some locations (14% yield response) when plant density increased from 24,000 to 96,000 plants ha^{-1} (e.g., Garden City), but no yield responses were observed at other locations (Pidaran, 2012). In Missouri, Conley et al. (2005) reported a yield-density linear-plateau response, with yields improving from 6.3 to 7.3 Mg ha^{-1} when plant density doubled from 73,600 to 147,300 plants ha^{-1}, but yield benefits from increasing plant density plateaued through 368,000 plants ha^{-1}. Under dryland systems in Texas, Welch et al. (1966) concluded that the optimal plant density for increasing yields was around 100,000 to 150,000 plants ha^{-1}. Additionally, in dryland conditions in Oklahoma, data collected from 2001 through 2004 showed yield differences between plant densities, but optimal densities were much lower than what Welch et al. (1966) observed (Fig. 11). It was found that yields increased when the seeding rate increased from 100,000 through 210,000 seeds ha^{-1}, but no further significant yield increases were noted. However, increasing the seeding rate above 210,000 seeds ha^{-1} did not hinder yields, especially in non-limiting environments.

Conversely, in Texas, Fernandez et al. (2012) did not report any differences in yield when the plant density ranged from 124,000 to 235,000 plants ha^{-1}. Similarly, Schnell et al. (2014) did not observe significant yield differences in response to different plant densities (69,000–245,000 plants ha^{-1}) across five locations in central

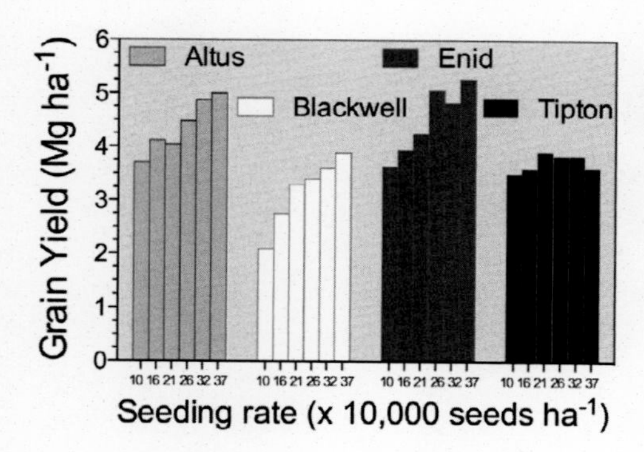

Fig. 11. Sorghum grain yield response to seeding rate at four diverse sorghum production environments in Oklahoma. Data provided by Drs. R. Kochenower, J. Lofton, and B. Arnall.

Texas; however, the yield for plant densities above 200,000 plants ha^{-1} was more sensitive to above- or below-normal precipitation.

The response to plant density is not consistent across sites, years, row configuration, and hybrids. The ability of the sorghum plant to compensate in non-optimal plant densities can be summarized in the association between the number of fertile panicles per plant and the final plant density (Fig. 12). Diverse sorghum genotypes with low tillering capacity may present a more consistent yield response to plant density than phenotypes having higher tillering ability, which can compensate for lower plant densities with tillers and thus result in more fertile panicles per plant (Fig. 12).

Geometry and Spatial Arrangement

Past work on geometry and spatial arrangement in sorghum was undertaken with the purpose of reducing water usage (Blum and Naveh, 1976). Yield improvement for skip-row (plant two rows, skip three rows) planting configurations was associated with water availability at the water peak demand: the boot to full-bloom stages. In diverse semiarid and dryland regions around the world (i.e., Australia and Ethiopia), skip-row configuration is a regular practice (Routley et al., 2003; Mesfin et al., 2010). Recent studies (2009–2011) performed in the Great Plains region of the United States tested alternative planting geometries for corn and sorghum (Haag, 2013). Five geometries were evaluated: (i) conventional—a row spacing of 76 cm with plants equidistantly separated within the rows; (ii)

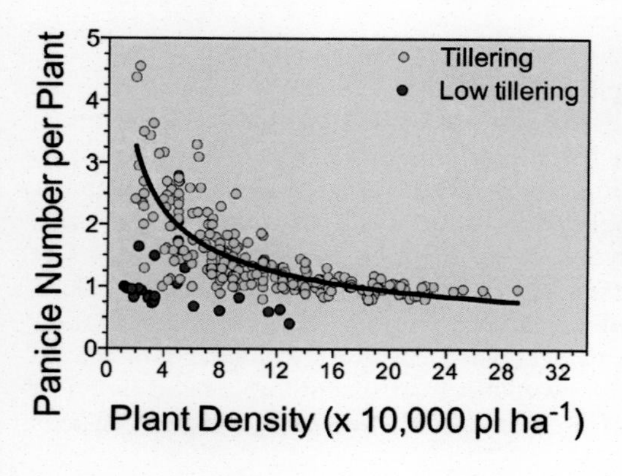

Fig. 12. Panicle (fertile tillers) number per plant versus plant density. Summary of information from Kansas, 1971–1996 (Vanderlip, 1972; Gerik and Neely, 1987; Koo Kim et al., 2010; Diawara, 2012; Fernandez et al., 2012; Pidaran, 2012; Haag, 2013; Ciampitti, unpublished data 2014; and Schnell et al., 2014).

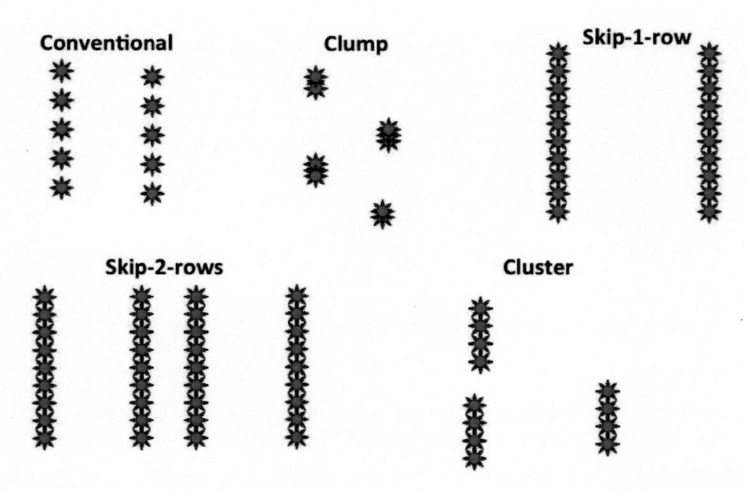

Fig. 13. Geometrical planting configuration for grain sorghum. Adapted from Haag, 2013.

cluster—six plants planted but alternating between rows (planted in row 1 and then a gap, planted in row 2, and planted again in row 1); (iii) clump—plants planted in group of three plants; (iv) skip one row and plant one row; and (v) skip two rows and plant two rows (Fig. 13). In these studies, three seeding rates were evaluated: 30,000; 40,000; and 51,000 plants ha^{-1}. Overall, sorghum planted either in the conventional (8.0 Mg ha^{-1}), clump (7.7 Mg ha^{-1}), or cluster (7.6 Mg ha^{-1}) geometries resulted in similar final grain yields, which was also related to comparable biomass and water use efficiency values. Clump planting shows great potential for stabilizing or increasing yields in low-yield environments. The skip-row configurations resulted in lower yields (the average for both configurations was 5.6 Mg ha^{-1}), primarily in high-yield environments, due to reduced light interception. In conclusion, alternative planting geometries such as cluster and clump appear to have fewer disadvantages than the skip-row geometry in dryland environments (Haag, 2013). Further research should be conducted on the potential G×E×M interactions of planting geometry with all farming-systems factors involved in sorghum production in dryland environments.

Rotation and Tillage Effects

The most common dryland cropping system in the semiarid Great Plains for decades was wheat-fallow (WF) using conventional tillage (CT). However, as early as the 1960s, research showed that more intensive crop rotations are suitable when tillage is reduced (Phillips, 1969). Later research also showed that a wheat-sorghum-fallow (WSF) rotation with reduced (RT) or no tillage (NT) was feasible in western Kansas (Norwood et al., 1990; Norwood, 1994; Schlegel et al., 1999). Norwood et al. (1990) reported on the effects of cropping and tillage effects in long-term studies at Garden City and Tribune, KS. They found that RT increased WF yields at both locations, whereas WSF wheat yields were increased at Tribune. Similarly sorghum yields were more consistently increased by RT at Tribune. Wheat yields were usually similar for WF and WSF at both locations, whereas average continuous wheat yields were 50% or less than those of WF or WSF at Tribune. The profiles of available soil water at wheat planting were similar for WF and WSF within the same tillage system. For sorghum in a WSF rotation, available soil water at planting was greater with RT than CT at Tribune but not at Garden City.

Not all studies have shown advantages to using RT or NT. Thompson and Whitney (1998) evaluated three tillage systems (CT, RT, and NT) in a WSF rotation near Hays, KS. Similar yields were documented with CT and RT and lower yields with NT, a result they attributed to poorer stands, increased weed competition, and drier soils, particularly for the wheat. Schlegel et al. (1999) compared WF and WSF under both RT and NT along with NT continuous winter wheat (WW). Wheat yields were usually similar for WF and WSF, with both being about twice as great as for WW. Wheat yields were similar with both RT and NT. Average sorghum yields were 23% greater with NT than RT. Available soil water at wheat planting was similar for WF and WSF regardless of tillage system and about twice as great as for WW. Tillage intensity did not have an effect on the available soil water at sorghum planting.

Norwood and Currie (1998) compared WSF with wheat-corn-fallow (WCF) in southwest Kansas under four tillage practices. From this research, the authors concluded that corn yielded more than sorghum in 12 of 16 tillage× year combinations, primarily due to above-average precipitation. Corn yields were larger under RT or NT in 3 of 4 yr, while sorghum yields were larger in only 1 yr. In a later study, Norwood (1999) compared the yields and water usage of CT and NT corn, sorghum, sunflower, and soybean. The grain yields of all the crops were increased with NT compared with CT. On average, corn yields were greater than sorghum yields, and sunflower yields were greater than soybean yields. Sorghum and sunflower removed the water that was deeper in the profile and removed more total soil water than either corn or soybean.

Other studies have evaluated crop rotations that were more intensive than two crops in 3 yr. As stated earlier, WW has been found to yield only about 50% that of WF or WSF (Norwood et al., 1990; Schlegel et al., 1999). Four-year rotations (3 yr of crop and 1 yr of fallow) involving wheat and grain sorghum were evaluated in western Kansas (Schlegel et al., 2002), with rotations of wheat-wheat-sorghum-fallow (WWSF) and wheat-sorghum-sorghum-fallow (WSSF). Wheat yields were 43% (1.2 Mg ha^{-1}) greater following sorghum than wheat, whereas sorghum yields were 36% (1.5 Mg ha^{-1}) greater following wheat than following sorghum. In eastern Colorado, Peterson et al. (1999) found that annualized grain production for 3- (WSF or WCF) and 4-yr (wheat-corn-millet-fallow and WSSF) were similar. More recently, 4-yr rotations involving wheat, grain sorghum, and corn were evaluated (Schlegel, 2014). The rotations of wheat-sorghum-corn-fallow and wheat-corn-sorghum-fallow were compared with WSF and WCF. Corn and grain sorghum yields (6-yr average) were about twofold greater following wheat than following corn or grain sorghum. Stone et al. (2002) evaluated the water depletion depth of grain sorghum and sunflower near Tribune, KS and found that the depth of water depletion was greater with sunflower (3.1 m) than with sorghum (2.5 m). Also, the end-of-season rooting depth was larger for sunflower (3.0 m) than sorghum (2.5 m).

Since crop production in the semiarid Great Plains is primarily limited by lack of water, the efficient use of water is critical for sustainable production.

Norwood (1994) reported that profile-available soil water at planting of sorghum in a WSF rotation was 38 mm greater with NT than CT, with 24 mm of this increase stored below 0.9 m. Stone and Schlegel (2006) evaluated the yield response of grain sorghum and wheat in relation to stored soil water and precipitation from studies conducted near Tribune, KS from 1973 to 2004. The authors found that grain yield was related to available soil water at emergence and increased 221 kg ha^{-1} cm^{-1} for sorghum and 98 kg ha^{-1} cm^{-1} for wheat. Growing-season precipitation increased sorghum yield 164 kg ha^{-1} cm^{-1} and wheat yield 83 kg ha^{-1} cm^{-1}. Available soil water at planting along with growing season precipitation accounted for 63% of the yield variation for sorghum and 70% for wheat. When the data was sorted by tillage system, the yield response to water supply (available soil water at planting plus growing-season precipitation) was greater with NT than CT. For sorghum, NT produced 184 kg ha^{-1} cm^{-1} compared with 129 kg ha^{-1} cm^{-1} for CT (a 42% increase). Similarly for wheat, NT produced 138 kg ha^{-1} cm^{-1} compared with 86 kg ha^{-1} cm^{-1} for CT (a 60% increase).

In a review of the economics of dryland cropping systems in the Great Plains, Dhuyvetter et al. (1996) found that in seven of eight studies net returns were greater from a more intensive crop rotation than WF when RT or NT were used following wheat harvest and prior to planting of a summer crop. Schlegel et al. (1999) reported that the sorghum yields had to be 3500 kg ha^{-1} or greater to make WSF more profitable than WF but that conclusion depended on relative grain prices, yield, and production costs. In wheat–summer-crop–fallow rotations, corn yields were 28% greater with NT than CT, increasing the net return 69%, whereas sorghum yields were only 11% greater, with no effect on net returns (Norwood and Currie, 1997). Norwood and Currie (1998) also reported that returns were greater for WCF than WSF in 10 of 16 tillage × year combinations. Schlegel et al. (2002) reported that second-year sorghum yields in a WSSF rotation had to be 3.5–4.0 Mg ha^{-1} and second-year wheat yields in a WWSF rotation had to be 2.5–3.0 Mg ha^{-1} to make 4-yr rotations more profitable than a 3-yr WSF rotation. Peterson et al. (1999) reported that opportunity cropping (growing a crop every year) maximized production but was less profitable than WCF.

The NUE Concept for Sorghum: A Physiological Perspective

Understanding changes in NUE and physiological strategies for sorghum will facilitate improvements in plant N uptake and their related efficiencies (N recovery and physiological efficiencies), which can consequently impact crop productivity. The study of grain N sources—herein understood as (i) reproductive-stage shoot N remobilization (remobilized N): the quantity of N remobilized from the stover to the grain after flowering; (ii) reproductive-stage whole-plant N uptake (reproductive N): the new N taken up by the plant and allocated in the grain fraction from flowering until maturity; and (iii) vegetative-stage whole-plant N uptake (vegetative N): the quantity of N taken up by the whole plant from emergence until flowering—is a novel approach for investigating NUE changes from a plant physiology viewpoint. A scientific summary for sorghum was performed with the goal of comparing NUE and its related efficiencies (Ciampitti and Prasad, 2016). A similar historical review and analysis was developed for corn (Ciampitti and Vyn, 2012, 2013, 2014). For sorghum, the yield-to-N-uptake relationship fitted an envelope function (Fig. 14).

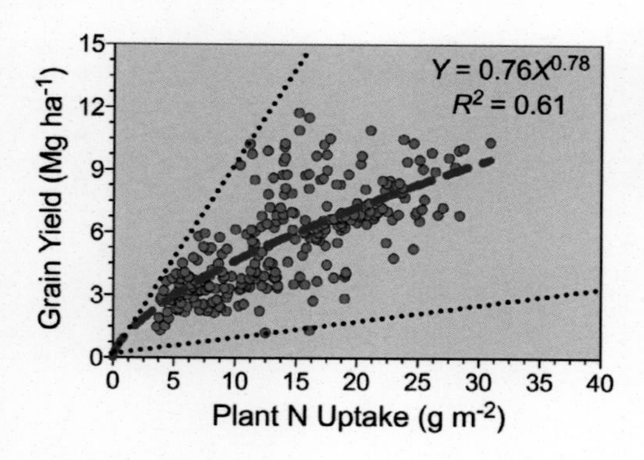

Fig. 14. Yield and plant N uptake (including grain + stover) both determined at the end of the season. Research trials from 1965 to 2014; adapted from Ciampitti and Prasad, 2016.

To shed some light on the NUE concept for sorghum, a deeper and more comprehensive evaluation was needed. This analysis considers the contribution of two main types of N sources within the plant: remobilized N and reproductive N. Briefly, high remobilization of N is a highly desirable plant trait, but it depends on the timing. Early-reproductive N remobilization is favorable under stress conditions that constrain grain yield, but it is a not a desirable trait in high-yielding sorghum systems (>12 Mg ha^{-1}). For corn and wheat, a trade-off was observed between remobilized N and reproductive N, suggesting that high reproductive N can only be achieved under low remobilized N, with a longer functional stay-green trait (Kichey et al., 2007; Bogard et al., 2010). Therefore, late-reproductive N remobilization is a highly desirable trait for concomitantly increasing plant N uptake and yield. A review paper on corn written by Ciampitti and Vyn (2013) documented a strong association between the quantity of N taken up before flowering and the remobilized N factor, suggesting a strong dependency of remobilized N on the plant N status achieved by flowering time. A similar relationship was documented for the data collected in this summary for sorghum (Fig. 15). Thus, the quantity of N remobilized after flowering is highly dependent on plant N status at flowering. For wheat and corn, a lower N remobilization efficiency was related to lower grain N demand (Pan et al., 1995; Bancal, 2009). Two scenarios are possible: (i) under high-yield environments, remobilized N is mainly driven by vegetative N (a strong positive relationship), and (ii) if the grain demand is affected by biotic or abiotic stress conditions, the association between

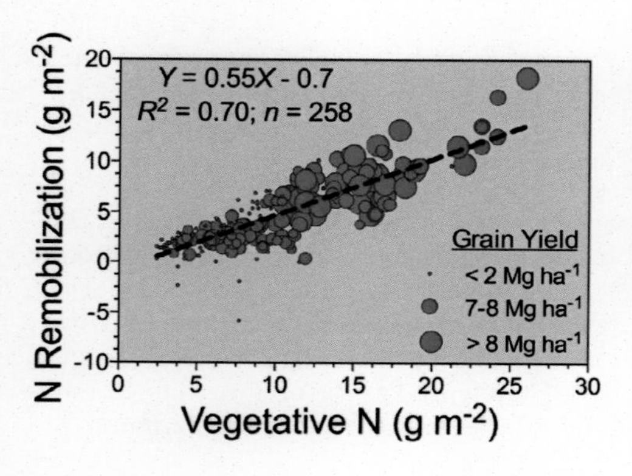

Fig. 15. Reproductive-stage shoot N remobilization (remobilized N), from flowering until the end of the season, versus vegetative-stage plant N uptake, including both leaf and stem fractions (vegetative N). Bubble sizes correspond to yields ranging from <2 to >8 Mg ha^{-1}. Adapted from Ciampitti and Prasad, 2016.

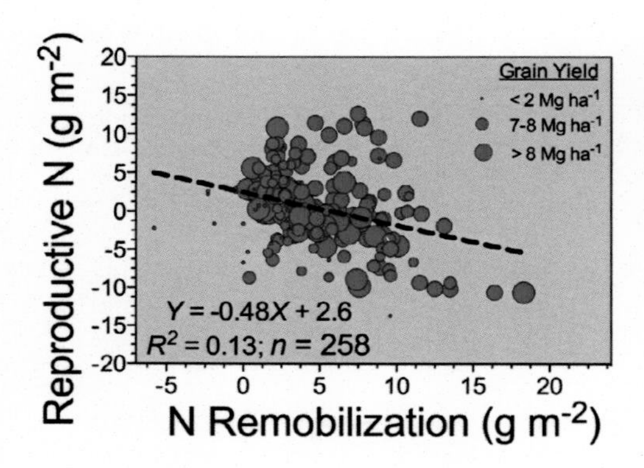

Fig. 16. Reproductive-stage whole-plant N uptake (reproductive N) versus shoot N remobilization (remobilized N). Bubble sizes correspond to yields ranging from <2 to >8 Mg ha^{-1}. Adapted from Ciampitti and Prasad, 2016.

remobilized N and vegetative N could be impacted (the lower the sink strength, the lower the requirement for remobilized N).

Increasing N remobilization can jeopardize the quantity of N taken up after flowering, thus decreasing the supply of N available for the grain. As documented for corn and wheat, a trade-off between remobilized N and reproductive N was documented also for sorghum in this synthesis-analysis (Fig. 16). Superior N remobilization was associated with lower reproductive N, but this relationship presented greater variability, as reflected in a low R^2 coefficient. Within the variability presented in this association, high yield values were related to high reproductive N and moderate to high remobilization N (large bubble size). The latter observation does not follow the trade-off previously documented, suggesting that sufficient variation exists for selecting genotypes that can simultaneously increase both reproductive N and remobilization N for maximizing sorghum yields. Remobilization of N might be related to a "late-season" nutrient remobilization, which can still have longer and functional reproductive N via the stay-green plant trait. Increasing NUE would require maximizing both processes during the reproductive stage in a balanced approach.

This synthesis-analysis for sorghum NUE provides valuable information on plant responses under diverse plant N content (Ciampitti and Prasad, 2016). In summary, the main points to be noted in this synthesis are (i) that a correlation and dependency exists between the quantity of N remobilized by the plant and the plant N status at flowering, and (ii) that an evident trade-off was shown for the quantity of N taken up after flowering and the shoot N remobilization during the same period. Continued NUE improvement should focus on investigating the physiological implications of this relationship and the G×E×M factors related to the trade-off mechanism (Ciampitti and Prasad, 2016).

Summary

Understanding sorghum yield response under diverse G×E×M scenarios (on-farm scale) should be pursued in depth. Information for modern hybrids in relation to all interactions with production factors and other yield-limiting components (e.g., weeds, insects, and disease pressure) needs to be generated for properly dissecting the G×E×M interaction. For example, the information on nutrient uptake and partitioning from Vanderlip (1972) is still being used as the preferential reference for sorghum on this topic. New research studies are currently being summarized in an effort to update the seasonal nutrient dynamics in sorghum (B.M. McHenry, P.V.V. Prasad, and I.A. Ciampitti, unpublished data, 2014–2015).

In the future, sorghum is a likely alternative in environments with a high probability of water stress because of the crop's drought tolerant traits: high water-use efficiency under low water availability, crop plasticity for delaying or "avoiding" water-stress conditions, and a better ability to tolerate high temperatures. Additionally, further market opportunities related to food, feed, and biofuel sources could increase sorghum demand and production in diverse regions of the United States.

Acknowledgments

This is contribution no. 17-076-B from the Kansas Agricultural Experiment Station. The authors gratefully acknowledge Dr. Richard Vanderlip for providing a helpful review of an early version of this manuscript.

References

Anda, A., and L. Pinter. 1994. Sorghum germination and development as influenced by soil temperature and water content. Agron. J. 86:621–624. doi:10.2134/agronj1994.00021962008600040008x

Assefa, Y., and S.A. Staggenborg. 2010. Grain sorghum yield with hybrid advancement and changes in agronomic practices from 1957 through 2008. Agron. J. 102:703–706. doi:10.2134/agronj2009.0314

Assefa, Y., and S.A. Staggenborg. 2011. Phenotypic changes in grain sorghum over the last five decades. J. Agron. Crop Sci. 197:249–257. doi:10.1111/j.1439-037X.2010.00462.x

Bancal, P. 2009. Decorrelating source and sink determinism of nitrogen remobilization during grain filling in wheat. Ann. Bot. (Lond.) 103:1315–1324. doi:10.1093/aob/mcp077

Barber, J., C. Trostle, and B.E. Warrick. 2007. Recommended last planting date for grain sorghum in the Texas Low Rolling Plains. http://sanangelo.tamu.edu/extension/agronomy/agronomy-publications/grain-sorghum-production-in-west-central-texas/recommended-last-planting-date-for-grain-sorghum-in-the-texas-low-rolling-plains/ (accessed 5 Oct. 2016).

Blum, A., and M. Naveh. 1976. Improved water-use efficiency in dryland grain sorghum by promoted plant competition. Agron. J. 68:111–116. doi:10.2134/agronj1976.00021962006800010029x

Bogard, M., V. Allard, M. Brancourt-Hulmel, E. Heumez, J.M. Machet, M.H. Jeuffroy, P. Gate, P. Martre, and J. Le Gouis. 2010. Deviation from the grain protein concentration–grain yield negative relationship is highly correlated to postanthesis N uptake in winter wheat. J. Exp. Bot. 61:4303–4312. doi:10.1093/jxb/erq238

Census of Agriculture. 2007. http://www.agcensus.usda.gov/Publications/2007/

Ciampitti, I.A., and P.V.V. Prasad. 2016. Historical synthesis-analysis of changes in grain nitrogen dynamics in sorghum. Front. Plant Sci. doi:10.3389/fpls.2016.00275

Ciampitti, I.A., and T.J. Vyn. 2012. Physiological perspectives of changes over time in maize yield dependency on nitrogen uptake and associated nitrogen efficiencies: A review. Field Crops Res. 133:48–67. doi:10.1016/j.fcr.2012.03.008

Ciampitti, I.A., and T.J. Vyn. 2013. Grain nitrogen source changes over time in maize: A review. Crop Sci. 53:367–377.

Ciampitti, I.A., and T.J. Vyn. 2014. Understanding global and historical nutrient use efficiencies for closing maize yield gaps. Agron. J. 106:2107–2117. doi:10.2134/agronj14.0025

Conley, S.P., W. Stevens, G. Dunn, and D. David. 2005. Grain sorghum response to row spacing, plant density, and planter skips. Crop Manage. 4. doi:10.1094/CM-2005-0718-01-RS

Cronholm, G., A. Knutson, R. Parker, and B. Pendelton. 2007. Managing insect and mite pests of Texas sorghum. Texas Coop. Ext. Serv. Bull. B-1220. Texas A&M Univ., College Station, TX:. http://lubbock.tamu.edu/files/2011/11/sorghum_guide_2007.pdf (accessed 5 Oct. 2016)

Dhuyvetter, K.C., C.R. Thompson, C.A. Norwood, and A.D. Halvorson. 1996. Economics of dryland cropping systems in the Great Plains: A review. J. Prod. Agric. 9:216–222. doi:10.2134/jpa1996.0216

Diawara, B. 2012. Effect of planting date on growth, development, and yield of grain sorghum hybrids. M.S. thesis. Kansas State Univ., Manhattan.

Duvick, D.N. 1999. Heterosis: Feeding people and protecting natural resources. In: J.G. Coors and S. Pandey, editors, The genetics and exploitation of heterosis in crops. ASA-CSSA-SSSA, Madison, WI. p. 25.

Fernandez, C.J., D.D. Fromme, and W.J. Grichar. 2012. Grain sorghum response to row spacing and plant populations in the Texas Coastal Bend region. Int. J. Agron. Article ID 238634. doi:10.1155/2012/238634

Food and Agriculture Organization of the United Nations—Statistics Division (FAOSTAT). 2012. http://faostat3.fao.org/home/E (accessed 5 Oct. 2016)

Fromme, D.D., C.J. Fernandez, W.J. Grichar, R.L. Jahn. 2012. Grain sorghum response to hybrid, row spacing, and plant populations along the Upper Texas Gulf Coast. Int. J. Agron. Article ID 930630, 5 pages, 2012. doi:10.1155/2012/930630

Gerik, T.J., and C.L. Neely. 1987. Plant density effects on main culm and tiller development of grain sorghum. Crop Sci. 27:1225–1230. doi:10.2135/cropsci1987.0011183X002700060027x

Haag, L.A. 2013. Ecophysiology of dryland corn and grain sorghum as affected by alternative planting geometries and seeding rates. Ph.D. diss., Kansas State Univ., Manhattan.

Khan, N.A. 2000. Simulation of wheat growth and yield under variable sowing date and seedling rate. M.S. thesis, Dep. of Agronomy. Univ. of Agriculture, Faisalabad, Pakistan.

Kichey, T., B. Hirel, E. Heumez, F. Dubois, and J. Le Gouis. 2007. In winter wheat (*Triticum aestivum*L.), postanthesis nitrogen uptake and remobilization to the grain correlates with agronomic traits and nitrogen physiological markers. Field Crops Res. 102:22–32. doi:10.1016/j.fcr.2007.01.002

Koo Kim, H., E.V. Oosterom, M. Dingkuhn, D. Luquet, and G. Hammer. 2010. Regulation of tillering in sorghum: Environmental effects. Ann. Bot. 106:57–67. doi:10.1093/aob/mcq079

Maiga, A. 2012. Effects of planting practices and nitrogen management on grain sorghum production. Ph.D. diss., Kansas State Univ., Manhattan.

Martin, V.L., and R.L. Vanderlip. 1997. Sorghum hybrid selection and planting management under moisture limiting conditions. J. Prod. Agric. 10:157–163. doi:10.2134/jpa1997.0157

Mason, S.M., D. Kathol, K.M. Eskridge, and T.D. Galusha. 2008. Yield increase has been more rapid for maize than for grain sorghum. Crop Sci. 48:1560–1568. doi:10.2135/cropsci2007.09.0529

Mesfin, T., B. Tesfahunegn, C.S. Wortmann, M. Mamo, and O. Nikus. 2010. Skip-row planting and tie-ridging for sorghum production in semiarid areas of Ethiopia. Agron. J. 102:745–750. doi:10.2134/agronj2009.0109

Miller, F.R., and Y. Kebede. 1984. Genetic contributions to yield grains in sorghum, 1950 to 1980. In: W. R. Fehr, editor, Genetic contributions to yield grains of five major crop plants. CSSA Spec. Publ. 7. CSSA, Madison, WI. p. 1–14.

Naeem, M. 2001. Growth, radiation use efficiency and yield of new cultivars of wheat under variable nitrogen rates. M.S. thesis, Dep. of Agronomy, Univ. of Agriculture. Faisalabad, Pakistan.

Norwood, C.A. 1994. Profile water distribution and grain yield as affected by cropping system and tillage. Agron. J. 86:558–563. doi:10.2134/agronj1994.00021962008600030019x

Norwood, C.A. 1999. Water use and yield of dryland row crops as affected by tillage. Agron. J. 91:108–115. doi:10.2134/agronj1999.00021962009100010017x

Norwood, C.A., and R.S. Currie. 1998. An agronomic and economic comparison of the wheat-corn-fallow and wheat-sorghum-fallow rotations. J. Prod. Agric. 11:67–73. doi:10.2134/jpa1998.0067

Norwood, C.A., and R.S. Currie. 1997. Dryland corn vs. grain sorghum in western Kansas. J. Prod. Agric. 10:152–157. doi:10.2134/jpa1997.0152

Norwood, C.A., A.J. Schlegel, D.W. Morishita, and R.E. Gwin. 1990. Cropping system and tillage effects on available soil water and yield of grain sorghum and winter wheat. J. Prod. Agric. 3:356–362. doi:10.2134/jpa1990.0356

Pan, W.L., J.J. Camberato, R.H. Moll, E.J. Kamprath, and W.A. Jackson. 1995. Altering source-sink relationships in prolific maize hybrids: Consequences for nitrogen uptake and crop science. Crop Sci. 35:836–845. doi:10.2135/cropsci1995.0011183X003500030034x

Peterson, G.A., D.G. Westfall, F.B. Peairs, L. Sherrod, D. Poss, W. Gangloff, K. Larson, D.L. Thompson, L.R. Ahuja, M.D. Koch, and C.B. Walker. 1999. Sustainable dryland agroecosystem management. Tech. Bull. TB99-1. Colorado State Univ. and Agric. Exp. Stn., Ft. Collins, CO.

Phillips, W.M. 1969. Dryland sorghum production and weed control with minimum tillage. Weed Sci. 17:451–454.

Pidaran, K. 2012. Effect of planting geometry, hybrid maturity, and population density on yield and yield components in sorghum. M.S. thesis, Kansas State Univ., Manhattan.

Prasad, P.V.V., M. Djanaguiraman, R. Perumal, and I.A. Ciampitti. 2015. Impact of high temperature stress on floret fertility and individual grain weight of grain sorghum: Sensitive stages and thresholds for temperature and duration. Front. Plant Sci. doi:10.3389/fpls.2015.00820

Routley, R., I. Broad, G. McLean, J. Whish, and G. Hammer. 2003. The effect of row configuration on yield reliability in grain sorghum: 1. Yield, water use efficiency and soil water extraction. In: Solutions for a better environment. Proceedings of the 11th Australian Agronomy Conference, Geelong, Victoria. 2–6 Feb. 2003. Australian Society of Agronomy, Gosford, Australia. www.regional.org.au/au/asa/2003/c/9/routley.htm (accessed 22 Aug. 2015).

Sanabria, J., J.F. Stone, and D.L. Weeks. 1995. Stomatal response to high evaporative demand in irrigated grain sorghum in narrow and wide row spacing. Agron. J. 87:1010–1017. doi:10.2134/agronj1995.00021962008700050040x

Schlegel, A.J. 2014. Large-scale dryland cropping systems. SWREC Field Day 2014. KSU AES and CES Report of Progress 1106:41–42.

Schlegel, A.J., K.C. Dhuyvetter, C.R. Thompson, and J.L. Havlin. 1999. Agronomic and economic impacts of tillage and rotation on wheat and sorghum. J. Prod. Agric. 12:629–636. doi:10.2134/jpa1999.0629

Schlegel, A.J., T.J. Dumler, and C.R. Thompson. 2002. Feasibility of four-year crop rotations in the central High Plains. Agron. J. 94:509–517. doi:10.2134/agronj2002.5090

Schnell, R.W., R. Collet, J. Gersbach, R. Sutton, and T. Provin. 2014. Optimizing grain sorghum seeding rates in Central Texas. Stiles Farm Foundation Research Report. Soil and Crop Science. Texas A&M AgriLife Extension. p. 44–46. http://stilesfarm.tamu.edu/files/2015/10/2014_Stiles_Booklet.pdf (accessed 14 Oct. 2016).

Staggenborg, S., B. Gordon, R. Taylor, S. Duncan, and D. Fjell. 1999. Narrow-row grain sorghum production in Kansas. Publ. MF-2388. Kansas State Univ. Agric. Exp. Stn. Coop. Ext. Serv. https://www.bookstore.ksre.ksu.edu/pubs/MF2388.pdf (accessed 5 Oct. 2016)

Steiner, J.L. 1986. Dryland grain sorghum water use, light interception, and growth responses to planting geometry. Agron. J. 78:720–726. doi:10.2134/agronj1986.00021962007800040032x

Steiner, J.L. 1987. Radiation balance of dryland grain-sorghum as affected by planting geometry. Agron. J. 79:259–265. doi:10.2134/agronj1987.00021962007900020017x

Stone, L.R., and A.J. Schlegel. 2006. Yield-water supply relationships of grain sorghum and winter wheat. Agron. J. 98:1359–1366. doi:10.2134/agronj2006.0042

Stone, L.R., D.E. Goodrum, A.J. Schlegel, M.N. Jaafar, and A.H. Khan. 2002. Water depletion depth of grain sorghum and sunflower in the central High Plains. Agron. J. 94:936–943. doi:10.2134/agronj2002.9360

Thompson, C.A., and D.A. Whitney. 1998. Long-term tillage and nitrogen fertilization in a west central Great Plains wheat-sorghum-fallow rotation. J. Prod. Agric. 11:353–359. doi:10.2134/jpa1998.0353

Trostle, C., and D. Fromme. 2011. South & Central Texas production handbook. United Sorghum Checkoff Program, Lubbock, TX.

Unger, P.W., and R.L. Baumhardt. 1999. Factors related to dryland grain sorghum yield increases: 1939 trough 1997. Agron. J. 91:870–875. doi:10.2134/agronj1999.915870x

United Stated Department of Agriculture National Agricultural Statistics Service (USDA-NASS). 2013. Agricultural Statistics 2013. US Gov. Printing Office, Washington, DC. http://www.nass.usda.gov/Publications/Ag_Statistics/2013/Agricultural_Statistics_2013.pdf (accessed 5 Oct. 2016).

Vanderlip, R.L. 1972. How a sorghum plant develops. Kansas Ext. Circular C-447.

Welch, N.H., E. Burnett, and H.V. Eck. 1966. Effect of row spacing, plant population, and nitrogen fertilization on dryland grain sorghum production. Agron. J. 58:160–163.

The Biology and Control of Sorghum Diseases

Christopher R. Little* and Ramasamy Perumal

Abstract

Sorghum [*Sorghum bicolor* (L.) Moench] is among the most important cereal crops in the world. It is a source of animal feed and fodder, traditional and processed foods and beverages, and an advanced biofuel. Biotic and abiotic stressors reduce global productivity of the sorghum crop. Among the biotic stressors, fungi, bacteria, viruses, nematodes, and parasitic plants cause diseases that compromise sorghum yield and quality. As a result, this review seeks to provide a framework to appreciate the causal organisms of the various sorghum diseases. Because fungi cause the bulk of the diseases, three disease classes are discussed for this pathogen group. These include (i) seedling, stalk, and root diseases, (ii) foliar diseases, and (iii) panicle and grain diseases. Throughout, integrated pest management strategies that emphasize reduced pathogen pressure, optimized plant growth conditions, and the utilization of disease resistance (or tolerance) are highlighted. To move forward in our understanding, each disease class or pathogen group discussion includes a summary of broad challenges for future work as perceived by the authors.

The Sorghum Crop

Sorghum (*Sorghum bicolor* (L.) Moench) is an important cereal grain crop for feed and food, but this highly diverse plant is also used as a fodder source and has potential in the biofuel industry (Smith and Frederiksen, 2000; Leslie, 2002; Tariq et al., 2012; Mocoeur et al., 2015). In the United States, average annual planting area ranged from 5.3 to 8.1 million acres per year during the period from 2009 to 2014 (USDA, NASS, 2015). During the same time, an average of 8.5 million metric tons of grain sorghum was produced per year. In 2013, worldwide, 61.4 million tons of sorghum were produced with an average yield of 14.6 thousand hectograms per hectare. Although the greatest proportion of sorghum was produced on the African continent (41.9%), the United States, Nigeria, and Mexico were the top three producers (FAO STAT, 2014).

Abbreviations: JgMV, Johnsongrass Mosaic virus; MCDV, maize chlorotic dwarf virus; MDMV, maize dwarf mosaic virus; MSpV, maize stripe virus; MSV, mastrevirus; PCR, polymerase chain reaction; QTL, quantitative trait loci; SCMV, sugar cane mosaic virus; SDM, sorghum downy mildew; SgCSV, sorghum chlorotic spot virus; SrMV, sorghum mosaic; VIGS, virus-induced gene silencing.

C.R. Little, Department of Plant Pathology, Kansas State University, Manhattan, KS 66506; R. Perumal, Agricultural Research Center- Hays, Kansas State University, Hays, KS 66506. *Corresponding author (crlittle@ksu.edu)

doi:10.2134/agronmonogr58.2015.0073

Sorghum is a traditional cereal crop in semiarid regions of Africa and is considered a "food security crop" (Strange and Scott, 2005). Productivity can be compromised due to a suite of perennial abiotic and biotic problems. In some regions of the developing world, many of the traditional, preferred landrace varieties might not be disease resistant. Fungi, bacteria, viruses, nematodes, and parasitic plants cause sorghum diseases (Frederiksen and Odvody, 2000; Smith and Frederiksen, 2000; Leslie, 2002; Little et al., 2012; Tesso et al., 2012). Some sorghum pathogens are cosmopolitan in nature; they cause disease every year, and only cause minimal or indistinguishable losses, although regional epidemics may occur. Examples of such diseases include zonate leaf spot (*Gloecercospora sorghi*), sooty stripe (*Ramulispora sorghi*), and bacterial leaf diseases (various genera and species) (Frederiksen and Odvody, 2000; Smith and Frederiksen, 2000; Leslie, 2002; Brady et al., 2011). Other diseases may be quite acute and have caused severe, epidemic-level losses. Such diseases include anthracnose (*Colletotrichum sublineola*), sorghum downy mildew (*Peronosclerospora sorghi*), and ergot (*Claviceps africana, Claviceps sorghi*) (Frederiksen and Odvody, 2000; Smith and Frederiksen, 2000; Leslie, 2002; Little et al., 2012; Tesso et al., 2012). Yet, other diseases cause perennial yield drag and require much ongoing research and attention such as Fusarium stalk rot (*Fusarium* spp.), charcoal rot (*Macrophomina phaseolina*), and grain mold (numerous fungal spp.) (Little et al., 2012; Tesso et al., 2012). In epidemics, yield losses can be significant, widespread, and economically impactful. In general, host resistance varies for many of these diseases, which is often worsened by the fact that pathogen populations, especially those undergoing sexual reproduction on a regular basis, can shift their genetic structure, and new pathotypes or races can occur on a relatively short timescale (McDonald and Linde, 2002; Stukenbrock and McDonald, 2008). Therefore, cultural controls continue to constitute best management practices for certain diseases including many of the root and stalk rots, which are predominantly initiated and worsened when plants are exposed to environmental stress (Tesso et al., 2012).

Sorghum Disease Control

Control of sorghum diseases is best achieved using integrated pest management strategies that function to reduce pathogen pressure, optimize plant growth conditions, and make use of highly adapted disease-resistant (or tolerant) sorghum hybrids. In general, these constitute the sides of the often-cited "plant disease triangle." In principle, removal or reduction of any component of the triangle (i.e., a virulent pathogen, conducive environment, susceptible host) will result in some level of disease control. However, the control measure should always be implemented before the establishment of an epidemic (Scholthof, 2007; Yuen and Mila, 2015). For example, pathogen pressure may be reduced if infested crop residue is removed via tillage, disease-free or treated seed is planted, or pesticides are applied in a preventative and timely manner (Ciancio and Mukerji, 2007). Sorghum growing conditions may be optimized if a proper seedbed is available for the germinating seedling, accurate and appropriate herbicide and insecticide applications are performed, or planting is timed to ensure that favorable conditions can be exploited and optimum yields achieved (Vanderlip and Reeves, 1972). The major goal of breeding programs in Kansas and other sorghum-producing regions is to develop breeding lines (and hybrids) that have good adaptation to environmental stress, but also confirmed resistance (or tolerance) to pathogens

(Kapanigowda et al., 2013a, 2013b; Perumal et al., 2015). This latter strategy has the greatest long-term impact for sorghum producers, as it is often difficult to modulate the pathogen and environment sides of the plant disease triangle.

Disclaimer: It is the responsibility of individual producers, extension specialists, and other professionals to be aware of their environmental and site constraints, choose the most logical and profitable cultural practices, and select the best genetic material for their production system. Additionally, accurate disease identification and diagnosis is critical for making correct pest management decisions. It is not the goal of this review chapter to replace the advice of highly trained university diagnosticians, experienced field consultants, or seed dealers.

Seedling, Stalk, and Root Diseases

Among the most understudied diseases of sorghum are those that affect seedlings, roots, and stalks. Historically, this has been due to an observation bias on the part of the pathologist, diagnostician, and breeder, that is, "out of sight, out of mind." These pathogens act on underground plant structures, or within stalks, and symptoms are often not visually apparent or if they are, may appear ambiguous compared to the abiotic stressors that often predispose these diseases (Frederiksen and Odvody, 2000; Leslie, 2002; Tesso et al., 2012).

Seedling, root, and stalk diseases have the potential to cause more damage than any other. Of these, charcoal rot, caused by *Macrophomina phaseolina* (Tassi) Goid., is likely the most damaging (Leslie, 2002; Tesso et al., 2012; Bandara et al., 2015). However, epidemics of Fusarium stalk rot and root rot in the Midwestern United States have brought attention to a suite of *Fusarium* spp. that remain largely uninvestigated and still require further study (Leslie et al., 1990; Tesso et al., 2010; Adeyanju, 2014). In both charcoal rot and Fusarium stalk rot, stalks may be spongy and show discoloration and shredding of the pith. The plants often have a grayish-green hue, do not produce a good yield, and may lodge. These fungi survive in crop residue and are capable of infecting seedlings early, but may progress through the host development period. Several environmental stressors cause root and stalk diseases to manifest themselves. Therefore, management strategies that mitigate plant stress include proper nutrition (especially N/K balance), avoiding excessive plant populations, and controlling weeds that compete with the plants (Pande et al., 1990; Cloud and Rupe, 1994; Leslie, 2002; Sweeney et al., 2011). As will be explored below, the use of resistant hybrids and rotation away from sorghum are only marginally effective strategies due to the ongoing search for resistance and the long period that pathogen propagules (e.g., microsclerotia and chlamydospores) may survive in the soil or in host debris.

Seed Rots and Seedling Diseases

Seed rots are those diseases that result in the failure of the seed to germinate in the soil (Swarup et al., 1962). Seedling diseases, including blights, may cause both pre- and post-emergent damping-off. These diseases are common throughout all sorghum production regions and are caused by both soilborne and seedborne pathogens (Frederiksen and Odvody, 2000). These pathogens include *Rhizoctonia solani* J.G. Kühn, *Fusarium* spp. (seedling blights), *Pythium* spp., *Macrophomina phaseolina* (Tassi) Goid (seedling root rot), *Colletotrichum sublineola* Henn ex Sacc. & Trotter, and several fungi that cause seed rot or damping-off due to improper seed storage prior to planting (Christensen and Kaufman, 1965; Pratt and Janke, 1980; Sumner and Bell, 1982; McLaren and Rijkenberg, 1989; Leslie et al., 1990; Jardine and Leslie, 1992; Frederiksen and Odvody, 2000; Leslie, 2002; Sharma et al., 2012). *R. solani, M. phaseolina,* and *Pythium* spp. are largely soilborne and debris-borne, whereas some *Fusarium* spp. also survive in the environment, but some species may be introduced to the developing seedling by a seedborne route (Singleton et al., 1992; Little et al., 2012). Arif and Ahmad (1969) and Narasirnhan and Rangaswami (1969) showed that *Alternaria* spp. could reduce germination when sorghum kernels were inoculated externally with conidia. *Alternaria alternata* caused post-emergent seedling death (Konde and Pokharjar, 1979). These pathogens also affect young, and older plant roots, as well as other plant parts and it may be difficult to differentiate root rots caused by these organisms as they share similar symptoms.

Some *Pythium* species cause seedling disease in sorghum and include *P. aphanidermatum* (Edson) Fitzp. (Freeman et al., 1966). *P. arrhenomanes* Drechsler (Forbes et al., 1987), *P. graminicola* Subr. (Pratt and Janke, 1980), *P. myriotylum* Drechsler (Wang et al., 2003), and *Pythium ultimum* var. *ultimum* Trow (Davis and Bockus, 2001). *P. arrhenomanes, P. graminicola,* and *P. myriotylum* are related species and fall within the same phylogenetic clade B1. However, technologies such as loop-mediated isothermal amplification (LAMP) PCR can distinguish and allow for detection of specific *Pythium* species within this group (Fukata et al., 2014).

Fusarium thapsinum Klittich, J.F. Leslie, P.E. Nelson, & Marasas, once classified as *F. moniliforme* Sheld. (sensu lato), infects sorghum florets, invades the developing kernels, reduces seed germination, and causes seedling blight (Castor, 1981; Frederiksen et al., 1982; Leslie, 2002; Little and Magill, 2003; Little et al., 2012). *F. thapsinum* is more pathogenic on sorghum seedlings than *F. verticillioides* (Sacc.) Nirenberg or other members of the classical *Gibberella fujikuroi* (Sawada) Wollenw. species complex and causes reduced germination and seedling vigor (Little and Magill, 2003). According to Castor (1981), "seedlings, which develop from fungal colonized grain, may be less vigorous than under normal conditions."

Rhizoctonia solani AG (anastamosis group) 1–1A causes sheath blight in sorghum and presumably this *R. solani* variant is more adapted to above-ground plant tissues than root tissues (Pascual et al., 2000; Frederiksen and Odvody, 2000). However, *R. solani* AG-2 has been found in sorghum debris, and *R. solani* AG-3 has been isolated from Johnsongrass (Ruppel, 1985; El Bakali et al., 2000). It is not currently known what the major seedling-infecting anastomosis group is for sorghum. However, geographical location, crop rotations, seed treatment use, and other agroecosystem characteristics may impact the *R. solani* populations or anastamosis group communities that predominate on sorghum seedlings.

Regardless of whether the pathogen gains entry to the host seedling via a soilborne or seedborne route, or due to improper storage of seed, the result is weakened or killed seedlings. Thin, uneven stands characterized by frequent row skips may lead to increased replanting costs. Those seedlings that do emerge are often weak or stunted compared to healthy seedlings, especially those that are infected by *Fusarium* spp. (Little and Magill, 2003). Further, if the rate of emergence is reduced, season-wide plant development is slowed, yield potential is reduced, and plants may become more susceptible to other diseases and environmental stressors (Debaeke et al., 2006; Kapanigowda et al., 2013a).

Management of seedling diseases is accomplished using high quality, disease-free seed that are treated with a broad-spectrum fungicide, planted at proper depths and at proper plant populations, planted into fertile soils of moderate pH (6.0 to 6.5) and at warmer temperatures (> 70°F), and planted into well-draining soil beds (Frederiksen and Odvody, 2000; Mughogho, 1984; Wu and Cheng, 1990; McGee, 1995; Debaeke et al., 2006). Proper pH and temperature at planting ensure rapid seedling germination, emergence, and development, which allow the seedling to avoid (i.e., "outgrow") seedling pathogens. However, combining seedling cold tolerance with pathogen resistance has been a major challenge where producers wish to take advantage of early season moisture and extend the overall growth period to realize greater season-wide yield potentials (Kapanigowda et al., 2013a). Davis and Bockus (2001) reported that damping-off caused by *P. ultimum* var. *ultimum* was most severe at cooler temperatures (10 to 15°C) compared to warmer temperatures (20 to 25°C), although some levels of damage were also observed at the higher temperatures. Conversely, some species found from sorghum, including *P. aphanidermatum* and *P. myriotylum*, exhibit greater growth at higher temperatures (Ishiguro et al., 2013). Some storage fungi, including *Aspergillus*, *Penicillium*, and *Rhizopus*, were shown to cause seedling blights and damping-off symptoms some decades ago. Leukel and Martin (1943) found that *Aspergillus niger* and *Rhizopus stolonifer* caused seedling blights and damping-off at lower soil temperatures (15°C) than *Penicillium* spp., which Edmunds et al. (1970) indicated as a seedling blight pathogen at 20 to 25°C.

Low pH soils (< 5.2) tend to favor seedling blights caused by *Fusarium* spp. For example, Davis et al. (1994) found that *Fusarium graminearum* Schwabe was the dominant species isolated from sorghum seedlings in south central Kansas. Under both wheat-grain sorghum and grain sorghum-grain sorghum rotation conditions, *F. graminearum* was isolated from seedling roots at higher levels under lower pH. Likewise, lower pH reduced the rate of secondary root growth. Other pathosystems involved *Rhizoctonia spp.* For example, *R. solani* caused less disease in *Agrostis stolonifera* L. under normal nitrogen levels at pH 8.5 and 10.0, compared to lower pH values (Bloom and Couch, 1960). Kumar et al. (1999) showed that *R. solani* AG-8 from wheat roots and AG-11 from lupin hypocotyls exhibited optimal growth on potato dextrose agar (PDA) at pH 6.0 to 7.0, however, AG-11 growth significantly decreased at pH 8.0 and 9.0, whereas AG-8 growth decreased significantly at pH 4.0 and 5.0. Based on these examples, clear trends are not obvious for *R. solani* regarding pH whether in the soil or in vitro.

In general, resistance in sorghum to *Pythium* spp., *Fusarium*, and *R. solani* is lacking or unknown. However, other management strategies can be employed to mitigate the poor vigor and yield reductions that occur with these diseases (de Milliano et al., 1992). Seedling diseases may be controlled using disease-free seed that is treated with combinations of broad-spectrum fungicides for both true fungi (e.g., *Fusarium* spp. and *R. solani*) and oomycetes (e.g., *Pythium* spp.) (Mueller et al., 2013). Information in the older literature suggested that various fungicide combinations including sulfur compounds (thiram), carboxamides (carboxin), PCNB (Terraclor, Olin), thiadiazoles (etridiazole), benzimidazoles (carbendazim), phthalimides (Captan), and dithiocarbamates (maneb) could control the primary fungal pathogens involved with seedling diseases in sorghum. However, due to the toxic nature of some of these active ingredients, many are no longer used. Davis and Bockus (2001) indicated that use of metalaxyl (73 g a.i./100 kg seed) increased grain sorghum yields by 24% in *Pythium*-infested fields. More recently, in 2010 and 2014, respectively, BASF expanded its label for Stamina F3

HL fungicide seed treatment and Priaxor (fluxapyroxad + pyraclostrobin) for sorghum. In 2010, Syngenta labeled Dynasty seed treatment fungicide for sorghum in addition to its previously labeled products, Apron XL and Maxim 4FS (Plant Management Network, 2010a, 2010b; Plant Management Network, 2014). However, seed treatments may not be recommended or available for sorghum in every production area.

Root Rots

Root rots in sorghum are understudied diseases that still require significant research, but many of the lessons learned for seedling and stalk rot diseases apply to their management. Numerous common soilborne fungi such as *Fusarium* spp., *R. solani*, *Pythium* spp., and *Macrophomina phaseolina* are associated with sorghum roots and form a "root rot complex" (Mughogho, 1984; Leslie et al., 1990; Frederiksen and Odvody, 2002; Idris et al., 2007; Idris et al., 2008).

A success story for sorghum disease management is the long-term control of milo disease caused by *Periconia circinata* (L. Mangin) Sacc. & D. Sacc, which causes a root and crown rot in sorghum. This disease has been managed for over 80 yr due to the discovery of sorghum mutants in the 1930s with resistance to the host-specific *peritoxin*. Thus, commercial sorghum hybrids grown today have used the same recessive gene (*pc*) for decades with no threat of the pathogen overcoming its durability (Frederiksen and Odvody, 2000; Smith and Frederiksen, 2000; Tesso et al., 2012; Nagy and Bennetzen, 2008). However, many root rot complexes caused by *Fusarium* spp., for example, are not so simple because resistance is suspected to be quantitatively inherited, there are numerous *Fusarium* spp. associated with roots, and/or the diseases are complicated by factors in the soil environment.

Root rots, along with seedling and stalk rot diseases, are present every year wherever sorghum is grown. Jardine and Leslie (1992) reported that crop losses to root rots ranged from 5 to 10% due to the poor grain fill, weakened peduncles, and plant lodging that is observed with these diseases. Since there are numerous organisms associated with the root rot complex, symptoms are largely dependent on the pathogen that caused the disease. Multiple pathogens may also attack sorghum roots simultaneously. However, the predominance of one pathogen in a production environment will present a characteristic set of symptoms. For example, *Fusarium* spp. causes elongated to circular, pink, red, or purple discoloration on roots. Other pathogens such as *Pythium* spp. and *R. solani* will cause root necrosis characterized by brown, watersoaked areas (Frederiksen and Odvody, 2000; Davis and Bockus, 2001). *M. phaseolina* causes blackening of the roots and form microsclerotia in senescing, dried out tissues (Frederiksen and Odvody, 2000; Leslie, 2002; Tesso et al., 2012). In most cases, the pathogen moves into the lower internodes of the plant and causes discoloration there. In cases of severe disease, it is thought that either *Fusarium* spp. or *M. phaseolina* may grow into the stalk of the plant. Above-ground, root rot pathogens cause similar symptoms including stunting in the apical regions of the plant, leaf chlorosis and necrosis, wilting, and plants that are uprooted with relative ease compared to their healthy counterparts due to reduced root development (Frederiksen and Odvody, 2000; Idris et al., 2007; Idris et al., 2008).

As with symptomology, different fungi are problematic under varying environmental conditions. For example, *Pythium* spp. prefer cool, wet conditions, whereas *Fusarium* causes diseases during periods that have been wet, then dry, or alternately so. *M. phaseolina* causes disease in hot, dry environments (Mughogho, 1984; Cloud and Rupe, 1994; Davis and Bockus, 2001). Along with these conditions, improper soil fertility, poor soil drainage, insect damage, and late planting may favor fungal progression into and through the root and crown tissues (Davis and Bockus, 2001; Leslie, 2002). The root rotting pathogens are necrotrophs and hemibiotrophs and will enter through wounds due to insect or mechanical damage and natural openings. Limited information exists for *Rhizoctonia* diseases of sorghum roots. However, more information is available for the *R. solani* anastomosis group 1-IA that causes banded leaf and sheath blight on above ground foliage (Pascual et al., 2000). According to Frederiksen and Odvody (2000), the strains that cause root rot are likely not the same as those causing banded leaf and sheath blight.

One of the best control strategies for root rot diseases is the use of locally adapted hybrids that withstand the range of stressors associated with a given production environment (de Milliano et al., 1992). Resistance to root rot is partial in most cases and heavily influenced by the environment. Host tolerance to adverse abiotic conditions, coupled with cultural management strategies, is necessary for control (Frederiksen and Odvody, 2000; Diourte et al., 2007). In some cases, rotation with other crops is a good strategy to reduce pathogen populations. For example, rotation of maize with wheat resulted in significantly reduced root rot compared to continuous corn (Govaerts et al., 2007). However, many of the pathogens survive in the soil for years, cause disease in multiple crop species, and crop rotation may not be economically viable for all producers. Oospores, chlamydospores, and microsclerotia serve as soil overseasoning structures for *Pythium*, *Fusarium*, and *Macrophomina*, respectively (Singleton et al., 1992). Instead of crop rotation, planting in warm soils (> 65°F/18°C) at a proper depth and seeding rate, as suggested for seedling disease control, helps to prevent undue seedling stress and fosters vigorous plant development (Kapanigowda et al., 2013a). Seedling injury should be avoided through the proper use of herbicides, insecticides, and fertilizer treatments. For example, nitrogen rates greater that 26.9 kg ha^{-1} using polymer coated urea resulted in seedling injury (Nelson et al., 2008). It is certain that treatments that lead to seedling damage may allow these pathogens to colonize weakened plants and cause subsequent root rot diseases. Azoxystrobin (Quadris) is labeled for in-furrow use to protect seedlings and roots against *R. solani* and *Pythium* spp. Additionally, this product has been labeled for protection of seedlings against *M. phaseolina*, which may also protect the developing roots of younger plants (Mueller et al., 2013).

Charcoal Rot

Charcoal rot is caused by the soilborne pathogen, *Macrophomina phaseolina* (Frederiksen and Odvody, 2000; Leslie, 2002; Cruz Jimenez, 2011; Tesso et al., 2012; Bandara et al., 2015). This fungus is a cosmopolitan generalist necrotroph and causes disease on numerous crops and wild species (Islam et al., 2012). Plants infected with this fungus exhibit poor grain filling and may lodge during the season (Tesso et al., 2012). Often, disease foci are scattered in the field where portions of the field may be completely lodged, and the areas outside these foci are normal. Lodging is due to internal decomposition of the pith at the crown level and above by the fungus. Often, the vessels remain, but the structural integrity of the stalk is compromised due to pith loss (Mughogho, 1984). Close inspection of diseased stalk materials reveal pepper-like microsclerotia on the remaining vascular strands, which are lacking if the stalk rot is caused by *Fusarium*, *Pythium*, or *Colletotrichum* (Khune et al., 1984; Tesso et al., 2012). If the plant has lodged, the development of microsclerotia will be greatest at the point where stalk damage is greatest. Lodged stalks, whether due to *M. phaseolina* or *Fusarium* spp. infection, cause severe yield losses, increased harvest difficulty, and are reduced in fodder quality (Trimboli, 1982; Khune et al., 1984; Omar et al., 1985; Seetharama et al., 1987; de Milliano et al., 1992; Mayek-Perez et al., 2001; Tesso et al., 2005). Normally, there is a lack of red discoloration of the stalk in charcoal rot disease, as this symptom is characteristic of plants with Fusarium stalk rot. Stalks infected by *M. phaseolina* exhibit an ash-gray or black color (Adeyanju, 2014). However, in plants that have red (*PPq^r^q^r^*) or purple (*PPQQ*) pigmentation, some accumulation of pigments in the stressed areas of living pith tissue may be observed. Often, chlorosis of the panicle and supporting stalk is the aerial symptom of the disease.

Microsclerotia serve as the primary inoculum source for *M. phaseolina* and produce hyphae that grow through the soil and infect roots of the developing plant (Cloud and Rupe, 1994; Odvody and Dunkle, 1979). In the southcentral United States and other arid regions of dryland production, this disease manifests itself when environmental conditions are the most stressful (Diourte et al., 2007; Odvody and Dunkle, 1979). Therefore, high soil temperatures (30 to 35°C) and low moisture levels (-1.3 to -413 MPa) enhance pathogen growth, and after flowering, may weaken host defenses and promote the manifestation of charcoal rot disease (Odvody and Dunkle, 1979). If grain filling occurs during stress, the physiological demand (first sink) in addition to the pathogen demand (second sink) have a significant combined effect on yield (Bandara et al., 2017). High plant populations, leaf diseases, hail damage, mechanical damage, high nitrogen fertilizer, and insect-feeding damage may stress the plant and increase disease severity (Williams et al., 1980; Pande et al., 1990; de Milliano et al., 1992; Russin et al., 1995). In general, light, shallow soils are more prone to drought stress, and plants in these soils will be more sensitive to high temperature and prolonged water deficit stress (Mahalakshmi and Bidinger, 2001; Tesso et al., 2012).

Although not always feasible, decreasing moisture stress during the post-flowering period has been shown to be a primary means of reducing the impacts of charcoal rot (Tesso et al., 2012; Bandara et al., 2016). Utilization of the "stay-green" trait, post-flowering non-senescence associated with drought tolerance, has been the primary mechanism to accomplish this in the host (Rosenow et al., 1977; Rosenow and Clark, 1981; Mughogho, 1984; Borrell et al., 2000). The maintenance of photosynthetically active leaves after anthesis is a stay-green trait that has been observed in numerous monocotyledonous crops in addition to sorghum, including maize, rice, wheat, and oats. Four QTLs (*stg*1, *stg*2, *stg*3, and *stg*4) are associated with this trait in sorghum (Xu et al., 2000). Stay-green provides a physiological means to avoid stress and tolerates *M. phaseolina* through reduced lodging, better grain fill under high temperature and drought stress, increased stem carbohydrates during and after flowering, maintenance of increased green leaf area during post-flowering drought, increased leaf nitrogen content, and improved transpiration efficiency (Rosenow and Clark, 1981; Mughogho, 1984; Borrell and Douglas, 1997; Rosenow et al., 1997; Borrell et al., 2000; Borrell and Hammer, 2000).

Other methods that reduce plant stress include avoiding excessive plant populations in non-irrigated environments, balancing N and K nutrient levels (i.e., avoid high N and low K), controlling weeds, and minimizing insect feeding (Frederiksen and Odvody, 2000; Adeyanju, 2014). Crop rotation is not a successful control strategy since more than 500 species of plants, including most legumes and large-grain cereals, are hosts for *M. phaseolina* (Short et al., 1980; Cruz, 2011). Cotton and wheat, which are relatively poor hosts of *M. phaseolina*, may serve as successful rotation crops (Zaki and Gaffar, 1988; Cruz, 2011). Saleh et al. (2010) found that there may be some host specialization in *M. phaseolina*. However, there is no evidence to suggest that the degree of host specialization or preference is sufficient between local or regional populations of the fungus to sufficiently differentiate between hosts and result in reduced pathogenicity and reproduction on any given rotation host. Clearly, this is an area for further research.

Management of crop residue may have mixed results since this material contains large numbers of microsclerotia that are disseminated in the soil when residue decomposes if the preceding crop had been infected. However, residue accumulation also promotes greater soil water content and lower soil temperatures that in turn reduce stress and charcoal rot disease incidence (Cook et al., 1973; Dhingra and Sinclair, 1975; Short et al., 1980; Baird et al., 2003; Mengistu et al., 2007).

Sorghum hybrids vary in resistance to charcoal rot, but it is recommended to plant hybrids with good stalk strength and standability and later maturity to avoid the highest temperature and moisture stress periods of the growing season and thus escape the disease (Vidyabhushanam et al., 1989). Disease resistance screening and release of highly adapted germplasm with charcoal rot tolerance or resistance requires persistence. Likewise, identification of the genetic loci and molecular mechanisms associated with charcoal rot resistance is a high research priority for this disease. Recent work by Adeyanju et al. (2015) has identified SNP loci and linked genes that associate with charcoal rot using a genome-wide association study (GWAS). For example, one SNP on chromosome 9 of *S. bicolor* is linked to a gene that encodes an ROP GTPase that has intimate involvement in hormone-mediated plant disease resistance.

Dynasty (azoxystrobin) seed treatment has been labeled for use on sorghum (Plant Disease Management Reports, 2010b) along with other products including ApronXL (mefenoxam) and Maxim 4FS (fludioxonil); however, it is not clear what the effect of these active ingredients will be on early colonization of sorghum seedlings by *M. phaseolina* or the impact on the health of the adult sorghum plant. In other studies, both mefenoxam and azoxystrobin were effective in reducing charcoal rot disease incidence in melon when used as a soil treatment (Cohen et al., 2012).

Fusarium Stalk Rots

Fusarium stalk rot is caused by numerous *Fusarium* spp. including *F. thapsinum*, *F. proliferatum* (Matsush.) Nirenberg ex Gerlach & Nirenberg, *F. subglutinans* (Wollenw. & Reinking) P.E. Nelson, Toussoun & Marasas, and *F. andiyazi* Marasas, Rheeder, Lampr., K.A. Zeller, & J.F. Leslie and is found in the same geographic areas as charcoal rot (Leslie et al., 1990; Tesso et al., 2010; Tesso et al., 2012). In many ways, the effects of Fusarium stalk rot and charcoal rot are similar, e.g., premature plant death, stalk discoloration (Fig. 1A & B), a shredded vascular system (although not to the same extent as charcoal rot), lodging (Fig. 1C), and leaves

Fig. 1. Examples of common sorghum diseases encountered in the Midwestern United States. A. Sorghum line ('SC599') resistant to Fusarium stalk rot infection by *Fusarium thapsinum* (*white arrow* = artificial inoculation point); B. Sorghum line susceptible to Fusarium stalk rot infection (*F. thapsinum*) (*white arrow* = artificial inoculation point; note pith discoloration due to bidirectional progress of the lesion); C. Lodging symptoms associated with fungal stalk rots of sorghum; D. Interveinal striping pattern caused by systemic infection by the sorghum downy mildew pathogen, *Peronosclerospora sorghi*; E. Adaxial foliar lesions caused by the sooty stripe pathogen, *Ramulispora sorghi* (note the broad chlorotic halos = *white arrow*); F. Abaxial microsclerotia produced by the sooty stripe pathogen, *R. sorghi*; G. Adaxial rust pustules (uredinia) produced by *Puccinia purpurea*; H. Honeydew exudate from sorghum florets produced by the ergot pathogen, *Claviceps africana*; I. Sorghum kernels (line 'SC170', moderately susceptible to grain mold) resulting from various artificial inoculations by sterile-distilled water (control, left panel), *F. thapsinum* (middle panel), and *Curvularia lunata* (right panel); J. Bacterial leaf stripe caused by *Pseudomonas andropogonis*; K. Red leaf symptoms caused by cool nighttime temperatures after infection by maize dwarf mosaic virus.

with a scorched, "grayed out," or freeze-damaged appearance (Tesso et al., 2005; Adeyanju, 2014; Adeyanju et al., 2015; Tesso et al., 2012; Bandara et al., 2015). However, plants affected by Fusarium stalk rot are often randomly dispersed in the field, rather than occurring in foci that spread out from a central infection area as in charcoal rot. Also, microsclerotia are not produced on the diseased tissue, the outer stalk remains green during the infection process, and the infected pith area is often discolored pink, red, or maroon to purple due to the production of pigments by the plant such as 3′-deoxyanthocyanins and other flavonoid-derived anthocyanin pigments (Adeyanju, 2014).

Like charcoal rot, post-flowering drought stress promotes Fusarium stalk rot. However, the disease manifests itself to its greatest extent when there has been sufficient moisture and moderate temperatures after panicle initiation, which occurs relatively early in plant development (Rosenow et al., 1977; Adeyanju, 2014). When drought stress occurs after this period, similar compounding factors as seen in charcoal rot worsen the disease, for example, high plant populations, high N and low K fertility, insect feeding, other foliar diseases, and hail damage (Trimboli and Burgess, 1983; Mughogho, 1984). Continuous cropping may increase the incidence of Fusarium stalk rot as the various *Fusarium* spp. overseason in crop debris and a build-up of pathogens in the soil increase the probability that propagules will germinate and enter plants through natural wounds or other means of damage at the crown of the plant and the root system (Mughogho, 1984; Frederiksen and Odvody, 2000; Leslie, 2002; Idris et al., 2008). Since such a wide array of *Fusarium* spp. may be associated with sorghum roots, crown tissues, and stalks, numerous grass hosts may act as reservoirs.

Similar cultural strategies may be employed to mitigate the effects of Fusarium stalk rot as in charcoal rot. Avoiding stress is the primary cultural control, and this may be accomplished through irrigation at panicle initiation, reduced tillage, improved soil fertility, reducing plant populations in dryland environments, and controlling plant damage by insects and other causes (Seetharama, 1987; Frederiksen and Odvody, 2000; Adeyanju, 2014). Another cultural management practice, *ecofallow,* combines rotation and no-till practices to efficiently manage Fusarium stalk rot in grain sorghum. Doupnik and Boosalis (1980) indicated that sorghum Fusarium stalk rot incidence was decreased from 39 to 11% and yields increased from approximately 2700 to 3800 kg ha^{-1} compared to conventional tillage, respectively, using an ecofallow approach.

Hybrid resistance varies. Therefore, to avoid lodging problems hybrids with good stalk strength characteristics should be grown. However, limited information is available in this regard. Plants with the stay-green phenotype exhibit physiological tolerance to *Fusarium* spp. The difference between the responses of stay-green lines to *Fusarium* spp. versus *M. phaseolina* remains unclear (Tesso et al., 2012). However, as detected by SPAD readings, Bandara et al. (2016) found that the Fusarium stalk rot resistant and staygreen line, SC599, exhibited negative senescence after inoculation from soft dough to physiological maturity, whereas susceptible hybrids and lines senesced. Some lines show clear resistance to *Fusarium,* different from lines showing resistance to *M. phaseolina,* and acts independently of the major QTLs for stay-green in sorghum (Adeyanju, 2014; Adeyanju et al., 2015; Bandara et al., 2015).

Acremonium Wilt

Acremonium wilt of sorghum is a disease caused by the fungus, *Acremonium strictum* (W.) Gams (syn = *Sarocladium strictum* (W. Gams) Summerbell) (Summerbell et al., 2011). The fungus lives as a saprophyte in the environment, but is a pathogen on numerous plant hosts including a wide range of monocots and dicots (Leslie, 2002). Acremonium wilt was first reported in Texas and Georgia, and then in the sorghum-producing regions of Mexico (Natural et al., 1982). The fungus is also an opportunistic pathogen of humans as it causes a hyalohyphomycosis that can be fatal in immunosuppressed patients (Guarro et al., 1997; Novicki et al., 2003). Taxonomic placement of this species has been difficult, but Saleh et al. (2004) found that *A. strictum* is a distinct species allied with related clades of ascomycete fungi included in the *Acremonium–Cephalosporium* and *Gaeumannomyces–Harpophora* species complexes. Subsequently, phylogenetic analysis of this complex group of fungi reclassified *A. strictum* and *S. strictum* based on ribosomal small subunit DNA sequences (Summerbell et al., 2011).

The development of asymmetrical necrosis along the midrib of mature sorghum leaves characterizes Acremonium wilt where one side of the blade dies, and the other remain as healthy. As the disease spreads, the entire leaf dies and gives the foliage a bleached appearance (Natural et al., 1982; Leslie, 2002). As with many biotic stresses of sorghum, the upper leaves are affected first. In the case of Acremonium wilt, the lower foliage may survive the disease as the upper leaves desiccate. If the plant wilts, a cross-section of the stem at the crown will show necrotic vascular bundles. Purple or red wound-associated pigments, which are largely 3′-deoxyanthocyanins and related flavonoid-derived molecules, accumulate at the sites of vascular necrosis. Pigment accumulation may occur in the affected leaves as well. As far as it is known, *A. strictum* does not produce a translocated toxin, and the pathogen itself may be isolated from any plant tissues showing disease symptoms (Natural et al., 1982). The pathogen moves systemically through susceptible hosts and has been reported to be seedborne (Bandyopadhyay et al., 1987).

Natural et al. (1982) indicated that Acremonium wilt could reduce grain yield by 50%. Plants with wounded roots that contact high amounts of soilborne inoculum appear to be more susceptible to the disease. The same authors indicated that lines exist that were resistant to natural infection in the field.

Anthracnose Stalk Rot

Anthracnose, caused by *Colletotrichum sublineola*, affects all parts of the sorghum plant and occurs in up to four phases: (i) seedling root rot, (ii) foliar, (iii) stalk, and (iv) grain. Anthracnose stalk rot occurs later in the season as the fungus moves from the foliage and into the stalk where it causes reddish discoloration and rot, which can lead to lodging (Tesso et al., 2012; Burrell et al., 2015). If grain develops during warm, humid periods, grain yield may be reduced through reduced translocation. Quality may also be reduced if the fungus colonizes the grain itself (Hagan et al., 2014). Colonization of the stalk may lead to the production of microsclerotia that can survive on the soil surface in crop residue (Tesso et al., 2012).

Chemical controls are ineffective because the pathogen has often progressed too far in diseased plants by the time symptoms are observed. Therefore, resistance is the best way to control anthracnose. Recently, Burrell et al. (2015) found a

QTL on the distal end of chromosome 5 that was consistent for foliar, stalk, and grain stages of anthracnose. In sweet sorghum, Hagan et al. (2014) found that 'Sugar Drip' and 'Dale' were more resistant to anthracnose compared to 'M81E'. Forage sorghums were not as resistant as the sweet, silage, or grain types tested. For additional information and management strategies for anthracnose, please see the section concerning the foliar phase of the disease.

Milo Disease

In the early twentieth century, Milo disease, caused by *Periconia circinata* (L. Mangin) Sacc., was one of the first major disease threats to emerge in sorghum in the United States (Leukel, 1948; Churchill et al., 2001). As previously mentioned in this section, *P. circinata* produces a root and crown rot disease. The fungus grows slowly in culture, but persists in the soil and can survive and grow over a wide range of temperatures (10 to 40°C) (Leukel, 1948; Odvody et al., 1977; Churchill et al., 2001). Leukel (1948) reported that *Pythium arrhenomanes* was originally thought to be the causal agent of Milo disease, but conducted experiments in which sterile soils were inoculated with various candidate pathogens isolated from diseased roots and reproduced Milo disease when *P. circinata* was used as the inoculum.

Normally, *P. circinata* grows as a saprophyte. However, isolates of the pathogen that produce the polyketide-peptide host-selective toxin, peritoxin, are pathogens (Churchill et al., 2001). Non-pathogenic isolates of the fungus exist and constitute most isolates found in the field. Odvody et al. (1977) found that only 13% of isolates from soil and 34% of isolates from susceptible roots could cause seedling death in their assays. Also, although the conidia of non-pathogenic isolates germinated, produced infection structures, and generated small cortical lesions on sorghum roots, they did not produce the toxin and therefore, could not incite disease. Furthermore, as is the case with host-selective toxins, sensitivity to the toxin conditions disease susceptibility in the plant. Peritoxin can cause cell death in susceptible sorghum genotypes at a level of 1 ng mL^{-1}. Pringle and Scheffer (1966) isolated at least one additional toxin from *P. circinata* in addition to peritoxin.

Mutations from toxin sensitivity to resistance in the semi-dominant *Pc* allele results in plants that are disease resistant (Frederiksen and Odvody, 2000; Churchill et al., 2001). Fortunately, this mutation occurs at a relatively high frequency. In the 1930s, resistance was discovered by selecting such spontaneously resistant genotypes, and toxin resistance has remained durable to this day.

Pokkah Boeng

Pokkah boeng is the Javanese name for "malformed" or "distorted top." This disease is also referred to as "twisted top." The causal agent of pokkah boeng in sorghum is *Fusarium subglutinans* (Wollenw. & Reinking), however *F. proliferatum*, *F. sacchari* Nirenberg, and *F. verticillioides* have been attributed to this disease in sugarcane where the disease is a more serious concern (Lin et al., 2014; Das et al., 2015). The disease is of minor importance in the United States, but has become endemic in most Indian sorghum growing areas. In the case of the *Fusarium* spp. mentioned above, all survive on host residue. For this disease, the spread of airborne conidia and subsequent infection and colonization of leaves leads to movement into the leaf sheaths, stalk, and growing point of the plant. There is some evidence to suggest that these *Fusarium* pathogens may be seed transmitted.

Pokkah boeng causes wrinkled, chlorotic leaves to form at the plant growing point. When the disease is less severe, infected leaves may resemble those infected by a virus causing a mosaic symptom. However, when disease is severe, internodes and nodes become twisted, stalks bend, and transverse indentations may form in the stalk rind (i.e., the "knife-cut" symptom), which facilitate stalk breakage under high wind conditions. Dry conditions followed by a period of wet weather and high humidity facilitate infection and disease development. In India, the disease is more prevalent in the post-rainy season.

Some Challenges for Seedling, Root, and Stalk Diseases

1. Resistance to seedling, stalk rot, and root diseases remains lacking. Since genetic resistance is difficult to find, tolerance to these diseases and the abiotic stressors that promote them is a priority.
2. Understanding the etiology of seedling blights, Fusarium stalk rot, and the root rot complex requires further investigation. More work is needed to catalog the array of *Fusarium* spp. that causes these diseases versus those that are casually associated with the host.
3. The infection process and resulting molecular-plant microbe interactions of the charcoal rot fungus, *M. phaseolina,* remain unclear for sorghum.
4. Although post-flowering drought tolerance, that is, stay-green, is a well-known trait in sorghum, the discovery, and deployment of pre-flowering drought tolerance is also important. The contribution of pre-flowering drought-tolerance to reduced charcoal rot and Fusarium stalk rot disease tolerance is presumed, but largely unknown.

Foliar Diseases

Sorghum is susceptible to several foliar diseases caused by bacterial, fungal, and viral pathogens, but only a few pathogens cause widespread damage. However, "micro-epidemics" do occur and can be quite severe under highly conducive environmental conditions. In this section, the foliar fungal pathogens of sorghum will be discussed. In temperate production systems, foliar pathogens attack leaves in the summer through the spread of fungal spores via rain splash and air currents among the plant canopy and then survive in crop residue after harvest. If the greatest damage occurs on the upper foliage of plants during the grain filling period, then yields can be greatly affected by foliar diseases. In general, many fungal pathogens attack the lower, older leaves first, although this is not always the case. Management strategies for foliar diseases include crop rotation, weed control, resistant hybrids, crop debris management, and use of labeled foliar fungicides where such practices are economically or practically viable (Ratnadass et al., 2012; Mueller et al., 2013).

Anthracnose

Colletotrichum sublineola causes anthracnose of sorghum, occurs to some degree throughout the world wherever sorghum is grown, and is considered one of the most damaging foliar diseases of grain sorghum (Harris and Sowel, 1970; Berquist, 1973; Sutton, 1980; Cardwell et al., 1989; Cole and Hoch, 1991; Ali and Warren, 1992). A draft genome for *C. sublineola* was published in 2014. As a result, the genome size is predicted to be 46.75 Mbp, contains 12,699 protein-encoding genes, and secretes 168 proteins that are uniquely different from those produced by the closely related corn pathogen, *C. graminicola* (Baroncelli et al., 2014). In the United States, this disease is common and a significant economic factor in the southeastern and southcentral grain sorghum production areas including Arkansas and the humid, Gulf Coastal region of Texas (Williams et al., 1980; Moore et al., 2008). The disease may also damage sweet, forage, and biomass sorghums in the southern United States (Hagan et al., 2014).

Colletotrichum sublineola is variable, and numerous pathotypes exist (Ali and Warren, 1987; Cardwell et al., 1989; Prom et al., 2012). Due to the pathotype shifts and the development of new pathotypes that occur in *C. sublineola*, hybrids that are resistant in one year may show susceptible reactions within years after that (Leslie, 2002; Prom et al., 2012). Also, there are pathogenic and non-pathogenic isolates that exist in nature, and there is a tendency for isolates to exhibit host preferences and differ between geographical locations. *Colletotrichum graminicola* is a closely related species that causes anthracnose of corn, ryegrass, and orchardgrass (Baroncelli et al., 2014).

During the foliar phase of anthracnose, *C. sublineola* produces small elliptical lesions with distinct margins on the leaf blade and midrib and may be visible as early as boot formation (Cole and Hoch, 1991; Mathur et al., 1997; Frederiksen and Odvody, 2000). Within these lesions, acervuli are formed, which are diagnostic for the disease. They contain melanized setae that protrude from the structures and are visible in the field using a hand lens (Williams et al., 1980; Frederiksen and Odvody, 2000; Gwary et al., 2004). Peduncle infections occur on the most susceptible hybrids and result in necrotic, sunken lesions. Under humid conditions, both foliar and peduncle lesions enlarge and coalesce to form large areas of dead tissue (Frederiksen and Odvody, 2000). Stalk infections are characterized by marbled brick red and white discolorations of the pith (Coleman and Stokes, 1954). In general, stalk and peduncle infections inhibit translocation of assimilates to the developing seed, which results in poor grain development (Dodd, 1980). Extensive lesion development on leaves results in defoliation and yield reduction if the infection is severe in the upper canopy of the plant. If stalk infections are severe enough, lodging may occur (Tesso et al., 2012). Seed infections are characterized by premature senescence, shriveling, and the presence of the black, setae-laden acervuli on the surface of the seed.

Colletotrichum sublineola invades nearly all tissues of the plant: stems, peduncles, rachis and rachis branches, grain, and leaves (Williams et al., 1980; Ali and Warren, 1987; Frederiksen and Odvody, 2000; Katilé, 2007). Anthracnose causes disease in four phases: seedling root rot, foliar, stalk rot, and seed mold, and these can occur within a single season (Coleman and Stokes, 1954; Williams et al., 1980; Casela and Frederiksen, 1993; Frederiksen and Odvody, 2000; Tesso et al., 2012). The seedling phase is characterized by infection of seeds as they germinate; this is thought to occur in infested soil (Basu Chaudhary and Mathur, 1979; Ali and Warren, 1987). However, it is possible that seedborne *C. sublineola* propagules that are present may be the result of the grain mold phase if the seed was produced in a conducive environment with epidemic levels of anthracnose and if it could serve as the source of inoculum for seedling infections (de Milliano et al., 1992). The foliar phase is often the most common, dramatic, and damaging to yield and begins when the pathogen colonizes the first true leaf and extends to panicle emergence (Ali and Warren, 1987; Thomas et al., 1996). During this phase, the foliar symptoms manifest themselves from the late vegetative to the early reproductive phase of the plant. The foliar phase is the most predominant in the humid southeastern and southcentral sorghum production areas of the United States. The stalk rot phase occurs when the pathogen has developed to an extent where conidia produced during the foliar phase germinate and infect above the uppermost leaf and invade the stalk interior (Mughogho, 1984; Ali and Warren, 1987). Stalk rot occurs in a top-down direction and appears to be more predominant in arid production areas. If the environment is conducive, e.g., high humidity, and sporulation is heavy, then *C. sublineola* may subsequently colonize the developing seed and cause grain mold disease during the final phase. Hybrid resistance plays a significant role in determining whether this final phase will occur or not; that is, highly susceptible varieties may produce large amounts of inoculum and are more likely to exhibit seed infection symptoms (Coleman and Stokes, 1954; Tesso et al., 2012).

As alluded to above, humid and rainy weather are necessary for the full manifestation of anthracnose (Prom et al., 2015b). However, high humidity alternating with periods of dry weather increases disease severity in both the foliar and stalk phases of the disease (Edmunds et al., 1970; Cardwell et al., 1989). Anthracnose is not a great concern in many of the drier production areas in the Great Plains of the United States. Splashing rain and wind move conidia during the foliar phase and rain and/or dew move conidia onto the leaf sheaths to initiate the stalk phase of anthracnose (Ali and Warren, 1987; de Milliano et al., 1992). The feeding activities of stem borers create wounds and may carry the fungus with them into the stem. Additionally, the movement of equipment through the field when leaves are wet leads to increased disease severity due to movement of the pathogen.

The best control for anthracnose is the use of resistant hybrids and resistance is controlled by several different dominant and recessive genes (Coleman and Stokes, 1954; Jones, 1979; Mathur et al., 1997; Boora et al., 1998; Tenkouano et al., 1998; Tesso et al., 2012). Interestingly, resistance may not occur for all of the disease phases. For example, plants may express foliar or stalk resistance to *C. sublineola*, but more particularly, there are lines that exhibit leaf blade resistance but lack midrib resistance (Coleman and Stokes, 1954; Jones, 1979; Erpelding, 2007). Most inbred parental lines were developed in the southern United States, but local pathogen populations outside of this area are an important consideration when deploying resistance (Prom et al., 2012; Tesso et al., 2012).

Ferreira and Warren (1982) reported that sorghum seedlings were not susceptible to inoculation before five weeks of age. In this case, it did not matter if plants were resistant or susceptible to anthracnose. Recent work by Ibraheem et al. (2010) indicates that the *y1* (*yellow seeded1*) gene, which "encodes a MYB transcription factor that regulated phlobaphene biosynthesis" is linked and/or co-segregates with phytoalexin-based anthracnose resistance. Perumal et al. (2009) identified two closely linked markers (*Xtxa6227* and *Xtxp549*) at the distal end of chromosome 5 for the anthracnose resistance locus *Cg1*, a dominant gene for resistance in cultivar SC748–5 and validated it in advanced breeding lines. Further, using a different breeding population, a QTL marker was identified by Burrell et al. (2015) on the same distal end of chromosome 5 that was consistent for resistance to the foliar, stalk, and grain stages of anthracnose. These markers could facilitate marker-assisted selection in breeding for anthracnose resistance and map-based cloning of resistance gene(s).

The anthracnose fungus overwinters in the soil, crop residue, leaf debris, infected seed, and via mycelium and small sclerotia (Ali and Warren, 1987; Basu Chaudhary and Mathur, 1979; Casela and Frederiksen, 1993; Tarr, 1962; Vizvary and Warren, 1982). However, tillage and crop rotation have been suggested as a method of control if some level of residue management is followed. For example, in a field where erosion control is not an issue, it is possible to plow under residue and reduce the chance of newly formed acervuli of sporogenic microsclerotia producing conidial masses. However, evidence that this approach is effective for this disease is currently lacking. Besides the tillage component, infested fields should be rotated with a non-related host before returning to sorghum (Williams et al., 1980). Cotton and various leguminous species, such as chickpea, lentils, and common beans, are effective non-host rotation crops for *C. sublineola*. Reservoir hosts, such as Johnsongrass, should be removed from the field and field perimeters (Ali and Warren, 1987; Tarr, 1962).

Some fungicides have been labeled for foliage protection against anthracnose in the United States. These include azoxystrobin (Quadris), azoxystrobin + propiconazole (Quilt), picoxystrobin (Aproach), and pyraclostrobin (Headline), all of which are QoI fungicides, with the exception of the propiconazole active ingredient in Quilt. Quadris and Quilt are labeled for forage sorghums, whereas Quilt, Headline, and Aproach are labeled for grain sorghum. Aproach may be used for other *Sorghum* spp. as well.

Leaf Blight

Although several saprophytic and facultatively pathogenic fungi can be associated with leaf blight-type lesions on sorghum leaves, *E. turcicum* (Pass.) Leo. & Suggs (*Setosphaeria turcica* (Luttrell) Leonard & Suggs) causes northern corn leaf blight on sorghum (Leonard and Suggs, 1974; Frederiksen et al., 1975). Leaf blight is a widespread disease that has been attributed to 45% yield losses in sorghum in India and 22% in the Philippines (Chenulu and Hora, 1962; Elazegui, 1971; Sharma et al., 2012). In Africa, leaf blight is considered a widespread and major disease of productive cultivars (Ngugi et al., 2001; Ngugi et al., 2002).

High levels of loss occur if the disease establishes itself before the boot stage. At harvest, the grain will appear shrunken or "pinched" due to large amounts of leaf area that may be damaged by lesions. From time-to-time, epidemics of leaf blight occur. For example, this was prevalent in Texas and sporadic across the sorghum-producing areas of the United States and Mexico in the mid-1990s (Leslie, 2002). As the disease name suggests, *E. turcicum* also causes leaf blight on sweet and field corn, as well as pearl millet (Frederiksen and Odvody, 2000). Another similar fungal pathogen, *Bipolaris maydis* (Y. Nisik. & C. Miyake) Shoemaker, causes southern corn leaf blight, but is confined to susceptible corn hybrids as exemplified by the widespread epidemics that occurred in corn hybrids with Texas male-sterile cytoplasm backgrounds. In this case, the host-selective toxin produced by *B. maydis* race T, T-toxin, targeted the URF-13 protein in the mitochondria of these plants, which resulted in necrosis of leaves and major yield losses (Levings, 1990). No such host-selective mechanism of toxin action has been described for the northern corn leaf blight pathogen in sorghum.

Symptoms of leaf blight include long, cigar-shaped necrotic lesions that may be 2 to 5 cm long or longer. Monocerin is a non-host specific polyketide toxin that is produced by *E. turcicum* and is responsible for lesion development (Robeson and Strobel, 1982; Axford et al., 2004). The lesions are watersoaked at first and then become tan or gray with a border that corresponds to the coloration of the plant genotype (i.e., *ppqq*, tan; *PPq^{r}q^r*, red, *PPQQ*, purple) (Smith and Frederiksen, 2000). In severe cases, lesions may coalesce and result in necrotic foliage that resembles prolonged water deficit stress. In warm, humid weather, conidiation occurs within the lesion and is dark gray to olive in appearance. If the pathogen establishes itself on a susceptible host prior to panicle exsertion, yield may be greatly reduced (Sifuentes Barrera and Frederiksen, 1994; Frederiksen and Odvody, 2000). In the case of seedling infections, red, tan, or purple lesions are common (Williams and Frederiksen, 1978). *Exserohilum turcicum* is capable of infecting plants at all stages throughout the growing season if conditions are conducive (e.g., warm, humid) (Berquist, 1986; Carson, 1995). Most lesions appear within a week after infection on older leaves and then move to younger tissue in susceptible plants (Julian et al., 1994). Sporulation occurs on lesions approximately one to two weeks after lesions appear and are spread by rain splash and wind (Parker et al., 1995; Ferrandino and Elmer, 1996). *Exserohilum turcicum* also infects volunteer sorghum plants, johnsongrass, sudangrass, teosinte, and gamagrass, which are reservoir hosts for the pathogen (Levy, 1984; Esele, 1995). *E. turcicum* can survive in seed, crop residue, and as chlamydospores in the soil (Frederiksen and Odvody, 2000).

Resistance and tolerance to northern corn leaf blight in sorghum has been identified (Boora et al., 1999). In sorghum there are six, NB-LRR type resistance genes (*St* genes) that reside as paired loci on chromosome 5. When these genes were silenced using a VIGS (virus-induced gene silencing) approach, resistance was significantly reduced compared to wild-type (Martin et al., 2011).

Rotation is not an effective control strategy if infected grasses or volunteer sorghum persists in and around production fields. If crop rotation is used, rotations should be performed with non-hosts of *E. turcicum,* such as dicotyledonous crops (Ratnadass et al., 2012). If there is a history of the disease at a location, and susceptible hybrids are to be grown, the reservoir hosts should be destroyed. Since the conidia and hyphae of the fungus survive in crop debris and are seedborne, proper field sanitation (e.g., burying sorghum debris via tillage) and the use of pathogen-free seed is advisable (Frederiksen and Odvody, 2000; Leslie, 2002). In a study to examine the application of host genotype mixtures (i.e., hybrid mixtures) as a disease management tool for controlling a sorghum disease, Sifuentes Barrera and Frederiksen (1994) found that incorporation of leaf blight–resistant hybrids improved yield and reduced the disease severity of the susceptible hybrids in the same mixture. Further, Ngugi et al. (2001) found that intrarow mixtures of non-host maize and sorghum could delay leaf blight and anthracnose infection and reduce season-wide disease progress, which has implications for small-holder intercropping strategies. Foliar fungicides have been shown to protect against *E. turcicum* infection. However, it is not clear if these products result in a yield benefit per se (Mueller et al., 2013). Quilt (azoxystrobin + propiconazole) and Headline (pyraclostrobin) have been labeled to control leaf blight in sorghum.

Sorghum Downy Mildew

As the most damaging oomycete disease in the crop, sorghum downy mildew (SDM) caused by *Peronosclerospora sorghi* (W. Weston & Uppal) C.G. Shaw gained widespread attention in the 1960s after epidemics caused severe losses in the Gulf Coast region of Texas where the pathogen is now endemic. The pathogen subsequently spread to Alabama, Arkansas, Georgia, Louisiana, Kansas, Missouri, eastern New Mexico, Oklahoma, Tennessee, and the panhandle of Texas in the late 1960s. In the 1970s, the pathogen spread further to Florida, Illinois, Indiana, Kentucky, and Nebraska (Frederiksen, 1980). In several of these areas, losses to this disease were significant and was under control by metalaxyl fungicide treatment. More recently, a second outbreak of SDM caused by a metalaxyl-resistant strain of pathotype P3, has occurred in Wharton County, TX. This discovery raised broader concerns about this pathogen's ability to develop fungicide resistance and the potential for it to cause new outbreaks after being largely under control since the early 1980s (Frederiksen, 1980; Isakeit and Jaster, 2005).

Symptoms in young, systemically-infected plants exhibit light green, yellowish, or white stripes that run lengthwise along the leaf blades (Fig. 1D). While the tissue is green, downy mildew sporulation may occur on the undersides of the leaves on the reverse sides of the lighter areas. Such systemically infected plants do not produce panicles and remain sterile (Schuh et al., 1986). When these leaves senesce, shredding often occurs, which gives plants the appearance that they've been damaged by hail or high winds. Seedling infections lead to chlorosis and death. Local lesion infection also occurs (Tesso et al., 2012).

Soilborne oospores cause systemic infection of seedlings (de Milliano et al., 1992). Oospores germinate, infect roots, colonize the seedling, and the sporangia (called "conidia") that are produced on the undersides of leaves are disseminated in an airborne manner (Radwan et al., 2011; Tesso et al., 2012; Prom et al., 2015c). Other plants near the systemically infected plant may be infected by the airborne conidia and these infections usually result in local lesions. Johnsongrass may be infected as a reservoir host and sorghum planted after sudangrass is especially vulnerable to the pathogen. Infections of highly susceptible grain and forage sorghum types can lead to build-up of *P. sorghi* soil populations (Schuh et al., 1987; Tesso et al., 2012).

Although most sorghum hybrids vary in their level of susceptibility to *P. sorghi*, selection of hybrids with resistance is the best control strategy for SDM. The most common pathotypes of the pathogen are P1 and P3, and resistance to these strains are used in the sorghum industry (Frederiksen, 1980; Isakeit and Jaster, 2005; Radwan et al., 2011). Pathotype P2 strains are less commonly encountered, however resistance exists in lines RTx430, SC414–12E, and QL-3 (Prom et al., 2015c). Hybrids that are grown in the Gulf Coast region of Texas, where the pathogen is endemic, should have resistance to both P1 and P3. The use of treated seed to prevent systemic infection of sorghum seedlings is achieved using metalaxyl and mefanoxam. The emergence of metalaxyl-resistant strains in *P. sorghi* (i.e., pathotype P6) is a concern for the long-term efficacy of this strategy (Isakeit and Jaster, 2005; Radwan et al., 2011). Since this new pathotype overcame resistance in some hybrids, development of new sources of genetic resistance for the endemic regions of southern Texas and northern Mexico has remained an important goal. Screening exotic African sources from Mali and Gambia provides viable alternative resistance sources, and the trait can be further introgressed into adapted breeding lines for enhanced resistance in hybrids (Prom et al., 2015c).

The oomycetes that cause downy mildews of cereals are obligate parasites that cannot be grown in pure culture. Consequently, relatively little genetic information is available. Restriction length fragment polymorphisms, amplified fragment length polymorphisms, and polymerase chain reaction (PCR) amplification of the ribosomal ITS-2 region have been shown to differentiate *Peronosclerospora* spp., providing support for the classic taxonomic delineation of the newly emerged SDM pathotypes (Perumal et al., 2006). Microsatellite primer sets based on *P. sorghi* sequences will continue to be useful as molecular diagnostic probes to monitor the emergence of new pathotypes in the future (Perumal et al., 2008).

Some research exists concerning the use of crop rotation for control of SDM (Tuleen et al., 1980), which is determined by the long-term survival of oospores in the soil. In general, rotations of greater than two years with crops such as cotton, wheat, soybeans, or non-sorghum-related forages along with the management of sorghum residue via tillage is an adequate strategy where there has been a perennial problem with SDM (Radwan et al., 2011; Tesso et al., 2012). Corn-sorghum rotations should be avoided since corn is another host for this pathogen. Also, controlling johnsongrass, shattercane, sudangrass, and other reservoirs near problematic fields is a key component of an integrated pest management strategy for SDM (Frederiksen and Odvody, 2000; Radwan et al., 2011).

Zonate Leaf Spot

Zonate leaf spot is caused by the fungal pathogen *Gloeocercospora sorghi* Bain & Edgerton ex Deighton and infects corn and millet in addition to sorghum (Lutrell, 1954). Although zonate leaf spot is of low to moderate importance, it is a common disease in most sorghum growing regions of the United States. This disease is more important in the southeast United States with significant losses in both yield and quality on forage and sweet sorghums (Hunter and Anderson, 1997; Frederiksen and Odvody, 2000).

Zonate leaf spot produces large, irregular-shaped lesions that exhibit a bullseye appearance characterized by alternating tan and reddish-purple bands. This pattern is sometimes confused with physiological spotting and lesion appearance is also hybrid dependent. Lesions progress up the plant from older to younger tissues and may coalesce and form a whole leaf blight symptom. During warm and wet periods, pink to salmon-colored conidia form a gelatinous matrix in the lesions (Rawla, 1973). Heavy rains, constant cloud cover, and high humidity provided conducive conditions for an unusually severe outbreak of zonate leaf spot recently at the Texas A&M AgriLife Research Farm (Burleson County, TX) in 2015 (Prom et al., 2015a). Sclerotia may also form in a manner like those produced by *Ramulispora sorghi* (see "Sooty stripe" section). The fungus overwinters in sclerotia in soil and plant debris and is endemic in south Texas (Coley-Smith and Cooke, 1971; Odvody and Madden, 1984; Brady et al., 2011). Conidia are dispersed by rain and water, and disease increases during periods of high rainfall, high humidity, and moderate to high temperatures like that of sooty stripe caused by *R. sorghi* (Rajasab et al., 1989; de Milliano et al., 1992). However, leaf wetness requirements for this pathogen are not known. The pathogen may be spread by the movement of equipment through the field.

Resistant hybrids, crop rotation, burial of crop debris, and tillage are practical approaches to control zonate leaf spot (de Milliano et al., 1992). However, it is not known how long *G. sorghi* can survive in the soil. Johnsongrass, bermudagrass, bentgrass, napiergrass, broomcorn, sugarcane, corn, sudangrass, and sudangrass-sorghum hybrids are reservoir hosts for *G. sorghi* (Sprague, 1950; Alvarez, 1976; Chiang et al., 1989; Lenne, 1990; Roane and Roane, 1997). Fungicides (e.g., Quilt Xcel, azoxystrobin + propiconazole) have been labeled for use in the United States to control this disease, although there is no clear yield benefit by doing so (Mueller et al., 2013).

Sooty Stripe

Sooty stripe is caused by the fungal pathogen, *Ramulispora sorghi* (Ellis & Everh.) L.S. Olive & Lefebvre (Brady et al., 2011). Along with a zonate leaf spot epidemic in 1998, epidemics of sooty stripe were observed in the Midwest United States during 1997 and 1998 due to the warm, moist conditions that favored both diseases. However, unlike zonate leaf spot, this disease is prevalent in Kansas, Missouri, Oklahoma, and Nebraska, but not so in the southern and south-central United States (Leslie, 2002).

Broad chlorotic halos typically surround sooty stripe lesions (Fig. 1E). The pathogen produces elongated, tan elliptical lesions characterized by the production of numerous rows of microsclerotia within the abaxial sides of the lesions (Fig. 1F). The older leaves are infected first and are the most extensively colonized in susceptible hybrids. Sooty stripe may cause a severe leaf blight disease when lesions coalesce and cover the leaf. Yield impacts of 10 to 26% have been associated with two or more lesions per leaf (de Milliano et al., 1992; Jardine and Gordon, 1997).

Conidia are spread by wind and rain splash, and infection may occur at any plant stage. The pathogen is thought to overseason in crop debris on or below the soil surface as microsclerotia. Microsclerotia are sporogenic, meaning they can directly germinate and produce conidia. Brady et al. (2011) found a positive relationship between the number of microsclerotia produced within a lesion and their size. Furthermore, although the local population of *R. sorghi* was found to be genetically uniform, the levels of host susceptibility governed microsclerotium size, that is, more susceptible varieties often produced larger microsclerotia with greater potential to sporogenically conidiate. Johnsongrass acts as a reservoir host for this pathogen (Sprague, 1950; Odvody et al., 1973; Lenne, 1990). Destruction of leaf debris and crop rotation helps to reduce *R. sorghi* inoculum. Although hybrid resistance varies for this disease, use of resistance is the best control strategy.

Rust

Sorghum rust is caused by the obligate fungal pathogen, *Puccinia purpurea* Cooke (Hooker, 1985; White et al., 2012). Another closely related species, *P. sorghi*, only causes significant disease in corn (Hooker, 1985). In general, forage sorghums are more susceptible to *P. purpurea* than grain sorghum hybrids (Sharma et al., 2012). Rust can be economically damaging in warm, moist environments where late-planted or ratoon sorghum grows. However, the disease often appears late in the season when it is less impactful on yield. If infection occurs early, yield losses are more noticeable (White et al., 2012).

Sorghum rust is characterized by the production of small, brown uredinia pustules that are raised above the leaf surface (Fig. 1G). Uredinia may occur on the upper and lower surfaces of the leaf and often occur on older plant tissues. The pathogen also infects the peduncle where uredinia take on a more linear appearance than the oval uredinia on leaves. Each uredinium is filled with orange-brown urediniospores that are released and spread by the wind when the peridium ruptures. Under epidemic conditions, 10 to 14 d are required for the formation of new pustules after urediniospores infect healthy tissue. After flowering, as the host senesces, so do the uredinia, which are converted into telia that contain dark black-brown teliospores that function as the overseasoning spore for the rust. Spermagonia and aecia are formed on *Oxalis corniculata* L. (creeping woodsorrel) as part of the sexual reproduction phase of the pathogen (Pavgi, 1972). However, overwintering uredinia on johnsongrass are likely to provide the bulk of the inoculum needed to initiate new infections on sorghum and propagate the disease cycle (White et al., 2014).

Control of rust may be achieved using resistant hybrids. Rotations with noncereals and control of weedy grasses are prudent as the pathogen can perpetuate a secondary cycle in this manner if green tissue is available and if temperatures are conducive for uredinia formation (de Milliano et al., 1992; Leslie, 2002; White et al., 2014). The contribution of the alternate host is not entirely understood as far as epidemics are concerned, although the distribution of *O. corniculata* is relatively wide across the United States. Fungicides (e.g., Aproach, picoxystrobin; Headline SC, pyraclostrobin) have been labeled for use in the United States. No significant yield advantage has been observed after using fungicides (Mueller et al., 2013).

Gray Leaf Spot

The fungal pathogen *Cercospora sorghi* Ellis & Everh. causes gray leaf spot and can be economically damaging in moist environments (Leslie, 2002). The disease is characterized by long, rectangular lesions that take on a grayish cast when the pathogen produces conidia within the lesion (Frederiksen and Odvody, 2000). Resistant hybrids and crop rotation are advised to control this disease if it has been a significant problem in corn or sorghum at a production site (Ward and Nowell, 1998). Fungicides (e.g., Aproach, picoxystrobin; Quadris, azoxystrobin; Quilt Xcel, azoxystrobin + propiconazole) have been labeled for use in the United States. No significant yield advantage has been observed after using fungicides (Mueller et al., 2013).

Some Challenges for Foliar Diseases

The following list highlights some important research areas for sorghum foliar diseases:

1. Pathotype and strain diversity for major fungal pathogens of sorghum require constant monitoring while those for minor fungal pathogens require further elucidation.
2. The environmental requirements for foliar disease (e.g., leaf wetness duration) and survival (e.g., the manner, length, and structure of overwintering inoculum) require further study for certain pathogens where this information is not known.
3. More information concerning the benefits, if any, of residue management, control of reservoir hosts, crop rotation, and the efficacy of registered fungicides are required.

Panicle and Grain Diseases

Important panicle and grain diseases include smuts, ergot, and grain mold. These diseases significantly impact both yield and quality of grain. Quality deterioration impacts nutritional and feed value. Most of the sorghum varieties cultivated in African countries lack resistance to panicle and grain diseases and include landrace varieties and some exotic or improved varieties (Malvick, 2008).

Head Smut

Head smut is caused by the fungal pathogen, *Sporisorium reilianum* (J.G. Kühn) Langdon & Full. *S. reilianum* is a soilborne, facultatively biotrophic basidiomycete smut fungus that is prevalent in all parts of the world where sorghum is grown and may cause significant yield loss. In the United States, the disease is most prevalent in the Gulf Coast and central regions of Texas, but also occurs in the Midwest. The pathogen readily mutates and is characterized by several races (Oh et al., 1994; Prom et al., 2011; Little et al., 2012).

Head smut disease is characterized by dark, brown smut sori that emerge in place of the panicle due to the active destruction of the reproductive tissues by the pathogen. The sorus (or gall) is covered by a white membrane (peridium) when immature, but disintegrates as the tissues age, which allows the teliospores contained within to be scattered by wind and water to the soil (Little et al., 2012). After the sorus disintegrates, the vascular strands of the plant remain and are a diagnostic symptom of the disease. Some hybrids dwarf or tiller profusely, and if tillering occurs, those heads will also be smutted (Little et al., 2012).

Infection occurs at the seedling stage but is not evident until plants produce panicles (Mehta, 1965). The germinated teliospore forms a germ tube that penetrates the exterior of the plant either in roots or mesoderm to reach the meristematic tissue in the sorghum seedling. Teliospore germination is high in moist environments when temperatures range between 27 and 31°C. The fungus requires actively growing meristematic tissue for development (Little et al., 2012). A dry soil with cool to moderate temperatures until the plants reach the 3 to 4 leaf stage is considered ideal for infection. The sorus that is produced in place of the panicle produces millions of teliospores that become soilborne and initiate systemic infection of seedlings (Frederiksen and Odvody, 2000). The teliospores germinate in the spring as the seed germinates and the mycelium invades the nodal region of the shoot apex. Teliospores remain viable in the soil for a decade at a time and hence disease-free or chemically treated seed may not prevent further infection (Little et al., 2012).

Since different races of *S. reilianum* exist, not every hybrid is resistant in every location due to the race-specific, vertical resistance that is required for this pathogen (Prom et al., 2011; Little et al., 2012). However, new sources of resistance have been found and should be utilized where possible as host resistance is the primary control for head smut. Chemical controls, seed treatments, and crop rotation have not been effective in controlling this disease (Smith and Frederiksen, 2000). However, some cultural management strategies should be adopted in problematic fields to reduce head smut incidence. These include: (i) avoid continuous sorghum cropping, (ii) avoid hybrids with compact heads, (iii) rogue and eradicate infected plants prior to sorus rupture to prevent additional teliospores returning to the soil, and (iv) avoid sowing hybrids in environments that will promote systemic infection of seedlings and manifestation of disease in the adult plant.

Covered Kernel Smut

Covered kernel smut is caused by the fungus *Sporisorium sorghi* Ehrenb. ex Link (Frederiksen and Odvody, 2000). This disease is also referred to as "grain smut" and occurs in the United States at very low levels, but is responsible for high levels of yield loss in Africa (Nzioki et al., 2000). This is the most common and popularly recognized smut by farmers in West Africa (Gwary et al., 2007).

Covered kernel smut destroys the grain in the panicle and replaces them with a cone-shaped sorus that is ~1 cm in length (Esele, 1995; Leslie, 2002; Ngugi et al., 2002). The disease may affect the entire panicle or portions of the panicle. At harvest, sori are broken open and teliospores contaminate healthy seed. Teliospores may become seedborne and germinate within seedlings and infect the growing point of the plant (Esele, 1995).

Chemical treatment of seed and the use of disease-free or "certified" seed has reduced the frequency and severity of covered kernel smut (de Milliano et al., 1992). Like head smut, infected plants may be rogued and eradicated to reduce teliospore levels. In general, host resistance has been the best option for disease management (Nzioki et al., 2000).

Loose Smut

Loose kernel smut is caused by the fungal pathogen, *Sporisorium cruentum* (J.G. Kühn) Vánky (Frederiksen and Odvody, 2000). The disease occurs on sorghum worldwide. Emechebe et al. (2010) reported head and loose smuts to be more prevalent in years that had little rainfall in Nigeria. The disease is one of most destructive smuts known (Kutama et al., 2013). Plants affected by *S. cruentum* are often stunted, have thin stalks, and their heads are said to emerge earlier than in healthy plants (Aggarwal and Mehrotra, 2003). Infected plants may also produce abundant tiller branches (Williams et al., 1978; Singh, 1998). Infected panicles are characteristically looser, bushier, and a darker green color than normal panicles, due to the induction of hypertrophic glumes by the pathogen (Malvick, 2008). Usually, the spikelets of an infected panicle are smutted, but some may escape or proliferate (Tarr, 1962; Williams et al., 1978). The panicle is always bushy because the covering membrane of the sorus ruptures very early releasing the dark teliospores (Singh, 1998). For this reason, plants infected by *S. cruentum* often head up to two weeks before normal as the fungus accelerates the growth cycle of the stem (Tarr, 1962). On infected plants, tillers are short, slender and abundant (Kutama et al., 2013). Plant height, the number of leaves, days to flowering, heading, and grain yield are all affected in plants that are infected by *S. cruentum* (Kutama et al., 2013).

The pathogen infects sorghum, sudangrass, and johnsongrass (Stoll et al., 2005). Teliospores may be carried on seed and germinate soon after the seed is planted and will infect and colonize the developing seedling. The mycelium grows quiescently in plants until panicle formation and exsertion when florets are replaced by smut galls (Frederiksen and Odvody, 2000).

In the developed world, problems with this smut do not exist, since the pathogen has been under long-term control since sorghum hybrids are resistant. However, if there is a problematic field, sanitation, rotation with non-cereals, grassy weed control, and the use of disease-free, fungicide treated seed can provide control (Leslie, 2002; Gwary et al., 2007).

Ergot

Ergot is caused by the fungal pathogens, *Claviceps africana* Frederickson, Mantle, & de Milliano and *Claviceps sorghi* Kulkarni, Seshadri & Hegde. However, this section will focus primarily on *C. africana*. *Claviceps africana* is most important in Africa, South America, and North America, whereas *C. sorghi* originated in Asia (Frederickson et al., 1991; Isakeit et al., 1998; Pažoutová and Bogo, 2002). A third ergot pathogen, *Claviceps sorghicola* Tsukib., Shiman., & Uematsu, will not be discussed here. Since its initial spread in the mid-1990s, *C. africana* is found worldwide wherever sorghum is grown. The pathogen is now endemic in locations ranging from Texas to Australia and may observed as far north in the United States as Kansas and Nebraska (Bandyopadhyay et al., 1998; Tooley et al., 2000; Bandyopadhyay and Frederiksen, 2006). The pathogen can be of moderate risk in the major sorghum growing regions of the United States, but usually does not cause significant problems. The risk is moderate to low in the southeastern United States, but some areas have seen occasional epidemics. In general, *C. africana* is more of a problem in seed production fields, especially if the parental lines being used flower asynchronously, for example, flower too late or too early for timely pollination, or tiller excessively after pollination is complete. These situations provide a window of opportunity for *C. africana* conidia to infect and colonize unfertilized florets. This pathogen is biologically interesting because once fertilization occurs it cannot establish an infection (Frederickson et al., 1993; Bandyopadhyay et al., 1998; Isakeit et al., 1998).

Ergot is characterized by the production of sugary exudates ("honeydew") from infected florets that turn a faint orange to pink color as it is exuded (Frederickson et al., 1991; Frederickson et al., 1993; Zummo et al., 1998; Tooley et al., 2000) (Fig. 1H). Residual ergot honeydew on healthy seed can interfere with harvest as it adheres to combine augers, headers, and sieves. *Cerebella* spp., *Alternaria* spp., and other saprophytic fungi can colonize *C. africana* honeydew, giving contaminated seed a moldy appearance (Bandyopadhyay et al., 1998; Leslie, 2002). However, this is not to be confused with grain mold or weathering (see next section). The fungus itself produces a bloom of white, powdery conidia on the honeydew surface under cool, humid conditions. The powdery growth also occurs wherever honeydew has dripped such as on other seeds, leaves, or the soil surface. Florets that are infected by *C. africana* conidia before fertilization are replaced with a fungal body instead of producing a seed. (Komolong et al., 2003; Little et al., 2012). These sclerotia resemble those that are produced by *C. purpurea*, the causal agent of ergot of rye. *C. africana* sclerotia are generally larger than the grain, are dark colored, but do not exhibit the same toxicity to livestock (or humans) as the rye pathogen. However, the sclerotia of *C. sorghi* are purple to black in color, are larger than those produced by *C. africana*, and produce several alkaloid toxins (Frederickson et al., 1991; Bandyopadhyay et al., 1998; Tooley et al., 2000; Leslie, 2002; Pažoutová and Bogo, 2002).

Claviceps africana survives in dried honeydew on the soil surface, plant residue, as sphacelia and sclerotia in seed lots, and via reservoir infection of johnsongrass, shattercane, and other sorghum-related grassy hosts (Pažoutová et al., 2000; Alderman et al., 2004; Pažoutová and Frederickson, 2005). These sources produce primary conidia, which may then germinate secondary conidia that may also be dispersed by wind, rain, insects, animals, humans, and machines (Frederickson et al., 1993; Bandyopadhyay et al., 1998; Isakeit et al., 1998; Pažoutová and Bogo, 2002). Infection of unfertilized florets results in the production of a white mycelial mass that replaces the reproductive tissues (a "sphacelium") and produces copious amounts of honeydew as mentioned above. Over time, the tissues of the sphacelium are converted into a sclerotium, which is a hard, compact mass of mycelium that is much darker in color than the sphacelium (Frederickson et al., 1991; Frederickson et al., 1993; Alderman et al., 1999). The sclerotium not only functions as an overseasoning structure, but also germinates stromata on stipes crowned with a knob-like capitulum that contains numerous perithecial cavities that are lined with asci containing sexually-derived ascospores (Bandyopadhyay et al., 1998; Alderman et al., 1999). In many *Claviceps* spp., these ascospores function as the primary inoculum, but in *C. africana*, sclerotia are not always produced, and the role for ascospores has not been established. It is thought that windborne secondary conidia from southern regions in the United States serves as the primary inoculum for production systems in the southern and central Great Plains (Frederickson et al., 1993; Bandyopadhyay et al., 1998; Tooley et al., 2000).

If sorghum pollinates during a period of higher temperature that is not favorable for the iterative germination of secondary conidia from primary conidia, the effects of *C. africana* may be reduced. High humidity and low temperatures favor the spread of ergot since pollination is often impaired or slowed at these temperatures (McLaren and Wehner, 1992; Frederickson et al., 1993; Leslie, 2002). This is one reason that the disease could be more serious in cooler production regions in the Great Plains if more inoculum were available for infection. When temperatures dip to 15 to 16°C (60°F), pollination is dramatically slowed, but pathogen activity is not decelerated to the same extent (McLaren, 1997). Consecutive days of low temperatures and high relative humidity during flowering are sufficient to observe significant floret infections by the fungus (McLaren and Wehner, 1992). A period of cooler than normal nighttime temperatures for two to three weeks before flowering, and during pollen formation, reduces pollen viability and will lead to increased ergot infection if the pathogen is present and the environment is conducive to disease (McLaren, 1997; Bandyopadhyay et al., 1998). Rain during flowering, for example, reduces pollen movement (de Milliano et al., 1992; McLaren and Wehner, 1992; Bandyopadhyay et al., 1998). Rain splash and wind can move inoculum through the field.

In hybrid production fields, male-sterile plants (e.g., seed parents) require pollination to produce seed. If male and female plants have vastly asynchronous flowering periods, there will be an increased chance that *C. africana* conidia may infect unfertilized florets (Alderman et al., 1999; Frederickson et al., 2003). Likewise, there is increased vulnerability in forage and hay sorghums due to higher levels of male-sterility. Sorghums that tiller late or exhibit reduced fertilization due to later tillering will exhibit greater amounts of ergot infection (Bandyopadhyay et al., 1998; Isakeit et al., 1998).

Hybrids with complete genetic resistance to ergot are not known. However, some genotypes have been identified with reduced susceptibility (McLaren, 1992; Tegegne et al., 1994; Reed et al., 2002). Also, currently available hybrids that exhibit cold tolerance during panicle and pollen formation and flowering may provide tolerance as well. These materials are promising sources of abiotic stress tolerance that retain high pollen viability and thus, reduced susceptibility to ergot. In hybrid production fields, the choice of parental lines that flower synchronously is necessary to avoid infection. Thus, choosing hybrids that produce large amounts of pollen and pollinate over a short period is favorable. Additionally, sowing times should be chosen that will maximize pollen production and viability. In the United States, propiconazole has been labeled for the control of ergot (Prom and Isakeit, 2013).

Grain Mold and Weathering

Grain mold, also called "head mold" of sorghum, is caused by a complex of fungi that infect the internal tissues of the developing spikelet. Infection occurs between anthesis and physiological maturity. Most of the fungi in the grain mold complex are saprophytes that live epiphytically on plant surfaces such as the seed coat or dead floral tissues (Prom and Erpelding, 2009; Little et al., 2012). However, they become facultative pathogens when exposed to developing seed tissues within the protective microclimate of the spikelet and floret. Many of the same fungi colonize the outer surfaces of grain before and after physiological maturity, and in this case, are referred to as grain weathering fungi (Little et al., 2012). Damage caused by this fungal disease complex includes a reduced yield due to loss of seed mass, decreased grain density, and decreased germination (Ibrahim et al., 1985; Katilé et al., 2009). Many of the grain mold and weathering fungi cause post-harvest storage deterioration and may produce harmful mycotoxins (Husseini et al., 2009; Sharma et al., 2010; Yassin et al., 2010). Some species colonize the sorghum peduncle, rachis, and rachis branches (rachilla) in a disease known as "head blight" (Little et al., 2012).

In grain mold and weathering, fungi grow on glumes and grain surfaces giving a moldy appearance (Fig. 1I). Kernels that have a blackened appearance have often been colonized by *Alternaria, Bipolaris,* or *Curvularia* spp. (Husseini et al., 2009; Rossman et al., 2009; Yassin et al., 2010). However, this dark discoloration is not to be confused with smut diseases (see previous sections). Kernels with white, pink, or peach to orange colored growth have often been colonized by *Fusarium* spp. (Frederiksen and Odvody, 2000; Leslie, 2002). Kernels that are covered in small black dots that are visible to the naked eye, or using a hand lens, have often been colonized by *Phoma* (i.e., pycnidia) or *Colletotrichum* (i.e., acervuli) (Williams and Rao, 1981; Navi et al., 2005; Little et al., 2012). If colonization by these organisms is superficial in the case of grain weathering, then damage may be limited to reduced market value due to grain discoloration. However, if damage has occurred within the kernel, grain nutritional quality may be reduced along with a reduction in grain mass and density (Little et al., 2012). Grain that is being produced in hybrid systems that exhibit grain mold or weathering symptoms may display reduced germination and seedlings grown from such grain may display reduced vigor (Girish et al., 2004; Little and Magill, 2003; Noll et al., 2010).

Although both the grain mold and grain weathering complexes result in quality and yield reductions, grain mold is thought of as a disease that takes place when infection occurs before black layer deposition of the grain, whereas grain weathering takes place after black layer deposition (Little et al., 2012). However, the situation in the field is rarely this clear cut. Further, some grain mold fungi also act as grain weathering fungi as infections and colonization may take place throughout the growing season if rainy, humid weather persists before and after grain maturity. However, infection in the strictest sense of grain mold occurs when fungi infect early at or near anthesis and quickly colonize the flower parts (glumes, lemma, palea, and lodicule) (Castor, 1981). From there, fungi may colonize the pedicel base and disrupt translocation of water, macronutrients, and photoassimilates through the transfer tissue into the developing kernel. Castor

(1981) found that *Curvularia lunata* (Wakker) Boedijn induced a premature formation of the black layer, that is, senescence of these transfer cells that normally signal the end of translocation, and resulted in shriveled, underdeveloped grain.

A prolonged warm, wet, humid (> 95% R.H.) environment during grain development is the primary factor that contributes to grain mold and weathering problems in the field (Williams et al., 1978; Garud et al., 2000; Navi et al., 2005). In fact, if grain is left in the field late in the season due to delayed harvest, grain weathering fungi will have a greater impact. Additionally, grain damage due to birds, insects (head bugs, worms) produce feeding sites that act as entry points for facultatively pathogenic fungi (Leslie et al., 2002). Insect feeding behaviors have been associated with high degrees of grain mold in Africa (Ratnadass et al., 2003).

Control of grain mold may occur through host resistance. However, no immunity to grain mold exists. Several physical traits associated with panicle architecture, floret structure (e.g., glume coverage), and composition of the developing seed may reduce grain mold (Brown et al., 2006; Little et al., 2012). For example, lines and hybrids should be avoided that have highly compact head architecture as this favors high humidity within the panicle and fungal conidia may be easily transferred from one spikelet to another via microcurrents of air (Leslie et al., 2002; Little et al., 2012). Tannins are often associated with grain mold resistance, but should be avoided in feed and food usage grain sorghum since they interfere with protein digestibility, reduce animal weight gain, and interfere with food utilization characteristics such as palatability (Reichert et al., 1980). Planting should be timed so that flowering occurs during a period that is less conducive to grain mold infection. Such periods include those with less frequent rain and lower relative humidity, but maintain an optimum environment for pollen viability and a sufficient grain filling period.

Some Challenges for Panicle Diseases

Grain mold and weathering:

1. The diversity of fungal pathogens associated with grain mold and grain weathering has been well established. However, more work should be done to address diversity within species at the molecular and pathogenic levels.
2. New species or isolates that are regionally associated with grain mold and weathering or widespread remain to be discovered. These unidentified species or isolates are members of the seed mycobiome that requires deeper investigation. Tools such as "metagenomics" may help reveal the complete fungal community associated with sorghum seeds, whether molded, weathered, or healthy, within different production environments.
3. More information is needed concerning the geographic distribution of mycotoxigenic species of grain mold, weathering, and storage fungi associated with sorghum seed. Foodways and cultural control practices that protect the safety of harvested grain should remain a priority especially for the newly industrialized and low income countries. This issue is also important for the dietary markets in the industrialized world.

Ergot:

4. Cooler temperatures prevent optimum pollination and increase risk for ergot infection. Thus, germplasm with cool temperature tolerance at flowering is of interest to mitigate infection of unfertilized ovules by ergot conidia.

5. There are no known sources of complete resistance to ergot. A thorough screen of a wide array of sorghum germplasm may reveal materials with complete genetic resistance in addition to the discovery of additional materials with reduced susceptibility.
6. Isolates of *C. africana* and *C. sorghi* should be screened for resistance to propiconazole and other active ingredients of commonly available fungicides.

Bacterial Diseases

As with fungal diseases, resistance to bacterial diseases of sorghum is the most economical control strategy. For example, resistance to bacterial leaf stripe (*Pseudomonas andropogonis* Smith; Fig. 1J), the most common bacterial foliar disease of sorghum, is available in several breeding lines including SC414, BTx378, B35–6, and Tx2862 (Muriithi and Claflin, 1997). Many bacterial pathogens of sorghum, including bacterial leaf spot (*Pseudomonas syringae* Van Hall), bacterial top and stalk rot (*Dickeya dadantii;* Samson et al., 2005), and yellow leaf blotch (*Pseudomonas* sp.), do not cause significant disease losses and therefore extensive screening and breeding activities have not occurred. In some cases, cultural controls are warranted for bacterial diseases in sorghum. For example, in the cases of bacterial leaf stripe, bacterial leaf streak [*Xanthomonas campestris* pv. *holcicola* (Elliot) Dye], and bacterial leaf blight, the pathogens are seedborne. Thus, the production of seed in an environment free of the bacterium is necessary as is high quality seed produced in a disease-free environment (Karunakar et al., 1995). The bacterial leaf stripe and leaf streak pathogens survive in leaf debris. Therefore, tillage operations that reduce or destroy sorghum residue will prevent bacterial spread.

Challenges for Bacterial Diseases

1. The bacterial pathogens of sorghum are an understudied group of organisms. The risk of an emerging bacterial pathogen in sorghum, such as has been seen in maize with the wilt pathogen, *Xanthomonas vasicola* pv. *vasculorum*, suggests a need to screen sorghum germplasm for resistance to a wide array of major and minor bacterial pathogens and their various strains.
2. Where environmental conditions are more conducive to bacterial infections, a bank of resistant germplasm to a wide array of potential bacterial pathogens should be established and well-documented for these regions.

Virus Diseases

The largest and most important group of plant pathogenic viruses on sorghum are the potyviruses. There are over 150 species in the *Potyvirus* genus (ICTV, 2015). All are characterized by a flexuous rod virion and a positive-sense, single-stranded RNA genome that is on the order of 10,000 kb in length. These viruses produce unique spindle-like structures within host cells that are visible in transmission electron microscope imagery of infected cells. Sorghum is infected by several species of *Potyvirus* including *Maize dwarf mosaic virus* (MDMV), *Johnsongrass mosaic virus* (JgMV), *Sugarcane mosaic virus* (SCMV), *Sorghum mosaic virus* (SrMV), and *Zea mosaic virus* (ZeMV) (Seifers et al., 2012). Johnsongrass serves as a perennial reservoir host for MDMV and JgMV, but SCMV and SMV cannot infect *Sorghum halepense.* Potyviruses (MDMV, SCMV) are vectored by aphid species including *Schizaphis graminum* (Rondani) and *Rhopalosiphum maidis* (Fitch).

Potyviruses, especially MDMV, produce a wide variety of symptoms, but the most recognizable on sorghum are the (i) red leaf symptom, which occurs following low overnight temperatures (< ~60°F) (Fig. 1K), (ii) concentric, necrotic rings, (iii) mosaic patterns, and (iv) necrosis of the rachis branches that results in a "little seed" symptom. In general, red leaf symptoms are considered more severe and are associated with greater yield losses than are mosaic type symptoms, which are less severe manifestations of systemic infection (Alexander et al., 1984).

Other viruses infect sorghum to one degree or another, but many do not cause serious or even observable losses on sorghum, although some cause serious losses on related hosts such as maize (Frederiksen and Odvody, 2000). The genera *Fijivirus* (*Fiji disease virus*, FDV; *Maize rough dwarf virus*, MRDV; *Mal de Rio Cuarto virus*, MRCV), *Furovirus* (*Sorghum chlorotic spot virus*, SgCSV; *Peanut clump virus*, PCV), *Mastrevirus* (*Maize streak virus*, MSV), *Nucleorhabdovirus* (*Sorghum stunt mosaic virus*, SSMV), *Tenuivirus* (*Maize stripe virus*, MSpV), and *Waikavirus* (*Maize chlorotic dwarf virus*, MCDV). Other species including *Barley yellow dwarf virus* (Luteovirus), *Panicum mosaic virus* (Panicovirus), and *Brome mosaic virus* (Bromovirus) have also been reported on sorghum, but appear to be quite minor.

In general, like the potyviruses mentioned above, most plant virus genomes are composed of positive-sense, single-stranded RNA. Such is the case for several viruses that infect sorghum including MCDV, SgCSV, and PCV (Stewart et al., 2014). Two nucleorhabdoviruses, SSMV and MMV, and the tenuivirus MSpV, have genomes composed of negative-sense, single-stranded RNA (Creamer et al., 1997). The fijivirus MRDV and the geminivirus MSV possess genomes consisting of double-stranded RNA and circular ssDNA, respectively. Further, the genomes may be complete (*monopartite*; MCDV, MSV, SSMV, MMV), or subdivided into two segments (*bipartite*; SgCSV, PCV) or multiple segments (MSpV, MRDV) (ICTV, 2015). The virions of these viruses vary from rod-shaped (SgCSV, PCV), bullet-shaped (SSMV, MMV), geminate (MSV), and icosahedral (MCDV, MRDV) (ICTV, 2015).

Insects vector many of these sorghum viruses. Although aphids are the primary vectors for the potyviruses, plant- and leafhoppers vector the bulk of the viruses discussed in this section. *Peregrinus maidis* Ashmead, the corn planthopper, is a persistent, transovarial vector of MSpV and MMV (Singh and Seetharama, 2008). However, MRDV may be transmitted by several delphacid planthoppers. *Graminella* spp. are leafhoppers that transmit MCDV (*G. nigrifons*, *G. sonora*) and SSMV (*G. sonora*) (Creamer et al., 1997). *Cicadulina* leafhoppers persistently transmit MSV. Alternatively, two sorghum furoviruses, SgCSV and PCV, are transmitted by *Polymyxa graminis* L., a soilborne plasmodiophorid pathogen that colonizes roots, and which is most well-known for transmission of the soilborne wheat mosaic virus (SBWMV) (Kanyuka et al., 2003).

Distinguishing between viruses in sorghum based on symptomology alone is a challenging task, since numerous symptoms may be shared between viruses on the same plant and co-infection can occur (Singh and Seetharama, 2008). For example, MSpV, SgCSV, SSMV, MMV, and MSV produce chlorotic stripes, spots, mottling, and streaks on leaf tissue. MSpV, MCDV, and MSV produce leaf banding, vein clearing, and white streaks. Necrotic lesions are produced by SSMV and MSV. Stunting is a very common symptom in MSpV, MCDV, SSMV, and MMV. Moreover, and most importantly, reductions in yield, via low seed production, failure to produce a panicle, and head distortion and/or sterility may occur after

SSMV, MMV, MSpV, and MRDV virus infection. Therefore, because these symptoms often occur simultaneously, if they are observed at all, and may vary based on host genotype, virus strain, and genus, serological (ELISA) and molecular (PCR, RT-PCR) tools are necessary for definitive diagnosis.

Some Challenges for Virus Diseases

1. Numerous viruses and confounding symptoms occur (and co-occur) in sorghum, but further characterization of sorghum virus species and strains is required.
2. Determine the distribution of reservoir hosts for viruses and insect vectors as part of integrated management strategies to reduce epidemics of disease.
3. Understand the genetic basis for virus resistance in sorghum germplasm.
4. Development of pathogen-derived resistance to *Potyvirus* spp. in sorghum.
5. Diagnostician training, technologies, and diagnostic tools for virus differentiation and detection, especially in regions where the sorghum crop is most vulnerable to virus diseases.

Nematodes

In general, symptoms of nematode damage may resemble drought stress, nutrient deficiency, root diseases, and insects. Such symptoms often include stunting, chlorosis, wilt, and generalized root dysfunction. There are a wide range of nematodes that cause diseases on sorghum. These include lesion, stunt, needle, cyst, and root knot nematodes.

Lesion Nematodes

Lesion nematodes are migratory, endoparasites that damage plant roots through constant feeding and movement within root tissues. Several species of lesion nematodes (*Pratylenchus* spp.) are found on sorghum. These include *P. brachyurus* (Godfrey), *P. coffeae* Goodey, *P. crenatus* Loof, *P. hexincisus* Taylor and Jenkins, *P. loosi* Loof, *P. thornei* Sher & Allen, and *P. zeae* Graham (Leslie, 2002). Of these, *P. coffeae* and *P. crenatus* appear to have low virulence on sorghum (Motalaote et al., 1987). The others may cause intermediate to severe damage to sorghum.

Of the species that are known to cause damage on sorghum, *P. zeae* can cause the greatest disease severity. Endo (1959) indicated that sorghum is a susceptible host for *P. zeae*. Although nematode densities are highly dependent on extraction method, soil type, and soil volume, several authors have found that various soil populations of *P. zeae* can affect sorghum growth. Motalaote et al. (1987) found that 600 nematodes per plot reduced plant height, shoot weight, and root weight; Cuarezma-Teran and Trevathan (1984) found that 500 nematodes per plot could reduce sorghum growth in Mississippi; Ayala and Bee-Rodriguez (1978) showed that *P. zeae* was responsible for sorghum seedling death; and Bee-Rodriguez and Ayala (1977) found that 1500 nematodes per 20 cm plot reduced above- and belowground growth. However, McDonald and van den Berg (1993) showed that 9000 nematodes per plot had no effect on sorghum growth.

Other *Pratylenchus* spp. have been shown to cause damage as well. *Pratylenchus hexincisus* reduced sorghum growth by 20% at 65 nematodes per 100 cm^3 soil (Norton, 1958). *P. brachyurus* caused disease in more than 70% of sweet sorghum cultivars tested (Sharma and De Souza Meideros, 1982) and many varieties in West Africa were also susceptible (Beujard, 1994). However, Endo (1959) found that sorghum was not highly susceptible to *P. brachyurus*.

Lesion nematodes interact with fungal root and stalk pathogens, which may cause increased damage compared to nematode or fungal pathogen alone. Norton (1958) found that *P. hexinus* interacted with *M. phaseolina* to produce more severe root disease, especially under drought conditions. Similarly, *P. zeae* was found to interact with *C. lunata*, *Fusarium* spp., *M. phaseolina*, and *R. solani* to decrease sorghum growth (Bee-Rodriguez and Ayala, 1977). Francl and Wheeler (1993) demonstrated that *P. zeae* and *Acremonium strictum* interacted to produce necrosis, wilt, and reduced plant growth.

Stunt Nematodes

Stunt nematodes are ectoparasites that feed on root parenchymal cells. These feeding activities result in swollen and stubby roots due to a decrease in root elongation and increase in root diameter (Frederiksen and Odvody, 2000). Several species of stunt nematodes have been reported to be pathogens of sorghum. These species come from the genera *Tylenchorhynchus* and *Quinisulcius*. Many species are responsible for growth and yield suppression in sorghum (de Milliano et al., 1992). For example, *Tylenchorhynchus nudus* Allen can cause 10 to 56% plant growth reduction and *T. germanii* Germani & Luc reduced root and shoot weight by 46 and 23%, respectively (Frederiksen and Odvody, 2000). Studies in West Africa have shown that *T. nudus* and *T. germanii* can reduce root and shoot weight by 31 to 53% and 1.7 to 23%, respectively. Further, all *Tylenchorynchus* spp., except for *T. vulgaris*, exhibited greater nematode reproduction on sorghum compared to peanut, pearl millet, and cowpea (Leslie, 2002). In Colorado and Kansas, *Merlinius*, *Quinisulcius*, and *Tylenchorhynchus* were common stunt nematodes in sorghum agroecosystems (Todd et al., 2014). In Louisiana, *T. annulatus* (Cassidy) Golden suppressed root and panicle dry weight as populations increased. The nematode was sensitive to antagonism by *Mesocriconema xenoplax* Raski, another plant pathogenic nematode that largely feeds on fruit species, and by the charcoal rot pathogen, *M. phaseolina*, which suppressed nematode reproduction (Wenefrida et al., 1997).

Plant rhizospheres from symptomatic plants yield high numbers of stunt nematodes. Nematicides and insecticide treatments, such as terbofos, aldicarb, and phorate, provide little yield response but may reduce nematode numbers. Rotations resulted in improved yields compared to monocropping, but soil type had a major influence (Trevathan and Robbins, 1995).

Needle Nematodes

At least two species of *Longidorus* are capable of parasitizing sorghum. These include *L. africanus* Merny and *L. breviannulatus* Norton & Hoffman. Several *Longidorus* spp. have also been noted as virus vectors. Kolodge et al. (1987) reported *L. africanus* on sorghum cv. G-499 GBR and indicated that sorghum was a more susceptible to this nematode than barley, bermudagrass, corn, wheat, or oats. In Leslie (2002), Kollo reported that *L. africanus* could accumulate as many as 50 nematodes per 100 cm^3 of soil and resulted in reduced sorghum growth. However, Koenning et al. (1999) reported *L. africanus* on lettuce in California and *L. breviannulatus* on corn from several states including Delaware, Iowa, Illinois, Indiana, and Michigan, but not from sorghum. This latter finding may be attributed to host crop distribution rather than strictly host preference of the nematode.

Root-knot Nematodes

Root-knot nematodes (*Meloidogyne* spp.) are sedentary, endoparasites that infect numerous crops. Several species of root-knot nematode are capable of infecting sorghum. These include, but are not limited to, *M. acronea* Coetzee, *M. inocognita* (Kofoid & White) Chitwood, *M. javanica* Treub, *M. graminicola* Golden & Birchfield, and *M. naasi* Franklin. Fortnum and Curren (1988) suggested that sorghum was a poor host for *Meloidogyne*. These authors tested several cultivars and races and found that the nematodes could not produce more than one to two egg masses per root on average. In their studies, sorghum was also relatively resistant to *M. arenaria* (races 1 and 2), *M. incognita* (race 3), and *M. javanica*.

In contrast to the finding by Fortnum and Currin (1988), Thomas and Murray (1987) found that *M. incognita* race 3 could cause up to 45% yield loss in sorghum. Other studies have shown sorghum damage caused by *Meloidogyne* spp. as well. For example, *M. acronea* caused delayed flowering and 56% yield loss. *M. graminicola* caused reduced sorghum growth at 2000 stage II juveniles/pot. Where *M. incognita* is involved, sorghum damage is greater after cotton. *M. naasi* race 5 reduces sorghum growth, and both species cause galling on sorghum roots (Leslie, 2002).

Some Challenges for Nematode Control

1. Understanding yield losses and field economic thresholds for the nematode pests of sorghum.
2. Interactions between soilborne nematodes themselves and with the fungal pathogens of seedlings, roots, and stalks.
3. High-throughput screening of sorghum germplasm against the most consequential nematode pests such as *Pratylenchus* spp.
4. Consistency in cultivars, inoculum densities, and environments for race and cultivar screening in the field.

Parasitic Plants

Witchweeds (*Striga* spp.) are flowering plants that parasitize many monocots. *Striga hermonthica* (Delile) Benth., the purple-flowered species, occurs across Africa and, in addition to sorghum, can parasitize maize, millet, rice, and sugarcane (Abbasher et al., 1998; Yoshida et al., 2010). *S. asiatica* (L.) Kuntze, the red-flowered species, occurs in sub-Saharan Africa, Asia, and has been introduced into other sorghum growing regions of the world including Australia and the United States (Cochrane et al., 1997). Like *S. hermonthica, S. asiatica* can parasitize sorghum, maize, rice, and sugarcane.

Witchweed is especially damaging in areas where sorghum is grown under poor soil nutrition and quality. Above-ground symptoms of witchweed infection include chlorosis, stunting, and wilting (Doggett, 1965; Spallek et al., 2013). Before the emergence of the witchweed plants from the ground, host symptoms may resemble that of drought stress, nutrient deficiency, or severe nematode infestation. Observation of clumps of witchweed adjacent to the sorghum plant occurs several weeks after initial infection and indicate that heavy damage has already occurred to the roots, which will lead to partial or up to 100% yield loss (Musselman, 1980; Spallek et al., 2013).

Like many parasitic plants, the production of tiny, numerous seeds is a key strategy to improve the probability of host infection and the establishment of the next witchweed generation. Each *Striga* plant can produce thousands to hundreds of thousands of "dust" seeds that are brown or black in appearance (Berner et al., 1995; Atera and Itoh, 2011; Spallek et al., 2013). These seeds can survive in the soil for extended periods of time, up to years or decades (Atera and Itoh, 2011).

Preexisting soilborne witchweed seeds are induced to germinate by host seedling root exudates. Root exudates containing strigolactones, (+)-strigol and sorgolactone, are stimulatory to *S. asiatica* seeds (Matsutova et al., 2005). Witchweed seedlings germinate exploratory roots to locate suitable host roots. Chang et al. (1986) found that initial recognition between parasite and host at the sorghum root surface is dependent on the production of the quinone, 2,6-dimethoxybenzoquinone. The haustorium uses mechanical force and enzymatic degradation to penetrate the host root. *Striga*'s haustorium interfaces with the host xylem via several lignified oscula, which are used to "perforate the host vascular system." (Dörr, 1997). After haustorium attachment to the host, witchweed seedlings continue to develop beneath the soil surface for several weeks. At approximately seven weeks, the plants will emerge and flower.

Witchweed management strategies have focused on preventing seed germination, inducing premature seed germination, that is, "fooling" the seed into germinating in the absence of a host, preventing haustorium attachment to the host root via infection-site-specific hypersensitive reaction, and preventing the movement and/or introduction of witchweed seeds to new sorghum production areas (Musselman, 1980; Maiti et al., 1984; Olivier et al., 1991; Arnaud et al., 1999; Spallek et al., 2013). Trap-crops, including cotton, sesame, tobacco, various grain and forage legumes, and other weedy plants, have been used to induce witchweed seed germination since these plants produce stimulatory molecules, but are not susceptible to haustorial infection (Cook et al., 1983; Ramaiah et al., 1983; Khan et al., 2006; Cordosa et al., 2011). The use of ethylene can induce witchweed seed germination, which results in seedling death if no host roots are available for infection(Cordosa et al., 2011).

Some Challenges for *Striga* Control

1. Effective crop rotation, intercropping, and trap and/or repellent crops are necessary that are economically feasible for various production systems.
2. Improved genetic resistance that prevents *Striga* attachment or development.
3. Improved host tolerance that enables plants to have adequate production even in the presence of *Striga* infestation.
4. Development of herbicide seed dressings that are cost-effective, efficient at killing witchweed seedlings, and can be balanced with herbicide resistance programs in sorghum.
5. Biocontrol agents, such as *Fusarium oxysporum* f. sp. *strigae* Elzein & Thines, that can attack all witchweed plant parts and can provide durable control over the long term.

Conclusions

Most sorghum diseases can be managed through a combination of resistance and cultural control strategies. There are some practical measures that can be taken to control sorghum diseases, and if one or more is followed, many problems can be avoided:

1. If available, plant disease resistant (or disease tolerant) germplasm that is well adapted to the local growing environment.
2. Plant high quality, disease-free (fungicide-treated) seed.
3. Plant in fertile (balanced N and K) soils at pH 6.0 to 6.5.
4. Plant when soil temperatures are > 65°F (averaged daily).
5. Prepare fields with adequate drainage to prevent flooding.
6. Control potential reservoir hosts such as grassy weeds (especially Johnsongrass and related *Sorghum* spp.) near production fields.
7. Plant hybrids with good stalk strength and sturdy structural characteristics.
8. Mitigate plant stress through the management of weeds, insects, and plant populations. Typically, dense plant populations in dryland production systems should be avoided.
9. Rotate sorghum with non-cereals on a regular basis to reduce the accumulation of soilborne plant pathogen propagules.

Although the general practices listed above will achieve sorghum disease control at some level, it is important to note that challenges still exist. In some cases, such as anthracnose, ergot, and grain mold, a great deal is known about the life cycle of the pathogens involved. However, in many situations, such as the virus diseases, bacterial leaf diseases, nematode pathogens, and many of the minor fungal foliar diseases, knowledge of etiological agents and their interaction with the sorghum host and the environment remain lacking.

For the next century, substantial global climate change has been forecast for most agricultural areas, which are predicted to become warmer and drier (Christensen et al., 2007). Under these changing conditions, it seems likely that the frequency or severity of epidemics caused by plant pathogens will be altered (Burdon et al., 2006). However, the directionality of such alterations may vary by pathosystem and agricultural ecosystem and remain to be determined as more empirical evidence is gathered. The unpredictable perennial weather changes due to global warming warrants continuous evaluation of germplasm to identify potential sources of resistance to major and minor diseases and integration with abiotic stress tolerance.

Acknowledgments

This manuscript is contribution no. 16-360-B from the Kansas Agricultural Experiment Station, Manhattan.

References

Abbasher, A.A., D.E. Hess, and J. Sauerborn. 1998. Fungal pathogens for biological control of *Striga hermonthica* on sorghum and pearl millet in West Africa. Afr. Crop Sci. J. 6:179–188.

Adeyanju, A. 2014. Genetic study of resistance to charcoal rot and Fusarium stalk rot diseases of sorghum. Ph.D. dissertation, Kansas State University. Manhattan, KS.

Adeyanju, A., C. Little, J. Yu, and T. Tesso. 2015. Genome-wide association study on resistance to stalk rot diseases in grain sorghum. G3: Genes, Genomes. Genetics 5:1165–1175.

Aggarwal, A., and R.S. Mehrotra. 2003. Fundamentals of plant pathology, 2nd Ed. Tata McGraw-Hill Publishing Company, New Dehli, India.

Alderman, S., D.E. Frederickson, G. Milbraith, N. Montes, J. Narro, and G.N. Odvody. 1999. A laboratory guide to the identification of *Claviceps purpurea* and *Claviceps africana* in grass and sorghum seed samples. Oregon Department of Agriculture. Salem, OR.

Alderman, S.C., R.R. Halse, and J.F. White. 2004. A reevaluation of the host range and geographical distribution of *Claviceps* species in the United States. Plant Dis. 88:63–81.

Alexander, J.D., R.W. Toler, and L.M. Giorda. 1984. Correlation of yield reductions with severities of disease symptoms in grain sorghum (*Sorghum bicolor* (L.) Moench) infected with sugarcane mosaic or maize dwarf mosaic viruses. (Abstr.). Phytopathology 74:794.

Ali, M.E.K., and H.L. Warren. 1987. Physiological races of *Colletotrichum graminicola* on sorghum. Plant Dis. 71:402–404.

Alvarez, M.G. 1976. Primer catalogo de enfermedades de plantas Mexicanas. Fitofilo 71:1–169.

Arif, A.G., and M. Ahmad. 1969. Some studies on fungi associated with sorghum seeds and sorghum soils and their control: Part I. Flora of sorghum seeds and seed treatment. W. Pakistan. J. Agric. Res. 7:102–117.

Arnaud, M.C., C. Veronesi, and P. Thalouarn. 1999. Physiology and histology of resistance to *Striga hermonthica* in *Sorghum bicolor* var. Framida. Aust. J. Plant Physiol. 26:63–70.

Atera, E., and K. Itoh. 2011. Evaluation and ecologies and severity of *Striga* weed on rice in Sub-Saharan Africa. Agric. Biol. J. North Am. 2:752–760.

Axford, L.C., T.J. Simpson, and C.J. Willis. 2004. Synthesis and incorporation of the first polyketide synthase free intermediate in monocerin biosynthesis. Angew. Chem. 116:745–748.

Bain, D.G., and C.W. Edgerton. 1943. The zonate leaf spot, a new disease of sorghum. Phytopathology 33:220–226.

Baird, R.E., C.E. Watson, and M. Scruggs. 2003. Relative longevity of *Macrophomina phaseolina* and associated mycobiota on residual soybean roots in soil. Plant Dis. 87:563–566.

Bandara, Y.M.A.Y., R. Perumal, and C.R. Little. 2015. Integrating resistance and tolerance for improved evaluation of sorghum lines against Fusarium stalk rot and charcoal rot. Phytoparasitica 43:485–499.

Bandara, Y.M.A.Y., D.K. Weerasooriya, T.T. Tesso, and C.R. Little. 2016. Stalk rot fungi affect leaf greenness (SPAD) of grain sorghum in a genotype- and growth-stage-specific manner. Plant Dis. 100:2062–2068.

Bandara, Y.M.A.Y., D.K. Weerasooriya, T.T. Tesso, P.V.V. Prasad, and C.R. Little. 2017. Stalk rot fungi affect grain sorghum yield components in an inoculation stage-specific manner. Crop Prot. 94:97–105.

Bandyopadhyay, R., D.E. Frederickson, N.W. McLaren, G.N. Odvody, and M.J. Ryley. 1998. Ergot: A new disease threat to sorghum in the Americas and Australia. Plant Dis. 82:356–367.

Bandyopadhyay, R., and R.A. Frederiksen. 2006. Contemporary global movement of emerging plant diseases. Ann. N. Y. Acad. Sci. 894:28–36.

Bandyopadhyay, R., L.K. Mughogho, and M.V. Satyanaryana. 1987. Systemic infection of sorghum by *Acremonium strictum* and its transmission through seed. Plant Dis. 71:647–650.

Baroncelli, R., J.M. Saenz-Martin, G.E. Rech, S.A. Sukno, and M. Thon. 2014. Draft genome of *Colletotrichum sublineola,* a destructive pathogen of cultivated sorghum. Genome Announc. 2:E00540–E14.

Basu Chaudhary, K.C., and S.B. Mathur. 1979. Infection of sorghum seeds by *Colletotrichum graminicola* 1. Survey, location in seed, and transmission of the pathogen. Seed Sci. Technol. 7:87–92.

Bebawi, F.F., R.E. Eplee, and R.S. Norris. 1984. Effects of seed size and weight on witchweed (*Striga asiatica*) seed germination, emergence, and host-parasitization. Weed Sci. 32:202–205.

Bee-Rodriguez, D., and A. Ayala. 1977. Interaction of *Pratylenchus zeae* with four soil fungi on sorghum. J. Agric. Univ. P R. 61:501–506.

Berner, D.K., J.G. Kling, and B.B. Singh. 1995. *Striga* research and control- A perspective from Africa. Plant Dis. 79:652–660.

Bergquist, R.R. 1973. *Colletotrichum graminicola* on *Sorghum bicolor* in Hawaii. Plant Dis. Rep. 57:272–275.

Beujard, P. 1994. Nématicides, nématodes phytoparasitaires et rendements des cultures pluviales dans la zone sahélienne de l'Afrique de l'ouest. Afro-Asian Journal of Nematology 4:129–146.

Bloom, J.R., and H.B. Couch. 1960. Influence of environment on diseases of turf-grasses. I. Effect of nutrition, pH, and soil moisture on Rhizoctonia brown patch. Phytopathology 50:532–535.

Boora, K.S., R.A. Frederiksen, and C.W. Magill. 1998. DNA-based markers for a recessive gene conferring anthracnose resistance in sorghum. Crop Sci. 38:1708–1709.

Boora, K.S., R.A. Frederiksen, and C.W. Magill. 1999. A molecular marker that segregates with sorghum leaf blight resistance in one cross is maternally inherited in another. Mol. Gen. Genet. 261:317–322.

Borrell, A.K., and G.L. Hammer. 2000. Nitrogen dynamics and the physiological basis of stay-green in sorghum. Crop Sci. 40:1295–1307.

Borrell, A.K., G.L. Hammer, and R.G. Henzell. 2000. Does maintaining green leaf area in sorghum improve yield under drought? II. Dry matter production and yield. Crop Sci. 40:1037–1048.

Borrell, A.K., and A.C.L. Douglas. 1997. Maintaining green leaf area in grain sorghum increased nitrogen uptake under postanthesis drought. International Sorghum and Millets Newsletter 38:89–91.

Bowden, R.L., M.K. Kardin, J.A. Percich, and L.J. Nickelson. 1984. Anthracnose of wild rice. Plant Dis. 68:68–69.

Brady, C.R., L.W. Noll, and C.R. Little. 2011. Disease severity and microsclerotium properties of the sorghum sooty stripe pathogen, *Ramulispora sorghi*. Plant Dis. 95:853–859.

Bramel-Cox, P.J., J.S. Stein, D.M. Rodgers, and L.E. Claflin. 1988. Inheritance of resistance to *Macrophomina phaseolina* (Tassi) Goid and *Fusarium moniliforme* Sheldon in sorghum. Crop Sci. 28:37–40.

Brown, P.J., P.E. Klein, E. Bortiri, C.B. Acharya, W.L. Rooney, and S. Kresovich. 2006. Inheritance of inflorescence architecture in sorghum. Theor. Appl. Genet. 113:931–942.

Burdon, J.J., P.H. Thrall, and L. Ericson. 2006. The current and future dynamics of disease in plant communities. Annu. Rev. Phytopathol. 44:19–39.

Burrell, A.M., A. Sharma, N.Y. Patil, S.D. Collins, W.F. Anderson, W.L. Rooney, and P.E. Klein. 2015. Sequencing of an anthracnose-resistant sorghum genotype and mapping of a major QTL reveal strong candidate genes for anthracnose resistance. Crop Sci. 55:790–799.

Cardwell, K.F., P.R. Hepperly, and R.A. Frederiksen. 1989. Pathotypes of *Colletotrichum graminicola* and seed transmission of sorghum anthracnose. Plant Dis. 73:255–257.

Casela, C.R., and R.A. Frederiksen. 1993. Survival of *Colletotrichum graminicola* sclerotia in sorghum stalk residues. Plant Dis. 77:827.

Castor, L.L. 1981. Grain mold histopathology damage assessment, and resistance screening within *Sorghum bicolor* (L.) Moench lines. Ph.D. dissertation. Texas A&M University, College Station, TX.

Chang, M. 1986. The haustorium and the chemistry of host recognition in parasitic angiosperms. J. Chem. Ecol. 12:561–579.

Chenulu, V.V., and T.S. Hora. 1962. Studies on losses due to Helminthosporium blight of maize. Indian Phytopathol. 15:235–237.

Chiang, M.Y., C.G. van Dyke, and K.J. Leonard. 1989. Evaluation of endemic foliar fungi for potential biological control of Johnsongrass (*Sorghum halepense*): Screening and host range tests. Plant Dis. 73:459–464.

Christensen, C.M., and H.H. Kaufman. 1965. Deterioration of stored grains by fungi. Annu. Rev. Phytopathol. 3:69–84.

Christensen, J.H., B. Hewitson, A. Busuioc, A. Chen, X. Gao, I. Held, R. Jones, R.K. Kolli, T.W. Kwon, R. Laprise, V. Magana Rueda, L. Mearns, C.G. Menendez, J. Raisanen, A.

Rinke, A. Sarr, and P. Whetton. 2007. Regional climate projections. In: S. Solomon, D. Qin, M. Manning, Z. Chen, M. Marquis, K.B. Averyt, M. Tignor, and H.L. Miller, editors, Climate change: The physical science basis. Contributions of working group I to the fourth assessment report of the introgovernmental panel on climate change. Cambridge Univ. Press, New York. p. 892–896.

Christine, K.Y.Yu., K. Springob, J. Schmidt, R.L. Nicholson, I.K. Chu, W.K. Yip, and C. Lo. 2005. Stilbene synthase gene (SbSTS1) is involved in host and nonhost defense responses in sorghum. Plant Physiol. 138:393–401.

Churchill, A.C.L., L.D. Dunkle, W. Silbert, K.J. Kennedy, and V. Macko. 2001. Differential synthesis of peritoxins and precursors by pathogenic strains of the fungus *Periconia circinata*. Appl. Environ. Microbiol. 67:5721–5728.

Cianco, A., and K.G. Mukerji, editors. 2007. General concepts in integrated pest and disease management. Springer, Dordrecht, The Netherlands.

Cloud, G.L., and J.C. Rupe. 1994. Influence of nitrogen, plant growth stage, and environment on charcoal rot of grain sorghum caused by *Macrophomina phaseolina* (Tassi) Goid. Plant Soil 158:203–210.

Cochrane, V., and M.C. Press. 1997. Geographical distribution and aspects of the ecology of the hemiparasitic angiosperm *Striga asiatica* (L.) Kuntze: A herbarium study. J. Trop. Ecol. 13:371–380.

Cohen, R., N. Oman, A. Porat, and M. Edelstein. 2012. Management of Macrophomina wilt in melons using grafting or fungicide soil application: Pathological, horticultural, and economical aspects. Crop Prot. 35:58–63.

Cole, G.T., and H.C. Hoch, editors. 1991. The fungal spore and disease initiation in plants and animals. Springer, New York.

Coleman, O.H., and I.E. Stokes. 1954. The inheritance of resistance to stalk red rot in sorghum. Agron. J. 46:61–63.

Coley-Smith, J.R., and R.C. Cooke. 1971. Survival and germination of fungal sclerotia. Annu. Rev. Phytopathol. 9:65–92.

Cook, G.E., M.G. Boosalis, L.D. Dunkle, and G.N. Odvody. 1973. Survival of *Macrophomina phaseoli* in corn and sorghum stalk residue. Plant Dis. Rep. 57:873–875.

Cook, C.E., L.P. Whichard, B. Turner, and M.E. Wall. 1966. Germination of witchweed (*Striga lutea* Lour) isolation and properties of a potent stimulator. Science 154:1189–1190.

Cordosa, C., C. Ruyter-Spira, and H.J. Bouwmeester. 2011. Strigolactones and root infestation by plant-parasitic *Striga*, *Orobanche* and *Phelipanche* spp. Plant Sci. 180:414–420.

Creamer, R., X. He, and W.E. Styer. 1997. Transmission of sorghum stunt mosaic rhabdovirus by the leafhopper vector *Graminella sonora* (Hemiptera: Cicadelidae). Plant Dis. 81:63–65.

Crouch, J.A., and M. Tomaso-Peterson. 2012. Anthracnose disease of centipedegrass turf caused by *Colletotrichum eremochloae*, a new fungal species closely related to *Colletotrichum sublineola*. Mycologia 104:1085–1096.

Cruz Jimenez, D.R. 2011. Influence of soils, nutrition, and water relations upon charcoal rot disease processes in Kansas. M.S. thesis, Kansas State University, Manhattan, KS.

Cuarezma-Teran, J.A., and L.E. Trevathan. 1984. Nematodes associated with sorghum in Mississippi. Plant Dis. 68:1083–1085.

Dahlberg, J., R. Bandyopadhyay, B. Rooney, G. Odvody, and P. Madera-Torres. 2001. Evaluation of sorghum germplasm used in U.S. breeding programs for sources of sugary disease resistance. Plant Pathol. 50:681–689.<

Das, I.K., S. Rakshit, and J.V. Patil. 2015. Assessment of artificial inoculation methods for development of sorghum pokkah boeng caused by *Fusarium subglutinans*. Crop Prot. 77:94–101.

Davis, M.A., and W.W. Bockus. 2001. Evidence for a *Pythium* sp. as a chronic yield reducer in a continuous grain sorghum field. Plant Dis. 85:780–784.

Davis, M.A., D.J. Jardine, and T.C. Todd. 1994. Selected pre-emergent herbicides and soil pH effect on seedling blight of grain sorghum. J. Prod. Agric. 7:269–276.

Debaeke, P., J.-M. Nolot, and D. Raffaillac. 2006. A rule-based method for the development of crop management systems applied to grain sorghum in south-western France. Agric. Syst. 90:180–201.

de Milliano, W.A.J., R.A. Frederiksen, and G.D. Bengston, editors. 1992. Sorghum and millets diseases: A second world review. ICRISAT Press, Patancheru, India.

Desai, S.A. 1998. Evaluation of advanced sorghum varieties against charcoal rot disease in Karnataka. Karnataka J. Agric. Sci. 11:104–107.

Dörr, I. 1997. How *Striga* parasitizes its host: A TEM and SEM study. Ann. Bot. (Lond.) 79:463–472.

Dhingra, O.D., and J.B. Sinclair. 1975. Survival of *Macrophomina phaseolina* sclerotia in soil, effects of soil moisture, carbon:nitrogen ratios, carbon sources, and nitrogen concentrations. Phytopathology 65:236–240.

Diourte, M., J.L. Starr, M.J. Jeger, J.P. Stack, and D.T. Rosenow. 2007. Charcoal rot (*Macrophomina phaseolina*) resistance and the effects of water stress on disease development in sorghum. Plant Pathol. 44:196–202.

Doupnik, B., and M.G. Boosalis. 1980. Ecofallow– A reduced tillage system– and plant disease. Plant Dis. 64:31–35.

Dodd, J.L. 1980. The photosynthetic stress translocation balance concept of sorghum stalk rot. In: Proceedings of International Workshop of Sorghum Diseases, Hyderabad, India. 11-15 Dec. 1978. Texas A&M University/ICRISAT Press, College Station, TX. p. 300-305.

Doggett, H. 1965. *Striga hermonthica* on sorghum in East Africa. J. Agric. Sci. 65:183–193.

Du, M., C.L. Schardl, E. Nuckles, and L.J. Vaillancourt. 2005. Using mating-type gene sequences for improved phylogenetic resolution of *Colletotrichum* species complexes. Mycologia 97:641–658.

Edmunds, L.K., M.C. Futrell, and R.A. Frederiksen. 1970. Sorghum diseases. In: J.S. Wall and W.M. Ross, editors, Sorghum production and utilization. AVI Publishing, Westport, CT. p. 200-234.

Edmunds, L.K., and N. Zummo. 1975. Sorghum diseases in the United States and their control. USDA Handbook No. 468. Washington, DC, U.S.A. U.S. Government Publishing Office, Washington, D.C.

Elazegui, F.A. 1971. Helminthosporium leaf spot of sorghum in the Philippines. M.S. thesis, University of the Philippines at Los Banos, Laguna, Philippines.

El Bakali, M.A., M.P. Martin, F.F. Garcia, B.A. Moret, and P.M. Nadal. 2000. First report of *Rhizoctonia solani* AG-3 on potato in Catalonia (NE Spain). Plant Dis. 84:806.

Emechebe, A.M., A.S. Kutama, and B.S. Aliyu. 2010. Incidence and distribution of head and loose smuts of sorghum (*Sorghum bicolor* L. Moench) in Nigerian Sudan savanna. Bayero Journal of Pure and Applied Sciences 3:142–147.

Endo, B.Y. 1959. Responses of root-lesion nematodes, *Pratylenchus, Brachyurus*, and *P. zeae* to various plants and soil types. Phytopathology 49:417–421.

Erpelding, J.E. 2007. Inheritance of anthracnose resistance for the sorghum cultivar Redlan. Plant Pathol. J. 6:187–190.

Esele, J.P. 1995. Foliar and head diseases of sorghum. Afr. Crop Sci. J. 3:185–189.

FAO STAT. 2014. Food and Agriculture Organization of the United Nations, Statistics (http://faostat.org/). (Accessed 01 Dec 2014).

Ferreira, A.S., and H.L. Warren. 1982. Resistance of sorghum to *Colletotrichum graminicola*. Plant Dis. 66:773–775.

Forbes, G.A., O. Ziv, and R.A. Frederiksen. 1987. Resistance in sorghum seedling disease caused by *Pythium arrhenomanes*. Plant Dis. 71:145–148.

Fortnum, B.A., and R.E. Currin, III. 1988. Host suitability of grain sorghum cultivars to *Meloidogyne* spp. Annals of Applied Nematology 2:61–64.

Francl, L.J., and T.H. Wheeler. 1993. Interactions of plant-parasitic nematodes with wilt-inducing fungi. In: W. Khan, editor, Nematode interactions. Chapman and Hall, London. p. 79–103.

Frederiksen, R.A. 1980. Sorghum downy mildew in the United States: Overview and outlook. Plant Dis. 64:903–908.

Frederiksen, R.A., L.L. Castor, and D.T. Rosenow. 1982. Grain mold, small seed, and head blight: The *Fusarium* connection. p. 26-36 In: Proceedings of the 37th Annual Corn and Sorghum Industry Research Conference. American Seed Trade Association, Washington, D.C.

Frederiksen, R.A., and G.N. Odvody. 2000. Compendium of sorghum diseases, 2nd ed. APS Press, St. Paul, MN.

Frederiksen, R.A., D.T. Rosenow, and D.M. Tuleen. 1975. Resistance to *Exserohilum turcicum* in sorghum. Plant Dis. Rep. 59:547–548.

Frederickson, D.E., P.G. Mantle, and W.A.J. De Milliano. 1991. *Claviceps africana* sp. nov.; the distinctive ergot pathogen of sorghum in Africa. Mycol. Res. 95:1101–1107.

Frederickson, D.E., P.G. Mantle, and W.A.J. De Milliano. 1993. Windborne spread of ergot disease (*Claviceps africana*) in sorghum A-lines in Zimbabwe. Plant Pathol. 42:368–377.

Freeman, T.E., H.H. Luke, and D.T. Sechler. 1966. Pathogenicity of *Pythium aphanidermatum* on grain crops in Florida. Plant Dis. Rep. 50:292–294.

Fukata, S., R. Takahashi, S. Kuroyanagi, Y. Ishiguro, N. Miyake, H. Nagai, H. Suzuki, T. Tsuji, R. Hashizume, H. Watanabe, and K. Kageyama. 2014. Development of loop-mediated isothermal amplification assay for the detection of *Pythium myriotylum*. Lett. Appl. Microbiol. 59:49–57.

Garud, T.B., S. Ismail, and B.M. Shinde. 2000. Effect of two mold-causing fungi on germination of sorghum seed. International Sorghum and Millets Newsletter 41:54.

Gaskin, T.A., and M.P. Britton. 1956. A new host for *Sclerophthora macrospora*. Plant Dis. Rep. 40:830.

Ghorade, R.B., V.B. Shekar, B.D. Gite, and B.D. Sakhare. 1997. Some general combiners for grain mould resistance in sorghum. Journal of Soils and Crops 7:8–11.

Girish, A.G., S. Deepti, V.P. Rao, and R.P. Thakur. 2004. Detection of seedborne grain mold fungi in sorghum and their control with fungicidal seed treatment. International Sorghum and Millets Newsletter 45:31–33.

Govaerts, B., M. Fuentes, M. Mezzalama, J.M. Nicol, J. Deckers, J.D. Etchevers, B. Figueroa-Sandoval, and K.D. Sayre. 2007. Infiltration, soil moisture, root rot and nematode populations after 12 years of different tillage, residue and crop rotation managements. Soil Tillage Res. 94:209–219.

Guarro, J., W. Gams, I. Puhhol, and J. Gene. 1997. *Acremonium* species: New emerging fungal opportunists—in vitro antifungal susceptibilities and review. Clin. Infect. Dis. 25:1222–1229.

Gwary, D.M., A. Obida, and S.D. Gwary. 2007. Management of sorghum smuts and anthracnose using cultivar selection and seed dressing fungicide in Maiduguri, Nigeria. Int. J. Agric. Biol. 9:326–328.

Gwary, D.M., T.D. Rabo, and A.B. Anaso. 2004. The development of anthracnose on sorghum genotypes in the Nigerian savanna. J. Plant Dis. Prot. 111:96–103.

Hagan, A.K., K.L. Bowen, and J. Jones. 2014. Nitrogen rate and variety impact diseases and yield of sorghum for biofuel. Agron. J. 106:1205–1211.

Hansing, E.D., and A. Hartley. 1962. Sorghum seed fungi and their control. Proceedings of the Association of Official Seed Analysts 52:143–148.

Harris, H.B., and G. Sowel. 1970. Incidence of *Colletotrichum graminicola* on *Sorghum bicolor* introductions. Plant Dis. Rep. 54:60–62.

Harris, H.B., B.J. Johnson, J.W. Dobson, and E.S. Luttrell. 1964. Evaluation of anthracnose on grain sorghum. Crop Sci. 4:460–462.

Hazra, S., R.P. Thakur, and R.P. Devi. 1999. Pathogenic and molecular variability among twelve isolates of *Colletotrichum graminicola* from sorghum. J. Mycol. Plant Pathol. 29:176–183.

Hooker, A.L. 1985. Corn and sorghum rusts. In: A.P. Roelfs and W.R.B.A. Press, editors, The cereal rusts. Diseases, distribution, epidemiology, and control. Vol. 2. Academic Press, San Diego, CA. p. 208–223.

Horvath, B.J., and J.M. Vargas. 2004. Genetic variation among *Colletotrichum graminicola* isolates from four hosts using isozyme analysis. Plant Dis. 88:402–406.

Huang, L.D., and D. Backhouse. 2006. Analysis of chitinase isoenzymes in sorghum seedlings incoulated with *Fusarium thapsinum* and *F. proliferatum*. Plant Sci. 171:539–545.

Hunter, E.L., and I.C. Anderson. 1997. Sweet sorghum. Hortic. Rev. (Am. Soc. Hortic. Sci.) 21:40–73.

Husseini, A.M., A.G. Timothy, H.A. Olufunmilayo, A.S. Ezekiel, and H.O. Godwin. 2009. Fungi and some mycotoxins found in mouldy sorghum in Niger State, Nigeria. World J. Agric. Sci. 5:5–17.

Ibraheem, K., I. Gaffor, and S. Chopra. 2010. Flavonoid phytoalexin-dependent resistance to anthracnose leaf blight requires a functional yellow seed1 in *Sorghum bicolor*. Genetics 184:915–926.

Ibrahim, O.E., W.E. Nyquist, and J.D. Axtell. 1985. Quantitative inheritance and correlations of agronomic and grain quality traits of sorghum. Crop Sci. 25:649–654.

Idris, H.A., N. Labuschagne, and L. Korsten. 2007. Screening rhizobacteria for biological control of root and crown rot of sorghum in Ethiopia. Biol. Control 40:97–106.

Idris, H.A., N. Labuschagne, and L. Korsten. 2008. Suppression of *Pythium ultimum* root rot of sorghum by rhizobacterial isolates from Ethiopia and South Africa. Biol. Control 45:72–84.

International Committee on Taxonomy of Viruses (ICTV). 2015. Virus taxonomy: 2015 Release. International Committee on Taxonomy of Viruses. (accessed 9 June 2016).

Isakeit, T., and J. Jaster. 2005. Texas has a new pathotype of *Peronosclerospora sorghi* the cause of sorghum downy mildew. Plant Dis. 89:529.

Isakeit, T., G.N. Odvody, and R.A. Shelby. 1998. First report of sorghum ergot caused by *Claviceps africana* in the United States. Plant Dis. 82:592.

Ishiguro, Y., T. Asano, K. Otsubo, H. Suga, and K. Kageyama. 2013. Simulataneous detection by multiplex PCR of the high-temperature-growing *Pythium* species: *P. aphanidermatum*, *P. helicoides* and *P. myriotylum*. J. Gen. Plant Pathol. 79:350–358.

Islam, M.S., M.S. Haque, M.M. Islam, E.M. Emdad, A. Halim, Q.M.M. Hossen, M.Z. Hossain, B. Ahmed, S. Rahim, M.S. Rahman, M.M. Alam, S. Hou, X. Wan, J.A. Saito, and M. Alam. 2012. Tools to kill: Genome of one of the most destructive plant pathogenic fungi *Macrophomina phaseolina*. BMC Genomics 13:493.

Jardine, D.J., and W.B. Gordon. 1997. Assessment of grain sorghum yield loss by sooty stripe (*Ramulispora sorghi*). (Abstr.). Phytopathology 87:S48.

Jardine, D.J., and J.F. Leslie. 1992. Aggressiveness of *Gibberella fujikuroi* (*Fusarium moniliforme*) isolates to grain sorghum under greenhouse conditions. Plant Dis. 76:897–900.

Jones, E.M. 1979. The inheritance of resistance to *Colletotrichum graminicola* in grain sorghum, *Sorghum bicolor*. Ph.D. Dissertation, Purdue University, West Lafayette, IN.

Kanyuka, K., E. Ward, and M.J. Adams. 2003. Polymyxa graminis and the cereal virus it transmits: A research challenge. Mol. Plant Pathol. 4:393–406.

Kapanigowda, M.H., R. Perumal, R.M. Aiken, T.J. Herald, S.R. Bean, and C.R. Little. 2013a. Analysis of sorghum [*Sorghum bicolor* (L.) Moench] lines and hybrids in response to early-season planting and cool conditions. Can. J. Plant Sci. 93:773–784.

Kapanigowda, M.H., R. Perumal, M. Djanaguiraman, R.M. Aiken, T. Tesso, P.V.V. Prasad, and C.R. Little. 2013b. Genotypic variation in sorghum [*Sorghum bicolor* (L.) Moench] exotic germplasm for drought and disease tolerance. Springerplus 2:650.

Karunakar, R.I., S. Pande, and R.P. Thakur. 1995. Effects of host resistance, temperature and duration of wetness on leaf blight development of grain sorghum. Indian Journal of Phytopathol. 23:146–151.

Katilé, S.O. 2007. Expression of defense genes in sorghum grain mold and tagging and mapping a sorghum anthracnose resistance gene. Ph.D. dissertation. Texas A&M University, College Station, TX.

Katilé, S.O., R. Perumal, W.L. Rooney, L.K. Prom, and C.W. Magill. 2009. Expression of pathogenesis-related protein PR-10 in sorghum floral tissues in response to inoculation with *Fusarium thapsinum* and *Curvularia lunata*. Mol. Plant Pathol. 11:93–103.

Kaufmann, P.J., and G.J. Weidemann. 1996. Isozyme analysis of *Colletotrichum gloeosporoides* from five host genera. Plant Dis. 80:1289–1293.

Khaleeque, M.I., A. Alam, M. Inam-ul-Haq, and M. Ahmad. 1995. Reaction of ten sorghum cultivars against grain smut under semi-arid conditions of Bahawalpur. Pakistan. J. Phytopathol. 7:90–91.

Khan, Z.R., J.A. Pickett, L.J. Wadhams, A. Hassanali, and C.A.O. Midega. 2006. Combined control of *Striga hermonthica* and stemborers by maize-*Desmodium* spp. intercrops. Crop Prot. 25:989–995.

Khune, N.N., D.E. Kurhekar, J.G. Raut, and P.D. Wangikar. 1984. Stalk rot of sorghum caused by *Fusarium moniliforme*. Indian Phytopathol. 37:316–319.

Koenning, S.R., C. Overstreet, J.W. Noling, P.A. Donald, J.O. Becker, and B.A. Fortnum. 1999. Survey of crop losses in response to phytoparasitic nematodes in the United States for 1994. J. Nematol. 31:587–618.

Kolodge, C., J.D. Radewald, and F. Shibuya. 1987. Revised host range and studies on the life cycle of *Longidorus africanus*. J. Nematol. 19:77–81.

Komolong, B., S. Chakrabotry, M. Ryley, and D. Yates. 2003. Ovary colonization by *Claviceps africana* is related to ergot resistance in male-sterile sorghum lines. Plant Pathol. 52:620–627.

Konde, B.K., and B.R. Pokharkar. 1979. Seedborne fungi of sorghum. Seed Res. 7:54–57.

Kumar, S., K. Sivasithamparam, J.S. Gill, and M.W. Sweetingham. 1999. Temperature and water potential effects on growth and pathogencity of *Rhizoctonia solani* to lupin. Can. J. Microbiol. 45:389–395.

Kutama, A.S., M.I. Auyo, S. Umar, and M.L. Umar. 2013. Reduction in growth and yield parameters of sorghum genotypes screened for loose smuts in Nigerian Sudan savanna. World J. Agric. Sci. 1:185–192.

Lenne, J.M. 1990. World list of fungal diseases of tropical pasture species. Phytopathological Papers 31:1–162.

Leonard, K.J., and E.G. Suggs. 1974. *Setosphaeria prolata*, the ascigerous state of *Exserohilum prolatum*. Mycologia 66:281–297.

Leukel, R.W. 1948. *Periconia circinata* and its relation to Milo disease. J. Agric. Res. 77:201–222.

Leslie, J.F., editor. 2002. Sorghum and millets diseases. Iowa State Press, Ames, IA.

Leslie, J.F., C.A.S. Pearson, P.E. Nelson, and T. Toussoun. 1990. *Fusarium* spp. from corn, sorghum, and soybean fields in the central and eastern United States. Phytopathology 80:343–350.

Leukel, R.W., and J.H. Martin. 1943. Seed rot and seedling blight of sorghum. U.S. Department of Agriculture Technical Bulletin No. 839. USDA, Washington, D.C.

Levings, C.S. 1990. The Texas cytoplasm of maize: Cytoplasmic male sterility and disease susceptibility. Science 250:942–947.

Levy, Y. 1984. The overwintering of *Exserohilum turcicum* in Israel. Phytoparasitica 12:177–182.

Lin, Z., X. Shiqiang, Y. Que, J. Wang, J.C. Comstock, J. Wei, P.H. McCord, B. Chen, R. Chen, and M. Zhang. 2014. Species-specific detection and identification of *Fusarium* species complex, the causal agent of sugarcane pokkah boeng in China. PLoS One 9:E104195.

Little, C.R., and C.W. Magill. 2003. Reduction of sorghum seedling vigor by inoculation with *Fusarium thapsinum* and *Curvularia lunata* at anthesis. International Sorghum and Millets Newsletter 44:112–113.

Little, C.R., R. Perumal, T. Tesso, L.K. Prom, G.N. Odvody, and C.W. Magill. 2012. Sorghum pathology and biotechnology- A fungal disease perspective: Part I. Grain mold, head smut, and ergot. The European Journal of Plant Science and Biotechnology 6:10–30.

Lutrell, E.S. 1954. Diseases of pearl millet in Georgia. Plant Dis. Rep. 38:507–514.

Mahalakshmi, V., and F.R. Bidinger. 2001. Evaluation of stay-green sorghum germplasm lines at ICRISAT. Crop Sci. 42:965–974.

Maiti, R.K., K.V. Ramaiah, S.S. Bisen, and V.L. Chidley. 1984. A comparative study of the haustorial development of *Striga asiatica* (L.) Kuntze on sorghum cultivars. Ann. Bot. (Lond.) 54:447–457.

Malvick, D.K. 2008. Sorghum smuts. Integrated Pest Management, Reports on Plant Disease, No. 208. University of Illinois Extension, Woodstock, IL.

Marley, P.S., and D.A. Aba. 1996. Varietal resistance of *Sorghum bicolor* (L.) Moench germplasm to *Sporisorium sorghi* Link in Nigeria. African Journal of Plant Protection 6:83–89.

Marley, P.S., R. Bandyopadhyay, R. Tabo, and O. Ajayi. 2001. Reactions of sorghum genotypes to anthracnose and gray leaf spot diseases under Sudan and Sahel savanna field conditions of Nigeria. J. Sustain. Agric. 18:105–116.

Martin, T., M. Biruma, I. Fridborg, P. Okori, and C. Dixelius. 2011. A highly conserved NB-LRR encoding gene cluster effective against *Setosphaeria turcica* in sorghum. BMC Plant Biol. 11:151.

Mathur, K., R.P. Thakur, and V.P. Rao. 1997. Intrapopulation variability in *Colletotrichum sublineolum* infecting sorghum. Journal of Mycology and Plant Pathology 27:310.

Matsusova, R., R. Kumkum, F.W.A. Verstappen, M.C.A. Franssen, M.H. Beale, and H.J. Bouwmeester. 2005. The strigolactone germination stimulants of the plant parasitic *Striga* and *Orobanche* spp. are derived from the carotenoid pathway. Plant Physiol. 139:920–934.

Mayek-Perez, N., C. Lopez-Castaneda, M. Gonzalez-Chavira, R. Garcia-Espinosa, J. Acosta-Gallegos, O.M. Vega, and J. Simpson. 2001. Variability of Mexican isolates of *Macrophomina phaseolina* based on pathogenesis and AFLP genotype. Physiol. Mol. Plant Pathol. 59:257–264.

McDonald, B.A., and C. Linde. 2002. The population genetics of plant pathogens and breeding strategies for durable resistance. Euphytica 124:163–180.

McDonald, A.H., and E.H. van den Berg. 1993. Effect of watering regimen on injury to corn and grain sorghum by *Pratylenchus* species. J. Nematol. 25:654–658.

McGee, D.C. 1995. Epidemiological approach to disease management through seed technology. Annu. Rev. Phytopathol. 33:445–466.

McKenzie, E.H.C., and G.C.M. Latch. 1984. New plant disease records in New Zealand: Graminicolous fungi. N. Z. J. Agric. Res. 27:113–123.

McLaren, N.W. 1992. Quantifying resistance of sorghum genotypes to the sugary disease pathogen (*Claviceps africana*). Plant Dis. 76:986–988.

McLaren, N.W. 1997. Changes in pollen viability and concomitant increase in the incidence of sorghum ergot with flowering date and implications in selection for escape resistance. J. Phytopathol. 145:261–265.

McLaren, N.W., and F.H.J. Rijkenberg. 1989. Efficacy of fungicide seed dressings in the control of pre- and post-emergent damping-off and seedling blight of sorghum. S. Afr. J. Plant Soil 6:167–170.

McLaren, N.W., and F.C. Wehner. 1992. Pre-flowering low temperature predisposition of sorghum to sugary disease (*Claviceps africana*). J. Phytopathol. 135:328–334.

Mehta, B.K. 1965. Variation in pathogenicity of the sorghum head smut fungus. Ph.D. dissertation, Texas A&M University, College Station, TX.

Mehta, P.J., S.D. Collins, W.L. Rooney, R.A. Frederiksen, and R.R. Klein. 2000. Identification of different sources of genetic resistance to anthracnose in sorghum. International Sorghum and Millets Newsletter 41:51–54.

Mengistu, A., J.D. Ray, J.R. Smith, and R.L. Paris. 2007. Charcoal rot disease assessment of soybean genotype using a colony-forming unit index. Crop Sci. 47:2453–2461.

Mocoeur, A., Y.-M. Zhang, Z.-Q. Liu, X. Shen, L.-M. Zhang, S.K. Rasmussen, and H.-C. Jing. 2015. Stability and genetic control of morphological and biomass and biofuel traits under temperate maritime and continental conditions in sweet sorghum (*Sorghum bicolor*). Theor. Appl. Genet. 128:1685–1701.

Montes-Belmont, R., H.E. Flores-Moctezuma, and R.A. Nava-Juárez. 2003. Alternate hosts of *Claviceps africana* Frederickson, Mantle and de Milliano, causal of sorghum "ergot" in the state of Morelos, Mexico. Rev. Mex. Fitopatol. 21:63–66.

Moore, J.W., M. Ditmore, and D.O. Tebeest. 2008. Pathotypes of *Colletotrichum sublineolum* in Arkansas. Plant Dis. 92:1415–1420.

Motalaote, B., J.L. Starr, R.A. Frederiksen, and F.R. Miller. 1987. Host status and susceptibility of sorghum to *Pratylenchus* species. Revue de Nématologie 11:65–74.

Mueller, D.S., K.A. Wise, N.S. Dufault, C.A. Bradley, and M.I. Chilvers. 2013. Fungicides for field crops. APS Press, St. Paul, MN. p. 112.

Mughogho, L.K., ed. 1984. Proceedings of a consultative group discussion on research needs and strategies for control of sorghum root and stalk diseases. ICRISAT Press, Patancheru, India. p. 267

Muriithi, L.M., and L.E. Claflin. 1997. Genetic variation of grain sorghum germplasm for resistance to *Pseudomonas andropogonis*. Euphytica 98:129–132.

Musselman, L.J. 1980. The biology of *Striga, Orobanche*, and other root parasitic weeds. Annu. Rev. Phytopathol. 18:463–489.

Nagy, E.D., and J.L. Bennetzen. 2016. Pathogen corruption and site-directed recombination at a plant disease resistance gene cluster. Genome Res. 18:1918–1923.

Narasirnhan, K.S., and G. Rangaswami. 1969. Influence of mold isolates from sorghum grain on viability of the seed. Curr. Sci. 38:389–390.

Natural, M.P., R.A. Frederiksen, and D.T. Rosenow. 1982. *Acremonium* wilt of sorghum. Plant Dis. 66:863–865.

Navi, S.S., R. Bandyopadhyay, R.K. Reddy, R.P. Thakur, and X.B. Yang. 2005. Effects of wetness duration and grain development stages on sorghum grain mold infection. Plant Dis. 89:872–878.

Nelson, K.A., P.C. Sharf, L.G. Bundy, and P. Tracy. 2008. Agricultural management of enhanced-efficiency fertilizers in the North-Central United States. Crop Management. doi:10.1094/CM-2008-0730-03-RV

Ngugi, H.K., S.B. King, G.O. Abayo, and Y.V.R. Reddy. 2002. Prevalence, incidence, and severity of sorghum diseases in western Kenya. Plant Dis. 86:65–70.

Ngugi, H.K., S.B. King, J. Holt, and A.M. Julian. 2001. Simultaneous temporal progress of sorghum anthracnose and leaf blight in crop mixtures with disparate patterns. Phytopathology 91:720–729.

Noll, L.W., D.N. Butler, and C.R. Little. 2010. Tetrazolium violet staining of naturally and artificially molded sorghum (*Sorghum bicolor* (L.) Moench) caryopses. Seed Sci. Technol. 38:741–756.

Norton, D.C. 1958. The association of *Pratylenchus hexicinus* with charcoal rot. Phytopathology 48:355–358.

Novicki, T.J., K. LaFe, L. Bui, U. Bui, R. Geise, K. Marr, and B.T. Cookson. 2003. Genetic diversity among clinical isolates of *Acremonium strictum* determined during an investigation of a fatal mycosis. J. Clin. Microbiol. 41:2623–2628.

Nzioki, H.S., L.E. Claflin, and B.A. Ramundo. 2000. Evaluation of screening protocols to determine genetic variability of grain sorghum germplasm to *Sporisorium sorghi* under field and greenhouse conditions. Int. J. Pest Manage. 46:91–95.

Odvody, G.N., and L.D. Dunkle. 1979. Charcoal rot of sorghum: Effect of environment on host-parasite relations. Phytopathology 69:250–254.

Odvody, G.N., L.D. Dunkle, and L.K. Edmunds. 1977. Characterization of the *Periconia circinata* population in a milo disease nursery. Phytopathology 67:1485–1489.

Odvody, G.N., L.D. Dunkle, and M.G. Boosalis. 1973. The occurrence of sooty stripe of sorghum in Nebraska. Plant Dis. Rep. 57:681–683.

Odvody, G.N., and G. Forbes. 1984. Pythium root and seedling rots. In: L.K. Mughogho, editor, Sorghum root and stalk rots: A critical review. International Crops Research Institute for the Semi-Arid Tropics, Patancheru, India. p. 31–35.

Odvody, G.N., and D.B. Madden. 1984. Leaf sheath blights of *Sorghum bicolor* caused by *Sclerotium rolfsii* and *Gleocercospora sorghi* in South Texas. Phytopathology 74:264–268.

Oh, B.J., R.A. Frederiksen, and C.W. Magill. 1994. Identification of molecular markers linked to head smut resistance (*Shs*) in sorghum by RFLP and RAPD analyses. Phytopathology 84:830–833.

Olivier, A., N. Benhamou, and G.D. Leroux. 1991. Cell-surface interactions between sorghum roots and the parasitic weed *Striga hermonthica*-cytochemical aspects of cellulose distribution in resistant and susceptible host tissues. Can. J. Bot. 69:1679–1690.

Omar, M.E., R.A. Frederiksen, and D.T. Rosenow. 1985. Collaborative sorghum disease studies in Sudan. Sorghum Newsletter 29:93.

Oosterom, E.J., R. Jaychandran, and F.R. Bidinger. 1996. Diallel analysis of stay green trait and its components in sorghum. Crop Sci. 36:549–555.

Pande, S., L.K. Mughogho, and R.J. Karunakar. 1990. Effect of moisture stress, plant population density, and pathogen inoculation on charcoal rot of sorghum. Ann. Appl. Biol. 116:221–232.

Pascual, C.B., A.D. Raymundo, and M. Hyakumachi. 2000. Resistance of sorghum line CS621 to *Rhizoctonia solani* AG1-IA and other sorghum pathogens. J. Gen. Plant Pathol. 66:23–39.

Pastor-Corrales, M.A. 1980. Variation in pathogenicity of *Colletotrichum graminicola* (Cesati) Wilson and in symptom expression of anthracnose of *Sorghum bicolor* (L.) Moench. Ph.D. Dissertation, Texas A&M University, College Station, TX. p. 122.

Pavgi, M.S. 1972. Morphology and taxonomy of the *Puccinia* species of corn and sorghum. Mycopathol. Mycol. Appl. 47:207–212.

Pažoutová, S., R. Bandyopadhyay, D.E. Frederickson, P.G. Mantle, and R.A. Frederiksen. 2000. Relations among sorghum ergot isolates from the Americas, Africa, India, and Australia. Plant Dis. 84:437–442.

Pažoutová, S., and A. Bogo. 2002. Recovery of *Claviceps sorghi* (Ascomycotina: Clavicipitaceae) in India. Mycopathologia 153:99–101.

Pažoutová, S., and D.E. Frederickson. 2005. Genetic diversity of *Claviceps africana* on sorghum and Hyparrhenia. Plant Pathol. 54:749–763.

Perumal, R., T. Isakeit, M.A. Menz, S. Katilé, E.G. No, and C.W. Magill. 2006. Characterization and genetic distance analysis of isolates of *Peronosclerospora sorghi* using AFLP fingerprinting. Mycol. Res. 110:471–478.

Perumal, R., M.A. Menz, P.J. Mehta, S. Katilé, L.A. Gutierrez Rojas, R.R. Klein, P.E. Klein, L.K. Prom, J.A. Schlueter, W.L. Rooney, and C.W. Magill. 2009. Molecular mapping of Cg1, a gene for resistance to anthracnose (*Colletotrichum sublineolum*) in sorghum. Euphytica 165:597–606.

Perumal, R., P. Nimmakayala, S. Erraitamuthu, E.G. No, U.K. Reddy, L.K. Prom, G.N. Odvody, D.G. Luster, and C.W. Magill. 2008. Simple sequence repeat markers useful for sorghum downy mildew (*P. sorghi*) and related species. BMC Genet. 9:77. doi:10.1186/1471-2156-9-77

Perumal, R., T. Tesso, K.D. Kofoid, P.V.V. Prasad, R.M. Aiken, S.R. Bean, J.D. Wilson, T.J. Herald, and C.R. Little. 2015. Registration of nine grain sorghum seed parent lines. J. Plant Reg. 9:244–248.

Plant Management Network. 2010a. BASF announces expanded label for Stamina F3 HL fungicide seed treatment at ASTA Seed Expo 2010. Plant Management Network. https://www.plantmanagementnetwork.org/pub/php/news/2010/StaminaF3HL/ (accessed 13 April 2015).

Plant Management Network. 2010b. Dynasty seed treatment fungicide now registered for use on sorghum. Plant Management Network. http://www.plantmanagementnetwork.org/pub/php/news/2010/Dynasty/ (accessed 13 Apr. 2015).

Plant Management Network. 2014. Sercadis fungicide receives EPA registration for disease control in rice. Plant Management Network. http://www.plantmanagementnetwork.org/pub/php/news/2014/SercadisFungicide/ (accessed 13 April 2015).

Pratt, R.G., and G.D. Janke. 1980. Pathogenicity of three species of *Pythium* to seedlings and mature plants of grain sorghum. Phytopathology 70:766–771.

Pringle, R.B., and R.P. Scheffer. 1964. Host specific plant toxins. Annu. Rev. Phytopathol. 2:133–156.

Prom, L.K., and J.E. Erpelding. 2009. New sources of grain mold resistance among sorghum accessions from Sudan. Trop. Subtrop. Agroecosyst. 10:457–463.

Prom, L.K., and T. Isakeit. 2003. Laboratory, greenhouse, and field assessment of fourteen fungicides for activity against *Claviceps africana*, causal agent of sorghum ergot. Plant Dis. 87:252–258.

Prom, L.K., T. Isakeit, H. Cuevas, W.L. Rooney, R. Perumal, and C.W. Magill. 2015a. Reaction of sorghum lines of zonate leaf spot and rough leaf spot. Plant Health Prog. doi:10.1094/PHP-RS-15-0040

Prom, L.K., R. Perumal, S.R. Erattaimuthu, J.E. Erpelding, N. Montes, G.N. Odvody, C. Greenwald, Z. Jin, R. Frederiksen, and C.W. Magill. 2011. Virulence and molecular genotyping studies of *Sporisorium relianum* isolates in sorghum. Plant Dis. 95:523–529.

Prom, L.K., R. Perumal, S.R. Erattaimuthu, C.R. Little, E.G. No, J.E. Erpelding, W.L. Rooney, G.N. Odvody, and C.W. Magill. 2012. Genetic diversity and pathotype determination of *Colletotrichum sublineolum* isolates causing anthracnose in sorghum. Eur. J. Plant Pathol. 133:671–685.

Prom, L.K., R. Perumal, T. Isakeit, G. Radwan, W.L. Rooney, and C.W. Magill. 2015b. The impact of weather conditions on response of sorghum genotypes to anthracnose (*Colletotrichum sublineolum*) infection. Am. J. Exp. Agric. 6:242–250.

Prom, L.K., R. Perumal, N. Montes-Garcia, T. Isakeit, G.N. Odvody, W.L. Rooney, C.R. Little, and C.W. Magill. 2015c. Evaluation of Gambian and Malian sorghum germplasm against the downy mildew pathogen, *Peronosclerospora sorghi,* in Mexico and the USA. J. Gen. Plant Pathol. 81:1–8.

Radwan, G.L., Perumal, R., Isakeit, T., Magill, C.W., Prom, L.K., and Little, C.R. 2011. Screening exotic sorghum germplasm, hybrids, and elite lines for resistance to a new virulent pathotype (P6) of *Peronosclerospora sorghi* causing downy mildew. Plant Health Progress. doi:10.1094/PHP-2011-0323-01-RS.

Rajasab, A.H., and A. Ramalingam. 1989. Splash dispersal in *Ramulispora sorghi* Olive and Lefebreve, the causal agent of sooty stripe in sorghum. Proc. Indiana Acad. Sci. 99:335–341.

Ramaiah, K.V., C. Parker, M.J. Vasudeva Rao, and L.J. Musselman. 1983. *Striga* identification and control handbook. Information Bulletin No. 15. ICRISAT, Patancheru, India. p. 52.

Ratnadass, A., P.S. Marley, M.A. Hamada, O. Ajayi, B. Cissé, F. Assamoi, I.D.K. Atokple, J. Beyo, O. Cisse, D. Dakouo, M. Diakite, S. Dossou-Yovo, B. Le Diambo, M.B. Vopeyande, I. Sissoko, and A. Tenkouano. 2003. Sorghum head-bugs and grain molds in West and Central Africa: I. Host plant resistance and bug-mold interactions on sorghum grains. Crop Prot. 22:837–851.

Ratnadass, A., P. Fernandes, J. Avelino, and R. Habib. 2012. Plant species diversity for sustainable management of crop pests and diseases in agroecosystems: A review. Agron. Sustain. Dev. 32:273–303.

Rawla, G.S. 1973. Gloeocercospora and Ramulispora in India. Trans. Br. Mycol. Soc. 60:283–292.

Reed, J.D., B.A. Ramundo, L.E. Claflin, and M.P. Tuinstra. 2002. Analysis of resistance to ergot in sorghum and potential alternate hosts. Crop Sci. 42:1135–1138.

Reichert, R.D., S.E. Fleming, and D.J. Schwabe. 1980. Tannin deactivation and nutritional improvement of sorghum by anaerobic storage of H2O-, HCl-, or NaOH-treated grain. J. Agric. Food Chem. 28:824–829.

Roane, C.W., and M.K. Roane. 1997. Graminicolous fungi of Virginia: Fungi associated with genera *Echinochloa* to *Zizania*. Va. J. Sci. 48:11–46.

Robeson, D.J., and G.A. Strobel. 1982. Monocerin, a phytotoxin from *Exersohilum turcicum* (º *Dreschlera turcica*). Agric. Biol. Chem. 46:2681–2683.

Rodriguez-Herrera, R., W.L. Rooney, D.T. Rosenow, and R.A. Frederiksen. 2000. Inheritance of grain mold resistance in sorghum without a pigmented testa. Crop Sci. 40:1573–1578.

Rogerson, C.T. 1956. Diseases of grasses in Kansas. Plant Dis. Rep. 40:388–397.

Rosenow, D.T., and L.E. Clark. 1981. Drought tolerance in sorghum. p. 18-31 In: H.D. Loden and D. Wilkinson, editors, Proceedings of the 36th Annual Corn and Sorghum Research Conference, Washington D.C. American Seed Trade Association. Alexandria, VA.

Rosenow, D.T., J.W. Johnson, R.A. Frederiksen, and F.R. Miller. 1977. Relationship of nonsenescence to lodging and charcoal rot in sorghum In: 1977 Agronomy abstracts. ASA, Madison, WI. p. 69.

Rosenow, D.T., G. Ejeta, L.E. Clark, M.L. Grilbert, R.G. Henzell, A.K. Borrell, and R.C. Muchow. 1997. Breeding for pre- and post-flowering drought stress resistance in sorghum. In: International Conference on Genetic Improvement of Sorghum and Pearl Millet, Lubbock, TX, 22-27 Sept. 1996, p. 400-411.

Rossman, A.Y. 2009. The impact of invasive fungi on agricultural ecosystems in the United States. Biol. Invasions 11:97–107.

Rothrock, C.S. 1992. Tillage systems and plant disease. Soil Sci. 154:309–315.

Ruppel, E.G. 1985. Susceptibility of rotation crops to a root rot isolate of *Rhizoctonia solani* from sugar beet and survival of the pathogen in crop residues. Plant Dis. 69:871–873.

Russin, J.S., C.H. Carter, and J.L. Griffin. 1995. Effects of grain sorghum (*Sorghum bicolor*) herbicides on charcoal rot fungus. Weed Technol. 9:343–351.

Saleh, A.A., and J.F. Leslie. 2004. *Cephalosporium maydis* is a distinct species in the Gaeumannomyces-Harpophora species complex. Mycologia 96:1294–1305.

Samson San Martin, F., P. Lavin, A. Garcia, and G. Garcia. 1997. Anamorphic states of *Claviceps africana* and *Claviceps fusiformis* (Ascomycetes, Clavicipitaceae) associated with different grasses in Tamaulipas, Mexico. Revista Mexicana Micologia 13:52–57.

Scholthof, K.B. 2007. The disease triangle; pathogens, the environment, and society. Nat. Rev. Microbiol. 5:152–156.

Schuh, W., R.A. Frederiksen, and M.J. Jeger. 1986. Analysis of spatial patterns in sorghum downy mildew with Morisita's index of dispersion. Phytopathology 76:446–450.

Schuh, W., M.J. Jeger, and R.A. Frederiksen. 1987. The influence of soil temperature, soil moisture, soil texture, and inoculum density on the incidence of sorghum downy mildew. Phytopathology 77:125–128.

Seetharama, N. 1987. Effect of pattern and severity of moisture deficit stress on stalk rot incidence in sorghum: I. Use of line source irrigation technique, and the effect of time of inoculation. Field Crops Res. 15:289–308.

Seetharama, N.R., C. Sachan, A.K. Huda, K.S. Gill, K.N. Rao, and F.R. Bidinger. 1991. Effect of pattern and severity of moisture deficit stress on stalk rot incidence in sorghum. II. Effect of source/sink relationships. Field Crops Res. 26:355–374.

Seetharaman, K., R.D. Waniska, and L.W. Rooney. 1996. Physiological changes in sorghum antifungal proteins. J. Agric. Food Chem. 44:2435–2441.

Seifers, D.L., R. Perumal, and C.R. Little. 2012. New sources of resistance in sorghum (*Sorghum bicolor*) germplasm are effective against a diverse array of *Potyvirus* spp. Plant Dis. 96:1775–1779.

Sharma, R., R.P. Thakur, S. Senthilvel, S. Nayak, S.V. Reddy, V.P. Rao, and R.K. Varshney. 2010. Identification and characterization of toxigenic fusaria associated with sorghum grain mold complex in India. Mycopathologia 171:223–230.

Sharma, R.D., and A.C. de Souza Medeiros. 1982. Reaços de Alguins de genotipos de sorgho sacarino. Aos nematoides, *Meloidogyne javanica* e *Pratylenchus brachyrus*. Pesquisa Agropecu. Bras. 17:697–701.

Sharma, R., H.D. Upadhyaya, S.V. Manjunatha, V.P. Rao, and R.P. Thakur. 2012. Resistance to foliar diseases in a mini-core collection of sorghum germplasm. Plant Dis. 96:1629–1633.

Short, G.E., T.D. Wyllie, and P.R. Bristow. 1980. Survival of *Macrophomina phaseolina* in soil and residue of soybean. Phytopathology 70:13–17.

Sifuentes Barrera, J.A., and R.A. Frederiksen. 1994. Evaluation of sorghum hybrid mixtures for controlling sorghum leaf blight. Plant Dis. 78:499–503.

Singh, B.U., and N. Seetharama. 2008. Host-parasite interactions of the corn planthopper, *Peregrinus maidis* Ashm. (Homoptera: Delphacidae) in maize and sorghum agroecosystems. Arthropod-Plant Interact. 2:163–196.

Singh, R.S. 1998. Plant Diseases, 7th ed. Oxford and IBH Publishing Company, New Delhi, India.

Singh, S.D., P. Sathiah, and K.E.P. Rao. 1994. Sources of rust resistance in purple-colored sorghum. International Sorghum and Millets Newsletter 35:100–101.

Singleton, L.L., J.D. Milhail, and C.M. Rush, editors. 1992. Methods for research on soilborne phytopathogenic fungi. APS Press, St. Paul, MN.

Smith, C.W., and R.A. Frederiksen, editors. 2000. Sorghum: Origin, history, technology, and production. John Wiley & Sons, New York.

Spallek, T., M. Musembu, and K. Shirasu. 2013. The genus *Striga*: A witch profile. Mol. Plant Pathol. 14:861–869.

Sprague, R. 1950. Diseases of cereals and grasses in North America. Ronald Press Company, New York. p. 538.

Stewart, L.R., R. Teplier, J.C. Todd, M.W. Jones, B.J. Cassone, S. Wijeratne, A. Wijeratne, and M.G. Redinbaugh. 2014. Viruses in maize and Johnsongrass in southern Ohio. Phytopathology 104:1360–1369.

Stoll, M., D. Begerow, and F. Oberwinkler. 2005. Molecular phylogeny of *Ustilago, Sporisorium,* and related taxa based on combined analyses of rDNA sequences. Mycol. Res. 109:342–356.

Strange, R.N., and P.R. Scott. 2005. Plant disease: A threat to global food security. Annu. Rev. Phytopathol. 43:83–116.

Stuckenbrock, E.H., and B.A. McDonald. 2008. The origins of plant pathogens in agro-ecosystems. Annual Review of Plant Pathology 46:75–100.

Summerbell, R.C., C. Gueidan, H.J. Schroers, G.S. de Hoog, M. Starink, Y. Archeta Rosete, J. Guarro, and J.A. Scott. 2011. Acremonium phylogenetic overview and revision of *Gliomastix, Sarocladium,* and *Trichothecium*. Stud. Mycol. 68:139–162.

Sumner, D.R., and D.K. Bell. 1982. Root diseases induced in corn by *Rhizoctonia solani* and *Rhizoctonia zeae*. Phytopathology 72:86–91.

Sutton, B.C. 1980. The coelomycetes: Fungi imperfecti with pycnidia, acervuli, and stromata. Commonwealth Mycological Institute, Kew, London. p. 696.

Swarup, G., E.D. Hansing, and C.T. Rogerson. 1962. Fungi associated with sorghum seed in Kansas. Trans. Kans. Acad. Sci. 65:120–137.

Sweeney, D.W., J.L. Moyer, D.J. Jardine, and D.A. Whitney. 2011. Nitrogen, phosphorus, and potassium effects on grain sorghum production and stalk rot following alfalfa and birdsfoot trefoil. J. Plant Nutr. 34:1330–1340.

Tariq, A.S., Z. Akram, G. Shabdir, M. Gulfraz, K.S. Khan, M.S. Iqbal, and T. Mahmood. 2012. Character association and inheritance studies of different sorghum genotypes for fodder yield and quality under irrigated and rainfed conditions. Afr. J. Biotechnol. 11:9189–9195.

Tarr, S.A.J. 1962. Diseases of sorghum, sudan grass, and brown corn. Commonwealth Mycological Institute, Kew, Surrey, UK.

Tegegne, G., R. Bandyopadhyay, T. Mulatu, and Y. Kebede. 1994. Screening for ergot resistance in sorghum. Plant Dis. 78:873–876.

Tenkouano, A., F.R. Miller, R.A. Fredericksen, and R.A. Nicholson. 1998. Ontogenic characteristics and inheritance of resistance to leaf anthracnose in sorghum. Afr. Crop Sci. J. 6:249–258.

Tenkuano, A., F.R. Miller, R.A. Frederiksen, and D.T. Rosenow. 1993. Genetics of nonsencescence and charcoal rot resistance in sorghum. Theor. Appl. Genet. 85:644–648.

Tesso, T., L.E. Claflin, and M.R. Tuinstra. 2005. Analysis of stalk rot resistance and genetic diversity among drought tolerant sorghum genotypes. Crop Sci. 45:645–652.

Tesso, T., N. Ochanda, C.R. Little, L. Claflin, and M.R. Tuinstra. 2010. [*Sorghum bicolor* (L.) Moench] Analysis of host-plant resistance to multiple *Fusarium* species associated with stalk rot disease in sorghum. Field Crops Res. 118:177–182.

Tesso, T., R. Perumal, C.R. Little, A. Adeyanju, G. Radwan, L.K. Prom, and C.W. Magill. 2012. Sorghum pathology and biotechnology- A fungal disease perspective: Part II. Anthracnose, stalk rot, and downy mildew. The European Journal of Plant Science and Biotechnology 6:31–44.

Thaung, M.M. 2008. Biodiversity survey of coelomycetes in Burma. Australas. Mycologist 27:74–110.

Thomas, S.H., and L. Murray. 1987. Yield reductions in grain sorghum associated with injury by *Meloidogyne incognita* race 3. J. Nematol. 19:559.

Thomas, M.D., I. Sissoko, and M. Sacko. 1996. Development of leaf anthracnose and its effect on yield and grain weight of sorghum in West Africa. Plant Dis. 80:151–153.

Todd, T.C., Appel, J.A., Vogel, J., and Tisserat, N.A. 2014. Survey of plant-parasitic nematodes in Kansas and Eastern Colorado wheat fields. Plant Health Progress. doi:10.1094/PHP-RS-13-0125.

Tooley, P.W., N.R. O'Neill, E.D. Goley, and M.M. Carras. 2000. Assessment of diversity of *Claviceps africana* and other *Claviceps* species by RAM and AFLP analyses. Phytopathology 90:1126–1130.

Tosic, M., R.E. Ford, D.D. Shukla, and J. Jilka. 1990. Differentiation of sugarcane, maize dwarf, johnsongrass, and sorghum mosaic viruses based on reactions of oat and some sorghum cultivars. Plant Dis. 74:540–552.

Trevathan, L.E., and J.T. Robbins. 1995. Yield of sorghum and soybean, grown as monocrops and in rotation, as affected by insecticide and nematicide applications. Nematropica 25:125–134.

Trimboli, D.S., and L.W. Burgess. 1982. The fungi associated with stalk and root rot of grain sorghum in New South Wales. Sorghum and Millets Newsletter 25:105.

Trimboli, D.S., and L.W. Burgess. 1983. Reproduction of *Fusarium moniliforme* basal stalk rot and root rot of grain sorghum in the greenhouse. Plant Dis. 67:891–894.

Tuleen, D.M., R.A. Frederiksen, and P. Vudhivanich. 1980. Cultural practices and the incidence of sorghum downy mildew in grain sorghum. Phytopathology 70:905–908.

USDA NASS. 2015. National Agricultural Statistics Service, Data & Statistics. USDA-NASS. https://www.nass.usda.gov. (Accessed 15 Jun 2015)

Vanderlip, R.L., and H.E. Reeves. 1972. Growth stages of sorghum (*Sorghum bicolor* (L.) Moench). Agron. J. 64:13–16.

Velesquez-Valle, R., J. Narro-Sanchez, R. Nora-Nolasco, and G.N. Odvody. 1998. Spread of ergot of sorghum (*Claviceps africana*) in central Mexico. Plant Dis. 82:447.

Vidyabhushanam, R.V., B.S. Rana, and B.V.S. Reddy. 1989. Use of sorghum germplasm and its impact on crop improvement in India. In: Collaboration on genetic resources: Summary proceedings of a joint ICRISAT/NBPGR (ICAR) Workshop on Germplasm Exploration and Evaluation in India. ICRISAT, Patancheru, India. p. 85-89.

Vizvary, M.A., and H.L. Warren. 1982. Survival of *Colletotrichum graminicola* in soil. Phytopathology 72:522–525.

Walulu, R.S., D.T. Rosenow, D.B. Wester, and H.T. Nguyen. 1994. Inheritance of stay green trait in sorghum. Crop Sci. 31:1691–1694.

Wang, P.H., C.Y. Chung, Y.S. Lin, and Y. Yeh. 2003. Use of polymerase chain reaction to detect the soft rot pathogen, *Pythium myriotylum*, in infected ginger rhizomes. Lett. Appl. Microbiol. 36:116–120.

Ward, J.M.J., and D.C. Nowell. 1998. Integrated management practices for the control of maize gray leaf spot. Integrated Pest Management Reviews 3:177–188.

Wenefrida, L.E., E.C. McGawley, and J.S. Russin. 1997. Interrelationships among *Macrophomina phaseolina*, *Criconemella xenoplax*, and *Tylenchorhynchus annulatus* on grain sorghum. J. Nematol. 29:199–208.

White, J.A., M.J. Ryley, D.L. George, G.A. Kong, and S.C. White. 2012. Yield losses in grain sorghum due to rust infection. Australas. Plant Pathol. 41:85–91.

White, J.A., M.I. Ryley, D.L. George, and G.A. Kong. 2014. Optimal environmental conditions for infection and development of *Puccinia purpurea* on sorghum. Australas. Plant Pathol. 43:447–457.

Williams, R.J., and K.N. Rao. 1981. A review of sorghum grain moulds. Trop. Pest Manage. 27:200–211.

Williams, R.J., R.A. Frederiksen, and J.-C. Girard. 1978. Sorghum and pearl millet disease identification handbook. ICRISAT Information Bulletin, No. 2. ICRISAT, Hyderabad, India.

Williams, R.J., R.A. Frederiksen, and L.K. Mughogho, editors. 1980. Sorghum diseases: A world review. International Crops Research Institute for the Semi-Arid Tropics, Andhra Pradesh, India.

Williams, R.J., and K.N. Rao. 1981. A review of sorghum grain molds. Trop. Pest Manage. 27:200–211.

Wu, W.S., and K.C. Cheng. 1990. Relationships between seed health, seed vigour and the performance of sorghum in the field. Seed Sci. Technol. 18:713–719.

Xu, W., P.K. Subudhi, O.R. Crasta, D.T. Rosenow, J.E. Mullet, and H.T. Nguyen. 2000. Molecular mapping of QTLs conferring stay-green in grain sorghum (*Sorghum bicolor* L. Moench). Genome 43:461–469.

Yassin, M.A., A.-R. El-Samawaty, A. Bahkali, M. Moslem, K.A. Abd-Elsalam, and K.D. Hyde. 2010. Mycotoxin-producing fungi occurring in sorghum grains from Saudi Arabia. Fungal Divers. 44:45–52.

Yoshida, S., S. Maruyama, H. Nozaki, and K. Shirashu. 2010. Horizontal gene transfer by the parasitic plant *Striga hermonthica*. Science 328:1128.

Yuen, J., and A. Mila. 2015. Landscape-scale disease risk quantification and prediction. Annu. Rev. Phytopathol. 53:471–484.

Zaki, M.J., and A. Ghaffar. 1988. Inactivation of sclerotia of *Macrophomina phaseolina* under paddy cultivation. Pak. J. Bot. 20:245–250.

Zummo, N., L.M. Gourley, L.E. Trevathan, M.S. Gonzalez, and J. Dahlberg. 1998. Occurrence of ergot (sugary disease) incited by a *Sphacelia* sp. on sorghum in Mississippi in 1997. Plant Dis. 82:590.

Weed Competition and Management in Sorghum

Curtis R. Thompson,* J. Anita Dille, and Dallas E. Peterson

Abstract

Weed management in grain sorghum has been and continues to be a production challenge for growers. Weed species can be very competitive with sorghum and reduce grain yields. In 15 experiments in Kansas, sorghum grown with populations of kochia (*Kochia scoparia* L. Shrad.), Palmer amaranth (*Amaranthus palmeri* S. Watson), velvetleaf (*Abutilon theophrasti* Medik.), and morningglory [*Ipomoea hederacea* Jacq.] yielded 58% less than sorghum treated with herbicides providing 90% or better weed control. Annual grasses are a challenge as only herbicides applied preemergence provide adequate control. A three- to four-week weedfree period following sorghum planting will minimize negative effects of weeds on sorghum yield. Sorghum in rotation with other crops and fallow periods allows for an integrated weed management approach utilizing a variety of planting dates, crop competition, multiple herbicides having different sites of action, varied timings of herbicide applications and possibly tillage. Atrazine (1-Chloro,-3-ethylamino-5-isopropylamino-2,4,6-triazine) remains the base for most herbicide programs in sorghum. Atrazine with acetamide herbicides provide broad spectrum broadleaf and grass weed control. Adding mesotrione or saflufenacil in herbicide combinations provide additional control of pigweeds (*Amaranthus* sp.), kochia, and large seeded weeds like velvetleaf and common cocklebur (*Xanthium strumarium* L.). New in 2016, herbicide-tolerant sorghum developed with conventional breeding allows nicosulfuron applied postemergence to control annual grasses in sorghum. The use of preeemergence followed by postemergence herbicides will be required for adequate broad spectrum weed control. With problem-resistant weeds, grower dependence on postemergence only herbicide programs will likely lead to failed sorghum.

Weeds have had a historical presence and have been an annual problem for sorghum growers (Stahlman and Wicks, 2000). These weeds are plants existing with the sorghum crop that intercept light, utilize moisture and nutrients, and adversely affect sorghum growth, development, and ultimately grain yield. Thus, weeds are unwanted and need to be controlled to have optimum performance from the grain sorghum crop.

Abbreviations: ALS, acetolactate synthase; COC, crop oil concentrate; CPWC, critical period of weed control; GMO, genetically modified organism; HPPD, hydroxyphenylpyruvate dioxygenase inhibitor; IWM, integrated weed management; PPO, protoporphyrinogen oxidase; WSSA, Weed Science Society of America.

Kansas State University, Department of Agronomy, Manhattan KS 66506. * Corresponding author (cthompso@ksu.edu)

Contribution no. 17-145-B from the Kansas Agricultural Experiment Station.

doi:10.2134/agronmonogr58.2014.0071

Integrated weed management (IWM) in sorghum encompasses a series of practices across the entire life cycle of the sorghum crop that affect weed emergence, growth and development. A combination of a sequence of practices during the sorghum growing season ultimately results in some adequate level of weed management, thus the term "Integrated Weed Management". Integrated weed management includes a systems approach that involves sorghum being planted within a rotation of crops over a period of seasons or years. Each production practice prior to and during the lifecycle of the sorghum will have an effect on weed populations and their management in grain sorghum. When cultivation was used with an herbicide program in grain sorghum, more consistent weed control was achieved compared to either method alone, which suggested the benefit of IWM (Burnside and Wicks, 1967). In a review of IWM in sorghum in countries other than the United States, authors concluded that a third of the potential sorghum production is lost to weeds, but with an IWM approach, a blend of systematic processes could achieve an acceptable level of weed control and increase productivity of grain sorghum (Vijayakumar et al., 2014). The intent in this chapter is to discuss various aspects of weed management.

When sorghum growers in the United States have been surveyed regarding concerns about sorghum production, weed management is frequently at or near the top of the growers' concerns (Franke et al., 2009; K. Al-Khatib, personal communication). Unlike many modified crops glyphosate (N-(phosphonomethyl) glycine) cannot be used postemergence on grain sorghum. Rather the crop must be planted into a weed-free seedbed to avoid early season weed competition. Many weed species can be a problem in grain sorghum, but the most problematic are weedy-sorghum relatives, including Johnsongrass [*Sorghum halepense* (L.) Pers.] and shattercane [*Sorghum bicolor* (L.) Moench ssp. *arundinaceum* (Desv.) de Wet & Harlan] (Defelice, 2006). Herbicides currently available for managing weeds in sorghum do not control these sorghum relatives. As a result, it is recommended that fields heavily infested with Johnsongrass or shattercane should be planted to crops other than sorghum. Some of the most common broadleaf weeds causing problems for sorghum growers include Palmer amaranth (*Amaranthus palmeri* S. Watson), waterhemp [*Amaranthus tuberculatus* (Moq.) J.D.Sauer], redroot pigweed (*Amaranthus retroflexus* L.), tumble pigweed (*Amaranthus albus* L.), kochia (*Kochia scoparia* L. Shrad.), Russian thistle (*Salsola tragus* L.), velvetleaf (*Abutilon theophrasti* Medik.), common cocklebur (*Xanthium strumarium* L.), morningglory (*Ipomoea spp.*), field bindweed (*Convolvulus arvensis* L.), and devil's claw [*Proboscidea louisianica* (Mill.) Thell.]. To date, there are strategies and tools available to manage these broadleaf weeds; however, with the development of herbicide resistance the challenge to attain effective weed control is increasing.

In a survey conducted in Oklahoma, 66% of the growers indicated that weeds were of major concern in their grain sorghum production system while only 1% indicated that weeds were not a concern (Franke et al., 2009). In this survey, 54% of respondents listed *Amaranthus sp.* as the most frequent weed problem in their fields and 34% indicated that field bindweed was the second most frequent problem. The most frequent grass problems were large crabgrass [*Digitaria sanguinalis* (L.) Scop.] and Johnsongrass at 28% followed by field sandbur (*Cenchrus spinifex* Cav.) at 23%. In a survey conducted across the sorghum growing areas of the United States in 2006, growers were asked to rank and prioritize several areas of research. New technologies for controlling grass weeds to increase yield potential were ranked first among the areas of research listed (K. Al-Khatib, personal communication).

Weed Competition with Sorghum

Fifteen research experiments were conducted during 2004 through 2012 near Tribune and Manhattan, KS (Thompson, personal communication). The predominant weeds in Tribune were kochia and *Amaranthus* spp. and in Manhattan were Palmer amaranth, velvetleaf, and ivyleaf morningglory [*Ipomoea hederacea* Jacq.]. All sorghum was planted in 76 cm rows. In these 15 different experiments, the average sorghum yield in the weedy checks (no weed control strategies utilized) was 2.95 t ha^{-1}. When the imposed herbicide treatments controlled more than 90% of the predominant weeds, sorghum yielded an average of 7.09 t ha^{-1} (15 out of 15 experiments). When 60 to 80% of the predominant weeds were controlled, averaged sorghum yields were 5.45 t ha^{-1} (9 of 15 experiments). When the control of the predominant weed species dropped below 60%, the average sorghum yield was reduced to 3.89 t ha^{-1} (3 of 15 experiments). These data show the economic importance for growers to implement effective weed control strategies in their grain sorghum crop.

Kochia density of 1.1 plants per m of sorghum row (spaced 51 cm) prevented grain production when sorghum was grown under dryland conditions (Phillips, 1958). The presence of kochia continues to be a serious concern for many grain sorghum producers.

Sorghum grain and fodder yields were reduced as densities of waterhemp increased (Feltner et al., 1969a). The natural waterhemp infestation that was allowed to compete season long reduced sorghum yield 80% compared to the sorghum yield in the hand weeded check. Similarly, yellow foxtail [*Setaria pumila* (Poir.) Roem. & Schult.] reduced sorghum yield as density increased, however, the natural yellow foxtail infestation only reduced the sorghum yield 33% compared to sorghum grown weed free (Feltner et al., 1969b).

Palmer amaranth reduced sorghum grain yield by 1.8% to 3.5% for each plant per 15 m of row (Moore et al., 2004). As Palmer amaranth density increased, grain moisture and foreign material increased leading to additional harvest losses. Palmer amaranth reduced grain sorghum yield from 4% with only 1 weed per 4 m of row to up to 40% if there were 6 plants per m of row (Unruh, 2013). Redroot pigweed (*Amaranthus retroflexus* L.) that emerged when sorghum was at the 1- to 2.6-leaf stage caused significant crop yield losses at low weed densities. But, if redroot pigweed did not emerge until after the 5.5-leaf stage of sorghum, no significant yield losses occurred across the weed densities (Knezevic et al., 1997). Maximum estimated yield losses varied from 3% to 46%, depending on experimental location, pigweed density, and time of weed emergence relative to the growth stage of the sorghum.

Annual grass weeds can be a serious problem in sorghum, with preemergence soil-applied herbicides being the most effective method of control. Barnyardgrass [*Echinochloa crus-galli* (L.) P. Beauv.], large crabgrass, and Texas panicum [*Urochloa texana* (Buckley) R.D. Webster] reduced sorghum yield at a rate of 3.6% loss for each week of weed interference regardless of weed species (Smith et al., 1990). The authors did observe a row width interaction as the effect of grass competition on narrow row sorghum (61 cm) was less than the effect on wide row sorghum (91 cm). Fabrizius (1998) found that season-long competition of longspine sandbur [*Cenchrus longispinus* (Hack.) Fernald] reduced sorghum yield by 42% compared to yield of sorghum maintained free from weed competition.

Critical Period of Weed Control in Sorghum

Critical period of weed control (CPWC), as defined by Swanton and Weise, (1991) involves two separate components. First is the critical duration of weed interference, a time period that weeds can exist with sorghum and not affect yield. The second component is the critical weed-free period, a time period from planting to when later emerging weeds will not affect sorghum yield. The results from these two sets of experiments are used to determine the CPWC, also defined as the window during the grain sorghum lifecycle when it must be kept weed free to prevent unacceptable yield losses, often set at 5% loss threshold (Knezevic et al., 2002). Several citations refer to the requirement of the first three to four weeks maintained weed free following grain sorghum planting to prevent sorghum yield losses (Burnside and Wicks, 1967; Everaarts, 1993; Fabrizius, 1998). This assumes sorghum is planted into adequate moisture for normal sorghum emergence. This CPWC also will be influenced by weed species and weed density.

Goosegrass [*Eleusine indica* (L.) Gaertn.], southern sandbur (*Cenchrus echinatus* L.) and a spurge [*Croton hirtus* L'Her.] grown in two different experiments did not cause grain yield loss if they emerged 30 days after planting grain sorghum (Everaarts, 1993). In the two experiments it was concluded that 19 (goosegrass and southern sandbur) and 22 (*Croton hirtus*) weed-free days after planting were needed to prevent grain sorghum yield loss.

The critical period of longspine sandbur control was four weeks, even though there was a 10-fold difference in weed density between the two years studied (Fabrizius, 1998). If longspine sandbur was allowed to compete with grain sorghum for four, five, and six weeks after crop emergence, yields were reduced by 27%, 31%, and 42% respectively, compared to the weed-free sorghum yield (Fabrizius, 1998).

Growers need to control all weeds during the CPWC, to protect sorghum yield potential.

Sorghum in Crop Rotations

Crop rotation can be an essential tool for preventing weed problems (Liebman and Dyck, 1993). Conversely, the worst scenario is for a farmer to raise the same crop on the same parcel of ground year after year. A continuous sorghum system will lead to weed problems primarily if the weed species are left uncontrolled (Wiese et al., 1985). Examples of problem weeds might be shattercane, Johnsongrass, or longspine sandbur. When a problem exists with any of these grass weed species, it suggests that sorghum is being used too frequently in the rotation and that some crop other than sorghum should be planted to allow tools to be used that control these weed species.

Using multiple crops in a cropping system allows for a diverse set of weed control practices which include variation in the timing of preplant tillage and burndown with herbicides, variation in planting dates, and the use of different chemistries for weed control in the different crops. The more diverse the weed control tools utilized in the entire system, the more difficult it is for a weed species to adapt and become a significant problem (Swanton and Weise, 1991).

In the Central and Southern Great Plains region of the United States, sorghum frequently is planted in rotation with winter wheat. Sorghum planting often follows a fallow period between wheat harvest and sorghum planting that requires good weed management. It is essential that weeds be controlled in the wheat crop and during the fallow period following wheat harvest to minimize moisture loss, maximize moisture accumulation for the sorghum crop, and reduce the potential weed seed bank and subsequent potential weed problems that adversely affect the sorghum crop. Any uncontrolled weeds during this fallow period will produce seed which can increase the potential weed problems in sorghum.

It is important to review herbicide labels and their plant-back restrictions to grain sorghum to prevent unacceptable crop injury from herbicide carry over. Precipitation increases in the Central and Southern Great Plains as you move from west to east in the sorghum growing areas (Figure 1). In areas with sufficient rainfall to allow continuous cropping, sorghum often is planted following crops other than wheat. When planting after corn, acetolactate-synthase (ALS)-inhibitor herbicides such as thiencarbazone methyl {methyl 4-[(3-methoxy-4-methyl-5-oxo-1,2,4-triazole-1-carbonyl)sulfamoyl]-5-methylthiophene-3-carboxylate}, flumetsulam [N-(2,6-difluorophenyl)-5-methyl-(1,2,4)triazolo(1,5-a)pyrimidine-2-sulfonamide], and nicosulfuron {2-[({[4,6-dimethoxy-(2-pyrimidinyl)amino]carbonyl}amino)sulfonyl)-N,N-dimethyl-3-pyridinecarboxamide} can carry over, especially when soil pH is above 7.5 (Shaner, 2014). When sorghum is planted following soybeans, herbicides containing chlorimuron {ethyl 2-[(4-chloro-6-methoxypyrimidin-2-yl)carbamoylsulfamoyl]benzoate}, cloransulam {methyl 3-chloro-2-[5-ethoxy-7-fluoro(1,2,4) triazolo(1,5-c)pyrimidin-2-ylsulfonamido]benzoate}, fomesafen {5-[2-chloro-4-(trifluoromethyl)phenoxy]-N-methylsulfonyl-2-nitrobenzamide}, imazethapyr [5-ethyl-2-(4-methyl-5-oxo-4-propan-2-yl-1H-imidazol-2-yl)pyridine-3-carboxylic acid], imazamox [5-(methoxymethyl)-2-(4-methyl-5-oxo-4-propan-2-yl-1H-imidazol-2-yl)pyridine-3-carboxylic acid], or imazaquin {3-Quinolinecarboxylic acid, 2-[4,5-dihydro-4-methyl-4-(1-methylethyl)-5-oxo-1H-imidazol-2-yl]-} may have plant-back restrictions that prohibit planting sorghum the following year. Restrictions may depend on soil pH, texture, and/or rainfall; thus, it is important to review the herbicide labels. In the Southern Great Plains it's common that sorghum is planted after cotton. After reviewing rotational restrictions in herbicide labels, herbicide active

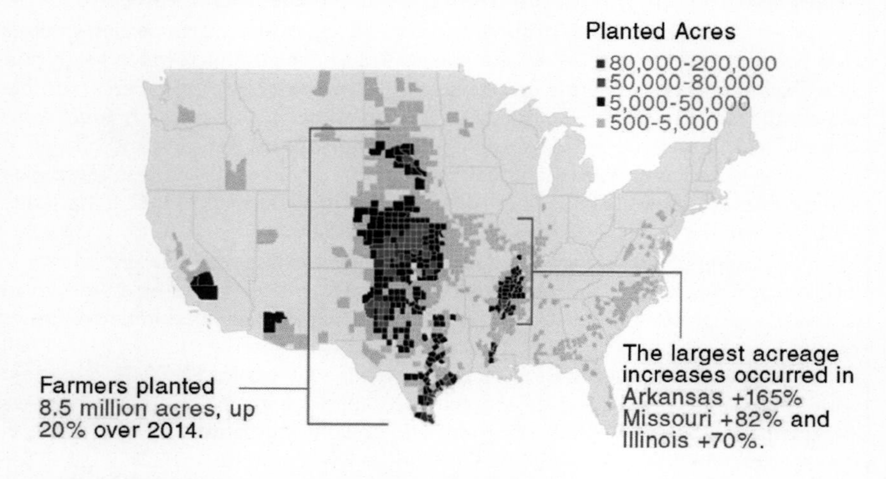

Figure 1. Map of the United States showing the sorghum growing areas in 2014. (Source: United Sorghum Checkoff Program).

ingredients that may be of concern to sorghum planted after cotton include norflurazon {4-chloro-5-(methylamino)-2-[3-(trifluoromethyl)phenyl]pyridazin-3-one}, prometryn [6-methylsulfanyl-2-N,4-N-di(propan-2-yl)-1,3,5-triazine-2,4-diamine], pyrithiobac–sodium [sodium;2-chloro-6-(4,6-dimethoxypyrimidin-2-yl)sulfanylbenzoate] or trifloxysulfuron–sodium {sodium;(4,6-dimethoxypyrimidin-2-yl) carbamoyl-[3-(2,2,2-trifluoroethoxy)pyridin-2-yl]sulfonylazanide}. When cotton is planted following sorghum, atrazine is often the concern, as propazine [6-chloro-2-N,4-N-di(propan-2-yl)-1,3,5-triazine-2,4-diamine] can be used and followed with cotton.

Tillage Systems for Grain Sorghum

Frequently, sorghum is no-till planted into wheat residue. Weed seeds produced in no-till systems generally remain on the soil surface which can leave seeds exposed to insect predation and weather elements. In some cases, there is inadequate moisture on the surface for seed germination and plant establishment. Clements et al. (1996) looked at vertical distribution of weed seeds in different tillage systems, finding weed seeds predominately on the surface in a no-till system. Small seeded broadleaf and grass weeds will continue to germinate and survive in no-till residue; however, large seeded broadleaf weeds like cocklebur, velvetleaf, and morningglory are at a disadvantage simply because there often is insufficient soil-to-seed contact and a lack of sufficient moisture for imbibition to allow proper establishment. Thus over time these large-seeded species tend to decline in a no-till system (Nice et al., 2007). This is a generalization as switching from a conventional-till system to a no-till system will change the predominate weed species and their frequency, but whether frequency increases or declines will vary from one species to another and the environment (Vencill and Banks, 1994). Weeds such as field bindweed often can be better managed in a no-till system over time than in a conventionally tilled system. Glyphosate and auxin herbicides applied between crops, in combination with postemergence herbicides in sorghum and other crops, can be effective to manage field bindweed.

No-till is a difficult environment to control perennial warm-season grasses that become established, especially species that have some tolerance to glyphosate. Problem grasses include bermudagrass [*Cynodon dactylon* (L.) Pers.], tumble windmillgrass (*Chloris verticillata* Nutt.), tumblegrass [*Schedonnardus paniculatus* (Nutt.) Trel.], sand dropseed [*Sporobolus cryptandrus* (Torr.) A. Gray] and others. In many cases, tillage may be the only option to effectively control these warm-season perennial grasses. The lack of tillage in no-till systems presents a challenge with perennial grasses that are tolerant to glyphosate.

Narrowing of crop row spacing and increasing seeding rates can provide some weed suppression. Sorghum produced significantly greater yields when grown in 45-cm rows than 60- and 90-cm row spacing, and sorghum planted in narrow rows resulted in fewer weeds than when planted in wide row spacing (Bishnoi et al., 1990). In 25-cm spaced rows, growth of emerged weeds was reduced by 24% compared with 51-cm spaced rows, and by 45% when compared with the 76-cm spaced rows (Staggenborg et al., 1999). Grain sorghum was equally competitive in 38- and 76-cm row spacing when weed pressure was low, but as weed pressure increased, grain sorghum was more competitive in 38-cm row spacing (Limon-Ortega et al., 1998).

Herbicides for Weed Management in Grain Sorghum

Fewer herbicides are available for weed management in grain sorghum compared with those available for corn and soybeans (Thompson et al., 2017). As a result, it is very important to plan a two-pass weed control system for a sorghum crop. Using preemergence herbicides followed by postemergence applied herbicides provide the most effective control of weeds, and often better than preemergence-only or postemergence-only programs.

In this section, discussion will center on herbicide active ingredients simply because herbicide trade names change over time, change within and among countries, concentrations change, and active ingredient combinations continue to change (Shaner, 2014). It is essential to know that not all herbicides and active ingredients are registered in all states or countries. Thus, it is essential to read the herbicide labels when developing a weed control program for sorghum with the various chemistries. Much of the information provided has come from the Herbicide Handbook published by Weed Science Society of America (Shaner, 2014). Within the discussion, a number in parenthesis following an active ingredient identifies the respective WSSA group designation for the herbicide site of action.

The triazine (1,3,4-Triazine) herbicides (Group 5) are photosystem II site A inhibitors that were among the first soil-active herbicides registered for weed control in sorghum. During the 1960s atrazine was being used in corn and atrazine use for weed management in grain sorghum was subsequently discussed (Bridges, 2008). The effectiveness of atrazine applied preemergence or postemergence in combination with several different herbicides has led to a heavy dependence on atrazine. This dependence on atrazine has continued into the current age of weed management in sorghum. Most herbicides used in sorghum are more effective when applied with atrazine than when applied alone. Atrazine controls many broadleaf weeds and suppresses some grass weed species. Propazine controls small-seeded broadleaf weeds and is safer than atrazine on grain sorghum. Very little propazine is currently used in sorghum due to less postemergence activity with propazine than with atrazine, higher cost than atrazine, and longer plant-back restrictions to soybean ("MiloPro" by Albaugh, LLC). However, propazine can be used in rotations back to cotton. Triazines can cause significant crop injury if used preemergence on sorghum planted on sandy soils. Labels of herbicides containing triazines prohibit preemergence use on sandy soils with low organic matter and high pH when planting sorghum. Postemergence applications of triazines are generally allowed after a specific stage of sorghum growth specified in the label. When atrazine is used postemergence, an adjuvant such as crop oil concentrate (COC) is usually recommended to enhance the postemergence activity. This will depend on the tank mix partner as some herbicides prohibit inclusion of COC. Without the inclusion of COC, atrazine will have less foliar activity and may not provide adequate weed control. Most atrazine labels prohibit use of atrazine after sorghum reaches 30 cm tall. An additional triazine used for broad spectrum weed control when applied preemergence or postemergence is terbuthylazine. Not registered currently in the United States, it is labeled in other countries and is used in Europe.

Triazines, especially atrazine, continue to be scrutinized for the potential adverse effect on the environment. Consequently, the maximum atrazine use rates have been reduced over the years in an attempt to maintain concentrations below maximum contamination levels in the environment. Weed management in sorghum would be severely compromised with the loss of atrazine registration. To date, an atrazine replacement for sorghum is very unlikely.

The acetamide herbicides (Group 15) are soil-active herbicides that provide control of annual grasses and small-seeded broadleaf weeds, including *Amaranthus* spp. It is essential that sorghum seed be treated with the current safener, fluxofenim "Concep III" (Syngenta) in order to prevent unacceptable sorghum injury from the acetamide herbicides. This warning applies to all herbicides containing acetamide chemistry. These herbicides currently are the most effective way to manage annual grass weeds in grain sorghum. Herbicide active ingredients in this group registered in sorghum include acetochlor [2-Chloro-N-(ethoxymethyl)-N-(2-ethyl-6-methylphenyl)acetamide], dimethenamid-P [(RS)-2-Chloro-N-(2,4-dimethyl-3-thienyl)-N-(2-methoxy-1-methylethyl)acetamide],metolachlor [(RS)-2-Chloro-N-(2-ethyl-6-methyl-phenyl)-N-(1-methoxypropan-2-yl) acetamide], and S-metolachlor [(RS)-2-Chloro-N-(2-ethyl-6-methyl-phenyl)-N-(1-methoxypropan-2-yl)acetamide]. Often marketed and applied in mixtures with atrazine, this combination provides broad-spectrum grass and broadleaf weed control. Specific herbicides containing acetochlor, dimethenamid-P, metolachlor, or S-metolachlor may be tank mixed with a postemergence herbicide program to extend the soil residual control of small seeded broadleaf and grass weeds. These herbicides only have soil activity and do not provide any foliar activity on emerged weeds. This type of application may become increasingly important as weeds develop resistance to postemergence-applied herbicides. In some countries, only the early postemergence application of the acetamide herbicides is registered. This does increase the risk of grass weed escapes, however.

Two additional acetamide herbicides (Group 15), propachlor (2-Chloro-N-isopropyl-N-phenylacetamide) and alachlor [2-Chloro-N-(2,6-diethylphenyl)-N-(methoxymethyl) acetamide] were commonly used in sorghum during the 1970s and 1980s. These herbicides have fallen out of favor in most states in the United States due to the development of other acetamide herbicides. The use of these products has been discontinued on sorghum in most U.S. states; however, these herbicides may continue to be used in other countries around the world that produce sorghum. Propachlor and alachlor will also control some small seeded broadleaf weeds and grasses.

The triketone herbicide, mesotrione {2-[4-(Methylsulfonyl)-2-nitrobenzoyl] cyclohexane-1,3-dione} (Group 27) is a hydroxyphenyl-pyruvatedioxygenase (HPPD)-inhibitor included in several herbicide products registered for preemergence application in grain sorghum. Mesotrione aids in the control of broadleaf weeds including *Amaranthus* spp., kochia, velvetleaf, and others. Herbicides containing mesotrione frequently contain S-metolachlor, which also provides very good preemergence grass control. Mesotrione is synergized when applied with atrazine (Abendroth et al., 2006), so when possible, the combination with atrazine and mesotrione should be used. Herbicides containing mesotrione (Group 27), S-metolachlor (Group 15) and atrazine (Group 5) are some of the most effective preemergence herbicides on the market for broad spectrum weed management in grain sorghum, despite the frequency of temporal injury to grain sorghum. Herbicides containing mesotrione should not be applied post-emergence on grain sorghum as mesotrione may cause unacceptable sorghum injury.

The isoxazole herbicide pyrasulfotole [(5-Hydroxy-1,3-dimethylpyrazol-4-yl) (a,a,a-trifluoro-2-mesyl-p- to1yl)methanone] (Group 27) is also an HPPD inhibitor that is premixed with bromoxynil (3,5-dibromo-4-hydroxybenzonitrile) (Group

6) for postemergence broadleaf weed control in grain sorghum. Pyrasulfotole [(5-Hydroxy-1,3-dimethylpyrazol-4-yl)(a,a,a-trifluoro-2-mesyl-p-tolyl)methanone] and bromoxynil will provide best control when tank mixed with atrazine. Generally, a surfactant and a nitrogen source, such as spray grade ammonium sulfate or urea ammonium nitrate, are added into the spray mixture to optimize activity. This herbicide effectively controls many broadleaf weed species including *Amaranthus* spp., kochia, velvetleaf, morningglory species, cocklebur, devilsclaw, and others, (Reddy et al., 2013). Ideally, weeds should be treated when they are small (10 cm or less, species dependent) for optimal control. Pyrasulfotole and bromoxynil often will cause leaf burn and some chlorosis on sorghum but rarely does this herbicide reduce grain sorghum yield (Reddy et al., 2013; Fromme et al., 2012).

The aryl-triazinone herbicide carfentrazone (ethyl 2-chloro-3-{2-chloro-5-[4-(difluoromethyl)-3-methyl-5-oxo-1,2,4-triazol-1-yl]-4-fluorophenyl}propanoate) (Group 14), the N-Phenylphthalimide herbicide flumioxazin {7-flouro-6-[(3,4,5,6-tetrahydro)phthalimido]-4-(2-propynyl)-1,4-benzoxazin-3-(2H)-o} (Group 14) and the pyrimidinedione herbicide saflufenacil (N′-{2-Chloro-4-fluoro-5-[1,2,3,6-tetrahydro-3-methyl-2,6-dioxo-4-(trifluoromethyl)pyrimidin-1-yl] benzoyl}-N-isopropyl-N-methylsulfamide) (Group 14) are protoporphyrinogen oxidase (PPO) inhibitors or cell membrane disruptors: herbicides which provide broadleaf weed control in sorghum. Carfentrazone (ethyl 2-chloro-3-{2-chloro-5-[4-(difluoromethyl)-3-methyl-5-oxo-1,2,4-triazol-1-yl]-4-fluorophenyl}propanoate) is effective postemergence primarily for control of velvetleaf and common lambsquarters (*Chenopodium album* L.) and often is tank-mixed with other broadleaf herbicides for broad-spectrum broadleaf weed control. Saflufenacil has excellent foliar activity and soil activity on broadleaf weeds but must be applied preemergence to sorghum. For broad-spectrum weed control, saflufenacil should be applied with a chloroacetamide (Group 15) and atrazine (Group 5). Saflufenacil will help control the large-seeded broadleaf weeds, common cocklebur, weedy sunflower, velvetleaf, and morningglory, along with other small seeded broadleaf weeds. Flumioxazin can be used 30 days before planting sorghum but must receive 2.5 cm rainfall during the 30-day period or delay planting sorghum until the rainfall has been received to minimize crop injury (see Valor label). Flumioxazin has good activity on *Amaranthus* spp, velvetleaf, and many other broadleaf weeds.

The synthetic auxin herbicides (Group 4) provide broadleaf weed control when applied post-emergence. The active ingredients in this group include 2,4-D, dicamba (3,6-Dichloro-2-methoxybenzoic acid), fluroxypyr {[(4-Amino-3,5-dichloro-6-fluoro-2-pyridinyl)oxy]acetic acid}, and quinclorac (3,7-Dichloro-8-quinolinecarboxylic acid). Also called growth regulators, these herbicides have different effectiveness on broadleaf species. For example, quinclorac has excellent postemergence activity on field bindweed and morningglory and also has activity on some annual grasses. This is the only Group 4 herbicide that can give some grass control when applied postemergence. Fluroxypyr and dicamba are the main growth regulator herbicides used to control kochia. Fluroxypyr and quinclorac do not provide adequate control of *Amaranthus* spp. 2,4-D is used to control many broadleaf weeds and is the growth regulator that is most effective on Palmer amaranth. The synthetic auxin herbicides are commonly tank-mixed with other herbicides including atrazine which broadens the spectrum of broadleaf weed species controlled.

The nitrile herbicide bromoxynil (Group 6) is a photosystem II site B inhibiting herbicide that provides control of broadleaf weed species only. Bromoxynil is premixed or tank-mixed with other chemistries to increase the spectrum of weeds controlled. Bromoxynil is most effective when applied to small broadleaf weeds, 5 cm or less. Applying bromoxynil alone may result in unacceptable weed control.

The benzothiadiazole herbicide bentazon [3-Isopropyl-1H-2,1,3-benzothiadiazin-4(3H)-one 2,2-dioxide] (Group 6) is a photosystem II site B herbicide. Frequently, bentazon is tank-mixed with other chemistries, especially atrazine, for postemergence control of broadleaf weeds. Bentazon also has activity on yellow nutsedge (*Cyperus esculentus* L.).

The acetolactate synthase (ALS)-inhibitor herbicides (Group 2) provide control of many broadleaf weed species, however, the frequency of ALS-inhibitor resistant weeds requires that these herbicides be tank mixed with other herbicides having a different site of action and having broadleaf activity. ALS active ingredients include halosulfuron {methyl 3-chloro-5-[(4,6-dimethoxypyrimidin-2-yl)carbamoylsulfamoyl]-1-methylpyrazole-4-carboxylate}, metsulfuron (2-{[(4-methoxy-6-methyl-1,3,5-triazin-2-yl)amino]-oxomethyl]sulfamoyl}benzoic acid methyl ester), and prosulfuron {1-(4-methoxy-6-methyl-1,3,5-triazin-2-yl)-3-[2-(3,3,3-trifluoropropyl)phenyl]sulfonylurea}. Herbicides containing halosulfuron also have activity on yellow nutsedge. These herbicides are frequently applied with a growth regulator herbicide and a nonionic surfactant. Halosulfuron is weak on some *Amaranthus* spp and kochia. Metsulfuron is used postemergence in sorghum and must be applied with 2,4-D to reduce metsulfuron injury to sorghum. Sorghum

Figure 2. Grain sorghum competing with large Palmer amaranth treated with dicamba and atrazine that is not providing adequate control, (Source: Thompson field experiment, 2015).

may be stunted and heading may be delayed from a metsulfuron and 2,4-D treatment. Prosulfuron can be used preemergence or applied postemergence to grain sorghum after it reaches 13 cm tall. Always mix a nonionic surfactant or crop oil concentrate with prosulfuron. The tank-mix partner may limit the adjuvant system used in the mixture. The length of residual may prohibit the use of prosulfuron in some rotations. Wheat can be safely planted following prosulfuron use in sorghum. An additional sulfonylurea herbicide used in Europe is tritosulfuron {1-[4-methoxy-6-(trifluoromethyl)-1,3,5-triazin-2-yl]-3-[2-(trifluoromethyl)phenyl]sulfonylurea} used to manage broadleaf weeds in cereals.

A new technology has been developed that will allow control of annual grasses with the ALS inhibiting herbicide nicosulfuron (Group 2) applied postemergence (Thompson et al., 2014). Kansas State University sorghum breeders transferred a gene from shattercane collected from southwest KS into sorghum lines using traditional breeding techniques (Tesso et al., 2011). This shattercane population had developed resistance to ALS grass corn herbicides. Hybrids with the ALS-inhibitor resistant trait will be marketed as "Inzen" sorghum hybrids in the United States. Nicosulfuron is most effective applied to annual grasses when they are small. Size of adequately controlled grass will vary with grass species and growing conditions (Hennigh et al., 2010a, 2010b). Shattercane and Johnsongrass if ALS susceptible, will be controlled with nicosulfuron, however, currently are not listed as weeds controlled in DuPont's "Zest" herbicide label. It will be essential to tank-mix nicosulfuron with broadleaf herbicides as many broadleaf weeds have developed resistance to the ALS inhibitors. Additional herbicides containing rimsulfuron (Group 2) and thifensulfuron (Group 2) likely will be registered in the near future for use preemergence to Inzen sorghum.

The phenyl-urea herbicides linuron [3-(3,4-dichlorophenyl)-1-methoxy-1-methylurea] and diuron (Group 7) are photosystem II site A inhibitors. Labeled only in certain sorghum growing states, these herbicides are used most commonly in a postdirected application when sorghum is at least 38 cm tall and weeds are small. Linuron and diuron will cause severe burn if applied to the sorghum leaves. This is often a rescue type of treatment.

Sorghum often is grown in dryer climates, which results in a sorghum grower's reluctance to use "expensive" preemergence herbicides. Growers that depend entirely on postemergence herbicide programs often make applications to large sorghum and large weeds. Research conducted over several years at Kansas State University suggests that herbicide applications to large sorghum and large weeds do not provide adequate control of the weeds, but can reduce the competitiveness of the weeds to result in increased sorghum yield when compared to untreated sorghum (Figure 2). Ten postemergence herbicide programs evaluated over a 4- to 6-year period found that the most effective treatment for broadleaf weed control was a pyrasulfotole and bromoxynil premix plus atrazine (Table 1). These data represent the effectiveness of rescue treatments and not a typical recommended practice. In addition, most of the applications would not comply with the herbicide labels based on crop and weed size.

Summary

Weed management in sorghum is a challenge and will become even more difficult with continued development of herbicide-resistant weeds. It is important to use effective preemergence applied herbicides followed by the use of effective

Table 1. Comparison of 10 "salvage" herbicide programs on grain sorghum yield, visual plant injury 14 to 18 d after application, and percent visual weed control within 14 to 28 d after application from a 6-yr (2009 and 2011 to 2015) experiment near Manhattan, KS. Treatments were applied to 30 to 40 cm tall sorghum and 10 to 50 cm tall weeds

Treatments‡	Rate¶	Yield	% Injury	Percent control 14 to 28 DAA§			
	g ha^{-1}	t ha^{-1}	14 to 18 DAA	AMAPA	IPOHE	ABUTH	HELAN
Atrazine + COC	1680 + 1% v/v	2.84 cd†	0 d	55d e	63 de	45 d	68 c
Pyrasulfotole & bromoxynil + atrazine + NIS+AMS	43 & 245 + 1120 + 0.25%v/v + 1% w/v	5.296 a	6 cd	86 a	84 abc	98 a	95 a
Carfentrazone + NIS	8.75 + 0.25% v/v	1.58 de	21 a	35 f	52 e	95 ab	34 d
Carfentrazone + atrazine + NIS	8.75 + 1680 + 0.25% v/v	3.06 bcd	20 ab	54 e	67 d	93 ab	64 c
2,4-D LV ester	280	3.64 abc	11 bc	68 bc	86 ab	83 ab	88 ab
2,4-D LV ester + atrazine	280 + 627	4.59 ab	8 cd	71 b	86 a	83 b	88 ab
Metsulfuron + 2,4-D amine	2.1 + 280	3.20 abcd	14 abc	60 cde	73 cd	65 c	70 c
Halosulfuron & dicamba + atrazine + COC	53 & 231 + 1120 +1% v/v	4.61 ab	2 d	64 bcd	68 d	68 c	78 bc
Dicamba + atrazine	315 + 594	4.41 abc	7 cd	67 bc	74 bcd	67 c	78 bc
Untreated		0.73 e					
HSD#		1.83	9	10	12	13	15

† Within columns, means that share the same letter are not significantly different ($p = 0.05$).

‡ The use of "&" between active ingredients indicates a premix herbicide, thus only a single use rate is provided

§ DAA, days after application; AMAPA, *amaranthus palmeri*; IPOHE, *Ipomoea hederacea*; ABUTH, *Abutilon theophrasti*; HELAN, *Helianthus annuus.*

¶ v/v, volume per volume; w/v, weight per volume.

HSD is minimum difference between two treatments used to declare they are significantly different using Tukey's Honest Significant Difference Test.

postemergence applied herbicides to minimize weed competition with sorghum. It is also essential to incorporate sorghum production into rotation with other crops to facilitate diversity in weed control strategies. IWM remains a key component to successful weed management in sorghum.

The future of weed management with chemicals in sorghum will continue to progress slower than that of other major crops. The lack of total profit to chemical companies from herbicides used in sorghum compared to herbicides used in major crops provides a significant barrier to herbicide development for sorghum. Because of limited sorghum acres relative to other crops, herbicides are not developed specifically for sorghum. Thus, it is very unlikely that a new herbicide will be developed specifically for sorghum. The United Sorghum Checkoff Program and the individual state Sorghum Commissions have provided funding to enhance weed control options in grain sorghum. Market potential and the risk for gene transfer to wild relatives will likely limit the development of herbicide resistant genetically modified sorghum.

References

Abendroth, J.A., A.R. Martin, and F.W. Roeth. 2006. Plant response to combinations of mesotrione and photosystem II inhibitiors. Weed Technology 20:267-274.

Bishnoi, U.R., D.A. Mays, and M.T. Fabasso. 1990. Response of no-till and conventionally planted grain sorghum to weed control method and row spacing. Plant Soil 129:117–120. doi:10.1007/BF00032403

Bridges, D.C. 2008. Benefits of triazine herbicides in corn and sorghum production. In: H.M. LeBaron, J.E. McFarland, and O.C. Burnside, editors, The Triazine herbicides 50 years revolutionizing agriculture. Elsevier, p. 163–174. doi:10.1016/B978-044451167-6.50016-7

Burnside, O.C., and G.A. Wicks. 1967. The effect of weed removal treatments on sorghum growth. Weeds 15:204–207. doi:10.2307/4041203

Clements, D.R., D.L. Benoit, S.D. Murphy, and C.J. Swanton. 1996. Tillage effects on weed seed return and seedbank composition. Weed Sci. 44:314–322.

Defelice, M.S. 2006. Shattercane, *Sorghum Bicolor* (L.) Moench Ssp. *Drummondii* (Nees ex Steud.) De Wet ex Davidese – Black sheep of the family. Weed Technol. 20:1076–1083. doi:10.1614/WT-06-051.1

Everaarts, A.P. 1993. Effects of competition with weeds on the growth, development and yield of sorghum. J. Agric. Sci. (Cambridge) 120:187–196. doi:10.1017/S0021859600074220

Fabrizius, C.H. 1998. Studies on the control and interference of longspine sandbur in corn and grain sorghum. MS Thesis, Kansas State University, Manhattan, KS. p. 23-45.

Feltner, K.C., H.R. Hurst, and L.E. Anderson. 1969a. Tall waterhemp competition in grain sorghum. Weed Sci. 17:214–216.

Feltner, K.C., H.R. Hurst, and L.E. Anderson. 1969b. Yellow foxtail competition in grain sorghum. Weed Sci. 17:211–213.

Franke, T.C., K.D. Kelsey, and T.A. Royer. 2009. Pest management needs assessment for Oklahoma grain sorghum producers. Oklahoma Coop. Exten. Service EPP-7082.

Fromme, D.D., P.A. Dotray, W.J. Grichar, and C.J. Fernandez. 2012. Weed control in grain sorghum (*Sorghum bicolor*) Tolerance of pyrasulfotole plus bromoxynil. Int. J. Agron. doi:10.1155/2012/951454

Hennigh, D.S., K. Al-Khatib, and M.R. Tuinstra. 2010a. Postemergence weed control in acetolactate synthase-resistant grain sorghum. Weed Technol. 24:219–225. doi:10.1614/WT-D-09-00014.1

Hennigh, D.S., K. Al-Khatib, R.S. Currie, M.R. Tuinstra, P.W. Geier, P.W. Stahlman, and M.M. Claassen. 2010b. Weed control with selected herbicides in acetolactate synthase-resistant sorghum. Crop Protection 29:879-883.

Knezevic, S.Z., M.J. Horak and R.L. Vanderlip. 1997. Relative time of redroot pigweed (*Amaranthus retroflexus* L.) emergence is critical in pigweed-sorghum (*Sorghum bicolor* (L.) Moench) competition. Weed Sci. 45:502–508.

Knezevic, S.Z., S.P. Evans, E.E. Blankenship, R.C. Van Acker, and J.L. Lindquist. 2002. Critical period for weed control: the concept and data analysis. Weed Sci. 50:773–786. doi:10.1614/0043-1745(2002)050[0773:CPFWCT]2.0.CO;2

Liebman, M., and E. Dyck. 1993. Crop rotation and intercropping strategies for weed management. Ecol. Appl. 3:92–122 Ecological society of America, John Wiley & Sons, Ltd., Washington, D.C. doi:10.2307/1941795

Limon-Ortega, A., S.C. Mason, and A.R. Martin. 1998. Production practices improved grain sorghum and pearl millet competitiveness with weeds. Agron. J. 90:227–232. doi:10.2134/agronj1998.00021962009000020020x

Moore, J.W., D.S. Murray, and R.B. Westerman. 2004. Palmer amaranth effects on the harvest yield of grain sorghum. Weed Technol. 18:23–29. doi:10.1614/WT-02-086

Nice, G., B. Johnson, and T. B. Gauman. 2007. Weed control in no-till systems. Coop. Extension in Agric. and Home Econ. State of Indiana, Purdue University and U.S. Department of Agriculture Cooperating. Rev 12/2007.

Phillips, W.M. 1958. Weed control in sorghum. Cir. 360. Fort Hays Branch, Kansas Agricultural Experiment Stations, Fort Hays, KS.

Reddy, S.S., P.W. Stahlman, P.W. Geier, C.R. Thompson, R.S. Currie, A.J. Schlegel, B.L. Olson, and N.G. Lally. 2013. Weed control and crop safety with premixed pyrasulfotole and bromoxynil in grain sorghum. Weed Technol. 27:664–670. doi:10.1614/WT-D-13-00005.1

Shaner, D.L., editor. 2014. Herbicide handbook. Weed Science Society of America Tenth Edition, 2014. Weed Science Society of America, Lawrence, KS.

Smith, B.S., D.S. Murray, J.D. Green, W.M. Wanyahaya, and D.L. Weeks. 1990. Interference of three annual grasses with grain sorghum (*Sorghum bicolor*). Weed Technol. 4:245–249.

Staggenborg, S.A., D.L. Fjell, D.L. Devlin, W.B. Gordon, and B.H. Marsh. 1999. Grain sorghum response to row spacing and seeding rates in Kansas. J. Prod. Agric. 12:390–395. doi:10.2134/jpa1999.0390

Stahlman, P.W., and G.A. Wicks. 2000. Weeds and their control in sorghum. In: C.W. Smith and R.A. Fredricksen, editors, Sorghum: Origin, history, technology, and production. John Wiley & Sons, New York. p. 535–590.

Swanton, C.J., and S.F. Weise. 1991. Integrated weed management: The rationale and approach. Weed Technol. 5:648–656.

Tesso, T., K. Kershner, N. Ochanda, K. Al-Khatib, and M. Tunstra. 2011. Registration of 34 sorghum germplasm lines resistant to acetolactate synthase-inhibitor herbicides. J. Plant Reg. 5(2):215–219. doi:10.3198/jpr2010.03.0184crg

Thompson, C.R., D.E. Peterson, W.H. Fick, P.W. Stahlman, and J.W. Slocombe. 2017. 2017 chemical weed control for field crops, pastures, rangeland and noncropland. SRP1132. Kansas State Univ. Agric. Exp. Station and Coop. Extension Service. Manhattan, KS.

Thompson, C.R., R.S. Currie, P.W. Stahlman, A.J. Schlegel, G. Cramer, D.E. Peterson, and J.L. Jester. 2014. Using Inzen z sorghum to manage annual grasses postemergence. Proc. NCWSS 69:204.

Unruh, B.J. 2013. Influence of nitrogen on weed growth and competition with grain sorghum. MS thesis, Kansas State University, Manhattan, KS. p. 66

Vencill, W.K., and P.A. Banks. 1994. Effects of tillage systems and weed management on weed populations in grain sorghum (*Sorghum bicolor*). Weed Sci. 42:541–547.

Vijayakumar, M., C. Jayanthi, R. Kalpana, and D. Ravisankar. 2014. Integrated weed management in sorghum [*Sorghum bicolor* (L.) Moench] – A review. Agri. Reviews 35(2):79–91. doi:10.5958/0976-0741.2014.00085.3

Wiese, A.F., P.W. Unger, and R.R. Allen. 1985. Reduced tillage in sorghum. In: A.F. Wiese, editor, Weed control in limited tillage systems. Weed Science Society of America, Champaign, IL. p. 51–60.

Irrigation of Grain Sorghum

Danny H. Rogers,* Alan J. Schlegel, Johnathon D. Holman, Jonathan P. Aguilar, and Isaya Kisekka

Abstract

Grain sorghum is used as a food, livestock feed, or both depending on location. It has a reputation as being a water-thrifty crop, but that characteristic is related to its ability to produce grain under harsh environmental conditions. Grain sorghum does respond to irrigation; nevertheless, the crop is often grown under conditions of limited or deficit irrigation because of insufficient water supply for full irrigation or producer preferences related to cultural practices and yield response compared with corn. Here we review the water requirements and response to various irrigation practices of grain sorghum.

Although grain sorghum is a staple food source in some parts of the world, especially in Africa (Dicko et al., 2006), it is more commonly used as a livestock feed source (United Sorghum Checkoff Program, 2016). The crop has a reputation of being water thrifty, but this characteristic is more related to its ability to produce yields in harsh environments, therefore making it an important dryland crop option in many semiarid production areas of the world. Although grain sorghum is an important crop, the worldwide production of corn, rice, and wheat exceeds that of grain sorghum by factors of 16, 7, and 11, respectively. The United States is a leading world producer of grain sorghum and the leading supplier of grain sorghum to the export market, providing about two-thirds of the world exports (US Grains Council, 2015). Most grain sorghum grown in the United States comes from Kansas, Texas, Oklahoma, Nebraska, and Missouri.

Grain sorghum acreage in the United States peaked in 1957. Total production peaked in 1985, as the yield increase offset the reduction of planted area (Fig. 1). Since peak production in 1985, there has been a general downward trend in total production and planted area. US yields increased at a rate of 0.045 t ha^{-1} yr^{-1} since 1929, but after the large yield increases of the late 1950s and early 1960s, the rate slowed to a more modest increase of 0.016 t ha^{-1} yr^{-1}. From 1974 to 2009, annual grain sorghum production in Kansas was reported either for the irrigated

---

Abbreviations: ET, evapotranspiration; SDI, subsurface drip irrigation.

D.H. Rogers, Dep. Bio and Ag Engineering, Kansas State Univ., 129 Seaton Hall, 920 North 17th St., Manhattan, KS 66502-2906; A.J. Schlegel (schlegel@ksu.edu), Kansas State Univ., Southwest Research-Extension Center, 1474 State Highway 96, Tribune, KS 67879; J.D. Holman (jholman@ksu.edu), J.P. Aguilar (jaguilar@ksu.edu), and I. Kisekka (ikisekka@ksu.edu), Kansas State Univ., Southwest Research and Extension Center, 4500 East Mary St., Garden City, Kansas 67846. *Corresponding author (drogers@ksu.edu).

Contribution no. 17-106-B from the Kansas Agricultural Experiment Station.

doi:10.2134/agronmonogr58.2014.0072

 Sorghum: State of the Art and Future Perspectives, Ignacio Ciampitti and Vara Prasad, editors. Agronomy Monograph 58.

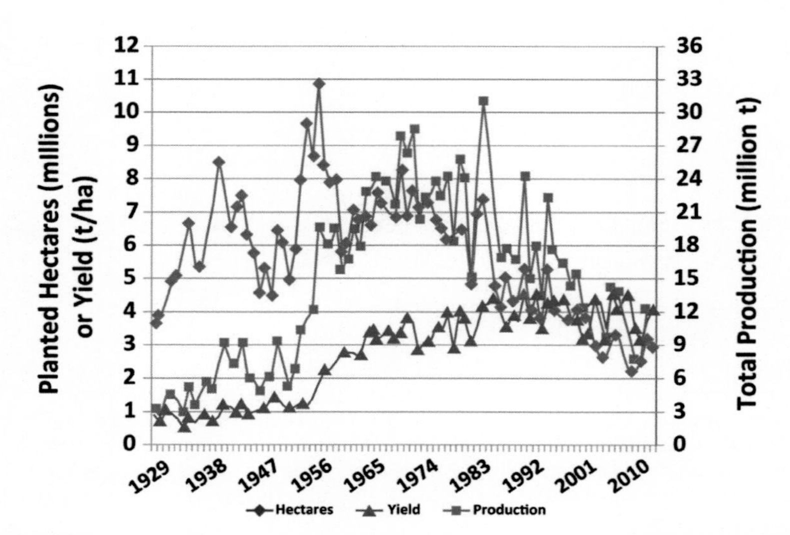

Fig. 1. Planted area, production, and yield trends for sorghum in the United States. From USDA National Agricultural Statistics Service Quick Stats (https://quickstats.nass.usda.gov/).

or the dryland crop (Fig. 2). In Kansas, yields from irrigated grain sorghum are higher than those from dryland grain sorghum (Fig. 2), but the annual trend for grain sorghum yield is 0.036 t ha^{-1} yr^{-1} for irrigated land, whereas the dryland yield trend is 0.047 t ha^{-1} yr^{-1}. The two rates are statistically significantly greater than zero but not significantly different from each other. Yields from irrigated sorghum tend to be less variable than those from dryland sorghum because of greater water availability. Even though most irrigated grain sorghum in Kansas is deficit irrigated, meaning that the crop will normally experience some water-use stress, the irrigation amount assures that the crop will be able to have some degree of yield potential. In Kansas, irrigated grain sorghum tends to be located in the drier areas of the state, where, in the early years of production, grain sorghum was irrigated with flood irrigation systems. Those systems tended to deliver greater amounts of water than center-pivot irrigation systems, which are used today; therefore, when yields of flood-irrigated sorghum are compared with yields of dryland sorghum that include significant production occurring in higher rainfall areas, the yield differential may have been greater during the early years of irrigation than in the later years. Surface or flood irrigation systems that introduced water in the furrow between the rows at an application depth of 3 or 4 inches would need to be applied so that the water would advance from the top of the field to the bottom. Flood irrigation systems tended to have low irrigation efficiency because of water loss resulting from deep percolation (water infiltrating beyond the crop root zone) and tailwater (irrigation water leaving the field at the bottom of the furrow), whereas center-pivot sprinkler systems can efficiently apply water at a shallower irrigation depth, which allows the irrigation water to be better targeted to the crop's water needs. Both irrigated and dryland production areas have also benefited from a shift to more conservation tillage

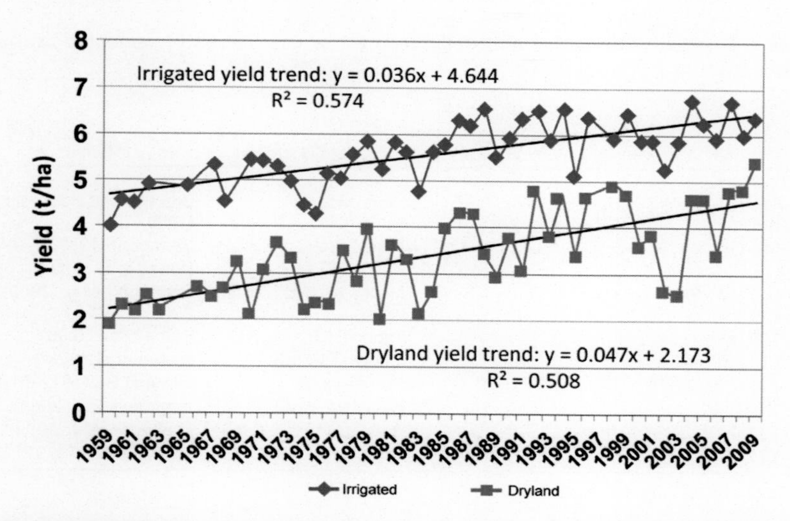

Fig. 2. Yield trend of grain sorghum in Kansas. From Kansas Farm Facts, Kansas Dep. of Agric., from 1958 to 2010.

practices that conserve soil water and may contribute to increased yield potential of dryland grain sorghum grown in wetter areas of the state. Conservation tillage practices can impact the water budget, and therefore the irrigation requirements, in several ways. Increased residue on the surface may allow more of the precipitation that falls on a field to infiltrate the soil. Higher residue covering also offers more protection for the soil surface from the impact of rainfall or sprinkler droplets, which can degrade the infiltration rate of the soil surface. Residue on the soil surface can also help limit soil-water evaporation. The benefits of residue on water conservation can occur throughout the year.

In the five-state production area of Kansas, Texas, Oklahoma, Nebraska, and Missouri, only about 5% of the total irrigated acreage is devoted to grain sorghum. About 40% of the grain sorghum in the five-state area is irrigated. Most (83%) of the irrigated grain sorghum is grown in Texas (USDA-NASS, 2007).

Grain Sorghum Water-Use Characteristics

Seasonal requirements for crop water use range from 410 to 780 mm (16 to 31 inches) for grain sorghum, as shown in Table 1 for various locations. This is similar to the range indicated by Wani et al. (2012) of 450–750 mm for 110- to 130-d sorghum crops. Crop water use is also referred to as crop evapotranspiration (ET) (Rogers and Alam, 2007; Rogers et al., 2015). Stone et al. (1996) reported an average water use of 663 mm for grain sorghum during a 14-yr study at Tribune, KS that was conducted using small irrigation basins and deep silt loam soils for the most well-watered treatment. The total amount of irrigation water needed depends on the season and the amount of soil water stored in the root zone. Dry-year-irrigation estimates (USDA-NRCS, 2015) for grain sorghum range from about 380 mm (15 inches) in southwest Kansas to less than 180 mm (7 inches) in

Table 1. Ranges of seasonal and daily crop water use values for selected crops for grain sorghum.†

Seasonal crop water use	Generalized and reported maximum daily peak crop water use	Study location	Reference
mm (inches)			
450–650 (17.72–25.59)	—	Worldwide	Brouwer and Heibloem (1986)
401–560 (16–22)	10 (0.40)	US Central Great Plains	Shawcroft (1989)
	10 (0.40)	Bushland, TX	Howell et al. (1997)
450–520 (17.76–20.59)	—	Manhattan and Tribune, KS	Hattendorf et al. (1988)
540–780 (21.2–30.6)	—	Tribune, KS	Stone et al. (1996)
460–510 (18.3–22.8)	—	Garden City, KS	Klocke et al. (2014)
550–710 (21.5–28.0)	13 (0.51)	Bushland, TX	Tolk and Howell (2001)
588 (23.1)	—	Zaragoza, Spain	Farré and Faci (2006)
526–580 (20.7–22.8)	—	Bari, Italy	Mastrorilli et al. (1999)
691–721 (27.2–28.4)	—	Albacete, Spain	López-Urrea et al. (2016)

† Adapted from Rogers et al., 2015.

southeast Kansas. Irrigation estimates for years with average rainfall are from about 330 mm (13 inches) in the west to 100 mm (4 inches) in the east. Average annual precipitation in Kansas varies from about 380 mm (15 inches) in the west to over 1000 mm (40 inches) in the east. The seasonal (April–October) pan evaporation ranges from about 1720 mm (68 inches) in western Kansas to 1080 mm (43 inches) in the eastern part of the state (Farnsworth and Thompson, 1982). The climate changes from semiarid to humid continental (Thornthwaite, 1931). The irrigation estimates are for well-watered conditions; however, irrigation is normally practiced in the semiarid areas of Kansas. The growing-season water-use requirement for grain sorghum can be met by in-season rainfall, irrigation applications, and stored root-zone soil water. Irrigation requirements would be larger in production areas with less in-season precipitation or for grain sorghum grown on shallow soils, soils with layers restricting root growth, or sandy soils with low soil-water-storage capacity.

Grain sorghum has a reputation for drought tolerance (Krieg and Lascano, 1990; Kebede et al., 2001; Howell et al., 2007; CGIAR, 2015), making it adaptable to marginal rainfed climates (Wani et al,. 2012), but also making it a choice for some irrigators with either low-capacity wells or otherwise limited water supplies because improved or more reliable yields are possible with low water application. The capacity of the irrigation system is the application depth of water that a field would receive if the entire field was irrigated in one day. Typically, large-scale irrigation systems, such as center-pivot irrigation systems, complete an irrigation event across multiple days and apply an amount of water in excess of the daily water use of the crop. The excess irrigation is stored in the crop root zone until used by the crop. Systems with irrigation capacities near or exceeding the daily crop water-use rate would be considered high-capacity systems. However, even if it has a capacity less than the peak water-use rate of the crop, an irrigation system can be considered high capacity if sufficient soil water reserves in the crop root zone are normally available to cover crop water needs when use rates exceed irrigation capacity (Rogers, 2009; Lamm and Rogers, 2015). For example, a system irrigating a crop grown on a field composed of a high water-holding-capacity soil, such as a silt loam, has a high probability of preventing yield-limiting water stress with an irrigation capacity of 0.10 cm d^{-1} (0.25 inches d^{-1}), whereas a system irrigating a field with sandy soils would need a greater irrigation capacity to have the same reliability. Irrigation systems servicing crops grown in regions with no in-season precipitation would also need an irrigation capacity that matches the long-term peak water-use rate to prevent yield-limiting water stress. The total volume of water available for irrigation can also be a limitation; for example, a surface-water storage impoundment that has a volume of water that is insufficient for a land area to be irrigated. Irrigation limitations can also comprise agency or governmental restrictions on the volume of water available relative to the potential irrigated area.

Grain sorghum is generally the last of the full-season summer crops to be planted in areas like the Central Great Plains of the United States, an area that has a continental climate. This timing allows for the soil profile to accumulate water prior to planting and often means that the reproductive stage begins after the hottest weather of the summer has passed. Water use varies by stage of growth, being low when the crop is small, increasing as the crop becomes larger and covers more of the surface area, and peaking when the canopy reaches full cover (meaning that little, if any, sunlight reaches the soil surface). The use rate will stay high until the leaves begin to senescence, as the plant approaches maturity (Rogers et al., 2015). Water-use rates for the various growth stages of grain sorghum are shown in Fig. 3. Average peak water-use rates are about 7.6 mm d^{-1} (0.3 inches d^{-1}), although occasionally a single-day peak use might approach 12.7 mm (0.5 inches), similar to the peak use rate of any field crop at full cover and active growing conditions.

Grain sorghum and corn are comparable feed grains, but sorghum initiates yield at a much lower level of water use than corn. Grain sorghum tends to be grown in areas with lower rainfall or with more limited irrigation than corn. These aspects are illustrated in Fig. 4. The water use needed by a crop to initiate

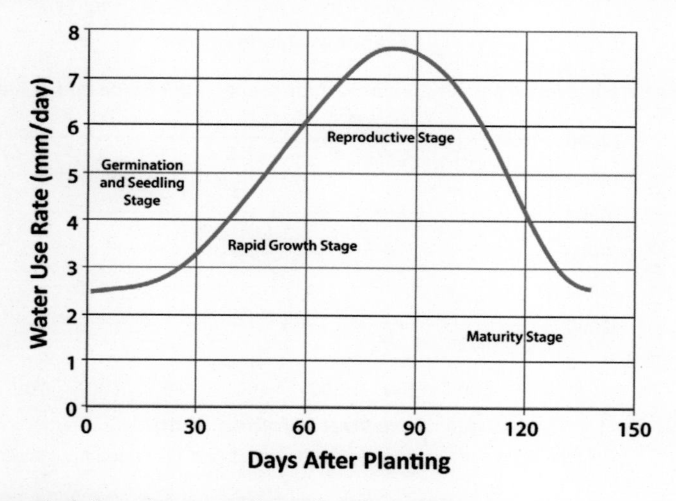

Fig. 3. Characteristic water-use pattern of grain sorghum.

yield is known as the threshold ET value for yield initiation. Of the crops shown, only sunflower initiates yield at a lower level of water use than grain sorghum.

Figure 5 shows the yearly yield plotted against ET for sorghum grown under six irrigation treatment levels in a multi-year study at Garden City, KS. The annual and growing season rainfall varied from above normal to extreme drought. Figure 6 shows the same yield data plotted against the applied irrigation amount and includes the average irrigation water-use response curves for the growing seasons with higher precipitation and for the two drought years. The water–yield-response curve is much flatter in the wetter years and steeper in the drought years. The amount of irrigation water needed to achieve the highest yield for the wet years was in the range of 150–200 mm, which was only about half the amount needed in the drought years.

Grain sorghum develops an extensive root system, which can extend deep into a friable soil. Stone et al. (2002) reported end-of-season rooting depths of 2.54 m (8.33 ft) for sorghum grown on Ulysses silt loam soils near Tribune, KS. Stone et al. (2001) reported root depths of 1.85 m for grain sorghum grown on Eudora silt loams near Manhattan, KS. Musick and Sletten (1966) indicated that grain sorghum water extraction was limited below 1.2 m on Pullman clay loams. Irrigation

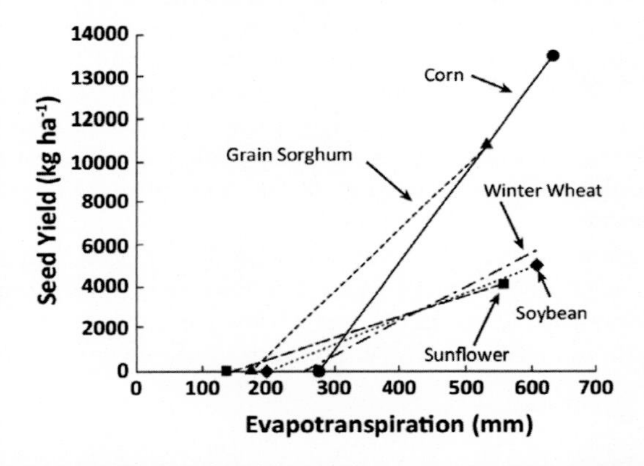

Fig. 4. Yield versus evapotranspiration for various crops. From Stone and Schlegel, 2006.

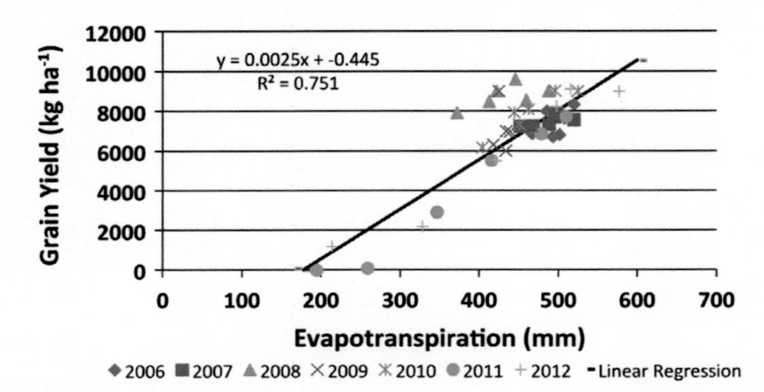

Fig. 5. Sorghum grain yield versus crop evapotranspiration for 2006–2012. From Klocke et al., 2014.

scheduling usually accounts for only the upper 1 m (3 ft) of the root zone, since most of the water extraction will occur in this region. About 75% of water use will occur in the upper half of the root zone. Under stress conditions, when the upper zone becomes water-limited, the crop will use significant deep water, as illustrated in Table 2. Because of the ability of grain sorghum to extract water and nutrients deeper in the profile than other crops such as corn, a crop rotation that alternates grain sorghum (deep-rooted) with a corn (shallow-rooted) could allow the grain sorghum to recover water and nutrients that move below the rooting depth of the corn. Or the grain sorghum can be used in combination with corn within a field to allow irrigation to be concentrated on corn while using grain sorghum as a limited-irrigation crop.

Irrigation Management

Grain sorghum is a crop that is adaptable to a limited irrigation-scheduling program (Stone et al., 1996; Stone and Schlegel, 2006; Howell et al., 2007; Klocke et al., 2012, 2014; Lamm et al., 2014). Stewart et al. (1983) chose grain sorghum as the crop to test a limited-irrigation dryland farming system using furrow

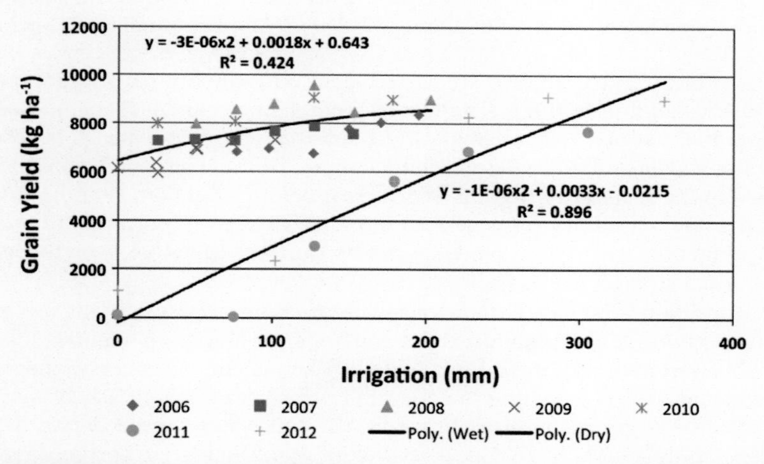

Fig. 6. Sorghum grain yield versus irrigation for years with above- and below-average precipitation. Hail on wheat in 2007 caused sorghum grain yield response to resemble that in wet years. From Klocke et al., 2014.

Table 2. Water extraction patterns under different soil water conditions, Garden City, KS.†

Depth	Normal (no stress)	Moderate stress	Moderate to serve stress
m (ft)		%	
0–0.30 (0–1)	31.4	25.3	7.5
0.30–0.61 (1–2)	23.2	18.9	7.3
0.61–0.91 (2–3)	18.4	19.9	14.8
0.91–1.22 (3–4)	13.4	17.9	24.9
1.22–1.52 (4–5)	7.6	11.7	24.4
1.52–1.83 (5–6)	6.0	6.3	21.0

† From Musick and Grimes, 1961.

irrigation and obtained favorable results. For high-water-holding-capacity soils, like medium-textured silt loams or heavier clay loams, limited water applications during the growing season of about half the full irrigation requirement for well-watered conditions—15–20 cm (6–8 inches)—will often produce 80–90% of the full yield potential.

Review of research trials in western Kansas demonstrates the utility of grain sorghum as a limited-irrigation crop (Table 3). In general, one or two irrigation applications, which were generally large (100 or 150 mm [4 or 6 inches]), provided near-maximum yield potential compared with treatments using three or four in-season irrigations. These trials were on deep silt loams, with the crop grown in small basins that were flood irrigated. Although most of these trials included a preplant irrigation, preplant irrigation is not recommended if any in-season irrigation is planned. In most years, sufficient rainfall is available to recharge the upper root zone, making preplant irrigation an inefficient use of water. For the studies in Garden City, KS from 1976 to 1978 and from 1976 to 1982, the treatment using 50% soil-water depletion as the irrigation trigger resulted in high yields. In the 1982–1985 Colby, KS study, treatments were scheduled using a percentage of the crop water use based on ET estimates instead of stage of growth. A long-term irrigation study of grain sorghum (2001–2008) at Tribune, KS used three annual irrigation treatments of about 130, 250, or 380 mm (5, 10, or 15 inches) of water (Table 4). The 130-mm treatment's average yield was 76 per cent of the 380 mm treatment. These studies seem consistent with the limited irrigation study of Musick and Dusek (1971). The results from a single 10-cm (4-inch) in-season irrigation at various growth stages indicated that applying water at heading or the milk stage resulted in the most efficient use of the irrigation water and was related to the severity of the soil water stress before irrigation. With limited irrigation water, stress during vegetative growth was less important, but as more total irrigation water was available, stress during vegetative growth became more important.

In addition to being able to extract water from a great depth in the soil profile, grain sorghum is also able to extract soil water at a lower percentage of available soil water without yield loss (Table 2). The general irrigation-management recommendation is to maintain soil water at or greater than 50% of available soil water. For grain sorghum, however, the soil water can be depleted to an average of 30–40% of available water before grain yields are severely reduced (Fig. 7). To ensure that soil water content is not depleted to yield-limiting levels before irrigation occurs, irrigation scheduling based on soil-water depletion or crop-water-use (ET) rate would be recommended when full irrigation of grain sorghum is intended.

Full and limited irrigation of grain sorghum on sandy soils require more-frequent and smaller irrigation applications, which matches the capability of center-pivot and subsurface drip irrigation (SDI) systems. Irrigation scheduling using ET or maintaining a given soil-water-depletion balance may be very useful in this condition, in which low water-holding capacity and restricted root zones present challenges to irrigation management. Under-irrigation can quickly result in yield-limiting stress. Single, large irrigation events can result in nutrient leaching and inefficient water use due to deep percolation. Such events also increase the potential for irrigation water redistribution on the soil surface within the field

Table 3. Summary of Kansas State University irrigated grain sorghum performance tests in western Kansas.

Irrigation treatment	Yield						
	Garden City 1954–1959	Colby 1970–1972	Tribune 1974–1977	Garden City 1976–1978	Garden City 1976–1982	Colby 1978–1979	Colby 1982–85
	kg ha^{-1} (bu acre^{-1})						
Unirrigated	2008 (32)	—	—	—	—	—	—
Timing							
Preplant (pre)	4833 (77)	6340 (101)	5900 (94)	6214 (99)	6905 (110)	5838 (93)	—
Pre + early vegetation (ev)	6403 (102)	6116 (107)	—	—	7782 (124)	—	—
Pre + ev + boot	7030 (112)	—	—	8097 (129)	7721 (123)	—	—
Pre + ev + boot + milk	7281 (116)	6716 (107)	6654 (106)	8097 (129)	—	6340 (101)	—
Pre + boot or bloom	—	6403 (102)	6465 (103)	7532 (120)	7405 (118)	6340, 6089 (101,97)	—
Pre + boot + head	—	6654 (106)	6277 (100)	—	—	6591 (105)	—
Pre + soft dough or milk	—	—	6528 (104)	—	—	6277 (100)	—
Pre + boot + milk	—	6591 (105)	6591 (105)	—	—	—	—
Pre + head + milk	—	6403 (102)	6340 (101)	—	—	—	—
Pre + 50% depletion	—		—	8035 (128)	8411 (134)	—	—
July	—	6654 (106)	—	—	6591 (105)	—	—
August	—	6465 (103)	—	—	—	—	—
July and August	—	7030 (112)	—	—	—	—	—
Amount							
1.4 × evapotranspiration (ET) (excess)	—	—	—	—	—	—	6591 (105)
1.2 × ET (excess)	—	—	—	—	—	—	6151 (98)
1.0 × ET (full)	—	—	—	—	—	—	6654 (106)
0.8 × ET (limited)	—	—	—	—	—	—	6214 (99)
0.6 × ET (limited)	—	—	—	—	—	—	5900 (94)
0.4 × ET (limited)	—	—	—	—	—	—	5335 (85)

Table 4. Average grain yield of grain sorghum as affected by irrigation amount at the Kansas State University Southwest Research Extension Center, Tribune for 2001–2008.

Irrigation amount	Seasonal water use	Yield
mm (inches)		kg ha^{-1} (bu acre^{-1})
127 (5)	484.89 (19.09)	5900 (94)
254 (10)	568.00 (22.40)	6967 (111)
381 (15)	646.17 (25.44)	7720 (123)

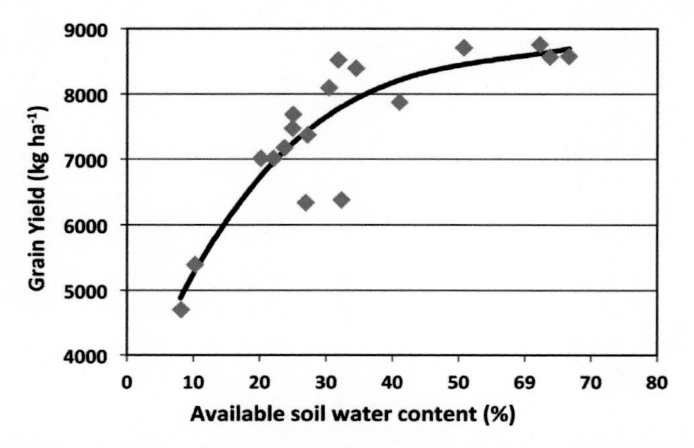

Fig. 7. The percentage of soil water available down to 1.2 m (4 ft) in depth prior to irrigation. Adapted from Jensen and Sletten, 1965.

and via runoff. Water redistribution within the field decreases the application uniformity and runoff water is a direct loss of irrigation efficiency.

Irrigation Systems

Grain sorghum is adaptable to water application by a variety of irrigation systems. The irrigation studies on water use cited above were conducted with both surface flood systems and pressurized sprinkler systems. Colaizzi et al. (2009) compared grain sorghum production for various types of sprinkler-nozzle packages and SDI. The sprinkler systems used included low-energy, precision applicators, low-elevation spray applicators, and mid-elevation spray applicators. The comparisons were made with four levels of irrigation, ranging from 25 to 100% of crop ET. Grain sorghum production differences were noted depending on the level of irrigation. There were, however, no statistically significant yield differences for the sprinkler types and SDI, except at the lowest irrigation rate, at which SDI had greater yields, indicating that grain sorghum production is robust over a wide range of irrigation system options.

Irrigation Summary

- Grain sorghum's water-use rate is similar to that of other summer crops and peaks at approximately 7.6 mm d^{-1} (0.3 inch d^{-1}). The peak use begins at initiation of the reproductive stage.

- The typical seasonal water need is 400–760 mm (16–30 inches).
- Grain sorghum has an extensive root system, which allows the crop to use soil water that is not available to other crops. Sorghum's drought tolerance makes it suitable for limited irrigation.

References

Brouwer, C., and M. Heibloem. 1986. Crop water needs. In: Irrigation water management: Irrigation water needs. Training manual no. 3/. FAO, Rome, Italy. p. 30–62. ftp://ftp.fao.org/agl/aglw/fwm/Manual3.pdf (accessed 29 July 2016).

CGIAR. 2015. Drought-tolerant crops for dryland. CGIAR, Montpellier, France. http://www.cgiar.org/www-archive/www.cgiar.org/pdf/drought_tolerant_crops_for_drylands.pdf (accessed 28 July 2016).

Colaizzi, P.D., S.R. Evett, T.A. Howell, and R.L. Baumhardt. 2009. Comparison of grain sorghum, soybean, and cotton production under spray, LEPA and SDI. In: Proceedings of the 2009 Central Plains Irrigation Conference, Colby, KS. 24–25 February. Kansas State Univ., Manhattan. p. 122–139. https://www.ksre.k-state.edu/irrigate/oow/cpic09.html (accessed 28 July 2016).

Dicko, M.H., H. Gruppen, A.S. Traore, A.G.J. Voragen, and W.J.H. van Berkel. 2006. Sorghum grain as human food in Africa: Relevance of content of starch and amylase activities. Afr. J. Biotechnol. 5(5):384–395 http://www.academicjournals.org/article/article1379753095_Dicko%20et%20al.pdf (accessed 29 July 2016).

Farré, I., and J.M. Faci. 2006. Comparative response of maize (*Zea mays* L.) and sorghum (*Sorghum bicolor* L. Moench) to deficit irrigation in a Mediterranean environment. Agric. Water Manage. 83:135–143. doi:10.1016/j.agwat.2005.11.001

Farnsworth, R.K., and E.S. Thompson. 1982. Mean monthly, seasonal, and annual pan evaporation for the United States. National Oceanic and Atmospheric Administration technical report NWS 34. Office of Hydrology, National Weather Service, Washington, DC.

Hattendorf, M.J., M.S. Redelfs, B. Amos, L.R. Stone, and R.E. Gwin, Jr. 1988. Comparative water use characteristics of six row crops. Agron. J. 80:80–85. doi:10.2134/agronj1988.00021962008000010019x

Howell, T.A., J.L. Steiner, A.D. Schnieder, S.R. Evett, and J.A. Tolk. 1997. Seasonal and maximum daily evapotranspiration of irrigated winter wheat, sorghum and corn. Trans. ASAE 40(3):623–634.

Howell, T.A., J.A. Tolk, S.R. Evett, K.S. Copeland, and D.A. Dusek. 2007. Evapotranspiration of deficit irrigated sorghum and winter wheat. In: A.J. Clemmens and S.S. Anderson, editors, The role of irrigation and drainage in a sustainable future. Proceedings of the USCID Fourth International Conference on Irrigation and Drainage, Sacramento, CA. 3–6 October.. U.S. Committee on Irrigation and Drainage, Denver, CO. p. 223–239

Jensen, M.E., and W.H. Sletten. 1965. Evapotranspiration and soil moisture-fertilizer interrelations with irrigated grain sorghum in the southern High Plains. Conservation Research Rep, 5. USDA-ARS, Washington, DC. doi:10.5962/bhl.title.65060

Kebede, H., P.K. Subudhi, and D.T. Tosenow. 2001. Quantitative trait loci influencing drought tolerance in grain sorghum (*Sorghum bicolor* L. Moench). Theor. Appl. Genet.103:266–276. doi:10.1007/s001220100541

Klocke, N.L., R.S. Currie, I. Kisekka, and L.R. Stone. 2014. Corn and grain sorghum response to limited irrigation, drought, and hail. Trans. ASABE 30(6):915–924.doi:10.13031/aea.30.10810.

Klocke, N.L., R.S. Currie, D.J. Tomsicek, and J.W. Koehn. 2012. Sorghum yield response to deficit irrigation. Trans. ASABE 55(3):947–955. doi:10.13031/2013.41526

Krieg, D.R., and R.J. Lascano. 1990. Sorghum. In: B.A. Stewart and D.R. Nielsen, editors, Irrigation of agricultural crops. Agron. Mongr. 30. ASA, CSSA, and SSSA, Madison, WI. p. 719–740.

Lamm, F.R., and D.H. Rogers. 2015. The importance of irrigation scheduling for marginal capacity systems growing corn. Appl. Eng. Agric. 31(2): 261–265.

Lamm, F. R., D. H. Rogers, J.P. Aguilar and I. Kisekka. 2014. Deficit irrigation of grain and oilseed crops. In: Proceedings of the 2014 Irrigation Association Technical Conference, Phoenix, AZ. 19–20 November. 2014 Technical Proceedings [CD]. Irrigation Assn., Falls Church, VA.

López-Urrea, R., L. Martinez-Molina, F. de la Cruz, A. Montor, J. Gonzalez-Piqueras, M. Odi-Lara, and J.M. Sanchez. 2016. Evapotranspiration and crop coefficients of irrigated biomass sorghum for energy production. Irrig. Sci. doi:10.1007/s00271-016-0503-y

Mastrorilli, M., N. Katerji, and G. Rana. 1999. Productivity and water use efficiency of sweet sorghum as affected by soil water deficit occurring at different vegetative growth stages. Eur. J. Agron. 11:207–215. doi:10.1016/S1161-0301(99)00032-5

Musick, J.T., and D.A. Duseck. 1971. Grain sorghum response to number, timing, and size of irrigation in the Southern High Plains. Trans. ASAE 14 (3): 401–404.

Musick, J.T., and D.W. Grimes. 1961. Water management and consumptive use by irrigated grain sorghum in western Kansas. Tech. bull. 113.Kansas Agric. Exp. Stn., Kansas State Univ., Manhattan.

Musick, J.T., and W.H. Sletten. 1966. Grain sorghum irrigation-water management on Richfield and Pullman soils. Trans. ASAE 9:369–371. doi:10.13031/2013.39981

Rogers, D.H. 2009. Irrigation scheduling using KanSched for a range of weather conditions. In: Proceedings of the 2009 Central Plains Irrigation Conference, Colby, KS. 24–25 February. Kansas State Univ., Manhattan. p. 66–73. https://www.ksre.k-state.edu/irrigate/oow/cpic09.html (accessed 28 July 2016)

Rogers, D.H., J. Aguilar, I. Kisekka, P.L. Barnes, and F.R. Lamm. 2015. Agricultural crop water use. Irrigation Management Series L934. Kansas State Univ. Agric. Exp. Stn. Res. and Coop. Ext. Serv., Manhattan.

Rogers, D.H., and M. Alam. 2007. What is ET? An evapotranspiration primer. Irrigation Management Series MF-2389 rev. Kansas State Univ. Agric. Exp. Stn. Res. and Coop. Ext. Serv., Manhattan.

Shawcroft, R.W. 1989. Crop water use. In: Proceedings of the Central Plains Irrigation Short Course, Colby, KS. 13–14 February. Kansas State Univ., Manhattan. p. 1–6.

Stewart, B.A., J.T. Musick, and D.A. Dusek. 1983. Yield and water use efficiency of grain sorghum in a limited irrigation-dryland farming system. Agron. J. 75:629–634. doi:10.2134/agronj1983.00021962007500040013x

Stone, L.R., D.E. Goodrum, M.M.N. Jaafar, and A.H. Khan. 2001. Rooting front and water depletion depths in grain sorghum and sunflower. Agron. J. 93:1105–1110. doi:10.2134/agronj2001.9351105x

Stone, L.R., D.E. Goodrum, A.J. Schlegel, M.N. Jaafar, and A.H. Khan. 2002. Water depletion depth of grain sorghum and sunflower in the central High Plains. Agron. J. 94:936–943. doi:10.2134/agronj2002.9360

Stone, L., and A. Schlegel. 2006. Crop water use in limited-irrigation environments. In: Proceedings of the 2006 Central Plains Irrigation Conference, Colby, KS. 21–22 February. Kansas State Univ., Manhattan. p. 173–184. https://www.ksre.k-state.edu/irrigate/oow/cpic06.html (accessed 29 July 2016).

Stone, L.R., A.J. Schlegel, R.E. Gwin, Jr., and A.H. Khan. 1996. Response of corn, grain sorghum, and sunflower to irrigation in the High Plains of Kansas. Agric. Water Manage. 30:251–259. doi:10.1016/0378-3774(95)01226-5

Tolk, J.A., and T.A. Howell. 2001. Measured and simulated evapotranspiration of grain sorghum with full and limited irrigation in three High Plains soils. Trans. ASAE 44(6):1553–1558.

Thornthwaite, C.W. 1931. The climates of North America: According to a new classification. Geogr. Rev. 21(4):633–655.

United Sorghum Checkoff Program. 2016. All about sorghum. United Sorghum Checkoff Program, Lubbock, TX. http://www.sorghumcheckoff.com/all-about-sorghum/.

USDA National Agricultural Statistics Service (NASS). 2007. Census of Agriculture. Table 27: Crops harvested from irrigated farms: 2008 and 2003. USDA-NASS, Washington, DC. http://www.agcensus.usda.gov/Publications/2007/Online_Highlights/Farm_and_Ranch_Irrigation_Survey/fris08_1_27.pdf (accessed 28 July 2016).

USDA National Resources Conservation Service (NRCS). 2015. Water requirements. Chap. 4 of Kansas Supplements—National Engineering Handb. Part 652: Irrigation guide. http://www.nrcs.usda.gov/wps/portal/nrcs/detail/ks/people/employees/?cid=nrcs142p2_033381 (accessed 29 July 2016).

US Grains Council. 2015. Sorghum . http://www.grains.org/ (accessed 16 Aug. 2016).

Wani, S.P., R. Albrizio, and N.R. Vajja. 2012. Soybean. In: P. Steduto, T.C. Hsaio, E. Fereres, and D. Raes, editors, Crop yield response to water. FAO Irrigation and Drainage Paper 66. Rome, Italy. pp. 124–131. www.fao.org/docrep/016/i2800e/i2800e.pdf (accessed 16 Aug. 2016).

Future Prospects for Sorghum as a Water-Saving Crop

David Brauer* and R. Louis Baumhardt

Abstract

Despite Earth's appearance from space, water is a minor component of the planet's mass, and relatively little water is present as freshwater available for crop production. Sorghum [*Sorghum bicolor* (L.) Moench] is ideally suited for grain and silage production in water-limited areas because of its ability to yield higher at lower levels of available water than other crops, including corn (*Zea mays* L.). In the United States, there are four areas in which sorghum is planted preferentially over other crops. Development of water policies in the 20th and 21st centuries and hydrology were examined in each of these four areas to provide insights into the possible future production trends for sorghum. In general, areas over the Ogallala Aquifer on the Southern High Plains are likely to see increases in sorghum production in the remainder of the 21st century, as water policies and hydrology will encourage the use of crops that produce more at lower levels of available water.

The initial images of Earth from space confirmed what humanity had known for centuries—a majority of the planet's surface, approximately 71%, is covered with water. However, these images of Earth are quite misleading in conveying the true abundance of water on the planet. If one assumes that the volume of water on Earth is approximately 1.386×10^9 cubic kilometers (Shiklomanov, 1993) and the planet's mass is approximately 6×10^{24} kilograms, then the relative abundance of water is only 0.02% of Earth's total mass. Therefore, the Earth is not a wet planet. Even a smaller percentage of the water is available for agriculture. Approximately 96.5% of the Earth's water is in oceans, seas, and bays (Shiklomanov, 1993) and thus is too saline for agricultural use except in rare circumstances where the cost of desalinization can be offset by cash receipts for high value crops. Another 1.7% of the planet's water supply is contained in ice caps, glaciers, and permanent snow (Shiklomanov, 1993). Groundwater accounts

Abbreviations: DFC, Desired Future Conditions; EAA, Edwards Aquifer Authority; GCDs, Groundwater Conservation Districts; GMA, Groundwater Management Areas; GMD, Groundwater Management Districts; HB, House Bill; HCUWCD, Hemphill County Underground Water Conservation District; HPWD, High Plains Underground Water Conservation District 1; IGUCA, Intensive Ground Water Use Control Areas; LEMA, Local Enhanced Management Area; NPGCD, North Plains Groundwater Conservation District; PACE, Property Assessed Clean Energy; PGCD, Panhandle Groundwater Conservation District; SB, Senate Bill; SWIFT, State Water Implementation Fund for Texas; TWDB, Texas Water Development Board; WAA, Water Appropriation Act.

D. Brauer and R.L. Baumhardt (R.Louis.Baumhardt@ars.usda.gov), Conservation and Production Research Lab., USDA-ARS, 2300 Experiment Station Dr., P.O. Drawer 10, Bushland, TX 79012. *Corresponding author (David.Brauer@ars.usda.gov).

doi:10.2134/agronmonogr58.2014.0073

 Sorghum: State of the Art and Future Perspectives, Ignacio Ciampitti and Vara Prasad, editors. Agronomy Monograph 58.

for another 1.7% of the planet's water supply, one-half of which is saline. The remaining 0.1% of the total water supply is in various other fresh surface water sources, including lakes, swamps, rivers, and the atmosphere. Therefore, the total water resource as freshwater groundwater, 0.8% of total, is several times greater than the surface freshwater supply.

Irrigation water from aquifers underlying cropland was used beginning in the second half of the 20th century to increase agricultural production. Without irrigation water withdrawn from aquifers, crop production would have been less in subhumid and humid areas in which seasonal drought lead to severe soil moisture deficits and in semiarid climates where annual rainfall is less than potential evapotranspiration. In many parts of the world, water withdrawals from aquifers have exceeded recharge, leading to net depletion of the groundwater resource. At the beginning of the 21st century, areas of severe groundwater depletion have been reported in Asia, Africa, Australia, and North America (Wada et al., 2010). The United States mirrors the global trend in which most of the liquid freshwater resource is groundwater. The United States has numerous aquifers (Fig. 1), many of which occur under the areas of intensive cultivation. Four American regions have experienced groundwater depletion in the 20th century: (i) the lower Mississippi alluvial aquifers, (ii) aquifers beneath southern Arizona, (iii) aquifers under the Central Valley of California, and (iv) the Ogallala Aquifer, also called the High Plains Aquifer, under a large portion of the central portion of the United States (Konikow, 2013). The Ogallala Aquifer is considered the largest groundwater resource of the United States, underlying more than 46,100,00 ha (178,000

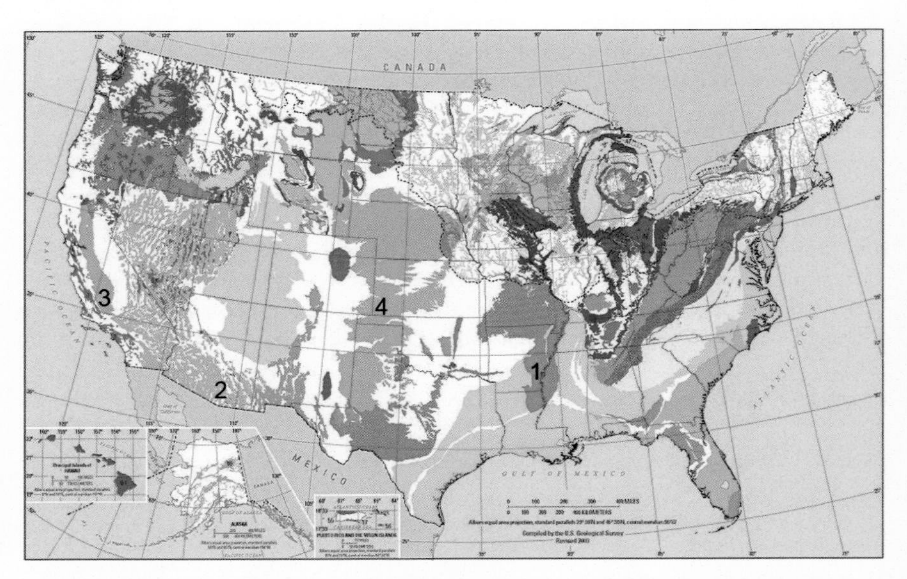

Fig. 1. Geographic distributions of aquifers in the United States. The four regions designated by numbers are places in which the aquifer(s) has experienced significant depletion during the 20th century. They are: (1) the lower Mississippi alluvial aquifers, (2) aquifers beneath southern Arizona, (3) aquifers under the Central Valley of California, and (4) the Ogallala Aquifer, or High Plains Aquifer. (Adapted from map by USGS.)

square miles) in eight states. Unfortunately, the relative abundance of the Ogallala Aquifer varied tremendously from state to state and location to location (McGuire, 2009 The main areas of groundwater depletion of the Ogallala Aquifer have occurred in Kansas and Texas, where withdrawals have exceeded recharge.

This groundwater depletion has the potential to limit agricultural productivity just at a moment when agricultural output needs to increase to adequately feed and clothe a growing worldwide population. The recognition of depleting groundwater resources comes at a time when other human activities are competing with agricultural production for the use of our limited freshwater resources, further attenuating problems of supplying agriculture with the levels of water to meet future food and fiber needs. New cropping systems will be necessary, especially in areas of the world where groundwater has become depleted, to produce enough food and fiber for future populations.

Sorghum as a Water-Smart Crop

Sorghum is ideally suited to provide feedstuff in areas with restricted water supplies for agriculture. Grain and forage–silage sorghum can readily substitute for corn in many of its uses as a livestock feed, human food component, and industrial-use crop. Therefore, it is quite natural to compare the water productivity of these crops. In addition, comparisons between corn and sorghum are not handicapped by their primary carbon fixation pathway, since both are C4 plants. An important aspect of water efficiency of sorghum relative to corn is presented in Fig. 2. The amount of water necessary for sorghum to produce the first increment

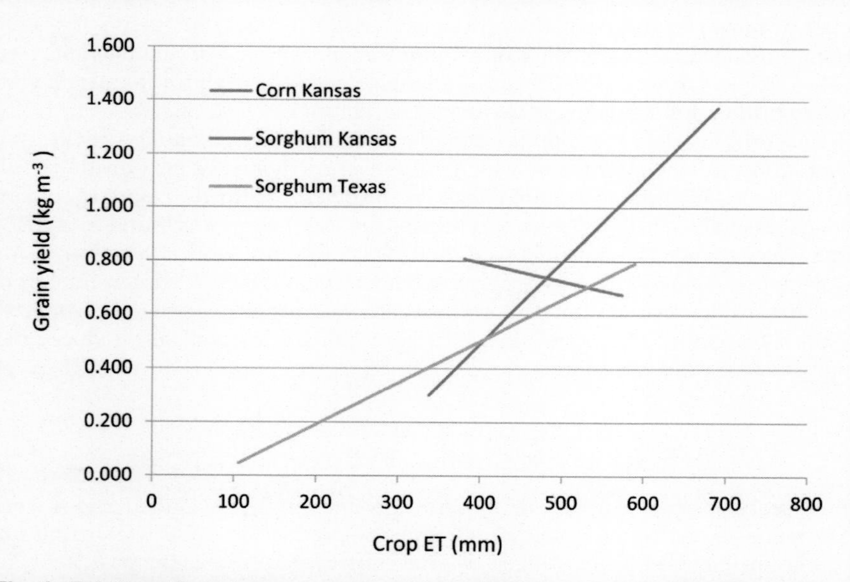

Fig. 2. Relationship between crop evapotranspiration and grain yield for corn and sorghum. Data are adapted from Klocke et al. (2011) and Klocke et al. (2012). Blue line is the data from Garden City, Kansas for the response of corn to water use. The red line is data from a similar experiment using sorghum from Garden City, KS. The data for sorghum in Texas was reported in Klocke et al. (2012) and was provided by USDA-ARS scientists in Bushland, TX.

of grain is significantly less than that of corn and grain crops with C3 metabolism like wheat (*Triticum aestivum* L.). At crop water use of <400 mm, sorghum produces more grain than corn. However, after that threshold, corn has the advantage in terms of grain production because it produces more grain per increment of additional water input than sorghum. In areas where water for irrigation is becoming limited, there appears to be a potential to substitute sorghum for corn, because of its similar uses and its lower water requirement to produce grain.

Most corn plants are monoecious, that is, with male and female flowers being borne on separate inflorescences on the same plant. Because of this attribute, corn tends to be cross pollinated. Cross fertilization or kernel set can be drastically reduced when either drought or high temperatures occur during flowering (Claassen and Shaw, 1970). Many factors have been attributed to decreased kernel set under drought and temperature stress, including failure of silks to elongate, decreased pollen viability, and decreased ability of the pistillate flower to produce seed (Herrero and Johnson, 1981; Jensen, 1971; Schoper et al., 1987). Sorghum, on the other hand, has inflorescence with spikelets that are perfect and fertile. Therefore, sorghum is almost always self pollinated in the field, and the fertilization process is not as vulnerable to extremes in water and heat stresses.

Sorghum for silage has been preferred over corn in areas with limited water supplies (Marsalis, 2011). For example, dairy farms have increased significantly between 2000 and 2010, with more than 150 dairies and 400,000 milking cows residing in western Texas and eastern New Mexico (Marsalis, 2011). A dairy cow's diet needs to be relatively high in fiber to maintain milk production and prevent excessive fat deposition (Morrison, 1957). Silage is an excellent source of fiber and energy; however, silage usually cannot be transported far from its site of production to the feeding site because of its high transportation cost relative to its feed value. When farmers are able to fully irrigate corn or sorghum for silage, corn silage production tends to exceed that of sorghum. However, when water is limited, sorghum tends to produce more dry matter per area than corn silage. In a 3-yr study in New Mexico in which crops were irrigated at two-thirds of full irrigation (~750 mm), sorghum silage produced from 5 to 10% more than corn silage (Marsalis, 2011). In Texas, sorghum silage production was found to use 20% less water than corn silage (Howell et al., 2008; Piccinni et al., 2009), but yields were approximately 20% less dry matter (Howell et al., 2008). A shorter time from planting to harvest was one factor affecting total crop water use (Piccinni et al., 2009). The relative efficiencies in which sorghum and corn silage produced biomass were similar, approximately 3.5 kg dry matter m^{-3} water (Howell et al., 2008).

Temporal and Spatial Trends for Sorghum roduction in the United States

Temporal and spatial trends in sorghum production in the United States were examined to determine if sorghum's superior production under water limiting conditions has been fully utilized by American agriculture in the past 50 yr. To date in the United States, corn production is more prevalent than sorghum production according to US agricultural census data (National Agricultural Statistics Service, 2014). Corn was harvested from approximately five times more acreage than sorghum in the United States in 1964 (Fig. 3A). The area of corn harvested for grain nearly doubled between 1964 and 2012 in the United States according

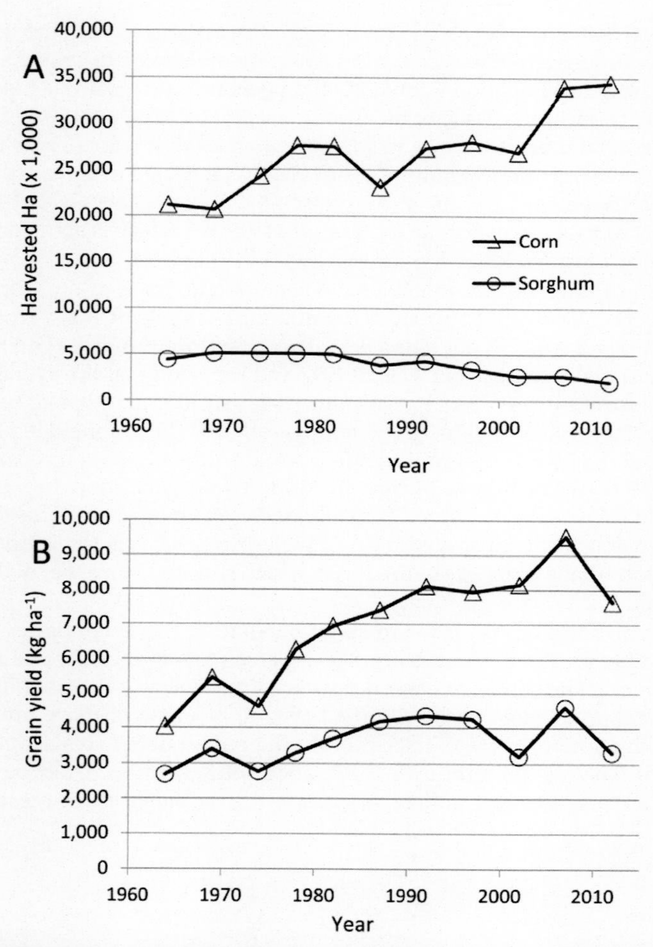

Fig. 3. US Agricultural Census Data comparing national production of corn and sorghum from 1964 to 2012. (a) Hectares of harvested corn (△) and sorghum (○); (b) national average grain yields.

to agricultural census data. In contrast, the area of sorghum harvested for grain remained unchanged between 1964 and 1992 and, since 1992, has tended to decline to one-half of the 1992 harvested area in 2012.

If corn and sorghum grains are so readily substituted for each other, why has sorghum area declined since the 1960s and corn area increased? One answer is the improvements in corn grain yields relative to sorghum grain yields. According to US agricultural census data, sorghum grain yields in 1964 were approximately two-thirds that of corn, 2680 versus 4030 kg ha^{-1} (Fig. 3B). Corn grain yields in each succeeding census have been higher than the preceding one, except for in 1974 and 2012. Decreases in national corn grain yields in 1974 may have been associated with changes in corn seed production immediately following the widespread *Helminthosporium maydis* (*Bipolaris maydis* or *Cochilobolus heterostrophus*) outbreak of the early 1970s (Ullstrup, 1972). Decreases in 2012 were associated with the widespread drought conditions experienced by a large portion of the central United States (The National Drought Mitigation Center, 2014). In general, corn grain yields increased 5500 kg ha^{-1} from 1964 to 2007, or an average annual increase of approximately 130 kg ha^{-1}. National grain sorghum yields have increased at a far slower pace since 1964. A national maximum grain sorghum yield, as reflected in agricultural census data, was also recorded in 2007 and equaled 4590 kg ha^{-1}. This yield change from 1964 to 2007 represents an annual increase of 44 kg ha^{-1}, which is about one-third of the increase in corn yields.

Dryland grain sorghum yields have been documented at the Conservation and Production Research Laboratory (Bushland, TX) since 1939 and from one uniformly managed experiment since 1958. Unger and Baumhardt (1999) analyzed variations in grain yields from 1939 to 1997 and reported that the average annual grain yield increase was 50 kg ha^{-1}, a value that is similar to the 44 kg ha^{-1} derived from agricultural census data. Unger and Baumhardt (1999) concluded that about one-third of the increased grain yields from a uniformly managed experiment during 1958 to 1997 could be attributed to improved genetics, while two-thirds of the yield increase was derived from changes in management. The component of management that contributed most to yield increases was the introduction of conservation tillage practices in the 1970s, which resulted in greater soil moisture at planting.

Although sorghum is grown in about one-half of the 48 contiguous states in the United States, the harvested acres as a percentage of the cropped land were concentrated in the southern central portion of the country in 2012 (Fig. 4). Four geographical areas have been identified (Fig. 4) as areas of more intensive sorghum production for further discussion in the remainder of this chapter. Of the hectares of harvested sorghum in 2012, about one-third occurred on the High Plains in areas over the Ogallala Aquifer, with the majority of these being in

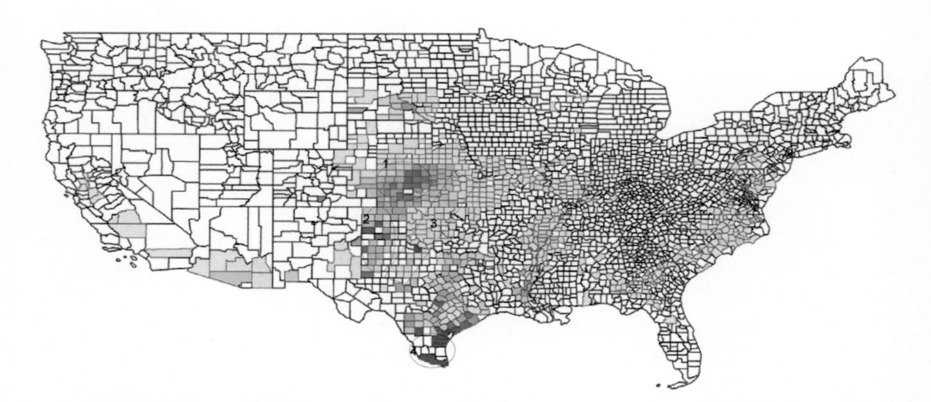

Fig. 4. Geographical distribution by county of sorghum production in the U.S. in 2012. Data from the U.S. agricultural census were used to create this figure. Sorghum acreage harvested as a percentage of the cropland was graphed for each county in the U.S. Counties with the highest percentages being redder in color. Four areas are delineated and discussed further in the text: (i) Kansas High Plains, (ii) Texas High Plains, (iii) Southern Rolling Plains, and (iv) Lower Rio Grande Valley and Coastal Bend regions of Texas.

Kansas and Texas (areas denoted by Regions 1 and 2, respectively, in Fig. 4). Because state laws in the United States have greater influence over water policies than federal laws, the geographic region of the High Plains will be further segregated into Kansas High Plains–Ogallala Aquifer region and Texas High Plains–Ogallala Aquifer regions. In both, Kansas and Texas, as well as Oklahoma, there are significant sorghum growing areas just east of the High Plains and the Ogallala Aquifer in the Rolling Plains portion of the Great Plains (Region 3 in Fig. 4). Some of the highest percentage of cropland being harvested as sorghum occurred in the lower Rio Grande Valley and Coastal Bend regions of southeastern Texas (Region 4 in Fig. 4). The future of sorghum in these regions will vary to some degree depending on future trends in water policies. In the following sections, water availability, water policies, and their possible influence on the amount of land cropped to sorghum will be explored.

Water Issues and Future Potential for Sorghum on the High Plains of Kansas

Annual rainfall in Kansas High Plains is greatest in the east, averaging 800 mm and declining to <400 mm at the western border with Colorado. Fortunately, the Ogallala Aquifer underlies much of the areas in the state of Kansas with annual rainfall less than 400 mm, the level that many consider minimum for dryland sorghum grain production (Fig. 1). Therefore, grain sorghum production persists in the region in part because of the ability to irrigate crops with water from the Ogallala Aquifer. Initial saturated thickness of the Ogallala Aquifer in Kansas varied tremendously with the greatest saturated thickness occurring in the southwestern corner of the state (Steward et al., 2013). Significant depletion in saturated thickness has been observed in most of the state, with an estimated 30% depletion of the water in the Ogallala Aquifer in Kansas by 2010 (Steward et al., 2013). Another 39% will be withdrawn by 2060. However, these are averages. In the southwestern portion of Kansas, the Ogallala Aquifer is likely to sustain irrigation for another 100 yr in several counties. Two hundred kilometers further north in central western Kansas, the saturated thickness of the Ogallala Aquifer was marginal to begin with and has been depleted sufficiently that many irrigation systems had ceased by 2010. The future for sorghum in the portion of the Kansas High Plains that has experienced significant depletion of the Ogallala Aquifer will be highly influenced by Kansas state laws regarding groundwater. Kansas state water policies will affect irrigated sorghum acres as well as dryland acreages. In times in which corn prices are high compared to sorghum, farmers may divert their water allotments to fewer hectares to guarantee high corn yields; thus, there will be opportunities for dryland sorghum production on those once irrigated acres.

Kansas water laws have evolved considerably during the second half of 20th century. Peck (1995) reviewed changes in Kansas water laws before 1995, which are briefly summarized here. Before 1945, groundwater use was governed by the absolute ownership doctrine that a landowner could withdraw and use water as desired, even if his withdrawal interfered with a neighboring landowner's use. Surface waters were dictated by the riparian doctrine: landowners adjoining a stream had rights to its use, whereas nonadjoining landowners had no rights. Under these water use doctrines, water was considered a personal property right that the state was reluctant to govern. In addition, disputes among water users were handled in courts on a one-off basis. Several laws were passed between 1880 and 1940 regarding water use and irrigation. In retrospect the most significant may have been the 1917 law that created the Water Commission. The name of the Water Commission was changed to the Division of Water Resources in 1927.

The years of 1944 and 1945 were highly significant in the development of current water laws in Kansas. In 1944, an Appellate Court in Kansas made it apparent that the current method for resolving water disputes was inefficient and contradictory in interpretation. In the aftermath of one such court decision, the presiding Kansas governor decided that the laws existing in 1944 were to be reviewed, and a committee was so established. By the summer of 1945, the committee had made their report, and the report became the basis of Water Appropriation Act (WAA) of 1945 that was passed by the state assembly and signed into law by the governor. The committee determined that the state could empower an administrative agency to appropriate and control the use of both surface and groundwater. The committee reached this conclusion from their interpretation of the 1917 law creating the Water Commission/Division of Water Resources. The 1945 WAA declared "all water within the state of Kansas....to the use of the people of the state, subject to control and regulation of the state." The WAA protected current users with vested rights, which were given priority over all others. New water users would have to obtain appropriated rights from the chief engineer, except in the case for domestic water use. The new water right could not conflict with existing use. The water right needed to be used, consumed, every 3 yr, or it was lost. The most important aspect for the future of water rights in Kansas was the dedication of water for public use. Initially, granted water rights were used to protect their owners' rights from conflicts or interference from other users.

It should be emphasized that WAA was passed in an era just before the rapid increase in irrigation in Kansas, especially over the Ogallala Aquifer. From 1945 to 1954, the chief engineer approved approximately 3500 water rights. In 1955 and 1956, during the waning years of a record drought in the central United States (Wessels Living History Farm, 2014), the chief engineer approved almost as many water rights (3200) as in the previous 9 yr (Peck, 1995).

The WAA was amended in two important ways in 1957. First, "water right" was defined as a real property right, meaning a landowner could sell or trade the water right independent of the land it flowed through or the groundwater under it. Second, domestic water use was limited to the water needed for humans, animals, and plants on 0.4 ha (1.0 acres).

An act was passed to establish Groundwater Management Districts (GMD) in 1972 in response to the understanding that withdrawals from the Ogallala Aquifer were exceeding recharge, leading to net depletion. The GMD Act authorized the establishment of five GMD, which cover most of the area over the Ogallala Aquifer and its closely related aquifers, corresponding to the High Plains Aquifer in Kansas. The purpose of the act was to preserve basic water law doctrine of the WAA while establishing the right and responsibility of local water users to determine their future groundwater use. The GMD were to adopt management practices based on the particular hydrological properties within their districts.

Several amendments were made to the WAA in the 1970s and 1980s. The 1977 amendment made it a criminal offense to use water without a permit, except for domestic water use. The 1978 amendment paved the way for the establishment of

Intensive Ground Water Use Control Areas (IGUCA). In areas of severe groundwater depletions, more restrictive water use plans could be prepared and submitted to the chief engineer as a means of slowing decreases in groundwater availability. An amendment in 1988 required water right owners to report annual water use.

The status quo remained in Kansas water law until the early 2010s when climatic conditions (a historical multi-year drought in western Kansas), better understanding of the relationship between water use and agricultural output, and political leadership created an environment for change. Prior to this, groundwater rights had evolved into a system in which a landowner was only allowed to withdraw a specified amount of water from a specified number of continuous hectares.

Climatic conditions of 2011 and 2012 were extremely difficult for Southern High Plain agriculture. In 2011, a record drought extended north from the Texas High Plain into southwestern Kansas (The National Drought Mitigation Center, 2014). Irrigation systems in these regions could not supply enough supplemental water to overcome deficits in rain, resulting in crop failures or extremely low yields. In 2012, the drought was less in terms of the reduction in rainfall from norms, but the geographic occurrence of drought was more widespread. The historical drought of western Kansas and Texas High Plains abated in 2014.

Agricultural economists have long debated which of two scenarios for the use of the Ogallala Aquifer is more favorable:

Scenario 1. Farmers irrigate as much as possible today, deplete the Ogallala Aquifer as fast as possible, and invest the enhanced income for future use and generations.

Scenario 2. Restrict current withdrawals from the Ogallala Aquifer, so that there will be more water available for future use.

In Scenario 1, when irrigated agriculture ceases to exist, all related agricultural industries (e.g., seed companies, irrigation equipment, fertilizer) will decline abruptly, and the only opportunity for future wealth is from investment of increased income while irrigating. In Scenario 2, farmers may lose some of the recent income, but there will be sufficient irrigation to sustain their farming operations and supporting enterprises and regional economies further into the future. A recent publication by Steward et al. (2013) supported Scenario 2 as a viable approach. Steward et al. (2013) assumed corn yields will increase during the next 60 yr as they have since World War II. Their analyses, which accounted for groundwater depletion, indicated that overall agricultural output during the next 60 yr will be greater if farmers decreased their water use by 10 to 20% now and continued to pump at those lower rates for a longer period of time.

However, predictions regarding future water use can be difficult to make. Kansas Water Office (1982) predicted that irrigated agriculture would decrease from 880,000 ha in 1977 to 308,000 ha in 2000 because of declining water availability and increasing energy cost to withdraw groundwater. Those predictions did not come true: irrigated hectares increased slightly to 935,000 ha in 2000 (Peterson and Bernardo, 2003). Peterson and Bernardo (2003) examined the assumptions of the 1982 study and found that two unpredicted conditions probably accounted for the differences between predictions and actual outcomes in irrigated area. First, significant water conservation occurred as furrow irrigation was replaced with sprinkler irrigation systems, and second, there were significant increases in the real prices for commodities.

In 2012, Governor Brownback of Kansas emphasized the need for water conservation. The state legislature followed his lead in making four changes to WAA. First, the use-or-lose provision of a water right was abolished. Second, laws were promulgated to create the Local Enhanced Management Area (LEMA). The IGUCA was rarely used because the chief engineer could modify water conservation plans without additional input from the submitting water right holders. Water right holders were unwilling to relinquish their control to the chief engineer. Under LEMA, the chief engineer has no means of modifying proposed water conservation plans. Third, multi-year water flexible accounts were modified. Water right owners had the ability to use more water in a year of need, on the condition that their water use over 5 yr did not exceed their total 5-yr water allotment. Under the law predating 2012, the water right owner was "taxed" a 10% of the extra water used as a conservation fee. This "water conservation tax" was eliminated in the 2012 amendments. Fourth, the "water conservation tax" for water placed in a virtual water bank for trading and selling was eliminated.

In 2013, Governor Brownback followed up the amendments to WAA by establishing a pathway to a long-term vision for the future water supply in Kansas. The second draft of the plan was presented in November 2014 at the Governor's Water Summit (Kansas Water Office, 2014). This vision had four guiding principles: (i) there should be locally driven solutions to water supply issues; (ii) new policies should not penalize those who have been good stewards of their water resource; (iii) water conservation efforts should be voluntary, incentive, and market-based; and (iv) action is required now to make a future reliable water supply. It is interesting to note that the vision currently stresses voluntary activities; however, if political will is great enough, the Kansas State Assembly with the approval of the governor has the power from WAA to dictate water conservation.

The second draft has a number of strategies that either directly mention sorghum as a part of the solution or may apply to sorghum production. One strategy articulated as "increase adoption of less water intensive crop varieties" states four points related to sorghum: (i) "form a collaborative stakeholder team to set sorghum research priorities and funding strategy"; (ii) "encourage producers to consider all aspects of agronomic management systems when trying to make water"; (iii) "implement sorghum research funding mechanism"; and (iv) "implement research to increase select pesticide resistance for sorghum." It would seem from these initiatives that Kansas is advancing the promotion of the United Sorghum Checkoff Program of "Sorghum: Rain or Shine."

There are indications that western Kansas agriculture is shifting in favor of sorghum production as water availability from the Ogallala Aquifer decreases. Scott County is in the central part of western Kansas. Saturated thickness of the Ogallala Aquifer in Scott County was less than a lot of other areas in the state before the development of irrigated agriculture, and by 2011 the saturated thickness had decreased sufficiently that irrigation could no longer be supported (Kansas Water Office, 2014). Agricultural census data indicated that there were more than 34,000 ha of sorghum harvested for grain in 1959 in Scott County (NASS, 2014). Harvested sorghum area declined rapidly between 1959 and 1974 when only 11,400 ha of sorghum were harvested in Scott County. Harvested area varied between 9000 and 14,000 ha from 1974 to 1992 in Scott County. Since 1992, hectares of sorghum harvested as grain have increased from 11,260 to more than 23,600 ha in 2007. Area harvested in 2012 was slightly less, but still almost twice that observed in the 1970s and 1980s. More than 80% of the acres in 2012 were irrigated. These census data support the hypothesis that Kansas farmers were managing a decreasing supply of irrigation water by increasing both irrigated and dryland sorghum production.

Water Issues and Future Potential for Sorghum on the High Plains of Texas

Although the Ogallala Aquifer resides under a vast area of the Texas High Plains, the amount of groundwater is considerably less than in other states, like Nebraska. Large areas of the Texas Ogallala Aquifer had less than 30 m of saturated thickness before the development of irrigation (McGuire, 2009). Two-thirds of the Ogallala Aquifer groundwater is in one-third of the area, north of the Canadian River. Musick et al. (1990) documented the trends in Texas High Plains irrigation from 1935 to 1989. The irrigated area increased rapidly from 0.2 million ha in the 1940s to a peak of 2.4 million ha in 1974. Irrigated area declined between 1974 and the mid-1990s and then stabilized with 1 to 1.3 million ha from 2000 to 2010 (Colaizzi et al., 2009). Groundwater extraction peaked in 1974, with a withdrawal of 10.0 km^3, and declined to approximately 5.6 km^3 in 1989. The declines in irrigated area and water withdrawals from 1974 to 1989 were associated with a change in irrigation method from furrow methods to sprinklers that consequently increased water use efficiency, that is, crop yield produced per unit of applied water. Decreases in irrigated area and groundwater withdrawals in the 1990s were associated with cropland being enrolled in the USDA Conservation Reserve Program, which pays farmers to return cropped land to native grasses to reduce or prevent wind and water induced soil erosion. Decreases in irrigated area are expected between now and 2030 in more areas in Texas over the Ogallala Aquifer as the aquifer becomes depleted to having less than 10 m of saturated thickness. It is obvious that hydrology will limit the future of irrigation to some extent on the Texas High Plains. But, in areas of more bountiful groundwater, Texas water policies may affect water availability and thus the preferences for corn or sorghum.

Surface water is owned by the state of Texas. However, rights to use that surface water can be assigned to a person or entity. Surface water rights in Texas follow the doctrine of prior appropriation; that is, water rights are based on the priority of water use, with the first water right owner having priority over others that came later. Texas Commission on Environmental Quality has been empowered to adjudicate surface water priorities and, as of April 1, 2012, to suspend water rights in the case of severe drought. Surface water rights have little bearing on farming on the Texas High Plains because the region lacks flowing rivers and streams on the Llano Estacado.

Current and future state laws regarding groundwater use may have a direct effect on farming practices on the Texas High Plains. Current groundwater rights in Texas, however, are quite different from most states, including Kansas. Modern groundwater law is derived from the 1904 decision in the Texas Supreme Court Case of Houston & Texas Central Railroad Company versus East (Drummond, 2014). East sued the railroad company because the East thought that the well drilled by the railroad company resulted in the family's well going dry during the drought of 1901–1902. In 1904, The Texas Supreme Court affirmed that county court ruling, quoting from an earlier New York state decision in its ruling: "An owner of soil may divert percolating water, consume or cut it off with impunity."

An oil and gas court case, Texas County versus Dougherty in 1915, affected future groundwater rulings (Drummond, 2014). The Texas Supreme Court ruled in this case that a landowner's right to access underground oil and gas resources is a private property right "of which may not be deprived without taking of private property."

Following droughts between 1910 and 1917, the Conservation Amendment to the Texas Constitution was adopted in 1917. This amendment empowered the state legislature to implement public policy to conserve and develop all of the natural resources of the state of Texas. As such, the amendment provided a mechanism by which laws and policies could be created to remedy water depletion and provided "promised stable water usage for the future." In some respects, the Conservation Amendment of 1917 provided the state of Texas with similar powers that the Kansas legislature provided in the same year when it created the Water Commission. However, Texas has not attempted to use the authority to the same extent as Kansas.

The private property of groundwater was confirmed in 1927 in the ruling on the case of Texas County versus Burkett. In addition, the 1927 ruling provided that a landowner had the absolute right to sell percolating ground water for industrial uses off site. Although the 1904 East ruling is often credited with the defining the term of "right of capture," that term was actually used first in Texas rulings in the 1935 ruling in the case of Brown versus Humble Oil and Refining Company. The Texas Supreme Court in a dispute regarding oil and gas rights stated that the "foregoing rule of ownership…should be considered in connection with the law of capture" (Drummond, 2014). The law of capture or right of capture was derived from Roman law in antiquity. The conditions of the rule of capture were further defined in another oil and gas ruling in the case of Elliff versus Texon Drilling Company of 1948. In this case, the Supreme Court ruled that "Each owner of land owns separately, distinctly, and exclusively all the oil and gas under his land." It would be more than 60 yr before the private property right to groundwater and rule of capture would be modified. Thus, from 1904 to the 21st century, Texas groundwater legal guiding principles would not be much different from those in Kansas before the passage of WAA in 1946.

Thirty years after the passage of the Conservation Act, Texas Legislature used its authority to conserve and develop the natural resource of groundwater. Groundwater Conservation Districts (GCDs) Act was passed in 1949 to provide the means by which locally governed GCDs could be formed (Drummond, 2014). The act states that the preferred method of groundwater management is via the GCDs. The primary legislatively mandated duties for the GCDs in the original act were to: (i) permit water wells, including governing well spacing; (ii) develop a comprehensive management plan; and (iii) adopt rules to implement management plans. A High Plains resident was quoted, "I favor no control, but if we must have it. Let it be local" (Drummond, 2014). A quote from the 1950s captures the objectives of GCDs: "Water district was not created to do away with the rights of the individual but rather to maintain those rights and provide for orderly development over the Ogallala Aquifer." A significant role in water conservation for the GCDs above the Ogallala Aquifer, however, would not occur during the 20th century.

Ruling in the case of Friendswood Development Company versus Smith-Southwest Industries, Inc. in 1978 led to the first major modification of the 1949 GCDs Act by allowing the creations of GCDs to control subsidence (Drummond 2014). The amendment also stated that landowners could be liable for the cause of subsidence of lands of others only when groundwater withdrawals were neglect, wasteful, or malicious. These conditions were similar to those imposed on negative effects of groundwater withdrawals on water uses by others.

Although not strictly a groundwater act, the Edwards Aquifer Authority (EAA) Act of 1993 has affected groundwater policies in Texas. To prevent federal interventions into water policies in Texas, the state legislature passed the EAA Act to restrict water withdrawals from the Edwards Aquifer to allow for continued nominal flows from springs and dependent streams as a means of habitat protection for endangered species. The EAA was authorized to adopt regulations and issue permits to limit groundwater production. The EAA Act was challenged in 1996 in the court case of Barshop versus Medina County Underground Water Conservation District. The GCD claimed that the Act constituted an unconstitutional deprivation of landowners' property right to the groundwater beneath their land. Barshop was a board member of the EAA and the State of Texas was joined as a necessary party. The Court ruled that any taking of private property was hypothetical in this case because no landowner had been denied the use of groundwater at the time of the ruling. Another 16 yr would pass before the court would have the opportunity in 2012 to rule on the taking of groundwater as a private property right.

Two significant water laws were developed in the aftermath of the drought of the 1990s. The first in 1995 was the passage of an act to create Groundwater Management Areas (GMA). The GMA were created to "provide for conservation, preservation, protection, recharging, and prevention of waste of the groundwater" and "to control subsidence caused by withdrawal of water from groundwater reservoirs" (Texas Water Development Board [TWDB], 2016a). Sixteen GMAs were delineated and adopted by 2002, two of which are over the Ogallala Aquifer (Fig. 5). The GMA provided a mechanism by which GCDs could coordinate their efforts of groundwater use and conservation for an aquifer that underlies multiple GCDs.

The other significant water-related act was Senate Bill 1 (SB 1), which provided for a regional water planning process (TWDB, 2016b). The Texas legislature passed SB 1 in 1997. The state of Texas was divided into 16 water planning areas, which do not necessarily coincide with the boundaries of GMAs. The water planning area were defined by county boundaries and therefore do not always create planning areas with defined hydrology. Two planning areas are located over the Ogallala Aquifer. The creation of the water plan is a bottom-up process that examines water supplies and demands over a 50-yr time horizon. In situations in which demand exceeds supply, the water plan tries to identify means by which the supply and demand can be balanced. These remedies, whether they are new sources of water or conservation efforts, are eligible for state support, primarily through low interest loans from the TWDB. All 16 GMA needed to develop and submit their first water plans to the TWDB in 2001. Plans are updated every 5 yr. Senate Bill 1 also increased the authority of locally controlled GCDs to establish

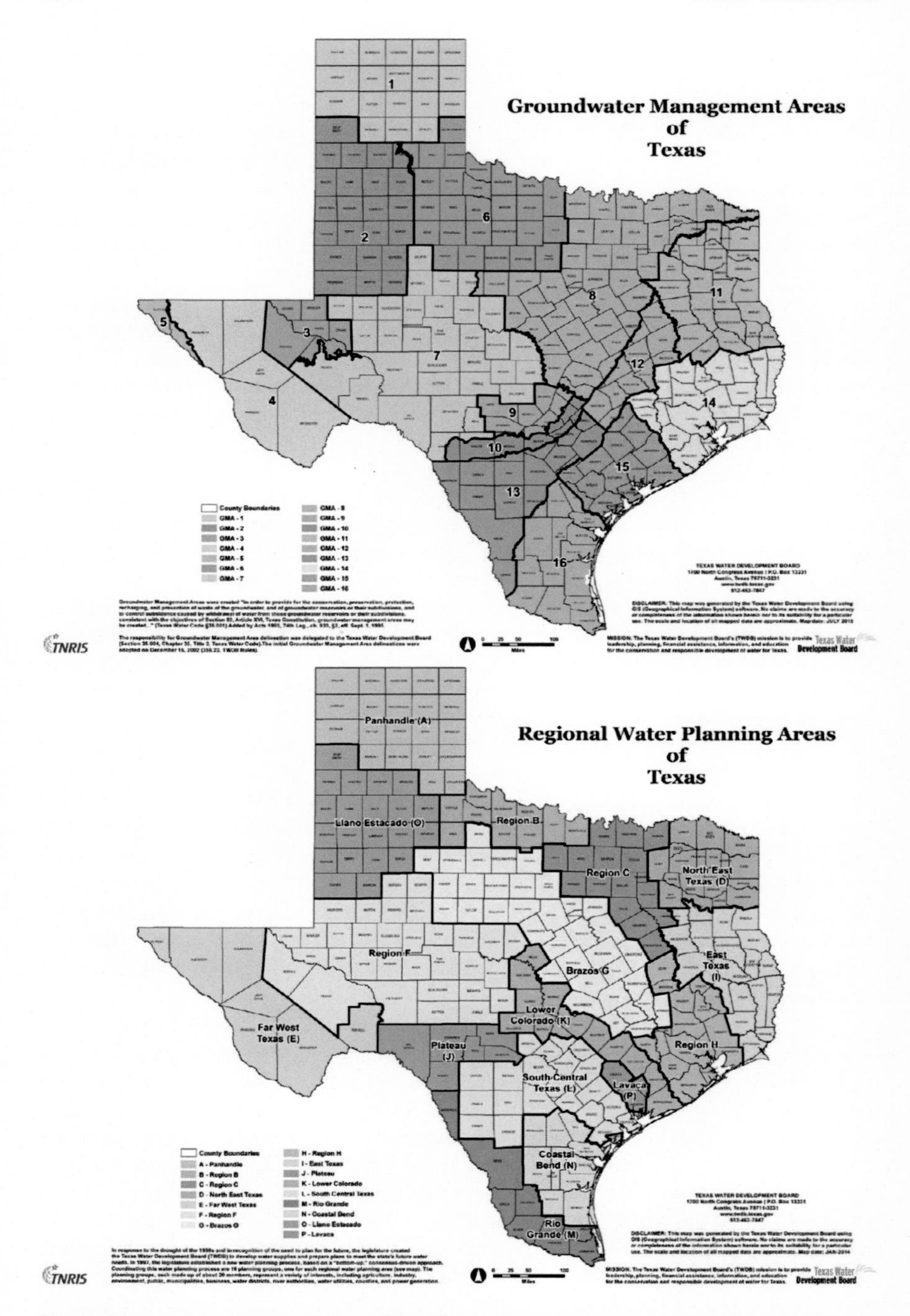

Fig. 5. Geographic representation of the Groundwater Management Areas and Regional Water Planning Areas in Texas. This figure was derived from images available from the Texas Water Development Board (TWDB, 2016a,b), used with permission of the Texas Water Development Board.

requirements for groundwater withdrawal permits and to regulate water transferred to locations outside of the district. The intent of this portion was to permit residents most affected by groundwater regulation to participate in the development of democratic solutions to groundwater issues.

The principle of absolute ownership of groundwater was affirmed in the 1999 case of Sipriano versus Great Spring Waters of America, Inc. (Drummond, 2014). Sipriano and another landowner sued the Great Spring Waters of America, Inc., or Ozarka, for groundwater depletion resulting from what Sipriano defined as "Ozarka's malicious and negligent withdrawals." Sipriano argued that the rule of capture should be abandoned. Unfortunately for Sipriano, the court ruled, "that the rule provides that, absent malice or willful waste, landowners have the right to take all the water they can capture under their land." One of the justices in the ruling of this case affirmed that Texas legislators had inadequately provided for the protection of groundwater and remedies for groundwater protection resided in that branch of state government by virtue of the 1917 Conservation Act.

Another significant groundwater legislative initiative was passed in 2005 as House Bill 1763 (HB 1763) (TWDB, 2016c). House Bill 1763 established a framework for regional collaboration among GCDs residing over a common aquifer within a GMA to set Desired Future Conditions (DFC). A DFC is the condition of an aquifer 50 yr in the future. The DFC for the Edward-Trinity Aquifer was set to meet minimum streamflows required for habitat for endangered aquatic species (TWDB, 2016d). For the Ogallala Aquifer, DFCs were set as a percentage of saturated thickness remaining in 50 yr, using 2010 as a baseline or a county average decrease in depth to water in the aquifer. Texas statute required that initial DFCs be submitted to TWDB by 1 Sept. 2010. The DFCs were to be incorporated into the 2007–2012 round of regional water planning because water left in aquifer to meet DFC would not account to the planning area's supply.

Groundwater Management Area 1 comprises the 47 million ha (17,000 square miles) of the 21 northern most counties of the Texas Panhandle and includes four GCD: Panhandle Groundwater Conservation District (PGCD), North Plains Groundwater Conservation District (NPGCD), Hemphill County Underground Water Conservation District (HCUWCD) and part of High Plains Underground Water Conservation District 1 (HPWD). The submitted statement defined three DFCs for three subareas within GMA 1. These subareas were defined by county boundaries. The DFC for the four northwestern counties in NPGCD was set at 40% of the water remaining in the Ogallala Aquifer in 50 yr. For the rest of NPGCD, and all of PGCD and HPWD, the DFC was set at 50% in 50 yr. For HCUWCD, DFC was 80% in 50 yr. These DFCs were submitted to TWDB by September 2010.

Mesa Water LP and G&J Ranch, Inc. filed a petition against the adoption of the DFCs. Hearings were held from November 2009 through May 2010 regarding the petition, in which both the petitioners and representatives of Region A Water Planning Group presented arguments opposing and in support of the proposed DFC, as well as about the method by which the DFCs were derived. Petitioners argued that the boundaries of the subareas for different DFCs within GMA 1 did not conform to the statute's description that specifies that subareas for DFC within GMA can be set because of hydrologically and geologically distinct subbasins within the aquifer or by unique geographic areas. Petitioners also argued that the proposed DFCs did not meet at least three other requirements of the law. The petitioners stated that 60% withdrawal in 50 yr within the subarea 1 was not physically or economically feasible. Petitioners said that the process lacked a socioeconomic impact assessment of the proposed DFCs for the entire GMA 1. Petitioners further argued that the proposed DFC were not fair and did not provide equal protection to both people and resources within GMA 1. Petitioner's final argument was that HB1763 envisioned that joint planning would occur within a GMA and that the proposed DFC was not the result of joint planning but rather each GCD acquiescing to the other's GCD proposed DFC. TWDB ruled against the petitioner's grievance, thus accepting the DFCs as submitted, and Region A Water Planning Group moved forward using the DFCs in their water plan to be submitted in 2012. Mesa Water LP and G&J Ranch, Inc. pursued further legal action, and a judgment in support of the Region A Water Planning Group was rendered. An important outcome from the court case was that TWDB's role in the water planning process was to certify that the methods used by the regional groups were consistent with HB 1763.

Although HB 1763 empowered GCDs to create rules for water conservation to meet DFCs, the following account indicates that conservation rules of GCDs require the will of members and water users to implement proposed conservation plans because of the local governance of GCDs. Acting on acceptance of DFC by TWDB, HPWD undertook the development of rules for a 16-county region, comprising the bulk of the Ogallala Aquifer south of Amarillo, TX and centered around Lubbock, TX. The governing board of HPWD proposed that all wells would be metered and annual volume of water pumped would have to be reported (Amarillo Globe News, 2011). In addition, allowable production of groundwater would be reduced from 5480 m^3 ha^{-1} (1.75 acre-feet) of continuously farmed land to 3920 m^3 ha^{-1} (1.25 acre-feet) in 2016 and beyond. These rules affected an estimated 40,000 irrigation wells that were not required to be metered in the past. There was vocal opposition to these rules from the water users in the district. The board postponed enforcement of the rules in 2012, 2013, and 2014. During these years, every member of the board of directors and the general manager of the HPWD retired.

The new board of directors of HPWD held hearings in 2014 that resulted in new rules for the district (HPWD, 2014). First of all, the requirement to have water meters was abolished in most instances. Second, most farmers will not have to report the volume of water pumped if the landowner certifies that only one irrigated crop per year is grown. Third, the annual water production was limited to 4700 m^3 ha^{-1} (1.5 acre-feet) of continuously farmed land. Farmers have several options to estimate water withdrawals in the absence of flow meters. The simplest is to certify that they are not double cropping any irrigated acres. Fourth, the new rule established a "water bank" in which water not used or pumped in 1 yr could be accumulated for future uses. This rule also specifies that water users have a starting allowance of 1700 m^3 ha^{-1} (6 acre-inches) in their "water banks." The earlier view demonstrated in the quote from a High Plain resident about preferring local governance of groundwater use appears to have prevailed in the 21st century. It will be interesting to see if these rules allow the water users in HPWD to meet the DFC.

In 2011, Texas state legislature made the first substantive change to groundwater ownership provision since the passage of GCD Act in 1949 (Drummond, 2014) when it passed SB 332. Before 2011, the pertinent portion of the state water code stated that "ownership and right of the owner of land and their lessee assigns in groundwater are hereby recognized." Interpretation of this portion of the water code had been problematic because of the lack of precision in the language. Senate Bill 332 clearly stated that the landowner owns the groundwater beneath the surface as a real property. However, the passed bill affirmed that the real property right of the groundwater does not: (i) prohibit a GCD from limiting or prohibiting drilling; (ii) affect the ability of GCDC to regulate withdrawals (production); and (iii) require rules adopted by GCD to proportionate a landowner's share of available groundwater based on the surface area. In addition, SB 332 requires GCDs to consider the public interest in conservation and protection in the development of management plans.

The most recent development in groundwater law in Texas resulted from the court case of EAA versus Day in 2012. Day and McDaniel purchased 150 ha (~375 acres) over the Edward Aquifer in 1994. The property had a functioning irrigation well from 1956 until 1983, when the well was removed. Day and McDaniel sought a permit from EAA for a replacement well to irrigate crops and pecans using a total of 860,000 m^3 (700 acre-feet) annually. The EAA provided a preliminary finding in support of the permit. Day and McDaniel spent $95,000 to drill the replacement well. Soon after the well's completion, EAA notified Day that the permit was denied. Through administrative appeal processes, Day and McDaniel were permitted to withdraw 17,000 m^3 (14 acre-feet) annually. Unsatisfied with that decision, Day sued EAA for taking property without compensation. The legal process eventually ended at the Texas Supreme Court. The Texas Supreme Court from the outset of its deliberations decided the legal principle in question was "whether land ownership includes an interest in groundwater in place that cannot be taken for public use with adequate compensation" (Drummond, 2014). This ruling would clarify the legal distinction between rule of capture and ownership in place with regard to groundwater. The Supreme Court stated, based on a similar ruling involving oil and gas rights, that the "landowner is not entitled to any specific molecules of groundwater." The court continued that it had never addressed the ownership of groundwater in place. The court ruled "each owner of land owns separately, distinctly and exclusively all the groundwater under his land." The court, however, expressed doubt that the EAA's action had denied Day and McDaniel all of the economically beneficial use of the property. The court did distinguish groundwater from being different from oil and gas because oil and gas are not recharged and groundwater may be. The court also distinguished differences in groundwater and surface water rights: non-use of groundwater conserves the resources, while non-use of appropriated surface water is equivalent to waste. In the end, The Texas Supreme Court referred the case back to lower courts. However, EAA settled the dispute with Day and McDaniel out of court. Thus, no substantial ruling was made as to whether or not EAA needed to compensate Day and McDaniel for taking private property.

The Day case has been heralded as a turning point in groundwater law in Texas because the private ownership of groundwater in place was affirmed, but this ownership is different from the concept of absolute ownership. In light of the Day rulings, the rule of capture as defined earlier in the East case confers ownership at the surface for groundwater and is the basis for disputes between different users of the groundwater, but has no effect on the ownership and use of groundwater in place. The Conservation Amendment empowers the State of Texas to regulate groundwater but that regulation has to be balanced against the

Texas Constitution's Taking Clause. In the Day case, the landowner was financially harmed by the actions taken by EAA because: (i) the landowner had spent considerable money to develop the groundwater, and (ii) the permitted groundwater withdrawal deprived the landowners of 98% of the requested withdrawal. Therefore, there is some uncertainty to the applicability of the rulings in the Day case to potential regulations of GCDs over the Ogallala Aquifer to reduce withdrawals by a smaller fraction, for example, by 20%. The results from Steward et al. (2013) that demonstrated that water conserved in the Ogallala Aquifer for future use will actually extend the life cycle productivity of the system adds further uncertainty to claims of taking private property by conservation reductions today. Currently other cases involving potential taking of property with regard to groundwater are working their way through the Texas court systems. As noted above, the local governance of GCD may preclude such cases being forwarded to the Texas Supreme Court for actions taken by GCDs over the Ogallala Aquifer.

In 2011, the Texas assembly passed Senate Bill 385 (SB 385) creating the Property Assessed Clean Energy (PACE) program. The primary purpose of this bill is to allow property owners to retrofit buildings and other infrastructure with more energy efficient devices. The program allows a property owner to repay expenses for these upgrades through increased payments of their property taxes, rather than via a traditional loan. Although the bill is named for its role to decrease energy use and expense, SB 385 allows similar repayment scheduling for retrofits for water conservation. It is conceivable that PACE could be used in the future for irrigators to buy equipment that increases water conservation.

In 2013, the Texas Assembly passed a series of bills leading to the creation of State Water Implementation Fund for Texas or "SWIFT" (TWDB, 2016e). SWIFT will utilize $2 billion from the state reserves to provide low interest loans and deferred loan payments to support water infrastructure projects. SWIFT will more than double the state's investment in water infrastructure. A project is eligible for funding if it was identified as a means of overcoming water deficit in one of the regional water plans. Twenty percent of the funds must be used for water conservation and reuse projects, and a minimum of 10% must be for projects assisting rural communities or agriculture. The first round of applications for funding were due on 1 Feb. 2015. The four major criteria for ranking applications are: (i) size of the population served, (ii) ability to assist a diverse urban and rural population, (iii) increasing regionalization, and (iv) meeting a high percent of water supply needs. Water supply projects to meet the irrigation needs on the Texas High Plains do not seem likely to rate high using the above four criteria. The future will be the judge of the importance of SWIFT for irrigation in Texas using water from the Ogallala Aquifer.

It appears certain that the hydrology of the Ogallala Aquifer, its past use in Texas, and the possible implementation of GCDs rules to meet DFCs will decrease the amount of water available for irrigation on the Texas High Plains. What role will sorghum play in maintaining crop production and farm income under these future scenarios? An examination of the strategies being offered by the Region A Water Planning Group to overcoming deficits between the water supply and demand for irrigation provide a basis for thinking that sorghum will be critical (Panhandle Water Planning Group, 2014). Shifting area from water intensive corn to less water intensive sorghum and cotton (*Gossypium hirsutum* L.) has the potential to reduce water withdrawals in Region A by an estimated 5000 km^3 of water (4 million acre-feet) between 2020 and 2070. Use of varieties with lower water use will save another additional 2500 km^3 of water over the 50 yr time span. Additional 5000 km^3 of water can be conserved through the application of irrigation scheduling techniques based on either evapotranspiration or crop water stress detected by on-pivot or remote sensors. Successful scheduling of irrigation applications to sorghum for optimum water efficiency by either ET (Howell et al., 2008; Piccinni et al., 2009) or plant stress sensors has already been documented (O'Shaughnessy et al., 2014). The Region A Water Planning Group is fairly confident that water conservation methods in which sorghum is a key component can be incorporated over the next 50 yr to maintain irrigated area within 5% of 2010 cropped area. The greatest unknown addressed by these water planners is the water conservation that may occur with the development and use of varieties with higher water use efficiency of water between 2030 and 2070 using advanced breeding techniques.

Recent production data indicate that sorghum is beginning to have an increased presence and influence on Texas High Plains agriculture. The dairy industry on the Texas High Plains has increased significantly since 2000 when there were fewer than 10,000 dairy cows in the region (Marsalis, 2011). Today the estimates of the Texas High Plain dairy herd exceed 150,000 cows. Milk production requires the cow's diet to be rich in fibrous materials (Morrison, 1957). Silage has traditionally been a major source of the fiber component in a dairy cow's diet (Morrison, 1957). Silage needs to be produced near its site of feeding since transportation is expensive due to its water content. In Bailey County (Texas), there were no reported areas of sorghum silage from 1964 to 2002 (NASS, 2014). Census data from 2007 and 2012 indicated the presence of 540 and 3100 ha of sorghum silage harvested in Bailey County, which corresponded to the growth of dairy operations in those years. A resurgence of sorghum for silage seems likely in areas where dairy operations are based, especially as the saturation thickness of the Ogallala Aquifer continues to decline.

Water Issues and Future Potential for Sorghum on the Rolling Plains of Texas, Oklahoma, and Kansas

In the south-central United States, sorghum is frequently produced in an area just to the east of the Ogallala Aquifer. This region is characterized by annual precipitation of approximately 500 to 600 mm with at least two-thirds of the rain occurring during the growing season. The annual precipitation in this region is ideal for optimum sorghum production (Fig. 2), but not adequate for reliable corn production (Fig. 2). This is also a region in which irrigation is not prevalent because of the lack of need due to adequate rainfall and lack of major, widespread aquifers. Water policies in this region, whether in Kansas, Oklahoma, or Texas, have little bearing on farmers choosing between sorghum and corn in their production system. The greatest unknown in this region is the possible effects of climate change on potential evapotranspiration. Increases in potential evapotranspiration will shift farmers' preferences further in the direction of sorghum because of its better growth under more adverse weather conditions.

Water Issues and Future Potential for Sorghum in Lower Rio Grande River Valley and Coastal Bend Region of Texas

Another area in the United States in which the frequency of planting sorghum is relatively high is in the lower Rio Grande River Valley and Coastal Bend Region in Texas, which correspond to Water Planning Regions M and N (TWDB, 2016b). The region's climate is characterized by a long growing season. Annual rainfall is quite variable, ranging from approximately 500 mm at the inland city of McAllen to 750 mm at the coastal town of Corpus Christi. Aquifers underlie much of the region; however, the aquifers in the lower Rio Grande Valley tend to be brackish. Those aquifers underlying the Coastal Bend Region tend to be fresh or slightly saline.

The production favors early planting and harvesting of sorghum. In the past, double cropping of sorghum, such as after a spring vegetable crop, occurred. In recent years, the presence of sugarcane aphids and other pests have prevented double-cropped sorghum. One reason for sorghum production in this area is its proximity to naval shipment centers for overseas transportation, with China being a leading purchaser in recent years.

In the Lower Rio Grande Valley, surface water is the dominant source for irrigation. However, the water operating systems guarantee the supply for municipal and industrial uses over that of agriculture. Population growth is anticipated in the Lower Rio Grande Valley, which will place more of a burden on the limited water supply. It is projected that irrigation's share of the water use will decline from 93% in 2010 to 42% by 2060. Part of this decrease will be from agricultural water conservation methods, both on-farm improvements and improved conveyance systems. The conservation strategies identified in the 2012 water plan included that of growing water-efficient crops and irrigation scheduling through evapotranspiration measurements (TWDB, 2012). As identified earlier, sorghum fits into both of these conservation strategies well. Significant savings in agricultural water use will occur as the growing population of the region transforms existing farm land into urban and suburban areas.

The future water supply in the Coastal Bend region looks more promising for continued agricultural production. Demand for water is projected to increase approximately 60% from 2000 to 2060. The Coastal Bend Region currently uses both surface and groundwater resources to meet its demands, with future development of the groundwater resources occurring to supply much of the water for increased demand. Groundwater tends to be used more than surface water for irrigation. Demand for irrigation water is expected to be constant from 2000 to 2060, while the demand for municipal, electricity generation, and manufacturing will increase. Kenedy County Groundwater Conservation District (2012) is the major authority that can regulate groundwater use in the region. Kenedy County Groundwater Conservation District requires permitting for well drilling, and once in operation annual reports of water production are required. The current DFC for the Gulf Coast Aquifer is a drawdown of 30 m between 2000 and 2060. Wells in the TWDB database have shown little or no drawdown since being drilled. Therefore, at present, it appears that the conditions that favor sorghum production in this area will persist into the foreseeable future.

Concluding Remarks

The world supply of freshwater is quite limited, and the supply of freshwater available for irrigation will probably be less in the future. Sorghum production tends to be more likely than corn under conditions of limited or restricted water availability because of sorghum's ability to produce grain at lower levels of water. Future availability of surface and groundwater will be a complex interaction between hydrology and water policies. This discussion has focused on the hydrology and water policies of four regions in the United States where sorghum production was common in 2012 to describe current water policies and envision how these policies may affect future sorghum production. For the two regions over the Ogallala Aquifer on the High Plains in Texas and Kansas, state and local water policies may greatly affect water usage by agriculture in the future in such a way to favor sorghum production over that of corn. In the Lower Rio Grande Valley of Texas, sustaining current levels of sorghum production may be difficult in the future as agricultural land is converted into suburban and urban areas with a concomitant redirection of the limited water supply from agricultural production to uses by municipalities and industries.

References

Amarillo Globe News. 2011. Water district votes to approve amendments. http://amarillo.com/news/local-news/2011-07-20/water-district-votes-approve-amendments (accessed 1 July 2016).

Claassen, M.M., and R.H. Shaw. 1970. Water deficit effects on corn. I. Grain components. Agron. J. 62:652–655. doi:10.2134/agronj1970.00021962006200050032x

Colaizzi, P.D., P.H. Gowda, T.H. Marek, and D.O. Porter. 2009. Irrigation in the Texas High Plains: A brief history and potential reductions in demand. Irrig. Drain. 58:257–274. doi:10.1002/ird.418

Drummond, D.O. 2014. Texas groundwater rights and immunities: From East to Day and beyond. Texas Water J. 5:59–94.

Herrero, M.P., and R.R. Johnson. 1981. Drought stress and its effect on maize reproductive systems. Crop Sci. 21:105–110. doi:10.2135/cropsci1981.0011183X002100010029x

High Plains Underground Water Conservation District 1. 2014. Rules of the High Plains Underground Water Conservation District No. 1. (as amended August 12, 2014). http://static.squarespace.com/static/53286fe5e4b0bbf6a4535d75/t/53ebc5efe4b0215d14909407/1407960559465/HPWD+Rules+as+amended+August+12+2014.pdf (accessed 1 July 2016).

Howell, T., S. Evett, J. Tolk, K. Copeland, P. Colaizzi, and P. Gowda. 2008. Evapotranspiration of corn and forage sorghum for silage. World Environmental and Water Resources Congress 2008, p. 1–14. doi:10.1061/40976(316)88

Jensen, S.D. 1971. Breeding for drought and heat tolerance in corn. Proc. 26th Ann. Corn Sorghum Res. Conf. 36:66–77.

Kansas Water Office. 1982. Ogallala Aquifer study in Kansas: L-P model. Kansas Water Office, Topeka, KS.

Kansas Water Office. 2014. A long-term vision for the future of water supply in Kansas, Draft 2. http://www.kwo.org/50_Year_Vision/Kansas%20Water%20Vision%20Draft%20II%2011.12.2014Final.pdf (accessed 1 July 2016).

Kenedy County Groundwater Conservation District. 2012. Rules of the Kenedy County Groundwater Conservation District. Available at: http://www.kenedygcd.com/Libraries/Rules/Adopted_RULE_AMENDMENTS_072512.sflb.ashx (accessed 1 July 2016).

Klocke, N.L., R.S. Currie, D.J. Tomsicek, and J.W. Koehn. 2011. Corn yield response to deficit irrigation. Trans. ASABE 54:931–940.

Klocke, N.L., R.S. Currie, D.J. Tomsicek, and J.W. Koehn. 2012. Sorghum yield response to deficit irrigation. Trans. ASABE 55:947–955.

Konikow, L.F. 2013. Groundwater depletion in the United States (1900–2008). USGS Sci. Invest. Rep. 2013-5079. http://pubs.usgs.gov/sir/2013/5079 (accessed 12 July 2016).

Marsalis, M.A. 2011. Advantages of forage sorghum for silage in limited input systems In: Proceeding of 2011 Western Alfalfa and Forage Conference, Las Vegas, NV. 11–13 Dec. 2011. UC Cooperative Extension, Plant Sciences Department, University of California, Davis, CA http://alfalfa.ucdavis.edu/+symposium/proceedings/2011/11-66.pdf (accessed 30 June 2016).

McGuire, V.L. 2009. Water-level changes in the High Plains aquifer, predevelopment to 2007, 2005–06, and 2006–07. USGS Invest. Rep. 2009-5019. http://pubs.usgs.gov/sir/2009/5019/ (accessed 12 July 2016).

Morrison, F.B. 1957. Feeds and feeding: A handbook for the student and stockman. 22nd ed. Morrison Publishing Company, Ithacha, NY.

Musick, J.T., F.B. Pringle, W.L. Harman, and B.A. Stewart. 1990. Long-term irrigation trends—Texas High Plains. Appl. Eng. Agric. 6:717–724. doi:10.13031/2013.26454

NASS (National Agricultural Statistics Service). 2014. Census of Agriculture, USDA. http://www.agcensus.usda.gov/ (accessed 12 July 2016).

O'Shaughnessy, S., S. Evett, P. Colaizzi, and T. Howell. 2014. Wireless sensor network effectively controls center pivot irrigation of sorghum. Appl. Eng. Agric. 29:853–864.

Panhandle Water Planning Group. 2014. Chapter 4. Identification, evaluation and selection of water management strategies based on needs. In: 2011 Adopted Water Plan. http://panhandlewater.org/2011_adopted_plan.html (accessed 12 July 2016).

Peck, J.C. 1995. The Kansas Water Appropriation Act—A fifty year perspective. Kansas Law Review 43(735). http://papers.ssrn.com/sol3/papers.cfm?abstract_id=2174074 (accessed 30 June 2016).

Peterson, J.M., and D.J. Bernardo. 2003. High Plains regional aquifer study revisited: A 20-year retrospective for western Kansas. Great Plains Res. 13:179–197.

Piccinni, G., J. Ko, T. Marek, and T. Howell. 2009. Determination of growth-state-specific crop coefficients (K_c) of maize and sorghum. Agric. Water Manage. 96:1698–1704. doi:10.1016/j.agwat.2009.06.024

Schoper, J.B., R.J. Lambert, B.L. Vasilas, and M.E. Westgate. 1987. Plant factors controlling seed set in maize: The influence of silk, pollen and ear-leaf water status and tassel heat treatment at pollination. Plant Physiol. 83:121–125. doi:10.1104/pp.83.1.121

Shiklomanov, S. 1993. World fresh water resources. In: Peter H. Gleick, editor, Water in crisis: A guide to the world's fresh water resources. Oxford Univ. Press, New York. p. 1–37.

Steward, D.R., P.J. Bruss, X. Yang, S.A. Staggenborg, S.M. Welch, and M.D. Apley. 2013. Tapping unsustainable groundwater stores for agricultural production in the High Plains Aquifer of Kansas, projections to 2110. Proc. Natl. Acad. Sci. 110(37):E3477–E3486. doi:10.1073/pnas.1220351110

Texas Water Development Board (TWDB). 2012. Water for Texas 20012 State Water Plan. http://www.twdb.texas.gov/waterplanning/swp/2012/index.asp (accessed 16 July 2016)

Texas Water Development Board (TWDB). 2016a. Groundwater management areas. http://www.twdb.texas.gov/groundwater/management_areas/index.asp (accessed 16 July 2016).

Texas Water Development Board (TWDB). 2016b.Water resources planning. http://www.twdb.texas.gov/waterplanning/index.asp (accessed 16 July 2016).

Texas Water Development Board (TWDB). 2016c. Desired future conditions. http://www.twdb.texas.gov/groundwater/management_areas/DFC.asp (accessed 16 July 2016).

Texas Water Development Board (TWDB). 2016d. Desired future conditions, Groundwater Management Area 13. http://www.twdb.texas.gov/groundwater/management_areas/dfc_mag/GMA_13_DFC.pdf (accessed 16 July 2016).

Texas Water Development Board (TWDB). 2016e. SWIFT. http://www.twdb.texas.gov/financial/programs/swift/index.asp (accessed 16 July 2016).

The National Drought Mitigation Center. 2014. U.S. Drought Monitor for August 7, 2012. http://droughtmonitor.unl.edu/MapsAndData/MapArchive.aspx (accessed 12 July 2016).

Unger, P.W. and R. L. Baumhardt. 1999. Factors related to dryland grain sorghum yield increases from 1939 through 1997. Agron. J. 91:870–875.

Ullstrup, A.J. 1972. The impacts of the southern corn leaf blight epidemics of 1970-1971. Annu. Rev. Phytopathol. 10:37–50. doi:10.1146/annurev.py.10.090172.000345

Wada, Y., L.P.H. van Beek, C.M. van Kempen, J.W.T.M. Reckman, S. Vasak, and M.F.P. Bierkens. 2010. Global depletion of groundwater resources. Geophys. Res. Lett. doi:10.1029/2010GL044571.

Wessels Living History Farm. 2014. Water issues in the 1950's. http://www.livinghistoryfarm.org/farminginthe50s/water_01.html (accessed 30 June 2016).

Sorghum: A Multipurpose Bioenergy Crop

P. Srinivas Rao,* K.S. Vinutha, G.S. Anil Kumar, T. Chiranjeevi, A. Uma, Pankaj Lal, R. S. Prakasham, H.P. Singh, R. Sreenivasa Rao, Surinder Chopra, and Shibu Jose

Abstract

Bioethanol and biodiesel produced from renewable energy sources are gaining importance in light of volatile fossil fuel prices, depleting oil reserves, and increasing greenhouse effects associated with the use of fossil fuels. Among several alternative renewable energy sources, energy derived from plant biomass is found to be promising and sustainable. Sorghum [*Sorghum bicolor* (L.) Moench] is a resilient dryland cereal crop with wide adaptation having high water, nutrient, and radiation use efficiencies. This crop is expected to enhance food, feed, fodder, and fuel security. Sweet sorghum is similar to grain sorghum but has the ability to accumulate sugars in the stalks without much reduction in grain production. Hence, it is used as a first-generation biofuel feedstock, where the grain and stalk sugars can be used for producing bioenergy, while energy sorghum or biomass sorghum is increasingly viewed as a potential feedstock for lignocellulosic biofuel production. Although the commercial use of sweet sorghum for bioethanol production has been demonstrated in China and India, the viability of large-scale lignocellulosic conversion of sorghum biomass to biofuels is yet to be demonstrated. This chapter dwells on sorghum feedstock characteristics, biofuel production models, sustainability indicators, and commercialization.

Sorghum (*Sorghum bicolor* (L.) Moench) originated in the Sahel region of Africa (Legwaila et al., 2003), is adapted to tropical and temperate cropping systems, and is one of the five most cultivated crop species globally. It is used for food, feed, fuel, fiber, brewing, and construction purposes (Srinivasarao et al., 2015). It is efficient in converting CO_2 into sugar (Schaffert and Gourley, 1982) and highly productive with tolerance to drought, water logging, and salinity (Almodares et

---

Abbreviations: *bmr*, brown midrib; CU, centralized unit; DCU, decentralized unit; GHG, greenhouse gas; IRR, internal rates of return; LCA, life cycle assessment; SWOT, strength–weakness–opportunities–threat.

P. Srinivas Rao, K.S. Vinutha (vinuthaks.mysore@gmail.com), and G.S.A. Kumar (anilkumargpb99@gmail.com), ICRISAT, Research Program on Dryland Cereals, Patancheru 502 324, India; T. Chiranjeevi (chiranjeevi.tulluri@gmail.com), and A. Uma, Jawaharlal Nehru Technological Univ., Dep. of Biotechnology, Hyderabad, India; P. Lal, Montclair State Univ., NJ (lalp@mail.montclair.edu); R.S. Prakasham, CSIR-Indian Inst. of Chemical Technology (IICT), Hyderabad 500007, India (prakashamr@gmail.com); H.P. Singh, Fort Valley State Univ., GA (singhh@fvsu.edu); R. Sreenivasa Rao, Inst. of Biological Environmental and Rural Sciences (IBERS), Aberystwyth Univ., Gogerddan, Aberystwyth, Ceredigion, SY23 3EE, UK (rsr@aber.ac.uk); S. Chopra, Dep. of Plant Sciences, Penn State Univ., PA (sic3@psu.edu); and S. Jose, Center for Agroforestry, Univ. of Missouri, Columbia, MO (joses@missouri.edu). *Corresponding author (psrao@ufl.edu; psrao72@gmail.com). P. Srinivas Rao and K.S. Vinutha contributed equally to this work.

doi:10.2134/agronmonogr58.2014.0074

 Sorghum: State of the Art and Future Perspectives, Ignacio Ciampitti and Vara Prasad, editors. Agronomy Monograph 58.

al., 2008; Promkhambut et al., 2010). It is commonly grown where maize (*Zea mays* L.) and other major cereals are less well adapted, especially because of soil and climate constraints. The plant height ranges from 1.5 to 3.0 m (energy sorghum up to 6.5 m); the optimum temperature for growth and photosynthesis is 32 to 34°C with a day length of 10 to 14 h; the optimum rainfall is 550 to 800 mm (Srinivasarao et al., 2009). With the availability of photoinsensitive genotypes, this crop is becoming popular for meeting food and fuel requirements. Hence, this review discusses the suitability of sorghum as an alternative feedstock for biofuel production and its value-chain sustainability. In the United States, sweet sorghum is primarily grown in the southeastern states and is usually planted between May and July. It was introduced to the United States in 1853 and cultivated for the sweet syrup extracted from its stalk and grain but is also used as a livestock grain and forage crop in the Great Plains states (Dweikat et al., 2012; National Sweet Sorghum Producers and Processors Association, 2015).

Types of Sorghum

Sorghum is a versatile species that is produced as a source of grain and forage. Based on the origin of the crop, five basic cultivated races reflect regional adaptation of sorghum, while 10 intermediate races and six spontaneous races were identified (House, 1985; Harlan and de Wet, 1972, 1971). However, genetic manipulation of grain sorghum has resulted in the development of improved cultivars (Prakasham et al., 2014) of sweet (sugars), low-lignin (*bmr*), and energy (biomass) sorghums. Please refer the Supplemental Material for further details.

Grain Sorghum

Grain sorghum contains either two or three dwarfing genes (Brown et al., 2008) and are 0.61 to 1.22 m tall. While there are several grain sorghum groups, most current grain sorghum hybrids have been developed by crossing *Milo* with *Kafir*. Other groups include *Hegari, Feterita, Durra, Shallu,* and *Kaoliang*. The grain is free-threshing, as the lemma and palea are removed during combining. The seed color is variable with yellow, white, brown, and mixed classes in the grain standards. Brown-seeded types are high in tannins, which lowers the palatability. High starch content, rapid liquefaction, and low viscosity during liquefaction was desired in the grain sorghum for ethanol production. Major factors adversely affecting the bioconversion process of sorghum are tannin content, low protein digestibility, and high mash viscosity.

Sweet Sorghum

Sweet sorghum, similar to grain sorghum except for the sugar and juice-rich stalk, has been grown for syrup and fodder for many centuries. Sweet sorghum is considered to be a potential bioethanol feedstock in addition to meeting food, feed, fodder, fuel, and fiber demands. Some sweet sorghum lines attain juice yields of 78% of total plant biomass, containing 15 to 23% soluble fermentable sugar (Srinivasarao et al., 2009; Vinutha et al., 2014). The sugar is composed mainly of sucrose (70–80%), fructose, and glucose. Most of the sugars are uniformly distributed in the stalk with about 2% in the leaves and inflorescences (Vietor and Miller, 1990), making the crop particularly amenable to direct fermentable sugar extraction. Sweet sorghum, with its structural sugars C_5 and C_6, obtained from cellulose and hemicellulose components of the bagasse could be used both for first- and second-generation biofuel. Ethanol is produced from sweet sorghum juice and is most similar to the molasses-based ethanol production process.

Forage Sorghum

Forage sorghum is used primarily as silage for livestock. It is harvested at the soft dough stage of development and ensiled. It contains 52 to 65% dry matter digestibility, 8 to 12% crude protein, 60 to 75% neutral detergent fiber, and 34 to 40% acid detergent fiber. Ensiled grain has a digestibility of about 90%. The sorghum silage contains less grain and is higher in fiber than corn silage. To obtain the optimum rate of gain for most livestock, sorghum silage must be supplemented with protein, minerals, and vitamins. Young plants contain an alkaloid, which releases hydrocyanic acid or prussic acid when hydrolyzed. This can be toxic to livestock; therefore, livestock feeding needs to be avoided during this stage. When the crop is cut and field-cured, the hydrocyanic acid dissipates. When it is ensiled, the hydrocyanic acid degrades slowly over 2 to 3 wk.

Brown Midrib Sorghum

Brown midrib (*bmr*) mutations are novel mutants in sorghum that usually contain less lignin (<6% in some lines) because of modifications in the phenylproponoid pathway (Saballos et al., 2008). In sorghum, *bmr* mutants were first developed at Purdue University via chemical mutagenesis, and the altered lignin content was characterized by brown vascular tissue (Porter et al., 1978). Introgression of several *bmr* alleles into high biomass and stay green lines was performed, and most of the *bmr* mutants resulted in increased yields of fermentable sugars followed by enzymatic saccharification albeit with varied background effects. Alleles of *bmr*12, *bmr*18, *bmr*26, and *bmr*6 have been characterized at the molecular level (Bout and Vermerris., 2003; Saballos et al., 2009; Sattler et al., 2010). Allelic genes *bmr*12 and *bmr*18 decrease caffeic acid *O*-methyltransferase activity and *bmr*6 has been linked to a decrease in cinnamyl alcohol dehydrogenase activity (Oliver et al., 2005).

Energy Sorghum

Energy sorghum is a forage sorghum bred for high biomass production. Elucidation of the flowering time gene regulatory network and identification of complementary alleles for photoperiod sensitivity enabled large-scale generation of energy sorghum hybrids for testing and commercial use in biofuel production. Energy sorghum hybrids with long vegetative growth phases were found to accumulate more than twice as much biomass as grain sorghum, owing to extended growing seasons, greater light interception, and higher radiation use efficiency. High biomass yield, efficient nitrogen recycling, and preferential accumulation of stem biomass with low nitrogen content contributed to energy sorghum's elevated nitrogen use efficiency. Several biomass sorghum hybrids have been developed and improved for the production of lignocellulosic sugar and as starch feedstock. Biomass yields vary between 15 and 25 t ha^{-1} but have been reported to be as high as 40 t ha^{-1} (Packer and Rooney, 2014).

Biomass Composition

The biomass composition performed with Fibertherm (C. Gerhardt Analytical Systems) has yielded a ballpark figure of the various components like cellulose, hemicellulose, lignin, and ash. The analysis was performed across grain, forage, sweet, brown midrib, and energy sorghum. The samples were collected from experiments conducted in the post–rainy season, 2013 to 2014 at ICRISAT. The biomass samples were dried, ground, and sieved (0.2 mm) to obtain uniform sized samples. The hemicellulose content ranged from 22.37 to 30.15%, cellulose from 34.32 to 39.16%, lignin from 4.63 to 16.82%, and ash from 1.83 to 3.2%. The hemicellulose content of *bmr* and energy sorghum are 29.2 and 30.15%, whereas the lignin content is lowest in *bmr* sorghum at 4.63%, and the energy sorghum has a lignin of 10.25%. Explicitly, the cellulose content is reported higher in *bmr* sorghum (39.16%) than in energy sorghum (37.55%) (Fig. 1).

Sweet sorghum plays a dual role in the bioethanol value chain. Once the juice is extracted for fermentation, the bagasse remaining can be used for the lignocellulosic conversion, thus increasing the ethanol yield and achieving complete consumption of raw material. The sweet sorghum has 24.87, 34.32, and 8.39% of hemicellulose, cellulose, and lignin, respectively. Thus, *bmr* and energy sorghum with high potential biomass yield can be used for bioethanol production. Sweet sorghum can also be promoted as a complementary feedstock in the sugar mill areas where sugarcane (*Saccharum officinarum* L.) is the popular feedstock (Srinivasarao et al., 2012).

Bioethanol from Sorghum: Current Scenario

Sweet sorghum is valued as one of the promising nonfood feedstocks for bioethanol production. Bioethanol from sweet sorghum is regarded as a 1.5-generation biofuel, that is, derived from a food crop part not used for consumption (Li et al., 2013; Wang et al., 2014a). The fuel produced from the sweet sorghum juice is relatively clean with a low production cost. An ethanol yield based on total sugar of

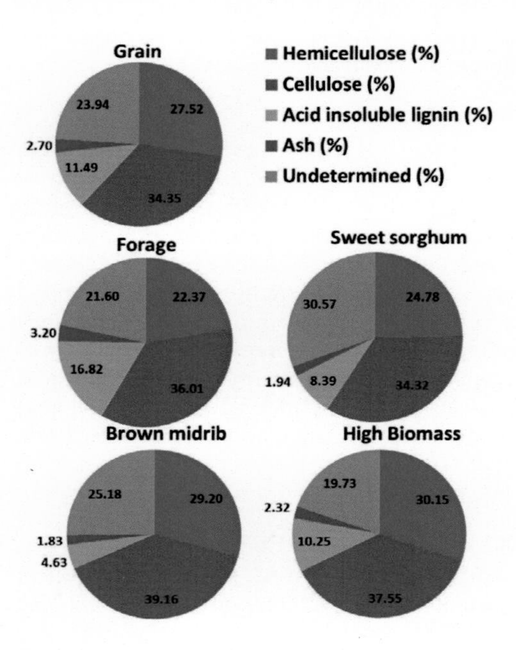

Fig. 1. Biomass composition of different types of sorghum analyzed in the Fibertherm.

480 g kg^{-1} was obtained after 24 h of fermentation using a mixed culture of organisms. This shows the potential for producing as much as 0.252 m^3 t^{-1} or 33 m^3 ha^{-1} ethanol using only the lignocellulose part of the sweet sorghum stalks. This yield is high enough to make the process economically attractive (Marx et al., 2014). The average ethanol productivity was ~220 g ethanol kg^{-1} of original dry stem of sweet sorghum, equivalent to 2465 L ethanol ha^{-1} (Cifuentes et al., 2014). It can be a supplementary feedstock to the sugar mills operating with the distilleries in the off season (2 mo), development of average revenue of US$3 million for a crushing rate of 6500 t d^{-1} can be achieved in dryland situations. Several studies have highlighted the potential of sweet sorghum for ethanol production (Srinivasarao et al., 2009), and it was reported as a viable feedstock for electricity production (Cutz et al., 2013).

Energy sorghums are being developed solely as an energy crop. Because of its height (3–6 m) and a long vegetative growth phase, it has with the advantage of greater light interception and higher radiation use efficiency. The biomass accumulated in energy sorghum is double that accumulated in grain sorghum. Theoretical ethanol yields of potential lines of energy sorghum are ~25% higher than the current yields. Biomass sorghums have the potential to produce high tonnages of C_5 and C_6 sugars and lignin. The biomass yield varies between 15 and 25 t ha^{-1}, and as high as 40 t ha^{-1} has been reported (Packer and Rooney, 2014). The yield of energy sorghum hybrids ranged from 27.2 to 32.4 t ha^{-1} with a wide variety of parental sources (Packer and Rooney, 2014).

The main dispute on second-generation bioethanol is the cost for the establishment of the enzymatic conversion of biomass to bioethanol (Giarola et al., 2012). The major hurdle in second-generation biofuel production is the intricate conversion of biomass to biofuel, pretreatment for removal of lignin, and the enzymes used for saccharification. During the conversion process of biomass to biofuel, the presence of lignin is not a hindrance for biohydrogen production by anaerobic fermentation. However, in the enzymatic conversion process to bioethanol, the effect of lignin is very significant (Prakasham et al., 2012). Hence, to reduce the pretreatment cost, conventional or mutational breeding methods can be used for developing lines possessing low lignin. In sorghum, *bmr* mutants are promising because of the defective genes in the phenylproponoid pathway resulting lower lignin accumulation in secondary cell walls (<6%). Deployment of introgression techniques for *bmr* genes into sweet or energy sorghum to develop biomass with low lignin content is being explored widely. These new introgressed lines are expected to catalyze the development of biofuel industries and make the conversion economically viable. The biomass yield and quality are of prime concern for the viability of commercial biofuel industries.

Sorghum as Raw Material for Industrial Products

Crystal Sugar

Sweet sorghum juice is a rich source of sucrose (85% sucrose, 9% glucose, and 6% fructose) that could effectively be used for the production of crystalline white sugar (Woods, 2000). However, the presence of several inorganic components and impurities need to be eliminated before starting crystalline sugar production. The juice is clarified with liming followed by saturation with carbonation to capture the impurities through precipitation of the lime milk. The thin purified juice obtained after the filtration process is thickened in a multieffect evaporator. The thin juice is diluted with water during extraction and purification and then passes to the evaporating station with an average sugar content of 15%. The thick juice leaving the evaporator contains ~70% sugar (Kangama and Rumei, 2005). White crystalline sugar production appears to be limited to lab scale but has not been commercialized anywhere.

Lipids

The sorghum bagasse remaining after juice extraction can be an excellent source for the production of various valuable bioproducts such as lipids produced by microbial fermentation. Some of the microbial candidates are referred to as oleaginous because of their high cellular lipid content (>20% w/w) accumulation (Ratledge and Wynn, 2002). Among different species, *Cryptococcus curvatus* is one of the most efficient candidates for microbial fermentation, which can accumulate storage lipids up to 60% of dry cell weight (Meng et al., 2009). This microbe can grow on a wide range of mono- and disaccharides (Glatz et al., 1984). Since sorghum straw is a good source for cellulose and hemicellulose sugars, it can be used for the production of single-cell oil.

Biodegradable Plastics

Biodegradable plastics could be made using lignin derived from sorghum biomass (Ashori, 2008). Lignin is a by-product of bioethanol production. Furthermore, sorghum cellulose fiber can be used for reinforcement of thermoplastic materials such as biodegradable wood composites (Ashori, 2008). Sorghum fiber, the residue that is obtained after pretreatment, can be used for the reinforcement of biodegradable composites. The term wood composite refers to any composite that contains plant fibers (wood and nonwood based) and thermosets or thermoplastics (Ashori, 2008). Wood composites are potential green materials, as nontoxic chemicals are employed in their manufacture. Furthermore, the dimensional stability of these materials is more advanced than the traditional wood products. The sorghum fibers can be mixed with polylactic acid, and the resulting composite can be used as a completely biodegradable matrix (Satyanarayana et al., 2009).

Beverage and Dietary Products

Sorghum is the primary food grain in parts of India and Africa, where it is mainly used in making bread, porridges, and opaque alcoholic beers (dregs) (Rooney, 1967; Serna-Saldivar et al., 1988; Bello et al., 1990; Mohammed et al., 1993). Though sorghum has been used for centuries to brew traditional (opaque) beer in Africa (Faparusi, 1970), it has only been recently that sorghum beer brewing has been established into a major industry. The types of beers differ from European (lager) in that lactic acid fermentation also occurs during sorghum beer processing. The sorghum alcoholic drink is consumed while still fermenting and contains large amounts of insoluble materials (Serna-Saldivar et al., 1988), which are mainly starch fragments and dextrins that are not digested during mashing and fermentation (Glennie and Wight, 1986). Ting is a spontaneously fermented sorghum food that is popular for its sour taste and unique flavor. Insight of the microbial diversity and population dynamics during sorghum fermentations is an essential component of the development of starter cultures for commercial production of ting.

Sorghum grains are gluten free and with a high potential to be used as an alternative to wheat flour for the celiac sprue market (Liu et al., 2012). In fact, sorghum grains are minor cereals that form the staple food for a large segment of the population in India and Africa. Use of sorghum for food is still mostly confined to the traditional consumers and populations in lower economic strata, partly because of nonavailability of these grains in ready-to-eat forms. Sorghum is not only nutritionally comparable but is also superior to major cereals with respect to protein, energy, vitamins, and minerals (McKevith, 2004). Moreover, it is rich in dietary fiber, phytochemicals, and micronutrients (Chakraborty et al., 2011).

Biofuel

The production of ethanol from the sugar-rich sweet juice is a common process and has been commercialized in Brazil for the past four decades. The alcohol concentration rises from 6 to 7% (v/v) in the last fermentation step where the temperature is kept between 33 and 35°C. Yeast cream is separated by centrifuges into holding tanks, and clarified juice from the separators is fed into the fermentation buffer tank. Ethanol is then recovered from the fermentation broth by distillation and dehydration. Anhydrous ethanol is realized by passing through a distillation column and a rectification column coupled with vapor-phase molecular sieves in which a mixture of nearly azeotropic water and ethanol is purified. The ethanol yield is 50 to 65 L t^{-1} from the sweet juice process.

The major components of sorghum bagasse are cellulose, hemicellulose, and lignin. Thus, like any other lignocellulosic biomass, sorghum biomass can also be converted to fuel ethanol. For this purpose, the key elements involved are pretreatment and enzymatic saccharification to liberate fermentable sugars. Several pretreatment strategies have been reported including sulfuric acid, hydrochloric acid, phosphoric acid, steam, dilute ammonia hydroxide, ammonia fiber explosion, and hot water (Menon and Rao, 2012). However, a conclusion has not been reached on the best pretreatment technology for industrial-scale ethanol production because of the high cost of downstream processing of fermentable sugars, cellulase enzymes, etc. The produced fermentable sugars can also be used for the production of biohydrogen by using various microbial consortia (Prakasham et al., 2012). The fermenting microbial candidates for the production of various potential industrial commodities are furnished in Table 1.

Paper Pulp

After the extraction of juice sugars in conditions adaptable to the industrial scale, sweet sorghum bagasse can be used for manufacturing paper pulp. The quality of the pulp obtained is similar as regular softwood used for paper making. Research reports suggest that the sorghum pulp exhibits a degree of cohesion higher than 80%; a low kappa number, indicating a good delignification; a high degree of polymerization; and exceptional physicomechanical properties. This allows the consideration of sweet sorghum as a major raw material for the paper industry in every region where it will be possible to grow it in association with sugarcane. These pulps can be used in sectors usually restricted to superior chemical pulps such as those obtained from softwood (Belayachi and Delmas, 1995).

Table 1. Fermenting microbial organisms used for production of various products using sorghum biomass (Prakasham et al., 2014).

Product	Microorganism	Reference
Bioethanol	*Saccharomyces cerevisiae*	Wu et al., 2010; Bridgers et al., 2011; Ostovareh et al., 2015
Acetone	*Bacillus acetobutylicum*	Cheng et al., 2008
Butanol	*Bacillus acetobutylicum*, *Clostridium acetobutylicum*	Cheng et al., 2008; Zhang et al., 2011
Biohydrogen	*Caldicellulosiruptor saccharolyticus*, *Ruminococcus albus*, *Clostridia* spp.	Antonopoulou et al., 2008; Ntaikou et al., 2010; Panagiotopoulos et al., 2010; Saraphirom and Reungsang 2010; Prakasham et al., 2012
Lactic acid	*Lactobacillus delbruckii*, *L. paracasei*, *L. plantarum*	Samuel et al., 1980; Richter and Berthold; 1998; Hetényi et al., 2010; Yadav et al., 2011
Lipids	*Chlorella protothecoides*, *Mortierella isabellina*, *Cryptococcus curvatus*, *Schizochytrium limacinum*	Economou et al., 2010; Gao et al., 2010; Liang et al., 2010, 2012
Methane	*Ubiquitous microflora*	Jerger et al., 1987; Klimiuk et al., 2010; Matsakas et al., 2014
Enzymes: cellulases, xylanases	*Trichoderma* spp., *Aspergillus* spp.	Uma, unpublished data, 2015

Biofuel Production Models

The Centralized and Decentralized Unit Ethanol Production Models

The biofuel production value-chain model for sweet sorghum encompasses the stalk production, transportation, crushing, ethanol production, and thus blending with gasoline. In this section two primary models are examined for the success of the value-chain production model of sorghum feedstock bioethanol. The fundamental capabilities are compared for a centralized unit (CU) model and a decentralized unit (DCU) model in Table 2.

Table 2. Characteristics of centralized unit (CU) and decentralized unit (DCU) biofuel production models (Belum et al., 2013).

Model	CU (Fig. 2)	DCU (Fig. 3)
Description	A constellation of villages closer to the crushing units are aimed for the cultivation of crops along with the transportation of the sweet sorghum stalks to the crushing units. In CU, the requirement of feedstock for 40 kL d^{-1} is 8000 ha crop area yr^{-1} for two seasons for ethanol distillery.	Involves the crushing of sweet sorghum stalks, extracting and boiling the juice for syrup production. The sweet sorghum is managed by a farmers group or microentrepreneurs under the DCU located in the villages that are more than 50 km distance from the distilleries.
Advantages	Supply of the sweet sorghum stalks directly from the farmer's field to the distillery takes little time, keeping the transportation costs minimal.	This model links farmers to input supply agencies including credit or financial institutions and output markets through capacity building. This model also assures links between the DCU and the distillery with a buyback agreement for syrup supply on a preagreed price.
Disadvantages	The farmer located more than 50 km radius has to bear high charges for transporting the feedstock. Delay in the supply of the feedstock because of a longer distance would further reduce the sugar stalk weight and juice yield. During postrainy cultivation season, it is difficult to find 4500 ha land with irrigation facilities.	For establishment of DCU, high initial investment is required. For procuring small quantities of syrup from DCU, assurance has to be given by distilleries alone. For industry payment, a schedule and price fixation criterion is syrup. Payments to farmers to be based on either stalk weight or syrup quality.

Sweet Sorghum Ethanol Value Chain: A Strength–Weakness–Opportunities–Threat Analysis of Centralized and Decentralized Unit Ethanol Production Models

Ethanol production from sweet sorghum has emerged as an alternative to fossil fuels. Sweet sorghum stalks are converted into juice and fermented to produce bioethanol. Two pilot models are tested for conversion of sweet sorghum into bioethanol. In the CU model (Fig. 2), the stalks are supplied directly to the distillery, and in DCU model (Fig. 3), the stalks are processed into syrup at the village level and then transported to the CU model distillery for ethanol production. A

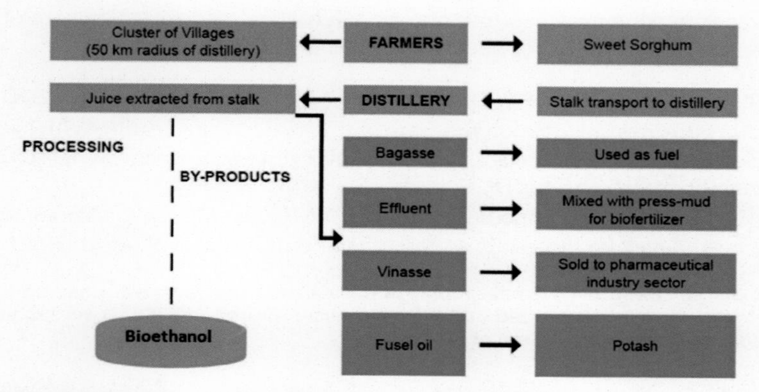

Fig. 2. Centralized unit of the sweet sorghum supply chain linking farmers to the distillery (Basavaraj et al., 2012).

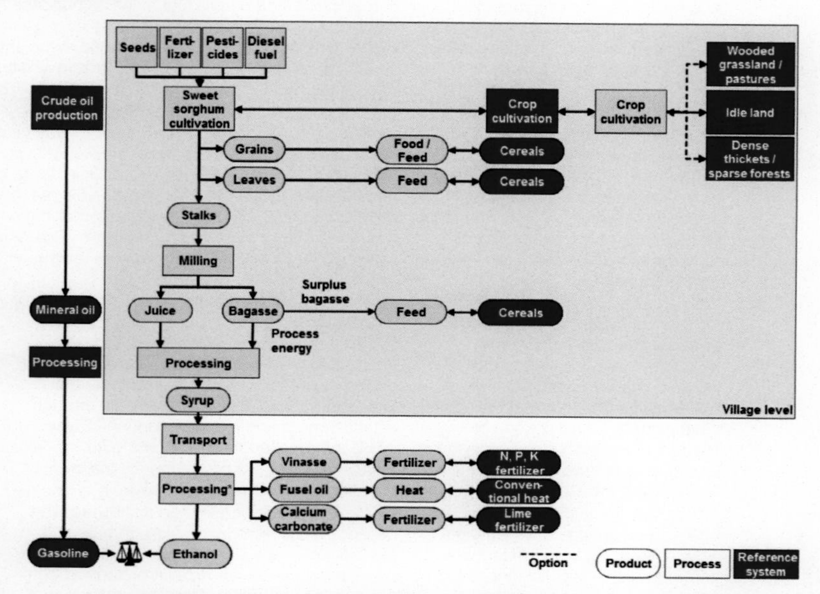

Fig. 3. Decentralized unit: A group of farmers or villages crush sweet sorghum stalks and produce syrup; the decentralized unit is linked with a centralized unit for ethanol production (Srinivasarao et al., 2013; Reinhardt and Cornelius, 2014).

strength–weakness–opportunities–threat (SWOT) analysis was performed on both CU and DCU models. Several attempts were made to use sweet sorghum for ethanol production in a CU model by crushing stalks for juice production at industry level (Srinivasarao et al., 2013). Because of the restriction of the availability of raw materials for a shorter time period, along with difficulties in timely transporting the harvested material from the farmer's field to the distillery, this model could not take off. But, sweet sorghum to ethanol value chain in DCU was also proved to have a better outlook on the conversion of the juice to syrup and further to ethanol in the CU (Basavaraj et al., 2013; Belum et al., 2013). During the implementation of the aforementioned two models, a number of issues have emerged in using sweet sorghum as an alternative feedstock. Thus, the SWOT analysis was performed for both models (Table 3).

Business Model for the Viability of Sweet Sorghum Decentralized Unit

The Agri-Business Incubation program of ICRISAT, in partnership with National Agriculture Innovation Project, has considered various business models based on the viability of the various DCU models (Table 4).

Cost is the major impedance component for the DC unit. Various models have been developed to establish the crushing unit that will offset investments.

Table 3. The strength–weakness–opportunities–threat analysis performed on the two developed models, centralized unit (CU) and decentralized unit (DCU), for biofuel production from sorghum (Belum et al., 2013).

Model	CU	DCU
Strengths	Targeted crop cultivation in the vicinity of CU improves juice recovery and minimizes the transportation cost of the stalks to the CU. Crushes the raw material in bulk quantities before in-stalk fermentation sets in.	Crushing of the stalks without delay resulting in good quality juice yield. Reduces transportation of feedstock. The by-product, such as bagasse, can be used for livestock feed and as an organic matter to enrich the soil. The produced syrup can be stored as long as 24 mo before its conversion to ethanol.
Weaknesses	The CU needs to be supplied with raw material throughout the year, which is neither feasible nor economical for the agroprocessors to run the processing unit to full capacity based on a single feedstock. Constraints for the availability of required land to produce raw material for whole year.	Lack of coordination and implementation of strategies. Broadcasting of information by way of initial training programs or awareness camps is expensive. Lack of management skills in operating and handling DCU. The costs involved in the establishment of DCU and further processing and operating is very high. Incompetence of DCUs in providing large quantities of syrup as raw material.
Opportunities	CU can enter into contractual agreements with the small farmers to overcome land constraints by contract farming or buyback agreements. Production activities must be seen as a part of the whole chain supply. Direct linkage of farmers to the markets will provide a long-term relationship between the farmers and the stakeholders.	Mechanize and standardize most of the processing activities. It adds value for other by-products that can be made available for alternative markets. This model provides opportunities for low-income group farmers to become microlevel entrepreneurs through establishment and management of DCU without any dependence on the partners. This model also makes possible the government's mandate of blending ethanol with gasoline.
Threats	The government can strengthen the efficiency of linkages by framing a policy benefiting both the farmers and the processors. It requires high volumes of stalk and transportation costs are high.	DCU is hugely dependent on distilleries. The cost of producing syrup is high, threatening the feasibility of a DCU. Syrup production can be affected due to labor. Lack of support from the government in implementing policies that would benefit the ethanol sector. A proper policy alignment is required for the production of fuel ethanol for the success of an ethanol program

Table 4. The five business models developed for the viability of sweet sorghum decentralized unit (DCU) (Belum et al., 2013).

Model parameters	Approach to business model	Model assumptions
1. DCU as a standalone unit	Operated by individuals or group.	Not sustainable, as negative return of investment at −7.9% was observed in ICRISAT study (Rao et al., 2013).
2. DCU established and managed by the centralized unit (CU)	As the distillery cannot be run beyond 4 mo, it may establish multiple DCUs and store the syrup to operate for the rest of the year.	All the DCUs will produce syrup consistently to feed the distillery.
3. DCU as an alternative for crushing unit of distillery outsourced to jaggery units	Crushing and syrup making is outsourced to local jaggery (traditional noncentrifugal cane sugar consumed in Asia and Africa) units and procured from them.	A viable option for the success of the DCU and CU.
4. DCU as a franchising miniethanol manufacturing unit	The sweet sorghum growers were organized into an association that functions as a united entity for crop cultivation, marketing, and management of DCU. Licensing mechanism may not be permitted in some countries.	Viable and useful at the village level with the miniprocessing plant. Easy management for the distillery and enhanced benefits to entrepreneurs.
5. DCU supplying syrup to alternative markets (food, ready-to-serve [RTS], confectionery, vinegar)	Setting up of processing units in micro, small, and medium levels for food, RTS, vinegar, and confectionary production.	Micro-, small-, and medium-level entrepreneurs will benefit by manufacturing nonethanol products.

Operating the DCU as a simple crushing unit and selling the syrup is one option, while outsourcing the crushing to other jaggery units might be considered to offset costs. Whichever option is selected, the distillery carries out the fermentation and distillation process to produce ethanol (Karuppanchetty and Selvaraj, 2013).

Sustainability

It is a well-known fact that many nations are facilitating biofuel production in a way that it must not compete with grain over land, it must not compete with food that population demands, it must not compete with feed for livestock, and it must not inflict harm on the environment, which are cornerstones of sustainability.

Environmental Assessment

End users are increasingly demanding sustainability and emissions accounting of sourced biomass used for biofuel production (Lal et al., 2015). In this backdrop, greenhouse gas (GHG) emissions and land-use change impacts are gaining traction as instruments for ensuring sustainability of bioenergy products and processes. This subsection briefly outlines environmental sustainability and associated complexities pertaining to sorghum-based biofuel production. The reduction of GHG emissions, like CO_2, CH_4, N_2O in land-use changes in terms of the use of grasslands or forests for biofuel feedstock production, is a critical factor in evaluating the agroenvironmental impact. Sorghum biomass as second-generation feedstock have a greater potential for positive environmental outcomes relative to sweet sorghum sugar or starch-based first-generation biofuel production. However, current production levels of second-generation biofuels are negligible.

Land Use, Land-Use Change and Carbon Storage

Economic policies and social pressures are used to assess the land-use competition for food or feedstock production for biofuel. Any crop performs better in fertile land than in marginal lands. Energy crops like cassava (*Manihot esculenta* Crantz), *Jatropha* L., and sweet sorghum can grow on marginal lands. Under better management practices, the sweet sorghum cultivation results in higher C stocks and a competition for food production. Thus, cultivation of sweet sorghum in the rainfed areas can provide food and income to farmers while they use their marginal land for bioethanol production (Basavaraj et al., 2013). In the last decade, the grain sorghum area cultivated during the rainy season in Maharashtra, India, has declined, as farmers chose to grow sugarcane. One positive example out of the limited contradictory studies on land-use change is that a study undertaken by the Interdisciplinary Center for Energy Planning, at the University of Campinas, São Paulo, concluded that Brazil could supply ethanol to substitute 5 and even 10% of its projected global gasoline use by 2025 without negatively affecting either the environment or food production (de Cerqueria Leite et al., 2009). Unfortunately, these kind of studies are not available for many countries. One has to mention, however, that using abandoned or marginal lands for bioenergy would avoid food impacts but not necessarily avoid the negative C impacts at times.

Biological Diversity and Soil Quality

The diversity in the agricultural system and the income security of small-scale farmers can be flourished by introducing sorghum in the existing cropping systems. Sorghum is pliable for various cropping systems: mono-, double, sequential, relay, strip, and intercropping or cultivation in fallow lands and crop ratooning. However, monocropping of sorghum, rather than integrating into existing diversified agricultural systems, may possibly lead to a loss of biodiversity and could be detrimental to ecosystems (Köppen et al., 2009). Further, in double cropping the input demands and the crop cycle is reduced, which might lead to lower agrobiodiversity. Whereas, by intercropping and crop rotation, eventual dwindling of the soil organic matter, and thus soil fertility, can be controlled. Integrated pest management, no-till, organic methods, and a general reduction of chemical inputs increase the agrobiodiversity of sorghum. This makes sweet sorghum cultivation more sustainable than other ethanol energy crop cultivation (Aziz et al., 2013). In tropical countries, sweet sorghum ratoons are harvested (Schaffert, 2007), reducing the need for field preparation and double use of machinery as crops other than sweet sorghum require. Hence, the nonsorghum cropping system has a higher impact on soil nutrient composition and soil erosion. Though frequent field work is essential for sorghum, as a dryland resilient crop, optimum soil conditioning with good arability helps to absorb the rain or irrigation water.

Greenhouse Gas and Other Emissions

Bioethanol production from sweet sorghum will alleviate the pressure on the use of the fossil fuel reserve and reduce the emission of GHGs in a standard scenario as explained by Köppen et al. (2009). All three products of sweet sorghum, grains, juice, and bagasse, are used for power generation (Fig. 4). Nevertheless, changes in cultivation practices, in general, to zero tillage, over intense mechanization and a shift to organic farming would further reduce the GHG emissions. The use of sweet sorghum juice from a hectare of crop, can save up to 2300 L of

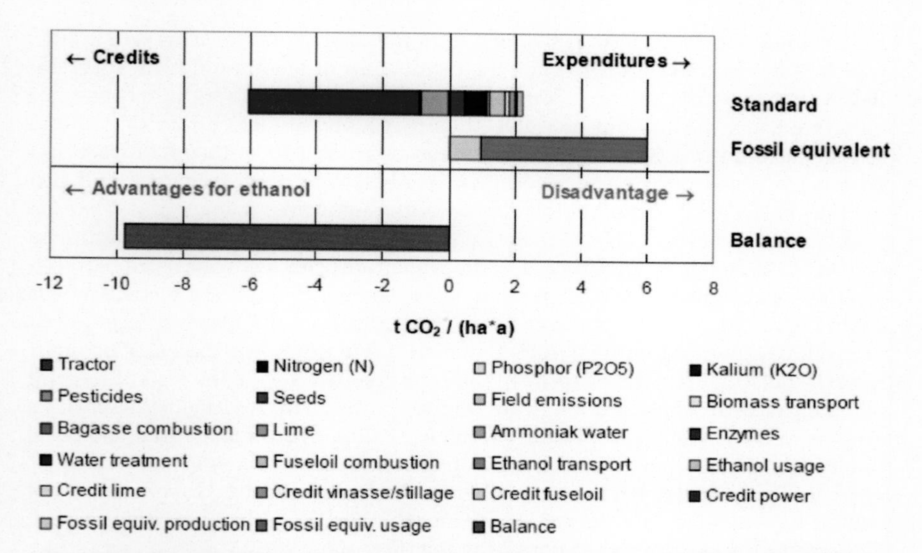

Fig. 4. Detailed greenhouse gas balance for the standard scenario as explained by Köppen et al. (2009).

crude oil and reduce GHG emissions by 1.4 to 22 kg of CO_2 (Köppen et al., 2014). Under the circumstances where the grain is used as food and in syrup production, the remaining bagasse can be used for power generation to compensate the GHG production. Cultivars of sweet sorghum with higher fermentable sugar and juice yield can be used more efficiently to produce ethanol when compared with corn and grain sorghum ethanol. A shift to biomass as a source for production of biofuel, as promoted by many countries, is to frontier the emission of GHG (Gerbens-Leenes et al., 2009). In comparison with the other feedstocks, like corn, the use of sweet sorghum has more competitiveness, though the gains are small in terms of agronomic and environmental benefits. Furthermore, the production of green electricity rather than biofuel from sweet sorghum bagasse is more advantageous.

Water Use Efficiency, Quality, and Footprint

Sorghum offers several advantages over traditional biofuel feedstocks, such as corn, owing to its lower input requirements for water and N fertilizers, its ability to withstand arid and drought-like environments, and its capability to grow in marginal conditions (Köppen et al., 2009). Furthermore, sorghum has an annual growth cycle, which is attractive for cultivators who are averse to long-term commitments in perennial feedstocks like energy grasses (Dweikat et al., 2012). The water requirements of biofuel production depend on the type of feedstock used and on geographic and climatic variables. These factors must be considered to determine water footprint and to identify critical scenarios as well as mitigation strategies. In India, sugarcane is mainly cultivated under irrigation, unlike Brazil, where rainfed sugarcane is in vogue, while sweet sorghum and energy sorghum are grown largely in rainfed conditions in most of the countries. The mechanism of better water use efficiency to convert unit amount of water to biomass has attributed to its drought-resistant and hardy characteristics. This will in turn lead to a lower water footprint for the production of bioethanol from sorghum than from sugarcane or corn. Compared to sugarcane, sweet sorghum consumes only one-third the amount of water, thus it is more suitable under rained and semiarid conditions. Sweet sorghum consumes around 1000–1500 L of water per crop cycle based on the water footprint studies of various feedstock and production pathways for biofuel, compared with sugarcane, switchgrass (*Panicum virgatum* L.), wheat (*Triticum aestivum* L.) straw, or corn under irrigated conditions (Wu et al., 2014; Fig. 5). Differences in the water footprint were observed between locations as a consequence of climatic variables even though the feedstock was common across regions. In the case of sweet sorghum, the blue water required to produce a liter of biofuel is 1000 L, which is five times less than that of sugarcane (Wu et al., 2014). The purpose of the crop does not decide the water footprint whether used for consumption or for energy production. A rice (*Oryza sativa* L.) crop has a lower water footprint in producing a unit amount of ethanol, biodiesel, or electricity than energy-dedicated crops like rapeseed (*Brassica napus* L.) or *Jatropha*. The principal debate as to whether food crops can be used for energy or not should be extended to a discussion about whether limited water resources are dedicated for food or for energy production (Roy et al., 2012).

Life Cycle Assessments and Emissions Impact

Life cycle assessments (LCAs) can be used to evaluate potential environmental impacts of biofuel production, resource levels, and public health. The LCA for energy ratio and emissions, resource level efficiency, and delineating public health impact can include all stages from feedstock production until the use of biofuel by the final consumer. This method determines the environmental suitability of a crop and associated emissions. The assessment for production of bioethanol from sweet sorghum stem juice in China revealed that energy efficiency and environmental performance show a positive net energy ratio of 1.56 and 8.37 MJ L^{-1} as the net energy gain from the production of biomass to its conversion to fuel (Wang et al., 2014b). If ethanol from sorghum is used supplemental to fossil fuels or completely substituted, 10 t ha^{-1} yr^{-1} of CO_2 equivalents of GHG

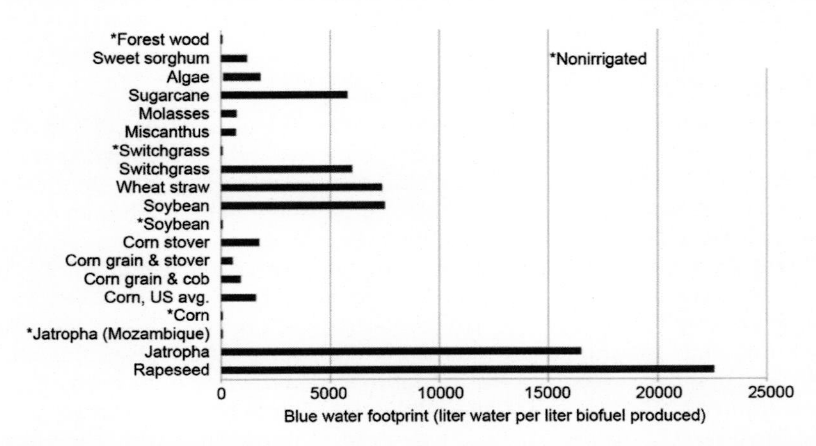

Fig. 5. Blue water foot print of various crops under irrigated and nonirrigated conditions (Wu et al., 2014).

can be saved. The GHG savings can be primarily attributed to high biomass yields and better cultivation practices with improved high-yielding varieties or hybrids. The fermentable sugars and juice yield is not highly influential.

Social Assessment

Tenures of Land for Bioenergy Production

The land tenure will be a criterion to check in a nation where land use is very well defined. Competition for fertile lands and increased productivity can be a major issue in the future as a result of the higher demand of biofuel feedstock production as farmers and smallholders tend to shift toward more remunerative options. Even if the cultivation of the energy crops are undertaken in low, marginal, or degraded lands, there is a risk that expansion of energy feedstocks may adversely impact smallholders and lease farmers. The social impact of the large-scale biofuel cultivation in developing countries and its impact on rural livelihoods has not been systematically studied. Shifting from an extensive to an intensive production system that requires highly specialized techniques leads to a technological migration that can adversely impact land occupancy patterns. The migration of small and medium farmers by abandoning agriculture and livestock or leasing out their fertile lands for sugarcane cultivation to biofuel has caused a rise in land prices in developing countries. Hence, sound land tenure policies, planning, and policy interventions will be crucial for future expansion of the bioenergy crops especially in the developing world (Janssen and Rutz, 2011).

Jobs in Bioenergy Sector and Change in Employment Impacts

Over the past decade, the growth of the bioenergy industry has garnered the interest of both public and private sectors, owing to its potential to spur economic activity, create jobs and enhance energy security through reduced dependence on imported fossil fuels, and the possibility of limiting environmental impacts (Lal et al., 2015). From an economic perspective, sorghum-based biofuels offer the potential to supplement a country's energy needs and reduce dependence on fossil fuel imports. Akin to other biofuel feedstocks, cultivation of sorghum is likely to create on-farm employment in rural regions. For example, the South African Biofuels Association aims to foster sustainable income generating opportunities, especially for the marginalized sections of society, through the promotion of its biofuel program (Southern African Biofuels Association, 2007). The rural economies can benefit by generating employment in the agriculture sector and by sharing improved technical knowledge on crop cultivation. Local impact can be improved by developing models such as this integrated food–energy system (Aziz et al., 2013).

Brazil as an example of a successful program, which is characterized by sustainability in the production system by using local renewable raw input while, in turn, enhancing rural employment prospects. Ethanol has environment-friendly characteristics over gasoline, and the production of sugarcane-derived ethanol provides a rural development benefit. The employment opportunities created by the bioenergy sector in developing countries, like India, are expected to be significant by the year 2020, generating ~838,780 jobs (Table 5). The deployment of bioenergy not only has the potential for the job creation but also industrial competitiveness, local development, and a resilient export industry (Domac et al., 2005).

Table 5. Global employment generation from different energy production units: A prediction model adopted from Domac et al. (2005).

Year	2005	2010	2020
Solar thermal heat	4,590	7,390	14,311
Photovoltaics	479	−1,769	10,231
Solar thermal electric	593	649	621
Wind onshore	8,690	20,822	35,211
Wind offshore	530	−7,968	−6,584
Small hydro	−11,391	−995	7,977
Bioenergy	449,928	642,683	838,780
Total	453,418	660,812	900,546

Bioenergy: Access to Modern Energy Services

Modern energy services are very essential constituents in policy-making decisions. Access to modern services is equally important in developing world context such as cooking, lighting, and transportation services. The lack of modern energy services in India, where 364 million people have no provision for electricity and 726 million people use traditional open-flame methods for cooking and heating, is an indicator of policy failure (Balachandra, 2011). The agricultural activities such as tilling, irrigation, and postharvest processing demand huge amounts of energy primarily met by human labor. The rural industry also has energy requirement for milling and processing the products. The agriculture feedstock-based energy production can meet the local demand from households such as cooking, lighting, and heating. Biofuel production could also increase access to energy services with positive effects on human welfare by expanding access to electricity and pumped potable water, improving health by reducing air pollution, and improving conditions of women and children in the developing world by weaning them away from wood fuel or charcoal-based energy production. Bioenergy, such as produced from sorghum, is more advantageous and provides tremendous opportunities to develop and access these modern services (Ejigu, 2008).

Economic Assessment

Economic Sustainability of Biomass and Productivity

Food and energy security demonstrated by sweet sorghum has been mentioned in earlier sections; it is also important to note that with the existing conversion technologies, the crop fits well in the biofuel production cycle with little modifications of machinery for crushing and processing. Further value-added products can also be produced with minimum inputs. The crop can be adjusted into any of the cropping system because of the availability of genotypes with different maturity period. The mechanization of sorghum in cultivation and postharvest handling of biomass and grains can be adapted to a greater extent, though affordability and availability of machinery are a constraint. In the Philippines provinces of Bukidnon (Mindanao), Tarlac, and Pampanga (Luzon, Philippines), large tracts of land that are suitable for producing sweet sorghum are available, thus distilleries around these areas are exploring the possibility of using sweet sorghum as a complementary feedstock (Belum et al., 2011).

Above all, the biomass productivity and conversion unit sustainability depends on financial returns from ethanol prices. The demand management of ethanol includes decisions to cap ethanol supply on considerations of fairness in distribution of ethanol to accommodate the needs of other sectors (potable and industrial). It was also recommended by the policymakers that the size of the ethanol blending program should be linked to the availability of feedstocks (Government of India, 2009; Basavaraj et al., 2012). Lower availability of molasses, and consequently higher prices in countries like India, has also affected the cost of ethanol production, putting ethanol blending programs at stake. Hence, sorghum can be chosen as the best alternative feedstock for biofuel production. Despite several advantages of sweet sorghum as a promising alternative crop for bioethanol production, national policies in developing nations on biofuel do not specify any clear road map for its commercialization and use (Basavaraj et al., 2013).

Profitability and Efficiency

Sorghum has a high net energy balance, and although the ethanol yield per unit weight of feedstock is lower for sweet sorghum than for sugarcane, the lower production costs and water requirement for this crop compensate the yield gap. Hence, sweet sorghum still ends up with a competitive cost advantage in the production of ethanol in countries like India (Rao et al., 2004). Either the price of the main product (grain or biomass) or the by-product (fuel, fusel oil, butanol, or any other product in the value chain) will decide the choice of the crop and the end product to be produced. Alternative uses of the feedstock play a key role in the decision-making process of farmers. If the price of fuel is lower than the other end products, the choice will be to not extract biofuel from the crop. Thus, long-term viability can be achieved from a biofuel system only if the profitability is enhanced by higher efficiency as is required for advanced biofuel system like lignocellulosic biofuel production.

Economic Equity and Net Energy Balance

Sorghum can be cultivated with very low financial resources. Farmers need agricultural land and seeds to grow the crop. The plant can easily be propagated by seed. However, good productivity and efficiency of the cultivation needs input such as human work, energy, fertilizers, and pesticides. Even if the feedstock production can be done at very low cost, considerable financial resources are needed for the further processing steps such as transport, milling, and conversion to ethanol. In general, it can be said that the larger the system is the larger the financial resources required. However, the availability of financial resources is often a key limiting factor especially in developing countries.

For instance, under the grain-to-food scenario, the value of grain significantly reduces the cost of ethanol production after due credit is given to grain value. With the ongoing debate on food vs. fuel, sweet sorghum as a feedstock is found to be economically promising when the grain is used for food and stalk is used for ethanol. Ethanol production using the syrup route is the most uneconomical, while biofuel production from the stalk-only scenario is competitive under the high case. For the syrup route, the extraction of syrup at village level is still not commercially viable, adding to the overall cost of production. In all cases, feedstock costs are the major contributor to the variable production costs of ethanol followed by processing costs, labor, maintenance, and operational costs (Fig. 6). Feedstock costs, however, tend to come down as we move from the low to high case. Ethanol production under the grain-to-food scenario is competitive in all cases (Reinhardt and Cornelius, 2014).

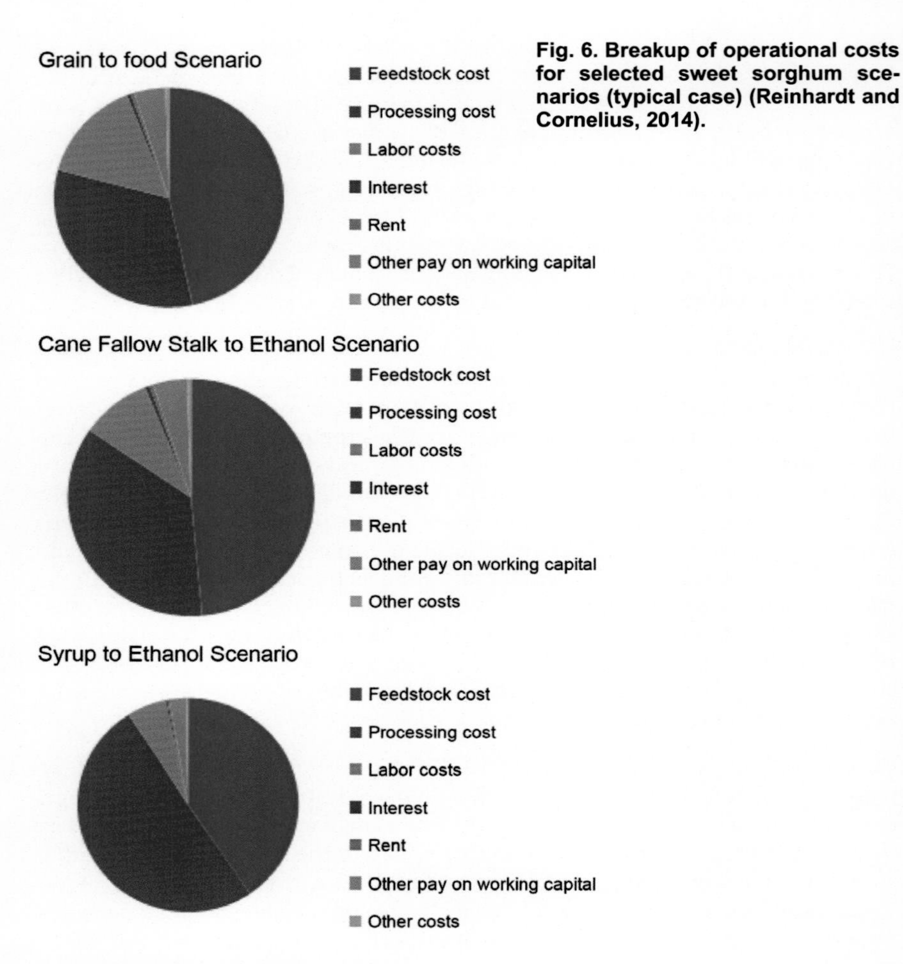

Fig. 6. Breakup of operational costs for selected sweet sorghum scenarios (typical case) (Reinhardt and Cornelius, 2014).

Competition with Food Crops

In comparison with current sugar and starch crops for bioethanol production, sorghum offers important benefits with respect to food security, as it can serve as a multiple-purpose crop used for food, feed, and fuel at the same time. Its seeds are valuable cereals and the leaves are high-value feed, thus contributing significantly to enhancing food supply and improving food security especially in rural areas of developing countries that are prone to food insecurity. In addition to the grain used for human or animal consumption, sweet sorghum accumulates sugars with little competition between grain and sugar production. The bagasse can be used as animal feed and it is reported to have a better nutritional value than the bagasse of sugarcane (Almodares and Hadi, 2009). The production of bioethanol based on traditional food crops may lead to increases of agricultural commodity prices, which negatively affects access to food particularly in net-food importing developing countries and for the poorest therein. Significant price increases have already occurred in major bioethanol feedstock markets such as corn and sugar. Thus, the food vs. fuel conflict can be resolved by framing and implementing proper policies that introduce sustainability criteria, standards, and best management practices. Overall, competition with food is a potentially significant concern when investing in biofuel. The issue is not entirely resolved with second-generation biofuel even if they use nonfood feedstock because of indirect land-use changes and because of the potentially huge market demand for renewable energy in comparison with agriculture. Regulatory approaches that include procedural rules, legislatively prescribed practices, reporting, monitoring, compliance, and enforcement (Ellefson et al., 2004; Lal et al., 2013) could contribute to mitigating this potential fuel vs. food conflict.

Economic Viability Assessments

The economic and financial viability analysis has shown that feasibility of ethanol production from sweet sorghum stalk depends on the ethanol and feedstock pricing in addition to the recovery rate of ethanol. As an illustration, with a marginal improvement in recovery to 4.9% from the current level of 4.5%, and feedstock price fixed in 2012 was at US$20 t^{-1} of stalk, ethanol production became attractive at 50 cents L^{-1} when the administered price of ethanol in India was 48 cents L^{-} (Basavaraj et al., 2012). The sweet-sorghum-to-ethanol scenarios, though positive, throws up mixed results. Stalk plus grain-to-ethanol (cane fallow 2020) and grain-to-food scenarios are economically most viable compared with stalk-only-to-ethanol scenario. For the stalk-plus-grain scenario, internal rates of return (IRR) of 70 and 148% are obtained under the typical (35 t ha^{-1}) and high productivity (>50 t ha^{-1}) scenarios, respectively. For the grain-to-food scenario, the IRRs are marginally lower under the typical and high cases. The stalk-only-to-ethanol scenario (cane fallow) is viable only under the high case with IRR of 25%. The syrup route to ethanol is the most unviable scenario, where syrup is produced at the village level and transported to the distillery for conversion to ethanol. This is because the syrup production at village level is small scale, leading to higher costs of production. By-products generated during crop production and processing stages make an important contribution to economic returns. Among the by-products from sweet sorghum processing for ethanol, the value of surplus bagasse used to generate electricity is the highest followed by excess power, calcium carbonate, and vinasse. For biomass sorghum to biogas the return on investment is positive with IRRs of 24, 44, and 57% under the three cases (low, typical, and high, respectively). Economic feasibility analysis of producing second-generation ethanol from sorghum biomass indicates that processing costs of second-generation ethanol determines its profitability, which in turn, depends on the enzyme price. Bringing down the enzyme price holds the key for the economic viability of second-generation ethanol (http://www.sweetfuel-project.eu/exploitable_results).

Initiatives on Bioenergy Sustainability: An Approach for Long-Term Viability

In recent years, intensive work has been done to develop alternative sources for fossil fuels. Bioenergy is one of the most promising solutions to the depleting fossil fuel reserves on the globe. A multitude of crops and technologies were studied for efficient biofuel production, providing an established model system for adoption and commercialization. While bioenergy is being accepted and the most investigated alternative to fossil fuel, sustainability is an essential requirement for biofuel long-term viability. With the growing adoption and introduction of new crops and technologies, it is imperative to evaluate environmental and socioeconomic impacts. While sorghum is used as a fuel source in industry, it also ranks poorly against fossil fuels in comparison with many other biofuels because of issues such as acidification, eutrophication, photochemical smog, and ozone depletion (Elbehri et al., 2013; Regassa and Wortmann, 2014). Numerous bioenergy sustainability initiatives have been developed over the past few years to address such issues associated with the production of biofuels, which includes regulatory frameworks, voluntary standards, certification schemes, and scorecards (Fig. 7). The value chain embracing these indicators would be ideal for maximizing the socioeconomic and environmental benefits to the regions of operation but specifically to poor small holders and marginal farmers (land holding of <2 ha).

Some sorghum-based bioenergy initiatives were taken up in India (Rusni Distilleries, Tata Chemicals Ltd), China (ZTE Ltd, Jilin Biofuel Ltd), Philippines (Sancarlos Bioenergy Inc., Ecofuels Ltd), Brazil (Embrapa and multinational seed companies working with sugar mills), Columbia (CLAYUCA), and the United States (Southeastern Biofuel Ltd, Chromatin Ltd, BioDimensions Inc., Ceres Energy Inc., etc.). Recently, some other countries, such as Mexico, Indonesia, Mali, Mozambique, South Africa, and Australia, are investing in sweet sorghum research and development. In India, both the initiatives stopped operation for lack of policy support and other concerns related to harvesting and crushing. Other areas of priority are (i) integration of the fermentation and distillation of sweet sorghum juice in corn ethanol plants, (ii) promotion of sweet sorghum as a bioethanol feedstock in existing sugar mills having a distillery, and (iii) lignocellulosic biofuel production from energy sorghum in the recently commercialized biofuel plants, which are dependent mostly on corn stover and agricultural residues.

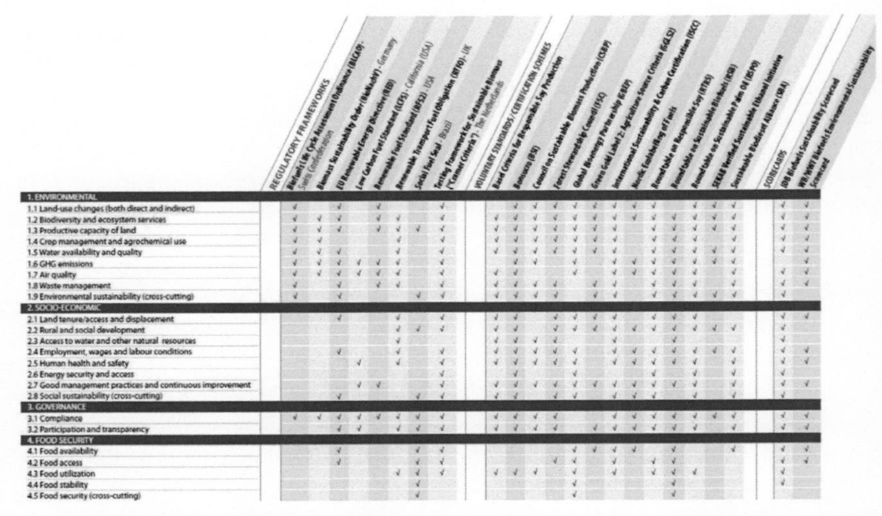

Fig. 7. Bioenergy sustainability initiatives addressing environmental, socioeconomic, governance, and food security aspects or issues (FAO, 2011).

Epilogue

Sorghum grows better than many other crops in marginal and arid climates and produces grain, food, feed, fodder, and biofuel, which makes sorghum the crop of crops as a climate-change-ready option enhancing global food, feed, fodder, and biofuel security potential. It also provides economic stability to farmers by diversifying and channelizing the use of various value-added end products from a single crop. It complies with various biotic and abiotic stress-prone marginal lands and is also a highly reliable crop that grows well in hot, dry environments. Hence, to meet the large-scale biofuel cultivation demand and supply the raw material required for industry the research and development, oriented activities have to be promoted. Development of genotypes suitable for different agroclimatic situations, biotic and abiotic stress tolerances, and photoperiod insensitivities (to make available the biomass during off season in sugarcane cultivating areas), are of prime importance. Optimization of processing technologies for biofuel production will play a key role in reducing input cost and maximizing the benefit with diversified end products. Transitioning sweet sorghum to a bioenergy crop is hindered by lack of technology for large-scale harvest, transport, and storage of the large quantities of biomass and juice produced in addition to in-stalk fermentation as a result of delayed crushing in the tropics. Generating ethanol from lignocellulosic material or energy sorghum through hydrolysis and fermentation has the potential to give very encouraging bioenergy yields in relation to the required fossil energy inputs, but the technology has yet to be fully deployed commercially. The policy support to farmers and processors from respective governments based on regional facts (feedstock production and supply costs, processing, and ethanol handling costs) has to be developed and implemented to derive advantageous biofuel production from a multipurpose crop like sorghum.

Acknowledgments

The authors acknowledge the financial support received from the Department of Biotechnology (DBT), Government of India through Indo–US Joint Clean Energy Development Centre (JCERDC) and development of low-lignin high-biomass sorghums suitable for biofuel production.

References

Almodares, A., and M.R. Hadi. 2009. Production of bioethanol from sweet sorghum: A review. Afr. J. Agric. Res. 4:772–780.

Almodares, A., R. Taheri, M. Chung, and M. Fathi. 2008. The effect of nitrogen and potassium fertilizers on growth parameters and carbohydrate contents of sweet sorghum cultivars. J. Environ. Biol. 29:849–852.

Antonopoulou, G., H. Gavala, I. Skiadas, K. Angelopoulos, and G. Lyberatos. 2008. Biofuels generation from sweet sorghum: Fermentative hydrogen production and anaerobic digestion of the remaining biomass. Bioresour. Technol. 99:110–119. doi:10.1016/j.biortech.2006.11.048

Ashori, A. 2008. Wood-plastic composites as promising green-composites for automotive industries. Bioresour. Technol. 99:4661–4667. doi:10.1016/j.biortech.2007.09.043

Aziz, I., T. Mahmood, and K.R. Islam. 2013. Effect of long term no-till and conventional tillage practices on soil quality. Soil Tillage Res. 131:28–35. doi:10.1016/j.still.2013.03.002

Balachandra, P. 2011. Modern energy access to all in rural India: An integrated implementation strategy. Energy Policy 39:7803–7814. doi:10.1016/j.enpol.2011.09.026

Basavaraj, G., P.P. Rao, C. Ravinder Reddy, A. Ashok Kumar, P. Srinivasa Rao, and B.V.S. Reddy. 2012. A review of the National Biofuel Policy in India: A critique of the need to promote alternative feedstocks. J. Biofuels 3:65–78. doi:10.5958/j.0976-4763.3.2.007

Basavaraj, G., P.P. Rao, L. Achoth, and C.R. Reddy. 2013. Assessing competitiveness of sweet sorghum for ethanol production: A policy analysis matrix approach. Agric. Econ. Res. Rev. 26:31–40.

Belayachi, L. and M. Delmas. 1995. Sweet Sorghum: A quality raw material for the manufacturing of chemical paper pulp. Biomass Bioenergy 8:411–417. doi:10.1016/0961-9534(95)00046-1

Bello, A.B., L.W. Rooney, and R.D. Waniska. 1990. Factors affecting quality of sorghum to a thick porridge. Cereal Chem. 67:20–25.

Belum, V.S.R., A.A. Kumar, C.R. Reddy, P.P. Rao, and J.V. Patil, editors. 2013. Developing a sweet sorghum ethanol value chain. Patancheru 502 324. International Crops Research Institute for the Semi-Arid Tropics, Andhra Pradesh, India.

Belum, V.S.R., H. Layaoen, W.D. Dar, P. Srinivasarao, and J.E. Eusebio, editors. 2011. Sweet sorghum in the Philippines: Status and future. ICRISAT, Patancheru, Andhra Pradesh, India.

Bout, S., and W. Vermerris. 2003. A candidate-gene approach to clone the sorghum *Brown midrib* gene encoding caffeic acid *O*-methyltranseferase. Mol. Genet. Genomics 269:205–214. doi:10.1007/s00438-003-0824-4

Bridgers, E.N., M.S. Chinn, M.W. Veal, and L.F. Stikeleather. 2011. Influence of juice preparations on the fermentability of sweet sorghum. Biol. Eng. Trans. 4:57–67. doi:10.13031/2013.38507

Brown, P.J., W.L. Rooney, C. Franks, and S. Kresovich. 2008. Efficient mapping of plant height quantitative trait loci in a sorghum association population with introgressed dwarfing genes. Genetics 180:629–637. doi:10.1534/genetics.108.092239

Chakraborty, S.K., D.S. Singh, B. Kumbhar, and S. Chakraborty. 2011. Process optimization with respect to the expansion ratios of millet- and legume (pigeon pea)-based extruded snacks. J. Food Process Eng. 34:777–791. doi:10.1111/j.1745-4530.2009.00433.x

Cheng, Y., S. Li, J. Huang, Q. Zhang, and X. Wang. 2008. Production of acetone and butanol by fermentation of sweet sorghum stalks juice. Trans. Chin. Soc. Agric. Eng. 24:177–180.

Cifuentes, R., R. Bressani, and C. Rolz. 2014. The potential of sweet sorghum: As a source of ethanol and protein. Energy Sustainable Dev. 21:13–19. doi:10.1016/j.esd.2014.04.002

Cutz, L., S. Sanchez-Delgado, U. Ruiz-Rivas, and D. Santana. 2013. Bioenergy production in Central America: Integration of sugar mills. Renew. Sustain. Energy Rev. 25:529–542. doi:10.1016/j.rser.2013.05.007

de Cerqueria Leite, R.C., M.R.L.V. Leal, L.A.B. Cortez, W.M. Griffin, and M.I.G. Scandiffio. 2009. Can Brazil replace 5% of the 2025 gasoline world demand with ethanol? Energy 34:655–661. doi:10.1016/j.energy.2008.11.001

Domac, J., K. Richards, and S. Risovic. 2005. Socio-economic drivers in implementing bioenergy projects. Biomass Bioenergy 28:97–106. doi:10.1016/j.biombioe.2004.08.002

Dweikat, I., C. Weil, S. Moose, L. Kochian, N. Mosier, K. Ileleji, and N. Carpita. 2012. Envisioning the transition to a next-generation biofuels industry in the US Midwest. Biofuels, Bioprod. Biorefin. 6:376–386. doi:10.1002/bbb.1342

Economou, C.N., A. Makri, G. Aggelis, S. Pavlou, and D.V. Vayenas. 2010. Semi-solid state fermentation of sweet sorghum for the biotechnological production of single cell oil. Bioresour. Technol. 101:1385–1388. doi:10.1016/j.biortech.2009.09.028

Ejigu, M. 2008. Toward energy and livelihoods security in Africa: Smallholder production and processing of bioenergy as a strategy. Nat. Resour. Forum 32:152–162 doi:10.1111/j.1477-8947.2008.00189.x

Elbehri, A., A. Segerstedt, and L. Pascal. 2013. Biofuels and the sustainability challenge: A global assessment of sustainability issues, trends and policies for biofuels and related feedstocks, Trade and markets division, Food and Agriculture Organization of the United Nations, Rome.

Ellefson, P.V., M.A. Kilgore, C.M. Hibbard, and J.E. Granskog. 2004. Regulation of forestry practices on private land in the United States: Assessment of state agency responsibilities and program effectiveness (No. 176). Dep. of Forest Resources, Univ. of Minnesota, St. Paul.

FAO. 2011. A compilation of bioenergy sustainability initiatives. Bioenergy and Food Security, Food and Agriculture Organization of the United Nations. http://www.fao.org/bioenergy/foodsecurity/befsci/62379/en (accessed 1 Feb. 2016).

Faparusi, S. 1970. Sugar changes during the preparation of Burukutu beer. J. Sci. Food Agric. 21:79–81. doi:10.1002/jsfa.2740210206

Gao, C., Y. Zhai, Y. Ding, and Q. Wu. 2010. Application of sweet sorghum for biodiesel production by heterotrophic microalga *Chlorella protothecoides*. Appl. Energy 87:756–761. doi:10.1016/j.apenergy.2009.09.006

Gerbens-Leenes, W., A.Y. Hoekstra, and T.H. van der Meer. 2009. The water footprint of bioenergy. Proc. Natl. Acad. Sci. USA 106:10219–10223. doi:10.1073/pnas.0812619106

Giarola, S., N. Shah and F. Bezzo. 2012. A comprehensive approach to the design of ethanol supply chains including carbon trading effects. Bioresour. Technol. 107:175–185.

Glatz, B.A., E.G. Hammond, K.H. Hsu, L. Baehman, N. Bati, W. Bednarski, D. Brown and M. Floetenmeyer. 1984. Production and modification of fats and oils by yeast fermentation. In: C. Ratledge, P. Dawson, and J. Rattray, editors, Biotechnology for the oils and fats industry. American Oil Chemists' Society, Champaign, IL. p. 163–176.

Glennie, C.W., and A.W. Wight. 1986. Dextrins in sorghum beer. J. Inst. Brew. 92:384–386. doi:10.1002/j.2050-0416.1986.tb04428.x

Government of India. 2009. Report of the working group on animal husbandry and dairying for the 11th five year plan (2007–2012), Government of India, Planning Commission, Delhi. http://planningcommission.nic.in/aboutus/committee/wrkgrp11/wg11_rpanim.pdf (accessed 1 Feb. 2016).

Harlan, J.R., and J.M. de Wet. 1971. Toward a rational classification of cultivated plants. Taxon 20:509–517. doi:10.2307/1218252

Harlan, J.R., and J.M.J. de Wet. 1972. A simplified classification of cultivated sorghum. Crop Sci. 12:172–176. doi:10.2135/cropsci1972.0011183X001200020005x

Hetényi, K., K. Gál, Á. Németh, and B. Sevella. 2010. Use of sweet sorghum juice for lactic acid fermentation: Preliminary steps in a process optimization. J. Chem. Technol. Biotechnol. 85:872–877. doi:10.1002/jctb.2381

House, L.R. 1985. A guide to sorghum breeding, 2nd ed. ICRISAT, Andhra Pradesh , India.

Janssen, R., and D.D. Rutz. 2011. Sustainability of biofuels in Latin America: Risks and opportunities. Energy Policy 39:5717–5725. doi:10.1016/j.enpol.2011.01.047

Jerger, D.E., D.P. Chynoweth, and H.R. Isaacson. 1987. Anaerobic digestion of sorghum biomass. Biomass 14:99–113. doi:10.1016/0144-4565(87)90013-8

Kangama, C.O., and X. Rumei. 2005. Production of crystal sugar and alcohol from sweet sorghum. Afr. J. Food Agric. Nutr. Dev. 5:1–5.

Karuppanchetty, S.M., and A. Selvaraj. 2013. Business models for viability of sweet sorghum decentralized crushing unit. In: B.V.S. Reddy, A.A. Kumar, C.R. Reddy, P.P. Rao, and J.V. Patil, editors, Developing a sweet sorghum ethanol value chain. ICRISAT. Andhra Pradesh, India. p. 193–196.

Klimiuk, E., T. Pokój, W. Budzynski, and B. Dubis. 2010. Theoretical and observed biogas production from plant biomass of different fibre contents. Bioresour. Technol. 101:9527–9535. doi:10.1016/j.biortech.2010.06.130

Köppen, S., H. Fehrenbach, S. Markwardt, A. Hannecke, U. Eppler, and U.R. Fritsche. 2014. Final report on implementing the GBEP indicators for sustainable bioenergy in Germany. Institut für Energie- und Umweltforschung (Heidelberg) and International Institute for Sustainability Analysis and Strategy, Darmstadt, Germany.

Köppen, S., G. Reinhardt, and S. Gärtner. 2009. Assessment of energy and greenhouse gas inventories of Sweet Sorghum for first and second generation bioethanol. In FAO Environmental and Natural Resources Service Series No. 30. FAO, Rome.

Lal, P., P. Burli, and J.R.R. Alavalapati. 2015. Policy mechanisms to implement and support biomass and biofuel projects in United States. In: S. Jose and T. Bhaskar, editors,

Biomass and biofuels: Advanced biorefineries for sustainable production and distribution. CRC Press, Boca Raton, FL. p. 279–301

Lal, P., T. Upadhyay, and J.R.R. Alavlapati. 2013. Woody biomass for bioenergy: A policy overview. In: J.M. Evans, R.J. Fletcher, J.R.R. Alavalapati, A.L. Smith, D. Geller, P. Lal, D. Vasudev, M. Acevedo, J. Calabria, and T. Upadhyay, editors, Forestry bioenergy in the Southeast United States: Implications for wildlife habitat and biodiversity. National Wildlife Federation, Merriðeld, VA. p. 248–255. http://www.nwf.org/~/media/PDFs/Wildlife/Conservation/NWF_Biomass_Biodiversity_Final.ashx (accessed 21 Mar. 2015).

Legwaila, G.M., T.V. Balole, and S.K. Karikari. 2003. Review of sweet sorghum: A potential cash and forage crop in Botswana. UNISWA J. Agric. 12:5–14.

Li, S., G. Li, L. Zhang, Z. Zhou, B. Han, W. Hou, J. Wang, and T. Li. 2013. A demonstration study of ethanol production from sweet sorghum stems with advanced solid state fermentation technology. Appl. Energy 102:260–265. doi:10.1016/j.apenergy.2012.09.060

Liang, Y., N. Sarkany, Y. Cui, J. Yesuf, J. Trushenski, and J.W. Blackburn. 2010. Use of sweet sorghum juice for lipid production by *Schizochytrium limacinum* SR21. Bioresour. Technol. 101:3623–3627. doi:10.1016/j.biortech.2009.12.087

Liang, Y., T. Tang, T. Siddaramu, R. Choudhary, and A.L. Umagiliyage. 2012. Lipid production from sweet sorghum bagasse through yeast fermentation. Renew. Energy 40:130–136. doi:10.1016/j.renene.2011.09.035

Liu, L., T.J. Herald, D. Wang, J.D. Wilson, S.R. Bean, and F.M. Aramouni. 2012. Characterization of sorghum grain and evaluation of sorghum flour in a Chinese egg noodle system. J. Cereal Sci. 55:31–36. doi:10.1016/j.jcs.2011.09.007

Marx, S., B. Ndaba, I. Chiyanzu, and C. Schabort. 2014. Fuel ethanol production from sweet sorghum bagasse using microwave irradiation. Biomass Bioenergy 65:145–150. doi:10.1016/j.biombioe.2013.11.019

Matsakas, L., U. Rova, and P. Christakopoulos. 2014. Evaluation of dried sweet sorghum stalks as raw material for methane production. BioMed Res. Int. 2014:1–7. doi:10.1155/2014/731731

McKevith, B. 2004. Nutritional aspects of cereals. British Nutrition Foundation. Nutr. Bull. 29:111–142. doi:10.1111/j.1467-3010.2004.00418.x

Meng, X., J. Yang, X. Xu, L. Zhang, Q. Nie, and M. Xian. 2009. Biodiesel production from oleaginous microorganisms. Renew. Energy 34:1–5. doi:10.1016/j.renene.2008.04.014

Menon, V., and M. Rao. 2012. Trends in bioconversion of lignocellulose: Biofuels, platform chemicals and biorefinery concept. Pror. Energy Combust. Sci. 38:522–550. doi:10.1016/j.pecs.2012.02.002

Mohammed, A.A., B.R. Hamaker, and A. Aboubacar. 1993. Effects of flour-to-water ratio and time of testing on sorghum porridge firmness as determined by a uniaxial compression test. Cereal Chem. 70:739–743.

National Sweet Sorghum Producers and Processors Association. 2015. NSSPPA, Cookeville, TN. http://www.nssppa.org/Sweet_Sorghum_FAQs.html (accessed 28 Sept. 2015).

Ntaikou, I., H.N. Gavala, and G. Lyberatos. 2010. Application of a modified anaerobic digestion model 1 version for fermentative hydrogen production from sweet sorghum extract by *Ruminococcus albus*. Int. J. Hydrogen Energy 35:3423–3432. doi:10.1016/j.ijhydene.2010.01.118

Oliver, A.L., J.F. Pedersen, R.J. Grant and T.J. Klopfenstein. 2005. Comparative effects of the sorghum *bmr*-6 and *bmr*-12 genes: I. Forage sorghum yield and quality. Crop Sci. 45:2234–2239. doi:10.2135/cropsci2004.0660

Ostovareh, S., K. Karimi, and A. Zamani. 2015. Efficient conversion of sweet sorghum stalks to biogas and ethanol using organosolv pretreatment. Ind. Crops Prod. 66:170–177. doi:10.1016/j.indcrop.2014.12.023

Packer, D.J., and W.L. Rooney. 2014. High-parent heterosis for biomass yield in photoperiod-sensitive sorghum hybrids. Field Crops Res. 167:153–158. doi:10.1016/j.fcr.2014.07.015

Panagiotopoulos, I.A., R.R. Bakker, T. de Vrije, E.G. Koukios, and P.A.M. Claassen. 2010. Pretreatment of sweet sorghum bagasse for hydrogen production by *Caldicellulosiruptor saccharolyticus*. Int. J. Hydrogen Energy 35:7738–7747. doi:10.1016/j.ijhydene.2010.05.075

Porter, K.S., J.D. Axtell, V.L. Lechtenberg, and V.F. Colenbrander. 1978. Phenotype, fiber composition, and *in vitro* dry matter disappearance of chemically induced *brown midrib* (*bmr*) mutants of sorghum. Crop Sci. 18:205–208. doi:10.2135/cropsci1978.0011183X001800020002x

Prakasham, R.S., P. Brahmaiah, D. Nagaiah, P. Srinivasarao, B.V.S. Reddy, R. Sreenivas, and P.J. Hobbs. 2012. Impact of low lignin containing *brown midrib* sorghum mutants to harness biohydrogen production using mixed anaerobic consortia. Int. J. Hydrogen Energy 37:3186–3190. doi:10.1016/j.ijhydene.2011.11.082

Prakasham, R.S., D. Nagaiah, K.S. Vinutha, A. Uma, T. Chiranjeevi, A.V. Umakanth, P. Srinivasarao, and N. Yan. 2014. Sorghum biomass: A novel renewable carbon source for industrial bioproducts. Biofuels 5:159–174. doi:10.4155/bfs.13.74

Promkhambut, A., A. Younger, A. Polthanee, and C. Akkasaeng. 2010. Morphological and physiological responses of sorghum (*Sorghum bicolor* L. Moench) to water logging. Asian J. Plant Sci. 9:183–193. doi:10.3923/ajps.2010.183.193

Rao, B.D., C.V. Ratnavathi, K. Karthikeyan, P.K. Biswas, S.S. Rao, B.S. Vijay Kumar, and N. Seetharama. 2004. Sweet Sorghum cane for biofuel production: A SWOT analysis in Indian context. National Research Center for Sorghum, Hyderabad, Andhra Pradesh, India.

Rao, P.P., G. Basavaraj, K. Basu, C.R. Reddy, A.A. Kumar and B.V.S. Reddy. 2013. Economics of sweet sorghum feedstock production for bioethanol. In: Developing a sweet sorghum ethanol value chain. ICRISAT, Patancheru, Andhra Pradesh, India. p. 99–109.

Ratledge, C., and J.P. Wynn. 2002. The biochemistry and molecular biology of lipid accumulation in oleaginous microorganisms. Adv. Appl. Microbiol. 51:11–51.

Regassa, T.H., and C.S. Wortmann. 2014. Sweet sorghum as a bioenergy crop: Literature review. Biomass Bioenergy 64:348–355. doi:10.1016/j.biombioe.2014.03.052

Reinhardt, G., and C. Cornelius. 2014. Environmental assessment of energy sorghum. Presented at the SWEETFUEL Regional Stakeholder Workshop, Hamburg, Germany. 26 June 2014. http://www.sweetfuel-project.eu/sweetfuel_events/sweetfuel_at_the_22nd_european_biomass_conference_exhibition (accessed 1 Feb. 2016).

Richter, K., and C. Berthold. 1998. Biotechnological conversion of sugar and starchy crops into lactic acid. J. Agric. Eng. Res. 71:181–191. doi:10.1006/jaer.1998.0314

Rooney, L.W. 1967. Properties of sorghum grain and new developments of possible significance to the brewing industry. Tech. Q. Master Brew. Assoc. Am. 6:277–282.

Roy, P., T. Orikasa, K. Tokuyasu, N. Nakamura, and T. Shiina. 2012. Evaluation of the life cycle of bioethanol produced from rice straws. Bioresour. Technol. 110:239–244. doi:10.1016/j.biortech.2012.01.094

Saballos, A., G. Ejeta, E. Sanchez, C. Kang, and W. Vermerris. 2009. A genome wide analysis of the cinnamyl alcohol dehydrogenase family in Sorghum [*Sorghum bicolor* (L.)Moench] identifies *SbCAD2* as the *brown midrib6* gene. Genetics 181:783–795. doi:10.1534/genetics.108.098996

Saballos, A., W. Vermerris, L. Rivera, and G. Ejeta. 2008. Allelic association, chemical characterization and saccharification properties of *brown midrib* mutants of sorghum (*Sorghum bicolor* (L.) Moench). Bioenerg. Res. 1:193–204. doi:10.1007/s12155-008-9025-7

Samuel, W.A., Y.Y. Lee, and W.B. Anthony. 1980. Lactic acid fermentation of crude sorghum extract. Biotechnol. Bioeng. 22:757–777. doi:10.1002/bit.260220404

Saraphirom, P., and A. Reungsang. 2010. Optimization of biohydrogen production from sweet sorghum syrup using statistical methods. Int. J. Hydrogen Energy 35:13435–13444. doi:10.1016/j.ijhydene.2009.11.122

Sattler, S.E., D.L. Funnell-Harris, and J.F. Pedersen. 2010. *Brown midrib* mutations and their importance to the utilization of maize, sorghum, and pearl millet lignocellulosic tissues. Plant Sci. 178:229–238. doi:10.1016/j.plantsci.2010.01.001

Satyanarayana, K.G., G.G.C. Arizaga, and F. Wypych. 2009. Biodegradable composites based on lignocellulosic fibers: An overview. Prog. Polym. Sci. 34:982–1021. doi:10.1016/j.progpolymsci.2008.12.002

Schaffert, R.E. 2007. Sweet Sorghum improvement and production in Brazil. Paper presented at the Global consultation on Pro-poor sweet sorghum development for bioethanol production and introduction to tropical sugar beet. Rome, Italy. 8–9 Nov. 2007

Schaffert, R.E., and L.M. Gourley. 1982. Sorghum as energy source. In: Proceedings of the International Symposium on Sorghum, 2-7 Nov. 1981. ICRISAT, Patancheru, Andhra Pradesh, India. p. 605–623.

Serna-Saldivar, S.O., D.A. Knabe, L.W. Rooney, T.D. Tanksley, Jr., and A.M. Sproule. 1988. Nutritional value of sorghum and maize tortillas. J. Cereal Sci. 7:83–94. doi:10.1016/S0733-5210(88)80062-6

Southern African Biofuels Association. 2007. Bio-ethanol from grains to increase food security? SABA, Johannesburg. http://www.ee.co.za/article/bio-ethanol-from-grains-to-increase-food-security.html (accessed 28 Sept. 2015).

Srinivasarao, P., C.G. Kumar, J. Malapaka, A. Kamal, and B.V.S. Reddy. 2012. Feasibility of sustaining sugars in sweet sorghum stalks during post-harvest stage by exploring cultivars and chemicals: A desk study. Sugar Tech. 14:21–25. doi:10.1007/s12355-011-0133-x

Srinivasarao, P., C.G. Kumar, R.S. Prakasham, A.U. Rao and B.V.S. Reddy. 2015. Sweet sorghum: Breeding and bioproducts. In: V.M.V. Cruz and D.A. Dierig, editors, Industrial crops. Springer, New York. p. 1–28.

Srinivasarao, P., C.G. Kumar, and B.V.S. Reddy. 2013. Sweet sorghum: From theory to practice. In: P.S. Rao and C.G. Kumar, editors, Characterization of improved sweet sorghum cultivars. Springer, India. p. 1–15.

Srinivasarao, P., S.S. Rao, N. Seetharama, and A.V. Umakath. P. SanjanaReddy, B.V.S. Reddy and C.L.L. Gowda. 2009. Sweet sorghum for biofuel and strategies for its improvement. Information Bulletin No. 77. ICRISAT, Patancheru, Andhra Pradesh, India.

Vietor, D.M., and F.R. Miller. 1990. Assimilation, partitioning, and nonstructural carbohydrates in sweet compared with grain sorghum. Crop Sci. 30:1109–1115. doi:10.2135/cropsci1990.0011183X003000050030x

Vinutha, K.S., L. Rayaprolu, K. Yadagiri, A.V. Umakanth, and P. Srinivasarao. 2014. Sweet sorghum research and development in India: Status and prospects. Sugar Tech. 16:133–143. doi:10.1007/s12355-014-0302-9

Wang, M., Y. Chen, X. Xia, J. Li, and J. Liu. 2014a. Energy efficiency and environmental performance of bioethanol production from sweet sorghum stem based on life cycle analysis. Bioresour. Technol. 163:74–81. doi:10.1016/j.biortech.2014.04.014

Wang, S., A. Hastings, S. Wang, G. Sunnenberg, M.J. Tallis, E. Casella, S. Taylor, P. Alexander, I. Cisowska, A. Lovett, G. Taylor, S. Firth, D. Moran, J. Morison, and P. Smith. 2014b. The potential for bioenergy crops to contribute to meeting GB heat and electricity. Global Change Biol. Bioenergy demands. 6:136–141.

Woods, J. 2000. Integrating sweet sorghum and sugarcane for bioenergy: Modelling the potential for electricity and ethanol production in SE Zimbabwe. Ph.D. thesis. King's College, London.

Wu, M., Z. Zhang, and Y.W. Chiu. 2014. Life-cycle water quantity and water quality implications of biofuels. Current Sustainable/Renewable Energy Rep. 1:3–10.

Wu, X., S. Staggenborg, J.L. Propheter, W.L. Rooney, J. Yu, and D. Wang. 2010. Features of sweet sorghum juice and their performance in ethanol fermentation. Ind. Crops Prod. 31:164–170. doi:10.1016/j.indcrop.2009.10.006

Yadav, A.K., N.K. Bipinraj, A.B. Chaudhari, and R.M. Kothari. 2011. Production of L(+) lactic acid from sweet sorghum, date palm, and golden syrup as alternative carbon sources. Starke 63:632–636. doi:10.1002/star.201100006

Zhang, T., N. Du, and T. Tan. 2011. Biobutanol production from sweet sorghum bagasse. J. Biobased Mater. Bioenergy 5:331–336. doi:10.1166/jbmb.2011.1158

Use of Grain Sorghum in Extruded Products Developed for Gluten-free and Food Aid Applications

Sajid Alavi,* Sue Ruan, Siva Shankar Adapa, Michael Joseph, Brian Lindshield, and Satyanarayana Chilukuri

Abstract

The use of alternative grains in foods is becoming common to address several concerns and trends such as nutrition, allergies, non-GMO, and sustainability. Sorghum is one such cereal crop that is increasingly being explored in applications targeting consumers who suffer from celiac disease (intolerance to wheat gluten). Another advantage of sorghum is its completely genetically modified organism (or GMO) free nature that can allow it to be used in foods designed for humanitarian aid in countries with GMO restrictions. This chapter focuses on two applications of sorghum in extruded foods, one commercial and related to gluten-free pasta, and the other non-commercial and related to non-GMO fortified blended foods used in government-sponsored aid programs. Pasta products such as spaghetti and macaroni are usually made from wheat (semolina). Gluten proteins in wheat have the unique property of forming an extensible, elastic, and cohesive mass when mixed with water, which imparts good strength, integrity of cooked product, and low cooking losses in pasta. Sorghum lacks this property and is inferior to wheat because it does not contain gluten-like proteins. Production of quality non-wheat pasta, using ingredients such as sorghum flour, is a challenge. This chapter describes a pilot-scale twin screw extrusion study on processing and evaluation of sorghum-based pre-cooked pasta in combination with rice. The sorghum-rice pasta was better in cooking quality as compared to its semolina counterpart, with the former having lower cooking loss (3.2–4.0% versus 10.1%), comparable or higher water uptake (111.1–130.1% versus 115.4%), and also considerably firmer cooked texture (246.0–375.9 g-f versus 117.5 g-f). In the second application of sorghum, processing of fortified blended foods (FBFs) for food aid applications using pilot-scale single screw extrusion was studied in comparison with corn-based FBFs, along with their quality and physicochemical characteristics. The sorghum-based FBFs were of superior quality with higher Bostwick flow (14.0– 20.0 cm/min) as compared with

---

Abbreviations: BD, rapid visco analyzer breakdown; CSB, corn soy blend; FAO, United Nations Food and Agriculture Organization; FAQR, Food Aid Quality Review; FBF, fortified blended food; GAO, Government Accountability Office; GMO, genetically modified organism; PT, Rapid visco analyzer peak time; PV, Rapid visco analyzer peak viscosity; TV, Rapid visco analyzer trough viscosity; FV, Rapid visco analyzer final viscosity; RVA, rapid visco analyzer; SB, Rapid visco analyzer setback; SSB, sorghum soy blend; USAID, United States Agency for International Development; USDA, United States Department of Agriculture.

S. Alavi, S. Ruan, and M. Joseph, Dep. of Grain Science and Industry, Kansas State University, Manhattan, KS 66506; S. Shankar Adapa and S. Chilukuri, College of Food Science & Technology, Acharya NG Ranga Agricultural University, Bapatla, Andhra Pradesh, India; B. Lindshield, Dep. of Food, Nutrition, Dietetics and Health, Kansas State University, Manhattan, KS 66506. *Corresponding Author (salavi@ksu.edu)

doi:10.2134/agronmonogr58.2018.0001

corn-based FBFs (6.5– 9.5 cm/min). Rapid viscoanlayzer (RVA) data was used to explain these differences and understand the interactions between various FBF components.

Introduction

Extrusion technology is commonly employed for processing and adding value to a range of grain-based raw materials, including corn, wheat, rice, sorghum, oats, and soybeans. These raw materials can either be in the form of whole grain or dehulled flours, starches, concentrated or isolated proteins from cereals and legumes, or bran. Extrusion has been used for industrial applications like rubber and plastics since the late 19th century, and only since the 1930s has it been applied to food products. In the past few decades, this technology has gained widespread use in the food industry because of its several benefits including economics and versatility (Alavi and Kingsly, 2016). Today extruded products comprise a multi-billion dollar market in the United States alone. These products include foods like breakfast cereal, ready-to-eat snacks, pasta, confectionery and meat-imitation products (or texturized vegetable proteins), and products for animal consumption including pet food, aquatic feed, and feed for farm animals like cattle and poultry.

As a multi-faceted and continuous processing technology, extrusion has several advantages over conventional batch cooking methods. Unlike batch-processing where several pieces of equipment might be needed to make the final product, the same extrusion equipment performs several functions including mixing and unitizing ingredients, cooking, forming of the product to the desired shape, expansion, texture alteration, sterilization and dehydration (due-to steam flash-off). By altering processing conditions, screw profile or the die, and by using different ingredients a wide-variety of products can be processed by the same equipment. Also extrusion allows better control over the process and product quality, and much greater processing capacity (ranging from a few hundred kilograms to several tons per hour).

This chapter describes two applications of extrusion processing for production of food products, both based on grain sorghum as a primary ingredient. The first application is gluten-free pasta as a commercial product, and the second is non-GMO fortified blended foods used in government-sponsored food aid programs.

Gluten-Free Pasta

A gluten-free diet is advised for individuals with celiac disease, which is a chronic enteropathy caused by consumption of prolamins present in wheat (gliadins), rye (secalins), barley (hordeins) and possibly oats (avidins) (Thompson, 2001). Intake of these proteins damages the villi lining the small intestine of celiac patients and interferes with absorption of food nutrients. Genetic predisposition, environmental factors, and immunogically-based inflammation are the main reasons for this ailment (Murray, 1999). The occurrence of celiac disease is higher than previously thought (Schober et al., 2005). Also it has higher prevalence rates in Europe and the United States than rest of the world. Grain sorghum is a recommended gluten-free food source (Mestres et al., 1993). Worldwide sorghum is the fifth largest crop in terms of production, and is especially common in Asia and Africa (Anglani, 1998; Dicko et al., 2006). It has traditionally been used in the United States for animal feed and ethanol production or exported in large quantities. Kansas is the leading state in production of sorghum, followed by Texas, Oklahoma, and Nebraska, because of suitable climate for this drought-tolerant crop (Smith, 2000). However, development of sorghum-based foods in the United States was rather neglected in the past.

Its use is becoming more common only recently in foods targeted at celiac patients and also health conscious consumers in general (Lovis, 2003; Devi et al., 2014; Licata et al., 2014; McCann et al., 2015; Mkandawire et al., 2015). These foods include breakfast cereal, ready-to-eat snacks, tortillas, couscous, porridges and baked goods including pancake mixes. This chapter describes a study on processing and quality attributes of gluten-free pasta with sorghum as one of the main ingredients.

Pasta products such as spaghetti and macaroni are frequently consumed and enjoyed in many countries. They are popular because they are versatile, natural, and wholesome, as well as made by a relatively simple manufacturing process (Kruger et al., 1996). Durum wheat semolina is the most common ingredient used in pasta. Semolina contains high quality and quantity of gluten and has the right particle size, which are attributes important for optimum processing, storage, and cooking of pasta. Besides semolina, sometimes flours of durum wheat or common wheat or a mixture of both are also used in pasta production. Elimination of wheat as an ingredient to enable production of gluten-free pasta presents significant challenges. The gluten proteins in wheat have a unique property of forming an extensible, viscoelastic, and cohesive mass when mixed with water. Pasta relies on this property of gluten to strengthen and retain its structure, maintain the integrity of cooked product, and reduce cooking losses. Sorghum lacks this property because it does not contain gluten-like proteins. A good quality pasta product cannot be produced with sorghum and other grains besides wheat when used alone (FAO, 1995).

Suhendro et al. (2000) reported a study on use of functional ingredients in the formulation of sorghum-based noodles to enhance their quality. Past work in our laboratory at Kansas State University has focused on addition of starch, gums, and egg in relatively small amounts to improve quality of pasta based on sorghum (Cheng et al., 2007). An important feature of that research was the use of extrusion, which led to full gelatinization of starch in the sorghum. The gelatinized starch in turn provided good binding properties in absence of gluten proteins. The pre-cooked nature of the final product also reduced preparation time in comparison to commercial pasta products. However, foods that use sorghum alone as the primary ingredient have protein of poor nutritional quality because of low content of essential amino acids, such as lysine and tryptophan (Anglani, 1998). The study described in this chapter utilized rice in addition to sorghum in the formulation for gluten-free pasta. Rice is an easily digested cereal and rice protein has one of the highest lysine content among all cereals (Luh, 1991). The nutritional value of rice has been found to significantly decrease at temperatures above 120 °C during extrusion processing (Eggum et al., 1986). However, the pasta extrusion process described here maintains temperatures below or not too much above 100 °C to produce a good quality dense product. Thus the use of a combination of sorghum and rice flours would be a good way to balance the nutritional profile in the resultant pre-cooked pasta product. Additionally, rice flour has a white color which can mitigate the tan color from sorghum flour, and its bland taste can potentially dilute any off flavors arising from sorghum.

Food Aid

The United Nations Food and Agriculture Organization report titled *The State of Food Insecurity in the World 2015* states that concerted global efforts to reduce hunger and malnutrition over the last decade and a half have resulted in decrease in number of undernourished people by 216 million, with a drop in prevalence of undernourishment from 18.6% to 10.9% worldwide and 23.3% to 12.9% for developing countries (FAO et al., 2015). Food aid can be an important tool in addressing certain food insecurity issues. International food aid programs supported by the United States are administered by the United States Agency for International Development (USAID) and United States Department of Agriculture (USDA) either as part of bilateral programs or through the United Nation's World Food Program. Fortified blended foods (FBFs) constitute a significant portion of U.S. food aid shipments. Fortified blended foods, designed for malnourished infants, children, pregnant women, and lactating mothers, are nutrition-rich, easy-to-prepare, high-protein blends of flours and other ingredients. They are often a combination of cereals with legumes and dried skimmed milk to increase the quality and quantity of proteins. The raw ingredients are processed before blending and also fortified with essential micronutrients (vitamins and minerals). Fortified blended foods are shipped in a dry form and consumed by the targeted individuals in the form of a porridge. An example of such a product is corn soy blend (CSB), which is used to treat moderate malnutrition and micronutrient deficiencies in underweight children and is the most commonly programmed specialized food in supplementary feeding programs (GAO, 2011).

It is clear that more effort is needed in the fight against hunger, as one in nine people around the world still lack the right nutrition for an active and healthy life. The USAID sponsored Food Aid Quality Review (FAQR) initiative resulted in a comprehensive review of food aid products and programs with recommendations for several changes (Webb et al., 2011). One of the recommendations was the development of new FBFs that incorporate grains other than the usual cereals such as corn. This would allow broadening of the basket of products available for food aid, and free up vendor initiative to explore the most cost-effective approaches for meeting defined nutritional characteristics and performance-based specifications. Sorghum is one such alternative cereal given its acceptability and traditional use in Africa and parts of Asia. Also its status as a non-GMO crop makes it more appealing for host governments. It is a low water usage crop and can even thrive in drought-like conditions (Ismail et al., 2003), which fits sustainable agriculture goals. Corn is known for its susceptibility to infestation by mold and resultant mycotoxins, whereas some varieties of sorghum are naturally mold-resistant (Pitt et al., 2013; Bandyopadhyay et al., 2003). For these reasons sorghum is well suited for use in FBFs. The use of extrusion for the development of fortified blended foods based on sorghum in combination with soy and the phyisco-chemical characterization of this new food aid product are another focus of this chapter.

One drawback of sorghum are anti-nutritional factors, especially phytic acid and tannins, which inhibit absorption of nutrients in the body (Saravanabavan et al., 2013; Awika et al., 2003). Also soybean contains trypsin inhibitors that interfere in the digestion of protein (Friedman and Brandon, 2001). Extrusion has been shown to break down or inactivate these anti-nutritional factors, while retaining key nutrients such as lysine (Joseph, 2016; Alonso et al., 2000; Awika et al., 2003; Nwabueze, 2007; Konstance et al., 1998). The various versions of FBFs procured and shipped by the U.S. government are typically not extruded (USDA, 2008; USDA, 2014). The study described in this chapter used extrusion for processing of sorghum and soy flours prior to fortification with other nutrients, not only to inactivate anti-nutrition factors but also to produce a fully cooked sorghum soy blend that would require much less preparation time before consumption as porridge as compared with traditional FBFs.

Process Description and Experimental Methods

This section describes pilot-scale extrusion processes for production of gluten-free pasta and fortified blended foods based on sorghum. Other details such as the experimental design used in the two studies and the methods for evaluation of product quality are also provided.

Gluten-free Pasta Based on Sorghum and Rice

A low shear extrusion cooking and forming process was used for production of the pasta based on sorghum and rice, and also a control product based on semolina for comparison purposes. White sorghum flour (Twin Valley Mills, Ruskin, Nebraska), long grain rice flour (Rivland, Houston, Texas) and durum semolina (Cargill, Minneapolis, Minnesota) were the main ingredients. Formulations are shown in Table 1.

For the gluten-free formulations, equal amounts of sorghum flours and rice flours were used. Addition of corn starch (Cargill, Minneapolis, Minnesota) at levels of 5 and 10% to the gluten free formulations was investigated for binding and structure. Guar gum (TIC Gums, Belcamp, Maryland) was also included in all formulations for improvement in texture. The hydrophilic component of guar gum can interact with proteins of sorghum and rice as a result of ionic charges and it can also interact with water-soluble starch, potentially leading to better binding (Raina et al., 2005). Monoglycerides (Danisco, New Century, Kansas) were added as a processing aid. They are also known to form complexes with the amylose molecules in starch, potentially leading to decreased swelling of granules and amylose leaching, and thus lower pasta disintegration and cooking losses (Mestres et al., 1993). Salt was also added to all formulations. Besides enhancing flavor, its hydrophilic nature aids in uniform distribution of moisture during drying, which is critical for pasta production.

A pilot-scale twin screw extruder (TX-52, Wenger Manufacturing, Sabetha, KS) was used for processing. The extruder screw diameter was 52 mm and L/D (length to diameter) ratio 25.5. All dry ingredients were premixed and metered into the extrusion system at a rate of 80 kg h^{-1}. Preconditioning with steam and water was done to achieve partial hydration and cooking before the material reached the extruder barrel. The net in-barrel process moisture was approximately 35% wet basis. Extruder screw speed was 150 rpm and the temperatures of the last four barrel segments were set at 30, 80, 75, and 70 °C to achieve optimal cooking while preventing the formation of any bubbles due to water vapor. An 'in-line' vacuum was applied after

Table 1. Formulations for pre-cooked pasta using pilot-scale twin screw extrusion.

	Semolina	Sorghum-Rice + 0% starch	Sorghum-Rice + 5% starch	Sorghum-Rice + 10% starch
Semolina (%)	97.0	–	–	–
Sorghum flour (%)	–	48.5	46	43.5
Rice flour (%)	–	48.5	46	43.5
Corn Starch (%)	–	–	5.0	10.0
Guar gum (%)	1.5	1.5	1.5	1.5
Salt (%)	1.0	1.0	1.0	1.0
Monoglycerides (%)	0.5	0.5	0.5	0.5

the cooking zone to achieve evaporative cooling and also removal of air trapped inside the dough before exit of the product from the extruder. A vertical rotini die with a pneumatically-driven rotary knife was used at the end of the extruder for shaping and cutting the pasta. The cut product was transferred pneumatically to a pilot-scale two-pass drying and one-pass cooling and conditioning system (Series 4800, Wenger Manufacturing, Sabetha, Kansas). Drying was achieved at 66 °C with retention time of 12 min and 28 min for the two passes, while cooling was done using room temperature air with a retention time of 10 min.

Optimum cooking time and cooking quality of pasta were measured according to AACC method 66–50 (AACC International, 2010a). Twenty five grams of dried pasta samples were put into a 500 mL beaker with 300 mL boiling distilled water and a timer was set to record the time. Two or three pieces of pasta were removed every 30 sec and squeezed between two clear glass Petri dishes to look for white cores. If there were no white centers, the samples were considered to be fully cooked and/or hydrated; the timer was stopped and the optimum cooking time was recorded. Two replicate cooks were conducted. About 25 g of additional samples were weighed and cooked, following the procedure and the time selected as described above. The cooked samples were drained, rinsed with 20 mL distilled water with draining for about 2 min, and weighed. Both the cooking and rinse waters were collected and evaporated in an air oven at 100 ± 1 °C for approximately 20 h. After cooling for 1 h in a desiccator, the remaining solids were weighed to determine the cooking loss. Three replicates were conducted. Cooking loss and water uptake were calculated as follows (Lai, 2001):

$$\text{Cooking Loss} = \frac{\text{weight of dried residue}}{\text{sample weight}} \times 100\%$$

$$\text{Water Uptake} = \frac{\text{weight of cooked pasta} - \text{sample weight}}{\text{sample weight}} \times 100\%$$

Pasta firmness was determined using a TA-XT2 texture analyzer (Stable MicroSystems, Godalming, United Kingdom) interfaced with a PC, following a procedure adapted from AACC method 66–50 (AACC International, 2010a; Suhendro et al., 2000). The pasta was cooked and drained after the optimum cooking time was reached, and then rinsed with distilled water for 30 sec before testing. A piece of the cooked pasta (about 1 inch long) was placed length-wise on the sample platform and a knife blade probe was used for the testing. The test speed was 0.5 mm s^{-1}, the pre-test and post-test speeds were both 5.0 mm s^{-1}, and trigger force was set at 0.1 N. The maximum cutting force in gram-force (g-f) was recorded. At least six replicates tests were conducted for each cooked product.

Color characteristics of both dried and cooked pasta were determined using a colorimeter (CR-210 Chroma-Meter, Minolta Camera Co. Ltd., Osaka, Japan). The measurements for the latter were performed 2 min after cooking. The instrument was calibrated using a standard white plate. Samples were placed in a glass Petri dish filled to the rim and leveled off with a plastic spatula. The two color parameters measured were L* (lightness, ranging from 0 for extreme dark to 100 for extreme light) and b* (+ve values for degree of yellowness and –ve values for degree of blueness) values. Measurements were replicated six times for each sample.

Fortified Blended Foods Based on Sorghum and Soybean

A high shear extrusion cooking and puffing process was used for production of FBFs based on sorghum and soybean, henceforth referred to as sorghum soy blends or SSB. For comparison FBFs based on corn and soybean, referred to as corn soy blends or CSB, were also studied. White sorghum flour (Nu Life Market, Scott City, Kansas), degermed corn meal (Agricor, Marion, IN) and defatted soy flour (Harvest Innovations, Indianola, IA) were the primary ingredients used in the study. As a first step, binary blends of cereal and/or legume flours in different ratios (Table 2) were prepared using a ribbon mixer.

The binary blends – sorghum/soy and corn/soy – were processed on a pilot-scale single screw extruder (X-20, Wenger Manufacturing, Sabetha, KS). The extruder screw diameter was 82.6 mm and L/D ratio 8.1. Steam and water were added in the preconditioner to achieve a discharge temperature of above 85 °C before the material entered the extruder barrel. The barrel temperatures were set at 45, 70, and 90 °C (inlet to discharge end) and the extruder screw speed at 500 to 570 rpm. A circular die was used with a 4.1-mm diam. opening. The expanded extrudates were cut at the die exit with a face-mounted rotary knife. The cut product was transferred pneumatically to a pilot-scale two-pass drying and one-pass cooling and conditioning system (Series 4800, Wenger Manufacturing, Sabetha, KS). Drying was done at 104 °C with retention time of five minutes each for the two passes, followed by cooling with a retention time of five minutes using room temperature air. The dried extrudates were ground to powder using a hammer mill (Schutte-Buffalo Hammermill, Buffalo, NY) fitted with a 3/64-inch (1190 μm) screen. To obtain the final fortified blended food, extruded and ground cereal and legume powder was blended with granulated white cane sugar (C&H Sugar, Crockett, California), soybean oil (Zeeland Farm Services, Zeeland, Michigan), and vitamin and mineral premixes (REPCO, Salina, Kansas) using a Hobart mixer (M802, Hobart Corporation, Troy, OH) according to the ratios shown in Table 3. The FBF composition was designed to be iso-protein (~19%), iso-fat (~10%) and iso-caloric (~ 400 kcal per 100 g).

Table 2. Formulations for binary cereal and legume blends processed in pilot-scale single screw extruder.

	SS64*	SS57	CS64	CS56
Corn meal (%)	–	–	64.0	56.0
Sorghum flour (%)	64.0	57.0	–	–
Soy flour (%)	36.0	43.0	36.0	44.0

*Number in the formulation code refers to the percentage of cereal flour in the extruded binary blend.

Table 3. Final composition of fortified blended foods.

	SSB (0% sugar)	SSB (10% sugar)	CSB (0% sugar)	CSB (10% sugar)
Extruded blend type	SS64	SS57	CS64	CS56
Extruded blend (%)	88.4	78.4	87.9	77.9
Sorghum or corn	56.4	44.4	55.9	43.9
Soy	32.0	34.0	32.0	34.0
Sugar (%)	–	10.0	–	10.0
Oil (%)	8.5	8.5	9	9
Mineral premix (%)	3.0	3.0	3.0	3.0
Vitamin premix (%)	0.1	0.1	0.1	0.1

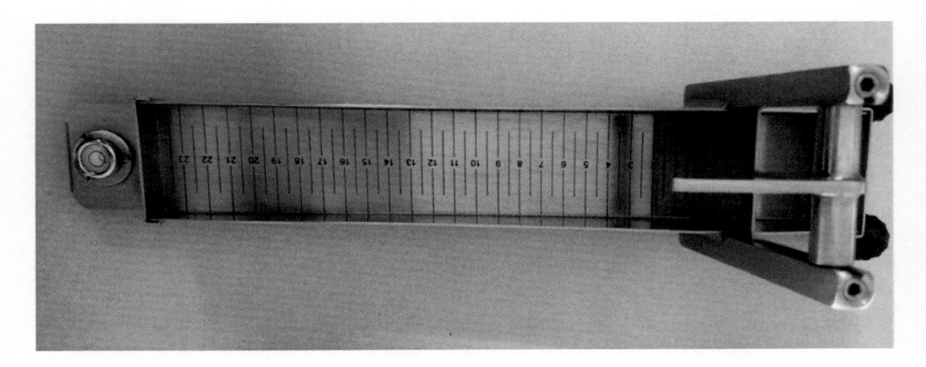

Fig. 1. Bostwick Consistometer (top view) for measuring flow of fortified blended food gruels.

Sugar is often not recommended in FBFs but was added in one SSB and CSB formulation each at 10% level to lower the consistency when hydrated into porridge form, and make the product easier to ingest and more palatable for infants and young children. One formulation each for SSB and CSB contained no sugar, in which case the level of cereal was increased to balance the calories.

The FBFs in powder form need to be hydrated in boiling water for consumption as porridge. Consistency of the porridge or gruels was tested at 20% solids, which represented a substantial increase over the 11.75 to 13.79% solids suggested traditionally (USDA, 2008; USDA 2014). Higher solid content was recommended in the FAQR for energy density given the limited gastric volume or capacity of infants (Webb et al., 2011). Gruels were prepared by adding 40 g of SSB or CSB to 160 mL of boiling distilled water with vigorous stirring using a fork for 1 min. The gruel was removed from heat and stirred for another 30 s, and then covered with an aluminum foil and cooled for 10 min in a water bath maintained at 30 °C. Distilled water was added to adjust for water loss through evaporation and bring the slurry back to the initial weight of 200 g. This was followed by further cooling in water bath at 30 °C for 1 h before testing. The Bostwick Consistometer (CSC Scientific Company, Fairfax, Virginia) was used to measure the consistency or flow of the gruel (Fig. 1). It consists of a long trough with 0.5-cm graduations along the base. The trough is separated at one end by a spring-loaded gate. This gate forms a chamber where the sample or cooled gruel was loaded. After allowing the gruel to settle for 30 s, the gate of the reservoir was opened and the distance of flow was recorded exactly after 1 min.

To understand the underlying causes for differences in porridge consistency, the pasting properties of extruded binary blends were examined using a rapid viscoanalyzer (RVA4, Newport Scientific, Warriewood, Australia) as per the standard AACC method 76–21.01 (AACC International, 2010b). The method relies on the change in rotary viscosity during hydration, swelling, and subsequent disintegration of starch granules in excess water while being stirred and heated, followed by possible realignment of molecules during cooling. Ground sample, 3.5 g in weight, was added to 25 mL of distilled water and tested on the RVA with continuous stirring for a total run time of 13 min. The following temperature profile was used: holding at 50 °C (00:00–01:00 min), heating at constant rate to 95 °C (01:00–04:42 min), holding at 95 °C (04:42–07:12 min), cooling at constant rate to 50 °C (07:12–11:00 min) and holding at 50 °C (11:00–13:00 min). Pasting parameters, including peak viscosity (PV), peak time (PT), trough viscosity (TV), breakdown (BD = PV– TV), final viscosity (FV) and setback (SB = FV– TV), were recorded.

Product Quality and Scientific Analyses

Comparison of Gluten-free Pasta with Semolina Pasta

Pictures of the dried semolina pasta and gluten-free pasta based on sorghum and rice are shown in Fig. 2. The optimum cooking times of the products were in the range of 2.0–3.5 min. This was substantially lower than the typical cooking time for ordinary commercial pasta (8–12 min), indicating the pre-cooked or pre-gelatinized nature of the extruded pasta described in this study. The color parameters of the extruded pasta are shown in Fig. 3. The dried gluten-free pastas were less bright (L^* = 45.4- 47.4) and also less yellow (b = 19.4 – 21.8) as compared with semolina control (L^* = 61.2 and b = 34.1). This was not unexpected, as sorghum flour is darker in color than

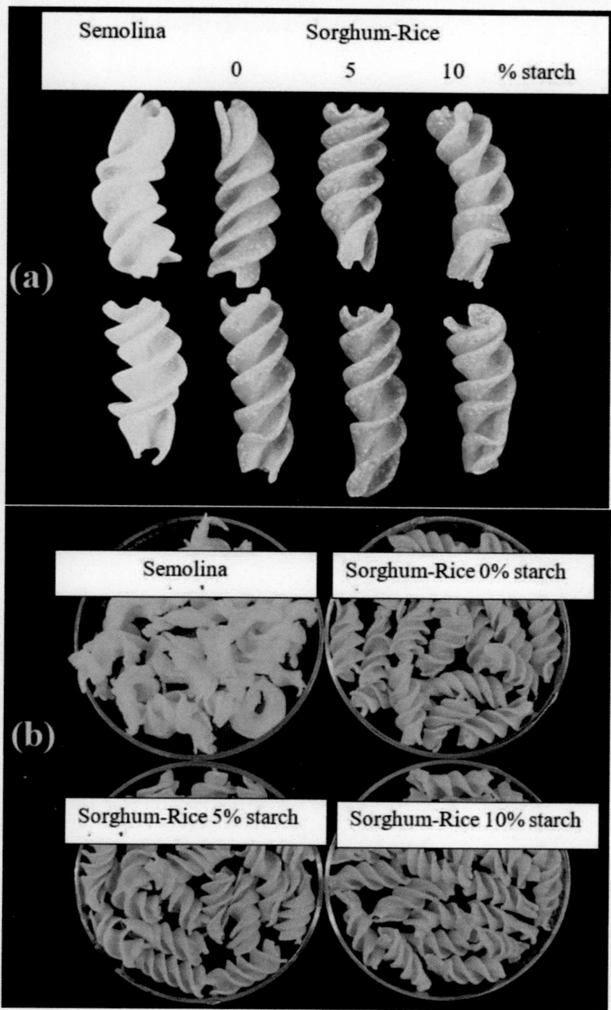

Fig. 2. Pre-cooked pasta processed using extrusion: a) dry products and b) after hydration or cooking. The two pictures have different magnifications.

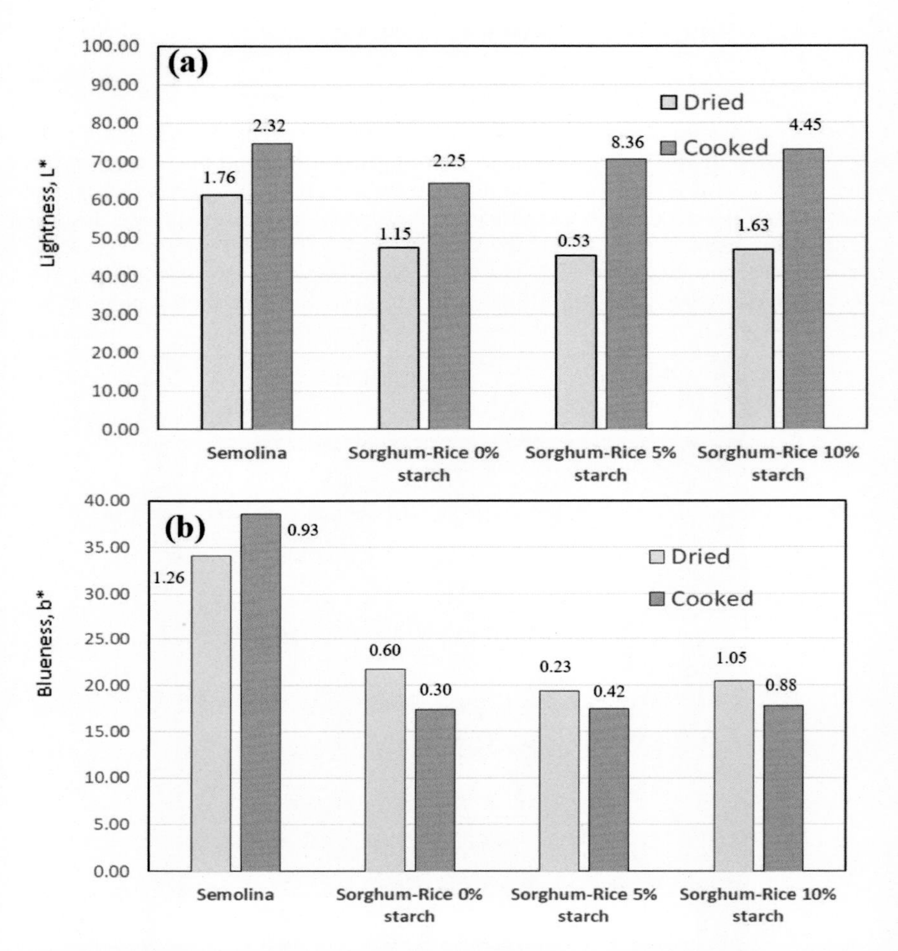

Fig. 3. Color values of pre-cooked pasta: a) lightness (L^*) and b) yellowness (b^*). Numbers on top of the bars represent standard deviation of the data.

semolina. Also the latter has a distinct yellow hue as compared to the dull greyish-white of sorghum flour and the white rice flour. The yellowness of the sorghum-rice product decreased slightly after cooking (b = 17.4– 17.8) and that of semolina pasta increased (b = 38.5), taking them further apart in appearance. On the other hand, the lightness of all the products increased after cooking. The improvement in lightness of cooked gluten-free pasta was substantial (L^* = 64.4 – 73.0), which could help with their visual appeal to consumers. Also higher starch levels in the gluten-free pasta increased the lightness of the cooked product. The product with 10% starch (L^* = 73.0) had almost the same brightness as semolina pasta (L^* = 74.8).

Cooking quality data of the pasta products are shown in Fig. 4 (cooking loss and water uptake) and Fig. 5 (firmness). The sorghum-rice pasta had much lower cooking loss (3.2–4.0% versus 10.1%), comparable or higher water uptake (111.1–130.1% versus 115.4%), and also considerably firmer cooked texture (246.0–375.9 g-f versus 117.5 g-f), as compared with the semolina pasta. This indicated that

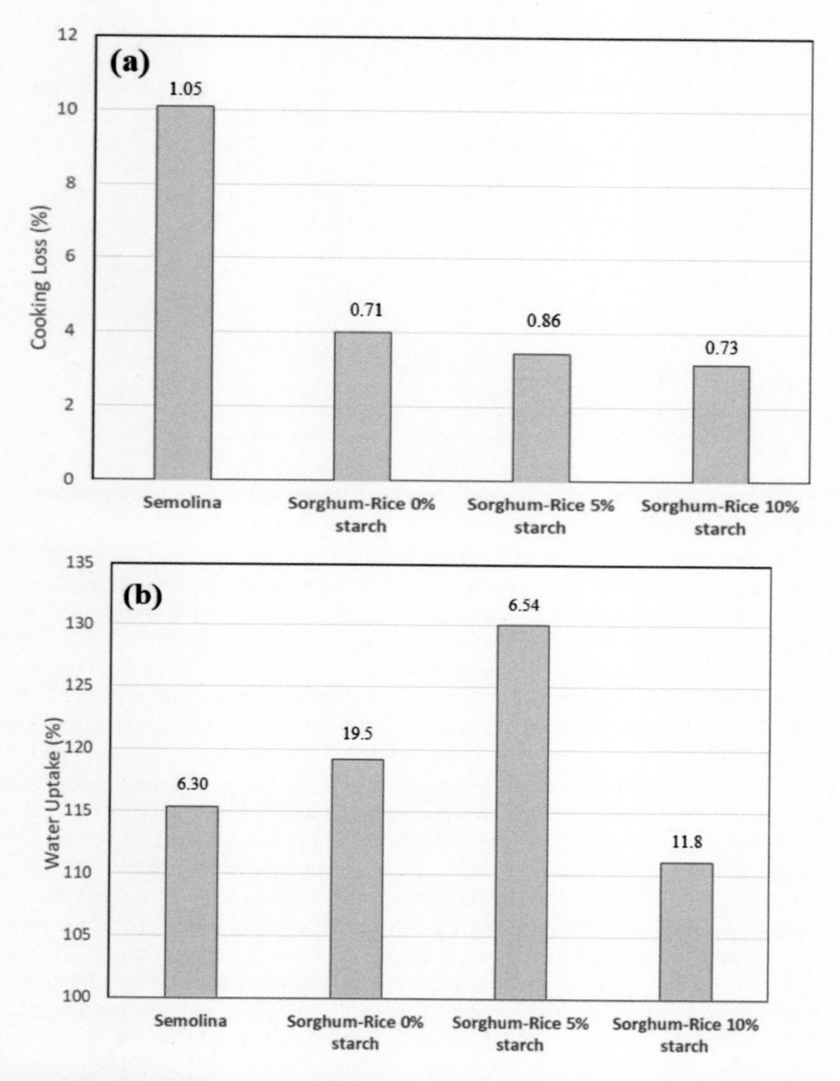

Fig. 4. Cooking loss (a) and water uptake (b) in pre-cooked pasta. Numbers on top of the bars represent standard deviation of the data.

the gluten-free pre-cooked pasta products were better in cooking quality as compared with their semolina counterpart. The cooking loss and water uptake of two commercial products, semolina-based rotini and rice-based gluten-free pasta, were also tested for comparison. The cooking loss of the extruded gluten-free products compared well with the commercial products (rotini 3.5% and rice 6.1%) but the water uptake of the latter was much higher (rotini 227.6% and rice 168.1%).

The most important differences between the sorghum-rice and semolina-based extruded pasta were the protein content and quality of the flours used, and also their particle size. The basic differences in the extrusion cooking process for

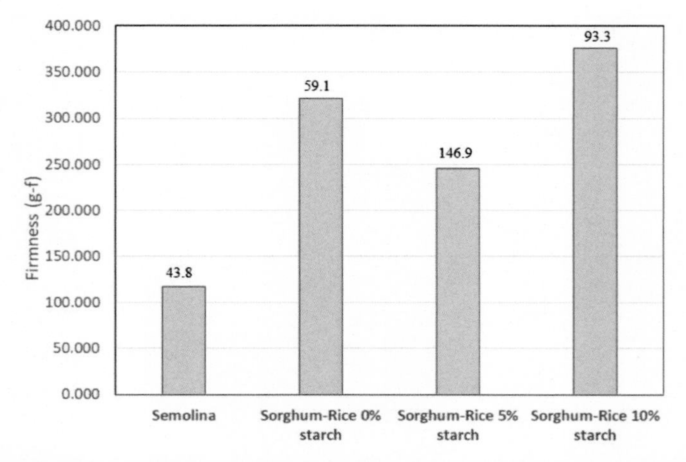

Fig. 5. Firmness of pasta after cooking. Numbers on top of the bars represent standard deviation of the data.

pre-cooked pasta as compared with the forming process for traditional pasta, and the resultant mechanisms of structure formation in the product matrices also need to be taken into account. In the latter process, temperatures are low (below 55 °C) and cooking is not involved. Semolina is hydrated and kneaded in a batch process or a continuous pasta press. The gluten in semolina forms an extensive continuous network that holds the product matrix together for a good quality pasta, and a high protein content (greater than 12%) aids in structure formation. In pre-cooked pasta using extrusion, as was the case in this study, the protein gets denatured due to shear forces and higher temperature involved in the process. The denatured protein loses its functionality and exists in a dispersed phase, but on the other hand, gelatinized starch forms the continuous network and binds the product structure. Lower protein actually is beneficial as there is less disruption of the matrix. This is one possible reason for the improved cooking quality of the gluten-free pasta based on sorghum flour (~11% protein) and rice flour (~7.5% protein). Also durum semolina has a mean particle size in the range of 300 to 450 mm, depending on its coarseness, whereas the typical mean particle size of sorghum flour ranges from 100 to 250 mm (Sacchetti et al., 2011; Dayakar Rao et al., 2016). The smaller particle size of sorghum and rice flours as compared with the semolina might have allowed better starch gelatinization and thus improved binding and cooking quality. Addition of 5 to 10% corn starch appeared to improve the product binding, lower the cooking loss, and increase the water uptake and firmness of the sorghum-rice pasta to some extent. Although the data trends in this respect were not consistent, it appears that the additional starch served to further counteract the disruptive effect of denatured protein leading to a more cohesive matrix in the pre-cooked pasta.

Comparison of Extruded Sorghum–Soy Blend with Corn–Soy Blend

For the food aid application, Bostwick flow measurements of gruels (20% solids) based on the SSB and CSB products are shown in Fig. 6. Two clear trends can be discerned for these data. The Bostwick flow of corn-based FBFs (6.5–9.5 cm min^{-1}) is much lower than sorghum-based FBFs (14.0–20.0 cm min^{-1}); and addition of 10% sugar in the formulation increased the Bostwick flow for both CSB and SSB, although the change is substantially higher in the latter. In fact, without addition of sugar the Bostwick flow of CSB was below the acceptable minimum of 9 to 11 cm min^{-1} (USDA, 2008; USDA, 2014). Obviously sugar had a plasticizing effect and helped in reducing the viscosity, which is especially critical, considering the much higher solids content of the gruels than traditionally recommended. Interestingly, the SSB product without any sugar still exhibited good flow and was well within the acceptable Bostwick range.

The above results can be better explained by understanding the interactions of sugar with starch, and the changes in the matrices of SSB and CSB as they undergo hydration, heating, and cooling during the gruel preparation process. Examination of RVA data for the extruded binary blends (Table 4) is helpful in this regard. It is clear that none of the extrudates exhibited significant cold water swelling, as can be seen from the peak time range of 1.8 to 2.2 min corresponding to a peak temperature range of 59.7 to 64.6 °C. Thus, it can be inferred that the starch granules were at least partly intact during the extrusion process. At the same time, starch in all the samples appeared to have been gelatinized during extrusion as they have maximum swelling and viscosity even before reaching the typical temperatures for completion of gelatinization (above 70 °C). Starch granules from corn in the CS blends exhibited greater hydration, swelling, and gel formation during testing and thus a higher peak viscosity (195.5–253.0 cP) than those from sorghum in the SS blends (PV = 168.5–241.0). The less accessible nature of sorghum starch might be one reason for this difference. All blends exhibited breakdown of the starch granules and gel structure leading to trough viscosity ranging from 60.5 to 112.5 cP.

Table 4. Rapid viscoanalyzer (RVA) data for extruded cereal/legume blends.

	SS64	SS57	CS64	CS56
Peak viscosity, PV (cP)	241.0 ± 2.8	168.5 ± 5.0	253.0 ± 18.4	195.5 ± 3.5
Peak time, PT (min)	2.4 ± 0.1	2.2 ± 0.7	2.2 ± 0.2	1.8 ± 0.2
Trough viscosity, TV (cP)	60.5 ± 0.7	73.5 ± 2.1	112.5 ± 3.5	86.0 ± 1.4
Final viscosity, FV (cP)	96.5 ± 0.7	85.5 ± 5.0	133.5 ± 0.7	149.5 ± 2.1

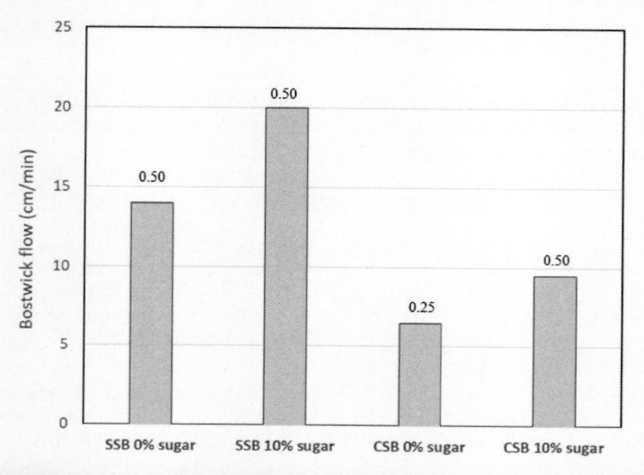

Fig. 6. Bostwick flow of gruels (20% solids) prepared from sorghum and corn based fortified blended foods. Numbers on top of the bars represent standard deviation of the data.

Sorghum proteins have a greater tendency for cross-linking and aggregation due to heat-induced denaturation or cooking (De Mesa-Stonestreet et al., 2012). Extruded blend SS57 had the highest breakdown, possibly due to interference of aggregated sorghum proteins and resultant weak gel structure. Similar interference might have caused a lower degree of starch reassociation and retrogradation in sorghum based blends during subsequent cooling as compared to corn based blends, leading to a low final viscosity (85.5 – 96.5 cP) for the former as compared with the latter (133.5–149.5 cP). This corresponded well with the high Bostwick flow observed for SSB as compared with CSB. Although percentage of cereal flour does not seem to have a big impact on FV of either the sorghum or corn based blends, the Bostwick flow of FBFs with lower cereal component (and 10% sugar) was much higher due to the plasticization effect of sugar and its retardation of starch retrogradation during cooling of gruels and subsequent gel formation (Wang et al., 2015).

It can be inferred from this study that the inherent properties of starch and protein in grain sorghum might make it more suitable as compared with corn for fortified blended foods of acceptable quality with optimum gruel consistency and palatability. It should also be noted that use of extrusion for processing of the cereal and legume components led to fully cooked SSB and CSB products, which required only 1 min of boiling for preparation of gruels as compared to 2 to 5 min for traditional CSB products (USDA 2008; USDA 2014). This can provide increased convenience and reduction of fuel costs for the care provider at home and other settings in food aid programs.

Conclusions

Grain sorghum can be used in various value-added food applications. This chapter provided an overview of research at Kansas State University that touched on two such sorghum-based foods– pre-cooked pasta as a gluten-free product and non-GMO fortified blended foods designed for humanitarian aid. The role of extrusion in the processing of these products, and its impact on the sorghum-based food matrix was described and related to product quality attributes. In both applications, the products based on sorghum was superior as compared to that based on more traditional grains; and use of extrusion led to pre-cooked products that were much more convenient to prepare than their traditional counterparts.

Acknowledgements

The authors would like to specially acknowledge the guidance for research on gluten-free pasta by the late Dr. Charles E. Walker (Department of Grain Science and Industry, Kansas State University). We also thank Mr. Eric Maichel, operations manager of the KSU extrusion lab, for his technical assistance and Wenger Manufacturing, Inc. (Sabetha, Kansas) for their continued support to the KSU extrusion lab.

References

Alavi, S., and A.R.P. Kingsly. 2016. Particulate flow and agglomeration in food extrusion. In: H. Merkus and G. Meesters, editors, Production, handling and characterization of particulate materials. Particle Technology Series 25. Springer, Cham. doi:10.1007/978-3-319-20949-4_5

Alonso, R., A. Aguirre, and F. Marzo. 2000. Effects of extrusion and traditional processing methods on antinutrients and in vitro digestibility of protein and starch in faba and kidney beans. Food Chem. 68:159–165. doi:10.1016/S0308-8146(99)00169-7

AACC International. 2010a. In: Approved methods of analysis, 11th ed. Method 66-50. Semolina, pasta, and noodle quality. AACC International, St. Paul, MN.

AACC International. 2010b. In: Approved methods of analysis, 11th ed. Method 76-21.01. General pasting method for wheat or rye flour or starch using the rapid visco analyzer. AACC International, St. Paul, MN.

Anglani, C. 1998. Sorghum for human food- a review. Plant Foods Hum. Nutr. 52:85–95. doi:10.1023/A:1008065519820

Awika, M.J., L. Dykes, L. Gu, L.W. Rooney, and R.L. Prior. 2003. Processing of Sorghum (Sorghum bicolor) and Sorghum products alters procyanidin oligomer and polymer distribution and content. J. Agric. Food Chem. 51:5516–5521. doi:10.1021/jf0343128

Bandyopadhyay, R., C.R. Little. R.D. Waniska, and D.R. Butler. 2003. Sorghum grain mold: Through the 1990s into the New Millennium. In: J. F. Leslie, editor, Sorghum and millets diseases. Iowa State Press, Ames, IA.

Cheng, E.M., S. Alavi, and S. Bean. 2007. Sorghum-based pre-cooked pasta utilizing extrusion processing. 25th Biennial Sorghum Research and Utilization Conference, 14–16 Jan 2007, Santa Ana Pueblo, NM.

Dayakar Rao, B., M. Anis, K. Kalpana, K.V. Sunooj, J.V. Patil, and T. Ganesh. 2016. Influence of milling methods and particle size on hydration properties of sorghum flour and quality of sorghum biscuits. LWT-. Food Sci. Technol. (Campinas) 67:8–13.

De Mesa-Stonestreet, N.J., S. Alavi, and J. Gwirtz. 2012. Extrusion-enzyme liquefaction as a method for producing sorghum protein concentrates. J. Food Eng. 108:365–375. doi:10.1016/j.jfoodeng.2011.07.024

Devi, N.L., S. Shobha, S. Alavi, K. Kalpana, and M. Soumya. 2014. Utilization of extrusion technology for the development of millet based complementary foods. J. Food Sci. Technol. 51:2845–2850. doi:10.1007/s13197-012-0789-6

Dicko, M.H., H. Gruppen, and A.S. Traore. 2006. Sorghum grain as human food in Africa: Relevance of content of starch and amylase activities. Afr. J. Biotechnol. 5:384–395.

Eggum, B.O., B.O. Juliano, M.G.B. Ibabao, and C.M. Perez. 1986. Effect of extrusion cooking on nutritional value of rice flour. Food Chem. 19:235–240. doi:10.1016/0308-8146(86)90073-7

FAO. 1995. Sorghum and millets in human nutrition. Food and Agriculture Organization of the United Nations, Rome, Italy.

FAO, IFAD, WFP 2015. The state of food insecurity in the world 2015: Meeting the 2015 international hunger targets: Taking stock of uneven progress. Rome, FAO.

Friedman, M., and D.L. Brandon. 2001. Nutritional and health benefits of soy proteins. J. Agric. Food Chem. 49:1069–1086. doi:10.1021/jf0009246

GAO. 2011. International food assistance: Better nutrition and quality control can further improve U.S. food aid. GAO11-491, U.S. Government Accountability Office, Washington, D.C.

Ismail, S., M. Immink, I. Mazar, and G. Nantel. 2003. Community-based food and nutrition programmes: What makes them successful– a review and analysis of experience. Food and Agricultural Organization of the United Nations, Rome, Italy.

Joseph, M.V. 2016. Extrusion, physico-chemical characterization and nutritional evaluation of sorghum-based high protein, micronutrient fortified blended foods. Ph.D. Thesis. Kansas State University, Manhattan, KS. p. 109.

Konstance, R.P., C.I. Onwulata, P.W. Smith, D. Lu, M.H. Tunick, M.D. Strange, and V.H. Holsinger. 1998. Nutrient-based corn and soy products by twin screw extrusion. J. Food Sci. 63:864–868. doi:10.1111/j.1365-2621.1998.tb17915.x

Kruger, J.E., R.B. Matsuo, and J.W. Dick. 1996. Pasta and noodle technology. In: J.E. Kruger, R.B. Matsuo, and J.W. Dick, editors, Pasta and noodle technology. AACC, St.Paul, MN. p. 95–120.

Lai, H.M. 2001. Effects of rice properties and emulsifiers on the quality of rice pasta. J. Sci. Food Agric. 82:203–216. doi:10.1002/jsfa.1019

Licata, R., J. Chu, S. Wang, R. Coorey, A. James, Y. Zhao, and S. Johnson. 2014. Determination of formulation and processing factors affecting slowly digestible starch, protein digestibility and antioxidant capacity of extruded sorghum–maize composite flour. Int. J. Food Sci. Technol. 49:1408–1419. doi:10.1111/ijfs.12444

Lovis, L.J. 2003. Alternatives to wheat flour in baked goods. Cereal Foods World 48:61–63.

Luh, B.S. 1991. Rice production. In: B.S. Luh, editor, Rice production. Van Nostrand Reinhold, New York. p. 402–403.

McCann, T., D. Krause, and P. Sanguansri. 2015. Sorghum-New gluten-free ingredient and applications. Food Aust. 67:24–26.

Mestres, C., P. Colonna, and M.C. Alexandre. 1993. Comparison of various processes for making maize pasta. J. Cereal Sci. 17:277–290. doi:10.1006/jcrs.1993.1026

Murray, J.A. 1999. The widening spectrum of celiac disease. Am. J. Clin. Nutr. 69:354–365.

Mkandawire, N.L., S.A. Weier, C.L. Weller, D.S. Jackson, and D.J. Rose. 2015. Composition, in vitro digestibility, and sensory evaluation of extruded whole grain sorghum breakfast cereals. LWT-. Food Sci. Technol. (Campinas) 62:662–667.

Nwabueze, T.U. 2007. Effect of process variables on trypsin inhibitor activity (TIA), phytic acid and tannin content of extruded African breadfruit–corn–soy mixtures: A response surface analysis. LWT. Food Sci. Technol. (Campinas) 40:21–29.

Pitt, J.I., M.H. Taniwaki, and M.B. Cole. 2013. Mycotoxin production in major crops as influenced by growing, harvesting, storage and processing, with emphasis on the achievement of food safety objectives. Food Contr. 32:205–215. doi:10.1016/j.foodcont.2012.11.023

Raina, C.S., S. Singh, and A.S. Bawa. 2005. Textural characteristics of pasta made from rice flour supplemented with proteins and hydrocolloids. J. Texture Stud. 36:402–420. doi:10.1111/j.1745-4603.2005.00024.x

Sacchetti, G., G. Cocco, D. Cocco, L. Neri, and D. Mastrocola. 2011. Effect of semolina particle size on the cooking kinetics and quality of spaghetti. Procedia Food Sci. 1:1740–1745. doi:10.1016/j.profoo.2011.09.256

Saravanabavan, S.N., M.M. Shivanna, and S. Bhattacharya. 2013. Effect of popping on sorghum starch digestibility and predicted glycemic index. J. Food Sci. Technol. 50:387–392.

Schober, T.J., M. Messerschmidt, and S.R. Bean. 2005. Gluten-free bread from sorghum: Quality differences among hybrids. Cereal Chem. 82:394–404. doi:10.1094/CC-82-0394

Smith, C.W. 2000. Sorghum production statistics. In: C.W. Smith and R.A. Frederiksen, editors. Sorghum: Origin, history, technology and production, John Wiley & Sons, New York, NY. p. 401–408.

Suhendro, E.L., C.F. Kunetz, C.M. McDonough, L.W. Roonet, and R.D. Waniska. 2000. Cooking characteristics and quality of noodles from food sorghum. Cereal Chem. 77:96–100. doi:10.1094/CCHEM.2000.77.2.96

Thompson, T. 2001. Wheat starch, gliadin, and the gluten-free diet. J. Am. Diet. Assoc. 101:1456–1459. doi:10.1016/S0002-8223(01)00351-0

USDA. 2014. USDA commodity requirements: CSBP2 corn soy blend plus for use in international food aid programs. United States Department of Agriculture, Farm Service Agency, Kansas City Commodity Office, Kansas City, MO. https://www.fsa.usda.gov/Internet/FSA_File/csbp2.pdf (accessed 13 Aug. 2018).

USDA. 2008. USDA Commodity requirements: CSB13 corn-soy blend for use in export programs. United States Department of Agriculture, Farm Service Agency, Kansas City Commodity Office, Kansas City, MO. http://www.fsa.usda.gov/Internet/FSA_File/csb13.pdf (accessed 13 Aug. 2018).

Wang, S., C. Li, L. Copeland, Q. Niu, and S. Wang. 2015. Starch retrogradation: A comprehensive review. Compr. Rev. Food Sci. Food Saf. 14:568–585. doi:10.1111/1541-4337.12143

Webb, P., B. Rogers, I. Rosenberg, N. Schlossman, C. Wanke, J. Bagriansky, K. Sadler, Q. Johnson, J. Tilahun, A.R. Masterson, and A. Narayan. 2011. Delivering improved nutrition: Recommendations for changes to U.S. Food Aid products and programs, Tufts University, Boston, MA.

Forage and Renewable Sorghum End Uses

Scott Staggenborg

Abstract

Sorghum is an important source of starch and fiber throughout the world. This chapter will focus on forage and biomass sorghum. Traits found in sorghum include photoperiod sensitivity and insensitivity, brown midrib, and brachytic dwarfism. Exceptional drought tolerance enables sorghum to produce high yields under limited water supply. The primary use for forage sorghum worldwide is animal feed, but sorghum may also be used as a dedicated energy crop for chemical production, direct combustion to produce heat, and anaerobic digestion to produce biogas. Sorghum's genetic diversity has not been fully exploited and could be the key to its future use in both traditional and new applications.

Sorghum is an important source of starch and fiber (cellulose, hemicellulose, and lignin) throughout the world. In the United States, grain sorghum is the prominent crop; however, each year a greater volume of forage sorghum seed than grain sorghum is purchased and planted. Interest in renewable feedstock crops has recently grown, and a wide range of annual and perennial species was suggested for such crops. Sorghum was quickly identified as one of the most appropriate dedicated energy crops. In addition, having scalable and well-understood production practices makes sorghum more acceptable than many lesser known perennial grasses. Relative high yields, especially in marginal environments, make it an excellent choice over grain-based annuals, and because it is generally not used for human food, it is of less concern in the food-versus-fuel debate (Staggenborg et al., 2008). Here I review the state of the art for sorghum as a renewable feedstock and its conversion to various biofuels and biochemicals.

Forage and Biomass Sorghum Types

The economic yield of sorghum can be defined as grain, forage and biomass, and sugars depending on the end use and the genetic composition of a particular hybrid. This section focuses on three groups: forage and the sweet sorghum hybrids, sorghum × sudangrass (So×Su), and sudangrass (Su×Su). Sorghum × sudangrass hybrids are known for high biomass yields and rapid regrowth, and

---

Abbreviations: AD, anaerobic digester; ADF, acid detergent fiber; BMR, brown midrib; ET, evapotranspiration; HI, harvest index; NDF, neutral detergent fiber; NREL, National Renewable Energy Laboratory; OD20, oxygen demand in 20 h; So×Su, sorghum × sudangrass; Su×Su, sudangrass × sudangrass; TS, total solids; VS, volatile solids.

Scott Staggenborg, 1301 East 50th, Lubbock, TX 79404 (sstaggenborg@chromatininc.com).

doi:10.2134/agronmonogr58.2014.0077.5

Su×Su hybrids are characterized by rapid early-season growth, rapid regrowth, fine stems, and lower prussic acid levels compared with forage sorghums and So×Su hybrids.

Plant Height and Maturity

These three groups can be further subdivided on the basis of hybrid maturity. Hybrid maturity is controlled by seven flowering genes: *Ma1* through, *Ma7*, which influence the duration of growth, or days to maturity (Quinby, 1974; Rooney et al., 2007; and Mullet et al., 2010). Manipulation of these genes can result in a forage sorghum, So×Su hybrid, or Su×Su hybrid that either produce a head after some variable vegetative stage (headed) or do not produce a head during a normal growing season in a temperate environment (photoperiod sensitive). Photoperiod-sensitive hybrids are those that require decreasing day length, a particular temperature, and a certain photoperiod to trigger floral initiation (Craufurd et al., 1999; Dingkuhn et al., 2008). In temperate environments, this photoperiod response results in floral initiation after a photoperiod of 12 h and 20 min, so panicles are rarely observed or are present late enough to produce no grain. Because these hybrids do not flower until very late in the year, they become very tall and produce very high biomass yields. They are traditionally higher in moisture at harvest because they do not head and trigger a natural drydown process like headed hybrids. From a forage-quality perspective, these hybrids produce no grain and are often considered to have low forage quality.

Sorghum height is controlled by four genes (dwarf) for height: dw_1, dw_2, dw_3, and dw_4. Genes dw_1 and dw_2 are known to affect internode length, dw_3 affects internode number, and dw_4 appears to affect panicle length (Goud and Vasudev Rao, 1977). Grain sorghum hybrids have been dwarfed to facilitate mechanized harvest and seldom exceed 2 m in height. Most grain sorghum hybrids developed in the United States are recessive at three height loci (3-dwarf) and their genotypes are generally $dw_1Dw_2dw_3dw_4$ (Quinby and Karper, 1954). One height gene, dw_3, is known only to reduce internodes on the lower part of the stalk and has little impact on panicle exsertion. From a forage perspective, plant height results in three distinct forage sorghum types: traditional, dual-purpose, and brachytic-dwarf forage hybrids. It is possible to find brachytic-dwarf So×Su hybrids.

Brown Midrib and Brachytic Dwarf Sorghum

Lignin decreases forage digestibility, so efforts to reduce lignin in forage sorghum hybrids have been conducted for decades. Brown midrib (BMR) mutations were introduced into sorghum by Porter et al. (1978). Fritz et al. (1990) selected three from this original population that were agronomically acceptable, and they have been the source of three genes: *bmr-6*, *bmr-12*, and *bmr-18*. BMR hybrids have noticeable brown pigment in the midrib of the leaves, the stem, the pith, and immature panicles (Porter et al., 1978). These hybrids have reduced lignin content and generally higher digestibility, yet they are often low in yield (Porter et al., 1978; Bucholtz et al., 1980; Hanna et al., 1981; Bean and McCollum, 2006; McCuistion et al., 2009; McCuistion et al., 2010).

Traditional forage sorghums have no recessive dwarf genes, resulting in hybrids that can exceed 4 m in height. While brachytic dwarf forage hybrids have a similar leaf number as traditional forage sorghum hybrids, they have brachytic mutations, which reduce internode length, resulting in a "stacking" appearance of the leaves. It is believe that because of their higher leaf-to-stem ratios, brachytic dwarf hybrids are more digestible compared with traditional forage sorghum hybrids.

Water Use

Sorghum tolerance to drought and heat stress has been reported by a large number of researchers. Forage sorghum is often compared with corn for improved water-use efficiency and for producing higher yields under water-limiting scenarios (Table 1). Marsalis et al. (2009) reported forage sorghum yields greater or equal to corn silage when the amount of applied irrigation water was reduced to two-thirds of what would have been applied to a fully irrigated corn crop. Howell et al. (2008) reported that water use in BMR sorghum was approximately 73% of that in corn water, as measured by a lysimeter, and Rajan and Maas (2007) reported higher daily water-use rates for corn compared with forage sorghum via eddy covariance and spectral crop coefficient methods in large-scale production fields.

McCuistion et al. (2009) compared the water use and yields of BMR forage sorghum, traditional forage sorghum, and photoperiod forage sorghum. Photoperiod forage sorghum had the highest yields and greatest water-use efficiency, 4.4 Mg ha^{-1} dry matter/100 mm of water. The water-use efficiency of headed forage sorghums was 2.1 Mg ha^{-1} dry matter/100 mm of water. Saeed and El-Nadi (1998) reported water-use efficiencies for forage sorghum ranging from 65 to 86 kg ha^{-1} mm^{-1}, with light, frequent irrigation resulting in the highest water-use efficiencies and dry matter yields. Ottman (2010) reported the greatest water-use efficiency at 75% of evapotranspiration (ET) replacement for forage sorghum. He reported an applied water-use efficiency of 78 kg ha^{-1} mm^{-1}. Hussein and Alva (2014) reported lower water-use efficiency for forage sorghum, 6 kg ha^{-1} mm^{-1}. They did not state the cultivar planted, so it is difficult to properly compare the data due to the diversity of forage sorghum types and yield potentials. Narayanan et al. (2013) reported water-use efficiencies ranging from 3.39 to 7.63 g dry matter/kg water from eight public biomass sorghum varieties. Their results illustrate the importance of biomass yield to improve water-use efficiency. Water use ranged from 218 to 256 kg m^{-2} during 2 yr, a relatively narrow range compared with yields that ranged from 796 to 1717 g m^{-2}.

Table 1. Corn, forage sorghum (FS), and brown midrib forage sorghum (BMR-FS) yields in New Mexico during two growing seasons under deficit irrigation.†

	2005	2006	2-yr avg.
Optimum harvest (60–65% moisture)		Mg ha^{-1}	
Corn	48.7	55.3	51.9
FS	61.5	59.0	60.3
BMR-FS	51.6	44.7	48.4
Late harvest (50–60% moisture			
Corn	57.3	56.3	56.8
Forage sorghum	54.8	57.0	56.1
BMR-FS	46.7	43.5	45.2
SEM	1.1	1.1	1.1

† Data from Marsalis et al., 2009; irrigation: 18–20 inches, or about two-thirds of full corn irrigation amounts.

Forages as Animal Feed

Forage sorghum, So×Su, and Su×Su hybrids have traditionally been included in rations for ruminant animals. These animals' unique digestive systems are capable of breaking down the complex carbohydrates into simple sugars, which are easily digested. Traditional harvesting and storage methods include grazing, greenchop, dry hay, or silage.

Direct Feeding and Dry Hay Methods

Direct feeding methods, grazing, and greenchop require additional management to reduce the potential of negative effects of prussic acid and nitrate poisoning. Prussic acid levels can be reduced by delaying grazing or greenchop harvest until plant heights exceed 45 cm. Prussic acid is concentrated in the leaf tissue, so delaying direct feeding and greenchop harvest until this height is achieved reduces the leaf-to-stem ratio. Nitrates accumulate in grasses that experience growth-reducing stress, drought being the most common. Reductions in dry matter because of stress result in higher concentrations of nitrates, which accumulate in the plant early in the growing season. Nitrate management in a direct-feeding scenario should be managed by testing the forage before feeding.

Sorghum × sudangrass and Su×Su hybrids are the primary hybrids used to produce dry hay. These hybrids can be harvested multiple times during the summer when adequate solar radiation and wind are available to reduce forage moisture content from 80 to 15% moisture (Hill, 1976). When forage sorghum hybrids are harvested for baled hay, additional crimping is required to break the large internodes and aid in the drying process (Hartley et al., 2011).

Silage

Sorghum silage is predominately produced from forage sorghum hybrids, although photoperiod-sensitive So×Su hybrids are occasionally used in a single-cut system to produce silage. Forage sorghum hybrids can often be divided into three categories: traditional, dual-purpose, and brachytic dwarf. The inclusion of BMR genes is common in traditional and brachytic dwarf hybrids.

Dual-purpose hybrids are dwarfed forage hybrids that resemble grain sorghum hybrids. Dual-purpose forage sorghums grow to between 1.5 and 2.5 m tall, depending on growing conditions, and will produce a significant amount of grain (Table 2). Lab analyses of silage from these hybrids will always result in high starch levels (~20%) because of the grain in the silage (Table 2). If, however, the plants are not harvested correctly, this starch availability may mean that the lab results are misleading for a nutrition (e.g., although the lab report may read 25% starch, only 15% may be available to the animal). In order for animals to digest all of this starch, harvest must occur between the milk and soft-dough stages. As the grain matures (proceeds from hard-dough to maturity), it becomes more difficult to digest because of grain hardness. At this point, the grain must be cracked or ground for the starch to be fully digested by the animal.

Full-season headed forage sorghum hybrids are preferred for sorghum silage production. These hybrids produce high biomass yields, contain some grain, and are typically drier at harvest than photoperiod-sensitive forage sorghum hybrids (Table 2). High biomass yields are accomplished through the production of long internodes and large leaves. Tall plants are often associated with lodging, so

Table 2. Yield and forage quality data for three Sorghum Partners forage sorghum hybrids grown over a 4-yr period.

Variable	Sorghum Partners NK 300	Sorghum Partners SS405	Sorghum Partners 1990
Harvest moisture†	62.1	65.2	70.0
Yield (Mg ha^{-1} @ 65% moisture)†	50.0	58.7	69.2
Grain (Mg ha^{-1})†	7.9	2.9	0.0
Crude protein (%)‡	9.76 ± 0.54§	8.01 ± 0.31	9.54 ± 1.36
Acid detergent fiber (%)‡	24.96 ± 0.06	34.63 ± 1.52	33.16 ± 3.61
Neutral detergent fiber (%)‡	39.50 ± 1.13	54.43 ± 2.5	54.54 ± 0.62
Total digestible nutrients (%)‡	65.24 ± 2.78	65.75 ± 0.35	61.26 ± 4.45
Starch (%)‡	33.64 ± 5.15	19.79 ± 9.73	5.93 ± 1.92
Crude fat (%)‡	2.25 ± 0.15	2.08 ± 0.52	0.84 ± 0.05
Ash (%)‡	6.68 ± 0.76	7.43 ± 0.41	8.65 ± 0.37
Net energy$_{lactation}$ (Mcal/kg)‡	0.77 ± 0.43	0.65 ± 0.02	0.67 ± 0.08

† From Bean and McCollum 2006.

‡ Chromatin internal data collected from different samples.

§ Standard deviation.

brachytic dwarf hybrids have become more popular in the industry. As mentioned previously, these hybrids have significantly smaller internode lengths, which reduce lodging. Forage hybrids also derive a significant yield from grain production, requiring proper harvesting to optimize silage digestion (Bell et al., 2015). While starch digestibility is optimized between the milk and soft-dough stages, forage moisture at harvest is also critical for optimizing silage quality. It is recommended that forages for silage be harvested at 35% moisture. Forages harvested at higher moisture contents can result in poor fermentation and excessive effluent drainage from the silage storage structure.

Forage Quality

From a livestock-feed perspective, forage composition focuses on nutritive value. Analyses for livestock nutrition emphasize digestibility of the fiber and use methods that resemble an animal rumen. The overall goal is to evaluate the carbohydrates, fat, protein, minerals, and vitamins available to animals for digestion and growth. Grass species are typically high in carbohydrates compared with legume forage sources. Neutral detergent fiber (NDF) is an estimation of the structural carbohydrates in forage (hemicellulose, cellulose, and lignin), and acid detergent fiber (ADF) is an estimate of the cellulose and lignin. Forage analyses also include in vitro digestibility estimation methods such as neutral detergent fiber digestibility and in vitro dry-matter digestibility (Martin and Barnes, 1980; Hatfield et al., 1994).

Attempts have been made to develop composite values that describe forage quality as a single number. Total digestible nutrients, net energy, relative feed value, and relative feed quality are all composite values that have been developed to assess feed quality (Undersander et al., 1993). A range of handbooks exist that illustrate livestock response to sorghum (Brouk and Bean, 2011; Brouk, 2012).

Dedicated Energy Sorghum

Energy sorghums have genetic roots similar to those used for livestock feed, are low in stalk juice and sugar contents, and are typically selected with a focus solely on yield rather than on both yield and animal nutritional quality, as is the case with forage sorghum. Our understanding of energy sorghum performance and current composition is based on forage sorghum performance measurements during the past 5 yr.

Sorghum biomass yields have been evaluated throughout the United States for almost three decades (Table 3). As expected, yields vary considerably on the basis of location and year. The lowest yields reported in the peer-reviewed literature were either rainfed or low-irrigation treatments in Texas (Bean and McCollum, 2006; Rooney et al., 2007) or Arizona (Ottman, 2010). The greatest yields with traditional forages were produced in high-yield environments with photoperiod-sensitive hybrids. The highest yield, 61 Mg ha^{-1} dry matter, was reported in Alabama in a narrow-row configuration with photoperiod-sensitive forage hybrid (Snider et al., 2012). Yields above 30 Mg ha^{-1} dry matter were reported with similar hybrids by Miller and McBee (1993), Venuto and Kindiger (2008), and Rocateli et al. (2012). The highest yields for energy sorghum was in Illinois, and that sorghum outyielded traditional forage hybrids by 35% (Maughan et al., 2012).

As with digestibility in ruminant animals, lignin reduces conversion of biomass to sugars and eventually ethanol or other chemicals (Dien et al., 2001). Because lignin is highly resistant to chemical cleaving and fills in the spaces between cellulose and hemicellulose, pretreatment processes to cleave lignin strands into more digestible segments have been evaluated in attempts to improve sugar recovery from cellulosic feedstocks (Yang and Wyman, 2004; Wyman et al., 2005; Pedersen et al., 2010; Hu et al., 2012). With the only commercial market currently being the livestock forage market, commercial-scale applications of BMR feedstocks into the biochemical markets are not widespread.

A range of in vitro lab procedures has been developed to estimate forage digestibility by ruminant animals; however, the classification of the carbohydrates as structural and nonstructural differs among methods. As interest in the use of crop biomass as a source of simple sugars increased, it became apparent that different analytical methods would be needed. Although estimations of cellulose and hemicellulose could be calculated from ADF and NDF, these methods were not adequate. Methods developed by the US DOE National Renewable Energy Laboratory (NREL) are the standards used in evaluating biomass samples for structural and nonstructural carbohydrates, lignin, and ash (Sluiter et al., 2005a; Sluiter et al., 2005b; Sluiter et al., 2011).

The composition of sorghum biomass has focused largely on the structural carbohydrates: cellulose, hemicellulose, lignin, and pectin. All four components are composed primarily of six sugars arranged in various complex and amorphous configurations (Heldt and Piechulla, 2011). Miller and McBee (1993) analyzed forage and grain sorghum for both structural and nonstructural carbohydrates (Table 4). As with mixed sorghum types, the wide range in starch values often indicates the presence of grain. Their values for hemicellulose and cellulose are similar to those from annual crops. Dien et al. (2009) reported similar data for a sample set that contained sorghum with the *bmr* genes. This is probably the

Table 3. Biomass sorghum yields across a range of years and US locations.

Source	Location	Years	Dry yield Max.	Min.	Mean	Description of entries†
			——— Mg ha^{-1} ———			
Caravetta et al., 1990	Lafayette, IN	1987–1988	15.9	8.3	11.8	PS
Miller and McBee, 1993	College Station, TX	1986–1987	28.0	11.4	17.7	Cultivar comparison
Hallam et al., 2001	Iowa	1988–1992	20.7	14.6	16.6	SWS and SS
Bean and McCollum, 2006	Bushland, TX	2000–2005	10.7	7.6	9.6	Irrigated PS, multiyear means
Rooney et al., 2007	College Station, TX	1985	8.5	5.1	6.7	FS nitrogen study
Venuto and Kindiger, 2008	El Reno, OK	2004–2006	33.1	23.6	29.4	PS and SS hybrids
Marsalis et al., 2009	Clovis, NM	2005–2006	21.5	19.2	20.4	FS hybrid
Marsalis et al., 2010	Clovis, NM	2007–2008			24.0	FS hybrid
Ottman, 2010	Maricopa, AZ	2009	15.1	7.6	12.1	FS irrigation study
Propheter et al., 2010	KS	2007–2008	26.8	12.7	19.7	FS, PS, and BMR
Dahlberg et al., 2011	Bushland, TX	2007	22.5	10.5	15.6	Irrigated SS and FS
Tamang et al., 2011	Lubbock, TX	2008–2009	20.0	10.8	15.1	Irrigated PS N study
Maughan et al., 2012	Illinois	2009–2010	39.9	13.5	24.5	ES N study, sum of two harvests
Rocateli et al., 2012	Shorter, AL	2008–2009	30.1	8.0	15.0	PS and FS hybrids
Snider et al., 2012	Alabama and Arkansas	2009–2010	61.0	15.0	31.8	PS hybrid

† FS, forage sorghum; PS, photoperiod sensitive; SWS, sweet sorghum; SS, sorghum × sudangrass; ES, energy sorghum.

reason for the low lignin values. Corredor et al. (2009) found that total structural carbohydrates in sorghum ranged from 48 to 80%, with lignin ranging from 11 to 20%. They also reported cellulose concentrations of 24 to 38% and hemicellulose concentrations of 18 to 22%. Additional work by Theerarattananoon et al. (2010) reported similar percentages of structural carbohydrates and also illustrated the range of compositions across different forage sorghum types. They reported lignin levels to be as low as 15% in a BMR forage sorghum compared with 18–20% in other forage and corn biomass samples. The BMR forage samples also had some of the highest glucan and xylan levels of the forages tested.

Dahlberg et al. (2011) conducted one of the most extensive evaluations of sorghum cultivars for yield and composition. They selected 22 cultivars that comprised both commercial forage and sweet sorghums (Table 4). The group included forage sorghum hybrids, So×Su hybrids, and sweet sorghum varieties. They also included subsets of BMR, non-BMR, photoperiod-sensitive, and non-photoperiod-sensitive cultivars. They found that carbohydrate composition varied widely among the different cultivars. Ash ranged from 7.3 to 11.3 g g^{-1}, lignin from 9.9 to 16.3 g g^{-1}, glucans from 18.7 to 35.2, and xylans from 12.6 to 19.9 g g^{-1}.

Sugar from Sorghum Biomass

Biomass to ethanol conversion, as with most plant-based biochemical production process, is driven by glucose and xylose, the two primary C6 and C5 sugars found in plant tissues. Pretreatments and extractions rates may differ with the method and potentially by feedstock species; however, if good analytical data can be obtained for the concentrations of the C6 and C5 sugars extracted from a given amount of biomass, then the ethanol yields will be predictable. The NREL currently cites the rate for the conversion of C6 and C5 sugars to ethanol as 0.51

Table 4. Forage sorghum structural composition data from four different research reports.

Source		Lignin	Cellulose	Hemicellulose	Starch	Ash	Description
		mg g^{-1}					
Miller and McBee, 1993	Max.	90.0†	344.0	251.0	52.0	NR‡	Forage sorghum
	Min.	53.0	226.0	184.0	15.0	NR	and grain types
	Mean	70.3	2663.8	204.8	29.7	NR	
Dien et al., 2009	Max.	148.8	256.0	203.0	39.0	67.6	Degrained
	Min.	100.3	240.0	183.0	19.0	51.6	samples, includes
	Mean	125.7	246.9	191.6	28.9	59.7	BMR§ hybrids
Corredor et al., 2009	Max.	204.7	387.2	224.8	229.1	108.7	PS-BMR, PS-non
	Min.	110.6	242.1	123.2	8.4	69.3	BMR, Forage
	Mean	163.3	341.2	182.0	96.7	92.7	BMR, and Forage Sorghum
Dahlberg et al., 2011	Max.	158.5	434.5	239.2	428.3	117.9	Forage, BMR,
	Min.	94.9	187.5	158.4	0.0	73.9	sorghum-
	Mean	123.6	274.3	188.7	154.1	92.1	sudangrass, and non-grain types

† Lignin determination method not listed.

‡ NR, not reported.

§ BMR, brown midrib.

kg ethanol/kilogram sugar (USDOE, 2006). These conversion factors take into account the addition of water during hydrolysis. This approach has been used by numerous authors to estimate ethanol yields. Dahlberg et al. (2011) estimated that ethanol yields from forage sorghum and sorghum-sudan hybrids ranged from 336 to 440 L Mg^{-1}, with a mean of 403 L Mg^{-1}. Theerarattananoon et al. (2010) reported sorghum composition data that resulted in ethanol estimates ranging from 486 to 521 L Mg^{-1}. Estimated ethanol yields for wheat, corn, big bluestem, and biomass sorghum were 473, 476, 449, and 471 L Mg^{-1}, respectively, based on composition data (Theerarattananoon et al., 2012). As previously mentioned, BMR sorghum hybrids typically have lower lignin concentrations and have been used in the livestock-feeding industry because of higher conversion rates. These characteristics carry over into cellulosic ethanol production. Dahlberg et al. (2011) reported that the highest conversion rate, 429 L Mg^{-1}, was from a BMR hybrid, but low yields across all the BMR hybrids evaluated did not result in higher ethanol yields per hectare compared with the other sorghum cultivars examined.

The production of organic acids, primarily acetic, propionic, butyric, and lactic acids, from biomass and grain has been under parallel development in the lignocellulosic ethanol production industry but has gained less public attention (Du and Yu, 2002; Zhan et al., 2002; Carole et al., 2004; Causey et al., 2004). These acids serve as precursors to plastics, fibers, and polyurethanes, which are currently produced primarily from petroleum substrates. While most of these developmental efforts have focused on a wide range of feedstocks, those with grain and sweet sorghum have met with a great deal of success (Samuel et al., 1980; Richter and Berthold, 1998; Richter and Träger, 1994; Zhan et al., 2003; Wee et al., 2006).

Combustion of Sorghum Biomass

The combustion of raw materials for the creation of heat has been a part of human existence for millennia. New technology in the late nineteenth century led to this energy source being harnessed to generate electricity. Coal dominated as the preferred source for heat and electricity generation until the late twentieth and early twenty-first centuries, when atmospheric CO_2 concentrations became a concern. Combustion of biomass became a topic of interest, with wood being the dominant source, and biomass from perennial and annual crops used on a smaller scale. More than 56 million megawatts of electricity are generated annually in the United States from biomass (USEIA, 2012), with wood being the dominant fuel; however, crop biomass and residues also have been used to generate electricity. The University of Minnesota, Morris generates electricity with a gasification system using corn residue, wood, and perennial grasses as fuel (Univ. Minnesota, Morris, 2013). Sorghum has been successfully combusted to generate electricity in California (Chromatin, 2012).

As a heat-generation fuel, sorghum has high heating values (HHV) that are comparable to those of other renewable resources (Table 5). Sorghum biomass HHVs have been reported to range from 16.00 to 18.32 MJ kg^{-1}. One of the challenges of using annual herbaceous crop biomass as a fuel source is that they are high in alkali metals (e.g., K and Na) and halides (e.g., Cl), with amounts often significantly exceeding the 2–3% of ash that is typically desired (Miles et al., 1995). High levels of these chemicals increase sintering and agglomeration in reactors and reactor beds (Ergudenler and Gahly, 1993; Miles et al., 1996; Jenkins et al., 1998; Nielsen et al., 2000; Fryda et al., 2008). The addition of kaolin, dolomite, and ammonium sulfate has been reported to reduce slagging and the corrosive effects of halides such as Cl (Davidsson et al., 2002; Ohman et al., 2004; Davidsson et al., 2008; Kassman et al., 2011).

Probably the most widely known use of sorghum for electricity generation is that of sweet sorghum bagasse at ethanol plants in India, Brazil, and China (Prasad et al., 2007; Reddy et al., 2007; Zhang et al., 2010). Cubuk et al. (2011) reported that co-combustion of sweet sorghum biomass and lignite coal resulted in similar NO*x* emissions but reduced C emissions. Cubuk and Heperkan (2004) noted that a similar mixture reduced pollutant concentrations in the emissions from lignite coal. Tillman (2000) pointed out that co-firing biomass with coal reduced the emission of NO*x*, SO*x*, fossil-fuel CO_2, and trace minerals, such as mercury; however, co-firing biomass with coal resulted in reduced boil efficiency in every test he reviewed.

The combustion of sweet sorghum bagasse will probably be quickly adopted because sugarcane bagasse has been used for centuries as a power source at sugarcane mills and also because it is readily available and eliminates the need for disposal. This system does not have the fouling and corrosion problems mentioned above because the K, Na, and Cl are physically removed from the bagasse during the sugar-extraction process along with the sugar. Das et al. (2004) and Jenkins et al. (1996) reported that nearly all (>92%) of the measured K, Na, and Cl were removed by leaching the biomass with water.

Anaerobic Digestion of Sorghum

Anaerobic digestion as a pathway to convert sorghum to biogas for heat and electricity or for transportation fuel is not widely used in the United States. Anaerobic digestion has been used for decades to reduce emissions from livestock waste systems by stimulating the biodegradable material to degrade and capturing the subsequent methane gas that is produced during degradations (Karellas et al., 2010). The inclusion of higher-carbon feedstocks that are also high in N are

Table 5. Fuel analyses for several plant biomass and fossil fuel sources from various reports.

Source	Crop	Fixed carbon	Volatile material	Ash	C	H	O	N	S	High heating value	SiO in ash	K_2O in ash	Cl in ash or plant material
		% (w/w) on dry wt. basis								MJ kg^{-1}	% of ash		
Fryda et al., 2008	Sweet sorghum bagasse	—	—	3.20	49.50	6.20	40.10	0.90	0.01	18.32	31.60	31.60	5.10
	Arundo donax L.	—	—	2.48	46.50	5.70	44.70	0.50	0.01	17.98	44.20	30.00	—
Hartley et al., 2011	Biomass sorghum	—	—	6.58	45.16	5.61	42.23	0.46	0.00	17.99	—	—	—
Ture et al., 1997	Sweet sorghum	—	—	2.18	44.04	6.26	—	0.22	0.08	16.83	—	—	—
	Corn stover	—	—	11.63	46.64	5.66	39.59	0.67	0.08	18.26	—	—	—
	Wheat straw	—	—	10.22	43.88	5.26	38.75	0.63	0.16	17.36	—	—	—
	Switchgrass	—	—	5.84	47.26	5.58	40.70	0.59	0.09	18.66	—	—	—
González et al., 2006	Sweet sorghum	10.30	61.30	2.70	34.00	4.50	60.23	0.80	0.02	16.00	—	—	0.45†
	Forest pellet	13.80	76.40	1.00	46.50	6.80	44.77	1.90	0.00	18.40	—	—	0.03†
	Arundo donax L.	11.40	58.40	2.20	40.30	5.30	53.13	0.40	0.07	17.40	—	—	0.80†
Channiwala and Parikh, 2002	Sugarcane bagasse	13.51	83.66	3.20	45.48	5.96	45.21	0.15	—	18.73	—	—	—
	Methane	—	—	—	74.85	25.15	—	—	—	55.35	—	—	—
	Bituminous coal	65.25	33.45	6.30	76.65	4.78	10.87	0.54	0.42	31.00	—	—	—
	Municipal solid waste	—	—	12.00	47.60	6.00	32.90	1.20	0.30	19.88	—	—	—
Jenkins et al., 1998	Yard waste	13.59	66.04	20.37	41.54	4.79	31.91	0.85	0.24	16.30	59.65	2.96	0.30
	Mixed paper	7.42	84.25	8.33	47.99	6.63	36.84	0.14	0.07	20.78	28.10	0.16	0.00

† Cl reported as % of plant dry matter.

more beneficial than the digestion of animal manure alone. Karellas et al. (2010) listed the advantages as (i) more flexibility in siting anaerobic digesters (ADs), (ii) reduction in AD volumes as well as construction and operating costs, (iii) lower mixing rates, which reduce parasitic electricity consumption, (iv) reduced cost of feedstock logistics because of higher-density materials, and (v) production of higher-quality fertilizer in the mineral effluents.

Biogas production rates are highly dependent on the composition of the feedstock. Total solids (TS) and volatile solids (VS) are two important components to be considered when evaluating a potential feedstock (Schievano et al., 2008; Karellas et al., 2010). Schievano et al. (2008 and 2009) reported that only lab measurements of VS and oxygen demand in 20 h (OD20) were needed to predict methane production in an AD. Angelidaki et al. (1999) developed a model to estimate biogas and methane production that accounted for substrate carbohydrate and N concentrations as well as the OD20, whereas Chandra et al. (2012) indicated that a C/N ratio between 20 and 30 was optimal for digestion.

Sorghum biomass is high in both TS and VS (Table 6), which is optimal for the fermentation process (Richards et al., 1991). As with animal nutrition, feedstocks with lower lignin levels result in greater digestion of TS (Herrmann et al., 2011). Herrmann et al. also found that ensiling had minimal impact on AD and that most of the dry matter lost during the ensiling process was converted to lactic and acetic acid, which are easily consumed in an AD. Also of interest is the use of thin stillage from grain-ethanol plants. Thin stillage from either corn or sorghum is quite limited in its ability to provide VS or organically available dry matter for digestion (Table 6). Sweet sorghum juice has similar properties as thin stillage in that is it low in both TS and VS. Probably both could be used as the liquid substrate in ADs to increase methane production while reducing the water use.

Addition of plant material as AD substrates increases yields over those of manure-only substrates. Zhou et al. (2011) reported a sixfold increase in biogas production when grass silage was digested compared with cow manure alone. Forage and sweet sorghum are feedstocks that produce high yields of biogas because of their elevated content of free sugars and VS (Table 6). High-biomass sorghum has been shown to be a good biogas feedstock. Claassen et al. (2004) and Antonopoulou et al. (2008) described systems in which sweet sorghum bagasse was used as a substrate by thermophilic bacteria to produce H_2 gas for biopower. High biomass yield per land area and high digestion rates result in high energy production per unit land area. Mahmood and Honermeier (2012) reported production of 8114 m^3 biogas ha^{-1} and 4333 m^3 methane ha^{-1} from sorghum. Jerger et al. (1987) reported more than 90% conversion of VS from sweet sorghum biomass to methane, with yields near 0.36 L g^{-1} VS.

Sweet Sorghum

Sweet sorghum cultivars are forage sorghum cultivars that have been selected for their high stem-juice yields and elevated sap-sugar concentrations. Sweet sorghum cultivars routinely have sugar concentrations above 15% (v/v) (Wu et al., 2010). Compared with grain sorghum, which produces a large panicle of grain containing complex carbohydrates, sweet sorghum stores nonstructural

Table 6. Anaerobic digester substrate composition analysis and methane production rate for a range of agricultural feedstocks.

Source	Feedstock	Total solids	Volatile solids	Methane
		g kg^{-1}	g kg^{-1} TS	L g VS^{-1}
Schievano et al., 2009	Cattle manure	18	799	0.058†
Schievano et al., 2009	Swine manure	30	602	0.126†
Schievano et al., 2009	Poultry litter	235	680	0.153†
Schievano et al., 2009	Maize silage	300	915	0.330†
Schievano et al., 2009	Sweet sorghum silage	200	905	0.290†
Richards et al., 1991	Sorghum silage	900	945	0.360
Lee et al., 2011	Corn thin stillage	73	64	0.630
Antonopoulou et al., 2008	Sweet sorghum juice	20	20	0.317‡
Miller and McBee, 1993	Forage sorghum	300	958	0.285
Jerger et al., 1987	Sweet sorghum (Rio)	343	945	0.400
Jerger et al., 1987	Grain sorghum (RS 610)	294	894	0.310
Jerger et al., 1987	High energy (ATx623 × Rio)	322	942	0.340

† Values estimated from model.

‡ Values estimated from H_2 yields.

carbohydrates in its stems (McBee et al., 1988). Sweet sorghum was originally used in the United States for human consumption as either crystal sugar or molasses.

Sweet sorghum production in the United States has largely been concentrated in the South and occupied enough acres to warrant attempts at improvement (Walton et al., 1938). It was estimated that nearly 190 million L of sweet sorghum syrup was produced in 1920. Cultivar improvement efforts were located at USDA facilities in Meridian, MS and Weslaco, TX from 1950 to 1970 (Coleman, 1970; Hipp et al., 1970). As a result, sweet sorghum cultivars were bred for either molasses production or crystal sugar production (Nathan, 1978). Sugar extracted from sweet sorghum, much like that from sugarcane, is composed primarily of sucrose, glucose, and fructose. Concentrations of each sugar vary among cultivars, with sucrose concentrations ranging from 65 to 93% (Corn, 2009; Kim and Day, 2011; Teetor et al., 2011; Wu et al., 2011). Glucose typically represents 75% of the remaining sugars, and fructose the balance. With increased interest in renewable fuels, research on sweet sorghum rose again in the early part of the twenty-first century, but largely on the same cultivars released from the 1950s through the 1980s (Miller and Ottman, 2010; Propheter et al., 2010; Wortmann et al., 2010; Teetor et al., 2011; Godsey et al., 2012).

Resent research indicates that sweet sorghum can attain similar fresh weights and fermentable sugars per unit of planted area as sugarcane (Table 7). The highest yields in the United States were reported in the humid, warm locations of Louisiana and South Texas and under irrigation in the desert Southwest. Yields were reduced by nearly 50% when sweet sorghum was grown under rainfed conditions in the southern and northern Great Plains, as reported by Tamang et al. (2011) and Wortmann et al. (2010). Sweet sorghum performance at locations that may be construed as the northern limit for sweet sorghum production in the United States—northeast Kansas (Propheter et al., 2010) and south central Nebraska (Wortmann et al., 2010)—indicate that water supply (stored soil water and in-season rainfall) can influence sweet sorghum yields, but these environments are also capable of producing relatively high yields.

Table 7. Sweet sorghum yields from various years and US locations.

Source	Site	Dry yields				Juice concentration	Ethanol†			
		Biomass	Stover‡	Grain‡	Juice§		Juice	Bagasse	Grain	Total
		Mg ha^{-1}				Brix	L ha^{-1}			
Propheter et al., 2010	KS	30.4	28.8	1.7	53.4	14.8	4390.1	13742.5	686.4	18819.0
Miller and Ottman, 2010	AZ	26.0	24.6	1.4	45.7		2472.2	11753.5	587.1	14812.7
Teetor et al., 2011	AZ	39.1	37.0	2.1	68.7	12.4	3324.5	17671.1	882.6	21878.2
Kim and Day, 2011	LA	30.0	28.4	1.6	52.7	10.5	3066.6	13561.7	677.4	17305.6
Corn, 2009	TX	43.7	40.2	3.5	74.7	13.7	3368.4	19228.3	1456.0	24052.8
Tamang et al., 2011	TX	14.9	14.1	0.8	26.2	11.3	1643.7	6735.6	336.4	8715.7
Tew et al., 2008	LA	28.6	27.1	1.6	50.3		4900.0	12942.4	646.4	18488.8
Wortmann et al., 2010	NE	9.8	9.2	0.5	17.2	9.9	1087.7	4418.8	220.7	5727.3
Wortmann et al., 2010	NE	15.9	15.1	0.9	28.0	11.2	2485.4	7208.0	360.0	10053.5

† Ethanol yields estimated from juice yields or dry yields using the following conversion rates: juice = 584.8 L ethanol per Mg sugar; bagasse = 478 L ethanol per Mg dry matter; grain = 416 L ethanol per Mg dry matter; sugar yield ha^{-1} = 0.95 × juice yield × 0.873 × (brix/100).

‡ Grain and stover yields may have been estimated based on total biomass yields, subtraction, and an assumed harvest index of 0.05.

§ Theoretical juice yield estimated as the difference between fresh weights and dry weights. If not reported, moisture content was assumed to be 65%.

Across all of the locations analyzed, juice concentration was less affected, which could have resulted from the situation that a majority of the data reported is from M81E, an open-pollinated sweet sorghum variety developed in 1981 at the USDA facility in Meridian, MS (Broadhead et al., 1981). It is expected that sugarcane sugar extraction technology will be used to extract sugar from the sweet sorghum stalks. This assumption allows for theoretical juice and sugar yields to be calculated on the basis of moisture content and sugar concentration. In many instances, sugar concentration is measured in degrees Brix, which is a measure of dissolved solids in a solution. For our analyses, sugar yields were estimated with Eq. [1], in which sugar and juice are measured in Mg ha^{-1} (Corn, 2009)

$$\text{Sugar yield} = 0.95 \times \text{juice yield} \times 0.873 \times \frac{\text{brix}}{100} \quad [1]$$

This relationship assumes that the expected sugar extraction rate from sweet sorghum stalks is approximately 95% (Bennett and Anex, 2009). The 0.873 is the percentage of fermentable sugars present (in degrees Brix) and is based on measurements made by Corn (2009).

Using a conversion rate for sugar to ethanol of 584 L Mg^{-1} sugar for sweet sorghum juice (Shapouri et al., 2006), 478 L Mg^{-1} biomass for sweet sorghum residual biomass (bagasse) (Dahlberg et al., 2011), and 416 L ethanol Mg^{-1} for sweet sorghum grain (Shapouri et al. (2006) resulted in ethanol estimates ranging from 5,727 to more than 24,000 L ha^{-1} (Table 7). On average, juice accounts for close to 20%, bagasse 77%, and grain 3% of these ethanol values. These estimates clearly illustrate the value of large amounts of bagasse as a feedstock. The fate of bagasse will be dictated by the value of the end energy product, which can include ethanol, electricity, or steam. This value proposition will also be dictated by the local infrastructure and energy needs.

Sweet sorghum hybrids and varieties are often overlooked as potential livestock feed. Their high yields of biomass and sugar make them ideal for forage products. In fact, many forage sorghum hybrids used for silage have high sugar contents but low juice yields. Adewakun et al. (1989) found that Brandes sweet sorghum resulted in similar average daily weight gain as corn silage and greater average daily weight gain compared with fescue hay. Organic matter digestibility, digestibility of crude protein and ether extracts, and gross energy digestibility were higher in steers fed sweet sorghum compared with corn silage. Caswell et al. (1983) suggested that the greater amounts of sugar and ether extracts in sweet sorghum provide more readily available energy for animal performance. Lance et al. (1964) reported similar milk yields from sweet sorghum feed compared with corn silage. Stefaniak et al. (2012) noted that sweet sorghum hybrids were higher in starch than other forage sorghum groups, suggesting that this could influence animal performance.

One advantage of sweet sorghum as a livestock feed is that the chemical producer is primarily interested in the sugar-containing juice. The by-product of this process would be the bagasse. Most studies have shown that there is very little loss of dry matter as a result of the juicing process and, in fact, that the biomass is macerated to into very fine pieces, potentially improving digestibility (Propheter et al., 2010). Being a by-product, bagasse also has the potential to reduce feed costs for the livestock producer and an additional income source for the chemical manufacturer.

Summary

Sorghum is one of the most genetically diverse crops used in agriculture. Forage sorghum is no exception, with the diversity increasing with the inclusion of So×Su and Su×Su hybrids. Current forage sorghum end users have the benefit of new traits, such as BMR and brachytic dwarfism, to improve digestibility and harvested yield. To date, most of these forages are used in the diets of ruminant livestock. In the past decade, an interest in using sorghum for renewable energy feedstocks has given rise to biomass sorghum, which comprises selections from forage sorghum hybrids with a focus on biomass yield and less on biomass quality. Attention to sorghum as a renewable feedstock also stimulated a renewed interest in sweet sorghum, which was developed by selecting forage sorghum hybrids with high stem-sugar content and juice yields. These hybrids have been positioned to extend the sugarcane growing season in some environments and have also been found to produce high levels of biogas in ADs. Biomass sorghum has also been proposed as a combustion feedstock, but slagging caused by high levels of K and Cl requires additional technology to be deployed to successfully use biomass sorghum as a heat source. Forage and biomass sorghum have proven to be reliable renewable food and fiber sources, especially in marginal environments in which water is limiting.

References

Adewakun, L.O., A.O. Famuyiwa, A. Felix, and T.A. Omole. 1989. Growth performance, feed intake and nutrient digestibility by beef calves fed sweet sorghum silage, corn silage, and fescue hay. J. Anim. Sci. 67:1341–1349.

Angelidaki, I., L. Ellegaard, and B.K. Ahring. 1999. A comprehensive model of anaerobic bioconversion of complex substrates to biogas. Biotechnol. Bioeng. 63:363–372. doi:10.1002/(SICI)1097-0290(19990505)63:3<363::AID-BIT13>3.0.CO;2-Z

Antonopoulou, G., H.N. Gavala, I.V. Skiadas, K. Angelopoulos, and G. Lyberatos. 2008. Biofuels generation from sweet sorghum: Fermentative hydrogen production and anaerobic digestion of the remaining biomass. Bioresour. Technol. 99:110–119. doi:10.1016/j.biortech.2006.11.048

Bean, B., and T. McCollum. 2006. Summary of six years of forage sorghum variety trials. Publ. SCS-2006-04. Texas A&M AgriLife, Texas A&M Univ., College Station.

Bell, J., Q. Xue, T. McCullom, P. Sirmon, T. Brown, and D. Pietsch. 2015. 2014 Texas Panhandle Sorghum Silage Trial. Texas A&M AgriLife, Texas A&M Univ., College Station.

Bennett, A.S., and R.P. Anex. 2009. Production, transportation and milling costs of sweet sorghum as a feedstock for centralized bioethanol production in the upper Midwest. Bioresour. Technol. 100:1595–1607. doi:10.1016/j.biortech.2008.09.023

Broadhead, D.M., K.C. Freeman, and N. Zummo. 1981. M 81E: A new variety of sweet sorghum. Info. Sheet 1309. Mississippi Agric. and Forestry Exp. Stn. Starkville.

Brouk, M.J. 2012. Sorghum in beef production feeding guide. United Sorghum Checkoff Program, Lubbock, TX.

Brouk, M.J., and B. Bean. 2011. Sorghum in dairy cattle production feeding guide. United Sorghum Checkoff Program, Lubbock, TX.

Bucholtz, D.L., R.P. Cantrell, J.D. Axtell, and V.L. Lechtenberg. 1980. Lignin biochemistry of normal and brown midrib mutant sorghum. J. Agric. Food Chem. 28:1239–1241. doi:10.1021/jf60232a045

Caravetta, G. J., J.H. Cherney, and K.D. Johnson. 1990. Within-row spacing influences on diverse sorghum genotypes: II. Dry matter yield and forage quality. Agron. J. 82:210–215.

Carole, T.M., J. Pellegrino, and M.D. Paster. 2004. Opportunities in the industrial biobased products industry. Appl. Biochem. Biotechnol. 115:871–885. doi:10.1385/ABAB:115:1-3:0871

Caswell, L.F., R.S. Kalmbacher, and F.G. Martin. 1983. Yield and silage fermentation characteristics of corn, sweet sorghum and grain sorghums. Proc. Soil Crop Sci. Soc. Fla. 42:139–142.

Causey, T.B., K.T. Shanmugam, L.P. Yomano, and L.O. Ingram. 2004. Engineering *Escherichia coli* for efficient conversion of glucose to pyruvate. Proc. Natl. Acad. Sci. USA 101:2235–2240. doi:10.1073/pnas.0308171100

Chandra, R., H. Takeuchi, and T. Hasegawa. 2012. Methane production from lignocellulosic agricultural crop wastes: A review in context to second generation of biofuel production. Renew. Sustain. Energy Rev. 16:1462–1476. doi:10.1016/j.rser.2011.11.035

Channiwala, S.A., and P.P. Parikh. 2002. A unified correlation for estimating HHV of solid, liquid and gaseous fuels. Fuel. 81:1051–1063. doi:10.1016/S0016-2361(01)00131-4

Chromatin. 2012. Constellation energy and chromatin announce partnership to test sorghum biomass as fuel to generate power. Press release, Chromatin, Inc., 21 Sept. 2011. http://www.chromatininc.com/news/Constellation-Energy-and-Chromatin-Announce-Partnership-to-Test-Sorghum-Biomass-as-Fuel-to-Generate-Power (accessed 9 May 2013).

Claassen, P.A.M., T. de Vrije, and M.A.W. Budde. 2004. Biological hydrogen production from sweet sorghum by thermophilic bacteria. Proceedings of the 2nd World Conference on Biomass for Energy, Industry, and Climate Protection, Rome Italy. 10–14 May. ETA-Renewable Energies, Florence, Italy, and WIP, Munich, Germany. p. 1522–1525.

Coleman, O.H. 1970. Syrup and sugar from sweet sorghum. In: J.S. Wall and W.M. Ross, editors, Sorghum production and utilization. AVI Publishing, Westport, CT. p. 416–441.

Corn, R.J. 2009. Heterosis and composition of sweet sorghum. Ph.D. diss. Texas A&M Univ., College Station. http://hdl.handle.net/1969.1/ETD-TAMU-2009-12-7409

Corredor, D.Y., J.M. Salazar, K.L. Hohn, S. Bean, B. Bean, and D. Wang. 2009. Evaluation and characterization of forage sorghum as feedstock for fermentable sugar production. Appl. Biochem. Biotechnol. 158:164–179. doi:10.1007/s12010-008-8340-y

Craufurd, P.Q., V. Mahalakshmi, F.R. Bidinger, S.Z. Mukuru, J. Chantereau, P.A. Omanga, A. Qi, E.H. Roberts, R.H. Ellis, R.J. Summerfield, and G.L. Hammer. 1999. Adaptation of sorghum: Characterisation of genotypic flowering responses to temperature and photoperiod. Theor. Appl. Genet. 99:900–911. doi:10.1007/s001220051311

Cubuk, M.H., and H.A. Heperkan. 2004. Investigation of pollutant formation of sweet sorghum–lignite (Orhaneli) mixtures in fluidised beds. Biomass Bioenergy 27:277–287. doi:10.1016/j.biombioe.2004.02.001

Cubuk, M.H., D.B. Ozkan, and O. Emanet. 2011. NO_x formation of co-combustion of sweet sorghum–ignite (Orhaneli) mixtures in fluidized beds. In: H. Gökçekus, U. Türker, and J.W. LaMoreaux, editors, Survival and sustainability. Springer, Berlin. p. 931–941.

Dahlberg, J., E. Wolfrum, B.W. Bean, and W.L. Rooney. 2011. Compositional and agronomic evaluation of sorghum biomass as a potential feedstock for renewable fuels. J. Biobased Mat. Bioenergy 5:507–513. doi:10.1166/jbmb.2011.1171

Das, P., A. Ganesh, and P. Wanikar. 2004. Influence of pretreatment for deashing of sugarcane bagasse on pyrolysis products. Biomass Bioenergy 27:445–457.

Davidsson, K.O., L.E. Åmand, B.M. Steenari, A.L. Elled, D. Eskilsson, and B. Leckner. 2008. Countermeasures against alkali-related problems during combustion of biomass in a circulating fluidized bed boiler. Chem. Eng. Sci. 63:5314–5329. doi:10.1016/j.ces.2008.07.012

Davidsson, K.O., J.G. Korsgren, J.B.B. Perrersson, and U. Jaglid. 2002. The effects of fuel washing techniques on alkali release from biomass. Fuel 81:137–142. doi:10.1016/S0016-2361(01)00132-6

Dien, B.S., G. Sarath, J.F. Pedersen, S.E. Sattler, H. Chen, D.L. Funnell-Harris, N.N. Nichols, and M.A. Cotta. 2009. Improved sugar conversion and ethanol yield for forage sorghum (*Sorghum bicolor* L. Moench) lines with reduced lignin contents. BioEnergy Res. 2:153–164. doi:10.1007/s12155-009-9041-2

Dien, B.S., N.N. Nichols, and R.J. Bothast. 2001. Recombinant *Escherichia coli* engineered for production of L-lactic acid from hexose and pentose sugars. J. Ind. Microbiol. Biotechnol. 27:259–264. doi:10.1038/sj.jim.7000195

Dingkuhn, M., M. Kouressy, M. Vaksmann, B. Blerget, and J. Chantereau. 2008. A model of sorghum photoperiodism using the concept of threshold-lowering during prolonged appetence. Eur. J. Agron. 28:74–89. doi:10.1016/j.eja.2007.05.005

Du, G., and J. Yu. 2002. Green technology for conversion of food scraps to biodegradable thermoplastic polyhydroxyalkanoates. Environ. Sci. Technol. 36:5511–5516. doi:10.1021/es011110o

Ergudenler, A., and A.E. Gahly. 1993. Agglomeration of silica sand in a fluidized bed gasifier operating on straw. Biomass Bioenergy 4:135–147. doi:10.1016/0961-9534(93)90034-2

Fritz, J.O., K.J. Moore, and E.H. Jaster. 1990. Digestion kinetics and cell wall composition of brown midrib sorghum × sudangrass morphological components. Crop Sci. 30:213–219. doi:10.2135/cropsci1990.0011183X003000010046x

Fryda, L.E., K.D. Panopoulos, and E. Kakaras. 2008. Agglomeration in fluidised bed gasification of biomass. Powder Technol. 181:307–320.

Godsey, C.B., J. Linneman, D. Bellmer, and R. Hunke. 2012. Developing row spacing and plant density recommendations for rainfed sweet sorghum production in the Southern Plains. Agron. J. 104:280–286. doi:10.2134/agronj2011.0289

González, J.F., C.M. González-García, A. Ramiro, J. Gañán, A. Ayuso, and J. Turegano. 2006. Use of energy crops for domestic heating with a mural boiler. Fuel Process. Technol. 87:717–726. doi:10.1016/j.fuproc.2006.02.002

Goud, J.V., and M. Vasudev. Rao. 1977. Inheritance of height in sorghum. Genet. Agrar. 31:39–51.

Hallam, A., I.C. Anderson, and D.R. Buxton. 2001. Comparative economic analysis of perennial, annual, and intercrops for biomass production. Biomass Bioenergy 21:407–424.

Hanna, W.W., W.G. Monson, and T.P. Gaines. 1981. In vitro dry matter digestibility, total sugars, and lignin measurements on normal and brown midrib (bmr) sorghums at various stages of development. Agron. J. 73:1050–1052. doi:10.2134/agronj1981.00021962007300060034x

Hatfield, R.D., J.G. Jung, J. Ralph, D.R. Buxton, and P.J. Weimer. 1994. A comparison of the insoluble residues produced by the Klason lignin and acid detergent lignin procedures. J. Sci. Food Agric. 65:51–58. doi:10.1002/jsfa.2740650109

Hartley, B.E., J.D. Gibson, J.A. Thomasson, and S.W. Searcy. 2011. Moisture loss and ash characterization of high-tonnage sorghum. Paper presented at ASABE Annual International Meeting, Louisville, KY. 7–10 August. Paper 1111528.

Heldt, H.W., and B. Piechulla. 2011. Plant biochemistry. 4th ed. Academic Press, Burlington, MA.

Herrmann, C., M. Heiermann, and C. Idler. 2011. Effects of ensiling, silage additives and storage period on methane formation of biogas crops. Bioresour. Technol. 102:5153–5161. doi:10.1016/j.biortech.2011.01.012

Hill, J.D. 1976. Predicting the natural drying of hay. Agric. Meteorol. 17:195–204. doi:10.1016/0002-1571(76)90055-8

Hipp, B.W., W.R. Cowley, C.J. Gerard, and B.A. Smith. 1970. Influence of solar radiation and date of planting on yield of sweet sorghum. Crop Sci. 10:91–92. doi:10.2135/cropsci1970.0011183X001000010033x

Howell, T.A, S.R. Evett, J.A. Tolk, K.S. Copeland, P.D. Colaizzi, and P.H. Gowda. 2008. Evapotranspiration of corn and forage sorghum for silage. Wetting Front 10 (1):3–11

Hu, G., C. Cateto, Y. Pu, R. Samuel, and A.J. Ragauskas. 2012. Structural characterization of switchgrass lignin after ethanol organosolv pretreatment. Energy Fuels 26:740–745. doi:10.1021/ef201477p

Hussein, M.M., and A.K. Alva. 2014. Growth, yield, and water use efficiency of forage sorghum as affected by N, P, K fertilizer and deficit irrigation. Am. J. Plant Sci. 5:2134–2140. doi:10.4236/ajps.2014.513225

Jenkins, B.M., R.R. Bakker, and J.B. Wei. 1996. On the properties of washed straw. Biomass Bioenergy 10:177–200. doi:10.1016/0961-9534(95)00058-5

Jenkins, B.M., L.L. Baxter, T.R. Miles, Jr., and T.R. Miles. 1998. Combustion properties of biomass. Fuel Process. Technol. 54:17–46. doi:10.1016/S0378-3820(97)00059-3

Jerger, D.E., D.P. Chynoweth, and H.R. Isaacson. 1987. Anaerobic digestion of sorghum biomass. Biomass 14:99–113.

Karellas, S., I. Boukis, and G. Kontopoulos. 2010. Development of an investment decision tool for biogas production from agricultural waste. Renew. Sustain. Energy Rev. 14:1273–1282. doi:10.1016/j.rser.2009.12.002

Kassman, H., M. Brostrom, M. Berg, and L. Amand. 2011. Measures to reduce chlorine in deposits: Application in a large-scale circulating fluidised bed boiler firing biomass. Fuel 90:1325–1334. doi:10.1016/j.fuel.2010.12.005

Kim, M., and D.F. Day. 2011. Composition of sugar cane, energy cane, and sweet sorghum suitable for ethanol production at Louisiana sugar mills. J. Ind. Microbiol. Biotechnol. 38:803–807. doi:10.1007/s10295-010-0812-8

Lance, R.D., D.C. Foss, C.R. Kruegar, B.R. Baumgardt, and R.R. Niedermeir. 1964. Evaluation of corn and sorghum silages on the basis of milk production and digestibility. J. Dairy Sci. 47:254–257. doi:10.3168/jds.S0022-0302(64)88635-5

Lee, P, J. Bae, J. Kim, and W Chen. 2011. Mesophilic anaerobic digestion of corn thin stillage: A technical and energetic assessment of the corn-to-ethanol industry integrated with anaerobic digestion. J. Chem. Technol. and Biotechnol. 86:1514–1520.

Mahmood, A., and B. Honermeier. 2012. Chemical composition and methane yield of sorghum cultivars with contrasting row spacing. Field Crops Res. 128:27–33. doi:10.1016/j.fcr.2011.12.010

Marsalis, M.A., S.V. Angadi, and F.E. Contreras-Govea. 2010. Dry matter yield and nutritive value of corn, forage sorghum, and BMR forage sorghum at different plant populations and nitrogen rates. Field Crops Res. 116:52–57. doi:10.1016/j.fcr.2009.11.009

Marsalis, M.A., S.V. Angadi, F.E. Contreras-Govea, and R.E. Kirksey. 2009. Harvest timing and byproduct addition effects on corn and forage sorghum silage grown under water stress. Bull. 799. New Mexico State Univ., Las Cruces.

Martin, G.C., and R.F. Barnes. 1980. Prediction of energy digestibility of forages with in vitro rumen fermentation and fungal enzyme systems. In: W.G. Pigden, C.C. Balch, and M. Graham, editors, Standardization of analytical methodology for feeds. Publ. IDR-134e. Intl. Res. Dev. Cent., Ottawa, ON.

Maughan, M., T. Voigt, A. Parrish, G. Bollero, W. Rooney, and D.K. Lee. 2012. Forage and energy sorghum response to nitrogen fertilization in central and southern Illinois. Agron. J. 104:1032–1040. doi:10.2134/agronj2011.0408

McBee, G.G., R.A. Creelman, and F.R. Miller. 1988. Ethanol yield and energy potential of stems from a spectrum of sorghum biomass types. Biomass 17:203–211. doi:10.1016/0144-4565(88)90114-X

McCuistion, K.C., B.W. Bean, and F.T. McCollum. 2009. Yield and water-use efficiency response to irrigation level of brown midrib, non-brown midrib, and photoperiod-sensitive forage sorghum cultivars. Forage Grazinglands 7. doi:10.1094/FG-2009-0909-01-RS

McCuistion, K. C., B. W. Bean, and F. T. McCollum. 2010. Nutritional composition response to yield differences in brown midrib, non-brown midrib, and photoperiod sensitive forage sorghum cultivars. Forage and Grazinglands 8. doi:10.1094/FG-2010-0428-01-RS

Miles, T.R., T.R. Miles, Jr., L.L. Baxter, R.W. Bryers, B.M. Jenkins, and L.L. Oden. 1995. Alkali deposits found in biomass power plants. A preliminary investigation of their extent and nature. Publ. NREL/TP-433-8142. Vol. 2. Sandia Natl. Lab., Albuquerque, NM.

Miles, T.R., T.R. Miles, Jr., L.L. Baxter, R.W. Bryers, B.M. Jenkins, and L.L. Oden. 1996. Boiler deposits from firing biomass fuels. Biomass Bioenergy 10:125–138. doi:10.1016/0961-9534(95)00067-4

Miller, A.N., and M.J. Ottman. 2010. Irrigation frequency effects on growth and ethanol yield in sweet sorghum. Agron. J. 102:60–70. doi:10.2134/agronj2009.0191

Miller, F.R., and G.G. McBee. 1993. Genetics and management of physiological systems of sorghum for biomass production. Biomass Bioenergy 5:41–49. doi:10.1016/0961-9534(93)90006-P

Mullet, J.E., W.L. Rooney, P.E. Klein, D. Morishige, R. Murphy, and J.A. Brady. 2010. Discovery and utilization of sorghum genes (*ma5/ma6*). U.S. Patent application 20100024065 A1. Published: 28 January.

Narayanan, S., R.M. Aiken, P.V.V. Prasad, Z. Xin, and J. Yu. 2013. Water and radiation use efficiencies in sorghum. Crop Sci. 105:649–656.

Nathan, R.A., ed. 1978. Fuels from sugar crops: Systems study for sugarcane, sweet sorghum, and sugar beets. Tech. Info. Cent. US DOE 71D-22781.US DOE, Washington, DC.

Nielsen, H.P., F.J. Frandsen, K. Dam-Johansen, and L.L. Baxter. 2000. The implications of chlorine-associated corrosion on the operation of biomass-fired boilers. Prog. Energy Combust. Sci. 26:283–298. doi:10.1016/S0360-1285(00)00003-4

Ohman, M., D. Bostrom, A. Nordin, and H. Hedman. 2004. Effect of kaolin and limestone addition on slag formation during combustion of wood fuels. Energy Fuels 18:1370–1376. doi:10.1021/ef040025+

Ottman, M.J. 2010. Water use efficiency of forage sorghum grown with sub-optimal irrigation. In: Forage and grain report. Univ. of Ariz. Coop. Ext. Publ. AZ1526. Univ. of Arizona, Tucson. p. 26–29. http://cals.arizona.edu/pubs/crops/az1526/az1526d.pdf

Pedersen, M., A. Vikso-Nielsen, and A.S. Meyer. 2010. Monosaccharide yields and lignin removal from wheat straw in response to catalyst type and pH during mild thermal pretreatment. Process Biochem. 45:1181–1186. doi:10.1016/j.procbio.2010.03.020

Porter, K.S., J.D. Axtell, V.L. Lechtenberg, and V.F. Colenbrander. 1978. Phenotype, fiber composition, and in vitro dry matter disappearance of chemically induced brown midrib (*bmr*) mutants of sorghum. Crop Sci. 18:205–208. doi:10.2135/cropsci1978.0011183X001800020002x

Prasad, S., A. Singh, N. Jain, and H.C. Joshi. 2007. Ethanol production from sweet sorghum syrup for utilization as automotive fuel in India. Energy Fuels 21:2415–2420. doi:10.1021/ef060328z

Propheter, J.L., S.A. Staggenborg, X. Wu, and D. Wang. 2010. Performance of annual and perennial biofuel crops: Yield during the first two years. Agron. J. 102:806–814. doi:10.2134/agronj2009.0301

Quinby, J.R. 1974. The genetic control of flowing and growth in sorghum. Adv. Agron. 25:125–162. doi:10.1016/S0065-2113(08)60780-4

Quinby, J.R., and R.E. Karper. 1954. Inheritance of height in sorghum. Agron. J. 46:211–216. doi:10.2134/agronj1954.00021962004600050007x

Rajan, N., and S. Maas. 2007. Comparative evaluation of actual crop water use of forage sorghum and corn for silage. Report as of FY2007 for 2007TX271B. Dep. of the Interior, USGS. http://water.usgs.gov/wrri/07grants/2007TX271B.html (accessed 1 June 2015).

Reddy, B.V.S., A. Ashok Kumar, and S. Ramesh. 2007. Sweet sorghum: A water saving bioenergy crop. Paper presented at the International conference on Linkages between Energy and Water Management for Agriculture in Developing Countries, ICRISAT campus, Hyderabad, India. 29–30 January.

Richards, B.K., R.J. Cummings, W.J. Jewell, and F.G. Herndon. 1991. High solid anaerobic methane fermentation of sorghum and cellulose. Biomass Bioenergy 1:47–53. doi:10.1016/0961-9534(91)90051-D

Richter, K., and C. Berthold. 1998. Biotechnological conversion of sugar and starchy crops into lactic acid. J. Agric. Eng. Res. 71:181–191. doi:10.1006/jaer.1998.0314

Richter, K., and A. Träger. 1994. L(+)-lactic acid from sweet sorghum by submerged and solid-state fermentations. Acta Biotechnol. 14:367–378. doi:10.1002/abio.370140409

Rocateli, A.C., R.L. Raper, K.S. Balkcom, F.J. Arriaga, and D.I. Bransby. 2012. Biomass sorghum production and components under different irrigation/tillage systems for the southeastern U.S. Ind. Crops Prods. 36:589–598. doi:10.1016/j.indcrop.2011.11.007

Rooney, W.L., J. Blumenthall, B. Bean, and J.E. Mullet. 2007. Designing sorghum as a dedicated bioenergy feedstock. Biofuels, Bioprod. Biorefin. 1:147–157. doi:10.1002/bbb.15

Saeed, I.A.M., and A.H. El-Nadi. 1998. Forage sorghum yield and water use efficiency under variable irrigation. Irrig. Sci. 18:67–71. doi:10.1007/s002710050046

Samuel, W.A., Y.Y. Lee, and W.B. Anthony. 1980. Lactic acid fermentation of crude sorghum extract. Biotechnol. Bioeng. 22:757–777. doi:10.1002/bit.260220404

Schievano, A., M. Pognani, G. D'Imporzano, and F. Adani. 2008. Predicting anaerobic biogasification potential of ingestates and digestates of a full-scale biogas plant using chemical and biological parameters. Bioresour. Technol. 99:8112–8117. doi:10.1016/j.biortech.2008.03.030

Schievano, A., B. Scaglia, G. D'Imporzano, L. Malagutti, A. Gozzi, and F. Adani. 2009. Prediction of biogas potentials using quick laboratory analyses: Upgrading previous models for application to heterogeneous organic matrices. Bioresour. Technol. 100:5777–5782. doi:10.1016/j.biortech.2009.05.075

Shapouri, H., M. Salassi, and J. Nelson. 2006. The economic feasibility of ethanol production from sugar in the United States. USDA, Washington, DC. http://www.usda.gov/oce/reports/energy/EthanolSugarFeasibilityReport3.pdf (accessed 6 Mar. 2016).

Sluiter, A., B. Hames, R. Ruiz, C. Scarlata, J. Sluiter, and D. Templeton. 2005a. Determination of ash in biomass, laboratory analytical procedure. Tech. Rep. 510-42622. National Renewable Energy Lab., Golden, CO.

Sluiter, A., B. Hames, R. Ruiz, C. Scarlata, J. Sluiter, D. Templeton, and D. Crocker. 2011. Determination of structural carbohydrates and lignin in biomass, laboratory analytical procedure. Tech. Rep. 510-42618. National Renewable Energy Lab., Golden, CO.

Sluiter, A., R. Ruiz, C. Scarlata, J. Sluiter, and D. Templeton. 2005b. Determination of extractives in biomass, laboratory analytical procedure. Tech. Rep. 510-42619. National Renewable Energy Lab., Golden, CO.

Snider, J.L., R.L. Raper, and E.B. Schwab. 2012. The effect of row spacing and seeding rate on biomass production and plant stand characteristics on non-irrigated photoperiod-sensitive sorghum (*Sorghum bicolor* (L.) Moench). Ind. Crops Prod. 37:527–535. doi:10.1016/j.indcrop.2011.07.032

Staggenborg, S.A., K.C. Dhuyvetter, and W.B. Gordon. 2008. Grain sorghum and corn comparisons: Yield, economic, and environmental responses. Agron. J. 100:1600–1604. doi:10.2134/agronj2008.0129

Stefaniak, T.R., J.A. Dahlberg, B.W. Bean, N. Dighe, E.J. Wolfrum, and W.L. Rooney. 2012. Variation in biomass composition components among forage, biomass, sorghum-sudangrass, and sweet sorghum types. Crop Sci. 52:1949–1954. doi:10.2135/cropsci2011.10.0534

Tamang, P.L., K.F. Bronson, A. Malapati, R. Schwartz, J. Johnson, and J. Moore-Kucera. 2011. Nitrogen requirement for ethanol production from sweet sorghum and photoperiod sensitive sorghums in the Southern High Plains. Agron. J. 103:431–440. doi:10.2134/agronj2010.0288

Teetor, V.H., D.V. Duclos, E.T. Wittenberg, K.M. Young, J. Chawhuaymak, M.R. Riley, and D.T. Ray. 2011. Effects of planting date on sugar and ethanol yield of sweet sorghum grown in Arizona. Ind. Crops Prod. 34:1293–1300. doi:10.1016/j.indcrop.2010.09.010

Tew, T.L., R.M Cobill, and E.P. Richard. 2008. Evaluation of sweet sorghum and sorghum × sudangrass hybrids as feedstocks for ethanol production. Bioenerg. Res. 1:147–152.

Theerarattananoon, K., X. Wu, S. Staggenborg, R. Propheter, R. Madl, and D. Wang. 2010. Evaluation and characterization of sorghum biomass as feedstock for sugar production. Trans. ASABE 53:509–525. doi:10.13031/2013.29561

Theerarattananoon, K., F. Xu, J. Wilson, S. Staggenborg, L. McKinney, P. Vadlani, Z. Pei, and D. Wang. 2012. Effect of pelleting conditions on chemical composition and sugar yield of corn stover, big bluestem, wheat straw and sorghum stalk pellets. Bioprocess Biosyst. Eng. 35:615–623. doi:10.1007/s00449-011-0642-8

Tillman, D.A. 2000. Biomass cofiring: The technology, the experience, the combustion consequences. Biomass Bioenergy 19:365–384. doi:10.1016/S0961-9534(00)00049-0

Türe, S., D. Uzun, and I. E. Türe. 1997. The potential use of sweet sorghum as a non-polluting source of energy. Energy 22:17–19. doi:10.1016/0360-5442(95)00024-0

Undersander, D., D.R. Mertens, and N. Theix. 1993. Forage analyses procedures. Natl. Forage Testing Assoc. Omaha, NE.

University of Minnesota, Morris. 2013. Renewable energy. University of Minnesota, Morris. https://www.morris.umn.edu/sustainability/renewable/ (accessed 1 Apr. 2016).

US Department of Energy (USDOE). 2006. Theoretical ethanol yield calculator. USDOE, Washington, DC. http://www1.eere.energy.gov/biomass/ethanol_yield_calculator.html (accessed 30 Apr. 2013).

US Energy Information Administration (USEIA). 2012. Electric power monthly. USDOE, Washington, D.C. http://www.eia.gov/electricity/annual/archive/03482012.pdf (accessed 30 Mar. 2016).

Venuto, B., and B. Kindiger. 2008. Forage and biomass feedstock production from hybrid forage sorghum and sorghum-sudangrass hybrids. Grassl. Sci. 54:189–196. doi:10.1111/j.1744-697X.2008.00123.x

Walton, C.F., Jr., E.K. Ventre, and S. Byall. 1938. Farm production of sorgo sirup. Farmers' Bull. 1791. USDA, Washington, DC.

Wee, J.Y., J.N. Kim, and H.W. Ryu. 2006. Biotechnological production of lactic acid and its recent applications. Food Technol. Biotechnol. 44(2):163–172.

Wortmann, C.S., A.J. Liska, R.B. Ferguson, D.J. Lyon, R.N. Klein, and I. Dweikat. 2010. Dryland performance of sweet sorghum and grain crops for biofuel in Nebraska. Agron. J. 102:319–326. doi:10.2134/agronj2009.0271

Wu, L., M. Arakane, M. Ike, M. Wada, T. Takai, M. Gau, and K. Tokuyasu. 2011. Low temperature alkali pretreatment for improving enzymatic digestibility of sweet sorghum bagasse for ethanol production. Bioresour. Technol. 102:4793–4799. doi:10.1016/j.biortech.2011.01.023

Wu, X.R., S.A. Staggenborg, J.L. Propheter, W.L. Rooney, J. Yu, and D. Wang. 2010. Features of sweet sorghum juice and their performance in ethanol fermentation. Ind. Crops Prod. 31:164–170. doi:10.1016/j.indcrop.2009.10.006

Wyman, C.E., B.E. Dale, R.T. Elander, M. Holtzapple, M.R. Ladisch, and Y.Y. Lee. 2005. Comparative sugar recovery data from laboratory scale application of leading pretreatment technologies to corn stover. Bioresour. Technol. 96:2026–2032. doi:10.1016/j.biortech.2005.01.018

Yang, B., and C.E. Wyman. 2004. Effect of xylan and lignin removal by batch and flowthrough pretreatment on the enzymatic digestibility of corn stover cellulose. Biotechnol. Bioeng. 86(1):88–98. doi:10.1002/bit.20043

Zhan, X., D. Wang, X.S. Sun, S. Kim, and D.Y.C. Fung. 2002. Lactic acid production from extrusion cooked grain sorghum. Trans. ASAE 46:589–593.

Zhan, X., D. Wang, M.R. Tuinstra, S. Bean, P.A. Seib, and X.S. Sun. 2003. Ethanol and lactic acid production as affected by sorghum genotype and location. Ind. Crops Prod. 18:245–255. doi:10.1016/S0926-6690(03)00075-X

Zhang, C., G. Xie, S. Li, L. Ge, and T. He. 2010. The productive potentials of sweet sorghum ethanol in China. Appl. Energy 87:2360–2368. doi:10.1016/j.apenergy.2009.12.017

Zhou, H., D. Löffler, and M. Kranert. 2011. Model-based predictions of anaerobic digestion of agricultural substrates for biogas production. Bioresour. Technol. 102:10819–10828. doi:10.1016/j.biortech.2011.09.014

Overview of Sorghum Industrial Utilization

Guangyan Qi, Ningbo Li, Xiuzhi Susan Sun, and Donghai Wang*

Abstract

Sorghum has a variety of uses, including food for human consumption (Africa, China, and India) and feed grain for livestock (America and Australia). Recently, sorghum has also gained interest as a new-generation bioenergy crop because of its multiple uses and wide adaptability to varied agroclimatic conditions. For example, more than 30% of grain sorghum is now used in fuel ethanol production in United States, which generates large amounts of distillers dried grains with solubles (DDGS) as a by-product. Therefore, the need for new value-added outlets for sorghum and its by-products has become crucial to maintaining the economic viability of the sorghum industry. In this chapter, we review the industrial applications of sorghum products, primarily sorghum flour, kafirin protein from sorghum flour, DDGS or brewing of lager beer, and sorghum wax. Sorghum flour has been used as a filler and extender in petroleum-based adhesives. Sorghum kafirin's unique properties, such as higher hydrophobicity and lower protein digestibility than other cereal prolamin protein, grant it great potential as biodegradable packaging materials. Many efforts have been made to investigate how kafirin film properties are affected by karirin extraction processes, film forming processes, and physical and chemical modifications, etc. Kafirin displays huge potential in preparing high-performance veneer adhesives. Kafirin protein also forms microparticles, which are useful in producing very thin plastic films, scaffolds, and encapsulating agents. Sorghum wax, a by-product of wet milling and ethanol production, shows potential as a source of bioplastic films and coatings for foods, mainly because of its hydrophobicity.

Sorghum, a grain, forage, or sugar crop, is among the most efficient crops in its conversion of solar energy and its use of water. Sorghum is also known as a high-energy, drought tolerant crop. Because of its wide uses and adaptation, "sorghum is one of the really indispensable crops" required for the survival of humankind (Harlan, 1971). The area planted to sorghum worldwide has increased by 66% over the past 50 yr, and the yield has increased by 244% (Stroade and

---

Abbreviations: DDG, dried distillers grains; DDGS, distillers dried grains with solubles; SPI, soy protein isolate.

Guangyan Qi (guangyan@ksu.edu) and Xiuzhi Susan Sun (xss@ksu.edu), Bio-Materials and Technology Laboratory, Dep. of Grain Science and Industry, 213 BIVAP Bldg., Kansas State Univ., Manhattan, KS 66506; Ningbo Li, 3622 Everett Dr. Manhattan, KS 66503; and Donghai Wang, Dep. of Biological and Agricultural Engineering, Kansas State Univ., 150 Seaton Hall, 920 N. 17th St., Manhattan, KS 66506. *Corresponding author (dwang@ksu.edu).

doi:10.2134/agronmonogr58.2014.0070

Boland, 2013). Currently, sorghum has a variety of uses, including food for human consumption in developing countries such as Africa, China, and India, and feed grain for livestock in developed countries, such as United States and Australia (Stroade and Boland, 2013; Queensland Government, 2012). Sorghum has also become a new-generation bioenergy crop because of its wide adaptability to varied agroclimatic conditions and its ability to minimize inputs such as water and nitrogen, which could be a key to sorghum's benefits as a bioenergy crop (Wallheimer, 2012; Rao et al., 2014). Kubecka (2011) found that sorghum could produce the same amount of ethanol per bushel as comparable feed grains while using up to one-third less water in the plant growth process. Approximately 30 to 35% of the sorghum crop in United States is used for fuel ethanol production, accounting for about 2% of U.S. fuel ethanol production (Kubecka, 2011). Distiller's dried grains with solubles is a by-product of the distillation and dehydration process during ethanol production (Bonnardeaux, 2009). Distiller's dried grains with solubles serves as an inexpensive source of protein (up to 30 to 40%) and is a non-animal-based livestock feed supplement. In 2009, 500 million bushels of sorghum were used for ethanol production in the United States, and generated more than 700 million pounds (317514 metric tons) of sorghum protein available from DDGS (Baker et al., 2010). In addition, brewing of lager beer from sorghum is becoming more widespread throughout Africa and generates large quantities of spent grain, which is rich in protein but high in moisture (Taylor and Taylor, 2009). Increased utilization of sorghum in ethanol fermentation and brewing of lager beer boosts the number of sorghum products available in the market, which means developing new outlets for coproducts has become crucial to maintaining the industry's economic viability.

Kafirin, a prolamin protein, accounts for approximately 70 to 90% of the total storage protein in sorghum (Hamaker et al., 1995). Kafirin shows homology similar to zein protein but with higher hydrophobicity (Wall and Paulis, 1978) and lower digestibility (Duodu et al., 2002). Those unique properties grant sorghum protein huge potential as a biopolymer for coating and films, with functional properties superior to zein protein film. The high hydrophobicity of kafirin also makes development of high-performance kafirin adhesives with excellent water resistance promising (Li et al., 2011). The purpose of this chapter is to give a brief summary of the industrial applications of sorghum and its coproducts from ethanol and lager beer production.

Sorghum Flour

A few decades ago, researchers reported that among sorghum products, flour was the most used in nonfood industrial applications (Anderson, 1969; Hahn, 1970). Gelatinized sorghum starch could be used in oil drilling as a fluid loss-control agent and in the paper industry as a coating and as beater adhesives. The binding properties of starch and protein in sorghum flour have also been exploited in the manufacture of building materials, such as insulation, gypsum board and ceiling tiles, foundry core binders, molding sand additives, low-grade taconite ore pellets, and charcoal briquettes (Anderson, 1969; Hahn, 1970).

Sorghum flour and brans have also been used as fillers or extenders in formaldehyde-based plywood adhesives for rollcoaters (Alexander and Krueger, 1978; Edler, 1981; Ramos et al., 1984), and more recently in inexpensive adhesives for wallboard and packaging materials (Rooney and Waniska, 2000; Hojilla-Evangelista and Bean, 2011). According to Ramos et al. (1984), the glue mixed with up to 50% sorghum flour extender still had good spreadability on veneer. Results demonstrated that the urea formaldehyde plywood adhesive extended with 20% sorghum flour had bond strength and water resistance equivalent to industrial-grade urea formaldehyde adhesive in terms of shear adhesion strength and wood failure. In the most recent study, Hojilla-Evangelista and Bean (2011) evaluated sorghum flour (0.2% db residual oil and 12.0% crude protein) as a protein extender in phenol formaldehyde-based plywood adhesive for sprayline coaters or foam extrusion. Soluble proteins in sorghum flour showed solubility and foaming properties that are desirable for the highly alkaline conditions in plywood glues for sprayline coaters. The adhesive containing sorghum flour as a protein extender had mixing properties and an appearance superior to those of the standard wheat flour–based plywood glue, but its viscosity and bond strength were markedly reduced. Doubling the amount of protein from sorghum flour in the glue mix significantly improved viscosity from 480 to 1104 cp and adhesion strength from 1.27 to 1.37 MPa. These results indicate that sorghum flour is a viable extender in plywood glues for sprayline coater or foam extrusion.

Sorghum Kafirin

Kafirin Films

Kafirin, the prolamin protein from sorghum, shows homology similar to zein in molecular weight, solubility, structure, and amino acid composition (Shull et al., 1991). Kafirin is reported to be more hydrophobic than zein (Wall and Paulis, 1978), is less easily digestible than zein after wet heat processing (Duodu et al., 2002), and is known to be nonallergenic (Ciacci et al., 2007). These unique properties make sorghum proteins of interest for industrial products such as edible and biodegradable protein films and adhesives.

Buffo et al. (1997) prepared films from laboratory-extracted sorghum kafirin, a by-product of wet milling, and demonstrated that sorghum kafirin could be used as a biopolymer for edible or nonedible film and coating applications for the first time. When plasticized with glycerol and polyethylene glycol (PEG400), kafirin film had tensile and water vapor barrier properties similar to commercial zein protein films derived from corn, and the kafirin films were more intensely colored than zein films (Table 1). Da Silva and Taylor (2005) made free-standing plasticized cast films from kafirin extracted from different dry milling fractions, including bran, the by-product of dry milling (or potentially in bioethanol production). Red and white sorghum flour were used as starting material in this research. The kafirin films showed much higher tensile strength, lower extensibility, and poorer water barrier properties than zein film. Compared with the study of Buffo et al. (1997), the kafirins prepared by Da Silva and Taylor (2005) were rich in β- and γ- kafirins with their tendency to cross-link at elevated temperatures; this protein composition difference was assumed to be responsible for the higher film tensile strength in the latter study. Also, the poorer water barrier properties of kafirin film were probably due to greater thickness or reduced flexibility, which might have caused microcracks. Films prepared from the bran kafirin were highly colored, less flexible, and had a less smooth surface texture than films made from flour, which was probably due to higher levels of

Table 1. Tensile strength, elongation at break, water vapor permeability, and HunterLab color values of kafirin and zein films.†

	Kafirin films	Zein films
Tensile strength, MPa	2.1 ± 0.3	2.6 ± 0.3
Elongation at break, %	106.1 ± 9.7	84.4 ± 3.5
Water vapor permeability‡	5.5 ± 0.2	5.7 ± 0.3
HunterLab color values§		
L	75.48 ± 2.85	94.03 ± 0.53
a	7.27 ± 1.85	−5.44 ± 0.47
b	48.10 ± 4.70	29.42 ± 4.74

† Data from Buffo et al. (1997). For all properties except water vapor permeability, means were significantly ($P < 0.05$) different between kafirin and zein films.

‡ g mm/m^2 h^{-1} kPa^{-1}.

§ $L = 0$ (black) to $L = 100$ (white); $-a$ (greenness) to $+a$ (redness); and $-b$ (blueness) to $+b$ (yellowness).

contaminants in the bran kafirins. The strong color of the kafirin bran films could limit their use in certain coating applications. Red kafirin would be better for coating red fruits such as litchis and plums, whereas yellow kafirin would be more suitable for coating nuts and pears.

Kafirin Extractants and Their Effects on Kafirin Films

Cast prolamin films are typically prepared by dissolving protein in aqueous ethanol at an elevated temperature, then pouring the solution onto a flat surface and allowing the solvent to evaporate (Taylor et al., 2006). The most efficient extractant for kafirin is aqueous *tert*-butyl alcohol with dithiothreitol (DTT), but both are expensive and not food compatible. Buffo et al. (1997) mentioned that kafirin extractability may be enhanced with the addition of FDA-approved reducing agents, such as sulfite, to the alcoholic extracting solution. Gao et al. (2005) also studied the effectiveness of extractants consisting of aqueous ethanol containing the food-compatible reducing reagent sodium metabisulfite, with or without sodium hydroxide, and then compared results with *tert*-butyl alcohol extraction. Addition of sodium hydroxide to aqueous ethanol extractant improved the yield and solubility of kafirin extracted from dry-milled whole-grain sorghum and enhanced the efficiency of the reducing regent (Table 2). The authors concluded that an appropriate procedure for the commercial extraction of kafirin is 70% (w/w) ethanol with 0.5% (w/w) sodium metabisulfite, and 0.35% (w/w) sodium hydroxide at 70°C, taking into consideration yield levels, film-forming properties, and food compatibility.

Ethanol use requires a government license, however, and some communities object to its use in food. Extracting solvents other than ethanol also may improve the kafirin isolation process; therefore, an alternative to ethanol for kafirin salvation was investigated to produce edible film. Taylor et al. (2005) examined several food-compatible solvents to replace aqueous ethanol for kafirin film casting and concluded that kafirin films cast from glacial acetic acid had the same tensile and water barrier properties as those cast from aqueous ethanol. Glacial acetic acid also had the advantage of casting films at a much lower temperature (25°C) than aqueous ethanol (70°C), which yielded films of more consistent quality.

Brewer's spent grain, the coproduct from the brewing of lager beer from sorghum, is rich in protein but high in moisture. Researchers have extracted the

Table 2. Characterization of kafirins prepared using different extraction processes. Data from Gao et al. (2005).

Extractant†	Extraction temp.	Drying conditions	Yield‡	Protein purity	Solubility§	FT-IR (powder) α-helix intermolecular β-sheet (1650 cm^{-1}; 1620 cm^{-1} peak intensity ratio)
			%	db	%	
tB/DTT	RT¶	Freeze-dried	32(3.7)a	82.4(1.63)a	100(0.1)d	1.39(0.02)e
tB/DTT	RT¶	40°C			95(1.6)c	0.90(0.03)a
Et/$Na_2S_2O_5$/NaOH	70°C	Freeze-dried	54(3.0)c	84.2(2.43)a	95(1.1)c	1.10(0.01)d
Et/$Na_2S_2O_5$/NaOH	70°C	40°C			93(0.2)c	0.96(0.02)b
Et/$Na_2S_2O_5$	70°C	Freeze-dried	38(0.8)b	83.4(0.49)a	90(1.2)b	1.00(0.01)c
Et/$Na_2S_2O_5$	70°C	40°C			80(2.0)a	0.90(0.01)a

† tB/DTT = 60% tert-butyl alcohol + 0.05% DTT. Et/$Na_2S_2O_5$/NaOH = 70% ethanol + 0.5% sodium metabisulfite with 0.35% sodium hydroxide. Et/$Na_2S_2O_5$ = 70% ethanol + 0.5% sodium metabisulfite without 0.35% sodium hydroxide.

‡ Figures in parentheses indicate standard deviations.

§ Dry weight basis. Values in the same column but with different letters are significantly different at the 95% level.

¶ Room temperature, approximately 25°C.

kafirin from brewer's spent grain and used it in biofilm preparation (Taylor and Taylor, 2009). Application of the ethanol solvent extraction technology also could be applied to brewer's spent grains, but the process requires optimization. Taylor and Taylor (2009) extracted 40% of total protein from sorghum brewer's spent grain with single extraction into glacial acetic acid after a 24-h presoak in 0.5% sodium metabisulphite. Because the spent grain had protein content as high as 30% (db), compared with 9% in sorghum whole grain, more protein was extracted from the former despite low extraction efficiency. In addition, the kafirin extracted from brewer's spent grain is more pure (85%) than from the whole grain (68%) because the water-soluble proteins were removed during the brewing process.

Wang et al. (2009) stated that extractants (acetic acid, HCl-ethanol, and NaOH-ethanol) affected the purity and physicochemical properties of extracted kafirin proteins from DDGS. Acetic acid extraction produced protein with the highest purity of 98.9% and extraction yields of 44.2%. An extraction yield of 56.8% was achieved by NaOH-ethanol but with a lower purity of 94.88%. The most inefficient solvent for kafirin extraction was HCl-ethanol, with only 24.2% yield and 42.3% purity. Size exclusion chromatography revealed that the protein extracted by acetic acid and HCl-ethanol solvents had a higher molecular weight than the NaOH-ethanol extracted. γ-Kafirins were found only in extracts from the NaOH-ethanol extraction method. Understanding the functional properties of kafirin proteins from DDGS, such as structural, molecular, and thermal properties, is critical to determining their applications.

Kafirin Film Modification

Although edible and biodegradable coatings and films have been studied intensively in recent years, only a few are used in commercial packaging systems because of performance limitations such as low tensile strength, low elongation, low oxygen permeability, and high water vapor permeability compared with synthetic plastic packaging (Rhim and Ng, 2007). Therefore, modification of protein film properties is necessary to meet a diverse range of packaging requirements (Krochta, 2002). Both physical and chemical methods have been used to improve the protein film properties. Physical methods are mainly heat treatment by microwave and irradiation. Chemical modifications usually introduce cross-linking agents and hydrophobic materials.

Microwave energy was used to thermally modify kafirin structure before casting films, and mechanical properties, barrier properties, digestibility, and biodegradability were altered accordingly (Byaruhanga et al., 2005). Microwave energy offers advantages over conventional heating methods because of rapid internal heat generation (Mudgett, 1982). Byaruhanga et al. (2005) found out that microwave heating at temperatures of 90 or 96°C with holding times of 1 to 2 min significantly increased tensile strength and Yong's modulus, whereas strain decreased by about one-third (Table 3). Water vapor permeability also decreased by about one-third. Formation of intermolecular disulfide and possibly nondisulfide cross-links is partly responsible for the changes in the functional properties of the films. However, for it to be effective, the kafirin must be wetted to increase its dielectric properties, and the kafirin must be close to its native state or undenatured state (aqueous tert-butanol-based solvent was used to extract kafirin at ambient temperature to minimize temperature-induced denaturation).

Emmambux et al. (2004) found that both tannic acid and sorghum-condensed tannins can bond to kafirin protein, thus decreasing protein chain mobility and free volume at a molecular level in the kafirin film. As a result, tensile stress increased by 50 to 100% and strain decreased three to four times as tannic acid and sorghum-condensed tannins contents increased by up to 20% of tannin compared with kafirin. Modification with tannins also reduced oxygen permeability by more than 50% but did not affect water vapor permeability.

To elucidate the molecular causes of the changes in tensile properties of the kafirin films with microwave heat and tannin complexation, Byaruhanga et al. (2006) utilized electrophoresis and vibrational spectroscopy to study their mechanisms. Results indicated that microwave heat induced cross-linking connected by disulfide bonds and changes in the secondary structure from α-helical to β-sheet conformation to form oligomers of kafirin, whereas tannin cross-linking was probably by hydrogen bonding. Both treatments increased the tensile strength of the cast kafirin bioplastic films. Gao et al. (2005) also found that the proportion of intermolecular β sheet structure increased during drying with elevated temperature as well. In their study, it was also found that a high proportion of the native α-helical structure of the protein remaining after extraction and drying is crucial to film properties of high mechanical strength and strain, low water vapor, and good sensory properties. Because α-helical structure could dissolve readily and form even, uniform films of consistent thickness, the study suggested that industrial-scale kafirin production must minimize protein aggregation and maximize the native α-helical structure of kafirin after extraction and drying to maintain even protein distribution in the solvent; then kafirin could be modified by heating or other treatments to form more β-sheet structures, thus increasing protein film strength.

Grafting of synthetic monomers has been used to make biopolymers thermoplastic and provide films with good properties (Reddy et al., 2012). Reddy et al. (2014) assumed that because dried distillers grains (DDG) are a mixture of proteins and carbohydrates, grafting, which is a milder process than acetylation,

Table 3. Tensile properties of films made from wet kafirin microwave-heated at 96°C and held for different time periods.†

Microwave Time, min	0	1		2		3		4		10	
Microwave temp, °C		90	96	90	96	90	96	90	96	90	96
Maximum tensile strength, MPa	1.37a (0.17)	2.43b (0.36)	2.65b (0.49)	2.38b (0.19)	3.04b (0.38)	2.79b (0.45)	3.13b (0.73)	2.38b (0.40)	2.74b (0.76)	2.16b (0.41)	2.57b (0.35)
Tensile strength at break, MPa	1.36a (0.30)	2.20ab (0.57)	2.48b (0.53)	2.08ab (0.4)	2.55b (0.42)	2.61b (0.25)	2.75b (0.72)	2.34b (0.87)	2.64b (0.87)	2.03ab (0.49)	2.45b (0.37)
Strain, %	142b (20.8)	131b (12)	105a (9)	112ab (11)	83a (16)	108a (10)	102a (17)	90a (9)	99a (20)	90a (12)	108a (10)
Young's modulus, MPa	21a (5.7)	46b (10)	75b (14)	56b (8)	101b (15)	54b (11)	104b (18)	37b (10)	92b (11)	44b (10)	75b (12)

† Data from Byaruhanga et al. (2005). Values followed by the same letter in the same row are not significantly different ($P < 0.05$). Values in parentheses are standard deviations of the means.

helps preserve the properties of the proteins and carbohydrates, therefore yielding thermoplastics with better properties. Three monomers, including methyl, ethyl, and butyl methacrylates, were used to graft on sorghum DDG and were then compression-molded into films. Results showed that at a grafting ratio of 40%, butyl methacrylate-grafted films had a dry strength of 4.8 MPa (elongation of 1.8%) and a wet strength of 3.1 MPa (elongation of 8.1%), indicating that the films had good strength and wet stability. This study also indicated that grafting appears to be a more viable approach to developing thermoplastics from DDG than other common types of chemical modification.

Di Maio et al. (2010) investigated conventional direct-melt mixing technology on zein and kafirin proteins to study the thermoplasticization process of the proteins with several plasticizers. Researchers avoided traditional lengthy procedures (forming the protein–solvent–plasticizer solution followed by drying or the protein–plasticizer emulsion followed by the precipitation of the extrudable resin) in particular, and the protein and plasticizer were fed directly into the mixer to obtain a plastic-like material. Thermoplasticized kafirin had tensile properties comparable to zein film, with moduli as high as 750 MPa, strength of 10 MPa, and strain at break of 2% at 25 % (w/w) of lauric acid, whereas the use of lactic acid led to more deformable films, with strain at break as high as 50%. Results indicated that a lower plasticizer amount was sufficient to produce equally flexible films, confirming the enhanced thermoplasticization efficiency of the mixing process compared with casting methods.

To improve food safety and reduce the use of chemical preservatives, the antimicrobial properties of kafirin film affected by natural antimicrobial were studied by Giteru et al. (2015). Incorporating the plant essential oil citral led to films with strong antimicrobial activity against *C. jejuni*, *L. monocytogenes*, and *P. fluorescens*, whereas control kafirin film had antimicrobial activity against only *L. monocytogenes*. This approach could improve food safety and quality for both food and nonfood applications. Addition of citral also decreased film tensile strength, increased elongation at break (Table 4), and significantly enhanced the oxygen barrier properties of the kafirin film. The water vapor transmission rate of films decreased with citral incorporation, but the water vapor permeability was not affected (Table 5). The authors mentioned that future research would be on

Table 4. Physical and mechanical properties of kafirin films. Data from Giteru et al. (2015).

Films†	Thickness	Moisture content	Solubility	Tensile strength	Elongation at break
	min	%	%	N mm^{-2}	%
P	0.07 ± 0.01a	14.0 ± 3.94a	29.1 ± 5.59a	3.48 ± 1.11b	79.7 ± 41.4a
C	0.09 ± 0.02b	13.5 ± 4.80a	24.7 ± 4.89a	1.89 ± 0.55a	141.0 ± 40.7b
C+Q	0.09 ± 0.02b	14.0 ± 2.52a	24.5 ± 2.99a	1.76 ± 0.50a	147.0 ± 48.8b
Q	0.08 ± 0.02ab	11.3 ± 3.31a	24.4 ± 3.53a	3.25 ± 0.23b	46.7 ± 44.5a

† P, plain film; C, P + 2.5% citral film; C+Q, P + 1.5% citral + 1% quercetin film; Q, P + 2% quercetin film. Values followed by the same letter in the same column are not significantly different ($P < 0.05$).

the evaluation of physical suitability and antimicrobial properties of these films in fresh food systems.

Kafirin Microparticles

Another application of kafirin is the preparation of microparticles (microspheres). They are mainly composed of small spherical particles (1 to 10 μm) with internal holes or vacuoles and very large surface area (Taylor and Taylor, 2009). The vacuolated spherical kafirin microparticles with a mean diameter of 5 μm can be prepared by dissolving kafirin in a primary solvent (glacial acetic acid) and then adding water; changing the solvent polarity leads to microparticle formation (Taylor et al., 2009a). Taylor and Taylor (2009) also suggested more work should be done on manipulation of extraction conditions of kafirin microparticles, which might result in different physicochemical properties, and on exploring additional applications in the food, biomedical, and pharmaceutical industries.

Kafirin microparticles made by phase separation from acetic acid were used to prepare very thin (<15 μm) free-standing films (Taylor et al., 2009a). Film formation involves controlled aggregation of kafirin microparticles then dissolution of the microparticles in acetic acid before drying into a cohesive film matrix. The kafirin microparticles film had excellent tensile strength but was not very extensible. Compared with conventional cast kafirin film, microparticle films had better water barrier properties and lower protein digestibility.

Anyango et al. (2011) investigated various treatments for kafirin microparticles to improve water stability and other related functional properties of the thin film (<50 μm). Glutaraldehyde was reported to induce the formation of covalent cross-linking between free amino groups of kafirin and carbonyl groups of aldehyde, resulting in up to 43% improvement in film tensile strength. Modified thin kafirin films cast from microparticles also proved to be more stable in water than conventional cast kafirin films.

Kafirin Microparticle Scaffolds and Encapsulating Agents

An alternative application for kafirin microparticles could be biomaterial tissue scaffolds, which need to be large structures with a high degree of interconnected porosity. Anyango et al. (2012) found that both wet heat treatment and glutaraldehyde treatment could enhance bone morphogenetic protein-2 binding to the kafirin microparticles. Bone morphogenetic protein-2 induces the formation of both cartilage and bone and plays a role in a number of nonosteogenic developmental processes, such as in neutral induction (Chen et al., 2004). Thus, these

Table 5. Water and oxygen barrier properties of kafirin films with different levels of citral and quercetin. Data from Giteru et al. (2015).†

Films	WVP	WVTR	OP × 10^7
	g mm/m^2/h/kPa	g/h m^2	g/m/day/Pa
P	0.66 ± 0.08a	31.5 ± 3.74b	6.24 ± 0.45c
C	0.69 ± 0.10a	25.6 ± 3.86a	4.86 ± 1.11b
C+Q	0.65 ± 0.12a	23.7 ± 4.38a	4.33 ± 0.34b
Q	0.74 ± 0.18a	28.9 ± 6.96ab	2.90 ± 0.46a

† P, plain film; C, P + 2.5% citral film; C+Q, P + 1.5% citral + 1% quercetin film; Q, P + 2% quercetin film. Same superscript letters in the same column are not significantly different ($P < 0.05$). WVTR, water vapor transmission rate; WVP, water vapor permeability of the films; OP, oxygen permeability of the films.

modified kafirin microparticles have potential as natural, nonanimal protein-derived bioactive scaffolds for hard or soft tissue repair. However, in the recent research, Taylor et al. (2014) reported that kafirin microparticles (5 μm in diameter) injected subcutaneously in mice induced chronic inflammation. This may have been caused by the release of hydrolysis products such as glutamate during the rapid degradation of microparticles. However, kafirin microparticle films (50 μm thick, folded into 1 cm^3) showed no abnormal inflammatory reaction in rats and were only partially degraded by Day 28. This difference can be attributed to the larger surface area of the microparticles, resulting in more rapid breakdown and a releasing of the toxic products. This study highlights the fact that even though kafirin films have potential as biomaterial, assessing the biocompatibility and biodegradability of prolamin-based materials by using implanted scaffolds should always be considered.

Kafirin also may have potential as an encapsulating agent because of its large and porous internal surface area. Antioxidants, catechin, and sorghum-condensed tannins have been encapsulated within kafirin microparticles, and the interactions between sorghum polyphenols and proteins were characterized in terms of antioxidant release profiles and cross-linking reaction by Taylor et al. (2009b). Over a period of 4 h, they reported that catechin and sorghum-condensed tannin-encapsulated kafirin microparticles released approximately 70 and 50% of total antioxidant activity, respectively. These findings demonstrated that encapsulated microparticles have potential as an effective way to deliver dietary antioxidants and enhance potential health benefits through controlled released of antioxidant activity.

Kafirin Veneer Adhesives

Because of the high hydrophobicity of kafirin, Li et al. (2011) assumed that the preparation of high adhesion performance of kafirin veneer glue is promising. The authors compared the adhesive performance of three types of sorghum proteins: acetic acid–extracted sorghum protein from DDGS (PI), aqueous ethanol-extracted sorghum protein from DDGS (PII), and acetic acid–extracted sorghum protein from sorghum flour (PF). The PI type had the best adhesion performance in terms of dry, wet, and soaked adhesion strength, followed by PF and PII. At a protein concentration of 12%, the wet strength of PI was 3.15 MPa, compared with 2.17 and 2.59 MPa for PII and PF, respectively (curing condition was 150°C for 10 min). According to the differential scanning calorimetry thermograms, PF protein contained a higher amount of carbohydrates than PI and PII; those nonprotein contaminates might be the reason for the low adhesion strength of PF. In addition, the authors also assumed that more hydrophobic amino acids may be aligned at the protein–wood interface than in PII, which would contributed to the better water resistance of PI. The effects of sorghum protein concentration and curing conditions on the adhesive properties of sorghum protein were also investigated in this study. The optimum sorghum protein concentration and cure temperature for best adhesion strength was 12% and 150°C. Compared with a native soy protein isolate, acetic acid–extracted PI from DDGS had a significantly higher wet strength of 3.15 MPa (1.63 MPa for soy protein). Amino acid composition analysis showed that PI had as high as 57% of hydrophobic amino acids, which was likely a key factor in the greater water resistance than soy protein (36% hydrophobic amino acids) (Table 6).

Sorghum Wax for Film Application

Sorghum wax can be obtained as a by-product of sorghum wet milling and sorghum fermentation for ethanol production. Wax is a lipid with long carbon chains that provides an excellent water barrier because of its high hydrophobicity. It has been reported that nonpolar solvents such as hexane, benzene, chloroform, light petroleum ether, or acetone can be used to extract plant and seed wax (Kim et al., 2002). The unrefined wax naturally contains some extraneous polar and hydrophilic materials and may be more miscible in an aqueous film-forming solution than refined wax. Therefore, unrefined sorghum wax can be used as an additive to improve water vapor barrier properties of protein films and as an ingredient in edible films and coatings.

Weller et al. (1998a) utilized sorghum wax as an ingredient in edible protective coating formulations for confections. Results showed that gelatin-based candies coated with medium-chain triglyceride oil combined with grain sorghum had reduced protein solubility in water and melting-induced expansion. Although sorghum wax-treated candies had less favorable surface reflection, clarity, chalkiness, off-flavor, and aftertaste sensory scores than carnauba wax-treated candies, the organoleptic properties of sorghum wax may likely improve through refining or by using an extracting organic solvent other than petroleum ether. Weller et al. (1998b) found that sorghum wax improved the water vapor barrier properties as effectively as carnauba wax, a substance commonly added

Table 6. Hydrophilicity properties of amino acids in different properties. Data from Li et al. (2011).

Amino acids	PI†	PII‡	PF§	DDGS	Sorghum	SPI¶
	% of total					
Hydrophobic#	57.32a††	58.49a	59.04a	53.79a	54.1a	36.89b
Hydrophilic‡‡	42.68b	41.51b	40.96b	46.21b	45.9b	63.11a

† PI, acetic acid–extracted sorghum protein from distillers dried grains with solubles (DDGS).

‡ PII, ethanol-extracted sorghum protein from DDGS.

§ PF, acetic acid–extracted sorghum protein from sorghum flour.

¶ SPI, soy protein isolates.

Hydrophobic amino acid = alanine, methinine, phenylalanine, isoleucine, leucine and proline.

†† Means in the same row followed by different letters are significantly different at $P < 0.05$.

‡‡ Hydrophilic amino acid = lysine, tyrosine, arginine, threonine, glycine, histidine, erine, glutamine, and asparagine.

to edible coating formulations for fruits and vegetables. Improved moisture-barrier ability was also accompanied by greater film extensibility and less intense film coloration.

Sorghum wax was also incorporated into soy protein isolate (SPI) to make the films (Kim et al., 2002, 2003). The SPI-sorghum wax paste films had better water barriers and physical properties than control films. In addition, unlike hexane, sorghum wax extracted with ethanol was miscible with film-forming solution. Compared with other refined lipid materials, sorbitol, glycerin, and unrefined sorghum wax had similar or better effects on the water barrier properties of SPI films. The film improved by sorghum wax had a higher tensile strength than film made with other lipid materials.

Summary

Sorghum, a millennia-old cereal grain, is mainly used as a human food and animal feed worldwide. Currently, sorghum has also become a new-generation bioenergy crop because of its multiple uses and wide adaptability to varied agroclimatic conditions. This quick boosting sorghum to fuel ethanol production generates a large amount of DDGS as a by-product. And the unique properties of kafirin protein (30–40%) available from DDGS such as higher hydrophobicity and lower protein digestibility than other cereal protein attract researchers' extensive attention to exploit their new value-added industrial application. A lot of effort has been undertaken to produce sorghum kafirin films and to improve the protein film properties through physical and chemical modification. Studies have shown that kafirin film has tensile and water vapor barrier properties similar to the films made from commercially produced zein. Kafirin protein also forms microparticles, which are useful in producing very thin plastic films, scaffolds, and encapsulating agents. Another by-product of wet milling and ethanol production, sorghum wax, has also shown potential as a source of bioplastic films and coatings for foods.

Furthermore, acetic acid–extracted kafirin from DDGS had significantly higher wet adhesion strength than soy protein, which is considered as the one of the most promising bio-based veneer adhesives. Therefore, more studies on sorghum kafirin used as veneer adhesive are strongly recommended. Future work may involve physical (i.e., adhesive blends with other protein-based adhesive or synthetic resin to have the synergic effect on the adhesive strength) or chemical treatment (i.e., grafting specific hydrophobic groups on kafirin protein and introducing certain cross-link agents) on sorghum kafirin to further improve its adhesion performance.

References

Alexander, R.J., and R.K. Krueger. 1978. Plywood adhesives using amylaceous extenders comprising finely ground cereal-derived high fiber by-product. US Patent 4070314. Date issued: 24 January.

Anderson, R.A. 1969. Producing quality sorghum flour. The Miller, October Issue.

Anyango, J.O., N. Duneas, J.R.N. Taylor, and J. Taylor. 2012. Physicochemical modification of kafirin microparticles and their ability to bind bone morphogenetic protein-2 (BMP-2), for application as a biomaterial. J. Agric. Food Chem. 60:8419–8426. doi:10.1021/jf302533e

Anyango, J.O., J. Taylor, and J.R.N. Taylor. 2011. Improvement in water stability and other related functional properties of thin cast kafirin protein gilms. J. Agric. Food Chem. 59:12674–12682. doi:10.1021/jf203273y

Baker, A., E. Allen, H. Lutman, and Y. Hamda. Feed Outlook/FDS-10i/September 14, 2010. Economic research service, USDA. http://usda.mannlib.cornell.edu/usda/ers/FDS/2010s/2010/FDS-09-14-2010.pdf. (accessed 14 Sep. 2010).

Bonnardeaux, J. 2009. Potential uses for distillers grains. http://www.agric.wa.gov.au/content/sust/biofuel/potentialusesgrains 042007.pdf (accessed 20 Nov. 2010).

Buffo, R.A., C.L. Weller, and A. Gennadios. 1997. Films from laboratory-extracted sorghum kafirin. Cereal Chem. 74:473–475. doi:10.1094/CCHEM.1997.74.4.473

Byaruhanga, Y.B., M.N. Emmammbux, P.S. Belton, N. Wellner, K.G. Ng, and J.R.N. Taylor. 2006. Alteration of kafirin and kafirin film structure by heating with microwave energy and tannin complexation. J. Agric. Food Chem. 54:4198–4207. doi:10.1021/jf052942z

Byaruhanga, Y.B., C. Erasmus, and J.R.N. Taylor. 2005. Effect of microwave heating of kafirin on the functional properties of kafirin films. Cereal Chem. 82:565–573. doi:10.1094/CC-82-0565

Chen, D., M. Zhao, and G.R. Mundy. 2004. Bone morphogenetic proteins. Growth Factors 22:233–241. doi:10.1080/08977190412331279890

Ciacci, C., L. Maiuri, N. Caporaso, C. Bucci, L. Del Giudice, D. Rita Massardo, P. Pontieri, N. Di Fonzo, S.R. Bean, B. Ioerger, and M. Londei. 2007. Celiac disease: In vitro and in vivo safety and palatability of wheat-free sorghum food products. Clin. Nutr. 26:799–805. doi:10.1016/j.clnu.2007.05.006

Da Silva, L.S., and J.R.N. Taylor. 2005. Physical, mechanical, and barrier properties of kafirin films from red and white sorghum milling fractions. Cereal Chem. 82:9–14. doi:10.1094/CC-82-0009

Di Maio, E., R. Mali, and S. Iannace. 2010. Investigation of thermoplasticity of zein and kafirin proteins: Mixing process and mechanical properties. J. Polym. Environ. 18:626–633. doi:10.1007/s10924-010-0224-x

Duodu, K.G., A. Nunes, I. Delgadillo, M.L. Parker, E.N.C. Mills, P.S. Belton, and J.R.N. Taylor. 2002. Effect of grain structure and cooking on sorghum and maize in vitro protein digestibility. J. Cereal Sci. 35:161–174. doi:10.1006/jcrs.2001.0411

Edler, F.J. 1981. Sulfite–waste liquor–urea formaldehyde resin plywood glue. US Patent 4244846. Date issued: 13 January.

Emmambux, M.N., M. Stading, and J.R.N. Taylor. 2004. Sorghum kafirin film property modification with hydrolysable and condensed tannins. J. Cereal Sci. 40:127–135. doi:10.1016/j.jcs.2004.08.005

Gao, C., J. Taylor, N. Wellner, Y.B. Byaruhanga, M.L. Parker, E.N. Clare Mills, and P.S. Belton. 2005. Effect of preparation conditions on protein secondary structure and biofilm formation of kafirin. J. Agric. Food Chem. 53:306–312. doi:10.1021/jf0492666

Giteru, S.G., R. Coorey, D. Bertolatti, E. Watkin, S. Johnson, and Z. Fang. 2015. Physicochemical and antimicrobial properties of citral and quercetin incorporated kafirin-based bioactive films. Food Chem. 168:341–347. doi:10.1016/j.foodchem.2014.07.077

Hamaker, B.R., A.A. Mohamed, J.E. Habben, C.P. Huang, and B.A. Larkins. 1995. Efficient procedure for extracting maize and sorghum kernel proteins reveals higher prolamin contents than the conventional method. Cereal Chem. 72:583–588.

Hahn, R.R. 1970. Dry milling and products of grain sorghum. In: J.S. Wall and W.M. Ross, editors, Sorghum production and utilization. AVI Pub. Co., Westport, CT. p. 590–595.

Harlan, J.R. 1971. Agricultural origins: Centers and noncenters. Science 174:468–474. doi:10.1126/science.174.4008.468

Hojilla-Evangelista, M.P., and S.R. Bean. 2011. Evaluation of sorghum flour as extender in plywood adhesives for sprayline coaters or foam extrusion. Ind. Crops Prod. 34:1168–1172. doi:10.1016/j.indcrop.2011.04.005

Kim, K.M., K.T. Hwang, C.L. Weller, and M.A. Hanna. 2002. Preparation and characterization of soy protein isolate films modified with sorghum wax. J. Am. Oil Chem. Soc. 79:615–619. doi:10.1007/s11746-002-0532-4

Kim, K.M., D.B. Marx, C.L. Weller, and M.A. Hanna. 2003. Influence of sorghum wax, glycerin, and sorbitol on physical properties of soy protein isolate films. J. Am. Oil Chem. Soc. 80:71–76. doi:10.1007/s11746-003-0653-9

Krochta, J.M. 2002. Proteins as raw materials for films and coatings: Definitions, current status and opportunities. In: A. Gennadios, editor, Protein based films and coatings. CRC Press, Boca Raton, FL. p. 1–41.

Kubecka, B. 2011. Sorghum plays role in ethanol's impact. http://www.ethanolproducer.com/articles/7408/sorghum-plays-role-in-ethanolundefineds-impact (accessed 1 Mar. 2016).

Li, N., Y. Wang, M. Tilley, S.R. Bean, X. Wu, X.S. Sun, and D. Wang. 2011. Adhesive performance of sorghum protein extracted from sorghum DDGS and flour. J. Polym. Environ. 19:755–765. doi:10.1007/s10924-011-0305-5

Mudgett, R.E. 1982. Electrical properties of foods in microwave processing. Food Technol. 36:109–115.

Queensland Government. 2012. Overview of the sorghum industry. https://www.daf.qld.gov.au/plants/field-crops-and-pastures/broadacre-field-crops/sorghum/overview (accessed date: 3 Mar. 2016)

Ramos, J.R., R.R. Cabral, and F.D. Chan. 1984. Utilization of sorghum flour as extender for plywood adhesive. Technol. J. 9:29–48.

Rao, P., S.W. Zegada-Lizarazu, D. Bellmer, and A. Monti. 2014. Prospect of sorghum as a biofuel feedstock. In: Y.H. Wang, H.D.Upadhyaya, and C. Kole, editors, Genetics, genomics and breeding of sorghum. Taylor & Francis, Boca Raton, FL. p. 303–330.

Reddy, N., E. Jin, L. Chen, X. Jiang, and Y. Yang. 2012. Extraction, characterization of components, and potential thermoplastic applications of camelina meal grafted with vinyl monomers. J. Agric. Food Chem. 60:4872–4879. doi:10.1021/jf300695k

Reddy, N., Z. Shi, L. Temme, H. Xu, L. Xu, X. Hou, and Y. Yang. 2014. Development and characterization of thermoplastic films from sorghum distillers dried grains grafted with various methacrylates. J. Agric. Food Chem. 62:2406–2411. doi:10.1021/jf405499t

Rhim, J.W., and PK. Ng. 2007. Natural biopolymer-based nanocomposite films for packaging applications. Crit. Rev. Food Sci. Nutr. 47 (4): 411–433.

Rooney, L.W., and R.D. Waniska. 2000. Sorghum food and industrial utilization. In: C.W. Smith and R.A. Frederiksen, editors, Sorghum: Origin, history, technology, and production. John Wiley & Sons, New York. p. 689–750.

Shull, J.M., J.J. Watterson, and A.W. Kirleis. 1991. Proposed nomenclature for the alcohol-soluble proteins (kafirins) of Sorghum bicolor (L. Moench) based on molecular weight, solubility, and structure. J. Agric. Food Chem. 39:83–87. doi:10.1021/jf00001a015

Stroade, J., and M. Boland. 2013. Sorghum profile. http://www.agmrc.org/commodities-products/grains-oilseeds/sorghum/sorghum-profile/ (accessed date: 6 June 2015).

Taylor, J., and J.R.N. Taylor. 2009. Some potential applications for brewers spend grains the potential rich coproduct, from sorghum lager beer brewing. The institute of Brewing & Distilling Africa Sect. 12th Scientific and Technical Convention 2009, March. London, UK.

Taylor, J., J.R.N. Taylor, M.F. Dutton, and S. De Kock. 2005. Identification of kafirin film casting solvents. Food Chem. 90:401–408. doi:10.1016/j.foodchem.2004.03.055

Taylor, J.R.N., T.J. Schober, and S.R. Bean. 2006. Novel food and non-food uses for sorghum and millets. J. Cereal Sci. 44:252–271. doi:10.1016/j.jcs.2006.06.009

Taylor, J., J.R.N. Taylor, P.S. Belton, and A. Minnaar. 2009a. Formation of kafirin microparticles by phase separation from an organic acid and their characterization. J. Cereal Sci. 50:99–105. doi:10.1016/j.jcs.2009.03.005

Taylor, J., J.R.N. Taylor, P.S. Belton, and A. Minnaar. 2009b. Kafirin microparticle encapsulation of catechin and sorghum condensed tannins. J. Agric. Food Chem. 57:7523–7528. doi:10.1021/jf901592q

Taylor, J., J.O. Anyango, M. Potgieter, K. Kallmeyer, V. Naidoo, M.S. Pepper, and J.R.N. Taylor. 2014. Biocompatibility and biodegradation of protein microparticle and film scaffolds made from kafirin (sorghum prolamin protein) subcutaneously implanted in rodent models. J. Biomed. Mater. Res. A. 103:2582–2590. doi:10.1002/jbm.a.35394

Wall, J.S., and J.W. Paulis. 1978. Corn and sorghum grain proteins. In: Y. Pomeranz, editor, Advances in cereal science and technology. Vol. II. Am. Assoc. of Cereal Chemists, St Paul, MN. p. 135–219.

Wallheimer, B. 2012. Researchers: Sorghum should be in the mix as a biofuel crop. http://www.purdue.edu/newsroom/research/2012/120619CarpitaSorghum.html (accessed 3 Mar. 2016).

Wang, Y., M. Tilley, S. Bean, X.S. Sun, and D. Wang. 2009. Comparison of methods for extracting kafirin proteins from sorghum distillers dried grains with solubles. J. Agric. Food Chem. 57:8366–8372. doi:10.1021/jf901713w

Weller, C.L., A. Gennadios, R.L. Saraiva, and S.L. Cuppett. 1998a. Grain sorghum wax as an edible coating for gelatine-based candies. J. Food Qual. 21:117–128. doi:10.1111/j.1745-4557.1998.tb00509.x

Weller, C.L., A. Gennadios, R.L. Saraiva, and S.L. Cuppett. 1998b. Edible bilayer films from zein and grain sorghum wax or carnauba wax. Lebenson. Wiss. Technol. 31:279–285. doi:10.1006/fstl.1997.9998

Domestic and International Sorghum Marketing

Daniel O'Brien*

Abstract

Commercially grown sorghum comprises species of grasses within the genus *Sorghum* (usually *S. bicolor*) and is adaptable to high altitudes, toxic soils, and a wide range of temperatures. Grain sorghum is native to tropical and subtropical regions of Africa and Asia, and is grown for both commercial and subsistence agriculture in warmer climates worldwide, such as grain, fiber, and fodder. Sorghum is known as a drought-resistant crop due to its large proportion of roots relative to leaf surface area, the waxy cuticle over its leaf area, its ability to survive drought by rolling its leaves to reduce water loss by transpiration, and its characteristic of going dormant rather than expiring during extreme drought conditions. Grain sorghum is categorized as a competitive coarse grain in world markets, and as a feedgrain in the United States, with protein levels commonly near 9%, and can be used in livestock feeding rations as well as for making starch-based ethanol in developed countries. In less developed countries and those with subsistence populations in the semiarid tropics of Asia and Africa, sorghum remains a primary source of human nutrition, providing energy, protein, vitamins, and minerals. The United States, Argentina, and Australia have been the major countries growing grain sorghum for primarily commercial use (i.e., for livestock feed, bioenergy, and other nonfood and industrial uses). China, Japan, and Mexico have been the largest importers of grain sorghum in recent years, while sub-Saharan Africa, South Asia, parts of South America, North Africa, and Central America use a substantial proportion of their sorghum production for subsistence-level human consumption. Changing supply-demand trends have driven world sorghum prices higher in recent years, especially in non-subsistence, predominantly commercial-use countries. A key issue is whether one recent year's (2014 to 2015) uptrend in Chinese demand for grain sorghum imports, with its strong positive impact on world and US grain sorghum prices and farm-production, will continue.

Grain sorghum is grown commercially in warmer climates worldwide. It is native to tropical and subtropical regions of Africa and Asia and is used for grain, fiber, and fodder. Commercially gown sorghum comprises species of grass within the genus *Sorghum* (usually *S. bicolor* L.), and is adaptable to high

---

Abbreviations: DT, drought tolerant; FSI, food, seed, and industrial; GMO, genetically modified organism; MY, marketing year.

Extension Agricultural Economist, Kansas State Univ., Northwest Research Extension Center, 105 Experiment Farm Road, Colby, KS 67701-0786 *Corresponding author (dobrien@ksu.edu).

doi:10.2134/agronmonogr58.2014.0079

 Sorghum: State of the Art and Future Perspectives, Ignacio Ciampitti and Vara Prasad, editors. Agronomy Monograph 58.

altitudes, toxic soils, and a wide range of temperatures. Sorghum is also known as a drought-resistant crop due to its large proportion of roots relative to leaf surface area, the waxy cuticle over its leaf area, its ability to survive drought by rolling its leaves to lessen water loss by transpiration, and its characteristic of going into dormancy rather than expiring during extreme drought conditions.

Grain sorghum can be used in developed countries as a primary input for making starch-based ethanol. In less developed countries and those with subsistence populations in the semiarid tropics of Asia and Africa, sorghum remains a primary source of human nutrition, providing energy, protein, vitamins, and minerals. The protein level in grain sorghum is commonly near 9%.

An average temperature of at least 25°C (77°F) is required for sorghum to produce maximum grain yields in a given year. Daytime temperatures of at least 30°C (86°F) are required to achieve maximum photosynthesis. If nighttime temperatures below 13°C (55.4°F) occur for more than a few days, the plant's production potential can be severely reduced. Soil temperatures of at least 17°C (62.6°F) are required for planting and successful germination of grain sorghum, as is a relatively long growing season of 90 to 120 d.

World grain sorghum production is estimated to have averaged 59.65 × 10^6 Mg during the past seven marketing years (i.e., 2009 to 2010 through 2015 to 2016), with a projection of 62.50 × 10^6 Mg for 2016 to 2017.[1] The United States is the largest producer, with an average of 9.47× 10^6 Mg during that 7-yr period, followed by Mexico (6.66 × 10^6 Mg), Nigeria (6.52 × 10^6 Mg), India (5.74 × 10^6 Mg), Argentina (4.03 × 10^6 Mg), Sudan (3.84 × 10^6 Mg), Ethiopia (3.61 × 10^6 Mg), China (2.47 × 10^6 Mg), Australia (1.92 × 10^6 Mg), Brazil (1.90 × 10^6 Mg), Burkina Faso (1.71 × 10^6 Mg), Niger (1.27 × 10^6 Mg), Mali (1.22 × 10^6 Mg), Cameroon (1.12 × 10^6 Mg), Tanzania (0.807 × 10^6 Mg), Chad (0.80 × 10^6 Mg), Egypt (0.77 × 10^6 Mg), European Union (0.67 × 10^6 Mg), Yemen (0.40 × 10^6 Mg), Uruguay (0.24 × 10^6 Mg), and all other combined countries (3.96 × 10^6 Mg) (USDA-FAS, 2017).

Whereas the United States, Argentina, Australia and some other countries on this list grow sorghum for largely commercial use (i.e., for livestock feed, bioenergy, and other nonfood and industrial uses), there are other countries for which a substantial proportion of their sorghum production is used for subsistence-level human-consumption purposes.

This chapter examines a number of primary grain-sorghum supply-demand, price, and profitability trends in the United States and in the state of Kansas (which is the primary grain-producing state in the western Corn Belt of the United States). It also examines trends that have occurred in world production, exports, imports, usage, and ending stocks of grain sorghum across time.

In summary, the US and world grain sorghum market has been dynamic and changing across time. Strong, predominantly supply-demand, trends have driven the broader US feed-grain and world coarse-grain market prices in general and grain sorghum market prices in particular, especially in non-subsistence, predominantly commercial-use countries. The key issue for the future is the degree to which one recent year's higher than historically normal Chinese demand for grain sorghum imports continues, with its strong positive impact on world and US grain sorghum prices and farmers' ongoing production decisions.

1 Although some variations exist, the sorghum marketing year is generally July through June for most countries except the United States, for which the marketing year is September through August. Data for the US 2015 to 2016 marketing year is preliminary in nature but is a reasonable projection by the USDA at the time of this writing.

US Grain-Sorghum Supply–Demand and Prices

Sorghum Acreage in the United States

Planted acreage of grain sorghum in the United States has averaged 7.65×10^6 acres over the 2000 to 2016 period, with a low of 5.37×10^6 acres in 2010, and a high of 10.25×10^6 acres (4.15×10^6 ha) in 2001 (Fig. 1). Planted acreage of US grain sorghum has trended lower, at a rate of 0.170×10^6 acre yr^{-1} (6885 ha yr^{-1})(annually during 2000 to 2016 and was projected to be 7.22×10^6 acres (2.92×10^6 ha) in 2016 USDA-NASS, 2017). Harvested acreage of grain sorghum for grain in the United States averaged 6.40×10^6 acres (2.59×10^6 ha) during 2000 to 2016, with a low of 3.95×10^6 acres (1.60×10^6 ha) in 2011 and a high of 8.58×10^6 acres (3.47×10^6 ha) in 2001 (Fig. 1). Harvested acreage of US grain sorghum for grain trended lower, at a rate of 0.101×10^6 acre yr^{-1} (40905 ha yr^{-1}) during 2000 to 2016. Harvested acreage of US grain sorghum for silage averaged 0.329×10^6 acre yr^{-1} (133245 ha yr^{-1}) during the same period, ranging from a low of 0.226×10^6 acres in 2011 to a high of 0.412×10^6 acres (0.17×10^6 ha) in 2008 (USDA-NASS, 2017).

The percentage harvested of planted acreage of grain sorghum for grain in the United States averaged 84.3% during 2000 to 2016, with a low of 72.4% in 2011 and a projected high of 92.8% in 2015 (Fig. 1). The percentage harvested of planted acreage of US grain sorghum for grain trended higher, at a rate of 0.39% yr^{-1} during 2000 to 2016. The percentage harvested of planted acreage of grain sorghum for grain and silage combined in the United States averaged 88.2% during 2000 to 2016, with a low of 76.5% in 2011 and a high of 96.4% in 2016. The percentage harvested of planted acreage of US grain sorghum for both grain and silage combined also trended higher during 2000 to 2016 at a rate of 0.33% yr^{-1}.

Kansas Grain Sorghum Acreage

Planted acreage of grain sorghum in Kansas averaged 3.06×10^6 acres (1.24×10^6 ha) during 2000 to 2016, with a low of 2.35×10^6 acres ($.95175 \times 10^6$ ha) in 2010 and a high of 4.00×10^6 acres (1.62×10^6 ha) in 2001 (Fig. 2). Planted acreage of Kansas grain sorghum trended lower, at a rate of 0.047×10^6 acre yr^{-1} (0.019035×10^6 ha yr^{-1}) during 2000 to 2016 (USDA-NASS, 2017).

Harvested acreage of grain sorghum for grain in Kansas averaged 2.76×10^6 acres (1.12×10^6 ha) during 2000 to 2016, with a low of 2.00×10^6 acres

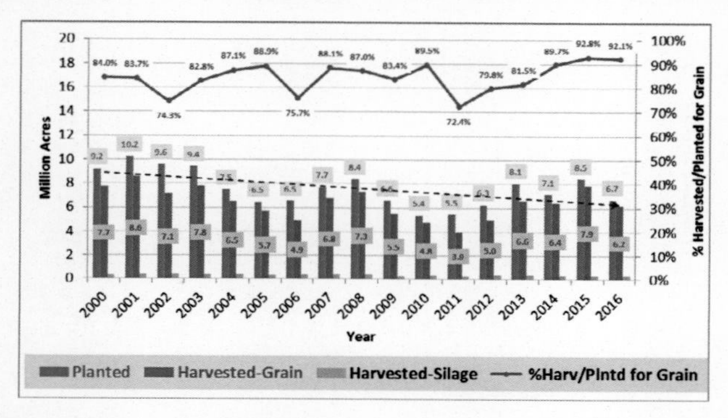

Fig. 1. US grain sorghum planted and harvested acreage, 2000–2016. Data from Quick Stats, USDA-NASS.

(0.81×10^6 ha) in 2011, and a high of 3.75×10^6 acres (1.52×10^6 ha) in 2001 (Fig. 2). Harvested acreage of Kansas grain sorghum for grain trended lower, at a rate of 0.033×10^6 acre yr^{-1} (0.01×10^6 ha) during 2000 to 2016. Harvested acreage of Kansas grain sorghum for silage averaged 0.078×10^6 acres yr^{-1} (0.03×10^6 ha yr^{-1}) during the same period, ranging from 0.040×10^6 acres (0.02×10^6 ha) in 2009 to 0.115×10^6 (0.05×10^6 ha) in 2002 (USDA-NASS, 2017).

The percentage harvested of planted acreage of grain sorghum for grain in Kansas averaged 90.3% during 2000 to 2016, with a low of 76.9% in 2011 and a high of 95.7% in 2010. The percentage harvested of planted acreage of Kansas grain sorghum trended higher, at a rate of 0.25% yr^{-1} during the same period. The combined percent harvested-of-planted acreage of grain sorghum for grain and silage in Kansas has averaged 92.9% during 2000 to 2016, with a low of 80.2% in 2011 and a high of 98.3% in 2010. The percentage harvested of planted acreage of US grain sorghum for both grain and silage combined has trended higher at a rate of 0.31% yr^{-1} during 2000 to 2016.

Sorghum Yields in the United States and Kansas

Grain sorghum yields in the United States averaged 64.4 bu acre^{-1} (4038.52 kg ha^{-1} during 1990 to 2016, with a low of 49.6 bu acre^{-1} (3110.42 kg ha^{-1}) in 2012 and an estimated high of 77.9 bu acre^{-1} (4885.12 kg ha^{-1}) in 2016 (Fig. 3). Yields of

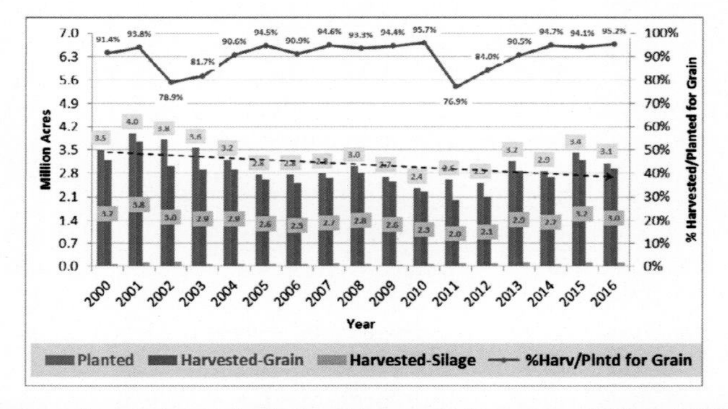

Fig. 2. Kansas grain sorghum planted and harvested acreage, 2000 to 2016. Data from Quick Stats, USDA-NASS.

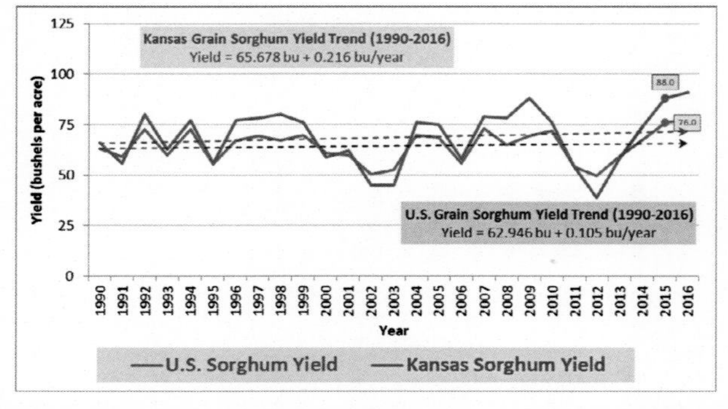

Fig. 3. US and Kansas grain sorghum yields, 1990 to 2016. Data from Quick Stats, USDA-NASS.

US grain sorghum have trended higher, at an annual rate of 0.105 bu acre^{-1} (6.58 kg ha^{-1}) during 1990 to 2016. However, over the most recent 8-yr period (i.e., 2009–2016), US grain sorghum yields averaged 65.2 bu acre^{-1} (4088.69 kg ha^{-1}) ranging from 49.6 to 77.9 bu acre^{-1} (3110.42 to 4885.11 kg ha^{-1} , while trending higher at an annual rate of 2.86 bu acre^{-1} yr^{-1} (2.86 kg ha^{-1} yr^{-1})(USDA-NASS, 2017).

Grain sorghum yields in Kansas averaged 68.7 bu acre^{-1} (4308.18 kg ha^{-1}) during 1990 to 2016, with a low of 39 bu acre^{-1} (2445.69 kg ha^{-1}) in 2012 and a high of 91 bu acre^{-1} (5706.61 kg ha^{-1}) in 2016 (Fig. 3). Yields of Kansas grain sorghum trended higher, at a rate of 0.216 bu acre^{-1} (13.55 kg ha^{-1}) annually during 1990 to 2016. Over the most recent 8-yr period (i.e., 2009–2016), Kansas grain sorghum yields averaged 68.7 bu acre $^{-1}$ (4308.18 kg ha^{-1}), ranging from 39.0 to 88.0 bu acre^{-1} (2445.69 to 5518.48 kg ha^{-1}, trending higher at a rate of 5.3 bu acre^{-1} (332.36 kg ha^{-1}) annually (USDA-NASS, 2017).

Grain Sorghum Production in the United States and Kansas

Grain sorghum production in the United States averaged 487 × 10^6 bu (12.37 × 10^6 Mg) annually during 1990 to 2016, with a low of 213 × 10^6 bu (5.41 × 10^6 Mg) in 2011 and a high of 875 × 10^6 bu (22.23 × 10^6 Mg) in 1992 (Fig. 4). Production of US grain sorghum trended lower, at a rate of 11.5 × 10^6 bu yr^{-1} (0.29 × 10^6 Mg yr^{-1}), during 1990 to 2016. However, during the most recent 8-yr period (i.e., 2009–2016), US grain sorghum production averaged 386 × 10^6 bu (9.80 × 10^6 Mg), ranging from 213 to 597 × 10^6 bu (5.41 to 15.16 × 10^6 Mg) and trending higher at a rate of 32.7 × 10^6 bu yr^{-1} (0.83 × 10^6 Mg yr^{-1}) (USDA-NASS, 2017).

Grain sorghum production in Kansas averaged 203 × 10^6 bu (5.16 × 10^6 Mg) annually during 1990 to 2016, with a low of 81.9 × 10^6 bu (2.08 × 10^6 Mg) in 2012 and a high of 354 × 10^6 bu (8.99 × 10^6 Mg) in 1996 (Fig. 4). Production of Kansas grain sorghum trended lower, at a rate of 1.3 × 10^6 bu yr^{-1} (0.03 × 10^6 Mg yr^{-1}) during 1990 to 2016. However, during the most recent 8-yr period (2009–2016), Kansas grain sorghum production averaged 188 × 10^6 bu (4.78 × 10^6 Mg) annually, ranging from 81.9 ′ 10^6 bu in 2012 to 282 ′ 10^6 bu in 2015 and trending higher at a rate of 14.6 bu yr^{-1} (USDA-NASS, 2017).

During 1990 to 2016, Kansas grain sorghum production averaged 42.8% of total US sorghum production, ranging from a low of 27.9% in 1992 to a high of 58.8% in 2009 and trending upward at a rate of 0.6% yr (Fig. 4). During the last 8 yr

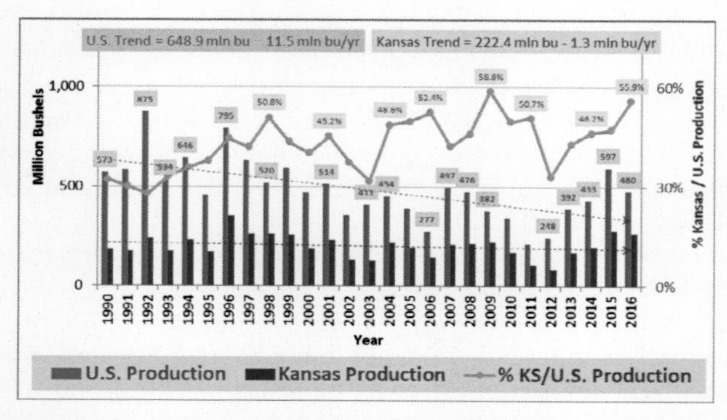

Fig. 4. US and Kansas sorghum production, 1990–2016. Data from Quick Stats, USDA-NASS.

(i.e., 2009 to 2016), Kansas grain sorghum production averaged 48.0% of total US sorghum production, ranging from a low of 33.1% in 2012 to a high of 58.8% in 2009 and trending downward at a rate of 0.4% yr^{-1}.

Competition with Corn Seed Technologies

Developments in competitive corn seed technologies have impacted US grain sorghum acreage and production in past years and have the potential to be a factor that limits US grain sorghum production in the future. Roundup Ready corn was introduced by Monsanto in 1998, greatly improving farmers' weed-control capability. The ability to spray glyphosate herbicide (i.e., Roundup) directly over and onto Roundup Ready corn crops in the field provided a competitive advantage compared with grain sorghum, especially in drought-prone parts of the US Central and Southern Plains.

Genetically modified organism (GMO) varieties of grain sorghum varieties with glyphosate resistance were not developed, in part because of concerns about introducing herbicide-resistant sorghum strains into existing US Great Plains cropping systems. The introduction of Roundup Ready corn in the United States is a major, if not the primary, factor in the decline in US grain sorghum acreage since 2000 and in US grain sorghum production since the 1990s.

With continued use of Roundup Ready seed technology, some significant weed pests have developed resistance to glyphosate herbicides. This resistance is now limiting the advantage that Roundup Ready corn had over non-GMO grain sorghum in drought-prone central and southern Great Plains growing environments.

More recently, several major seed companies have introduced corn hybrids specifically designed to provide enhanced drought tolerance. These corn hybrids were developed specifically for the generally drier western Corn Belt growing environments with the long-term goal of making more efficient use of increasingly scarce groundwater in irrigated crop enterprises and for enhancing feedgrain productivity under non-irrigated cropping systems. If successful, these drought-tolerant corn varieties—DT corn—will provide yet more competition for grain sorghum acreage and production in the US western Corn Belt and elsewhere. To date, DT corn productivity gains have shown promise, but not yet to the degree in drought-prone areas where there has been a large-scale changeover from grain sorghum to DT corn in Western Corn Belt growing regions.

Grain Sorghum Total Supplies in the United States

Total supplies of grain sorghum in the United States trended sideways to marginally lower during the 2000 to 2001 through estimated marketing year (MY) 2016 to 2017 (Fig. 5). During these 16 marketing years, US total grain sorghum supplies averaged 451 $\times$ 10^6 bu (11.46 $\times$ 10^6 Mg), with a median (middle 50th percentile) value of 454 $\times$ 10^6 bu (11.53 $\times$ 10^6 Mg), indicating a small amount of negative skew (i.e., presence of a few much-smaller-than-average observations) (USDA Office of the Chief Economist, 2017).

From MY 2000 to 2001 through MY 2016 to 2017, the minimum amount of US grain sorghum supplies was 241 $\times$ 10^6 bu (6.12 $\times$ 10^6 Mg) in MY 2011 to 2012, while the maximum amount of 620 $\times$ 10^6 bu (15.75 $\times$ 10^6 Mg) was in MY 2015 to 16. Total supplies of US grain sorghum trended lower, at a rate of 2.7 $\times$ 10^6 bu (.07 $\times$ 10^6 Mg) to MY during MY 2000 to 2001 through 2016 to 2017, with 518 $\times$ 10^6 bu

(13.16×10^6 Mg) projected for MY 2016 to 17 (USDA Office of the Chief Economist, 2017).

Beginning stocks of US grain sorghum have averaged 41.3×10^6 bu (1.05×10^6 Mg) since MY 2000 to 2001, ranging from a projected low of 15.2×10^6 bu (0.39×10^6 Mg) in MY 2013 to 2014 to a high of 65.7×10^6 bu (1.67×10^6 Mg) in MY 2006 to 2007 (Fig. 5). Beginning stocks have trended lower during the most recent 15-yr period, declining on average by 2.1×10^6 bu MY^{-1} ($.05 \times 10^6$ Mg MY^{-1}), with 37×10^6 bu (0.94×10^6 Mg) projected for MY 2016 to 2017. Imports into the United States of grain sorghum have been minimal from MY 2000 to 2001 through MY 2016 to 2017, with only 1 yr in which any appreciable amount of sorghum imports came into the United States, that is, 9.6×10^6 bu (0.24×10^6 Mg) in the drought-affected, short crop MY 2012 to 2013 (USDA-Office of the Chief Economist, 2017).

Grain Sorghum Usage in the United States

Total use of grain sorghum in the United States varied greatly from MY 2000 to 2001 through MY 2016 to 2017 (Fig. 6). During these 17 yr, US total usage of grain sorghum averaged 4.10×10^8 bu ($.10 \times 10^6$ Mg), ranging from a low of 218×10^6 bu (5.54×10^6 Mg) in MY 2011 to 2012 to a high of 583×10^6 bu (14.81×10^6 Mg) in MY 2015 to 2016. Total use of US grain sorghum has trended marginally higher, at a rate of 1.0×10^6 bu ($.03 \times 10^6$ Mg) per MY since MY 2000 to 2001. During the last 8 yr (MY 2009 to 2010 through

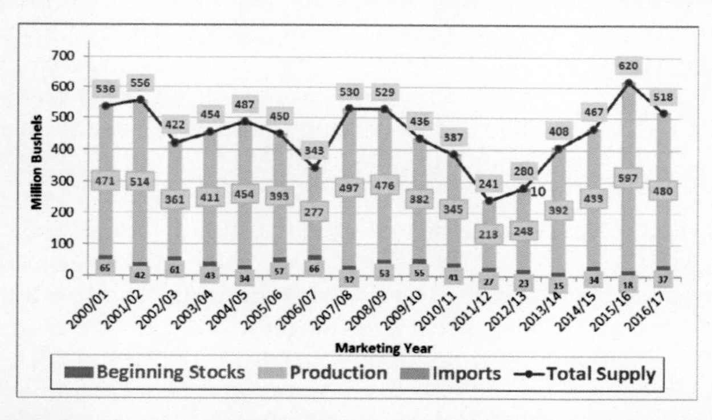

Fig. 5. US sorghum total supplies for 2000 to 2001 through 2016 to 2017 marketing years. Data from WASDE Report, 11 April 2017, USDA-OCE.

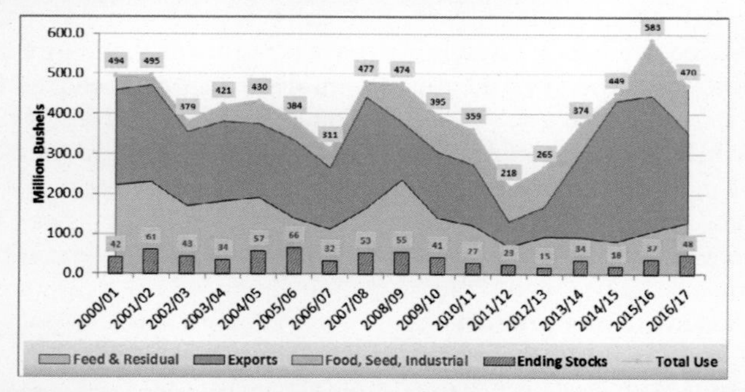

Fig. 6. Cumulative US sorghum sage for 2000 to 2001 through 2016 to 2017 marketing years. Data from WASDE Report, 11 April 2017, USDA-OCE.

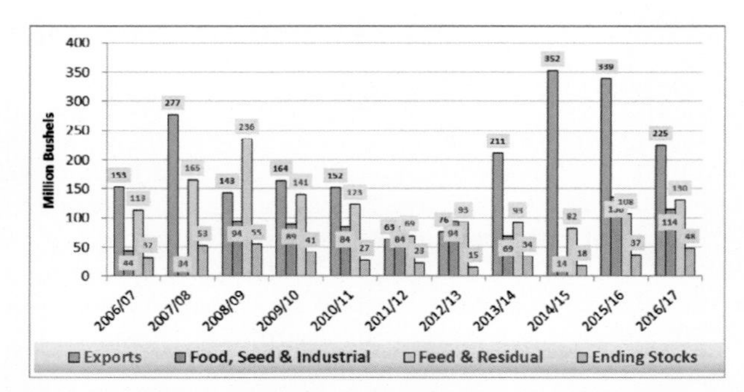

Fig. 7. US sorghum usage by category for 2006 to 2007 through 2016 to 2017 marketing years. Data from WASDE Report, 11 April 2017, USDA-OCE.

MY 2016 to 2017), US total use of grain sorghum initially dropped from 395 × 10^6 bu (10.03 × 10^6 Mg) in MY 2009 to 2010 down to 359 × 10^6 bu (9.12 × 10^6 Mg) in MY 2010 to 2011 and to 218 × 10^6 bu (5.54 × 10^6 Mg) in MY 2011 to 2012. Over the next four marketing years total use increased an average of 91.4 × 10^6 bu (2.32 × 10^6 Mg) annually—up to 583 × 10^6 bu (14.81 × 10^6 Mg) in MY 2015 to 2016—with a projected total use of 470 × 10^6 bu (11.94 × 10^6 Mg) in MY 2016 to 2017 (USDA-Office of the Chief Economist, 2017).

Food, seed, and industrial (FSI) use of US grain sorghum has been trending sideways to slightly lower from MY 2009 to 2010 through MY 2016 to 2017 (Fig. 6 and 7). During this 8-yr period, US total grain sorghum FSI use averaged 85.7 × 10^6 bu MY^{-1} (2.18 × 10^6 Mg MY^{-1}), with a minimum amount of 14.0 × 10^6 bu (0.36 in MY 2014 to 2015 and a maximum amount of 136 × 10^6 bu (3.45 × 10^6 Mg) in MY 2015 to 2016. United States' grain sorghum FSI use is projected to be 114 × 10^6 bu (2.90 × 10^6 Mg) in MY 2016 to 2017 (USDA Office of the Chief Economist, 2017).

The United States' FSI use has trended marginally higher since MY 2009 to 2010, increasing 2.3 × 10^6 bu (0.06 × 10^6 Mg) annually on average during MY 2009 to 2010 through MY 2016 to 2017. Grain sorghum used for ethanol production is included in the industrial aspect of this category. Usage of domestic-source grain sorghum for US ethanol declined temporarily in MY 2013 to 2014 through MY 2015 to 2016 due to a large increase in Chinese imports of US grain sorghum and to the higher price for grain sorghum feedstock that existed at the time compared with other feedgrains.

Under Title IX of the 2008 Farm Bill, the USDA recognized grain sorghum as an advanced feedstock eligible for payment in the Bioenergy Program for Advanced Biofuels. Then in 2012, the EPA published what is referred to as a "pathway" for grain sorghum under the Renewable Fuels Standard 2 (RFS2). The EPA pathway would allow ethanol plants to use biogas to provide the heat and electricity to be used in the process of turning grain sorghum feedstock into ethanol and distillers' grains co-products to produce what would qualify as an advanced biofuel. The expectation of this program was and is that advanced biofuel product from grain sorghum would qualify for an enhanced price over that of regular feedgrain-based ethanol products.

Initial attempts to produce grain sorghum in industry-scale plants have met with mixed success. However, as the technology continues to develop, production of grain sorghum as an advanced biofuel may eventually come to fruition on an industrial scale.

Exports of US grain sorghum were highly variable and trended higher during MY 2006 to 2007 through MY 2016 to 2017 (Fig. 6 and 7). During this 12-yr period, US total grain sorghum exports averaged 193 × 10^6 bu (4.90 × 10^6 Mg), ranging from 63.4 × 10^6 bu (1.61 × 10^6 Mg) in MY 2011 to 2012 to 352 × 10^6 bu (8.94 × 10^6 Mg) in MY 2014 to 2015. Exports of US grain sorghum trended higher, at a rate of 13.3 × 10^6 bu MY^{-1} (0.34 × 10^6 Mg MY^{-1}) for MY 2006 to 2007 through 2016 to 2017, with a projection of 225 × 10^6 bu (5.72 × 10^6 Mg) for MY 2016 to 2017 (USDA Office of the Chief Economist, 2017).

Combined livestock feed and residual use of US grain sorghum was also highly variable and trended slightly higher during MY 2006 to 2007 through MY 2016 to 2017 (Fig. 6 and 7). During this period, US total grain sorghum feed and residual use averaged 122 × 10^6 bu (3.10 × 10^6 Mg), ranging from a low of 69.2 × 10^6 bu (1.76 × 10^6 Mg^{-1}) in MY 2011 to 2012 to a high of 236.1 × 10^6 bu (6.00 × 10^6 Mg) in MY 2008 to 2009. Feed and residual use of US grain sorghum trended lower, at a rate of 9.4 × 10^6 bu MY^{-1} (0.24 × 10^6 Mg MY^{-1}) since MY 2006 to 2007, with a projection of 130 × 10^6 bu (3.30 × 10^6 Mg) for MY 2016 to 2017 (USDA Office of the Chief Economist, 2017).

Ending Stocks and Percent Ending Stocks-to-Use of Sorghum in the United States

The quantity of ending stocks of grain sorghum in the United States has been trending marginally lower from MY 1992 to 1993 through MY 2016 to 2017 (Fig. 7 and 8). During this 25-yr period, US grain-sorghum ending stocks averaged 39.8 × 10^6 bu (1.01 × 10^6 Mg^{-1}), ranging from a low of 15.2 × 10^6 bu (0.39 × 10^6 Mg^{-1}) in MY 2012 to 2013 to a high of 65.7 × 10^6 bu (1.67 × 10^6 Mg^{-1}) in MY 2005 to 2006. Ending stocks of grain sorghum have trended lower at a rate of 1.9 × 10^6 bu (0.05 × 10^6 Mg) per MY from MY 1992 to 1993 through MY 2016 to 2017 and are projected to be 48 × 10^6 bu (1.22 × 10^6 Mg) in MY 2016 to 2017 (USDA-OCE, 2017).

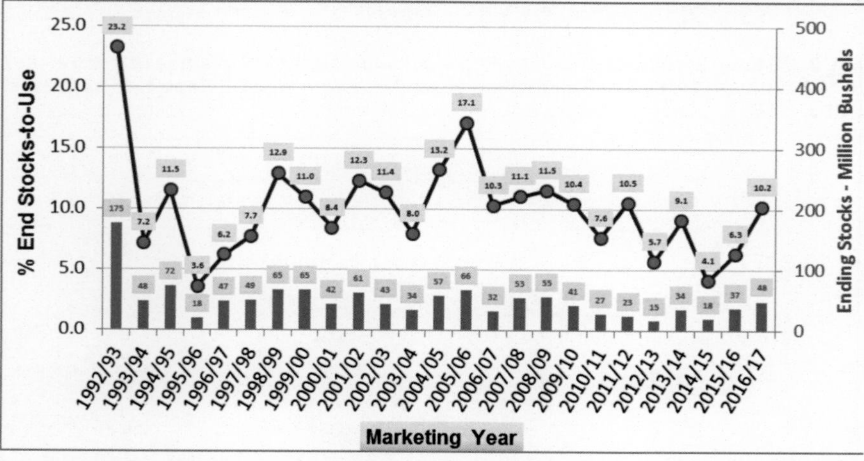

Fig. 8. US sorghum ending stocks and percent ending stocks-to-use for 1992 to 1993 through 2016 to 2017 marketing years. Data from WASDE Report, 11 April 2017, USDA-OCE

Percent ending stocks-to-use of grain sorghum in the United States also trended marginally lower from MY 1992 to 1993 through MY 2015 to 2016 (Fig. 8). United States' grain sorghum percent ending stocks-to-use averaged 9.83% over this period, with a median (middle 50th percentile) value of 10.38%, indicating that there were some relatively large percent ending stocks-to-use outcomes during this time. United States' grain sorghum percent ending stocks-to-use are projected to be 10.19% for MY 2016 to 2017. Percent ending stocks-to-use of US grain sorghum trended lower at a rate of 0.37% MY^{-1} from MY 1992 to 1993 through MY 2015 to 2016 and are projected to be 10.19% in MY 2016 to 2017 (USDA-OCE, 2017).

Sorghum Percent Ending Stocks-to-Use and Season Average Prices in the United States

Figures 9 and 10 show the relationship across time between US grain sorghum percent ending stocks-to-use and US grain sorghum season average prices from alternative perspectives.

Figure 9 illustrates this relationship for consecutive years from MY 1992 to 1993 through MY 2016 to 2017 and shows the annual pairing of percent ending stocks-to-use and prices, with the percent ending stocks-to-use for both US grain sorghum and US feedgrains overall (i.e., including sorghum, corn, barley, and oats).

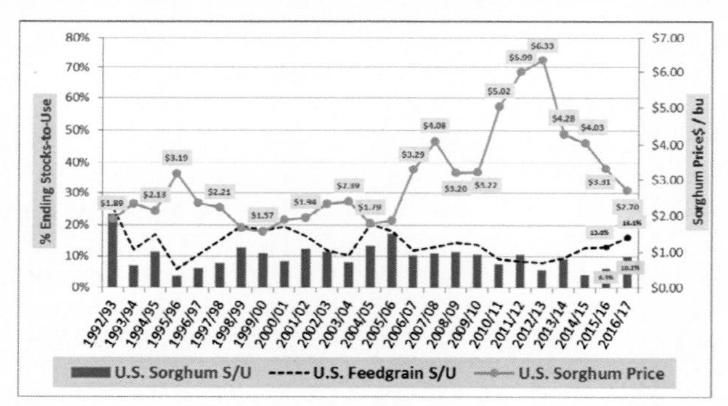

Fig. 9. US grain sorghum percent ending stocks-to-use and US average prices for 1992 to 1993 through 2016 to 2017 marketing years. Data from WASDE Report, 11 April 2017, USDA-OCE.

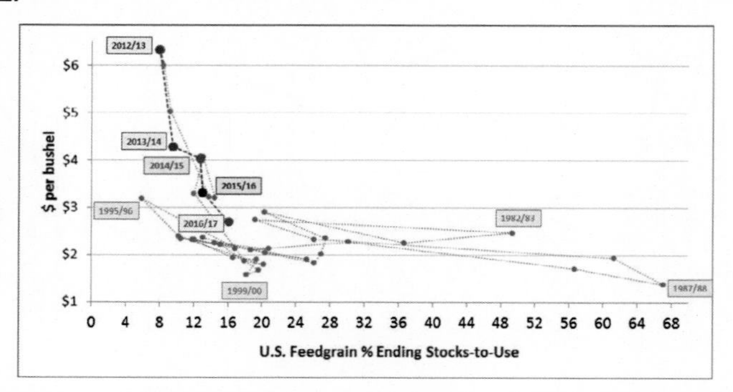

Fig. 10. US feedgrain percent ending stocks-to-use and US average sorghum prices for 1973 to 1974 through 2016 to 2017 marketing years. Data from WASDE Report, 11 April 2017, USDA-OCE.

The upward trend in US grain sorghum season average prices since MY 2005 to 2006 is shown in Figure 9. From MY 1992 to 1993 through MY 2005 to 2006, US grain sorghum season average prices ranged from \$1.57 to \$3.19 bu^{-1} (\$61.81 to \$125.58 Mg^{-1}). However, since that time—from MY 2006 to 2007 through MY 2015 to 2016, US season average prices ranged from lows of \$3.20 and \$3.22 bu^{-1} (\$125.98 and \$126.76 Mg^{-1}) during MY 2008 to 2009 and MY 2009 to 2010 to record highs of \$5.99 and \$6.33 bu (\$235.81 and \$249.20 Mg^{-1}) in MY 2011 to 2012 and MY 2012 to 2013, with a forecast price range of \$2.50 to \$2.90 (\$98.42 to \$114.17 Mg^{-1}) and a midpoint forecast of \$2.70 bu (\$106.29 Mg^{-1}) for MY 2016 to 2017 (USDA-OCE, 2017).

Figure 10 shows the intersection of these same factors (US feedgrain percent ending stocks-to-use and US grain sorghum season average prices) for the longer period since MY 1973 to 1974. Of these two charts, Figure 9 more clearly shows the annual evolution of prices across time through consecutive marketing years in response to changes in percent ending stocks-to-use for US feedgrains.

Figure 10, however, more clearly illustrates how US average grain sorghum prices are generally higher and more reactive to changes in supply-demand balances during times of relatively tight US feedgrain supplies than they are during periods of relatively abundant US feedgrain supply-demand balances. The category "US feedgrains" includes corn, grain sorghum, barley, oats, and millet.

This difference in price flexibility, or responsiveness of prices to changes in available supplies, is consistent with the body of economic theory in which commodity prices tend to respond in a more volatile, or "inflexible," manner to small changes in relative supplies when supplies or stocks of a commodity are more scarce. This situation is opposed to the one in which supplies of a commodity are relatively abundant, when changes in the relative amount of available supplies have a comparatively small impact on prices. This situation is representative of price responses to changes in quantity indicative of price flexibility.

In agricultural price analysis in general, and price forecasting applications in particular, information is often available about changes in supply for a particular commodity. Grain market participants are often interested in estimating how the prices of one commodity (such as grain sorghum) may change in response to a change in the supply of either that same commodity or in the aggregate supplies of a group of closely substitutable commodities (such as feedgrains).

The major types of feedgrains produced in the United States are corn, grain sorghum, barley, and oats. Other crops and grain-processing by-products, such as wheat and distillers grains, may also be potential competitive substitutes for grain sorghum in feed and industrial uses, and as a result, they may have tangible cross-market effects on grain sorghum prices. To assess these potential "own" and cross-commodity price impacts, measures of price flexibility are often used to address the issue of how grain sorghum prices may change in response either directly to the supply of grain sorghum, or indirectly to the broader supply of feedgrains and other potential feedgrain substitutes.

In economic theory, these types of quantified responses of commodity prices to changes in commodity supplies are associated with the concept of price flexibility of demand. By definition, "own price flexibility of demand" is calculated as the percentage change in prices of a commodity divided by the percentage change in quantity for that same commodity. In the absence of any competitive substitutes, price flexibility can be calculated as the inverse of price elasticity, that

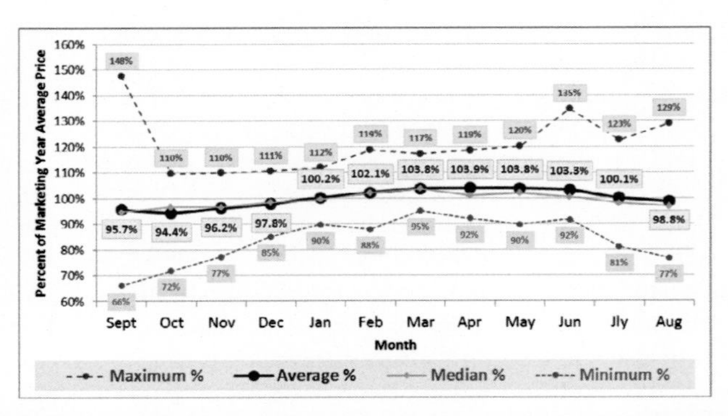

Fig. 11. Annual Kansas grain sorghum cost of production and net returns for 1998 through 2016. Data from Statewide Kansas Farm Management Association Non-irrigated Enterprise Data.

is, the percentage change in quantity demanded of a commodity divided by the percentage change in that same commodity's price.

However, in agricultural markets in general, and for grain sorghum in particular, competitive substitutes usually do exist. Therefore, the concept of cross price flexibility of demand is relevant to this discussion of grain sorghum price determination, being defined as the percentage change in prices of a commodity divided by the percentage change in quantity for an alternative commodity.

In the previous discussion of grain sorghum prices being more or less flexible, the intent is to illustrate how grain sorghum prices vary in their response to changes in supplies of grain sorghum and other substitute feedgrains. Figure 10 provides an example of cross price flexibility of demand as the response of US grain sorghum prices to changes in the broader class of US feedgrain supplies.

In Figure 10, more price inflexible, volatile price responses to changes in supply-demand balances are shown in the upper left of this graph. Conversely, more-price-flexible, less-volatile US grain sorghum price responses to changes in available US feedgrain supplies are shown.in the lower right. Since MY 2007 to 2008, US feedgrain ending stocks-to-use versus US grain sorghum marketing year average cash prices have generally been in the price-inflexible, volatile, high-price region of the upper left quadrant of Figure 10.

However, with the large projected increase in ending stocks and ending stocks-to-use in MY 2014 to 2015, MY 2015 to 2016 and MY 2016 to 2017, this price x percent stocks-to-use relationship has shifted to the right toward the middle of Figure 10, that is, toward the more economically flexible, moderate-to-lower priced, less-volatile areas of the graph.

Grain Sorghum Profitability in Kansas

The net profitability of grain sorghum production enterprises in Kansas generally trended higher during the 1998 through 2016 calendar year, although losses have occurred in the last couple of years (Fig. 11). Information in Figure 11 is based on the Kansas State University Kansas Farm Management Association enterprise records. These records which provide information on a per-bushel basis on (i) annual Kansas grain sorghum selling prices, (ii) revenue from all sources, including government payments, crop insurance, and crop sales, (iii) cost of production,

(iv) net returns to management, and (v) net returns to labor and management (KFMA, 2017; KSU Agric. Econ., 2017).

Of these key factors, both selling price and total revenues generally trended higher from 1998 to 2016, rising at an annual rate of \$0.13 bu^{-1} (\$5.12 Mg^{-1}). Total revenues for grain sorghum ranged from lows of \$2.25 bu^{-1} (\$88.58 Mg^{-1}) in 2004 and \$2.28 (\$89.76 Mg^{-1}) in 1999 to highs of \$6.84 bu^{-1} (\$269.28 Mg^{-1}) in 2011, and \$10.45 bu^{-1} (\$411.39 Mg^{-1}) in 2012, while falling to \$2.78 bu^{-1} (\$109.44 Mg^{-1}) in 2016. The cost of production has also trended higher, at an annual rate of \$0.11 bu^{-1} (\$4.33 Mg^{-1}), with lows of \$2.32 bu^{-1} (\$91.33 Mg^{-1}) in 2004, \$2.36 (\$92.91 Mg^{-1}) in 1999, \$2.81 (\$110.62 Mg^{-1}) in 2000, and \$2.88 (\$113.38 Mg^{-1}) in 2001. High costs per bushel harvested occurred in 2012 [\$9.36 (\$368.48 Mg^{-1})], 1998 and 2013 [both at \$5.43 (\$213.77 Mg^{-1})], 2011 [\$4.99 (\$196.45 Mg^{-1})], and 2010 [\$4.31 (\$169.68 Mg^{-1})], the with cost of production in 2016 at \$3.69 bu^{-1} (145.27 Mg^{-1}). High costs of production per bushel at least partly coincide with years in which bushels produced per acre are reduced.

Net returns to management trended marginally higher, at an annual rate of \$0.02 bu^{-1} (\$0.79 Mg^{-1}), while also averaging \$0.02 bu^{-1} produced during the 1998 to 2016 period. The low levels of net returns to management per bushel produced during this period were net losses of −\$0.92 (-\$36.22 Mg^{-1}) in 2016, −\$0.80 (-\$31.49 Mg^{-1}) in 2003, −\$0.63 (-\$0.63 Mg^{-1}) in 2005, −\$0.54 (-\$21.26 Mg^{-1}) in 2015, −\$0.36 (-\$14.17 Mg^{-1}) in 2001 and 2014, and −\$0.29 (-\$11.42 Mg^{-1}) in 2002. The highest levels of net returns to management per bushel produced during this period were net profits of \$1.85 (\$72.83 Mg^{-1}) in 2011, \$1.09 (\$42.91 Mg^{-1}) in 2012, \$0.89 (\$35.04 Mg^{-1}) in 2007, and \$0.63 (\$24.80 Mg^{-1}) in 2010. From 2007 forward, Kansas grain sorghum net returns to management have averaged \$0.28 bu^{-1} (\$11.02 Mg^{-1}) produced, ranging from −\$0.92 bu^{-1} (-\$36.22 Mg^{-1}) produced in 2015 to \$1.85 bu^{-1} (-\$72.83 Mg^{-1}) produced in 2011.

Net returns to management and labor also trended higher at an annual rate of \$0.02 bu^{-1} (\$0.79 Mg^{-1}) produced from 1998 through 2016, averaging \$0.57 bu^{-1} (\$22.44 Mg^{-1}) produced. The low levels of net returns to management and labor per bushel produced during this period were net losses of −\$0.46 (-\$18.11 Mg^{-1}) in 2016, −\$0.20 (-\$7.87 Mg^{-1}) in 2005, −\$0.19 (-\$7.48 Mg^{-1}) in 2003, and −\$0.05 (-\$1.97 Mg^{-1}) in 2015, and a small profit of \$0.10 (\$3.94 Mg^{-1}) in 2001. The highest levels of net

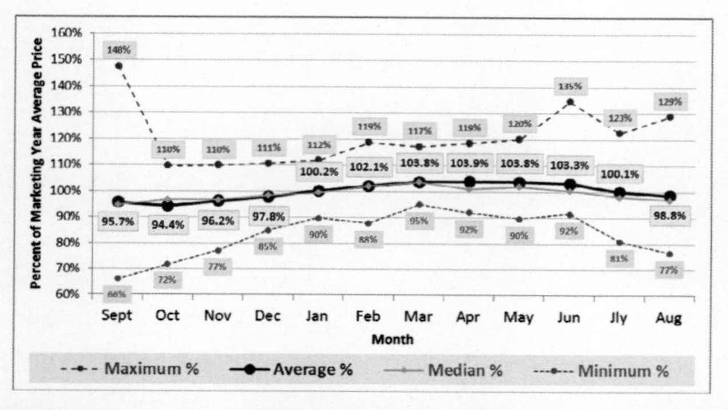

Fig. 12. Kansas grain sorghum monthly seasonal price index for 1992 to 1993 through 2015 to 2016 marketing years. Data from Kansas State University AgManager.info, KSU Agricultural Economics.

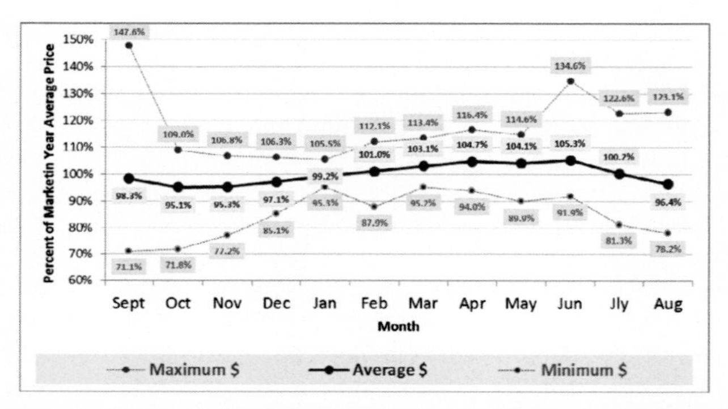

Fig. 13. Kansas grain sorghum monthly seasonal price index for 2007 to 2008 through 2015 to 2016 marketing years. Data from Kansas State University AgManager.info, KSU Agricultural Economics.

returns to management and labor per bushel produced during this period were net profits of \$2.37 (\$93.30 Mg^{-1}) in 2011, \$2.19 (\$86.22 Mg^{-1}) in 2012, \$1.45 (-\$57.08 Mg^{-1}) in 2007, \$0.68 (-\$26.77 Mg^{-1}) in 1998 and 2013, \$0.62 (-\$24.41 Mg^{-1}) in 2009, and \$0.56 (\$22.05 Mg^{-1}) per bushel in 2008. From 2007 forward, Kansas grain sorghum net returns to management and labor averaged \$0.89 bu^{-1} (\$35.04 Mg^{-1}) produced, ranging from −\$0.46 (-\$18.11 Mg^{-1}) in 2016 to \$2.37 bu^{-1} (\$93.30 Mg^{-1}) produced in 2011.

Grain Sorghum Seasonal Prices in Kansas

Cash prices for Kansas grain sorghum showed definite seasonal price patterns from MY 1992 to 1993 through MY 2015 to 2016. Focusing on an index of Kansas grain sorghum price movements across time, on average, prices tended to be lowest during what is typically the Kansas grain sorghum harvest period in October and proximate months and to be steadily higher through late fall, winter, and early spring, through March (Fig. 12).

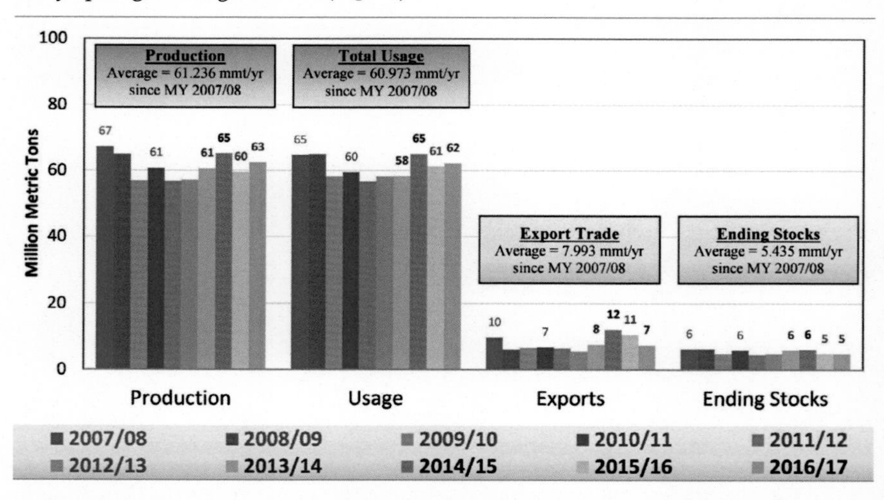

Fig. 14. World sorghum usage and ending stocks for 2007 to 2008 through 2016 to 2017 marketing years. Data from WASDE Report, 11 April 2017, USDA-OCE.

Kansas grain sorghum prices have tended to stay higher during spring and early summer (i.e., March–June), before beginning to decline in July and August (KSU Agric. Econ, 2017).

Periods of the lowest price volatility around the seasonal index trend occurred in January and March, while seasons of higher price variability occurred during summer (June, July, August) and early fall (September), that is, when the greatest uncertainty about the production of US grain sorghum and combined US feedgrains typically occurs. The distribution of US grain sorghum price indices from MY 1992 to 1993 through MY 2015 to 2016 is skewed negatively during the harvest period, that is, during October and November. This observation indicates that a small number of much-lower-than-average Kansas grain sorghum prices relative to their marketing-year average have occurred during harvest season.

During the recent period of MY 2007 to 2008 through MY 2015 to 2016, cash prices for Kansas grain sorghum continued to show pronounced seasonal price patterns. Focusing on the average and minimum-maximum range of movements in Kansas grain sorghum prices across this 9-yr period, the Kansas grain sorghum price index during the months of September through December averaged 98.3, 95.1, 95.3, and 97.1% of the seasonal average prices, respectively.

Kansas grain sorghum prices on average then trended upward, to 99.2% in January, 101.0% in February, 103.1% in March, 104.7% in April, 104.1% in May,

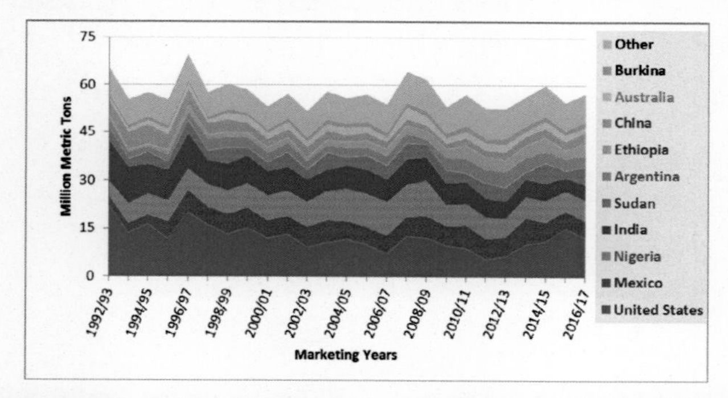

Fig. 15. World sorghum production by major countries for 1992 to 1993 through 2016 to 2017 marketing years. Data from PSD Online, USDA-FAS, and WASDE Report, 11 April 2017, USDA-OCE.

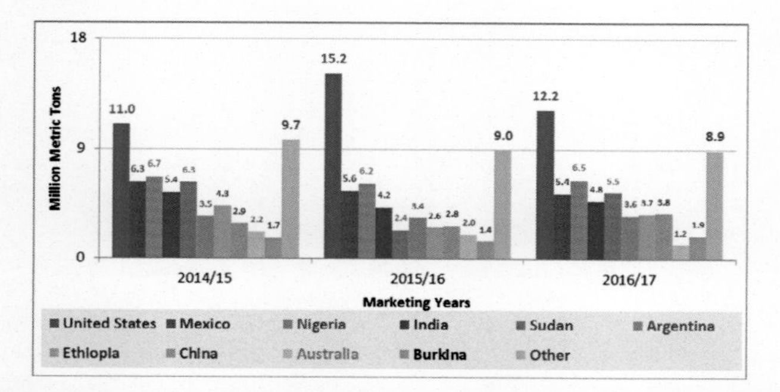

Fig. 16. Top world sorghum-producing countries, 2014 to 2015 through 2016 to 2017 marketing years. Data from PSD Online, USDA-FAS and WASDE Report, 11 April 2017, USDA-OCE.

and 105.3% in June. During July and August, Kansas grain sorghum prices have generally trended downward from early summer highs in June, to 100.2% of the marketing-year average price in July and 96.4% in August (Fig. 13). Patterns of seasonal price variability were similar between the longer (i.e., since MY 1992 to 1993) and intermediate (i.e., since MY 2007 to 2008) periods for Kansas grain sorghum prices (comparing Fig. 12 with Fig. 13).

International Grain-Sorghum Supply-Demand

World grain sorghum production was projected to average 61.2×10^6 Mg MY^{-1} for MY 2007 to 2008 through MY 2016 to 2017 (Fig. 14). Global grain sorghum production is projected to be 62.5×10^6 Mg in MY 2016 to 2017 (USDA-FAS, 2017).

In comparison, world grain sorghum total usage was forecast to average 62.0×10^6 Mg MY^{-1} since MY 2007 to 2008. Global grain sorghum usage was projected to be 62.2×10^6 Mg in MY 2016 to 2017.

Following production and usage, world grain-sorghum export trade was projected to average 7.93×10^6 Mg annually for MY 2007 to 2008 through MY 2016 to 2017 (7.50×10^6 Mg in MY 2016 to 2017), while world grain-sorghum ending stocks were expected to average 5.44×10^6 Mg MY^{-1} from MY 2007 to 2008 through MY 2016 to 2017 (4.86×10^6 Mg in MY 2016 to 2017).

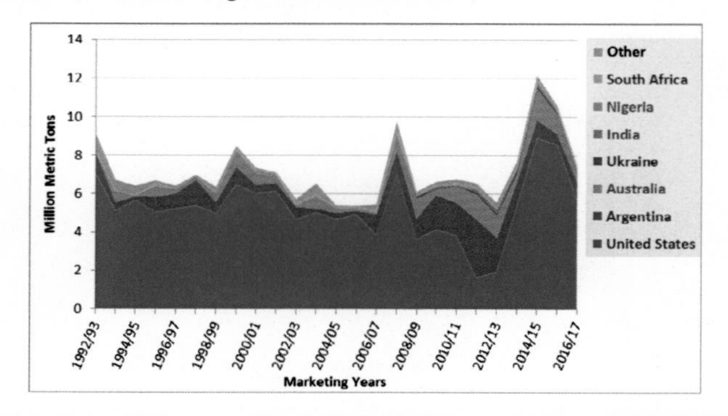

Fig. 17. World sorghum exports by major countries, 1992 to 1993 through 2016 to 2017 marketing years. Data from PSD Online, USDA-FAS, and WASDE Report, 11 April 2017, USDA-OCE.

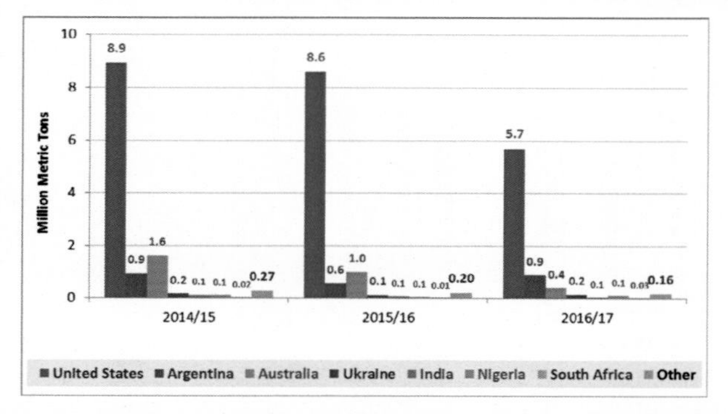

Fig. 18. Top world sorghum-exporting countries, 2014 to 2015 through MY 2016 to 2017. Data from PSD Online, USDA-FAS, and WASDE Report, 11 April 2017, USDA-OCE.

World Sorghum Production by Country

Figures 15 and 16 show world grain sorghum production by country. Figure 15 represents the position of major world grain sorghum–producing countries from a cumulative viewpoint since MY 1992 to 1993. Figure 16 focuses on grain sorghum production from selected major producing countries during the three most recent marketing years: MY 2014 to 2015 through MY 2016 to 2017 (USDA-FAS, 2017).

During the last 5 yr (2012–2016), the United States was projected to have produced an average of 10.9×10^6 Mg (430×10^6 bu and 19.6% of world total) of grain sorghum per marketing year. This is compared with 6.39×10^6 Mg for Mexico (11.5% of the world total), 6.38×10^6 Mg for Nigeria (11.4% of the world total), and 5.06×10^6 Mg for India. These countries are followed by 4.19×10^6 Mg for Sudan, 3.92×10^6 Mg for Argentina, 3.61×10^6 Mg for Ethiopia, 2.98×10^6 Mg for China, 1.79×10^6 Mg for Australia, 1.77×10^6 Mg for Burkina, and 9.29×10^6 Mg for all other countries (16.7% of the world total) (Fig. 15 and 16).

Since MY 2007 to 2008, these same countries have had varying trends in world grain sorghum production. In the United States grain sorghum production has trended up 0.17×10^6 Mg MY^{-1}. This was followed by Mexico which trended lower by 0.09×10^6 Mg annually; Nigeria, down 0.40×10^6 Mg annually; and India,

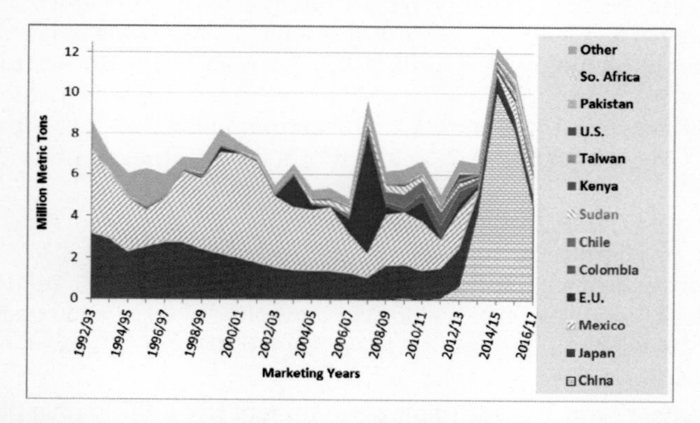

Fig. 19. World sorghum imports by major countries for 1992 to 1993 through 2016 to 2017 marketing years. Data from PSD Online, USDA-FAS, and WASDE Report, 11 April 2017, USDA-OCE.

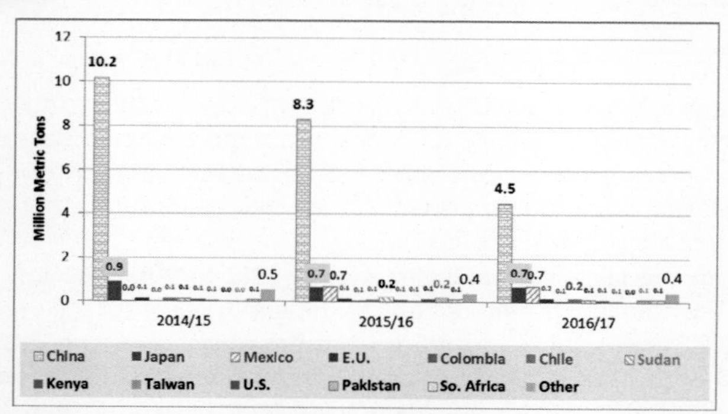

Fig. 20. Top world sorghum-importing countries, 2014 to 2015 through 2016 to 2017 marketing years. Data from PSD Online, USDA-FAS, and WASDE Report, 11 April 2017, USDA-OCE.

down 0.37 × 10^6 Mg annually. These countries were followed by Sudan, up 0.02 × 10^6 Mg annually; Argentina, up 0.11 × 10^6 Mg MY^{-1}; Ethiopia, up 0.09 × 10^6 Mg annually; China, up 0.19 × 10^6 Mg annually; Australia, down 0.16 × 10^6 Mg per year; Burkina, up 0.01 × 10^6 Mg MY^{-1}; and other countries, down 0.06 × 10^6 Mg annually.

Figures 15 and 16 show (i) the predominant roles and positions that North American countries of the United States and Mexico in world grain sorghum production, (ii) the important position of developing countries such as Nigeria, India, Sudan, and Ethiopia as world grain sorghum producers, and (iii) at least moderate growth in grain sorghum production from Argentina in recent years.

World Sorghum Exports by Country

Figures 17 and 18 show exports of grain sorghum by country. Figure 17 represents the position of major grain sorghum exporting countries relative to world grain sorghum exports from a cumulative viewpoint since MY 1992 to 1993. Figure 18 focuses on grain sorghum exports from selected major exporting countries during the most recent three marketing years: 2014 to 2015 through 2016 to 2017 (USDA-FAS, 2017).

During the most recent 5 yr, the United States is estimated to have exported an average of 6.11 × 10^6 Mg (241 × 10^6 bu, and 70.2% of world exports) of grain sorghum per marketing year. This is compared with 1.09 × 10^6 Mg from Argentina (12.5% of world exports), 0.92 × 10^6 Mg from Australia (10.5% of world exports), 0.16 × 10^6 Mg from Ukraine, 0.11 × 10^6 Mg from India, 0.07 × 10^6 Mg from Nigeria, 0.02× 10^6 Mg from South Africa, and 0.23 × 10^6 Mg from all other countries (Fig. 17 and 18).

Since MY 2007 to 2008 these same countries have had varying trends in world grain sorghum exports. The United States was up 0.313 × 10^6 Mg MY^{-1} while Argentina was down 0.083 × 10^6 Mg annually. Australia was up 0.009 × 10^6 Mg annually; Ukraine, up 0.013 × 10^6 Mg annually; India, unchanged; Nigeria, up 0.004 × 10^6 Mg annually; South Africa, down 0.002 × 10^6 Mg annually; and other countries, down 0.009 × 10^6 Mg annually.

Figures 17 and 18 show (i) the predominant role and position that the United States, followed by Argentina and Australia, has in world grain sorghum exports, and (ii) the participation of developing countries such as India and Nigeria in world grain sorghum export markets.

World Sorghum Imports by Country

Figures 19 and 20 show world grain sorghum imports by country. Figure 19 represents the position of major grain sorghum importing countries in world grain sorghum markets from a cumulative viewpoint since MY 1992 to 1993. Figure 20 focuses on grain sorghum imports from selected major importing countries since MY 2007 to 2008 (USDA-FAS, 2017).

During the last five marketing years—2012 to 2013 through projected MY 2016 2017—China is estimated to have averaged 5.548 × 10^6 Mg of grain sorghum imports (62.3% of total world grain sorghum imports during this period). This compared with 1.187 × 10^6 Mg of grain sorghum imports for Japan (13.5% of the world total), 0.803 × 10^6 Mg for Mexico (9.2% of the world total), 0.170 × 10^6 Mg for the European Union (1.9% of the world total), 0.270 × 10^6 Mg for Columbia (3.1% of the world total), 0.245 × 10^6 Mg for Chile (2.8% of the world total), and

0.139×10^6 Mg for Sudan (1.6% of the world total). Kenya, Taiwan, the United States, Pakistan, and South Africa have combined to import from 0.055 to 0.087×10^6 Mg of grain sorghum over the MY 2012 to 2013 through MY 2016 to 2017 period, averaging 4.15% of global grain sorghum imports.

Several other countries have consistently imported an average of 0.055 to 0.087×10^6 Mg MY^{-1} during the last five marketing years, totaling 5.5% of the world grain sorghum imports for this period. These countries include Kenya, Taiwan, the United States, Pakistan, and South Africa. All other countries combined have accounted for 0.536×10^6 Mg MY^{-1} on average, or 6.2% of the world sorghum imports for this period. These major grain-sorghum-importing countries have had varying higher and lower trends in the quantity imported each year since MY 2007 to 2008.

Chinese grain sorghum imports were negligible (never more than 0.083×10^6 Mg) in any year during from MY 2007 to 2008 through MY 2011 to 2012. However, Chinese grain sorghum imports then increased dramatically, to 0.631×10^6 Mg in MY 2012 to 2013, 4.16×10^6 Mg in MY 2013 to 2014, 10.16×10^6 Mg in MY 2014 to 2015, 8.284×10^6 Mg in MY 2015 to 2016, and to 4.50×10^6 Mg in MY 2016 to 2017. From MY 2007 to 2008 through MY 2016 to 2017, Chinese grain sorghum imports increased by an average of 0.980×10^6 Mg MY^{-1} over the MY 2012 to 2013 through MY 2016 to 2017 period, with the largest increases occurring from MY 2012 to 2013 through MY 2014 to 2015, as indicated above. The increased imports of China during that period and the continuation of historically high sorghum imports in succeeding years have been the major supportive demand factor in US and world grain sorghum markets during this same time.

Since MY 2007 to 2008, Japanese grain sorghum imports have trended lower by an average of 0.090×10^6 Mg MY^{-1}. Grain sorghum imports for Mexico have declined by an average of 0.214×10^6 Mg annually during this period. Since MY 1992 to 1993, world grain sorghum imports have increased by 0.057×10^6 Mg annually, while increasing at an accelerated rate of 0.260×10^6 Mg MY^{-1} since MY 2007 to 2008, with most of the recent growth due to Chinese imports.

Figures 19 and 20 show several predominant trends in global grain sorghum imports. First, the growth and dominant position of China as a world grain sorghum importer since MY 2012 to 2013 stands out. Second, recent declines in grain sorghum imports on the part of Japan and Mexico are also noteworthy. Third, the periodic entry of the European Union into the grain sorghum import market with sizable purchases is a major market facto, specifically in MY 2003 to 2004, MY 2007 to 2008, and MY 2010 2011. And fourth, there is an important role in global sorghum import markets for a number of relatively small but somewhat consistent import buyers such as Kenya, Chile, Sudan, Taiwan, South Sudan, Ethiopia, and Columbia, as well as other importing countries.

World Sorghum Usage by Type and Region of the World

Figure 21 represents the growth in global grain sorghum usage from MY 1992 to 1993 through MY 2016 to 2017. World grain sorghum total use averaged 0.060×10^6 Mg annually during this period, trending higher at a rate of 0.061×10^6 Mg MY^{-1}. Since MY 2007 to 2008, total use was estimated to average 61.0×10^6 Mg annually, trending lower at a rate of 0.105×10^6 Mg MY^{-1}. Global use of grain sorghum is projected to be 62.2×10^6 Mg in MY 2016 to 2017 (USDA-FAS, 2017).

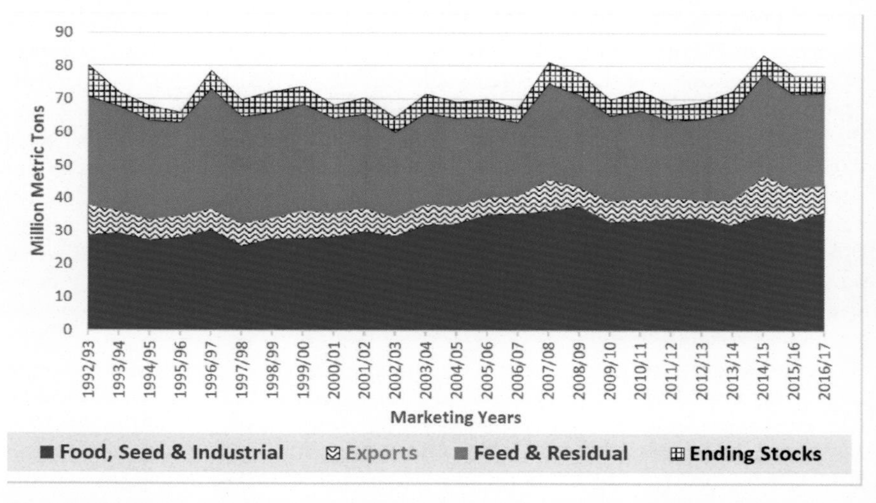

Fig. 21. World sorghum total use by type and ending stocks for 1992 to 1993 through 2016 to 2017 marketing years. Data from PSD Online, USDA-FAS, and WASDE Report, 11 April 2017, USDA-OCE.

By category, Food, Seed, and Industrial (FSI) use averaged 31.6 $\times$ 10^6 Mg annually since MY 1992 to 1993 (trending up by 0.36 $\times$ 10^6 Mg MY^{-1}). Since MY 2007 to 2008, world FSI use has averaged 34.4 $\times$ 10^6 Mg annually but has trended downward by 0.106 $\times$ 10^6 Mg MY^{-1}. Global grain sorghum FSI use is projected to be 36.5 $\times$ 10^6 Mg in MY 2016 to 2017.

Feed and residual use averaged 28.2 $\times$ 10^6 Mg from MY 1992 to 1993 through projected MY 2016 to 2017, while trending down by 0.262 $\times$ 10^6 Mg MY^{-1}. Since MY 2007 to 2008, world feed and residual use has averaged 26.5 $\times$ 10^6 Mg annually but has trended level to upward by 0.001 $\times$ 10^6 Mg MY^{-1}. Global grain sorghum feed and residual use is projected to be 25.7 $\times$ 10^6 Mg in MY 2016 to 2017.

Exports of grain sorghum in world markets have averaged 7.18 $\times$ 10^6 Mg during the MY 1992 to 1993 through projected MY 2016 to 2017 period, trending upward by 0.057 $\times$ 10^6 Mg annually. Since MY 2007 to 2008, world grain sorghum exports have averaged 7.93 $\times$ 10^6 Mg MY^{-1}, trending upward at a rate of 0.244 $\times$ 10^6 Mg annually. Global grain sorghum exports are projected to be 7.50 $\times$ 10^6 Mg in MY 2016 to 2017.

World ending stocks of grain sorghum have averaged 5.19 $\times$ 10^6 Mg from MY 1992 to 1993 through projected MY 2016 2017, trending upward marginally by 0.057 $\times$ 10^6 Mg MY^{-1}. Since MY 2007 to 2008, world grain sorghum ending stocks have averaged 5.44 $\times$ 10^6 Mg MY^{-1}, trending down at a rate of 0.094 $\times$ 10^6 Mg annually. Global grain-sorghum ending stocks are projected to be 4.86 $\times$ 10^6 Mg in MY 2016 to 2017.

Figure 22 represents global grain sorghum usage by major region for the MY 1992 to 1993 through the projected MY 2016 to 2017 period. During this time, world grain sorghum usage averaged 59.8 $\times$ 10^6 Mg annually, trending higher at a rate of 0.061 $\times$ 10^6 Mg annually, and is projected to be 62.2 $\times$ 10^6 Mg in MY 2016 to 2017.

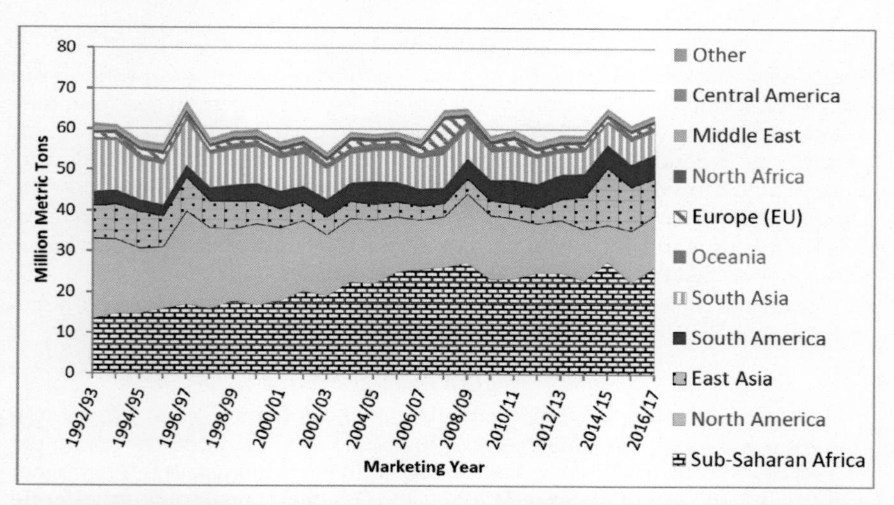

Fig. 22. World sorghum use by world region for 1992 to 1993 through 2016 to 2017 marketing year. Data from PSD Online, USDA-FAS, and WASDE Report, 11 April 2017, USDA-OCE.

Sub-Saharan Africa

Sub-Saharan Africa has been the largest major consumer of grain sorghum in the world over this 25-yr period, averaging 21.2 × 10^6 Mg of grain sorghum total use per marketing year (35.5% of the total world grain sorghum use during this period), trending upward by 0.541 × 10^6 Mg annually. From MY 2007 to 2008 through projected MY 2016 to 2017, average total use of grain sorghum in sub-Saharan Africa averaged 25.1 (41.1% of world grain sorghum total use), trending lower by 0.033 × 10^6 Mg annually in this later period. Sub-Saharan Africa use of grain sorghum is projected to be 26.5 × 10^6 Mg in MY 2016 to 2017 (USDA-FAS, 2017).

Countries included in sub-Saharan Africa are Angola, Burundi, Cape Verdi, Central African Republic, Chad, Congo, Djibouti, Equatorial Guinea, Ethiopia, the Gambia, Guinea, Kenya, Liberia, Madagascar, Mali, Mauritania, Nigeria, Rwanda, Senegal, Sierra Leone, Somalia, South Africa, South Sudan, Sudan, Tanzania, Uganda, Zambia, and Zimbabwe, among others.

North America

North America has been the second largest major grain-sorghum-using region in the world since MY 1992 to 1993, averaging 15.4 × 10^6 Mg of grain sorghum total use per marketing year, 25.8% of the total world grain sorghum use during this period. Usage has trended downward at a rate of 0.332 × 10^6 Mg MY^{-1}. Since MY 2007 to 2008, total use of grain sorghum in North America has averaged 13.1× 10^6 Mg, 21.4% of the world total use, and has trended downward at a rate of 0.421 × 10^6 Mg annually. North American use of grain sorghum is projected to be 12.3 × 10^6 Mg in MY 2016 to 2017 (USDA-FAS, 2017).

Countries included in North America are the United States, Canada, Mexico, Greenland, and St. Pierre and Miquelon.

South Asia

South Asia is estimated to be the third-largest major user of grain sorghum in the world during this 25-yr period, averaging 7.74 × 10^6 Mg grain sorghum total use per marketing year, or 12.9% of the total world grain sorghum use. Grain sorghum usage in South Asia has been trending lower at a rate of 0.254 × 10^6 Mg MY^{-1} since MY 1992 to 1993. Since MY 2007 to 2008, South Asia averaged 6.11 × 10^6 Mg per marketing year, 10.0% of the world total use, with a downward trend of 0.344 × 10^6 Mg annually during this same period. South Asia use of grain sorghum is projected to be 5.00 × 10^6 Mg in MY 2016 to 2017 (USDA-FAS, 2017).

Countries included in South Asia are Afghanistan, Bangladesh, Bhutan, India, Nepal, Pakistan, and Sri Lanka, among others.

East Asia

East Asia is estimated to be the fourth-largest major user of grain sorghum in the world since MY 1992 to 1993, averaging 6.22 × 10^6 Mg of grain sorghum per marketing year, or 10.4% of the total world grain sorghum use, trending higher at a rate of 0.013 Mg MY^{-1}. Since MY 2007 to 2008, the annual total use of grain sorghum for East Asia has averaged 6.61 Mg MY^{-1}, 10.8% of world total use, and has trended higher at a rate of 1.068 × 10^6 Mg annually. East Asian use of grain sorghum is projected to be 9.26 × 10^6 Mg in MY 2016 to 2017 (USDA-FAS, 2017).

Countries included in East Asia are China, Hong Kong, Japan, North and South Korea, Taiwan, Macau, Mongolia, Ryuku Island, and Nansei Island.

South America

South America is estimated to be the fifth-largest major user of grain sorghum in the world since MY 1992 to 1993, averaging 4.52 × 10^6 Mg of grain sorghum per marketing year, or 7.6% of total world grain sorghum use during this period, while trending higher at a rate of 0.104 × 10^6 Mg annually. Since MY 2007 to 2008, the total use of grain sorghum by South America averaged 5.26 × 10^6 Mg MY^{-1} (8.6% of world total use), trending higher at a rate of 0.087 × 10^6 Mg MY^{-1}. South American total use of grain sorghum is projected to be 5.31 × 10^6 Mg in MY 2016 to 2017 (USDA-FAS, 2017).

Countries included in South America are Argentina, Bolivia, Brazil, Chile, Columbia, Ecuador, French Guiana, Guyana, Paraguay, Peru, Uruguay, and Venezuela, among others.

Oceania

Oceania is estimated to be the sixth-largest major user of grain sorghum in the world since MY 1992 to 1993, averaging of 1.17 × 10^6 Mg of grain sorghum usage per marketing year, or 1.9% of the total world grain sorghum use during this period, with an uptrend of 6,000 Mg annually. Average total use was projected to equal 1.46 from MY 2007 to 2008 (2.4% of world total use), with a downtrend of 0.298 × 10^6 Mg MY^{-1}. Oceania is projected to use 0.88 × 10^6 Mg of grain sorghum in MY 2016 to 2017 (USDA-FAS, 2017).

Countries included in Oceania are Australia, French Polynesia, Fiji, Guam, New Caledonia, New Zealand, Papua New Guinea, Samoa, Solomon Islands, and Tonga, among others.

European Union

The European Union has been the seventh-largest major user of grain sorghum in the world since MY 1992 to 1993, averaging of 1.24×10^6 Mg of grain sorghum per marketing year, or 2.1% of total world grain sorghum use during this period, with an uptrend of 3,000 Mg annually. Since MY 2007 to 2008, the EU's grain sorghum usage has averaged 1.21×10^6 Mg MY^{-1} (2.0% of world total use), with a downtrend of 0.137×10^6 Mg per marketing year. The EU's use of grain sorghum is projected to be 0.805×10^6 Mg in MY 2016 2017 (USDA-FAS, 2017).

Countries included in the European Union are Austria, Belgium, Bulgaria, Croatia, Cyprus, Czech Republic, Denmark, Finland, France, Germany, Greece, Hungary, Ireland, Italy, Latvia, Lithuania, Luxembourg, Netherlands, Poland, Portugal, Romania, Slovakia, Slovenia, Spain, Sweden, and the United Kingdom, among others.

Middle East

The Middle East is estimated to be the eighth-largest major user of grain sorghum in the world since MY 1992 to 1993, averaging 0.837×10^6 Mg of grain sorghum per marketing year (1.4% of total world use during this period), with a downtrend of 5000 Mg annually. Since MY 2007 to 2008, total usage of grain sorghum in the Middle East has averaged 0.826×10^6 Mg per marketing year (1.4% of the world total use), with a downtrend of 0.017×10^6 Mg MY^{-1}. The Middle East's use of grain sorghum is projected to be 0.755×10^6 Mg in MY 2016 to 2017 (USDA-FAS, 2017).

Countries included in the Middle East are Bahrain, Iran, Iraq, Israel, Jordan, Kuwait, Lebanon, Oman, Qatar, Saudi Arabia, Syria, Turkey, United Arab Emirates, and Yemen, among others.

North Africa

North Africa is estimated to be the ninth-largest major user of grain sorghum in the world since MY 1992 to 1993, averaging of 0.798×10^6 Mg of grain sorghum per marketing year (1.3% of the total world grain sorghum use during this period), with a downtrend of 8000 Mg annually. Since MY 2007 to 2008, North Africa average total usage equals 0.786×10^6 Mg MY^{-1} (1.3% of the world total use), with a downtrend of 9,000 Mg per marketing year. North African use of grain sorghum is projected to be 0.775×10^6 Mg in MY 2016 to 2017 (USDA-FAS, 2017).

Countries included in North Africa are Algeria, Egypt, Libya, Morocco, and Tunisia, among others.

Central America

Central America is estimated to be the tenth-largest major user of grain sorghum in the world since MY 1992 to 1993, averaging 0.352×10^6 Mg of grain sorghum per marketing year (0.6% of the total world grain sorghum use during this period), with a downtrend of 5,000 Mg annually. Since MY 2007 to 2008, the total usage of grain sorghum in Central America has averaged 0.322×10^6 Mg MY^{-1} (0.5% of the world total use during this period), with a marginal uptrend. Central American usage of grain sorghum is projected to be 0.335×10^6 Mg in MY 2016 to 2017 (USDA-FAS, 2017).

Central America comprises the countries of Belize, Costa Rica, El Salvador, Guatemala, Honduras, Nicaragua, and Panama.

Figure 22 shows several important factors in global grain sorghum imports. First, it shows the importance across time of sub-Saharan Africa to world grain sorghum use, followed by North America, South Asia, East Asia, and South America. Second, the recent growth in grain sorghum use in East Asia is apparent, including China. Third, this graphic shows the relatively static amount of world grain sorghum use total use across time.

World Sorghum Ending Stocks and Percent Ending Stocks-to-Use

Figure 23 shows world grain sorghum domestic use and world ending stocks for the MY 1992 to 1993 through projected MY 2016 to 2017 period, and how global grain sorghum percent ending stocks-to-use has changed across time.

World grain-sorghum ending stocks have averaged 5.19 $\times$ 10^6 Mg MY^{-1} since MY 1992 to 1993, with a minimum of 2.89 $\times$ 10^6 Mg in MY 1995 to 1996, and a maximum of 9.51 $\times$ 10^6 Mg in MY 1992 to 1993. During this period, world grain-sorghum ending stocks have been nearly unchanged on average, increasing at a rate of 0.061 $\times$ 10^6 Mg per marketing year since MY 1992 to 1993. Since MY 2007 to 2008, world grain-sorghum ending stocks have averaged 5.44 $\times$ 10^6 Mg MY^{-1}, with a minimum of 4.56 $\times$ 10^6 Mg in MY 2011 to 2012 and a maximum of 6.33 $\times$ 10^6 Mg in MY 2008 to 2009. During the period since MY 2007 to 2008, world grain-sorghum ending stocks have trended lower on average, decreasing at a rate of 0.094 $\times$ 10^6 Mg MY^{-1}, and are projected to be 4.86 $\times$ 10^6 Mg in MY 2016 to 2017 (USDA-FAS, 2017).

Similarly, world grain-sorghum percent ending stocks-to-use have averaged 8.7% since MY 1992 to 1993, with a minimum of 5.1% in MY 1995 to 1996 and a maximum of 15.4% in MY 1992 to 1993. During this period, world grain-sorghum percent ending stocks-to-use has had no discernible positive or negative trend on average. Since MY 2007 to 2008, global grain sorghum percent ending stocks-to-use has averaged 8.9%, with a minimum of 7.8% projected for MY 2016 to 2017

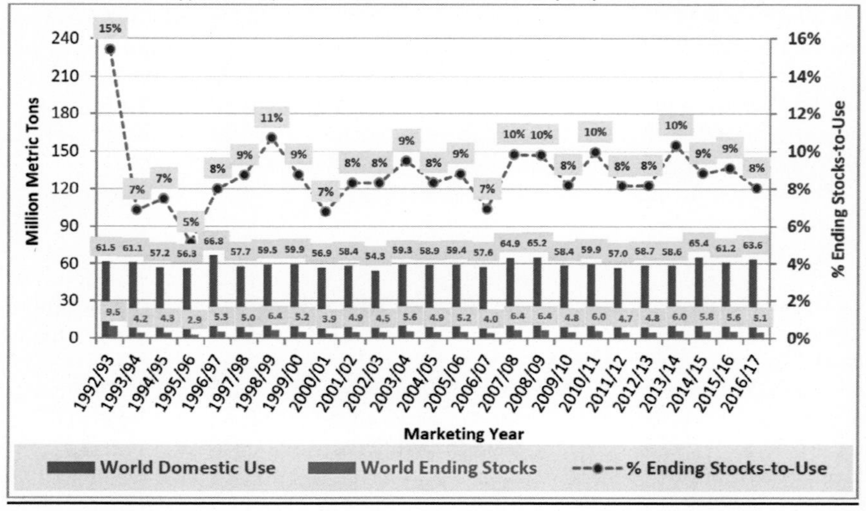

Fig. 23. World sorghum domestic use, ending stocks, and percent ending stocks-to-use for 1992 to 1993 through 2016 to 2017 marketing years. Data from PSD Online, USDA-FAS, and WASDE Report, 11 April 2017, USDA-OCE.

and a maximum of 10.2% in MY 2013 to 2014. Since MY 2007 to 2008, world grain sorghum percent ending stocks-to-use has trended marginally lower, decreasing at a rate of 0.1% MY^{-1}.

These data also indicate that since MY 2007 to 2008, world grain-sorghum ending stocks and percent ending stocks-to-use have displayed variability but have remained within ranges of 4.56 to 6.33 × 10^6 Mg, and 7.8% to 10.2% Mg, respectively. The strong increase in exports to China from the United States, Australia, and other major exporters had a notable impact on world ending stocks and percent stocks-to-use in MY 2014 to 2015 and MY 2015 to 2016. However, low prices for grain sorghum has helped to maintain global usage in MY 2016 to 2017 even though Chinese imports of grain sorghum have declined.

World Grain-Sorghum Percent Ending Stocks-to-Use versus US Sorghum Prices

Figure 24 shows the relationship between world coarse-grain percent ending stocks-to-use and US grain sorghum prices since MY 1992 to 1993. This chart is less instructive in terms of providing a direct, definable relationship between world grain-sorghum supply-demand balances and US prices than the domestic US stocks-to-use versus US price relationship in Figure 10.

However, it provides a more accurate representation of the broader world-market supply-demand balance information that influences the US price of grain sorghum than would world-percent ending stocks-to-use measure based on grain sorghum alone. In world grain markets, the title "coarse grains" refers to the aggregate amount of corn, grain sorghum, millet, oats, and mixed grains.

Shifts downward and to the right in Figure 24 of the points of world grain-sorghum percent ending stocks-to-use and US grain sorghum prices for the three most recent marketing years help to explain the reasons why US grain sorghum prices dropped in MY 2015 to 2016 and are projected to in MY 2016 to 2017. As

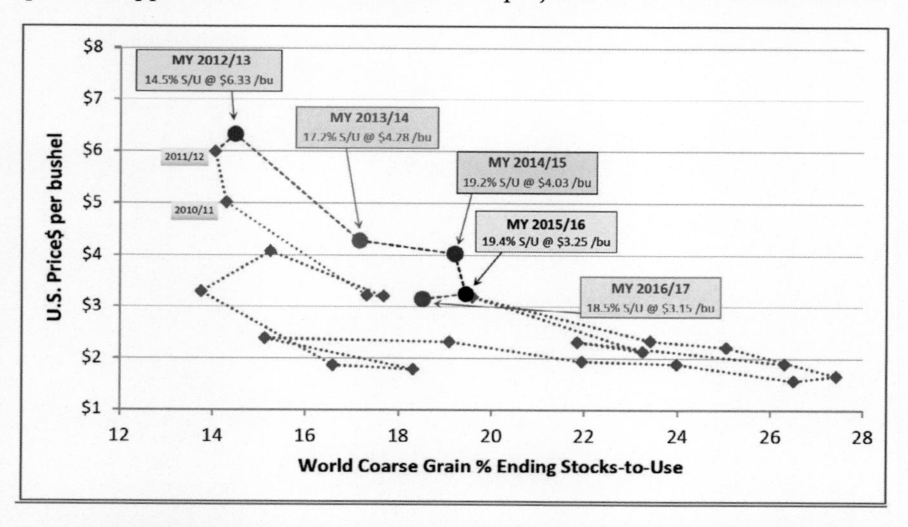

Fig. 24. US sorghum price vs. world coarse-grain percent stocks-to-use for 1992 to 1993 through 2016 to 2017 marketing years. Data from PSD Online, USDA-FAS, and WASDE Report, 11 April 2017, USDA-OCE.

world coarse grain percent ending stocks-to-use has grown in recent years, the downward pressure on US grain sorghum prices has increased.

References

Kansas Farm Management Association (KFMA). 2017. Enterprise reports. AgManager Info., Dep. of Agric. Economics, Kansas State Univ., Manhattan. https://agmanager.info/kfma/enterprise-reports (accessed 25 April 2017) (verified 15 May 2017).

Kansas State University Agricultural Economics. 2017. Seasonal cash grain prices. AgManager info. Kansas State Univ., Manhattan. http://www.agmanager.info/seasonal-cash-grain-prices (accessed 25 April 2017).

USDA, Foreign Agricultural Service (USDA-FAS). 2017. Production, supply, and distribution (PSD) market and trade data portal. USDA, Washington, DC. https://apps.fas.usda.gov/psdonline/app/index.html#/app/advQuery (accessed 25 April 2017) (verified 15 May 2017).

USDA National Agricultural Statistics Service (USDA-NASS). 2017. Quick stats. USDA, Washington, DC. https://www.nass.usda.gov/Quick_Stats/index.php (accessed 25 April 2017) (verified 15 May 2017).

USDA Office of the Chief Economist (USDA-OCE). 2017. World agricultural supply and demand estimates (WASDE). USDA, Washington, DC. https://www.usda.gov/oce/commodity/wasde/ (accessed 25 April 2017) (verified 15 May 2017).

The Sorghum Industry and Its Market Perspective

John Duff, Doug Bice, Ian Hoeffner, and Justin Weinheimer*

Abstract

The last 15 yr have seen significant change in the agricultural sector of the United States. The structure of markets is a prominent example, and nowhere have these changes been more pronounced than in the Sorghum Belt. We begin with an examination of historical trends in and challenges for the sorghum industry. Among those described are the acreage declines and yield stagnation caused by farm policy changes, an exodus of private industry investment, and low market liquidity and transparency. We continue by chronicling the rise of the ethanol industry and detailing its effect on the sorghum industry. Particular emphasis is placed on the transformational nature of the changes brought about by the ethanol industry. We conclude with an overview of the effect that export demand has had on the sorghum industry. Included in this analysis is a historical synopsis of sorghum export markets and an examination of the recent rise in demand from China.

During the last 15 yr, the rise of technology along with shifting global economic and political landscapes have combined to make agricultural markets some of the most complex in existence. This is compounded by market interconnectedness, the ascent of industries that cut across multiple market sectors, and an increase in the number of investment institutions and vehicles. It could well be said that agricultural markets have undergone more change in the last 15 yr than they did in the previous 50.

At the center of all this flux lies the US farmer, who faces many of the same challenges and opportunities encountered by previous generations. Yield is still the primary consideration that drives cropping and agronomic choices, with weed and pest control being necessary for a profitable operation, and weather the most important and uncontrollable external factor affecting farm profitability. Yet the complexity of the current agricultural marketing environment requires management practices far different than those used by the majority of farmers even a single generation ago, new practices that include a comprehensive knowledge of the global economy and the tools to analyze its impact on farm gate prices.

Abbreviations: ADM, Archer Daniels Midland; EISA, Energy Independence and Security Act; RFS, renewable fuel standard.

John Duff (john@sorghumcheckoff.com), Doug Bice (dougb@sorghumcheckoff.com), Ian Hoeffner (Ianh@sorghumgrowers.com), and Justin Weinheimer, Sorghum Checkoff, 4201 N Interstate 27, Lubbock, TX 79403. *Corresponding author (justinw@sorghumcheckoff.com).

doi:10.2134/agronmonogr58.2014.0080

Nowhere have these changes been more impactful than in the US Sorghum Belt, where farmers contend with some of the harshest environmental conditions in North American row-crop agriculture. Consolidation is taking place at a rapid pace in this region, and resources are becoming increasingly scarce. To be sure, the sorghum industry is on the forefront of change. In this chapter we will focus on historical trends in and challenges for the sorghum industry, the effect of the ethanol industry on sorghum, and the effect of export demand for sorghum.

Historical Trends in and Challenges for the Sorghum Industry

The historical evolution of modern sorghum began in the early twentieth century. Grown in the southwest due to its drought- and heat-tolerant nature, sorghum dominated much of the area it dominates today, which includes the Texas High Plains, western Kansas, and parts of Nebraska and southern South Dakota (see Fig. 1 for 2013 sorghum acreage by agricultural statistics district). Many areas further east also composed the Sorghum Belt of the day, and from the late 1920s through the 1930s, the industry acreage held consistent at about 10 million acres. Utilizing only conventional breeding practices and open-pollinated varieties, national yield averages hovered at about 12 bushels/acre during this time. Despite yields that would be deemed as very poor by today's standards, the sorghum industry saw a surge to 21.2 million acres in 1940, probably due to extended drought in the southwest United States. (See Fig. 2 and 3 for planted sorghum acres and sorghum yields over time.)

The next and probably most notable moment in the history of the sorghum industry came in the late 1950s, with the introduction of hybrids. Farmers in the United States fully adopted hybrid sorghum within 5 yr, raising the national average yield by about 50%, with the metric reaching 35.2 bushels/acre in 1958. On this new yield strength, the sorghum industry grew to 26.9 million planted acres in 1957, the highest ever recorded. This massive growth in production brought

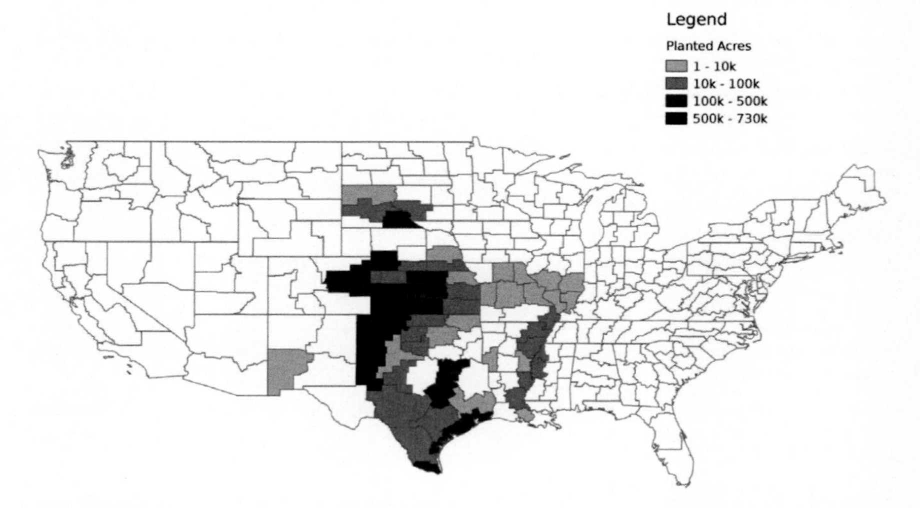

Fig. 1. Sorghum acreage planted in 2013. From USDA National Agricultural Statistics Service (https://quickstats.nass.usda.gov/).

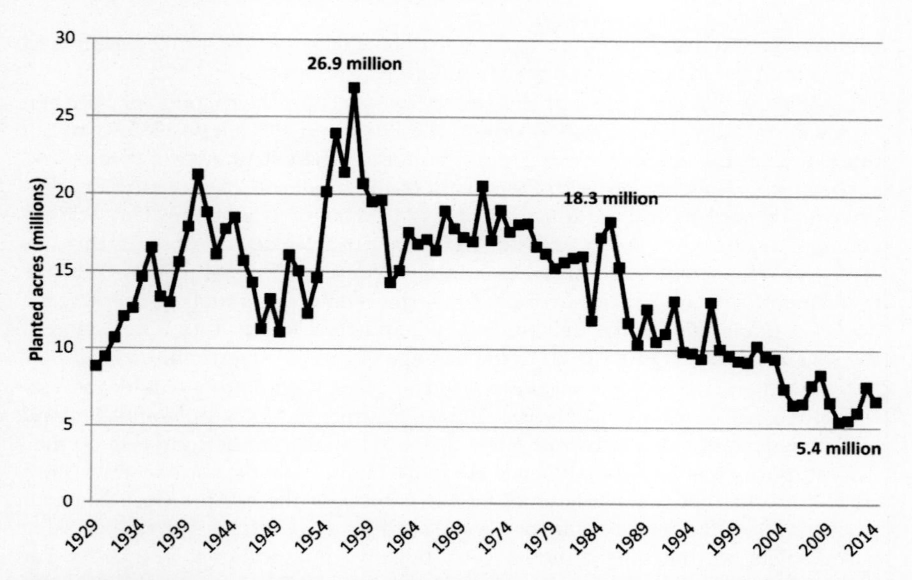

Fig. 2. Sorghum acres planted from 1929 to 2014. From USDA National Agricultural Statistics Service.

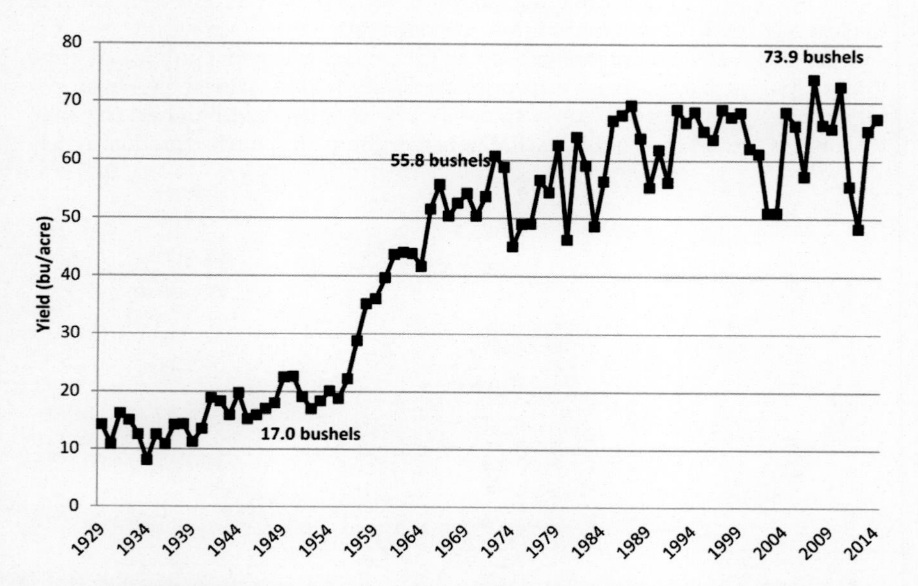

Fig. 3. Average national sorghum yields from 1929 to 2014. From USDA National Agricultural Statistics Service.

cattle feeders to the High Plains to capitalize on cheap, locally sourced grain and to help manage the problems associated with its oversupply.

After the acreage peaks of the late 1950s, sorghum held constant between 15 and 20 million planted acres through the 1960s and 1970s. The 1985 Farm Bill marked a significant shift away from this norm, when strong corn, cotton, and soybean policies, coupled with the advent of the modern Conservation Reserve Program, forced several million acres of sorghum out of production. These policy changes resulted in a contraction of the sorghum industry by about a third in only 2 yr. This event then served as a catalyst for an exodus of private-industry investment, resulting in fewer tools for farmers and perpetuating the acreage declines. (Note the decline in acres in Fig. 2 and refer to Fig. 4 for a depiction of the sorghum industry just prior to the passage of the 1985 Farm Bill.)

Making matters worse, sorghum faced significant marketing challenges. Historically, a crop's value has been driven by consumer needs, supply-and-demand interactions, and value substitutions within exchanges. In the coarse-grain markets, sorghum has traditionally been compared with the most similar starch crop, corn, and most local and regional market values for the grain reflect this comparison. The difference in value between sorghum and corn should, in theory, be based solely on the value differences of the two commodities to end users. For example, if beef cattle fed a ration of sorghum perform 95% as well as beef cattle fed a ration of corn, the value paid to the sorghum farmer should be 95% of the local corn price. Conversely, if a swine producer finds his animals' performance on sorghum is equal to their performance on corn, he should be willing to pay the sorghum farmer 100% of the local corn price.

However, the actual value that sorghum farmers have received for most of the last century has not reflected this economic rationality. There are several reasons for this discrepancy, and chief among them is the lack of a public exchange for transparent price discovery in sorghum. Prices paid to farmers have been low for decades because of the lack of transparency associated with public exchange trading. As a result of no public exchange for sorghum, the crop has predominately

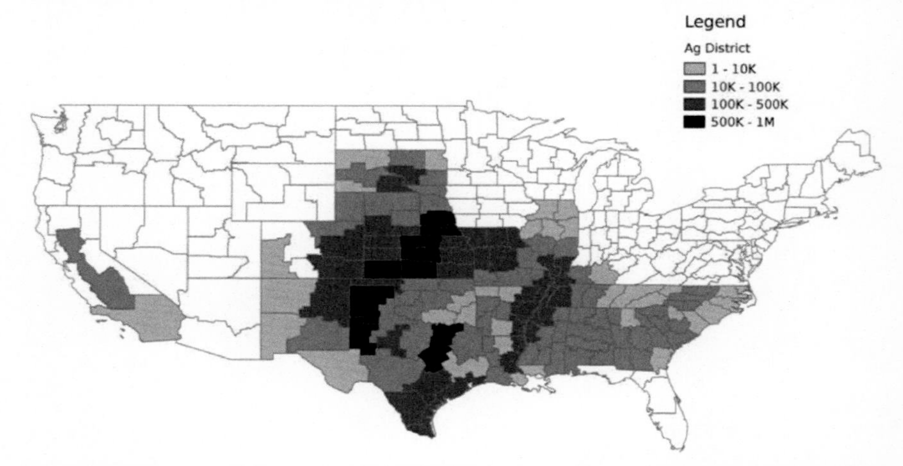

Fig. 4. Sorghum acreage planted in 1985. From USDA National Agricultural Statistics Service.

been traded at a discount to corn futures prices, with an adjustment for local supply and demand of both corn and sorghum (basis). Sorghum has been traded on a public exchange at least twice, during the 1970s and 1980s; however, the challenge of inadequate liquidity due to thin trading proved too much to overcome, and the contracts were discontinued each time. Low liquidity was in part due to mostly unpredictable yields and national production from year to year. But the lack of transparency creates a void of accountability that directly and negatively impacts the sorghum farmer. Not only has this void resulted in an inconsistent farmer basis, but it has also led to a lack of marketability for sorghum.

These problems are exacerbated by poor and inconsistent demand, which has caused problems for both farmers and handlers. Lack of demand from end users not only leads to low prices, but it also puts handlers at substantial risk from inconsistent marketability. Though this situation has changed significantly with the recent rise of demand from the ethanol industry and new export markets, these challenges have resulted in chronic apathy in the sorghum industry, leading to the very acreage declines that cause low liquidity, lack of transparency, and ultimately low marketability. Thus a vicious circle exists, and it is not confined to the marketing area of the sorghum industry. Acreage declines lead to less investment in crop improvement, which results in fewer tools for farmers. Fewer tools for farmers mean lower acres and even less liquidity, transparency, and marketability, and so the circle continues.

It is important to note many of these trends have seen reversals in recent years, with drought and high input prices joining market forces to usher in significant positive change for the sorghum industry. In addition to favorable market conditions, industry investment has increased considerably. For the first time in nearly three decades, the sorghum industry is seeing new R & D efforts in public and private breeding institutions. Originally driven by interest in sorghum improvement for next-generation biofuel production, investments in sorghum research related to yield, pest control, and even breeding technology are now occurring at a rapid pace. In addition to increased interest from established private seed industries such as DuPont Pioneer, investments by the emerging technology companies Chromatin Inc. and NexSteppe are yielding genetic gains as well. In the public sector, sorghum research stalwarts like the USDA-ARS, Kansas State University, and Texas A&M University have been joined in recent years by newly interested institutions such as the USDOE and East Coast universities like Clemson in moving the crop forward in support of private industry.

Effect of the Ethanol Industry on Sorghum

Grain ethanol has been used as a fuel source almost as long as internal combustion engines have been in existence. Early engines were designed to accommodate the fuel, as were early automobiles. This gave rise to a boom in ethanol production in the United States at the beginning of the twentieth century, and the fuel was promoted extensively in the Midwest as a panacea for the perennial oversupply of grain that plagued farmers and rural communities alike. Prohibition marked the end of this boom (and temporarily the entire industry), but the fortunes of ethanol would rise and fall a few more times over the next several decades, eventually giving way to the modern, comparatively mature, and markedly more stable industry thriving in the United States today.

Boom-and-bust cycles are all too familiar to stakeholders involved in the production, distribution, and use of agricultural commodities. As the history of its development clearly shows, the ethanol industry is no exception to this (it is probably subject to more volatility given that it uses one commodity to produce another commodity). However, because of the significant grain demand it creates, it is not surprising that ethanol has been looked to as a price stabilizer and a means to add value to struggling rural communities. Also not surprising are the players who helped build the modern industry and their maneuvers to stabilize its persistent volatility through political means.

The longtime grain handler and processor Archer Daniels Midland (ADM) entered the ethanol industry during a period of relative prosperity, several cycles removed from the bust ushered in by Prohibition. As did many of the farmer groups before and alongside it, ADM saw ethanol as a way to stabilize the price of its chief commodity, grain, and add value to its business as well. Seeing volatility in both input and output markets, though, ADM soon looked to policy mechanisms to assist the fledgling industry in gaining market access in the late 1970s and early 1980s. The political climate of the day lent itself well to this type of policy support, as energy crises and environmental disasters had made headlines in the United States throughout the 1970s. With newfound public and political support as well as a large institutional champion in ADM, the stage was set for the rise of the ethanol industry we know today.

Alongside ADM were numerous farmer groups and individuals interested in adding value to their farms and communities. While most of these groups were concentrated in Midwestern states with an overabundance of corn production, many were located squarely in the Sorghum Belt, from South Dakota to Kansas and even in eastern New Mexico. Clearly, oversupply and instability were national problems, as were difficult times in rural communities. Accordingly, ethanol projects and plants began increasing in number significantly throughout the 1980s and into the 1990s (but not without the familiar cyclical interruptions), with Sorghum Belt farmers and communities being key beneficiaries.

Ethanol production in the United States continued to increase into the early 2000s, and with this increase came the rise of supporting industries and infrastructure. Numerous construction companies, engineering groups, consulting firms, and other businesses emerged to provide the knowledge and skill necessary to move a small industry built largely out of necessity into the mainstream of US energy production. Ethanol marketing also became more sophisticated, and with the decline in use of lead and methyl tertiary butyl ether as octane agents, ethanol became the preferred method of boosting the octane content of gasoline and reducing particulate matter associated with petroleum-based aromatic gasoline additives such as benzene, toluene, and xylene.

While government still had an active and significant role in the ethanol industry, this role would be increased markedly with the passage of the Energy Policy Act in 2005. The act established the renewable fuel standard (RFS), which set ethanol blending requirements for the gasoline supply and its proprietors. These requirements would increase annually, with a goal of blending 7.5 billion gallons of ethanol into the US fuel supply by 2012. However, well before the RFS could reach maturation, Congress acted again, this time by passing the landmark Energy Independence and Security Act (EISA) in 2007. The EISA revamped the RFS (which would come to be known as the RFS2), accelerating the pace at which the annual blending requirements increased, thus setting a new goal of blending 36 billion gallons of all renewable fuels, including both grain ethanol and advanced biofuels, into the fuel supply by 2022 and adding significant environmental litmus tests to new fuels.

Though this seems like significant government intrusion into energy markets (and for better or worse, it is widely regarded as significant government intrusion into energy markets), ethanol offers an important value proposition not only to farmers but also to gasoline producers and marketers, regardless of policy. The positive impacts on grain prices and rural communities are obvious and well documented, but few understand the fuel-market dynamics that necessitate the presence of ethanol in the gasoline supply chain. Gasoline refiners today produce a lower octane fuel than that sold to consumers as gasoline, and this fuel is much cheaper to produce than its higher-octane alternative. After initial production, the product is blended with relatively cheaper ethanol to boost its octane content to a level suitable for sale as regular unleaded gasoline. This gasoline is sold for significantly more than it costs to produce because of these comparatively lower inputs costs, and refiners are still able to pass along meaningful savings to consumers. Clearly, with or without government policy, ethanol is valuable to farmers, rural communities, refiners, and ultimately consumers.

Much like their Midwestern counterparts, Sorghum Belt farmers benefitted from the added value both to their grain and to their communities. In the mid-1950s, the modern cattle feeding industry was built by High Plains farmers looking for ways to manage the chronic grain oversupply problems, with sorghum as a key component in rations due to its abundance resulting from the crop's recent hybridization. However, by the 1980s and 1990s sorghum acres had begun to decline precipitously, and the corresponding lack of supply led cattle feeders to look elsewhere for feedstuffs. The only viable option was Midwestern corn, and by the early 2000s, the cattle feeding industry had reequipped itself to feed grain sourced almost exclusively from the Corn Belt via rail.

The result of this shift on the part of cattle feeders was to exacerbate the oversupply and harvest-time price instability problems that had been hallmarks of grain production—particularly in the Sorghum Belt and Midwest—for decades, necessitating alternative opportunities and new demand drivers. Historically, foreign buyers had been key markets for sorghum produced in the United States, with Mexico leading the way. Thus export markets were one option, and while demand from those markets was significant, much of the price instability of the past remained, regardless of the amount of export demand present. Foreign competition also eroded US market share in some areas, and the inability of farmers to produce the quantity of sorghum needed to meet the needs of foreign buyers further hurt the US position. Fortunately for sorghum farmers, ethanol offered the answer to these problems.

After the passage of the EISA and subsequent introduction of the RFS2, ethanol production began to increase rapidly, at a rate much higher than had ever been seen in the industry. The number of ethanol plants operating rose dramatically, and many grain farmers in the Sorghum Belt and Midwest were involved in building, or at least exploring the possibility of building, a plant. The number of plants rose to more than 200 by 2008, when another downturn led to many plants becoming idle. However, the infrastructure was in place for the industry to accomplish what it was intended to accomplish, which was to add value to farmers' grain and rural communities. In retrospect, the industry has been very successful in both of these endeavors, and the effects can be seen from the conditions of farms and farmers' balance sheets to the growth of small communities scattered across the Sorghum Belt and Midwest whose very existences were once considered in grave jeopardy.

Today about a quarter of US ethanol production capacity is located in the Sorghum Belt. Of those roughly 50 ethanol plants, 12 are consistent sorghum users, with individual demand ranging from 10 to 100% of total grain usage, with all using the grain interchangeably with corn, sometimes even storing the two in the same bin. Each plant's usage level varies significantly depending on the time of year and availability of sorghum. Few changes must be made to a plant when transitioning from one to the other, and most of the 12 consistent sorghum users make no changes at all. Over the course of a year, sorghum meets an average of 50% of the needs of these plants, totaling roughly 120 million bushels. As of late, ethanol has consumed roughly a third of the annual US sorghum crop, with a third of sorghum ethanol production occurring in Texas and two-thirds occurring in Kansas.

Not surprisingly, a market effectively introduced within the last 15 yr that now consumes a third of the annual crop is having a significant market impact. For sorghum farmers, these impacts have been felt most immediately in a stabilization of harvest prices and a strengthening of overall prices relative to corn. Traditionally, sorghum prices in certain areas fell dramatically at harvest to reflect the lack of market options and then gradually adjusted back upward before and during the growing season, often rising above the price of corn before harvest. Consistent demand from ethanol plants has helped to minimize the occurrence of these large price movements, allowing farmers to take better advantage of profitable marketing opportunities.

Appreciation of price relative to corn has also been positive for sorghum farmers, particularly in areas where sorghum fits better from an agronomic or water management standpoint. On the High Plains, for example, because of price disadvantages, irrigated acres have traditionally been planted to cotton or corn regardless of how well sorghum might perform in limited irrigation environments. Ethanol, however, has helped close this gap, and farmers with limited irrigation capacity have been able to take advantage of sorghum's ability to better manage resource and other challenges. Figure 5 depicts ethanol production in millions of gallons along with average sorghum-to-corn price ratios over time. Note that prior to 2002 (the first year with ethanol production over two billion gallons in the United States) the average sorghum-to-corn price ratio was 91.3%, and after 2002 the average ratio was 95.3%. These ratios clearly show that ethanol has been beneficial for sorghum farmers.

In addition to opportunities in grain ethanol for sorghum farmers, significant opportunity exists in advanced biofuels. Advanced biofuels are produced with feedstocks other than corn starch that meet certain environmental requirements related to reducing CO_2 emissions. Grain sorghum ethanol producers can qualify to produce advanced biofuels—which can often command a premium in the marketplace under the RFS2—as can biofuel producers who use sweet and higher-biomass-type sorghums. Although considerable opportunity is available, little commercial progress has been made to date due to policy, political, and

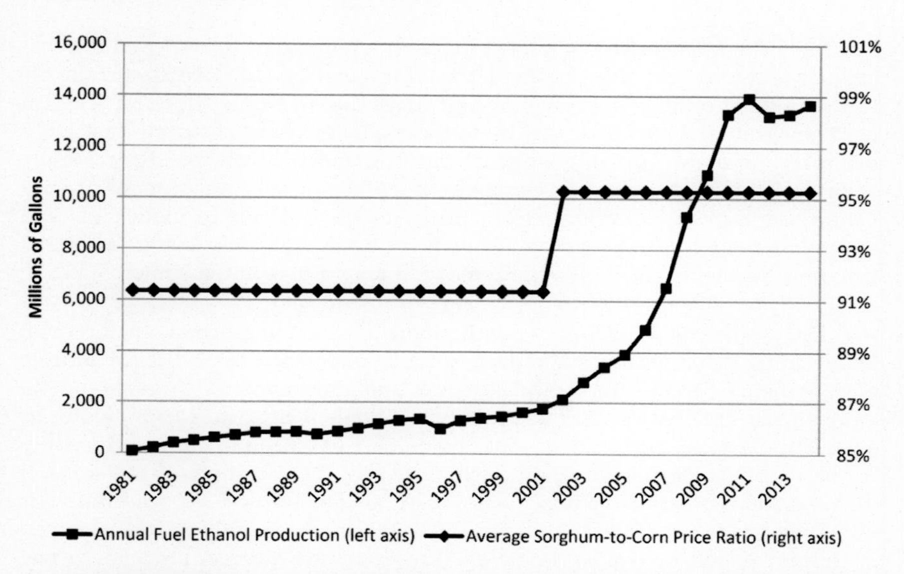

Fig. 5. Annual fuel ethanol production and average sorghum-to-corn price ratios for two periods between 1981 and 2014. From DOE Energy Information Administration and National Agricultural Statistics Service.

technology risks. Current fuel market dynamics, the state of financing, and technology costs have also proven to be difficult obstacles to overcome.

Regardless of government policy and despite the seemingly endless iterations of boom-and-bust cycles, the US ethanol industry has proven to have staying power. It offers farmers and rural communities a true value proposition, and in a short time it has grown to be the largest domestic market for sorghum. Looking ahead, the industry will continue to evolve, searching for ways to be more efficient and more competitive in the global energy economy. Undoubtedly, managing the grain supply more effectively will be a central tenet of this development, and sorghum will be well positioned given its growing importance in regional agricultural economies in the central United States.

Effect of Export Demand for Sorghum

It is often forgotten the founding member of the US Grains Council was the National Grain Sorghum Producers. Exports have long been a chief component of the US sorghum-demand portfolio, and the recent upswings in production have brought these markets back into focus. While China has captured most of the recent headlines, Mexico has historically been a pillar of US sorghum export demand. With about one-fourth of US sorghum production being concentrated in the Rio Grande Valley and along the Texas Gulf Coast, Mexican grain buyers have been central to maintaining workable stocks-to-use ratios on the US sorghum balance sheet. Other notable partners are Japan and the European Union, with the latter intermittently entering the market for sorghum and changing the world demand picture significantly. Without a doubt, though, China is currently top-of-mind for those with a stake in US sorghum exports.

Much has been made of China's burgeoning population growth and the challenge for a country roughly the size of the continental United States to sustainably feed a population that is more than four times that of North America. The Chinese government has not sat idly by as these changes have occurred. Instead, it has taken an active role in enhancing farming and livestock practices, supply-chain logistics, trade policy development, and other agricultural economic issues. China has made a concerted effort during the past decade to modernize and upgrade its agricultural markets through technological, policy, and infrastructure improvements not unlike those made in agriculture in the United States in the 1970s and 1980s with government support in kind.

The proliferation of China's population and associated urban advancement has recently come with a methodical, parallel expansion of policies that favor entrepreneurial and technological advances and a move toward larger and more automated farm operations. The country has made a clear shift from taxing its agricultural sector to subsidizing it, and as part of efforts to gain membership in the World Trade Organization during the past decade, it has eased some of its historical market-distorting policies. Also as part of these efforts China has eliminated its tax on farmers and introduced three new subsidy programs aimed at those engaged in grain production. These include a direct payment, a payment that subsidizes the use of improved seed technology, and a partial rebate for specific types of farm equipment. Since this original effort, agricultural insurance and county-based award programs have also been initiated as incentives to enhance farming practices (Gale, 2013). These along with other market and policy initiatives have seen robust results, with participation from many in the global agricultural sector.

China has been an exceptional market for US sorghum since 2013, providing a unique opportunity for growth in US sorghum exports. During the fall of 2013, the first-ever bulk shipment of US sorghum to China was offloaded at the Guangzhou port in China. The nearly 2.5-million-bushel shipment was designated as livestock feed, and the crop has been a major US export commodity ever since. Shipments to China have constituted nearly 90% of the total US sorghum exports during this time. Preceding this increase in exports, significant research on feeding sorghum to diverse livestock populations was conducted, and the crop was proven both cost- and quality-competitive with other grains. This was especially true in the context of least-cost ration formulation.

Research has shown that factors such as grain particle size and processing methodology (i.e., flaking or rolling) are key considerations for both ruminant and nonruminant livestock populations. Additionally, the constituent profiles associated with sorghum (e.g., amino and fatty acids, crude protein, phosphorous, linoleic acid, etc.) have been proven by several studies to make the grain an advantageous ration ingredient (United Sorghum Checkoff Program, 2014). This research has helped promote sorghum as a valuable alternative feedstuff in numerous areas, and the years of work in advance of the current sorghum export spike was very beneficial in positioning the crop for these opportunities. Emphasis on nutritional benefits for the swine and poultry industries as well as a presentation on various milling techniques were part of a recent promotional tour through Guangzhou, Nanning, and Shanghai that was sponsored by the Sorghum Checkoff and the US Grains Council and aimed at continuing the current momentum by highlighting sorghum's diverse uses and dietary benefits for these important markets.

Other key advantages for sorghum with regard to timely export shipments are the logistical and supply-chain benefits in the United States, and in particular, the Gulf Coast region. With respect to terminal storage and railway and port capacity, the United States is unmatched. All other factors being the same, this is why the United States has marked advantages over the competing sorghum-export countries of Argentina and Australia, two countries that have distinct geographic advantages over the United States. In part, the US advantages explain why Argentina and Australia have seen declines in sorghum exports of close to 50% since their recent 2012 export marketing-year highs. Terminal storage capacity in Kansas alone is well above 400 million bushels (which is more than the entirety of US sorghum production in 2013), and these facilities have direct links to port facilities through intricate and well-established shuttle infrastructure, which is a hallmark of the US grain industry. With regard to port capacity and travel expediency, Texas has more than 1000 channels and 16 available ports, and the port facilities at Corpus Christi and Houston are two of the largest in the nation in terms of grain storage. Having such a reliable logistical network has been an integral part of the recent and historical success of US sorghum export endeavors.

China's total purchases of US sorghum for the 2013–14 marketing year were over 200 million bushels, a milestone that was already reached by the end of the first quarter of the 2015 marketing year. There was a more than 400% increase in sorghum exports to China in this reporting period from the previous marketing year, and exporters sold more than 23 million bushels of US sorghum for the week ending 18 Dec. 2014, representing the largest weekly sale since 1995 and the second largest since 1990, with Chinese buyers accounting for 57% of the total. As of this writing, nearly 240 million bushels—55% of the projected 2014–15 US sorghum crop—are destined for export markets. The benefit to US sorghum farmers and exporters has been significant, with a 10% appreciation in the price of sorghum—to $253 per metric ton at the Gulf of Mexico (as of December 2014), from the same time a year earlier. Compared with average pricing from 2010 to 2013, this represents a net benefit to US sorghum farmers and exporters of $172 million.

Leading up to this surge in sorghum exports to China, corn imports to the country dwarfed those of sorghum four to one. However, for a myriad of reasons, the trend has been toward a more balanced approach to coarse grain imports and usage in general. Clearly, as the world's top exporter, pork producer and consumer (according to the US Grains Council, official swine inventories for China indicate the country accounts for half of the world total), dried distillers-grains importer, and the second-largest corn producer, China has significant influence on global supply-and-demand interactions in many commodity markets. This is in part due to limitations associated with current farming practices as well as marginal soil quality, resulting in an average corn yield in China about two-thirds of that in the United States The result is China's strong reliance on agricultural imports.

Looking forward to China's food-related needs during the next decade, as determined on the basis of projected urban growth and dependence on meat and protein for social advancement, the country will have to increase its yield and acreage to meet its needs. Even with favorable assumptions for both (i.e., corn yields and acreage), China would still fall almost 900 million bushels short of its projected consumption requirements (Gale et al., 2014). Furthermore, it is not probable that imported corn could fill even the void of this best-case scenario due to tariffs, quotas, value-added tax provisions, transgenic seed concerns, and excess inventory issues, just to name a few challenges that have recently thwarted the US corn export potential to China. Sorghum, on the other hand, whether intended to be used as a feed or food source, has not experienced such restrictions. The probable scenario that China's coarse-grain needs increase beyond this projected shortfall is all the more reason that the more than 50,000 sorghum producers in the United States work to increase sorghum yields and acres to meet China's significant needs.

The United States has welcomed the transitions and shifts with respect to China's agricultural policy. More specifically, entities such as the US Grains Council and Sorghum Checkoff have worked to position sorghum to cost-effectively support China with respect to its feed- and food-resource needs. Research expenditures from Sorghum Checkoff have more than doubled from the most recent prior year, and market development investments have increased by nearly a third over the prior 5-yr average.

With a continued interest in market development, production enhancement, competitive pricing, and logistical and quality considerations, the sorghum industry has more than stepped up to meet specific voids in the coarse-grain market in China. Also through sorghum's inherent benefits for lower water usage and input cost, strong livestock feed and nutritional food profiles, and diverse usage and hybrid portfolios, this often-overlooked grain continues to show that it has a key role to play in meeting the agricultural needs of many countries.

References

Gale, F. 2013. Growth and evolution in China's agricultural support policies. Economic Res. Rpt. ERR-153. USDA-ERS, Washington, DC.

Gale, F., M. Jewison, and J. Hansen. 2014. Prospects for China's corn yield growth and imports. Feed Outlook Rep. FDS-14D-01. USDA-ERS, Washington, DC.

United Sorghum Checkoff Program. 2014. Livestock nutrition. http://sorghumcheckoff.com/sorghum-markets/animal-nutrition/ (accessed 27 July 2016).

Printed and bound by CPI Group (UK) Ltd, Croydon, CR0 4YY

17/04/2025

14658856-0001